Essential Stem Cell Methods

Reliable Lab Solutions

Essential Stem Cell Methods

Reliable Lab Solutions

Edited by

Dr. Robert Lanza

Advanced Cell Technology
Worcester, Massachusetts

Dr. Irina Klimanskaya

Advanced Cell Technology
Worcester, Massachusetts

ELSEVIER

AMSTERDAM • BOSTON • HEIDELBERG • LONDON
NEW YORK • OXFORD • PARIS • SAN DIEGO
SAN FRANCISCO • SINGAPORE • SYDNEY • TOKYO
Academic Press is an imprint of Elsevier

Academic Press is an imprint of Elsevier
Linacre House, Jordan Hill, Oxford OX2 8DP, UK
30 Corporate Drive, Suite 400, Burlington, MA 01803, USA
525 B Street, Suite 1900, San Diego, CA 92101-4495, USA
32 Jamestown Road, London NW1 7BY, UK

First edition 2009

ISBN: 978-0-12-374741-9

For information on all Academic Press publications
visit our website at books.elsevier.com

Printed and bound in USA

09 10 11 12 10 9 8 7 6 5 4 3 2 1

CONTENTS

Contributors   xiii

Preface   xix

SECTION I   Organ-Derived Stem Cells

1. Neural Stem Cells: Isolation and Characterization

Rodney L. Rietze and Brent A. Reynolds

    1. Introduction   4
    2. Reagents and Instrumentation   5
    3. Methods   8
       References   21

2. Neural Stem Cells and Their Manipulation

Prithi Rajan and Evan Snyder

    1. Introduction   24
    2. Adult Niches for Stem Cells *In Vivo*   25
    3. *In Vitro* Manipulation of NSCs   27
    4. Conclusions and Projections   39
    5. Sample Protocols for the Culture and Characterization of NSCs   40
    6. Culture of Rodent/Murine NSCs   41
    7. Culture of human NSCs   43
    8. ICC of markers to identify stem cells and differentiation products   45
       References   45

3. Retinal Stem Cells

Thomas A. Reh, Joseph A. Brzezinski IV, and Andy J. Fischer

    1. Introduction   54
    2. Materials and methods   60
    3. Overview   68
       References   69

4. Dental Pulp Stem Cells

He Liu, Stan Gronthos, and Songtao Shi

1. Introduction	74
2. Identification of Dental Pulp Stem Cells (DPSCs)	75
3. Isolation of DPSCs	77
4. Differentiation of DPSCs	79
References	83

5. Mouse Spermatogonial Stem Cells: Culture and Transplantation

Jon M. Oatley and Ralph L. Brinster

1. Introduction	88
2. Spermatogonial Stem Cells (SSCs)	89
3. SSC Transplantation	91
4. SSC Culture	98
5. siRNA Transfection of Cultured Mouse SSCs	105
6. Implications	109
References	110

6. Stem Cells in the Adult Lung

Xiaoming Liu, Ryan R. Driskell, and John F. Engelhardt

1. Introduction	114
2. Anatomical and Cellular Diversity of the Adult Lung	115
3. Stem Cell Phenotypes and Niches in the Adult Lung	117
4. *In Vivo* Injury Models of the Lung	121
5. *Ex Vivo* Epithelial Tracheal Xenograft Model to Study Stem Cells Expansion in the Proximal Airway	128
6. *In Vitro* Colony-Forming Efficiency (CFE) Assay to Characterize Stem/Progenitor Cell Populations in Conducting Airway Epithelium	136
7. Models to Study Stem/Progenitor Cells of Airway SMG	141
References	143

7. Pancreatic Cells and Their Progenitors

Seth J. Salpeter and Yuval Dor

1. Introduction	150
2. Pancreas Development	151
3. Origins of Beta Cells During Postnatal Life	153
4. Preexisting Beta Cells	154
5. Ducts	155
6. Acini	155
7. Bone Marrow Cells	156
8. Adult Stem Cells	156

9. Dedifferentiation of Beta Cells   157
10. Summary and Perspective   157
11. Methods: Design of a Lineage Tracing Experiment in Mice   158
    References   160

8. Pluripotent Stem Cells from Germ Cells: Derivation and Maintenance

Candace L. Kerr, Michael J. Shamblott, and John D. Gearhart

1. Introduction   166
2. Germ Cell Development   167
3. EGC Derivation   168
4. Characterization of EG Cultures   177
5. EB Formation and Analysis   182
6. EB-Derived Cell Formation   183
    References   185

9. Pluripotent Stem Cells from Amniotic Fluid and Placenta

Dawn M. Delo, Paolo DeCoppi, Anna Milanesi, Minhaj Siddiqui, and Anthony Atala

1. Introduction   192
2. Amniotic Fluid and Placenta in Developmental Biology   192
3. Amniotic and Placental Cells for Therapy-A New Connection   193
4. Isolation and Characterization of Progenitor Cells   194
5. Differentiation of Amniotic and Placental-Derived Progenitor Cells   195
6. Conclusion   199
    References   199

10. Hematopoietic Stem and Progenitor Cells from Cord Blood

Hal E. Broxmeyer, Edward Srour, Christie Orschell, David A. Ingram, Scott Cooper,
P. Artur Plett, Laura E. Mead, and Mervin C. Yoder

1. Introduction: Cord Blood Transplantation   204
2. Methodologies for Assessing Hematopoietic Progenitor and Stem Cells,
    as Well as EPC, Present in Human Cord Blood   211
3. Uses of Above Assays   227
    References   229

11. Hematopoietic Stem Cells from Bone Marrow: Purification and Functional Analysis

K. K. Lin and M. A. Goodell

1. Introduction   238
2. Other HSC Surface Markers: Tie-2, Endoglin, and the SLAM
    Family Receptors   239
3. Fluorescent Dye Efflux in HSCs and the SP^{KLS} 239
4. Characteristics of HSCs from Different Purification Schemes   242

5. Protocol of HSC Sorting with Hoechst 33342 Staining (SP Population)   243
References   247

12. Microarray Analysis of Stem Cells and Their Differentiation

Howard Y. Chang, James A. Thomson, and Xin Chen

1. Introduction   250
2. Overview of Microarray Technology   250
3. Experimental Design   253
4. Confirmation Studies   270
5. Examples of Microarray Experiments for Stem Cell Biology
and Differentiation   271
6. Identification of "Stemness"   271
7. Differentiation   272
8. Stem Cell Niches   273
9. Future Directions   275
References   275

13. Tissue Engineering Using Adult Stem Cells

Daniel Eberli and Anthony Atala

1. Introduction   282
2. Biomaterials   284
3. Angiogenic Factors   286
4. Adult Stem Cells for Tissue Engineering   287
5. Conclusion   292
References   292

14. Tissue Engineering Using Mesenchymal Stem Cells

Jeremy J. Mao and Nicholas W. Marion

1. MSCs: Definition and Therapeutic Promise   298
2. Isolation and Expansion of MSCs   300
3. Multilineage Differentiation of MSCs   303
4. Clinical Translation of MSC-Based Therapies   310
5. Conclusions   312
References   313

SECTION II   Embryonic Stem Cells and Their Derivatives

15. Mouse Embryonic Stem Cells

Andras Nagy and Kristina Vintersten

1. Historical Overview   322
2. Factors Affecting the Efficiency of Mouse ES Cell Establishment   324

3. Factors Affecting the Contribution of Mouse ES Cell to Chimeric Embryos   325
4. Critical Events During Mouse ES Cell Establishment   326
5. Freezing of ES Cell Lines   330
6. Characterization   330
7. Protocols   332
    References   337

16. Human Embryo Culture

Amparo Mercader, Diana Valbuena, and Carlos Simón

1. Introduction   342
2. Human Embryo Development   342
3. Embryo Biopsy   345
4. Human Embryo Culture   347
5. Results   351
    References   354

17. Human Embryonic Stem Cells: Derivation and Maintenance

Hidenori Akutsu, Chad A. Cowan, and Douglas Melton

1. Introduction   360
2. Derivation of hES Cell Lines   360
3. Maintenance of hES Cells   365
4. Conclusion   368
    References   369

18. Human Embryonic Stem Cells: Characterization and Evaluation

Chunhui Xu

1. Introduction   374
2. Characterization of Undifferentiated hESCs   375
3. Conclusion   389
    References   389

19. Human Embryonic Stem Cells: Feeder-Free Culture

Michal Amit and Joseph Itskovitz-Eldor

1. Introduction   396
2. Methods for Feeder Layer-Free Culture of hESCs   401
    References   404

20. Neural Stem Cells, Neurons, and Glia from Embryonic Stem Cells

Steven M. Pollard, Alex Benchoua, and Sally Lowell

    1. Introduction   408
    2. Protocols   409
    3. Summary   421
    4. Media and Reagents   422
       References   424

21. Hematopoietic Cell Differentiation from Embryonic Stem Cells

Malcolm A. S. Moore, Jae-Hung Shieh, and Gabsang Lee

    1. Introduction   426
    2. Methods   436
       References   451

22. Cardiomyocyte Differentiation from Embryonic Stem Cells

X. Yang, X.-M. Guo, C.-Y. Wang, and X. Cindy Tian

    1. Introduction   460
    2. Embryoid Body Generation   461
    3. Cardiomyocyte Differentiation   470
    4. Enrichment of Cardiomyocytes   471
    5. Conclusion   473
       References   474

23. Insulin-Producing Cells from Mouse Embryonic Stem Cells

*Insa S. Schroeder, Sabine Sulzbacher, Thuy T. Truong, Przemyslav Blyszczuk,
Gabriela Kania, and Anna M. Wobus*

    1. Introduction   478
    2. Materials and Methods   485
    3. Results   493
    4. Summary   494
       References   496

24. Transgene Expression and RNA Interference in Embryonic Stem Cells

Holm Zaehres and George Q. Daley

    1. Retrovirus Expression Vectors and ESCs   502
    2. RNA Interference and ESCs   503
    3. siRNA Expression Vector Design   504
    4. Retrovirus Production   505

5. Retroviral and Lentiviral Gene Transfer into Mouse and Human ESCs   506
6. Transgene and siRNA Expression in Mouse and hESCs   507
7. Biotechnological and Medical Applications   511
   References   512

25. Lentiviral Vector-Mediated Gene Delivery into Human Embryonic Stem Cells

Michal Gropp and Benjamin Reubinoff

1. Introduction   519
2. Design of HIV-1 Based Vectors for Transduction of hESCs   523
3. Generation of Recombinant Viral Particles   527
4. Transduction of hESCs   529
5. Measurement of Transduction Efficiency   532
6. Enrichment for Transduced hESCs Expressing High Levels
   of the Transgene   533
7. Determination of Viral Titer   535
   References   536

26. Engineering Embryonic Stem Cells with Recombinase Systems

Frank Schnütgen, A. Francis Stewart, Harald von Melchner,
and Konstantinos Anastassiadis

1. Introduction   542
2. Site-Specific Recombination   543
3. Designing Substrates for Site-Specific Recombination   547
4. Generation of Conditional Alleles   553
5. Recombinase Mediated Cassette Exchange (RMCE)   558
6. Molecular Switches   559
7. Protocols   562
   References   572

27. Tissue Engineering Using Embryonic Stem Cells

Shahar Cohen, Lucy Leshanski, and Joseph Itskovitz-Eldor

1. Introduction   580
2. Special Considerations when using hESCs as the Cell Source for TE   581
3. Growing hESCs in Defined Animal-free Conditions   581
4. Obtaining the Desired Cell Population   582
5. Choosing the Right Scaffold   583
6. Scaling-Up a Regulable Bioprocess   583
7. hESC-Derived Connective Tissue Progenitors for TE   584
8. Culture and Maintenance of hESC on MEF Feeders   584
9. Harvesting Samples for Analyses   598
   References   593

Index   595

CONTRIBUTORS

Numbers in parentheses indicate the pages on which the authors' contributions begin.

Hidenori Akutsu (359), Department of Reproductive Biology, National Research Institute for Child Health and Development, Setagaya, Tokyo 157-8535, Japan

Michal Amit (395), Department of Obstetrics and Gynecology, Rambam Medical Center, Haifa, and The Bruce Rappaport Faculty of Medicine, Technion-Israel Institute of Technology, Haifa, Israel

Konstantinos Anastassiadis (541), Genomics, Technische Universitaet Dresden, BioInnovationZentrum, Am Tatzberg 47, 01307 Dresden, Germany

Anthony Atala (191, 281), Wake Forest Institute for Regenerative Medicine, Wake Forest University School of Medicine, Medical Center Boulevard, Winston-Salem, North Carolina 27157–1094, USA

Alex Benchoua (407), INSERM/UEVE U861861, ISTEM/AFM, 5 Rue H. Desbrueres, 91030 Evry, France, and Centre Development in Stem Cell Biology, Institute for Stem Cell Research, School of Biological Sciences, University of Edinburgh, Edinburgh EH9 3JQ, United Kingdom

Przemyslav Blyszczuk (477), Present Address: University Hospital, Experimental Medical Care, CH-4031 Basel, Switzerland

Ralph L. Brinster (87), Department of Animal Biology, School of Veterinary Medicine, University of Pennsylvania, Philadelphia, Pennsylvania 19104

Hal E. Broxmeyer (203), Departments of Microbiology and Immunology, Medicine, The Walther Oncology Center, Indiana University School of Medicine, Indianapolis, Indiana 46202, and the Walther Cancer Institute, Indianapolis, Indiana 46206

Joseph A. Brzezinski IV (53), Department of Biological Structure, University of Washington, Seattle, Washington

Howard Y. Chang (249), Program in Epithelial Biology, Stanford University School of Medicine, Stanford, California

Xin Chen (249), Department of Biopharmaceutical Sciences, University of California, San Francisco

Shahar Cohen (579), Stem Cell Center, Bruce Rappaport Faculty of Medicine, Technion—Israel Institute of Technology, Haifa, Israel 31096, Israel

Scott Cooper (203), Departments of Microbiology and Immunology, The Walther Oncology Center, Indiana University School of Medicine, Indianapolis, Indiana 46202, and The Walther Cancer Institute, Indianapolis, Indiana 46206

Chad A. Cowan (359), MGH Center for Regenerative Medicine, Cardiovascular Research Center, and Harvard Stem Cell Institute, Richard B. Simches Research Center, Boston, Massachusetts

George Q. Daley (501), Department of Biological Chemistry and Molecular Pharmacology, Harvard Medical School; Division of Pediatric Hematology/ Oncology, Children's Hospital Boston and Dana Farber Cancer Institute; Division of Hematology, Brigham and Women's Hospital; Harvard Stem Cell Institute, Boston, Massachusetts

Paolo DeCoppi (191), Wake Forest Institute for Regenerative Medicine, Wake Forest University School of Medicine, Medical Center Boulevard, Winston-Salem, North Carolina 27157–1094

Dawn M. Delo (191), Wake Forest Institute for Regenerative Medicine, Wake Forest University School of Medicine, Medical Center Boulevard, Winston-Salem, North Carolina 27157–1094

Yuval, Dor (149), Department of Cellular Biochemistry and Human Genetics, The Hebrew University-Hadassah Medical School, Jerusalem 91120, Israel

Ryan R. Driskell (113), Department of Anatomy and Cell Biology, College of Medicine, The University of Iowa, Iowa City, Iowa 52242

Daniel Eberli (281), Wake Forest Institute for Regenerative Medicine, Wake Forest University School of Medicine, Medical Center Boulevard, Winston-Salem, North Carolina 27157–1094, USA

John F. Engelhardt (113), Department of Anatomy and Cell Biology, College of Medicine, The University of Iowa, Iowa City, Iowa 52242

Andy J. Fischer (53), Department of Neuroscience, The Ohio State University, Columbus, Ohio

John D. Gearhart (165), Department of Obstetrics and Gynecology, Institute for Cell Engineering, Johns Hopkins University, Baltimore, Maryland

M. A. Goodell (237), Department of Pediatrics, Baylor College of Medicine, Houston, Texas, and Department of Immunology, Baylor College of Medicine, Houston, Texas, and Stem Cells and Regenerative Medicine Center, Baylor College of Medicine, Houston, Texas

Stan Gronthos (73), Mesenchymal Stem Cell Group, Division of Haematology, Institute of Medical and Veterinary Science, Adelaide, South Australia, Australia

Michal Gropp (517), The Hadassah Human Embryonic Stem Cell Research Center, The Goldyne Savad Institute of Gene Therapy and the Department of Gynecology, Hadassah University Medical Center, Jerusalem 91120, Israel

X.-M. Guo (459), Institute of Basic Medical Sciences, Tissue Engineering Research Center, Academy of Military Medical Sciences, Beijing, People's Republic of China

David A. Ingram (203), Department of Pediatrics, and The Wells Center for Pediatric Research, Indiana University School of Medicine, Indianapolis, Indiana 46202

Joseph Itskovitz-Eldor (395, 579), Department of Obstetrics and Gynecology, Rambam Medical Center, Haifa, Israel 31096, Israel, and Stem Cell Center, Bruce Rappaport Faculty of Medicine, Technion—Israel Institute of Technology, Haifa, Israel 31096, Israel

Gabriela Kania (477), Present Address: University Hospital, Experimental Medical Care, CH-4031 Basel, Switzerland

Candace L. Kerr (165), Department of Obstetrics and Gynecology, Institute for Cell Engineering, Johns Hopkins University, Baltimore, Maryland

Gabsang Lee (425), Moore Laboratory, Cell Biology Program, Memorial Sloan-Kettering Cancer Center, New York, NY, 10021

Lucy Leshanski (579), Stem Cell Center, Bruce Rappaport Faculty of Medicine, Technion—Israel Institute of Technology, Haifa, Israel 31096, Israel

He Liu (73), Peking University School of Stomatology, Beijing, China

Xiaoming Liu (113), Department of Anatomy and Cell Biology, College of Medicine, The University of Iowa, Iowa City, Iowa 52242

K. K. Lin (237), Department of Immunology, Baylor College of Medicine, Houston, Texas, and Stem Cells and Regenerative Medicine Center, Baylor College of Medicine, Houston, Texas

Sally Lowell (407), Centre Development in Stem Cell Biology, Institute for Stem Cell Research, School of Biological Sciences, University of Edinburgh, Edinburgh EH9 3JQ, United Kingdom

Jeremy J. Mao (297), Department of Biomedical Engineering, College of Dental Medicine, Columbia University, Fu

Nicholas W. Marion (297), Foundation School of Engineering and Applied Sciences, New York

Laura E. Mead (203), Department of Pediatrics, and The Wells Center for Pediatric Research, Indiana University School of Medicine, Indianapolis, Indiana 46202

Harald von Melchner (541), Department of Molecular Hematology, University of Frankfurt Medical School, Theodor-Stern-Kai 7, 60590 Frankfurt am Main, Germany

Douglas Melton (359), Department of Molecular and Cellular Biology, Howard Hughes Medical Institute, Harvard University, Cambridge, Massachusetts

Amparo Mercader (341), Instituto Universitario—Instituto Valenciano de Infertilidad, Plaza de la Policía local, 3. 46015, Valencia, Spain

Anna Milanesi (191), Wake Forest Institute for Regenerative Medicine, Wake Forest University School of Medicine, Medical Center Boulevard, Winston-Salem, North Carolina 27157–1094

Malcolm A. S. Moore (425), Moore Laboratory, Cell Biology Program, Memorial Sloan-Kettering Cancer Center, New York, NY, 10021

Andras Nagy (321), Mount Sinai Hospital, Samuel Lunenfeld Research Institute, Toronto, M5G 3H7, Canada, and Department of Medical Genetics and Microbiology, University of Toronto, Canada

Jon M. Oatley (87), Department of Animal Biology, School of Veterinary Medicine, University of Pennsylvania, Philadelphia, Pennsylvania 19104

Christie Orschell (203), Department of Medicine, Indiana University School of Medicine, Indianapolis, Indiana 46202

P. Artur Plett (203), Department of Medicine, Indiana University School of Medicine, Indianapolis, Indiana 46202

Steven M. Pollard (407), Wellcome Trust Centre for Stem Cell Biology, University of Cambridge, Cambridge CB2 1QR, United Kingdom

Prithi Rajan (23), Program in Stem Cells and Regeneration, Center for Neuroscience and Aging, Burnham Institute for Medical Research, La Jolla, California 92037

Thomas A. Reh (53), Department of Biological Structure, University of Washington, Seattle, Washington

Benjamin Reubinoff (517), The Hadassah Human Embryonic Stem Cell Research Center, The Goldyne Savad Institute of Gene Therapy, and the Department of Gynecology Hadassah University Medical Center, Jerusalem 91120, Israel

Brent A. Reynolds (3), Queensland Brain Institute, University of Queensland, Brisbane 4072, Australia

Rodney L. Rietze (3), Queensland Brain Institute, University of Queensland, Brisbane 4072, Australia

Seth J. Salpeter (149), Department of Cellular Biochemistry and Human Genetics, The Hebrew University-Hadassah Medical School, Jerusalem 91120, Israel

Frank Schnütgen (541), Department of Molecular Hematology, University of Frankfurt Medical School, Theodor-Stern-Kai 7, 60590 Frankfurt am Main, Germany

Insa S. Schroeder (477), *In Vitro* Differentiation Group, Leibniz Institute of Plant Genetics and Crop Plant Research (IPK), Gatersleben, Germany

Michael J. Shamblott (165), Department of Obstetrics and Gynecology, Institute for Cell Engineering, Johns Hopkins University, Baltimore, Maryland

Songtao Shi (73), Salivary Gland Disease Center and the Molecular Laboratory for Gene Therapy Capital Medical University School of Stomatology, Beijing, China

Jae-Hung Shieh (425), Moore Laboratory, Cell Biology Program, Memorial Sloan-Kettering Cancer Center, New York, NY, 10021

Minhaj Siddiqui (191), Wake Forest Institute for Regenerative Medicine, Wake Forest University School of Medicine, Medical Center Boulevard, Winston-Salem, North Carolina 27157–1094

Carlos Simón (341), Centro de Investigación Príncipe Felipe, Avda. Autopista del Saler, 16-3. 46013 Valencia, Spain, and Fundación IVI. Instituto Universitario—Instituto, Valenciano de Infertilidad Guadassuar, 1 bajo. 46015, Valencia, Spain

Evan Snyder (23), Program in Stem Cells and Regeneration, Center for Neuroscience and Aging, Burnham Institute for Medical Research, La Jolla, California 92037

Edward Srour (203), Department of Medicine, Indiana University School of Medicine, Indianapolis, Indiana 46202

A. Francis Stewart (541), Genomics, Technische Universitaet Dresden, BioInnovationZentrum, Am Tatzberg 47, 01307 Dresden, Germany

Sabine Sulzbacher (477), *In Vitro* Differentiation Group, Leibniz Institute of Plant Genetics and Crop Plant Research (IPK), Gatersleben, Germany

James A. Thomson (249), Wisconsin National Primate Research Center, Genome Center of Wisconsin and Department of Anatomy, Medical School, University of Wisconsin, Madison

X. Cindy Tian (459), Center for Regenerative Biology, University of Connecticut, Unit 4243, Storrs, CT 06269-4243

Thuy T. Truong (477), *In Vitro* Differentiation Group, Leibniz Institute of Plant Genetics and Crop Plant Research (IPK), Gatersleben, Germany

Diana Valbuena (341), Centro de Investigación Príncipe Felipe, Avda. Autopista del Saler, 16-3. 46013, Valencia, Spain

Kristina Vintersten (321), Mount Sinai Hospital, Samuel Lunenfeld Research Institute, Toronto, M5G 3H7, Canada

C.-Y. Wang (459), Institute of Basic Medical Sciences, Tissue Engineering Research Center, Academy of Military Medical Sciences, Beijing, People's Republic of China

Anna M. Wobus (477), *In Vitro* Differentiation Group, Leibniz Institute of Plant Genetics and Crop Plant Research (IPK), Gatersleben, Germany

X. Yang (459), Center for Regenerative Biology, University of Connecticut, Unit 4243, Storrs, CT 06269-4243

Chunhui Xu (373), Geron Corporation, 230 Constitution Drive, Menlo Park, California

Mervin C. Yoder (203), Department of Pediatrics, and The Wells Center for Pediatric Research, Indiana University School of Medicine, Indianapolis, Indiana 46202

Holm Zaehres (501), Department of Biological Chemistry and Molecular Pharmacology, Harvard Medical School; Division of Pediatric Hematology/Oncology, Children's Hospital Boston and Dana Farber Cancer Institute; Division of Hematology, Brigham and Women's Hospital; Harvard Stem Cell Institute, Boston, Massachusetts

PREFACE

Essential Stem Cell Methods is an abridgement and update of the three-volume stem cell series originally published in *Methods in Enzymology* in 2006: *Embryonic Stem Cells* (vol. 418), *Adult Stem Cells* (vol. 419), and *Stem Cell Tools and Other Experimental Protocols* (vol. 420). Stem cells are of great interest to scientists and clinicians due to their unique ability to differentiate into various tissues of the body. In addition to being a promising source of cells for regenerative medicine and drug discovery, they also serve as an excellent model of vertebrate development.

Growing stem cells in culture and differentiating them on demand requires specific skills and knowledge beyond basic cell culture techniques. We have tried to assemble the most robust and current techniques, including both "conventional" and novel methods in the stem cell field. The world's leading scientists with hands-on expertise were invited to write chapters on methods they are experts in or even established themselves. The book offers a variety of know-how from derivation to differentiation of stem cells, advances in differentiation of essential stem cells, and maintenance of their derivatives. Among them are derivatives of all three germ layers: cells of neural lineage, cardiomyocytes, hematopoietic, spermatogonial, pulmonary, and insulin-producing cells.

Each chapter is written as a short review of the field followed by an easy-to-follow set of protocols that would enable even the least experienced researchers to successfully establish the techniques in their laboratories. Such techniques include gene expression profiling, RNAi and gene delivery, embryo culture for human essential stem cell derivation, characterization and purification of stem cells, and tissue engineering using derivatives of stem cells.

We wish to thank the contributors for sharing their invaluable expertise in comprehensive and easy to follow step-by-step protocols which guide the reader through the challenges of stem cell isolation and differentiation.

Irina Kliminskaya, Ph.D.
Robert Lanza, M.D.

SECTION I

Organ-Derived Stem Cells

CHAPTER 1

Neural Stem Cells: Isolation and Characterization

Rodney L. Rietze and Brent A. Reynolds

Queensland Brain Institute
University of Queensland
Brisbane 4072, Australia

Abstract
1. Introduction
2. Reagents and Instrumentation
    2.1. Dissection equipment
    2.2. Tissue culture equipment
    2.3. Growth factors
    2.4. Media solutions
3. Methods
    3.1. Establishment of primary embryonic neurosphere cultures
    3.2. Establishment of primary adult neurosphere cultures
    3.3. Passaging neurosphere cultures
    3.4. Differentiation of neurosphere cultures
    3.5. Flow cytometric enrichment of adult NSCs
References

Abstract

Throughout the process of development and continuing into adulthood, stem cells function as a reservoir of undifferentiated cell types, whose role is to underpin cell genesis in a variety of tissues and organs. In the adult, they play an essential homeostatic role by replacing differentiated tissue cells "worn off" by physiological turnover or lost to injury or disease. As such, the discovery of such cells in the adult mammalian central nervous system (CNS), an organ traditionally thought to have little or no regenerative capacity, was most unexpected. Nonetheless, by employing a novel serum-free culture system termed the neurosphere assay (NSA),

Reynolds and Weiss demonstrated the presence of neural stem cells (NSCs) in both the adult (Reynolds and Weiss, 1992) and embryonic mouse brain (Reynolds and Weiss, 1996). Here, we describe how to generate, serially passage, and differentiate neurospheres derived from both the developing and adult brain, and provide more technical details that will enable one to achieve reproducible cultures, which can be passaged over an extended period of time.

1. Introduction

While originally debated, it is now clear that neurogenesis continues in at least two regions of the adult mammalian brain, namely, the olfactory bulb and hippocampal formation (Gross, 2000). This continuous and robust generation of new cells strongly argues for the existence of a founder cell with the ability to proliferate, self-renew, and ultimately generate a large number of differentiated progeny, that is, a stem cell (Potten and Loeffler, 1990). One of the difficulties in identifying and studying stem cells is their poorly defined physical nature, thereby affecting our ability to directly measure their presence and follow their activity. This problem has been overcome by defining stem cells based on a functional criterion such that stem cells, in general, are defined by what they do, not by what they look like. This creates a number of problems, both conceptual and practical, with the most obvious being that one must first force a stem cell to act in order to determine its presence, and hence, does the action of imposing an action accurately reflect the original or true nature of the cell in question. Clearly, what is needed is a specific selective positive marker that will allow us to definitely identify stem cells both *in vivo* and *in vitro*. In this review, we will discuss and detail a culture methodology that allows for the isolation, propagation, and identification of stem cells from the mammalian brain and provide practical advice on the use of flow cytometry to isolate a relatively pure population of putative stem cells.

While their presence was eluded to in a number of previous studies, the elucidation of the appropriate culture conditions which permitted the functional attributes of a stem cell to be demonstrated enabled the unequivocal demonstration of a neural stem cell, for the first time in 1992. To isolate and expand the putative stem cell from the adult brain, Reynolds and Weiss employed a serum-free culture system known as the Neurosphere assay (NSA) whereby the majority of primary differentiated CNS cells harvested would not be able to survive. While this system caused the death of the majority of cell types harvested from the periventricular region within 3 days of culture, it allowed a small population ($<0.1\%$) of epidermal growth factor (EGF)-responsive stem cells to enter a period of active proliferation, even at very low cell densities (Reynolds and Weiss, 1992). By using such a system, Reynolds and Weiss were able to demonstrate that a single adult CNS cell could proliferate to form a ball of undifferentiated cells they called a neurosphere, which in turn, could: (1) be dissociated to form more numerous secondary spheres, or (2) induced to differentiate, generating the three major cell types of the CNS. In doing so, they showed that the cell they had isolated exhibited the stem cell attributes of proliferation, self-renewal, and the ability to give rise to a number of differentiated, functional progeny (Hall and Watt, 1989; Potten and Loeffler, 1990). Subsequent studies have since demonstrated that by following a well-defined protocol, and using EGF, or basic fibroblast growth factor (bFGF), or both as mitogens, one could produce a consistent, renewable source of undifferentiated CNS precursors (a portion of which are stem cells), which could be expanded as neurospheres, or reliably differentiated into defined proportions of neurons, astrocytes, and oligo-dendrocytes (Gritti *et al.*, 1995, 1996, 1999; Reynolds and Weiss, 1996; Reynolds *et al.*, 1992; Weiss, 1996a,b).

More than 1000 citations to date which have employed the NSA, attest to the robust and reliable nature of the assay, its value in studying developmental processes, elucidating the role of genetic and epigenetic factors on the potential of CNS stem cells, and the determination of CNS phenotypes. While the method-ology seems relatively simple to carry out, a strict adherence to the procedures, described here, is required in order to achieve reliable and consistent results. Here, we describe in detail, the protocols for the isolation and culture of NSCs harvested from various regions of the embryonic and adult murine brain. These protocols assume a basic knowledge of murine brain anatomy. The reader is referred to O'Connor *et al.* (1998) for reference on this topic which is essential to perform the procedures for culturing murine NSCs outlined below.

2. Reagents and Instrumentation

2.1. Dissection equipment

Large scissors
Small fine scissors
Ultrafine spring microscissors (Fine Science Tools, Cat # 15396–01)
Small forceps (Fine Science Tools, Cat # 11050–10)
Small fine forceps (Fine Science Tools, Cat # 11272–30)
Ultrafine curved forceps (Fine Science Tools, 11251–35)
Bead sterilizer (Fine Science Tools, Cat # 250)
Dissection microscope

2.2. Tissue culture equipment

Flasks: 25 cm (Reynolds *et al.*, 1992) 0.2 μm vented filter cap (TPP Cat # 9026)
    75 cm (Reynolds *et al.*, 1992) 0.2 μm vented filter cap (TPP Cat # 90076)
    175 cm (Reynolds *et al.*, 1992) 0.2 μm vented filter cap (TPP Cat # 90151)
Tubes: 17 $\times$ 100 mm polystyrene, sterile (TPP Cat # 91015)
    50 ml polypropylene, sterile (TPP Cat # 91050)
    FACS tubes, sterile (Falcon Cat # 352054)
    Petri dishes: 100 mm, 35 mm (Nunc Cat # 351029, 174926)
Tissue sieve: 70 μm (Falcon Cat # 352350)

TC plates: 6-well, 24-well, 96-well (Falcon Cat # 353046, 353047, 353072)

8-well coated chamber slides: poly-d-lysine/laminin (BioCoat BD Cat # 35–4688)

8-well coated chamber slides: human fibronectin (Bio Coat BD Cat # 35–4631)

2.3. Growth factors

- EGF: human recombinant (Stem Cell Technologies Cat #02633). To a stock solution of 10 μg/ml, add 10 ml of hormone-supplemented neural culture media to each vial of EGF. Store as 100 μl aliquots at −20 °C.
- FGF2: human recombinant (Stem Cell Technologies Cat #02634). To a stock solution of 10 μg/ml, add 999 μl of hormone-supplemented neural culture media, 1 μl BSA to each vial of bFGF. Store as 100 μl aliquots at −20 °C.
- 0.2% Heparin: mix 100 mg heparin (Sigma Cat # H-3149) in 50 ml water. Filter sterilize. Store at 4 °C.

2.4. Media solutions

These cultures are extremely sensitive to contaminants present in water or glassware. If media is being made in the laboratory, use only tissue-culture-grade components. We strongly suggest you purchase as many components as possible, as this will minimize batch-to-batch inconsistencies and provide greater consistency of results, over all. Optimized reagents for the culture and differentiation of neurospheres are available from Stem Cell Technologies Inc. (www.stemcell.com).

2.4.1. Commercial media components

Phosphate buffered saline (D-PBS; Stem Cell Technologies, Cat # 37350)
Basal medium (NeuroCult™ NSC basal media, Cat # 05700)
10× hormone mix (NeuroCult™ NSC proliferation supplement, Cat #05701)
Differentiation medium (NeuroCult™ Differentiation supplement, Cat # 05703)
Preparation of complete NSC media is thoroughly described at http://www.stemcell.com/technical/manuals.aspx.

As with the in-lab preparation of media components described below, combining 450 ml of NeuroCult™ NSC basal media with 50 ml of NeuroCult™ NSC proliferation supplement will comprise the hormone-supplemented growth medium described below (Stock solutions—item 4). As below, complete NSC growth medium is achieved by the addition of EGF and/or bFGF.

2.4.2. Media preparation components

For the in-lab preparation of tissue culture media and hormone mix, a set of glassware to be used only for tissue cultures should be prepared. Bottles, cylinders, beakers, etc. should be properly rinsed several times with distilled water before being sterilized in an autoclave that is used for tissue culture purposes only. We strongly suggest that all media and stock solutions be prepared only in sterile disposable tubes and/or bottles, thereby avoiding contamination caused by cleaning solution residue or poor autoclaving techniques. Wherever possible, commercial stock solutions should also be employed.

1. **30% Glucose** (Sigma Cat # G-7021). Mix 30 g glucose in 100 ml distilled water. Filter, sterilize, and store at 4 °C.
2. **7.5% Sodium bicarbonate** (Sigma Cat # S-5761). Mix 7.5 g of NaHCO3 in 100 ml water. Filter, sterilize, and store at 4 °C.
3. **1 M HEPES** (Sigma Cat # H-0887). Dissolve 238.3 g HEPES in 1 l of distilled water. Store at 4 °C.
4. **3 mM Sodium selenite** (Sigma Cat #S-9133). Add 1.93 ml of distilled water to a 1 mg vial of sodium selenite. Mix, aliquot into sterile tubes, and store at −20 °C.
5. **2 mM Progesterone** (Sigma Cat # P-6149). Add 1.59 ml of 95% ethanol to a 1 mg vial of progesterone. Mix, aliquot into sterile tubes, and store at −20 °C.
6. **200 mM l-Glutamine** (Gibco Cat # 25030–024)
7. **Apotransferrin** (Serologicals Cat # 820056–1). Dissolve 400 mg of apotransferrin directly into 10× hormone mix solution.
8. **Insulin** (Roche Cat # 977–420). Dissolve 100 mg of bovine insulin in 4 ml of sterile 0.1N HCl, then add 36 ml of distilled water to this solution. Transfer entire volume to 10× hormone mix.
9. **Putrescine** (Sigma Cat # P-7505). Dissolve 38.6 mg of putrescine in 40 ml of distilled water. Transfer entire volume to 10× hormone mix.
10. **0.1% DNase-1** (Boehringer Mannheim Cat # 704159). Dissolve 100 mg of DNase-1 in 100 ml of HEM. Mix thoroughly, filter sterilize, aliquot into sterile tubes (1 ml/aliquot), then store −20 °C.
11. **Propidium iodide** (Sigma Cat # P4170)
12. **Trypsin** (Calbiochem Cat # 6502)
13. **Trypsin inhibitor** (Sigma T-6522). Combine 14 mg of trypsin inhibitor, 1 ml of 0.1% DNase-1, and 99 ml of HEM. Mix well, filter sterilize, and store at 4 °C for a maximum of 14 days.
14. **Minimum esssential medium** (Gibco Cat # 41500–018)

2.4.3. Stock solutions

1. **Preparation of 10× DMEM/F12:** Combine five 1-l packages of DMEM (Gibco]-Invitrogen Cat # 12100–046) and five 1-l packages of F12 powder (Gibco-Invitrogen Cat # 21700–075) in 1 l of water under gentle continuous stirring. Filter, sterilize, and store at 4 °C.

2. **Preparation of 10× hormone mix:** Combine individual components in the following order: (a) 300 ml ultrapure distilled water, (b) 40 ml of 10× DMEM/F12, (c) 8 ml of 30% glucose, (d) 6 ml of 7.5% NaHCO3, (e) 2.5 ml of 1 M HEPES.

Mix well, then add (a) 400 mg of apo-transferrin, (b) 40 ml of 2.5 mg/ml insulin stock, (c) 40 ml of 10 mg/ml putrescine stock, (d) 40 μL of 3 mM sodium selenite, and (e) 40 μL of 2 mM progesterone. Mix all components thoroughly, filter sterilize, then aliquot into 10 or 25 ml volumes in sterile tubes and store at -20 °C.

3. **Preparation of basal medium:** (for 450 ml) Combine individual components in the following order: (a) 375 ml of ultrapure distilled water, (b) 50 ml of 10× DMEM/F12 stock, (c) 10 ml of 30% glucose, (d) 7.5 ml of 7.5% NaHCO3, (e) 2.5 ml of 1 M HEPES, and (f) 5 ml of 20 nM L-Glutamine. Mix thoroughly, filter, sterilize, and store at 4 °C for a maximum of 3 months.

4. **Preparation of hormone-supplemented growth medium** (for 500 ml): Combine 50 ml of 10× hormone mix with 450 ml of basal medium, mix thoroughly, and store at 4 °C for a maximum of 1 week. Add 1 ml 0.2% heparin, 20 μl EGF, or/and 10 μl FGF2-stock (final concentration: 20 ng/ml EGF and 10 ng/ml FGF2).

5. **Preparation of complete NSC medium:** Add 2 μl of EGF for every 1 ml of hormone-supplemented growth medium and/or 1μl of bFGF and 1μl of heparin for every 1 ml of hormone-supplemented growth medium.

6. **Preparation of tissue dissociation medium:** (for 200 ml) Add 476 mg HEPES, 40 mg EDTA, 50 mg Trypsin, and 1 ml 0.1% DNase-1 to 200 ml Ca^{2+}/Mg^{2+} HBSS. Mix well, filter sterilize, then aliquot (3 ml/aliquot) and store at -20 °C.

7. **Preparation of Hanks Eagle Medium (HEM):** (for 8.75 l) Add contents of one 10l packet of Minimum Essential Medium to 3 l of distilled water in a 5-l flask. Combine 160 ml 1 M HEPES and 175 ml Penicillin/Streptomycin (1:50 dilution) to a separate flask containing 3 l of distilled water. Combine the contents of two flasks and set the pH to 7.2 with 10 M NaOH. Filter, sterilize, and aliquot into 100 ml portions. Store at 4 °C for maximum of 3 months.

2.4.4. Miscellaneous

10× PBS: without calcium, without magnesium, Gibco BRL, Cat. No. 14200–067

Penicillin/streptomycin: Gibco-BRL, Cat. No. 15140–114

Trypsin/EDTA: Sigma, Cat. No. E-6511

Matrigel: growth factor-reduced, Becton Dickinson, Cat. No. 40230

Laminin: Roche, Cat. No. 1 243 217

Poly-l-ornithine: Sigma, Cat. No. P-3655

Fetal bovine serum: Gibco BRL, Cat. No. 10106–151

3. Methods

3.1. Establishment of primary embryonic neurosphere cultures

Neurospheres have been generated from various regions of the embryonic CNS and from numerous strains of mice. As such, the protocol that we describe here has been made sufficiently broad so as to increase its applicability, yet most accurately reflects the methodology required to generate neurospheres from the lateral and medial ganglionic eminences of embryonic day 14 (E14) mice, as originally described by Reynolds *et al.* (1992).

3.1.1. Dissection of embryonic tissue

Mice (e.g., CD1 albino) are typically mated overnight, then separated the next morning, and checked for the presence of a gestational plug. This will count as embryonic day zero (E0). Alternatively, one can purchase time-pregnant animals from specialized animal care facilities. For the establishment of embryonic neurosphere cultures, we typically harvest pups at E14–E15 (note dissection of embryonic CNS is much easier at E15), sacrificing the mother in accordance with rules dictated by the animal ethics committee. Perform the dissection as quickly as possible (within 2 h), as tissue becomes soft and sticky over time and may be difficult to dissect. If you estimate that more than 2 h are required, remove and dissect 8–10 brains at a time, keeping the remaining embryos at 4 °C.

3.1.2. Setup

1. Add cold sterile HEM to two 100-mm sterile plastic Petri dishes.

2. Sterilize dissection tools immediately before use by using a glass bead sterilizer, or well in advance by autoclaving (120 °C for 20 min). Tools needed for the gross dissection include: large scissors, small pointed scissors, larger forceps, and small curved forceps. Ultrafine forceps and scissors will be used for the microdissection of CNS tissue.

3. Place gauze on the bottom of a small glass beaker, then fill with 70% ethanol. This is where forceps and scissors are stored during the dissection so as to reduce contamination.

4. Prepare a gross dissection area on a lab bench by laying several absorbent towels flat, then soaking the towels with 70% ethanol. Place gross dissection tools to the side.

5. Arrange the dissecting microscope, two Petri dishes containing HEM, and the ultrafine dissection tools within the laminar flow hood. As a precaution, keep some sterile Petri dishes and HEM ready at hand.

6. Warm up culture medium to 37 °C in a thermostatic water-bath.

3.1.3. Harvesting of embryonic brain tissue

1. Anesthetize the pregnant mother by an intraperitoneal injection of pentobarbital (120 mg/Kg), and upon deep anaesthesia, sacrifice mother by cervical dislocation.

2. Lay the pregnant mother on its back on the absorbent towels, and then liberally rinse the abdomen with 70% ethanol, so as to sterilize the area.

3. Grasp the skin above the genitalia using large forceps, then cut through the skin and fascia with large scissors, so as to expose the peritoneal cavity sufficiently to view the uteri.

4. Remove the uteri using small forceps and scissors and transfer them into a 100-mm dish, containing HEM. Typically, one can anticipate a litter size of 8–12 pups, however, only 2 or 3 are needed to establish a bulk culture (See notes 3 and 4). Ensure that tools are rinsed frequently in ethanol, so as to exclude fur. Upon completion of the dissection, dispose carcass immediately.

5. Transfer uterine tissue to a laminar flow hood, then rinse once or twice by placing them in 100-mm Petri dishes containing fresh sterile HEM.

6. Cut open the uterine horns then transfer the pups to a new 100-mm dish containing HEM using small forceps. At this point, check the age of the pups and discard those that appear malformed, or too small with respect to gestational age.

7. Separate the head(s) of the pups at the level just below the cervical spinal cord, discarding the skulls.

8. Transfer tissue culture dish to dissecting microscope, and under 10× magnification, begin to remove the brain by positioning the head side up and hold it from the caudal side at the ears using fine curved forceps. Use microscissors to cut a horizontal opening above the eyes and tease brain out of the opening by gently pushing on the head from the side opposite to the cut.

9. After removing all of the brains, increase the magnification (25×) then dissect out the desired brain region(s) to be used for establishing the culture. Typically the lateral and medial ganglionic eminences are removed, but refer to rodent brain atlas for details on how to dissect the specific areas.

10. Transfer harvested brain regions to a 15 ml falcon tube containing 2 ml of ice-cold HEM.

3.1.4. Establishing primary embryonic cultures

1. Several methods may be used to mechanically dissociate the dissected tissue, including fire polished glass pipette or 200 μl plastic tips together with a P200 Gilson pipette (which we routinely use). In either case, wet the plastic tip or glass pipette by sucking (and discarding) a small amount of sterile media, then proceed to triturate the tissue approximately 10 times, until a milky single cell suspension is achieved. Make sure to avoid generating air bubbles, as this reduces the number of viable cells and makes for an inefficient trituration. Also, the expulsion of cells during the trituration should not be too vigorous, as this will also significantly reduce viability.

2. If undissociated pieces of tissue are still present in the suspension following the initial trituration, wait 2 min, which will allow the undissociated cells and tissue to settle, then transfer the majority of the supernatant containing single cells into a

fresh tube leaving the undissociated tissue behind. Add an appropriate volume of complete NSC medium to the undissociated cells, so as to bring the total volume to 0.5–2 ml (depending on volume of tissue and method of dissociation). Repeat step 1.

3. Pool the two suspensions you have created, then centrifuge the resulting suspension at 800 rpm (110 g) for 5 min. Aspirate the supernatant, and then gently resuspend the cells to achieve a final 2 ml volume of complete NSC medium.

4. Combine a 10 μl-aliquot of the cell suspension with 90 μl of trypan blue in a microcentrifuge tube, mix, then transfer 10μl to a hemocytometer so as to determine the number of viable cells in the suspension.

5. For primary cultures, seed cells at a density of 2×10^6 cells per 10 ml (T-25 cm^2 flask) or 8×10 (Gritti *et al.*, 1995) cells in 40 ml media (T-175 cm^2 flask), in complete NSC Medium. Please note that the cell density for plating primary cells harvested directly from the E14 CNS is higher than that prescribed for subsequent subculturing conditions.

3.1.5. General comments

- Upon plating primary cells, individual cells will become hypertrophic and adhere to the substrate, while the majority of cells will either die or differentiate. Following 2–3 days in culture, proliferative cells will lift off the base of the tissue culture vessels. Aggregates of cells resembling neurospheres will most likely be observed within the first 48 h of culture. These should not be mistaken for primary spheres. The prevalence of aggregates is directly related to the amounts of debris and/or dead cells in the cultures. Typically, these pseudospheres are quite large, but are comprised of unusually small, phase-dark, and irregularly shaped cells.

- Bona fide neurospheres will appear phase bright and exhibit a somewhat spherical form to begin with, becoming more spherical as size increases. As shown in Fig. 1.1, small microspikes should be apparent on the outer surface of viable spheres at day 3.

- Primary neurospheres are often associated with cellular debris; however, subculturing will effectively select for proliferating precursor cells and remove cell aggregates, debris, and dead cells.

3.2. Establishment of primary adult neurosphere cultures

De novo neurogenesis has been reported to occur within discrete areas of the adult brain, namely the olfactory bulb, hippocampus, and the cortex. Here, we describe how to isolate adult murine NSCs and to establish continuous, stem cell lines by means of growth factor stimulation. This protocol can also be applied to rats, and implies the use of enzymatic predigestion, prior to mechanical dissociation. Note that while stem cells isolated from many different mice strains display

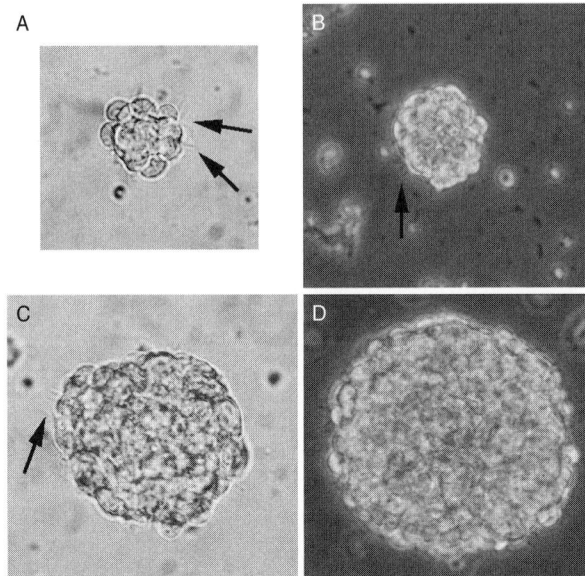

Figure 1.1 EGF-responsive murine neural stem cells, isolated from the E14 striatum were grown for 7 days in culture and then passaged. Small clusters of cells can be identified 2 days after passaging (A). The shape and opacity of the sphere, along with the presence of microspikes (arrows) assist in identifying a young, healthy neurosphere. Microspikes are still present in neurospheres after 3 DIV (B), and 4 DIV (C). By 6 DIV, the neurosphere is ready to be passaged. Magnification: 200×.

similar general features, differences regarding their growth rate and differentiation capacity may also be observed.

3.2.1. Setup

Sacrifice of animals, removal, and dissection of brain and/or spinal cord are performed outside the laminar flow hood. Particular caution should be exercised to avoid contamination. Have all the materials and instrumentation ready before starting the dissection procedure.

1. Add cold HEM to sterile plastic Petri dishes: one or two 100-mm dishes to hold tissue; several 60-mm dishes to wash tissues; some 35-mm dishes to hold dissected tissues.

2. Dissection tools can be sterilized in a hot bead sterilizer, in a preheated oven (250 °C for 2 h) or by autoclaving (120 °C for 20 min).

3. Select tools needed to remove brain and spinal cord (large scissors, small pointed scissors, large forceps, small curved forceps, and a small spatula) or for the tissue dissection (small forceps, curved fine forceps, small scissors, curved fine

scissors, scalpel). Immerse the two sets of tools in 70% ethanol in two beakers with gauze on bottom, to avoid spoiling the tips of the microforceps and scissors.

4. Warm culture medium and tissue dissociation medium to 37 °C in a thermostatic water-bath.

5. Begin the dissection.

3.2.2. Dissection of adult periventricular region

1. Anesthetize mice by intraperitoneal injection of pentobarbital (120 mg/Kg) and sacrifice them by cervical dislocation. Tissues from two or three mice (age: from 2 to 8 months) are generally pooled to start a culture.

2. Using large scissors cut off the head just above the cervical spinal cord region. Rinse the head with 70% ethanol.

3. Using small pointed scissors make a medial caudal-rostral cut and part the skin of the head to expose the skull. Rinse the skull with sterile HEM.

4. Using the skin to hold the head in place, place each blade of small scissors in orbital bone, so as to make a coronal cut between orbits of the eyes.

5. Using the coronal cut as an entry point, make a longitudinal cut through the skull along the sagittal suture. Be careful not to damage the brain by making small cuts ensuring the angle of the blades is as shallow as possible. Cut the entire length of the skull to the foramen magnum.

6. Using curved, pointed forceps grasp and peel the skull of the each hemisphere outward to expose the brain, then using a small wetted curved spatula, scoop the brain into a Petri dish containing HEM.

7. Repeat steps 1–6 until all of the brains have been harvested.

8. Wash brains twice by subsequently transferring them to new Petri dishes containing PBS.

9. To dissect the forebrain subventricular region, place the dish containing the brain under the dissecting microscope (10× magnification). Position the brain flat on its ventral surface and hold it from the caudal side using fine curved forceps placed on either side of the cerebellum. Use scalpel to make a coronal cut just behind the olfactory bulbs.

10. Following the removal of the olfactory bulbs, rotate the brain to expose the ventral aspect. Make a coronal cut at the level of the optic chiasm (Fig. 1.2A), discarding the caudal aspect of the brain.

11. Repeat steps 8–10 until all brains are sectioned.

12. Shift to a 25× magnification. Rotate the rostral aspect of the brain with the presumptive olfactory bulb facing downwards. Using fine curved microscissors, first remove the septum and discard, then cut the thin layer of tissue surrounding the ventricles, excluding the striatal parenchyma and the corpus callosum (Fig. 1.2B). Pool dissected tissue in a newly labeled 35-mm Petri dish.

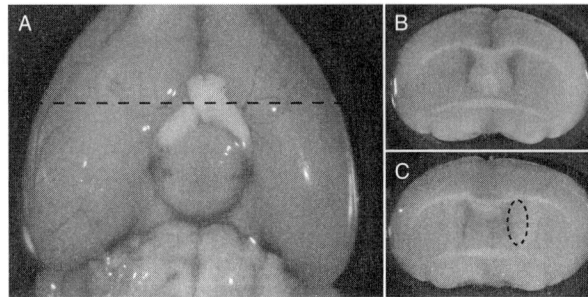

Figure 1.2 Ventral view of an adult C57Bl/6 mouse brain illustrating the rostral/caudal coordinate (dotted line) to section the brain coronally in order to harvest the rostral periventricular region of the lateral ventricles (A). Resulting coronal section when brain is sliced along the dotted line in panel A (B). Dotted line highlights the periventricular region that is harvested in a typical dissection following the removal of the septum (C).

14. Upon harvesting the periventricular regions from all brains, transfer dish to tissue culture laminar flow hood. Continue to use strict sterile technique.

3.2.3. Dissociation protocol

1. Using a scalpel blade, mince tissue for ~1 min until only very small pieces remain.

2. Using a filter tipped glass pipette and a total volume of 3 ml of tissue dissociation medium, transfer all of the minced tissues into the base of a 15 ml tube.

3. Incubate the tube for 7 min in a 37 °C water bath. Greater incubation times may be required, depending on the amount of tissue and on the overall size of the particles (larger pieces may be present due to inadequate mincing of the tissue).

4. At the end of the enzymatic incubation, return tube to hood, then add an equal volume of trypsin inhibitor (3 ml).

5. Avoiding the generation of air bubbles, mix well, then pellet the tissue suspension by centrifugation at 110 g for 7 min.

6. Discard virtually all of the supernatant overlaying the pellet, and then add an appropriate volume of HEM, so as to attain a final volume of 1 ml. Using a Gilson P1000 pipette (or similar) and a wetted 1000 μl filter-tip, begin to dissociate by triturating 1–2 times, then place the tip at the bottom of the tube, so as to restrict the flow of cells by ~50%, and continue triturating 5–7 times until the cell suspension takes on a milky or smooth appearance. Let the suspension settle for 3–4 min.

7. If many undissociated pieces of tissue are left, move cell suspension to a clean, labeled tube leaving about 100 μl behind. To the latter, add 900 μl of HEM and triturate again 5–7 times, until almost no undissociated pieces are left. Let the suspension settle down for 3–4 min. Transfer all but 100 μl of this tube to the labeled tube, thus pooling the cells from both trituration steps.

8. Bring the resulting cell suspension to a total volume of 14 ml by adding fresh HEM, then pass the suspension through a 70 μm sieve into a 15 ml tube, so as to remove debris or undissociated pieces, and then pellet the cells by centrifugation at 110 g for 7 min.

9. Remove virtually all of the supernatant, and resuspend the pellet in complete NSC culture medium, so as to bring the total volume of the resulting cell suspension to 0.5 ml.

10. Combine a 10 μl-aliquot from the cell suspension with 90 μl of Trypan blue in a microcentrifuge tube, mix, then transfer 10 μl to a hemocytometer so as to perform a cell count.

12. Seed cells at a density of 3500 viable cells/cm^2 in complete culture medium, in untreated 6-well tissue culture dishes (3 ml volume) or 25 cm^2-tissue culture flasks (5 ml volume).

13. Incubate at 37 °C, 5% CO2 in a humidified incubator.

14. Cells should proliferate to form spherical clusters that eventually lift off as they grow larger. These primary spheres should be ready for subculturing 7–10 days after plating, depending on the growth factors used.

Comments

• The 3 ml volume of tissue dissociation solution is sufficient for a good digestion of tissue from up to 8 mice. In the case of cell sorting where 8–16 mice are used, use a single 15 ml tube containing 3 ml of tissue dissociation solution for every 8 brains.

• In primary cultures from adult brain, a lot of debris is normally present, particularly in spinal cord cultures, together with adherent cells. To reduce debris, you may rinse tissue more frequently (steps 8–9). Generally, debris and adherent cells are eliminated after a couple of passages.

• Counting cells is sometimes difficult, due to the presence of debris, a large number of blood-derived cells and to the small number of CNS cells that can be isolated. In our experience, this protocol should yield about 5×10^4 cells from the subventricular region of one brain. Accurate quantification based on low cell counts of the CNS derived cells with a hemocytometer can be misleading. Thus, if quantification of the primary neural cell number is not to be carried out, a cell suspension derived from two mice may be plated in 4 dishes of a 6-well tissue culture dish, yielding an approximate final cell density of about 3500 cells/cm^2, or in one 25 cm^2-tissue culture flask, obtaining a final density of about 4000 cells/cm^2. Once competent with this procedure, one should expect to generate 400–600 neurospheres per mouse.

3.3. Passaging neurosphere cultures

As a rule of thumb, embryonic primary and passaged neurospheres should be ready for subculture between 4 and 5 days after plating, while adult primary and passaged neurosphere cultures should be ready for subculture 7–10 and 5–7 days after plating, respectively. However, one should monitor the cultures each day to ensure that neurospheres are not allowed to grow too large. Typically, a variety of diameters are apparent in a bulk culture. To determine whether spheres are ready to passage, the majority of neurospheres should equal 150 μm in diameter. If neurospheres are allowed to grow too large, they become difficult to dissociate and eventually, begin to differentiate *in situ*.

1. Observe the neurosphere cultures under a microscope to determine if the NSCs are ready for passaging. The average size of neurospheres across the culture should be ~150 μm. If neurospheres are attached to the culture substrate, forcefully strike the side of the tissue culture flask (attempting to minimize vessel movement by applying an equal force with the opposing hand).

2. Remove medium with suspended cells and place in an appropriate sized sterile tissue culture tube. If some cells remain attached to the substrate detach them by shooting a stream of media across the attached cells. Spin at 400 rpm (75 g), for 5 min.

3. Remove essentially 100% of the supernatant and resuspend cells using 1 ml of trypsin/EDTA, incubating at room temperature for 2 min in complete NSC medium (this volume allows for the most efficient trituration manipulations and is recommended for T-75 flasks). If more than 1 tube was used to harvest cultures, resuspend each pellet in 1 ml of trypsin/EDTA. If a T-175 flask is used, increase volume of trypsin/EDTA to 3 ml and incubate for 7 min.

4. Add an equal volume of trypsin inhibitor (as compared to trypsin/EDTA) to each tube, mix well, then centrifuge cell suspension(s) at 800 rpm (110 g) for 5 min.

5. Remove essentially 100% of the supernatant and resuspend cells by the addition of ~950 μL of complete NSC medium so as to produce a total volume of 1 ml. Using a Gilson P1000 pipette (or similar) and a wetted 1000 μl filter-tip, begin to dissociate by triturating 1–2 times, then place the tip at the bottom of the tube, so as to restrict the flow of cells by ~50%, and continue triturating 5–7 times until the cell suspension takes on a milky or smooth appearance.

6. Combine a 10 μl-aliquot from the cell suspension with 90μl of Trypan blue in a microcentrifuge tube, mix, then transfer 10μl to a hemocytometer so as to perform a cell count. If whole spheres appear, triturate cell suspension 2–3 times and recount.

7. Seed cells for the next culture passage in complete NSC medium at a density of 7.5×10 (Hall and Watt, 1989) cells/ml.

3.4. Differentiation of neurosphere cultures

When cultured in the presence of EGF and/or bFGF, NSCs and progenitor cells proliferate to form neurospheres which, when harvested at the appropriate time-point and using the appropriate methods as described here, can be passaged practically indefinitely. However, upon the removal of the growth factors and addition of a small amount of serum, neurosphere-derived cells are induced to differentiate into neurons, astrocytes, and oligodendrocytes (See Fig. 1.3). Overall, two methods have been described for the differentiation of neurospheres: as whole spheres cultured at low density (typically used to demonstrate individual spheres are multipotent) or as dissociated cells at high density (typically used to determine the relative percentage of differentiated cell types generated). The techniques for both methods are provided here.

3.4.1. Differentiation of whole neurospheres

If poly-l-ornithine coated coverslips are to be used, precoat glass slides by adding a sufficient volume of poly-l-ornithine (15 mg/ml) to completely cover the glass coverslip for a period of 2 h at 37 °C. Alternatively, 96-well plates can be

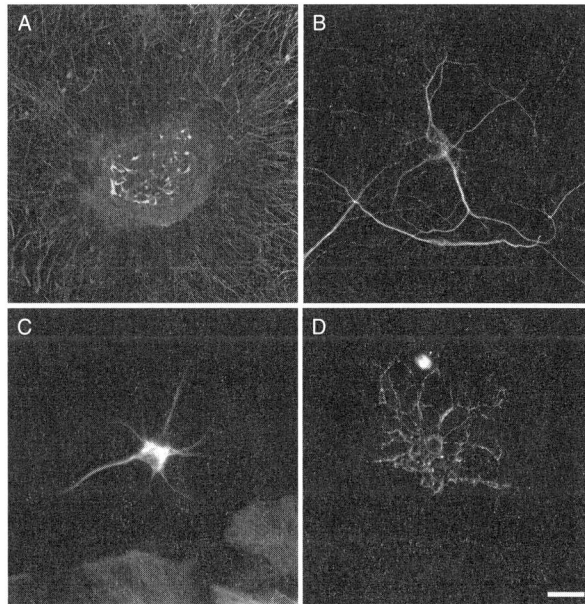

Figure 1.3 When transferred to differentiating conditions for 7 DIV, neurospheres will lose their spherical shape and flatten to essentially form a monolayer. The greatest concentration of cells will remain in the centre of the neurosphere (DAPI^{+ve} cells, blue), with astrocytes apparent throughout the sphere (GFAP, green), and neurons (B-tubulin, red) surrounding the core of the sphere lying on top of the astrocytes (A). Neurons are identified with a fluorescent label antibody raised against b-tubulin, a neuron specific antigen found in cell bodies and processes (B). Both protoplasmic and stellate astrocytes are identified with a fluorescent tagged antibody against the astrocyte specific protein GFAP (C). Oligodendrocytes are identified with an antibody against myelin basic protein (MBP) (D). Scale bar = 20 μm (B, C, D).

precoated with poly-l-ornithine. Aspirate poly-l-ornithine and immediately rinse 3 times (10 min each) with sterile PBS (do not allow coverslips or plate to dry). Remove PBS immediately prior to the addition of neurospheres and differentiation medium.

1. Once primary or passaged neurospheres reach 150 μm, (typically after 7–8 days *in vitro*), use percussion to remove adherent spheres, and then transfer contents of the flask to an appropriate sized sterile tissue culture tube. Spin at 400 rpm (75 g), for 5 min.

2. Aspirate essentially 100% of the growth medium, then gently resuspend (so as not to dissociate any neurospheres) with an appropriate volume of basal media + 1% sterile fetal calf serum. Note: an equal volume of commercially available NSC differentiation medium can also be used here (NeuroCult™ Differentiation supplement, Stem Cell Technologies, Cat # 05703).

3. Transfer neurosphere suspension to a 60-mm dish (or other sized vessel) to enable the harvesting/plucking of individual neurospheres with a disposable plastic pipette.

4. Transfer approximately 10 neurospheres using a sterile disposable plastic pipette or a Gilson P1000 pipette, and deposit into individual wells of 24- or 96-well tissue culture plate containing a poly-L-ornithine coated surface with NSC differentiation medium. Alternatively, commercially available, precoated chamber slides can be employed here.

5. After 6–8 days *in vitro*, individual neurospheres should have attached to the substrate and dispersed in such a manner, so as to appear as a flattened monolayer of cells.

6. Proceed to fix cells with the addition of 4% paraformaldehyde (in PBS, pH 7.2) for 10 min at room temperature and then process the adherent cells for immunocytochemistry as required.

3.4.2. Differentiation of dissociated cells

1. Once primary or passaged neurospheres reach 150 μm, (typically after 7–8 days *in vitro*), use percussion to remove adherent spheres, and then transfer contents of the flask to an appropriate sized sterile tissue culture tube. Spin at 400 rpm (75 g), for 5 min.

2. Remove essentially 100% of the supernatant and resuspend cells using 1 ml of trypsin/EDTA, incubating at room temperature for 2 min (this volume allows for the most efficient trituration manipulations). If more than 1 tube was used to harvest cultures, resuspend each pellet in 1 ml of trypsin/EDTA.

3. Add 1 ml of trypsin inhibitor to each tube, mix well, then centrifuge cell suspension(s) at 800 rpm (110 g) for 5 min.

4. Remove essentially 100% of the supernatant and resuspend cells by the addition of 1 ml of basal media +1% sterile fetal calf serum. Note: an equal volume of commercially available NSC differentiation medium can also be used here (NeuroCult™ Differentiation supplement, Stem Cell Technologies, Cat # 05703). Triturate cells until suspension appears milky and no spheres can be seen (~5–7 times).

5. Combine a 10 μl-aliquot from the cell suspension with 90 μl of Trypan blue in a microcentrifuge tube, mix, then transfer 10 μl to a hemocytometer, so as to perform a cell count.

6. Prepare the appropriate cell suspension in 1 ml of complete NSC differentiation media, so as to seed individual wells of 24-well tissue culture plate containing a poly-L-ornithine coated glass coverslip with 5×10^5 cells. Alternatively, commercially available, precoated chamber slides can be employed here, seeding wells at the same density.

7. After 4–6 days *in vitro*, neurosphere-derived cells will have differentiated sufficiently. Proceed to fix cells with the addition of 4% paraformaldehyde (in PBS, pH 7.2) for 10 min at room temperature and then process the adherent cells for immunocytochemistry as required.

3.5. Flow cytometric enrichment of adult NSCs

While approximately 1:300 cells harvested from the periventricular region of the adult mouse brain have the ability to form neurospheres, we have previously described a negative selection flow cytometric method by which NSCs can be greatly enriched (Rietze, 2001). This protocol essentially begins with the addition of Peanut Agglutinin (PNA) and Heat Stable Antigen (HSA or mCD24a) to a single cell suspension of adult cells, whose preparation is described in Section 3.2. This protocol has been established using CBA mice, but has been found to be applicable to many different mouse strains.

1. Harvest the periventricular region from 16 adult mice, processing as two separate samples (8 brains each), bringing both to a single cell suspension as described in Section 3.2. When combined, the total volume of the suspension should equal 400 μl.

2. Add 175 μl of complete NSC medium and 25 μl of the adult cell suspension to a total of four FACS tubes (labeled: (a) cells alone, (b) PI, (c) PNA-FITC, and (d) HSA-PE), these will serve as controls. Transfer the remaining 300 μl to a single FACS tube labeled "sort sample."

3. Add 2 μl of PNA-FITC to control tube (c), 1 μl of HSA-PE to control tube (d). Add 3 μl of PNA-FITC and 1.5 μl HSA-PE to the sort sample tube. Cap tubes and incubate on ice in the dark for 15 min.

4. Add 2.5 ml of NS to tubes (a), (c), and (d), while tube (b) receives 2.5 ml of PI rinsing solution. Add 5 ml of PI rinsing solution to the sort sample tube. Mix the contents of each tube using a pipette, then centrifuge at 110 g for 7 min.

5. Remove essentially 100% of the supernatant and resuspend each control pellet with 300 μl of complete medium, and the sort sample pellet with 2 ml of complete medium.

6. Bring FACS tubes to cytometer, using each of the control tubes to set the appropriate voltage and compensation. Voltages should be adjusted so that the forward versus side scatter pattern appear essentially as per Fig. 1.4A, and FITC/PE detectors as per Fig. 1.4C.

7. A triangle gate should be set first as shown in Fig. 1.4A, then a second gate set so as to exclude dead (PI-positive) cells from those included within the triangle gate (Fig. 1.4B).

8. NSCs are greatly enriched by selecting for the PNA^{lo}HSA^{lo} population as shown in Fig. 1.4C. Sorted cells should be collected in a 96-well plate containing 200 μl of complete NSC medium in each well. Given the low frequency of stem cells, a maximum of 20 wells are typically required to collect all of the PNA^{lo}HSA^{lo} population from the sort tube.

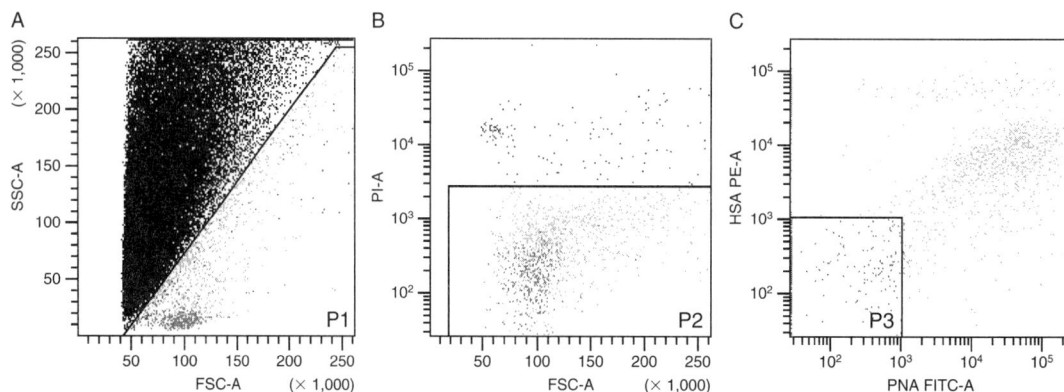

Figure 1.4 (A) Dot plot comparing the forward scatter (FSC-A) and side scatter (SSC-A) attributes of periventricular cells harvested from the rostral periventricular region. Selecting cells in population 1 (P1) excludes the majority of cellular debris without affecting the number of neurospheres generated. (B) Viable cells are distinguished from those cells contained within P1 in (A), by comparing FSC-A and propidium iodide intensity, and then gating for those cells within the propidium iodide negative population (P2). (C) Dot plot of viable periventricular cells comparing PNA and HSA staining intensities. Harvesting cells in the PNA^{lo}HSA^{lo} population (P3) will greatly enrich for stem cell activity.

Acknowledgments

The authors like to thank Dr. Preethi Eldi, Ms. Kristin Hatherley, and Dr. Dan Blackmore for their assistance in the preparation of this chapter.

References

Gritti, A., Cova, L., Parati, E. A., Galli, R., and Vescovi, A. L. (1995). Basic fibroblast growth factor supports the proliferation of epidermal growth factor-generated neuronal precursor cells of the adult mouse CNS. *Neurosci. Lett.* **185,** 151–154.

Gritti, A., Parati, E. A., Cova, L., Frolichsthal, P., Galli, R., Wanke, E., Faravelli, L., Morasutti, D. J., Roisen, F., Nickel, D. D., and Vescovi, A. L. (1996). Multipotential stem cells from the adult mouse brain proliferate and self-renew in response to basic fibroblast growth factor. *J. Neurosci.* **16,** 1091–1100.

Gritti, A., Frölichsthal Schoeller, P., Galli, R., Parati, E. A., Cova, L., Pagano, S, F., Bjornson, G. R., and Vescovi, A. L. (1999). Epidermal and fibroblast growth factors behave as mitogenic regulators for a single multipotent stem cell-like population from the subventricular region of the adult mouse forebrain. *J. Neurosci.* **19,** 3287–3297.

Gross, C. G., (2000). Neurogenesis in the adult brain: Death of a dogma. *Nat. Rev. Neurosci.* **1,** 67–73.

Hall, P. A., and Watt, F. M. (1989). Stem cells: The generation and maintenance of cellular diversity. *Development* **106,** 619–633.

O'Connor, T. J., Vescovi, A. L., and Reynolds, B. A. (1998). "Isolation and Propagation of Stem Cells from Various Regions of the Embryonic Mammalian Central Nervous System" pp. 149–153. Academic Press, London.

Potten, C. S., and Loeffler, M. (1990). Stem cells: Attributes, cycles, spirals, pitfalls and uncertainties. Lessons for and from the crypt. *Development* **110,** 1001–1020.

Reynolds, B. A., Tetzlaff, W., and Weiss, S. A. (1992). Multipotent EGF-responsive striatal embryonic progenitor cell produces neurons and astrocytes. *J. Neurosci.* **12,** 4565–4574.

Reynolds, B. A., and Weiss, S. (1992). Generation of neurons and astrocytes from isolated cells of the adult mammalian central nervous system. *Science* **255,** 1707–1710.

Reynolds, B. A., and Weiss, S. (1996). Clonal and population analyses demonstrate that an EGF-responsive mammalian embryonic CNS precursor is a stem cell. *Dev. Biol.* **175,** 1–13.

Rietze, R. L., *et al.* (2001). Purification of a pluripotent neural stem cell from the adult mouse brain. *Nature* **412,** 736–739.

Weiss, S., *et al.* (1996a). Is there a neural stem cell in the mammalian forebrain? *Trends Neurosci.* **19,** 387–393.

Weiss, S., *et al.* (1996b). Multipotent CNS stem cells are present in the adult mammalian spinal cord and ventricular neuroaxis. *J. Neurosci.* **16,** 7599–7609.

CHAPTER 2

Neural Stem Cells and Their Manipulation

Prithi Rajan and Evan Snyder

Program in Stem Cells and Regeneration
Center for Neuroscience and Aging
Burnham Institute for Medical Research
La Jolla, California 92037

Abstract
1. Introduction
2. Adult Niches for Stem Cells *In Vivo*
3. *In Vitro* Manipulation of NSCs
    3.1. Proliferation
    3.2. Survival/apoptosis
    3.3. Symmetric and asymmetric cell division
    3.4. Differentiation
4. Conclusions and Projections
5. Sample Protocols for the Culture and Characterization of NSCs
6. Culture of Rodent/Murine NSCs
    6.1. Preparation of dishes
    6.2. Isolation of NSCs
    6.3. Passaging of NSCs
    6.4. Freezing and thawing cells
    6.5. Reagents
7. Culture of human NSCs
8. ICC of markers to identify stem cells and differentiation products
References

Abstract

Extracellular signals dictate the biological processes of neural stem cells (NSCs) both *in vivo* and *in vitro*. The intracellular response elicited by these signals is dependent on the context that the signal is received, which in turn is decided by previous and concurrent signals impinging on the cell. A synthesis of signaling pathways which control proliferation, survival, and differentiation of NSCs *in vivo* and *in vitro* will lead to a better understanding of their biology, and will also permit more precise and reproducible manipulation of these cells to particular end points. In this review, we summarize the known signals which cause proliferation, survival, and differentiation in mammalian NSCs.

1. Introduction

Neural stem cells (NSCs) may be isolated from embryonic and adult brains, and are defined by their dual properties of self-renewal and their capacity to differentiate into the fates characteristic of the adult nervous system. The resident population of NSCs in the developing brain peaks prior to embryonic day 12–14 (E12–14) in the rat, and gradually diminishes due to differentiation into neurons initially, followed by astrocytes and then oligodendrocytes. Neurogenesis is maximal around E14, followed by gliogenesis, which peaks around E19 (Frederiksen and McKay, 1988; Caviness *et al.*, 1995). Neuronal architecture and glial differentiation continue to occur postnatally, especially synaptic pruning and myelination. NSCs are isolated with ease from all areas of the embryonic central nervous system, including cerebral cortex, hippocampus, striatum, mid-brain, including the substantia nigra, cerebellum, and spinal cord. In the adult there are structural zones to which these cells are restricted which will be discussed in relative detail below. NSCs have also been isolated from the neural crest and retina, and have been used as biological platforms to study the mechanisms of regulation of proliferation, survival, and differentiation. Neural crest stem cells (NCSCs) are isolated from chick, mouse, or rat neural tube explants and give rise to tissue derivatives of the neural crest, including neurons and glia from the peripheral nervous system and smooth muscle (Shah and Anderson, 1997). Retinal stem cells have been isolated from the ciliary margin zone of the adult eyes of amphibians and fish, and have recently been cultured from pigmented ciliary margin of mouse retinae too (Moshiri *et al.*, 2004; Tropepe *et al.*, 2000).

The most obvious practical advantage of NSCs is their potential to be an unrestricted source of neurons for replacement therapies. In addition to these transplantation therapies other important uses of stem cells lies in the creation of platforms for drug discovery (Rajan *et al.*, 2006). In order to make these processes more efficient by the successful manipulation of these cells, it is necessary to study the biology of NSCs in *in vitro* culture systems which are used to maintain and propagate them, and determine the signaling pathways which control the proliferation, differentiation, and survival of these cells in culture. Although the population of NSCs in the adult brain is appreciably less than in the embryonic brain, NSCs exist in localized regions called niches. The biology of the adult NSCs within their niches also need to be studied in detail for two reasons: in addition to replacement therapies where cells from allogeneic donors may be transplanted, resident stem cells may also be mobilized in order to harness their therapeutic potential. The use of fetal and adult NSCs for transplantation therapies may also benefit by knowledge of the *in vivo* survival and differentiation requirements of these cells.

In this review, we will attempt to summarize the signals controlling some aspects of NSC biology including the *in vivo* niches in which NSCs reside in the adult and during development in the embryo, and the signals which are known to orchestrate the proliferation, survival, and differentiation of NSCs *in vitro*. The inclusion of the details of each pathway is beyond the scope of this review, but may be found in several excellent reviews focusing on individual pathways (Bieberich, 2004; Blaise *et al.*, 2005; Cross and Templeton, 2004; Kleber and Sommer, 2004; Louvi and Artavanis-Tsakonas, 2006; McMahon, 2000; Polster and Fiskum, 2004; Sela-Donenfeld and Wilkinson, 2005; Stupack, 2005). Depending on the laboratory in which these studies were performed, the NSC cultures have been derived from rodent or human tissues and cultured under conditions, which vary in three-dimensional structure (monolayer or neurosphere) and culture additives (serum, B27, EGF, or LIF). Admittedly, these parameters confer significant differences in the NSC cultures generated. For example, neurospheres, which are NSC cultures which are maintained in suspended balls of cells, generate cultures of vastly different local density and extracellular matrix (ECM) composition when compared to NSCs generated in monolayers. In addition, NSC cultures prepared from tissue derived from identical brain regions but from different ages of embryos differ in their responses to growth factors attesting to the idea that the response elicited from a cell by an impinging ligand is affected by the context in which the signal is received, which in turn is dictated by cell-intrinsic and extracellular cues. For these contextual reasons, only those results which have been obtained in mammalian, and preferably neural-related, systems have been considered here. Finally, some conclusions have been drawn by over-expression of signaling proteins and transcription factors which may lead to spurious effects. However, in this review, we will consider data generated using all these paradigms in addition to *in vivo* studies to generate a heuristic model which will undoubtedly undergo refinements as our understanding of stem cell biology progresses.

2. Adult Niches for Stem Cells *In Vivo*

The adult mammalian brain has two prominent zones wherein stem cells reside. Stem cells are known to exist in the subventricular zone (SVZ) which extends anatomically around the ventricle in the cerebral cortex. These stem cells participate in the rostral-migratory stream (RMS) in rodents generating interneurons in the olfactory bulb; however, there is little evidence for the presence of a migrating stream similar to the RMS in humans (Sanai *et al.*, 2004). The second concentration of stem cells in the adult occurs in the subgranular zone (SGZ) in the hippocampus from which neurons are generated in the dentate gyrus

(Gage, 2000). Although the stem cells which reside in the SVZ were originally thought to comprise the layer of ependymal cells which line the ventricle (Johansson *et al.*, 1999), the current consensus suggests instead that astrocytes serve as functional stem cells (Alvarez-Buylla *et al.*, 2002), while the ependymal cells participate in regulating the niche which these cells occupy (Lim *et al.*, 2000). The current model suggests that the niche comprises three types of cells: the stem cell/astrocytes which contact the basal lamina and is capable of self-renewal (B cell), the preneuronal cells (C cell), and the young migrating neurons (A cells). B cells express GFAP, a mature astrocytic marker, while SSEA1 which is considered characteristic of rodent embryonic stem cells is present on a subset of astrocytes *in vivo* (Alvarez-Buylla and Lim, 2004). C cells may also be considered stem cells and respond to EGF as a mitogen *in vitro* to give rise to NSC cultures. A similar hierarchy of cells is present in the SGZ in the adult hippocampus. In this case, GFAP positive astrocytes give rise to neuronal precursors which mature into granule cells and populate the dentate gyrus.

In addition, there is some evidence for the presence of restricted progenitors distributed in the white matter of the cerebral cortex which have been designated white matter precursor cells (WMPCs) (Goldman and Sim, 2005). These are largely glial progenitors and less restricted multipotential progenitors which are scattered in the SVZ and throughout the parenchyma of the brain. WMPCs express PDGFRα and the A2B5 epitope and thus appear to be oligodendrocyte precursors (Scolding *et al.*, 1998, 1999), but have the capacity to generate all neural phenotypes when cultured *in vitro* (Nunes *et al.*, 2003). When sorted WMPCs were analyzed for their gene expression profiles they exhibited some characteristics of neural progenitor cells, such as HES1, musashi, doublecortin, and MASH1 (Goldman and Sim, 2005). Surprisingly, about 4% of the dissociated white matter from human surgical biopsies was shown to comprise these A2B5 positive cells.

Although the biology by which the WMPCs are maintained in the brain remains to be elucidated, a picture is emerging of the SVZ and SGZ niches, mostly by *in vivo* transgenic and gene ablation studies. The participation of ECM, basal lamina, blood vessels, and the paracrine effects of the cell themselves are important for the maintenance of these niches *in vivo*. The ligands, which are important for the maintenance, proliferation, differentiation, and motility of these cells include Notch/jagged, bone morphogenetic protein (BMP)/noggin, transforming growth factor alpha (TGFα), vascular endothelial growth factor 2 (VEGF2), Ephrins/ Ephs, sonic hedgehog (shh), and soluble amyloid precursor protein (sAPP), slit1, and slit2 (Alvarez-Buylla and Lim, 2004; Conti and Cattaneo, 2005; Hu *et al.*, 1996; Wu *et al.*, 1999). Most interestingly, noggin appears to be expressed by the ependymal cells, which inhibits BMP mediated astrocyte differentiation, thus maintaining the neurogenic niche. In the SVZ, the secreted fragment of β-amyloid precursor protein (sAPP) has been shown to be a mitogen. This is an interesting function of this protein, which is involved in the pathogenesis of Alzheimer's disease (Caille *et al.*, 2004). The polycomb transcription factor bmi1 and the forebrain restricted orphan nuclear receptor TLX are also thought to participate in the maintenance of the stem cell population in the developing, post natal, and adult brain (Leung *et al.*, 2004; Roy *et al.*, 2004; Shi *et al.*, 2004; Zencak *et al.*, 2005).

Not surprisingly, it is largely the same players which appear in studies of cultured NSCs. NSCs have been successfully cultured from various brain regions and ages. The aspiration of the field is that these cultures will be sufficiently reproducible and predictable across researchers that one may arrive at a desired end point, be it the large scale expansion of NSCs or a precursor, or the complete maturation of a particular subset of neurons or glia.

3. *In Vitro* Manipulation of NSCs

3.1. Proliferation

The successful culture of any cell type is a result of the processes of proliferation and survival. Proliferation may be further qualified as symmetrical or asymmetrical cell division, while the regulation of apoptosis may be included in cell survival (Fig. 2.1). Manipulation of the speed and length of the cell cycle appears to be critical for the nature of cells which result. Cell divisions which have an abridged G1 phase result in the proliferation of undifferentiated stem cells, while a progressive increase in the length of G1 facilitates the onset of differentiation (Calegari and Huttner, 2003).

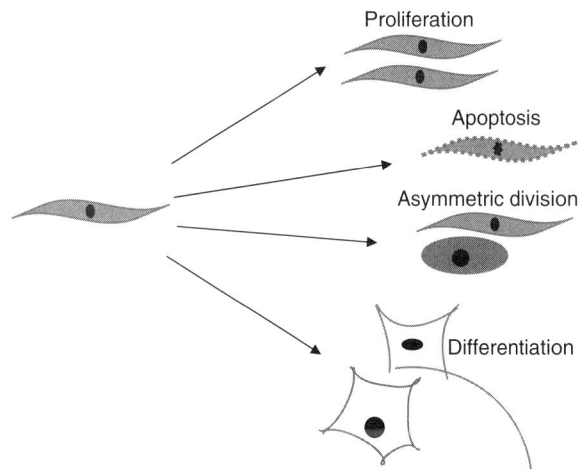

Figure 2.1 Processes which are regulated in NSC biology. The processes of proliferation, survival/apoptosis, and differentiation comprise the major biological processes of a NSC. Proliferation may lead to asymmetric division, which may result in the generation of two similar or different progeny, or death of one of the daughter cells. Proliferation and the signals involved are described in Figs. 2.2 and 2.3, while differentiation and its signals are described in Figs. 2.4 and 2.5.

Proliferation of NSCs in culture is most commonly controlled by the addition of basic fibroblast growth factor (bFGF) and epidermal growth factor (EGF). Implicit in the culture conditions is also the presence of ECM molecules, such as fibronectin, laminin, collagen, or vitronectin if serum is used, and ECM molecules which is secreted by the cell themselves. The self-secreted ECM is especially true in the case of neurospheres where no other extraneous ECM molecule is added. The involvement of FGF and EGF normally alludes to the activation of the mitogen activated protein kinase (MAPK) pathway. However, as seen in Fig. 2.2 there is evidence for the involvement of other protein and lipid signal mediators. In addition to MAPK activation, bFGF is also known to increase levels of active β catenin which is an important mediator of what is commonly called the canonical signaling pathway of the wnt ligand, and appears to maintain NSCs in a proliferative state when cultured as neurospheres (Israsena et al., 2004; Viti et al., 2003). Wnt, a signal upstream of β-catenin, has been shown to regulate the proliferation of a subset of neural progenitor cells in mouse forebrain and chick spinal cord, where the expression of Wnt-1 or stabilized β-catenin results in increased numbers of neural progenitors along with decreased neuronal differentiation (Chenn and Walsh, 2002; Megason and McMahon, 2002; Sommer, 2004). Along with wnt, notch, and shh also regulate NSC proliferation. The response elicited by notch signaling appears to be dependent on the levels of expression of notch and its ligands within the tissue. The mammalian ligands of notch include delta and jagged (Louvi and Artavanis-Tsakonas, 2006). The effects mediated by notch in response to its ligands are regulated by the relative expression of the ligand in adjacent cells. When there is high levels of delta in all cells, equal intensities of signaling between all the cells causes proliferation (Akai et al., 2005; Jessell and Sanes, 2000). However, an imbalance in notch signaling in adjacent cells, for example, by modulation of levels of delta expression in adjacent cells by intrinsic or extrinsic signals, leads to an instructive notch signal (Artavanis-Tsakonas et al., 1999). FGF causes induction of delta expression through the action of CASH in experiments performed in the chick (Akai et al., 2005), and thus precipitates notch function through Hes 1, Herp, and other similar transcriptional regulators. Lipid signaling is also an integral modulator of growth factor signaling and the ceramide GM1 is a cofactor of the FGF receptor (Rusnati et al., 2002). Although FGF appears to be an integral regulator of proliferation in culture, it is not frequently associated with proliferation in vivo, and may preferentially regulate the proliferation of these cells in culture. Similarly, EGF has been shown to be a mitogen in vitro for the "C" cells present in the SVZ (described in the previous section) (Doetsch et al., 2002). Although there is no evidence for the presence of EGF or bFGF in vivo, TGFα is present and it functions through the EGF receptor. It is possible that it is a ligand that mediates the in vivo proliferation of NSCs. In fact, the TGFα null mice do show the expected deficit of reduction of proliferation of NSCs in the SVZ (Tropepe et al., 1997).

Shh is thought to affect proliferation of precursors and in keeping with this observation is mutated frequently in medulloblastomas and gliomas

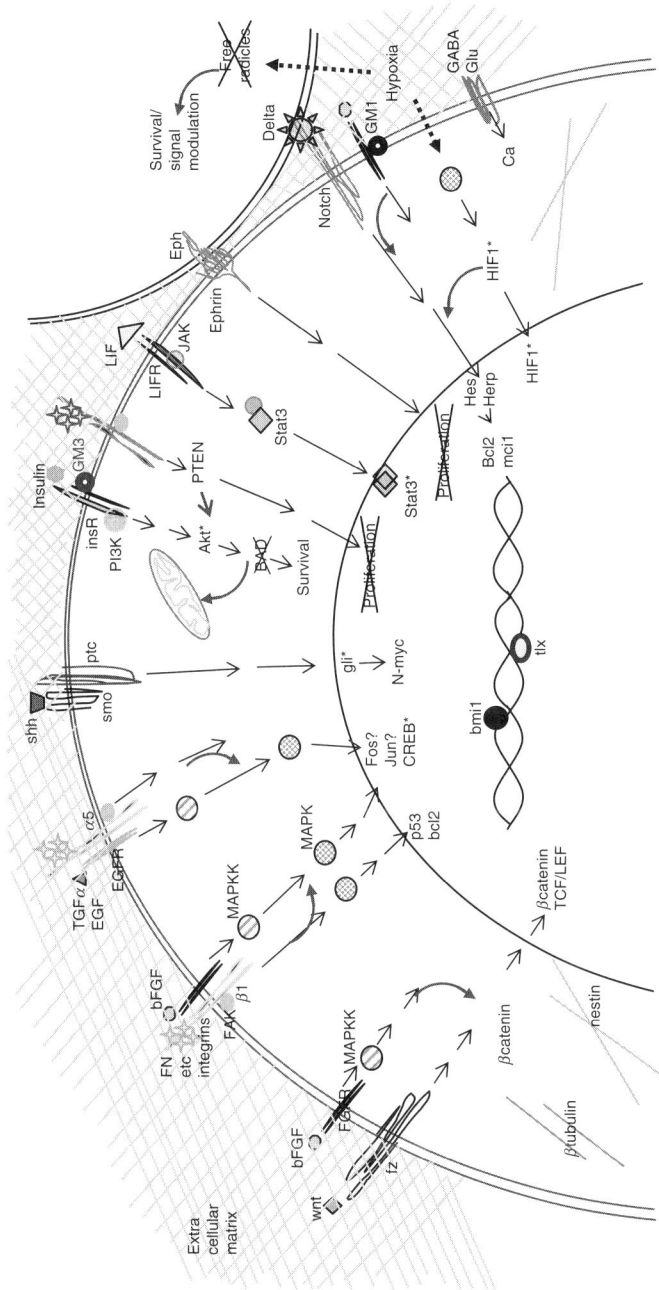

Figure 2.2 **Signals regulating proliferation of NSCs.** Extracellular signals which regulate the proliferation of NSCs may be soluble ligands, ECM molecules, and immobilized ligands present on neighboring cells. Interactions occur between intracellular signals generated by the growth factor receptor tyrosine kinases (RTKs) and the integrin receptors which respond to ECM molecules. Interactions also occur between the MAP kinase pathway and wnt signals, and between hypoxia induced signals and notch. Stat3 signals are activated by LIF, while sonic hedgehog possibly acts through the activation of myc in the nucleus. The epigenetic modifiers bmi1 and tlx are also involved in proliferation of NSCs. In addition to protein signaling intermediates, lipid signals are also involved especially at the level of receptors at the cell surface (FGF and insulin receptors). It is interesting to note that almost all the transcription factors which are invovled (TCF/LEF, Stat3, N-myc, p53, and bmi1), and proteins which are upregulated as a result of the signals (bcl2, mci) are associated with oncogenesis (* denotes an activated transcription factor. See text for details and references).

(Sanai *et al.*, 2005). Shh functions by activating Gli based transcriptional transactivation through a complex series of "double-negative" interactions with its receptor smoothened/patched (McMahon, 2000; Sanai *et al.*, 2005). It relieves transcriptional repression on Gli targets imposed by patched and causes activation of the same genes. shh effects the proliferation of precursors by inhibition of Rb, the retinoblastoma protein, and induction of N-myc expression (Kenney and Rowitch, 2000; Sjostrom *et al.*, 2005). Leukemia inhibitory factor (LIF) is also considered a mitogen in some NSC neurosphere culture systems, including NSCs derived from early mouse neurectoderm and human mesencephalon (Hitoshi *et al.*, 2004; Jin *et al.*, 2005). Surprisingly, it has been shown by several groups to cause astrocytic differentiation in rodent NSCs cultured in monolayers, reiterating the fact that the biological effects elicited by factors are context, and species, dependent. EphB1–3, EphA4 are present on cells of the SVZ and ephrinB ligands are associated with SVZ astrocytes. The Eph receptors signal via tyrosine kinase domains in their intracellular domains, and the ephrins themselves can signal through the tyrosine residues present in their intracellular domains (Holland *et al.*, 1996). This bi-directional signaling may be important for the demarcation of boundaries (Mellitzer *et al.*, 1999). Infusion of truncated EphB2 or ephrinB2 into the lateral ventricle resulted in an increase in cell proliferation particularly astrocytes, and blocks migration of cells from the SVZ (Conover *et al.*, 2000). Alternatively, ephrinA/EphA receptor regulates apoptosis in the developing brain and thus brain size and cell number (Depaepe *et al.*, 2005). EphrinA2 reverse signaling induced by EphA7 negatively regulates proliferation in the adult mouse brain, a disruption of this interaction causes increase proliferation of progenitors and increased neurogenesis (Holmberg *et al.*, 2005).

In addition to soluble and immobile factors discussed above, neurotransmitters are also mitogenic in the earlier stages of CNS development. GABA and glutamine appear before the onset of synapse formation during CNS development, and regulate proliferation of the precursor cell population depending on the stage of development, and the paradigm studied (Haydar *et al.*, 2000). GABA has been shown to inhibit the proliferative effect of bFGF on cortical progenitor cells (Antonopoulos *et al.*, 1997). These neurotransmitters modulate DNA synthesis and calcium concentration within the cytoplasm of NSCs (Haydar *et al.*, 2000; LoTurco *et al.*, 1995; Owens *et al.*, 2000). Although initially used in hematopoietic stem cell cultures, the biological basis for the culture of various stem cells including NSCs in low oxygen conditions is being increasingly understood. Hypoxia causes a general decrease in free radicals in the cell and thus decreases the amount of apoptosis which may occur in the culture (Cross and Templeton, 2004). It also causes the induction of a heterodimeric transcription factor hypoxia inducible factor 1 (HIF1) which causes an increase in the proliferation of undifferentiated cells (Gustafsson *et al.*, 2005). HIF1 interacts with the notch pathway to cause transcription of notch regulated genes which are involved in the maintenance of the undifferentiated state.

3.2. Survival/apoptosis

Apoptosis plays a central role in NSC biology. *In vivo*, the development of the mammalian CNS undergoes phases of regulated cell death including one occurring in early embryonic development which is thought to regulate the number of NSCs (De Zio *et al.*, 2005). Mechanistically, reduction in NSC apoptosis was evidenced by knocking out the expression of Apaf1, caspase 3–9, and EphA7 (Depaepe *et al.*, 2005; Kuan *et al.*, 2000; Kuida *et al.*, 1996). In addition, experiments with null mice suggest that the "intrinsic pathway'" of apoptosis is involved in cell death during synaptic development in the CNS (Chang *et al.*, 2002; Polster and Fiskum, 2004). This involves formation of active caspase 9 in the presence of Apaf1 and cytochrome C, which then leads to the activation of caspase 3. Although it has not been shown in stem cells per se, insulin mediates the survival of cells in culture via the Akt which phosphorylates and causes degradation of the proapoptotic protein BAD (Datta *et al.*, 1997, 1999). Akt also regulates the activity of the forkhead transcription factors and prevents it from transcribing genes which induce apoptosis including the FAS ligand (Brunet *et al.*, 1999). Interestingly, the GM3 ceramide is involved in down-regulating signals from the insulin receptor (Yamashita *et al.*, 2003). In addition, by extrapolation of experiments performed in related systems, it is possible that the MAPK pathway functions in NSC survival via the regulation of BAD (Bonni *et al.*, 1999). Notch is also involved in regulating survival of NSCs and appears to work by increasing the expression of the prosurvival genes bcl-2 and mcl-1 (Oishi *et al.*, 2004).

Integrins are also major effectors of cell survival, and are included in the "dependence receptors'" which, generally speaking, mediate survival in the presence of their respective ligand, and mediate apoptosis in its absence. The family includes the receptors DCC (deleted in colon cancer), RET (rearranged during transfection), and the integrin receptors (Stupack, 2005). There are 18α and 8β subunits which form at least 24 heterodimeric integrin receptors which mediate signals from a limited set of ECM molecules which form their cognate ligands (Hynes, 2002). The integrin receptor itself mediates MAPK signals via kinases such as focal adhesion kinase (FAK), but also mediates pathways related to cell survival including Akt activation, p53activation, and bcl2 expression (Matter and Ruoslahti, 2001; Stromblad *et al.*, 1996; Zhang *et al.*, 1995). In addition to eliciting these signals, the integrin receptors are also complexed to growth receptors in the cell, thus making EGF and PDGF signaling, among others, integrin dependent (Giancotti, 1997; Schneller *et al.*, 1997). The effects of integrin signaling are this far reaching, and may not be limited to survival alone. The $\beta 1$ intergrin subunit has been shown to be essential for NSC proliferation and survival (Leone *et al.*, 2005), possibly in combination with $\alpha 5$ and $\alpha 6$ subunits, respectively (Jacques *et al.*, 1998). Interestingly, it has recently been shown that PTEN a phosphatase which regulates Akt and functions downstream of integrins, negatively regulates NSC proliferation by affecting cell cycle entry at G0-G1 (Groszer *et al.*, 2006) again demonstrating the interaction between pathways related to cell proliferation and survival.

3.3. Symmetric and asymmetric cell division

Cell division in stem cells can result in two identical daughter cells which arise due to symmetric division, or two disparate progeny may be arrived at by asymmetric cell division. Symmetric division may give rise to two stem cells which are identical to the progenitor, or could give rise to two differentiated cells which are identical. Asymmetric division may give rise to one daughter cell which is differentiated while the other remains a multipotent stem cell, or may lead to the death of one cell thus maintaining total cell number within the environment. One of the first demonstrations of asymmetric cell division was by Chenn and McConnell where they elegantly demonstrated that notch is sequestered to the cells of the dividing SVZ during development of the CNS, while the other daughter cell becomes a neuron (Chenn and McConnell, 1995). Numb and Numb-like are PTB-domain proteins which are thought to negatively regulate notch signaling by causing receptor turnover of the notch protein. The loss of these proteins may cause an increase in the neuronal progenitor population due to an increase of symmetric cell divisions (Castaneda-Castellanos and Kriegstein, 2004; Li et al., 2003).

NSCs are derived from the neuroepithelial layer of the developing and developed neural tube, and are thus epithelial in nature. This implies the presence of a basement membrane and an apical surface for all cells. The most obvious manner of asymmetric division in such cells would entail a horizontal axis for cytokinesis, where one daughter cell inherits the basement membrane and all its associated characteristics, while the other daughter inherits the apical membrane. However, a large proportion of asymmetric divisions appear to undergo cytokinesis in the vertical axis which is not at perfectly perpendicular to the axis of the basement membrane of the cell, thus conferring one daughter cell with a larger portion of the apical membrane than the other (Huttner and Kosodo, 2005) (Fig. 2.3). This appears to be a method for the specification of radial glia fate (which inherits the apical membrane) versus neuronal fate in the mammalian CNS (Kosodo et al., 2004). The biochemical signals which are sequestered in the apical membrane are not clear, however, it is possible that lipid rafts are involved in anchoring intercellular components (Bieberich, 2004) in a manner similar to numb and prospero as shown in *Drosophila* (Clevers, 2005). In addition to these physical parameters, some genes have also been identified which are involved in the specification of polarity axes and positioning of the spindle. The mammalian homologue of the *Drosophila* lethal giant larvae gene (Lgl1) is essential for the localization of numb and lack of function leads to hyperproliferation of the neuroepithelium and radial glial cells, and tissue disorganization *in vivo* (Klezovitch et al., 2004). Similarly Lkb1 is a protein kinase which was originally isolated as a tumor suppressor gene and is known to regulate polarity in mammals (Baas et al., 2004). Nde1, a centrosome protein, and ASPM (abnormal spindle like microcephaly associated) are thought to function in the positioning of the mitotic spindle (Bond et al., 2003; Feng and Walsh, 2004). Loss of their respective functions lead to increased

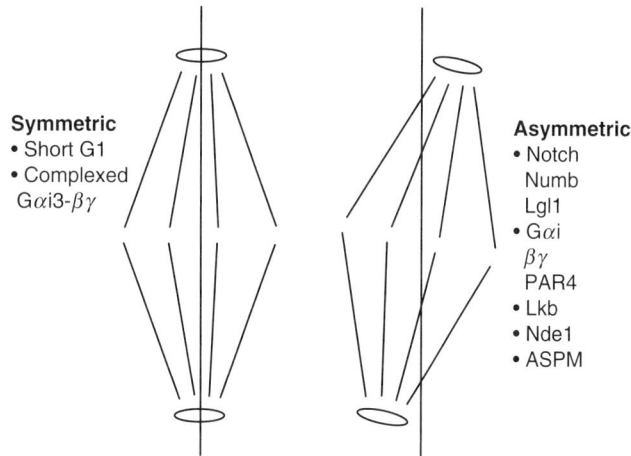

Figure 2.3 Regulation of symmetric and asymmetric division of NSCs. A schematic representation of the spindle apparatus of a dividing cell and its orientation relative to the longitudinal axis of the cells denotes the phenomena of symmetric and asymmetric cell division. The genes which have been shown to be involved in these processes in NSCs are listed.

horizontal/oblique axes of cell divisions causing a depletion of NSCs, and a concomitant increase of early neurogenesis.

Ceramide is also involved in signals which orchestrate asymmetric cell division, and the death of one of the progeny. While ceramide itself partitions normally in the dividing neural progenitor cells PAR4 segregates to only one of the progeny, PAR4 being a PKCζ inhibitor which sensitizes cells to ceramide induced apoptosis possibly through a p53-mediated mechanism (Bieberich *et al.*, 2003; Roussigne *et al.*, 2003; Wang *et al.*, 2005). Lipid signaling may also be involved in other aspects of interaction between nuclear and cytoskeletal proteins. Tubulin and other intermediary filaments such as nestin also localize asymmetrically during cell division as shown in embryoid body derived stem cells, and it is possible that interaction with ceramide is involved in these interactions (Bieberich, 2004). Since ceramide is present in membranes, it is likely that the intracellular membranes are in contact with the mitotic spindle and intermediary filaments thus alluding to a tertiary level of interactions during cell cycle which needs further investigation. G-protein signals are also intertwined with the lipid mediated signals described and may be involved in the positioning of the mitotic spindle. When $G\alpha_{i3}$-$\beta\gamma$ is present as a heterotrimer majority of the cleavage planes are vertical, while non-vertical cleavage planes result when $\beta\gamma$ is free to interact with downstream effectors (Sanada and Tsai, 2005). On the other hand $G\alpha_i$, in a manner similar to ceramide, binds tubulin via LGN and NUMA (Du and Macara, 2004). The exact mechanism by which these entities regulate the cleavage plane remains to be elucidated.

The length of the cell cycle is another parameter which is manipulated for manifestation of asymmetric cell division and the resulting differentiation. Cells which have a rapid pace of cell division, that is, a short G1, will not result in asymmetric cell divisions even if they have inherited cell components asymmetrically (Huttner and Kosodo, 2005). The lengthening of the cell cycle is required for the specification of progenitors and definitive neurons from NSCs (Cai *et al.*, 2002; Calegari and Huttner, 2003). One mechanism that has been proposed for the lengthening of G1 is the accumulation of ceramide in the cell during G0-G1. Ceramide accumulation may lead to quiescence or apoptosis in the extreme case possibly involving the Rb protein, but its levels may also be regulated to just cause an increase in the length of G1 to enable cell fate decisions (Bieberich, 2004).

3.4. Differentiation

NSCs can be differentiated into the three major cell types which constitute the adult CNS-namely neurons, astrocytes, and oligodendrocytes. NSCs may also be differentiated into cells of the neural crest lineages such as smooth muscle and neurons and glia of the peripheral nervous system (Rajan *et al.*, 2003; Sailer *et al.*, 2005) (Fig. 2.4). Fate choice decisions are orchestrated by the response of the cell to extracellular ligands which include soluble ligands and those which are immobilized by incorporation into either to the ECM or in the surface membrane of neighboring cells. As mentioned before, the end point of the differentiation is a result of the sum total of all signals impinging on the cell, and the biochemical state of the cell which is receiving the signal, that is, the qualitative result of any signal is entirely context-dependent. This is illustrated by the effect elicited by wnt 7a on NSC cultures: in cultures derived from embryos of age E13.5 wnt causes neuronal differentiation, while it causes proliferation in cultured derived from younger animals (Hirabayashi *et al.*, 2004), and BMP differentiation of NSCs into smooth muscle or astrocytes depending on basal levels of Stat3 activity (Rajan *et al.*, 2003).

Growth factors have been used extensively for NSC differentiation. The neuropoietic cytokines CNTF/LIF are robust inducers of the astrocytic fate (Johe *et al.*, 1996), and do so by the activation of the signal transducers and activators of transcription (Stat) proteins, particularly Stat1 and 3 (Bonni *et al.*, 1997; Rajan and McKay, 1998). The LIF receptor (which serves as the signaling moiety for both CNTF and LIF) activates the janus kinase/tyrosine kinase (JAK/TYK) family of kinases and also MAPK in the cytoplasm, which results in Stat activation. The inhibition of MAPK delays initiation of CNTF-mediated differentiation suggesting that CNTF-mediated MAPK activation is required for optimal differentiation (Rajan and McKay, 1998). BMPs also cause glial differentiation by the activation of Stat3, but do so under conditions which involve the FKBP12 rapamycin associated protein (FRAP) and dense cultures which have high levels of basal Stat3 activation. Thus in this case the high basal levels of active Stat3 becomes a context in which the active Stat signal generated by BMP via FRAP yields glia (Rajan *et al.*, 2003). The thyroid hormone T3 causes oligodendrocyte

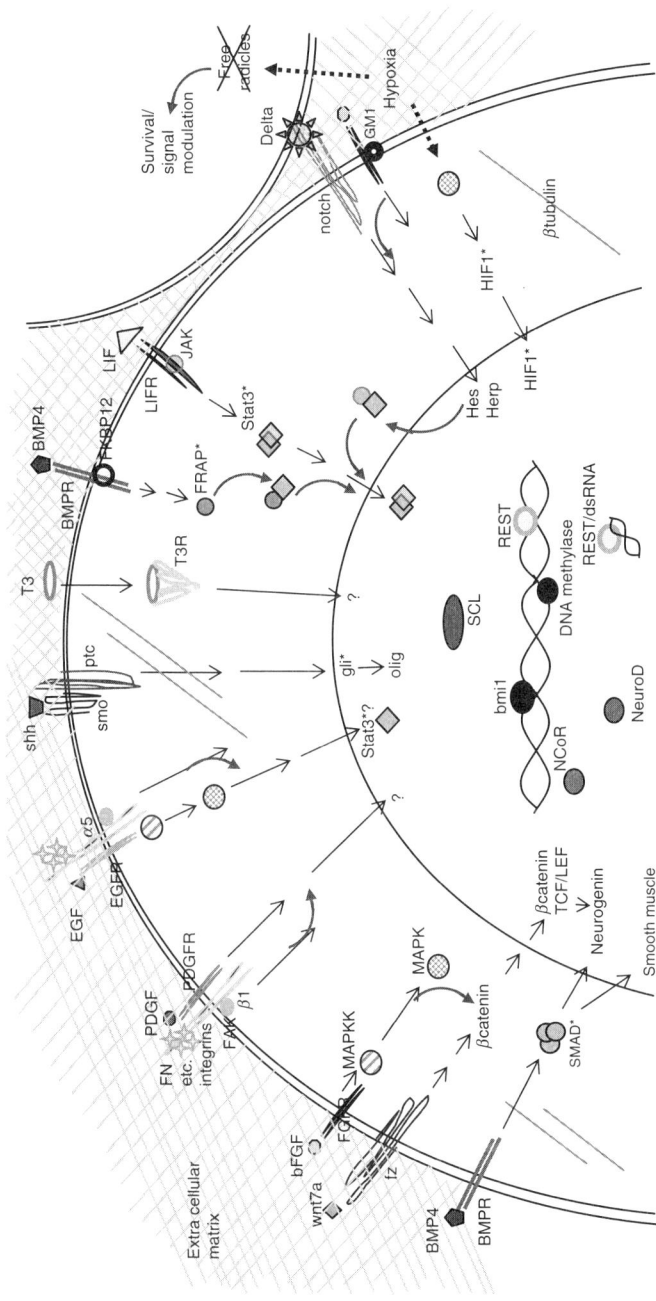

Figure 2.5 Signals regulating differentiation of NSCs. As seen in the case of proliferative signals extracellular signals which regulate the differentiation of NSCs may be soluble ligands, ECM molecules, and ligands present immobilized in neighboring cells. Almost all the ligands and receptor systems which were involved in proliferation are also involved in the regulation of differentiation. The one exception in its absence is the Eph/ephrin system. Also, the transcription factors activated by ligands during the induction of differentiation are distinct from the nuclear effects which are elicited during proliferation. There are interactions of the RTKs with the integrin receptors, however, in this case they signals lead to neuronal differentiation possibly through the activation of neurogenin. Wnt and BMP activation leads to neuronal differentiation possibly through the activation of neurogenin. PDGF appears to lead to the proliferation of a neuronal precursor. Interaction of the notch and BMP pathways lead to astrocytic differentiation, and so does LIF receptor activation by LIF or CNTF. Stat3 activation appears to be integrally involved in astrocytic differentiation. shh signals lead to oligodendrocyte differentiation, and so also signals originating from the triiodotyronine (T3) receptor. While hypoxia activates HIF1 which leads to specific differentiation effects, it also regulates free radical levels in the cell thus modulating signals in general. The nuclear component of differentiation is more complex than proliferation, and there are several more epigenetic and cofactor molecules involved. In addition helix–loop–helix proteins including neurogenin, SCL, and NeuroD are involved in fate choice specification. The complexity of possible transcription events during differentiation may perhaps be explained by the number of fates that a NSC can differentiate into (* denotes an activated transcription factor. See text for details and references).

astrocytic differentiation presumable by Hes and its targets. At the next level, one moves onto the "cross talk" or "signal integration" scenarios which may be illustrated by the mechanisms of Stat activation to yield glia from NSCs. There is a density mediated signal which causes the activation of Stat3 in NSCs, which was first documented by Rajan *et al.* (2003). Kamakura *et al.* subsequently showed that activation of notch (which could be density mediated) causes the complex formation between JAK and Stat3 which is facilitated by Hes causing Stat activation (Kamakura *et al.*, 2004). Thus it is the timing and summation of the effects of two or more antagonistic or complementary signals which will finally dictate the outcome.

In addition, there is extensive interaction of these transcription factors among themselves and with other constant or induced transcription factors, and the higher order transcription machinery. For instance SMADs, which are the major group of transcription factors activated by BMPs, are thought to interact with Stat3 on the GFAP promoter via p300 which serves as transcriptional coactivator which in addition to binding the transcription factors themselves, also possesses histone acetyltransferase activity which facilitates transcription (Nakashima *et al.*, 1999). Neurogenin appears to inhibit this interaction when it is bound to DNA by sequestering the p300/SMAD complex (Sun *et al.*, 2001). The other higher order transcriptional proteins which have been implicated are NCoR in glial differentiation (Hermanson *et al.*, 2002), and REST in glial maintenance by blocking neuronal genes in nonneuronal cells in most cases (Lunyak and Rosenfeld, 2005). In an added level of regulation, REST function is modulated by a double stranded RNA which appears to relieve repression of neuronally-related genes which are blocked by REST (Kuwabara *et al.*, 2004). Transcription of the GFAP promoter is influenced by methylation of a cytosine residue in the promoter (Takizawa *et al.*, 2001). It is thus possible that methylases are controlled based on the stage of proliferation/ differentiation of NSCs, which provides a further "context" to the cell receiving the signal. There is also increasing evidence that genes such as bmi1 which is one of the polycomb group of genes, an "oncogene" which regulates cell-cycle related genes such as INK4a and p16ARF, and is involved in the maintenance, proliferation, and differentiation states of NSCs. bmi1 forms part of a complex called the polycomb repressive complex 1 (PRC1) thought to maintain stable maintenance of gene expression by regulation of epigenetic chromatin modifications (Valk-Lingbeek *et al.*, 2004). Mice deficient in bmi1 display deficits in NSCs and cerebellar neurons (Leung *et al.*, 2004). Interestingly, they also show an increase in astrocytes (Zencak *et al.*, 2005). Finally, the bHLH protein SCL is thought to be involved in glial specification in the developing embryo (Muroyama *et al.*, 2005).

Differentiation may be brought about by inhibition of the cell cycle, or by the induction of differentiation per se without cell cycle inhibition. In the former case, inhibition of the cell cycle is brought about by the regulation of cdk inhibitors such as p27, which is described in Edlund and Jessell (Edlund and Jessell, 1999). Other cdk inhibitors which function in a similar manner are p21 and p15, these interaction mainly result in inhibition of phosphorylation of the Rb protein which would permit cells to re-enter the cell cycle (Weinberg, 1995). In the second case, regulation of differentiation occurs by proteins such as Id and Hes which are directly involved in the induction of differentiation regimes (Norton, 2000; Ohtsuka *et al.*, 2001).

4. Conclusions and Projections

Complex signaling networks regulate the life and death of a NSC. In this review, we have created a model for some of the permutations of the signals which are associated with NSC proliferation and differentiation, albeit a simplistic one. Our model is based largely on physical interaction of signaling entities and chemical modifications of enzymes and substrates, and implicit throughout the discussion is the assumption that temporal and spatial factors affect signal intensity and generation. "Context" is provided by prior or simultaneously occurring chemical, spatial, and temporal events, and is crucial for the outcome of signaling events, a balance of signaling events will decide whether a cell will divide, differentiate, or die. Computer modeling could be a further approach where networks of signals arising from the ligand-receptor interaction at the cell surface to transcription factor activation in the nucleus may be described along with regulatory motifs along the cellular networks (Ma'ayan *et al.*, 2005). The models provided in Figs. 2.2 and 2.5 depict pathways at their most basic level. An inclusion of all the "points of regulation" with detailed depictions of pathways and in four dimensions will exponentially increase complexity of the interactions.

A comparison of Figs. 2.2 and 2.5 will show that there are several apparently common signals involved in proliferation and differentiation. However, their interactions and the resulting end points are substantially different. Some striking examples are shh which regulates proliferation by the activation of N-myc, and differentiation by the activation of the olig transcription factor. The delta/notch system regulates proliferation possibly by genes which are activated by its responding transcription factors hes/herp, but activation of Stat proteins may be a signal which is involved in mediating the complex differentiation phenotype of notch. Another example is MAPK which although usually associated with proliferation can mediate differentiation by augmenting signals such as those mediated by wnt and CNTF. Integrins can modulate signals from several receptor tyrosine kinases (RTKs) inlcuding FGF, EGF, and PDGF.

Signals which regulate survival/apoptosis are more apparent and integral to proliferation than differentiation. Interestingly, several of the genes which are involved in these processes were identified as oncogenes or suppressor genes. These include signals which regulate bcl2, mci1, and p53. Among the transcriptional regulators of NSC proliferation are Stat3, N-myc, and bmi-1. Stat3 seems to be required for the proliferation of retinal progenitors, N-myc is responsible for proliferation in related systems and it functions downstream of shh. Bmi-1 is one of the polycomb group of proteins which is involved in epigenetic regulation of the genome, and along with tlx1 appears to be involved in NSC renewal. The Eph/

ephrin proteins appear to have important roles in the regulation of NSCs *in vivo*. It is possible that genes identified as oncogenes and tumor suppressor genes have as yet undefined roles in NSC proliferation and maintenance.

The number and complexity of nuclear signals which are associated with differentiation is much greater than those responsible for proliferation of NSCs. This is not surprising since there are several more outcomes which can occur due to differentiation, and possibly many avenues by which these fates can be arrived at, depending on the context in which the differentiation event is occurring. The determination of the cytoplasmic signals which regulate these nuclear proteins will enhance our understanding of the regulation of differentiation of NSCs, and the mechanisms by which cytoplasmic interactions occur. Our current knowledge indicates that BMPs, CNTF/LIF, notch, wnt, shh, and T3 are inducers of specific differentiation events, robust inducers like CNTF can induce differentiation independently, whereas others such as BMPs require a specific context to bring about a particular fate, and the RTKs appear to have a supporting role.

As mentioned in the text, some factors elicit different responses from NSCs derived from different species, for example, LIF causes differentiation in NSCs derived from E14 rat embryos and supports proliferation in NSCs derived from younger embryos, and in human NSCs. Due to such context considerations only those results which have been obtained in mammalian, and preferably neural-related, systems have been considered here. However, some extrapolations have been made from related systems, such as the case of insulin and integrins. The assumption of the involvement of Akt/BAD and integrin signaling in NSC culture is not unreasonable. Insulin is included in most cell culture systems including NSC cultures. Adherent monolayer cultures are dependent on a coating of matrix molecule such as fibronectin or laminin on the culture dish for survival and proliferation, while neurosphere cultures synthesize their own ECM. Ideas and practical methods to manipulate cell surface and cytoplasmic signals independently and in concert will permit appreciable control over NSCs and may even allow us to minimize differences which arise due to culture protocols. This may involve the concerted inhibition signals by inhibition of enzymes, manipulation of density, ECM, temporally regulated coactivation of pathways, and compartmentalized activation of signals especially for immobile ligands, and the configuration of intersecting gradients for immobile and soluble ligands.

5. Sample Protocols for the Culture and Characterization of NSCs

NSCs from various sources are capable of continued proliferation as undifferentiated cells and of then differentiating *in vitro* into either mixed populations of clonally-related neurons, astrocytes, or oligodendrocytes or of populations relatively enriched for one neural cell type versus another. Rodent NSCs are preferentially maintained in serum free medium supplemented with bFGF as mitogen. Differentiation towards enriched populations of various neural lineages is routinely accomplished by allowing the cells to exit the cell cycle by removing mitogens from the medium while simultaneously exposing them to various inducing agents including NT-3 for neurons, CNTF and serum for astrocytes, and T3 and PDGF for oligodendrocytes (Johe *et al.*, 1996). These cultures are characterized by immunocytochemistry (ICC) and/or reverse transcriptase-polymerase chain reaction (RT-PCR) with stem cell markers. Differentiated cultures are characterized with ICC markers, PCR, and whole cell patch clamp to confirm functionality of neurons. Here, we describe one of the several methods which are in existence and used to culture NSCs from mammalian tissue.

6. Culture of Rodent/Murine NSCs

6.1. Preparation of dishes

1. 2 days before start of culture: coat with poly-L-ornithine over night at 37 °C.
2. Day before: aspirate the poly-L-ornithine and wash twice with PBS; coat with fibronectin over night.
3. Next day aspirate the fibronectin, wash once with PBS, and plate cells.

6.2. Isolation of NSCs

1. Isolate required tissue from embryonic rat or mouse brain. Typically E14.5 rat or E12.5 mouse is used to derive cortical stem cells. Sacrifice the timed-pregnant mother with excess carbon dioxide, and remove the embryos from their sacs in a 10 cm culture dish in PBS. Place the cleaned embryos in a fresh 10 cm dish in HBSS (neutral pH), and place on ice.

2. Dissection of the embryonic brain is performed in 6 cm culture dishes in HBSS of neutral pH under a dissection microscope. Change dishes as frequently as needed, as debris will continuously accumulate. Cortex, striatum, and midbrain are easily isolated from embryos of this age, while younger embryos may be required for some mesencephalon stem cell cultures, and older ones for cerebellum stem cell cultures. Roughly a million cells will be obtained from the cortex of a rat embryo at age E14.5.

3. Collect the tissue from all the embryos in HBSS as the dissection progresses in a 15 ml tube on ice; move to tissue culture hood when the dissection is completed.

4. Aspirate medium until 0.5 ml HBSS remains in the tube; wash once with HBSS (about 10 ml) and once with sterile DMEM/F12/N2 medium (also about 10 ml).

5. Aspirate as much of the wash medium and add 1 ml of fresh medium. Triturate gently about 8–10 times with a P1000 pipette tip. One might create a smaller orifice by placing the pipette tip at the very bottom of the tube before sucking up. (Note: mechanical trituration is the method of choice rather than enzymatic dissociation, as this does not cause break up of the membranes of the meninges and result in fibroblast contamination of the stem cell culture).

6. Add another 5 ml of medium and triturate gently about 5 times with a 5 ml pipette.

7. Place the tube in the hood and allow the membranes and larger pieces of tissue to settle to the bottom of the tube. This may take about 5–10 min. During this period 10 μl of the cell suspension may be used for a cell count. Once the larger pieces settle, gently pipette the top 5 ml of the cell suspension into a 50 ml tube.

8. Count cells with a haemocytometer by mixing 10 μl of the cell suspension with 10 μl of a 0.2% suspension of trypan blue. Ideally over 80% of the cells should exclude trypan blue indicating viability.

9. After counting, dilute the cells to the required concentration with additional plating medium. Plate cells in 10 cm dishes at a concentration of a million cells per dish in a volume of 5 ml (scale down or up as needed). The cells are plated in DMEM/F12 medium supplemented with N2 and bFGF to a final concentration of 10 ng/ml. (This culture is designated P0 for passage 0).

10. The complete medium (+ bFGF) is replaced the next day to wash off the debris resulting from the primary plating. For subsequent days, the culture is fed with bFGF daily, and medium replaced every alternate day. In another variation, 50% of the medium may be replaced daily, along with the appropriate amount of bFGF for the entire volume.

6.3. Passaging of NSCs

1. Passage the culture onto new precoated dishes after 4–5 days (50–70% confluent). Wash culture 2–3 times with HBSS, then incubate cells for up to 45 min at 37 °C in 5–7 ml HBSS. Using a 10 ml pipette, gently spray the HBSS over the surface of the dish. If the cells remain stubbornly attached a cell lifter may be gently used. Pipette cells gently with a 10 ml pipette to disperse any clumps and pellet by centrifugation at 1000g for 5 min.

2. Aspirate the supernatant and resuspend the cells in 5–10 ml N2-medium, then count cells and assess the viability as above. Plate cells at about a million cells per 10 cm dish in 5 ml of N2 medium containing 10 ng/ml bFGF. The number of cells plated can be adjusted based on the desired density for the experiment. (P1)

3. Repeat steps 1 and 2 if a further passage is needed. (P2)

At this point, the culture should be relatively homogeneous and consist almost entirely of NSCs. In order to induce differentiation to neurons and glia remove the medium containing FGF and replace with DMEM/F12/N2 without FGF. Alternatively specific factors may be added to direct differentiation.

6.4. Freezing and thawing cells

1. Add 250 ml of propanol to the freezing container. Store at 4 °C for several hours. Keep DMSO at room temperature (RT).

2. Remove cells from the culture dish with HBSS. Count cells and dilute to a final concentration of 2–3 million cells per milliliter.

3. Add 100 μl of DMSO to the freezing vial and add 900 μl of cells to the same vial. Mix well by inversion and place in the freezing container overnight at −80 °C. Speed is of the essence.

4. Transfer the vials to liquid nitrogen storage.

5. Thawing: warm vial at 37 °C quickly. To wash off the DMSO as quickly as possible, add the cells to 10 ml of prewarmed medium in a 15 ml tube, and spin cells down at 1000 g for 5 min.

6. Resuspend cells in 5 ml of DMEM/F12/bFGF and plate in at 10 cm dish.

6.5. Reagents

Medium:
DMEM/F12/N2-recipe for 500ml:
DMEM/F12 medium
N2:
Apotransferrin 500 mg (Sigma T1147).
Insulin 12.5 mg (Dissolve in about 5 ml 0.01 N NaOH).
Putrescine 50 μl of stock (1 M Putrescine stock—Sigma, cat# P5780. Store aliquots at 20 °C).
Selenite 30 μl of stock (500 μM Sodium selenite stock—Sigma, cat# S5261. Store aliquots at 20 °C).
Progesterone 100 μl of stock (100 μM Progesterone stock—Sigma cat# P8783. Dissolve in ethanol; store aliquots at −20 °C).
Pen/strep to 1×
Adjust pH to 7.2. Filter. Store at 4 °C. Do not use if older than 2–3 weeks.
Poly-L-ornithine: (Sigma P4638). Final concentration of 15 μg/ml in sterile water. Make stock of 100× and store at −20 °C.
Fibronectin (Sigma F4759). Final concentration of 1 μg/ml in PBS. Make stock of 1000× at store at −20 °C.
Recombinant bFGF (R&D Systems). Final concentration of 10 ng/ml. Make 1000× stock in PBS + 0.1% BSA.
Freezing container (Mr. Frosty, Nalgene, Cat. No 5100–0001)

7. Culture of human NSCs

We have been successful in isolating human NSCs from the telencephalic ventricular zone of a 13-week gestation human fetal cadaver.

1. Briefly, we first establish a primary dissociated, stable serum-containing monolayer culture of fetal ventricular zone. Cultures that do not grow well or do not continue to proliferate are no longer pursued. A promising culture is then subjected to a 6–8 week sequential growth factor selection process based on growth parameters rather than on markers (as per Flax *et al.*, 1998). In fact, cells that form

clusters that are greater than 10 cell diameters and cannot be readily disaggregated are excluded.

2. Cells grown in serum are switched to serum-free conditions containing EGF + bFGF + LIF (bFGF; 20 ng/ml; human, recombinant; Calbiochem, EGF; 20 ng/ml; human, recombinant; Calbiochem, LIF; 10 ng/ml; human, recombinant; Chemicon). They are passaged once-per-week for 2 weeks.

3. Cells that successfully passaged are then switched to bFGF alone. They are similarly passaged once-per-week for 2 weeks.

4. Cell that successfully passaged in bFGF are then switched to EGF alone. They are similarly passaged once-per-week for 2 weeks.

5. Cells that successfully passaged and then switched back to bFGF and the 2 week selection process is continued.

6. Cells that successfully passaged in bFGF are then switched to bFGF + LIF.

7. Cultures that have successfully passaged over the previous 6–8 weeks and continue to maintain stem-like growth following this selection process are then subjected to an *in vitro* and *in vivo* functional screen. *In vitro*, the cells must be able to express undifferentiated markers yet, in response to induction, differentiate in a dish. The *in vivo* functional screen entails continuing to use only those cells that have the ability to engraft, migrate, and differentiate *in vivo* following implanatation into the ventricles and cerebella of newborn (P0) mice and yield olfactory bulb neurons or cerebellar granule neurons, respectively. After 3–4 weeks, the mice are sacrificed to determine which hNSCs yielded neurons in the olfactory bulb, glia in the cortex, and granule neurons in the cerebellum.

8. Based on this protracted screen, only a very small number of lines are ultimately selected for further use. These are stored as stable lines to be used for multiple experiments. They are not grown as floating clusters ("spheres") but rather as monolayers in serum-free medium in bFGF + LIF with the addition (1:1) of medium conditioned by hNSCs (i.e., 50% "self-conditioned medium"). No genetic manipulations are imposed on these cultures. Monolayers and any clusters are disaggregated frequently with trypsin or accutane. Karyotypes have remained stable. Most flasks are grown with the combination of bFGF and LIF with the addition (1:1) of medium self-conditioned by hNSCs and stored for future use. The hNSCs are predominantly adherent and grown as a monolayer, although there is a moderate percentage of floating cells. Tapping the flask and/or triturating easily dislodge the adherent cells to permit passaging and preclude terminal differentiation. The cells tend to form clusters both when floating and when attached. Frequent passaging excludes those cells that are not actively proliferative and hence allows their cell cycle and differentiation state to be more or less synchronized.

9. hNSCs are grown in 25cm^2 flasks (Falcon, "tissue culture-treated") containing 8ml of medium, fed and/or split 1:2 once-per-week. Concentrated splits do better than more dilute splits. hNSCs are dissociated into single cell suspensions

when any floating or adherent cellular clusters grow to a point where they can no longer readily be dissociated mechanically by simple trituration, or when they are too adherent to be dislodged by simple agitation, or when they are greater than 10 cell diameters in width—and prior to all transplantations. This generally coincides with our once-per-week passaging regime. Again, when passaging, we split only 1:2 each time. All cells—both floating and adherent—are passaged. For routine passaging, Accutase is used as the dissociating agent (Accutase; Innovative Cell Technologies). Accutase can be inactivated by simply diluting with fresh medium.

8. ICC of markers to identify stem cells and differentiation products

1. Fix the cells with the PAF solution (4% paraformaldehyde, 0.12 M Sucrose, in PBS, pH 7.2). The cells are washed once very quickly and gently with PBS at RT and fixed by adding PAF (0.5 ml/well for 24 well, 0.1 ml /well for 96 wells) for 20 min at RT.

2. Wash 10 min (min) with PBS. Repeat twice.

3. Saturate nonspecific sites and permeabilize with PBS containing 0.3% Triton and 10% normal goat serum (NGS) for 40 min at RT. Wash 5 min with PBS. Repeat twice.

4. Incubate the cells with primary antibodies for 1 h at RT: Dilute the antibodies in PBS containing 5% NGS. Wash 5 min with PBS. Repeat twice.

5. Add the secondary antibody for 1 h at RT. Dilute the antibody in PBS containing 5% NGS. Wash 5 min with PBS. Repeat twice.

6. Permeabilize the nuclei with 100% methanol ($-20\,°C$) for 5 min. Wash 5 min with PBS. Repeat twice.

7. Incubate the cells with bisBenzimide for 10 min at RT: Wash the cells 3 times with ultra pure water and mount them with AquaPolymount. Let it dry at store at 4 °C.

8. Some antibody sources: Mouse anti-MAP-2 (Sigma, dilution 1:250), Mouse anti-□Tubulin III (Covance, 1:600), neurofilament M (NF-m) was used (1:500, Chemicon), GFAP (1:500, Chemicon), GDNF (1:1,000, Chemicon), Nestin (1:100, chemicon), Alexa conjugated secondary antibodies (1:500, Invitrogen).

Acknowledgments

We like to acknowledge the financial support of the A-T Children's Project to PR and NIH to EYS, and Dr. J. F. Loring for suggestions on the manuscript.

References

Akai, J., Halley, P. A., and Storey, K. G. (2005). FGF-dependent Notch signaling maintains the spinal cord stem zone. *Genes Dev.* **19**, 2877–2887.

Alvarez-Buylla, A., and Lim, D. A. (2004). For the long run: Maintaining germinal niches in the adult brain. *Neuron* **41,** 683–686.

Alvarez-Buylla, A., Seri, B., and Doetsch, F. (2002). Identification of neural stem cells in the adult vertebrate brain. *Brain Res. Bull.* **57,** 751–758.

Anderson, D. J., Groves, A., Lo, L., Ma, Q., Rao, M., Shah, N. M., and Sommer, L. (1997). Cell lineage determination and the control of neuronal identity in the neural crest. *Cold Spring Harb. Symp. Quant. Biol.* **62,** 493–504.

Antonopoulos, J., Pappas, I. S., and Parnavelas, J. G. (1997). Activation of the GABAA receptor inhibits the proliferative effects of bFGF in cortical progenitor cells. *Eur. J. Neurosci.* **9,** 291–298.

Artavanis-Tsakonas, S., Rand, M. D., and Lake, R. J. (1999). Notch signaling: Cell fate control and signal integration in development. *Science* **284,** 770–776.

Baas, A. F., Smit, L., and Clevers, H. (2004). LKB1 tumor suppressor protein: PARtaker in cell polarity. *Trends Cell Biol.* **14,** 312–319.

Bieberich, E. (2004). Integration of glycosphingolipid metabolism and cell-fate decisions in cancer and stem cells: Review and hypothesis. *Glycoconj. J.* **21,** 315–327.

Bieberich, E., MacKinnon, S., Silva, J., Noggle, S., and Condie, B. G. (2003). Regulation of cell death in mitotic neural progenitor cells by asymmetric distribution of prostate apoptosis response 4 (PAR-4) and simultaneous elevation of endogenous ceramide. *J. Cell Biol.* **162,** 469–479.

Blaise, G. A., Gauvin, D., Gangal, M., and Authier, S. (2005). Nitric oxide, cell signaling and cell death. *Toxicology* **208,** 177–192.

Bond, J., Scott, S., Hampshire, D. J., Springell, K., Corry, P., Abramowicz, M. J., Mochida, G. H., Hennekam, R. C., Maher, E. R., Fryns, J. P., Alswaid, A., Jafri, H., *et al.* (2003). Protein-truncating mutations in ASPM cause variable reduction in brain size. *Am. J. Hum. Genet.* **73,** 1170–1177.

Bonni, A., Brunet, A., West, A. E., Datta, S. R., Takasu, M. A., and Greenberg, M. E. (1999). Cell survival promoted by the Ras-MAPK signaling pathway by transcription-dependent and-independent mechanisms. *Science* **286,** 1358–1362.

Bonni, A., Sun, Y., Nadal-Vicens, M., Bhatt, A., Frank, D. A., Rozovsky, I., Stahl, N., Yancopoulos, G. D., and Greenberg, M. E. (1997). Regulation of gliogenesis in the central nervous system by the JAK-STAT signaling pathway. *Science* **278,** 477–483.

Brunet, A., Bonni, A., Zigmond, M. J., Lin, M. Z., Juo, P., Hu, L. S., Anderson, M. J., Arden, K. C., Blenis, J., and Greenberg, M. E. (1999). Akt promotes cell survival by phosphorylating and inhibiting a Forkhead transcription factor. *Cell* **96,** 857–868.

Cai, L., Hayes, N. L., Takahashi, T., Caviness, V. S., Jr., and Nowakowski, R. S. (2002). Size distribution of retrovirally marked lineages matches prediction from population measurements of cell cycle behavior. *J. Neurosci. Res.* **69,** 731–744.

Caille, I., Allinquant, B., Dupont, E., Bouillot, C., Langer, A., Muller, U., and Prochiantz, A. (2004). Soluble form of amyloid precursor protein regulates proliferation of progenitors in the adult subventricular zone. *Development* **131,** 2173–2181.

Calegari, F., and Huttner, W. B. (2003). An inhibition of cyclin-dependent kinases that lengthens, but does not arrest, neuroepithelial cell cycle induces premature neurogenesis. *J. Cell Sci.* **116,** 4947–4955.

Castaneda-Castellanos, D. R., and Kriegstein, A. R. (2004). Controlling neuron number: Does numb do the math? *Nat. Neurosci.* **7,** 793–794.

Caviness, V. S., Jr., Takahashi, T., and Nowakowski, R. S. (1995). Numbers, time and neocortical neuronogenesis: A general developmental and evolutionary model. *Trends Neurosci.* **18,** 379–383.

Chang, L. K., Putcha, G. V., Deshmukh, M., and Johnson, E. M., Jr. (2002). Mitochondrial involvement in the point of no return in neuronal apoptosis. *Biochimie* **84,** 223–231.

Chenn, A., and McConnell, S. K. (1995). Cleavage orientation and the asymmetric inheritance of Notch1 immunoreactivity in mammalian neurogenesis. *Cell* **82,** 631–641.

Chenn, A., and Walsh, C. A. (2002). Regulation of cerebral cortical size by control of cell cycle exit in neural precursors. *Science* **297,** 365–369.

Cipolleschi, M. G., Dello Sbarba, P., and Olivotto, M. (1993). The role of hypoxia in the maintenance of hematopoietic stem cells. *Blood* **82,** 2031–2037.

Clevers, H. (2005). Stem cells, asymmetric division and cancer. *Nat. Genet.* **37,** 1027–1028.

Conover, J. C., Doetsch, F., Garcia-Verdugo, J. M., Gale, N. W., Yancopoulos, G. D., and Alvarez-Buylla, A. (2000). Disruption of Eph/ephrin signaling affects migration and proliferation in the adult subventricular zone. *Nat. Neurosci.* **3,** 1091–1097.

Conti, L., and Cattaneo, E. (2005). Controlling neural stem cell division within the adult subventricular zone: An APPealing job. *Trends Neurosci.* **28,** 57–59.

Cross, J. V., and Templeton, D. J. (2004). Thiol oxidation of cell signaling proteins: Controlling an apoptotic equilibrium. *J. Cell Biochem.* **93,** 104–111.

Datta, S. R., Brunet, A., and Greenberg, M. E. (1999). Cellular survival: A play in three Akts. *Genes Dev.* **13,** 2905–2927.

Datta, S. R., Dudek, H., Tao, X., Masters, S., Fu, H., Gotoh, Y., and Greenberg, M. E. (1997). Akt phosphorylation of BAD couples survival signals to the cell-intrinsic death machinery. *Cell* **91,** 231–241.

Depaepe, V., Suarez-Gonzalez, N., Dufour, A., Passante, L., Gorski, J. A., Jones, K. R., Ledent, C., and Vanderhaeghen, P. (2005). Ephrin signalling controls brain size by regulating apoptosis of neural progenitors. *Nature* **435,** 1244–1250.

De Zio, D., Giunta, L., Corvaro, M., Ferraro, E., and Cecconi, F. (2005). Expanding roles of programmed cell death in mammalian neurodevelopment. *Semin. Cell Dev. Biol.* **16,** 281–294.

Doetsch, F., Petreanu, L., Caille, I., Garcia-Verdugo, J. M., and Alvarez-Buylla, A. (2002). EGF converts transit-amplifying neurogenic precursors in the adult brain into multipotent stem cells. *Neuron* **36,** 1021–1034.

Du, Q., and Macara, I. G. (2004). Mammalian pins is a conformational switch that links NuMA to heterotrimeric G proteins. *Cell* **119,** 503–516.

Edlund, T., and Jessell, T. M. (1999). Progression from extrinsic to intrinsic signaling in cell fate specification: a view from the nervous system. *Cell* **96,** 211–224.

Feng, Y., and Walsh, C. A. (2004). Mitotic spindle regulation by Nde1 controls cerebral cortical size. *Neuron* **44,** 279–293.

Flax, J. D., *et al.* (1998). Engraftable human neural stem cells respond to developmental cues, replace neurons, and express foreign genes. *Nat. Biotechnol.* **16,** 1033–1039.

Frederiksen, K., and McKay, R. D. (1988). Proliferation and differentiation of rat neuroepithelial precursor cells *in vivo. J. Neurosci.* **8,** 1144–1151.

Gage, F. H. (2000). Mammalian neural stem cells. *Science* **287,** 1433–1438.

Gaiano, N., and Fishell, G. (2002). The role of notch in promoting glial and neural stem cell fates. *Annu. Rev. Neurosci.* **25,** 471–490.

Giancotti, F. G. (1997). Integrin signaling: Specificity and control of cell survival and cell cycle progression. *Curr. Opin. Cell Biol.* **9,** 691–700.

Goldman, S. A., and Sim, F. (2005). Neural progenitor cells of the adult brain. *Novartis Found. Symp.* **265,** 66–80; discussion 82–97.

Grandbarbe, L., Bouissac, J., Rand, M., Hrabe de Angelis, M., Artavanis-Tsakonas, S., and Mohier, E. (2003). Delta-Notch signaling controls the generation of neurons/glia from neural stem cells in a stepwise process. *Development* **130,** 1391–1402.

Groszer, M., Erickson, R., Scripture-Adams, D. D., Dougherty, J. D., Le Belle, J., Zack, J. A., Geschwind, D. H., Liu, X., Kornblum, H. I., and Wu, H. (2006). PTEN negatively regulates neural stem cell self-renewal by modulating G0-G1 cell cycle entry. *Proc. Natl. Acad. Sci. USA* **103,** 111–116.

Gustafsson, M. V., Zheng, X., Pereira, T., Gradin, K., Jin, S., Lundkvist, J., Ruas, J. L., Poellinger, L., Lendahl, U., and Bondesson, M. (2005). Hypoxia requires notch signaling to maintain the undifferentiated cell state. *Dev. Cell* **9,** 617–628.

Haydar, T. F., Wang, F., Schwartz, M. L., and Rakic, P. (2000). Differential modulation of proliferation in the neocortical ventricular and subventricular zones. *J. Neurosci.* **20,** 5764–5774.

Hermanson, O., Jepsen, K., and Rosenfeld, M. G. (2002). N-CoR controls differentiation of neural stem cells into astrocytes. *Nature* **419,** 934–939.

Hirabayashi, Y., Itoh, Y., Tabata, H., Nakajima, K., Akiyama, T., Masuyama, N., and Gotoh, Y. (2004). The Wnt/beta-catenin pathway directs neuronal differentiation of cortical neural precursor cells. *Development* **131,** 2791–2801.

Hitoshi, S., Seaberg, R. M., Koscik, C., Alexson, T., Kusunoki, S., Kanazawa, I., Tsuji, S., and van der Kooy, D. (2004). Primitive neural stem cells from the mammalian epiblast differentiate to definitive neural stem cells under the control of Notch signaling. *Genes Dev.* **18,** 1806–1811.

Holland, S. J., Gale, N. W., Mbamalu, G., Yancopoulos, G. D., Henkemeyer, M., and Pawson, T. (1996). Bidirectional signalling through the EPH-family receptor Nuk and its transmembrane ligands. *Nature* **383,** 722–725.

Holmberg, J., Armulik, A., Senti, K. A., Edoff, K., Spalding, K., Momma, S., Cassidy, R., Flanagan, J. G., and Frisen, J. (2005). Ephrin-A2 reverse signaling negatively regulates neural progenitor proliferation and neurogenesis. *Genes Dev.* **19,** 462–471.

Hu, H., Tomasiewicz, H., Magnuson, T., and Rutishauser, U. (1996). The role of polysialic acid in migration of olfactory bulb interneuron precursors in the subventricular zone. *Neuron* **16,** 735–743.

Huttner, W. B., and Kosodo, Y. (2005). Symmetric versus asymmetric cell division during neurogenesis in the developing vertebrate central nervous system. *Curr. Opin. Cell Biol.* **17,** 648–657.

Hynes, R. O. (2002). Integrins: Bidirectional, allosteric signaling machines. *Cell* **110,** 673–687.

Israsena, N., Hu, M., Fu, W., Kan, L., and Kessler, J. A. (2004). The presence of FGF2 signaling determines whether beta-catenin exerts effects on proliferation or neuronal differentiation of neural stem cells. *Dev. Biol.* **268,** 220–231.

Jacques, T. S., Relvas, J. B., Nishimura, S., Pytela, R., Edwards, G. M., Streuli, C. H., and ffrench-Constant, C. (1998). Neural precursor cell chain migration and division are regulated through different beta1 integrins. *Development* **125,** 3167–3177.

Jessell, T. M., and Sanes, J. R. (2000). Development. The decade of the developing brain. *Curr. Opin. Neurobiol.* **10,** 599–611.

Jin, G., Tan, X., Tian, M., Qin, J., Zhu, H., Huang, Z., and Xu, H. (2005). The controlled differentiation of human neural stem cells into TH-immunoreactive (ir) neurons *in vitro*. *Neurosci. Lett.* **386,** 105–110.

Johansson, C. B., Momma, S., Clarke, D. L., Risling, M., Lendahl, U., and Frisen, J. (1999). Identification of a neural stem cell in the adult mammalian central nervous system. *Cell* **96,** 25–34.

Johe, K. K., Hazel, T. G., Muller, T., Dugich-Djordjevic, M. M., and McKay, R. D. (1996). Single factors direct the differentiation of stem cells from the fetal and adult central nervous system. *Genes Dev.* **10,** 3129–3140.

Kamakura, S., Oishi, K., Yoshimatsu, T., Nakafuku, M., Masuyama, N., and Gotoh, Y. (2004). Hes binding to STAT3 mediates crosstalk between Notch and JAK-STAT signalling. *Nat. Cell Biol.* **6,** 547–554.

Kenney, A. M., and Rowitch, D. H. (2000). Sonic hedgehog promotes G(1) cyclin expression and sustained cell cycle progression in mammalian neuronal precursors. *Mol. Cell. Biol.* **20,** 9055–9067.

Kleber, M., and Sommer, L. (2004). Wnt signaling and the regulation of stem cell function. *Curr. Opin. Cell Biol.* **16,** 681–687.

Klezovitch, O., Fernandez, T. E., Tapscott, S. J., and Vasioukhin, V. (2004). Loss of cell polarity causes severe brain dysplasia in Lgl1 knockout mice. *Genes Dev.* **18,** 559–571.

Kosodo, Y., Roper, K., Haubensak, W., Marzesco, A. M., Corbeil, D., and Huttner, W. B. (2004). Asymmetric distribution of the apical plasma membrane during neurogenic divisions of mammalian neuroepithelial cells. *EMBO J.* **23,** 2314–2324.

Kuan, C. Y., Roth, K. A., Flavell, R. A., and Rakic, P. (2000). Mechanisms of programmed cell death in the developing brain. *Trends Neurosci.* **23,** 291–297.

Kuida, K., Zheng, T. S., Na, S., Kuan, C., Yang, D., Karasuyama, H., Rakic, P., and Flavell, R. A. (1996). Decreased apoptosis in the brain and premature lethality in CPP32-deficient mice. *Nature* **384,** 368–372.

Kuwabara, T., Hsieh, J., Nakashima, K., Taira, K., and Gage, F. H. (2004). A small modulatory dsRNA specifies the fate of adult neural stem cells. *Cell* **116,** 779–793.

Lee, H. Y., Kleber, M., Hari, L., Brault, V., Suter, U., Taketo, M. M., Kemler, R., and Sommer, L. (2004). Instructive role of Wnt/beta-catenin in sensory fate specification in neural crest stem cells. *Science* **303,** 1020–1023.

Leone, D. P., Relvas, J. B., Campos, L. S., Hemmi, S., Brakebusch, C., Fassler, R., Ffrench-Constant, C., and Suter, U. (2005). Regulation of neural progenitor proliferation and survival by beta1 integrins. *J. Cell Sci.* **118,** 2589–2599.

Leung, C., Lingbeek, M., Shakhova, O., Liu, J., Tanger, E., Saremaslani, P., Van Lohuizen, M., and Marino, S. (2004). Bmi1 is essential for cerebellar development and is overexpressed in human medulloblastomas. *Nature* **428,** 337–341.

Li, H. S., Wang, D., Shen, Q., Schonemann, M. D., Gorski, J. A., Jones, K. R., Temple, S., Jan, L. Y., and Jan, Y. N. (2003). Inactivation of numb and numblike in embryonic dorsal forebrain impairs neurogenesis and disrupts cortical morphogenesis. *Neuron* **40,** 1105–1118.

Lim, D. A., Tramontin, A. D., Trevejo, J. M., Herrera, D. G., Garcia-Verdugo, J. M., and Alvarez-Buylla, A. (2000). Noggin antagonizes BMP signaling to create a niche for adult neurogenesis. *Neuron* **28,** 713–726.

LoTurco, J. J., Owens, D. F., Heath, M. J., Davis, M. B., and Kriegstein, A. R. (1995). GABA and glutamate depolarize cortical progenitor cells and inhibit DNA synthesis. *Neuron* **15,** 1287–1298.

Louvi, A., and Artavanis-Tsakonas, S. (2006). Notch signalling in vertebrate neural development. *Nat. Rev. Neurosci.* **7,** 93–102.

Lunyak, V. V., and Rosenfeld, M. G. (2005). No rest for REST: REST/NRSF regulation of neurogenesis. *Cell* **121,** 499–501.

Ma'ayan, A., Jenkins, S. L., Neves, S., Hasseldine, A., Grace, E., Dubin-Thaler, B., Eungdamrong, N. J., Weng, G., Ram, P. T., Rice, J. J., Kershenbaum, A., Stolovitzky, G. A., *et al.* (2005). Formation of regulatory patterns during signal propagation in a mammalian cellular network. *Science* **309,** 1078–1083.

Mabie, P. C., Mehler, M. F., and Kessler, J. A. (1999). Multiple roles of bone morphogenetic protein signaling in the regulation of cortical cell number and phenotype. *J. Neurosci.* **19,** 7077–7088.

Matter, M. L., and Ruoslahti, E. (2001). A signaling pathway from the alpha5beta1 and alpha(v)beta3 integrins that elevates bcl-2 transcription. *J. Biol. Chem.* **276,** 27757–27763.

McMahon, A. P. (2000). More surprises in the Hedgehog signaling pathway. *Cell* **100,** 185–188.

Megason, S. G., and McMahon, A. P. (2002). A mitogen gradient of dorsal midline Wnts organizes growth in the CNS. *Development* **129,** 2087–2098.

Mellitzer, G., Xu, Q., and Wilkinson, D. G. (1999). Eph receptors and ephrins restrict cell intermingling and communication. *Nature* **400,** 77–81.

Moshiri, A., Close, J., and Reh, T. A. (2004). Retinal stem cells and regeneration. *Int. J. Dev. Biol.* **48,** 1003–1014.

Mungrue, I. N., Stewart, D. J., and Husain, M. (2003). The Janus faces of iNOS. *Circ. Res.* **93,** e74.

Muroyama, Y., Fujihara, M., Ikeya, M., Kondoh, H., and Takada, S. (2002). Wnt signaling plays an essential role in neuronal specification of the dorsal spinal cord. *Genes Dev.* **16,** 548–553.

Muroyama, Y., Fujiwara, Y., Orkin, S. H., and Rowitch, D. H. (2005). Specification of astrocytes by bHLH protein SCL in a restricted region of the neural tube. *Nature* **438,** 360–363.

Muroyama, Y., Kondoh, H., and Takada, S. (2004). Wnt proteins promote neuronal differentiation in neural stem cell culture. *Biochem. Biophys. Res. Commun.* **313,** 915–921.

Nakashima, K., Yanagisawa, M., Arakawa, H., Kimura, N., Hisatsune, T., Kawabata, M., Miyazono, K., and Taga, T. (1999). Synergistic signaling in fetal brain by STAT3-Smad1 complex bridged by p300. *Science* **284,** 479–482.

Norton, J. D. (2000). ID helix-loop-helix proteins in cell growth, differentiation and tumorigenesis. *J. Cell Sci.* **113**(Pt. 22), 3897–3905.

Nunes, M. C., Roy, N. S., Keyoung, H. M., Goodman, R. R., McKhann, G., II, Jiang, L., Kang, J., Nedergaard, M., and Goldman, S. A. (2003). Identification and isolation of multipotential neural progenitor cells from the subcortical white matter of the adult human brain. *Nat. Med.* **9,** 439–447.

Ohtsuka, T., Sakamoto, M., Guillemot, F., and Kageyama, R. (2001). Roles of the basic helix-loop-helix genes Hes1 and Hes5 in expansion of neural stem cells of the developing brain. *J. Biol. Chem.* **276,** 30467–30474.

Oishi, K., Kamakura, S., Isazawa, Y., Yoshimatsu, T., Kuida, K., Nakafuku, M., Masuyama, N., and Gotoh, Y. (2004). Notch promotes survival of neural precursor cells via mechanisms distinct from those regulating neurogenesis. *Dev. Biol.* **276,** 172–184.

Owens, D. F., Flint, A. C., Dammerman, R. S., and Kriegstein, A. R. (2000). Calcium dynamics of neocortical ventricular zone cells. *Dev. Neurosci.* **22,** 25–33.

Polster, B. M., and Fiskum, G. (2004). Mitochondrial mechanisms of neural cell apoptosis. *J. Neurochem.* **90,** 1281–1289.

Rajan, P., and McKay, R. D. (1998). Multiple routes to astrocytic differentiation in the CNS. *J. Neurosci.* **18,** 3620–3629.

Rajan, P., Panchision, D. M., Newell, L. F., and McKay, R. D. (2003). BMPs signal alternately through a SMAD or FRAP-STAT pathway to regulate fate choice in CNS stem cells. *J. Cell Biol.* **161,** 911–921.

Rajan, P., Park, K.-I., Ourednik, V., Lee, J.-P., Imitola, J., Mueller, F.-J., Teng, Y. D., and Snyder, E. (2006). Stem cell research and applications for human therapies. *In* "Drug Discovery Research in the Post Genomics Era." Wiley, New York.

Roussigne, M., Cayrol, C., Clouaire, T., Amalric, F., and Girard, J. P. (2003). THAP1 is a nuclear proapoptotic factor that links prostate-apoptosis-response-4 (Par-4) to PML nuclear bodies. *Oncogene* **22,** 2432–2442.

Roy, K., Kuznicki, K., Wu, Q., Sun, Z., Bock, D., Schutz, G., Vranich, N., and Monaghan, A. P. (2004). The Tlx gene regulates the timing of neurogenesis in the cortex. *J. Neurosci.* **24,** 8333–8345.

Rusnati, M., Urbinati, C., Tanghetti, E., Dell'Era, P., Lortat-Jacob, H., and Presta, M. (2002). Cell membrane GM1 ganglioside is a functional coreceptor for fibroblast growth factor 2. *Proc. Natl. Acad. Sci. USA* **99,** 4367–4372.

Sailer, M. H., Hazel, T. G., Panchision, D. M., Hoeppner, D. J., Schwab, M. E., and McKay, R. D. (2005). BMP2 and FGF2 cooperate to induce neural-crest-like fates from fetal and adult CNS stem cells. *J. Cell Sci.* **118,** 5849–5860.

Samanta, J., and Kessler, J. A. (2004). Interactions between ID and OLIG proteins mediate the inhibitory effects of BMP4 on oligodendroglial differentiation. *Development* **131,** 4131–4142.

Sanada, K., and Tsai, L. H. (2005). G protein betagamma subunits and AGS3 control spindle orientation and asymmetric cell fate of cerebral cortical progenitors. *Cell* **122,** 119–131.

Sanai, N., Alvarez-Buylla, A., and Berger, M. S. (2005). Neural stem cells and the origin of gliomas. *N. Engl. J. Med.* **353,** 811–822.

Sanai, N., Tramontin, A. D., Quinones-Hinojosa, A., Barbaro, N. M., Gupta, N., Kunwar, S., Lawton, M. T., McDermott, M. W., Parsa, A. T., Manuel-Garcia Verdugo, J., Berger, M. S., and Alvarez-Buylla, A. (2004). Unique astrocyte ribbon in adult human brain contains neural stem cells but lacks chain migration. *Nature* **427,** 740–744.

Schneller, M., Vuori, K., and Ruoslahti, E. (1997). Alphavbeta3 integrin associates with activated insulin and PDGFbeta receptors and potentiates the biological activity of PDGF. *EMBO J.* **16,** 5600–5607.

Scolding, N., Franklin, R., Stevens, S., Heldin, C. H., Compston, A., and Newcombe, J. (1998). Oligodendrocyte progenitors are present in the normal adult human CNS and in the lesions of multiple sclerosis. *Brain* **121**(Pt. 12), 2221–2228.

Scolding, N. J., Rayner, P. J., and Compston, D. A. (1999). Identification of A2B5-positive putative oligodendrocyte progenitor cells and A2B5-positive astrocytes in adult human white matter. *Neuroscience* **89,** 1–4.

Sela-Donenfeld, D., and Wilkinson, D. G. (2005). Eph receptors: Two ways to sharpen boundaries. *Curr. Biol.* **15,** R210–R212.

Shah, N. M., and Anderson, D. J. (1997). Integration of multiple instructive cues by neural crest stem cells reveals cell-intrinsic biases in relative growth factor responsiveness. *Proc. Natl. Acad. Sci. USA* **94,** 11369–11374.

Shi, Y., Chichung Lie, D., Taupin, P., Nakashima, K., Ray, J., Yu, R. T., Gage, F. H., and Evans, R. M. (2004). Expression and function of orphan nuclear receptor TLX in adult neural stem cells. *Nature* **427,** 78–83.

Sjostrom, S. K., Finn, G., Hahn, W. C., Rowitch, D. H., and Kenney, A. M. (2005). The Cdk1 complex plays a prime role in regulating N-myc phosphorylation and turnover in neural precursors. *Dev. Cell* **9,** 327–338.

Sommer, L. (2004). Multiple roles of canonical Wnt signaling in cell cycle progression and cell lineage specification in neural development. *Cell Cycle* **3,** 701–703.

Stromblad, S., Becker, J. C., Yebra, M., Brooks, P. C., and Cheresh, D. A. (1996). Suppression of p53 activity and p21WAF1/CIP1 expression by vascular cell integrin alphaVbeta3 during angiogenesis. *J. Clin. Invest.* **98,** 426–433.

Stupack, D. G. (2005). Integrins as a distinct subtype of dependence receptors. *Cell Death Differ* **12,** 1021–1030.

Sun, Y., Nadal-Vicens, M., Misono, S., Lin, M. Z., Zubiaga, A., Hua, X., Fan, G., and Greenberg, M. E. (2001). Neurogenin promotes neurogenesis and inhibits glial differentiation by independent mechanisms. *Cell* **104,** 365–376.

Takizawa, T., Nakashima, K., Namihira, M., Ochiai, W., Uemura, A., Yanagisawa, M., Fujita, N., Nakao, M., and Taga, T. (2001). DNA methylation is a critical cell-intrinsic determinant of astrocyte differentiation in the fetal brain. *Dev. Cell* **1,** 749–758.

Tropepe, V., Coles, B. L., Chiasson, B. J., Horsford, D. J., Elia, A. J., McInnes, R. R., and van der Kooy, D. (2000). Retinal stem cells in the adult mammalian eye. *Science* **287,** 2032–2036.

Tropepe, V., Craig, C. G., Morshead, C. M., and van der Kooy, D. (1997). Transforming growth factor-alpha null and senescent mice show decreased neural progenitor cell proliferation in the forebrain subependyma. *J. Neurosci.* **17,** 7850–7859.

Valk-Lingbeek, M. E., Bruggeman, S. W., and van Lohuizen, M. (2004). Stem cells and cancer; the polycomb connection. *Cell* **118,** 409–418.

Vicario-Abejon, C., Johe, K. K., Hazel, T. G., Collazo, D., and McKay, R. D. (1995). Functions of basic fibroblast growth factor and neurotrophins in the differentiation of hippocampal neurons. *Neuron* **15,** 105–114.

Viti, J., Gulacsi, A., and Lillien, L. (2003). Wnt regulation of progenitor maturation in the cortex depends on Shh or fibroblast growth factor 2. *J. Neurosci.* **23,** 5919–5927.

Wang, G., Silva, J., Krishnamurthy, K., Tran, E., Condie, B. G., and Bieberich, E. (2005). Direct binding to ceramide activates protein kinase Czeta before the formation of a pro-apoptotic complex with PAR-4 in differentiating stem cells. *J. Biol. Chem.* **280,** 26415–26424.

Weinberg, R. A. (1995). The retinoblastoma protein and cell cycle control. *Cell* **81,** 323–330.

Wu, W., Wong, K., Chen, J., Jiang, Z., Dupuis, S., Wu, J. Y., and Rao, Y. (1999). Directional guidance of neuronal migration in the olfactory system by the protein Slit. *Nature* **400,** 331–336.

Yamashita, T., Hashiramoto, A., Haluzik, M., Mizukami, H., Beck, S., Norton, A., Kono, M., Tsuji, S., Daniotti, J. L., Werth, N., Sandhoff, R., Sandhoff, K., *et al.* (2003). Enhanced insulin sensitivity in mice lacking ganglioside GM3. *Proc. Natl. Acad. Sci. USA* **100,** 3445–3449.

Zencak, D., Lingbeek, M., Kostic, C., Tekaya, M., Tanger, E., Hornfeld, D., Jaquet, M., Munier, F. L., Schorderet, D. F., van Lohuizen, M., and Arsenijevic, Y. (2005). Bmi1 loss produces an increase in astroglial cells and a decrease in neural stem cell population and proliferation. *J. Neurosci.* **25,** 5774–5783.

Zhang, Z., Vuori, K., Reed, J. C., and Ruoslahti, E. (1995). The alpha 5 beta 1 integrin supports survival of cells on fibronectin and up-regulates Bcl-2 expression. *Proc. Natl. Acad. Sci. USA* **92,** 6161–6165.

Zhu, L. L., Wu, L. Y., Yew, D. T., and Fan, M. (2005). Effects of hypoxia on the proliferation and differentiation of NSCs. *Mol. Neurobiol.* **31,** 231–242.

CHAPTER 3

Retinal Stem Cells

Thomas A. Reh,[*] **Joseph A. Brzezinski IV,**[*] **and Andy J. Fischer**[†]

[*]Department of Biological Structure
University of Washington
Seattle, Washington

[†]Department of Neuroscience
The Ohio State University
Columbus, Ohio

Abstract
1. Introduction
    1.1. Retinal stem/progenitor cells during development
    1.2. Retinal stem cells and the ciliary marginal zone
    1.3. Other sources of retinal stem cells
2. Materials and methods
    2.1. Primary cell culture of retinal stem/progenitors cells
    2.2. *In vivo* methods for the study of retinal/stem progenitors
3. Overview
    References

Abstract

During the embryonic development of the eye, a group of founder cells in the optic vesicle gives rise to multipotent progenitor cells that generate all the neurons and the Muller glia of the mature retina. In most vertebrates, a small group of retinal stem cells persists at the margin of the retina, near the junction with the ciliary epithelium. In fish and amphibians, the retinal stem cells continue to produce progenitors throughout life, adding new retina to the periphery of the existing retina as the eye grows. In birds, the new retinal addition is more limited, while it is absent in those mammals that have been analyzed. Nevertheless, cells from the retinal periphery and ciliary body of mammals can be isolated and grown *in vitro* for extended periods. Methods for the study of both embryonic progenitors and adult retinal stem cells, *in vitro* and *in vivo,* have led to a better understanding of retinal development, allowed for the screening of factors important in retinal growth and differentiation, and enabled the development of methods to direct stem and progenitor cells to specific fates. These methods may ultimately lead to the development of strategies for retinal repair.

1. Introduction

1.1. Retinal stem/progenitor cells during development

The vertebrate retina arises during development as an evagination, called the optic vesicle, of the diencephalon of the neural tube. The cells in the early optic vesicle express a unique complement of transcription factors, termed eye-field transcription factors (ETFs), including Pax6, Rx, Six3, and Chx10 (see Zuber *et al.*, 2003 for review). On the basis of this combination of transcription factors, the cells of the optic vesicle can be distinguished from the other cells of the neural tube. The optic vesicle cells undergo extensive divisions to generate all the neurons and the Muller glia of the adult retina. Analysis of the lineages of the proliferating cells from the early stages of eye development indicates that they undergo both symmetric and asymmetric divisions, and that many form clones containing very large numbers of cells (2795 cells; Fekete *et al.*, 1994). The majority of the clones obtained from labeling the dividing cells at early stages of retinal development contain multiple types of retinal neurons, as well as Muller glia. Thus, these cells are known as multipotent retinal progenitors.

Although the clonal analysis of retinal progenitor cells demonstrates a wide variety of clone size and composition, some patterns have emerged. First, ever since the first birthdating studies of Sidman (1961), it has been consistently found that the different types of retinal neurons are generated in a sequence, with ganglion cells, cone photoreceptors, amacrine cells, and horizontal cells generated during early stages of development, and most rod photoreceptors, bipolar cells, and Muller glia generated in the latter half of the period of retinogenesis. This has led to the hypothesis that retinal progenitor cells undergo a progressive change during development that constrains them to a smaller range of fates (Reh and Kljavin, 1989). However, an alternative model is that a changing environment directs the cells to progressively later fates, but the progenitor cells themselves remain competent to generate all retinal cell types throughout the period of retinogenesis (James *et al.*, 2003) There is experimental support for both models, and both the environment and intrinsic state of the cell are likely to be important factors in determining its ultimate fate (see Reh and Cagan, 1994; Livesey and Cepko, 2001 for review).

The first evidence that progenitor cells at a given stage of retina development are not all identical came from an analysis of their proneural gene expression. Retinal progenitor cells express at least one of the following members of this class: Ascl1, Ngn2, and NeuroD1. The bHLH transcription factor Ascl1 (formerly known as

Mash1 or Cash1) is expressed in only a subset of retinal progenitors (Jasoni and Reh, 1995; Jasoni *et al.*, 1994), and the proneural gene for the remaining progenitors appears to be either Ngn2 or NeuroD1 (Akagi *et al.*, 2004). More recently, another class of transcription factors, the Fox class, also reveals heterogeneity in the progenitor pool. Foxn4 is expressed in a subset of progenitors and specifically biases them to generate either amacrine or horizontal cells (Li *et al.*, 2004). Retinal progenitors can also be distinguished by their response to growth factors. Progenitors isolated from the early embryonic retina are stimulated to proliferate by FGF, but are only minimally responsive to epidermal growth factor (EGF) or TGFa (Anchan and Reh, 1996; Lillien and Cepko, 1992). By embryonic day 18 in the rat, the progenitors have now acquired a robust response to EGF (Anchan *et al.*, 1991). In addition, progenitor cells differ in their response to changes in intracellular cAMP. Postnatal progenitors are induced to differentiate by treatments that raise cAMP, while embryonic progenitors are not (Taylor and Reh, 1989). The evidence for progenitor heterogeneity is somewhat at variance with the indeterminate lineages of single progenitor cells. One possibility is that the different types of progenitor cells can interconvert. For example, loss of Ascl1 causes an expansion of the Ngn2 expressing progenitors (Akagi *et al.*, 2004). There is also evidence from other regions of the developing CNS that FGF-responsive neural stem cells can be converted to EGF-responsive stem cells (Ciccolini and Svendsen, 1998).

In summary, the majority of mitotically active cells in the embryonic retina are competent to generate multiple different types of retinal neurons, as well as Muller glia. Although these cells are typically referred to as multipotent progenitors, those isolated from early stages of retinogenesis could also be considered retinal stem cells on the basis of their potential to generate all retinal cell types. Moreover, the fact that they can generate very large clones indicates that many of their divisions are symmetric. In addition, several groups have shown the early progenitors retain the capacity to proliferate *in vitro* for extended periods of time, and can be cultured as "neurospheres"; a capacity that neural stem cells are known to possess (see e.g., Klassen *et al.*, 2004). As will be described below, the adult retina of some vertebrates continues to add new neurons and glia at the peripheral margin, and thus, true retina stem cells exist. Presumably, these cells were derived from a population of similar cells in the developing retina, but at this point, there is no definitive way to distinguish the stem cells from the progenitors during retinogenesis.

1.2. Retinal stem cells and the ciliary marginal zone

In many vertebrates, the development of the retina is not complete after the embryonic or neonatal period, but rather continues throughout life. This is most dramatically observed in teleosts; from the time of their hatching to when they reach their mature size, the eye of a teleost fish can grow 100-fold. The cellular mechanisms that enables new retinal neurons to be generated throughout the lifetime of the fish is found at the peripheral margin of the retina, where it joins with the ciliary epithelium. In this region, there is a small cluster of cells that forms a ring around the ciliary margin of the retina, the so-called ciliary marginal zone (CMZ; Hollyfield, 1968).

The cells of the CMZ in fish and frogs resemble the early progenitor cells of the eye, and possibly even the "founder" cells of the optic vesicle. Lineage tracing studies of the CMZ cells have shown that they can give rise to clones that contain all types of retinal neurons, including those that are generated both early and late in embryonic development (Wetts and Fraser, 1988). Most of the CMZ cells express the transcription factor profile of retinal progenitors, including the paired-class transcription factors, such as Rx, Chx10, and Pax6 (Perron *et al.*, 1998). The CMZ cells also express proneural transcription factors, like Ngn2 and Ascl1 (Harris and Perron, 1998), and at least some of them respond to mitogenic growth factors like their counterparts in the embryo (Mack and Fernald, 1993). Since these cells are capable of generating most of the retina of the mature frog (Reh and Constantine-Paton, 1983) or fish, and they continue to generate new retina throughout the lifetime of the animal, it is likely that this region contains a population of true retinal stem cells. This zone of cells is extremely productive in the fish and larval frog, but in other vertebrates, it is greatly reduced or absent. While it is relatively easy to identify the CMZ cells in fish and amphibians, it is not known how many cells in this zone represent true retinal stem cells and what proportion of them are progenitors.

In the eyes of amphibians and teleost fish, the retina continues to grow in parallel to the overall growth of the eye, whereas in birds most of the retina is generated *in ovo* with at least 90% of the retinal cells generated more than one week before hatching (Prada *et al.*, 1991). At the time of hatching, most birds have a fully functional retina, and it was generally assumed that the retina of postembryonic birds lacked the CMZ. However, a study nearly 30 years ago, first described the addition of newly generated cells to the peripheral edge of the postnatal chicken retina (Morris *et al.*, 1976). This work had gone largely unnoticed until several years ago when we demonstrated that new retinal neurons are generated at peripheral edge of the retina in chickens up to one month of age (Fischer and Reh, 2000). This CMZ-like zone has also been identified in the adult quail eye (Kubota *et al.*, 2002), and therefore may be a common feature of the avian eye. In addition to their potential to generate new retinal neurons, the chicken CMZ cells express a number of different genes that are also expressed by embryonic retinal progenitor cells. These genes include Pax6, Chx10, PCNA (Fischer and Reh, 2000), Notch1, cHairy (Fischer, 2005), transitin, the avian homologue of mammalian nestin (Fischer and Omar, 2005), Gli1, Gli3 (Moshiri *et al.*, 2005), and Cath5 when stimulated by the combination of insulin and FGF2 (Fischer *et al.*, 2002).

Although the zone of proliferating cells in the postnatal chicken eye is reminiscent of the CMZ of fish and amphibians, the CMZ of chickens appears to generate only bipolar and amacrine cells (Fischer and Reh, 2000). We fail to find evidence for the production of photoreceptor, horizontal, or ganglion cells in the untreated chicken CMZ, suggesting that the progenitors in the CMZ may be limited to producing particular neuronal cell types. However, we found that the combination of insulin and FGF2 stimulated to production of ganglion cells, suggesting that the types of cells produced in the avian CMZ are limited by local microenvironment rather than cell-intrinsic limitations that restrict the multipotency of CMZ progenitors (Fischer *et al.*, 2002). Unlike the CMZ progenitors in cold-blooded vertebrates, the CMZ progenitors in birds do not regenerate the retina when it has been damaged (see Moshiri *et al.*, 2004). Toxic doses of NMDA or kainic acid, that destroy numerous retinal neurons, do not stimulate the proliferation of CMZ progenitors (Fischer, 2005; Fischer and Reh, 2000).

The proliferation of cells in the normal posthatch chicken CMZ is relatively modest, but can be increased as much as 10-fold by intraocular delivery of growth factors (Fischer and Reh, 2000). Growth factors that stimulate the proliferation of CMZ progenitors include insulin, insulin-like growth factor I (IGF-I), EGF (Fischer and Reh, 2000, 2002), and sonic hedgehog (Shh) (Moshiri *et al.*, 2005), but not fibroblast growth factor (Fischer and Reh, 2000). Since the levels of proliferation are low in an untreated CMZ, it is possible that the factors that stimulate proliferation and neuronal differentiation are present in limiting quantities in the postnatal retina (Fischer and Reh, 2000). Alternatively, the proliferation of progenitors in the CMZ may be suppressed by factors produced by mature retinal neurons. For example, an unusual type of glucagon-expressing neuron within the retina produces neurites that project into and densely ramify within the CMZ and glucagon acts to suppress the proliferation of CMZ progenitors (Fischer *et al.*, 2005a). In addition to exogenous growth factors, the proliferation of CMZ progenitors can be increased by experimentally increasing rates of ocular growth. Visual deprivation nearly doubles the number of cells that are added to the edge of the retina (Fischer and Reh, 2000). The mechanisms that link postnatal ocular growth and the addition of cells to the peripheral edge of the retina may involve glucagon-expressing retinal neurons that are known to respond to growth-guiding visual cues (Fischer *et al.*, 1999) and influence the proliferation of CMZ progenitors (Fischer *et al.*, 2005a).

1.3. Other sources of retinal stem cells

1.3.1. The ciliary epithelium

The ciliary epithelia of the ciliary body, like the retina, are derived from the optic vesicle during embryonic development and have been shown to contain cells with neurogenic potential. For example, the nonpigmented epithelium (NPE) of the ciliary body is capable of producing neurons in the intact chicken eye (Fischer and Reh, 2003). Intraocular injections of growth factors (insulin, FGF2, and EGF) stimulate the proliferation and neuronal differentiation of NPE cells within the ciliary body (Fischer and Reh, 2003). Like the CMZ, the NPE cells express Chx10 and Pax6, but are found up to 3 mm anterior to the peripheral edge of the retina. This region of Pax6/Chx10-expressing cells in the NPE of the ciliary body coincides with the region where proliferating and newly generated neurons appear in response to growth factor-treatments. Newly generated neurons in the NPE express markers for amacrine cells, ganglion cells, and Müller glia, but not for bipolar cells or photoreceptors. The potential for the ciliary epithelium and adjacent iris to generate neurons may extend to the mammalian retina; forced expression of the paired-class homeodomain transcription factor Crx induces the expression photoreceptor genes in cells derived from the rat iris (Haruta *et al.*, 2001). Neuron-like cells have been identified in the NPE of adult nonhuman primates (Fischer *et al.*, 2001). Furthermore, a report from Zhao *et al.* (2002) has indicated that the blockade of BMP signaling interferes with the normal formation of the NPE of the ciliary body and promotes ectopic neural differentiation in the developing NPE of the rodent eye. Several groups have also reported that extended culture of both the pigmented and nonpigmented cells of the ciliary epithelium results in progenitor-like cells. These cells form neurospheres and can be passaged at least once. Moreover, the cells cultured from the pigmented ciliary epithelium express many of the markers of retinal progenitors and their progeny express proteins normally present in subtypes of retinal neurons. As a result of these characteristics, the cells have been termed retinal stem cells (Ahmad *et al.*, 2000; Coles *et al.*, 2004; Tropepe *et al.*, 2000). The relationship between the sphere-forming pigmented cells and the true retinal stem cells present in the CMZ of fish and frogs is not clear, since the latter are not thought to be pigmented. In addition, it is not clear how either of these cell types relate to the "founder" cells of the optic vesicle that produce all the progenitors of the retina. These issues may eventually be resolved by developing better markers that discriminate among the different types of "retina stem cells" (Fig. 3.1).

1.3.2. The pigmented epithelium

It has been known for over half a century that the pigmented cells of the eye are capable of acting as a source of retinal regeneration (or neural stem cells) (Coulombre and Coulombre, 1965; Orts-Llorca and Genis-Galvez, 1960, Reh *et al.*, 1987; Stone, 1950; Stone and Steinitz, 1957). The RPE is a well-known source of retinal stem cells in neotenic amphibians, larval anurans, and embryonic chicks (see Moshiri *et al.*, 2004 for review). Neurogenesis from RPE cells requires dedifferentiation, loss of pigmentation, and cell division (Stone, 1950, Stroeva and Mitashov, 1983). Collectively, this process has been named transdifferentiation (Okada, 1980). RPE cells that have been stimulated to transdifferentiate produce new neurons in a manner that resembles normal retinal histogenesis (Reh *et al.*, 1987, Sakaguchi *et al.*, 1997). In the embryonic chick and rodent, the ability of RPE cells to become retinal stem cells is lost during early stages of development (Park and Hollenberg, 1993; Pittack *et al.*, 1991; Zhao *et al.*, 1995). During the dedifferentiation process, the pigmented epithelial cells acquire a gene expression profile that resembles the retinal progenitor cells (Sakami *et al.*, 2005). It is possible that these cells go through a stage where they resemble stem or "founder" cells, because the RPE cells can regenerate the entire retina in some species, up to four complete times (Stone, 1957).

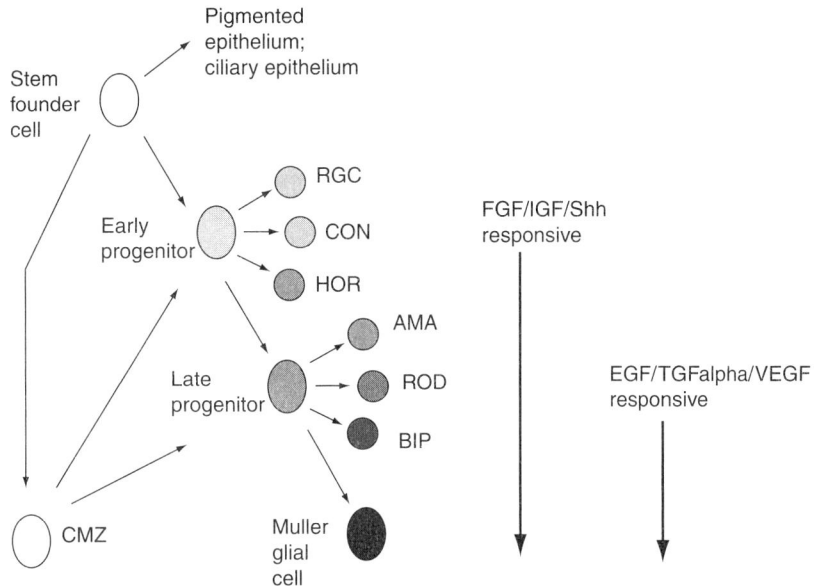

Figure 3.1 Diagram showing the possible relationships among the various types of retinal progenitor cells and retinal stem cells. The early optic vesicle is composed of cells that ultimately give rise to all the retinal cells, as well as the pigmented epithelium and ciliary epithelium. Progenitor cells in the retina generate the neurons and ultimately the Muller glia and in nonmammalian vertebrates there is a specialized zone of proliferating cells at the junction between the retina and the ciliary epithelium, called the CMZ. At least some cells of the CMZ or adjacent ciliary epithelium must be true retinal stem cells, since these cells are capable of generating all of the various types of retinal neurons and Muller glia and persist throughout the life of the animal.

1.3.3. Müller glia

Mature Müller glia, the major type of support cells in the retina, are capable of dedifferentiating into proliferating progenitor-like cells in the retinas of chickens (Fischer and Reh 2001a; Fischer *et al.*, 2002b), zebrafish (Yurco and Cameron, 2005), and rat (Ooto *et al.*, 2004). Under normal conditions, Müller glia are the predominant type of support cell in retina providing structural, nutritive, and metabolic support to retinal neurons. In response to sufficient retinal damage, or exposure to the combination of insulin and FGF2 without damage (Fischer *et al.*, 2002b), Müller glia reenter the cell cycle and express transcription factors found in embryonic retinal progenitors. These transcription factors include Ascl1, Pax6, Chx10 (Fischer and Reh, 2001a), and Six3 (Fischer, 2005). While Müller glia undergo only 1 round of division *in vivo*, these cells continue to proliferate and produce some new neurons when dissociated from the intact retina and are grown in culture (Fischer and Reh, 2001a). *In vivo*, the majority (about 80%) of cells that are generated by proliferating Müller glia remain as undifferentiated progenitor-like cells, while some differentiate into Müller glia and a few differentiate into amacrine or bipolar neurons. Destruction of ganglion cells, combined with insulin/ FGF2-treatment, stimulated the regeneration of a few ganglion cells (Fischer and Reh, 2002).

In the adult zebrafish, Yurco and Cameron (2005) have recently reported that acute lesions to the retina result in the reentry of Muller glia into the cell cycle. This study indicated that the proliferating Muller glia become progenitor-like and suggested that the glia may regenerate neurons in the damaged teleost retina.

In the rat retina, Müller glia have been shown to be a potential source of retinal regeneration (Ooto *et al.*, 2004; see Lamba *et al.*, 2008 for review). Similar to studies in the chicken retina, Ooto and colleagues used *N*-methyl-D-aspartate to induce excitoxicity and damage the adult rat retina. In response to damage, Muller glia were stimulated to proliferate and produce new neurons. Although these newly produced neurons were limited in number, numbers of regenerated neurons were increased by treatment with retinoic acid or the misexpression of basic helix-loop-helix and homeobox genes. These findings suggest the Muller glial cells are a potential source of neural regeneration in the adult mammalian retina, but may require stimulation (drugs and/or gene therapy) to regenerate significant numbers of neurons to treat retinal degenerative diseases

2. Materials and methods

2.1. Primary cell culture of retinal stem/progenitors cells

In order to establish and maintain cell cultures of retinal progenitors/stem cells, the optimal source is embryonic retina from either rodent or chick. The following methods work well for chick embryos from stages 25–35 (embryonic days 4–8), and for either rat or mouse from embryonic days 14.5 to birth and up to postnatal day 7 (Anchan *et al.*, 1991; Kelley *et al.*, 1994; Levine *et al.*, 1997; Reh and Kljavin, 1989). After postnatal day 7 in the rodent, there are few progenitor/stem cells, and the vast majority of cells that proliferate *in vitro* after this stage are likely Muller glia (Close *et al.*, 2005). We have also used the same methods for fetal human retina, up to day 70 post conception (Kelley *et al.*, 1995).

2.1.1. Harvest and culture of retinal stem/progenitor cells

1. The embryos are harvested or the pup is sacrificed in accordance with approved protocols for the institution.

2. The eyes are removed and placed in a sterile Petri dish containing cold (4 °C) Hanks's Balanced Salt Solution added with 3% D-glucose and 0.01 M HEPES, pH 7.4 (HBSS+).

3. The retinas are dissected from the extraocular tissue in this solution. The pigmented epithelium, lens, and scleral tissue are easily removed, and the retina is then transferred to a new sterile Petri dish containing (Calcium/Magnesium Free) CMF–HBSS using forceps or a Pasteur pipette.

4. The retinas are then transferred to a 15 ml centrifuge tube containing 0.25% trypsin, in HBSS–CMF (stock at 2.5%; 0.5 ml plus 4.5 ml HBSS–CMF).

5. The centrifuge tube containing the retinas is then put on a rocking platform in a 37 °C incubator and gently rocked for 5–15 min. As a guideline, embryonic chick retinas typically require only 5 min, while postnatal mouse retina can require up to 10 min for thorough dissociation. In all cases, however, care must be taken not to over-treat the retinas with trypsin, since this will result in low yields of viable progenitor cells.

6. The trypsin treatment is complete when the retinas are broken up into small pieces, but not yet into single cells. Add 0.5 ml FBS to the tube to inactivate the trypsin.

7. Centrifuge the tube 1500 rpm ($\sim750 \times g$) for 10 min to pellet the cells.

8. Carefully remove the supernatant (leaving a small amount of solution at the bottom, so that the pellet is not disturbed).

9. Resuspend the cell pellet cells in 2 ml of medium (see below) with gentle trituration using a (fire-polished) Pasteur pipette.

10. Determine the number of cells using a haemocytometer; trypan blue can be used to estimate the percentage viability.

11. Plate cells between 50,000 (low density) and 500,000 (high density) cells per well (for a 24 well plate) onto poly-D-lysine/Matrigel coated coverslips (see below). Cultures are maintained at 37 °C and 5% CO_2.

The culture medium contains DMEM/F12 (without glutamate or aspartate), 25 μg/ml insulin, 100 μg/ml transferrin, 60 μM putrescine, 30 nM selenium, 20 nM progesterone, 100 U/ml penicillin, 100 μg/ml streptomycin, 0.05 M HEPES, and 1% fetal bovine serum (Invitrogen). We prefilter all the stock solutions so there is no need to filter the final medium; however, if contamination is suspected, we have found that the medium can be filtered once without loss of potency. Once the retinal cells are established *in vitro*, one half of the medium is replaced every other day. The final medium is effective for $\sim$1 week when stored at 4 °C. The progenitor cells of chick or rodent retina can be maintained in this medium for up to 1 week and they retain their ability to generate neurons, as demonstrated by double labeling for BrdU/thymidine and neuronal-specific markers (see below). We typically use a serum concentration of 1%, but this can be reduced to 0.1% with little reduction in proliferation. The medium can also be supplemented with growth factors to stimulate the proliferation of the progenitor cells. We have also used the same medium for serum-free cultures, but under these conditions, the progenitor cells are more likely to differentiate (Fig. 3.2).

2.1.2. Substrate for adherent culture of retinal stem/progenitor cells: poly–lysine/matrigel

The retinal progenitor/stem cells proliferate in adherent cultures on glass coverslips coated with poly-D-lysine (PDL) and Matrigel. We have also maintained these cells on laminin coated coverslips and our recent experience indicates this is as

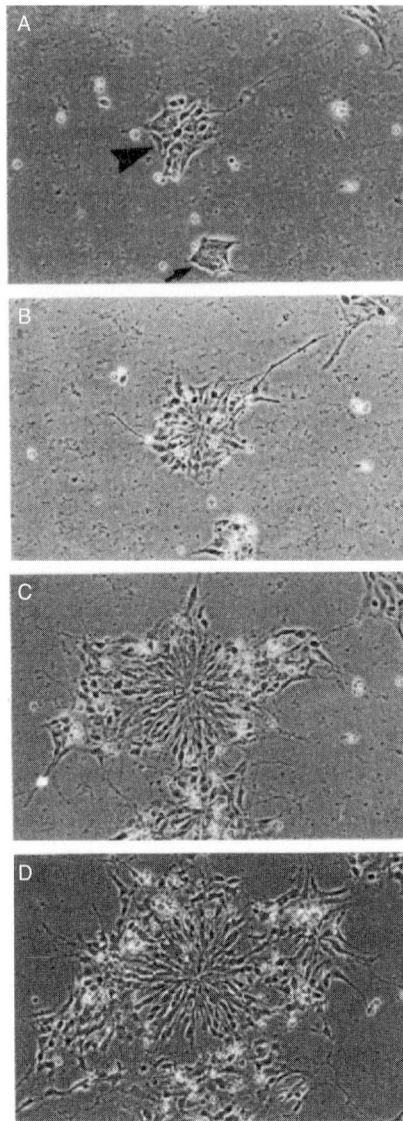

Figure 3.2 Stem/progenitor cells grown *in vitro* on adherent substrates and supplemented with TGF-alpha. The same cluster of cells was continuously monitored with time-lapse microscopy over several days. The arrowhead in (A) points to a small group of cells that eventually grows to a large rosette, containing both dividing cells and differentiated neurons (as determined by subsequent immunolabeling for retinal-neuron-specific antigens.

effective and much easier. However, the method we have used in previous publications is given here.

1. To prepare the coverslips, use high molecular weight PDL (30–70,000 MW). The PDL is dissolved in sterile water at a concentration of 0.5 mg/ml and aliquotted into sterile 15 ml conical tubes for storage at −20 °C.

2. Prior to use, a tube is thawed, and 9 ml of sterile water is added for a final concentration of 50 μg/ml. The coverslips used most frequently are 12 mm circular. They are placed into a small vial for sterilization with an autoclave.

3. Typically the retinal cells from a litter of E18 rats or from one E5–7 chick embryo are plated in 24 wells, and so 25–30 coverslips are placed in a large Petri dish for coating.

4. The PDL solution is added to the dish and care is taken to ensure all coverslips have sunk in the solution, and the coverslips are incubated 37 °C for 15–30 min.

5. The PDL solution is removed and the coverslips are washed in sterile water 3 times for 5 min each. Care is taken to wash the coverslips very well, mixing the coverslips with flamed forceps, since poly-L-lysine in solution can be toxic to cells. The coverslips can be dried and stored for up to 2 weeks in the Petri dish at 4 °C, or used immediately.

6. When ready to use, put one coverslip in each well of a 24 well plate using flamed forceps. Then proceed to coat them with Matrigel. Matrigel is supplied by the manufacturer as a frozen solution. Thaw the bottle slowly on ice (for several hours) to prevent gel formation. Small (200 μl) aliquots are made using precooled tubes (15 ml) on ice and a prechilled pipette. *If the Matrigel warms during the aliquotting, it will gel and not be effective for the cell cultures.* The aliquots are stored at −20 °C for up to 6 months.

7. To coat the coverslips, remove one aliquot of Matrigel from the −20 °C freezer and place on ice for 15–30 min to thaw (200 μl is used for one 24-well plate).

8. Add 10 ml of ice cold HBSS+ to the 15 ml tube containing 200 μl of thawed Matrigel. Mix gently.

9. Immediately, put 0.5 ml of the dilute Matrigel solution into each well of a 24-well plate in which you have already placed poly-lysine coated coverslips and place the plate in the incubator for 30 min at 37 °C.

10. Remove the plate from the incubator, and under the sterile hood, remove nearly all of the liquid from the wells, A small amount of the dilute Matrigel solution is left in the well just to cover the coverslips.

11. Let the plate dry in the hood uncovered for 15–30 min. The Matrigel will dry into a thin coating.

12. Plate cells onto the Matrigel, or store the plate at 4° for not more than 2 days prior to plating the cells.

2.1.3. N–cadherin substrates

An alternative substrate that works very well for adherent dissociated cultures of retinal stem/progenitor cells is N-cadherin. N-cadherin and other cell adhesion molecules can be applied to coverslips with the following protocol.

1. To coat glass with N-cadherin, first coat the coverslips (12 mm) in a 24-well plate with PDL, by incubating at 37 °C for ~30 min.
2. The coverslips are then washed 3 times with sterile water to remove excess (toxic) PDL.
3. Next, a small piece of nitrocellulose filter (~4 cm^2) is dissolved in 10 ml of MeOH (HPLC grade); this can be stored at 4 °C (up to ~1 month) in a 50 ml conical centrifuge tube.
4. To coat the coverglass with nitrocellulose solution, use a 200 μl Pipettman to apply 2 drops of solution to each coverslip.
5. Allow the nitrocellulose to "dry" (~1 min). Repeat this step.
6. Then dilute N-cadherin stock to 20 μg/ml with water. Stock should be 200 μg/ml in 0.1% BSA. Apply a dot (12–15 μl) of N-cadherin to the middle of the nitrocellulose coated coverslip.
7. Gently spread the dot out to cover more of the coverslips surface (do not scratch off the coating though).
8. Incubate the coverslips and plate at room temperature for ~15 min.
9. Wash coverslips once with culture media (DMEM/F12 works fine) to remove excess N-cadherin and nitrocellulose.
10. Block nitrocellulose with 10% FBS in DMEM/F12 for 30 min at 37 °C.
11. Wash coverslips with HBSS to remove excess serum. Finally, plate the cells.

2.1.4. Analysis of culture of retinal stem/progenitor cells

One of the primary methods used for the characterization of retinal stem/progenitor cells and their progeny is immunofluorescent labeling. With this technique, the overall number of stem/progenitor cells in the cultures can be estimated, and their potential for generation of the different types of retinal neurons can be evaluated. The techniques for labeling the dissociated retinal stem/progenitor cell cultures on coverslips are presented, because this is typically how we do the analysis; however, we have successfully used the same procedures when the cells have been directly cultured in the tissue culture wells without coverslips.

To label the cultures for stem/progenitor-specific antigens or retinal neuron-specific antigens, the coverslips are fixed in 4% paraformaldehyde in PBS (phosphate-buffered saline; 0.05 M sodium phosphate, 195 mM NaCl, pH 7.4) for 30 min. The fix is removed and replaced with PBS, and the plates are stored in the cold room for up to a week before staining. In preparation for immunofluorescent labeling, the coverslips are incubated with a "blocking solution" to prevent nonspecific binding of the antibody and reduce background fluorescence. The blocking solution is PBS with 0.3% Triton-X100 and 5% serum (goat serum for nongoat antibodies; FBS for goat antibodies.) The primary antibody is then diluted in the blocking solution and the coverslips are incubated in the primary antibody overnight. The primary antibody solution is then removed and the coverslips are washed three times with 10–15 min incubations in PBS. The secondary antibodies are diluted in PBS and 0.3% Triton-X100 and incubated for 1.5 h in the dark at room temperature. The coverslips are then rinsed three times for 10–15 min in PBS, and then finally in water to remove residual salt. They are then dried at room temperature and mounted, cell side down, on slides, with a drop of Fluoromount. To double label for two or more different antibodies, we incubate the coverslips with multiple primary antibodies simultaneously, as long as they have been raised in different species (e.g., one raised in rabbit and one raised in mouse or rat). Multiple secondary antibodies are also added at the secondary antibody incubation step.

There are many different primary antibodies that can be used to label retinal stem/progenitor cells, though none is definitive. Studies of the developing retina have shown that Pax6, Chx10, Sox2, Prox1, Six3, nestin (transitin in the chick; mab73b6 Dr. J. Weston, University of Oregon), mushashi, vimentin, Mash1, and Ngn2 are expressed by retinal stem/progenitor cells during development and regeneration (Anchan *et al.*, 1991; Belecky-Adams *et al.*, 1997; Burmeister *et al.*, 1996; Fischer and Omar, 2005; Fischer and Reh, 2000; Jasoni and Reh, 1996; Mathers *et al.*, 1997; Oliver *et al.*, 1995). This is by no means an exhaustive list, however, antibodies raised against these antigens have been used to label cells in various species, and double-labeling with BrdU has demonstrated that these proteins are expressed by stem/progenitor cells. One problem with these markers is that they are not only expressed by stem/progenitor cells, but most are also expressed in one or more types of differentiated retinal neurons. Another problem is that many are also expressed in Muller glia (albeit at lower levels; Blackshaw *et al.*, 2003), and so they cannot distinguish between stem/progenitor cells and glia (though Mash1 and Ngn2 may be the exception and are not expressed in Muller glia). Lastly, none of these markers has been used to discriminate between a stem cell and a progenitor. With these caveats in mind, the use of these markers can be informative, however, the only definitive way to demonstrate that there is a stem/progenitor in the culture is to show that the cells can generate neurons and glia of the retina.

Immunofluorescent analysis of stem/progenitors has been used to demonstrate that neurons are generated *in vitro*, when a neuron-specific antibody is used in conjunction with BrdU. In this method, BrdU is added to the culture, and the mitotically active stem/progenitor cells incorporate the nucleotide into their DNA during S-phase. After several days *in vitro*, the stem/progenitor cells give rise to new neurons, and by double-labeling with both an antibody raised against BrdU and a neuron-specific antibody, one can definitively demonstrate that neurogenesis is occurring *in vitro* (see e.g., Anchan *et al.*, 1991; Fischer and Reh, 2001; Kelley *et al.*, 1994; Levine *et al.*, 1997; Reh, 1992). There are many good retinal-neuron-specific antibodies that give reliable labeling when used in conjunction with BrdU. For rod photoreceptors, we have used monoclonal antibodies raised against rhodopsin (3A6 and 4D2 from Dr. Bob Molday, University of British Columbia). To label all photoreceptors (and a few bipolar cells), we have used anti-recoverin antibodies (Dr. Jim Hurley, University of Washington). Recoverin has the advantage of being expressed more quickly after the final mitotic division in the new rods and cones, whereas rhodopsin is not expressed for several days after the progenitor cells have generated the new rod. More recently, antibodies generated against photoreceptor-specific transcription factors have been generated and are likely to be ideal for *in vitro* studies, due to their nuclear localization. The use of multiple colocalized markers, along with BrdU, is the best way to ensure that the specific cell type is generated in the cultures, and that indeed retinal stem/progenitor cells are present. Other retinal cell types can also be identified *in vitro*, using antibodies that are frequently used in other regions of the nervous system. TuJ1, an antibody directed against neuron-specific tubulin, will label most inner retinal neurons, including ganglion cells, amacrine cells, horizontal cells, and bipolar cells (though less well). We have also used commercially available antibodies raised against Hu, NeuN, Brn3, calbindin, and calretinin. The reader is referred to earlier publications for details of the different antibodies used to characterize the various types of retinal neurons.

To label for proliferating cells, we add BrdU (1–10 g/ml final concentration) to the cultures prior to fixation. To identify the cells that have incorporated the BrdU, we pretreat the coverslips with 4N HCl for 7–8 min, rinse 3×5 min with PBS, and block nonspecific labeling with 5% goat serum in ∼0.1% Triton-X in PBS. Next we dilute rat-anti BrdU 1:250 in 5% goat serum in ∼0.2% Triton-X in PBS and incubate overnight. The coverslips are then rinsed 3×5 min in PBS, and the appropriate secondary antibody is applied as above. As described above, the coverslips to be labeled can be incubated with two or more primary antibodies simultaneously. When labeling for BrdU alone, it is important to wash the cells with Tx-100 before the acid treatment. The acid treatment is far less effective if the cells have not been permeablized. Also, double-labeling for BrdU and neuronal markers often requires that the neuronal marker and appropriate secondary are applied first, followed by a brief fixation (2% PFA for 15 min), and then the acid treatment. Most antigens loose their antigenicity with the acid treatment (Fig. 3.3).

Although many different secondary antibodies are available and work well, in general Alexafluor-conjugated secondaries (Invitrogen, Molecular Probes) provide the brightest and most photostable choices. Alexafluor-conjugated secondary antibodies include anti-rabbit, anti-rat, and anti-mouse diluted to 1:1000 in PBS plus 0.2% Triton X-100. For triple-labeling, provided that your microscope is equipped with the appropriate filter-sets and a camera capable of detecting far red wavelengths, it is best to combine the Alexa-488 and Alexa-568 with an Alexa-647 (far red), versus and Alexa-350 (blue) because the Alexa-647 is brighter and far more photo-stable than the Alexa-350.

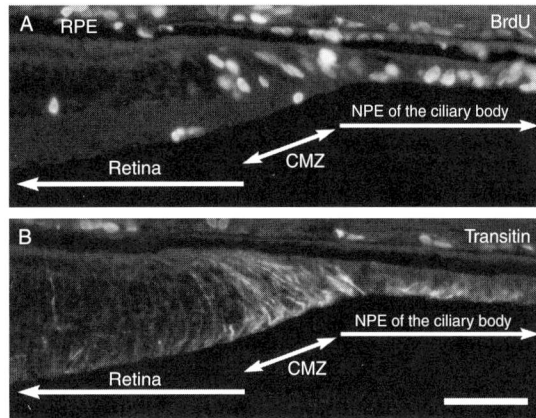

Figure 3.3 Photomicrographs of the CMZ of a posthatch chicken retina, showing (A) the mitotically active, BrdU-labeled cells in this zone after an intraocular injection of BrdU. Notice that cells in the ciliary epithelium (NPE) and the pigmented epithelium (RPE) are also labeled with BrdU. (B) The CMZ cells also label with transitin, a homolog to nestin, a neural stem, and progenitor marker.

2.2. *In vivo* methods for the study of retinal/stem progenitors

2.2.1. Intraocular injections

We have developed methods to study the retinal progenitors *in vivo*, both at the retinal margin, the ciliary epithelium and those derived from the Muller glia. The method described here details the intraocular injections and analysis of the tissue common to all these studies. The reader is referred to specific relevant publications for additional details relevant to specific experimental paradigms.

1. Prior to intraocular injections, posthatch chicks must be anesthetized. The simplest way to anesthetize a postnatal chicken is to place about 1 ml of a 4:1 mixture of mineral oil to halothane or isoflurane into the bottom of a large glass jar. Inhalation of the halothane or isoflurane vapors will render the chicks unconscious within 1 min.

2. Once removed from the jar, the animals will regain consciousness within 2 min. To keep the chickens unconscious for longer periods of time and to more precisely deliver anesthetic, machines equipped with the appropriate halothane/isoflurane vaporizer can be used.

3. Once the chickens are unconscious, the eyelids are swabbed with 70% ethanol in water or betadine solution to sterilize the area where needle will enter the eye.

4. Intravitreal injections can be made using standard 26G needles provided by Hamilton. The standard 26G needles are 55 mm long and are flexible, making it difficult to control during insertion into the eye and difficult to estimate whether the tip of the needle is in the desired location within the liquid vitreous. However, custom 22 mm 26G needles can be obtained.

5. The maximum volume per injection should not exceed 30 1. Larger injection volumes will result in backflow from the injection site. Puncturing the eye through the pars plana does not influence the proliferation of NPE cells or cells in the CMZ (Fischer and Reh, 2003).

6. Doses of growth factors can vary from 10 to 2000 ng, but growth factors such as IGF-I, EGF, BMP4, and FGF2 influence progenitors in the CMZ at 100 ng per dose delivered over 2–3 consecutive days. Most of the growth factors we have used in our studies have been obtained commercially and the reader is referred to the relevant publications for details.

2.2.2. Dissection and fixation of tissues

Our studies have relied heavily on the immunolabeling of the chicken retina prepared as cryosections or whole-mounts. The eyes of postnatal chickens are large and easy to dissect. The ease of dissection accommodates expeditious isolation of tissues and obviates the need for perfusion. To prepare tissues for immunolabeling, brief (less than 30 min) exposure to fixation is preferable. Although longer exposure to fixative may result in high-quality sections and better preservation of gross tissue morphology, the modification of basic side-chains in antigens with PFA often prevents interactions with antibodies. Enucleated eyes are hemisected equatorially with a fresh razor blade and the gel vitreous removed with forceps from the posterior eye cup. Eye cups are fixed (4% paraformaldehyde plus 3% sucrose in 0.1 M phosphate buffer, pH 7.4, 30 min at 20 °C), washed three times in PBS (phosphate-buffered saline; 0.05 M sodium phosphate, 195 mM NaCl, pH 7.4), cryoprotected in PBS plus 30% sucrose, immersed in embedding medium (OCT-compound; Tissue-Tek), and freeze-mounted onto sectioning blocks. Vertical sections, nominally 12 μm thick, are thaw-mounted onto SuperFrost Plus[tm] slides (Fisher Scientific), air-dried at 37 °C and stored at -20 °C until use. Depending on the quality of the fixation, the sections can be safely stored for several months without compromising immunoreactivity within the tissue.

For whole-mount preparations of the chicken retina, fixed retinas are dissected away from the pigmented epithelium, choroid and sclera, cryprotected in 30% sucrose in PBS and taken through 3 cycles of freezing (at -80 °C) and thawing (on a 37 °C hotplate) (Fischer and Stell, 1997; Fischer *et al.*, 1996).

The primary antibodies described in the previous section on *in vitro* immunofluorescence will also label the cells *in vivo*. In addition, the secondary antibodies used for labeling the sections or flatmount preparations are similar to those used for the cells in culture.

3. Overview

The development of *in vitro* and *in vivo* methods for the study of retinal progenitors and stem cells over the past 20 years has allowed rapid progress in our understanding of the factors that regulate the generation of retinal neurons and

Mears, A. J., Kondo, M., Swain, P. K., Takada, Y., Bush, R. A., Saunders, T. L., Sieving, P. A., and Swaroop, A. (2001). Nrl is required for rod photoreceptor development. *Nat. Genet.* **29**(4), 447–452.

Moshiri, A., Close, J., and Reh, T. A. (2004). Retinal stem cells and regeneration. *Int. J. Dev. Biol.* **48,** 1003–1014.

Moshiri, A., McGuire, C. R., and Reh, T. A. (2005). Sonic hedgehog regulates proliferation of the retinal ciliary marginal zone in posthatch chicks. *Dev. Dyn* .

Moshiri, A., and Reh, T. A. (2004). Persistent progenitors at the retinal margin of ptc+/-mice. *J. Neurosci.* **24**, 229–237.

Morris, V. B., Wylie, C. C., and Miles, V. J. (1976). The growth of the chick retina after hatching. *Anat. Rec.* **184**, 111–113.

Okada, T. S. (1980). Cellular metaplasia or transdifferentiation as a model for retinal cell differentiation. *Curr. Top. Dev. Biol.* **16,** 349–380.

Ooto, S., Akagi, T., *et al.* (2004). Potential for neural regeneration after neurotoxic injury in the adult mammalian retina. *Proc. Natl. Acad. Sci. USA* **101**(37), 13654–13659.

Oliver, G., Mailhos, A., Wehr, R., Copeland, N. G., Jenkins, N. A., and Gruss, P. (1995). Six3, a murine homologue of the sine oculis gene, demarcates the most anterior border of the developing neural plate and is expressed during eye development. *Development* **121**, 4045–4055.

Otteson, D. C., and Hitchcock, P. F. (2003). Stem cells in the teleost retina: Persistent neurogenesis and injury-induced regeneration. *Vision Res.* **43**, 927–936.

Park, C. M., and Hollenberg, M. J. (1989). Basic fibroblast growth factor induces retinal regeneration *in vivo. Dev. Biol.* **134**, 201–205.

Perron, M., Kanekar, S., *et al.* (1998). The genetic sequence of retinal development in the ciliary margin of the Xenopus eye. *Dev. Biol.* **199**(2), 185–200.

Pittack, C., Jones, M., and Reh, T. A. (1991). Basic fibroblast growth factor induces retinal pigment epithelium to generate neural retina *in vitro. Development* **113,** 577–588.

Prada, C., Puga, J., Perez-Mendez, L., Lopez, R., and Ramirez, G. (1991). Spatial and Temporal Patterns of Neurogenesis in the Chick Retina. *Eur. J. Neurosci.* **3**, 559–569.

Raymond, P. A., and Hitchcock, P. F. (1997). Retinal regeneration: Common principles but a diversity of mechanisms. *Adv. Neurol.* **72**, 171–184.

Reh, T. A. (1992). Cellular interactions determine neuronal phenotypes in rodent retinal cultures. *J. Neurobiol.* **23**, 1067–1083.

Reh, T. A., and Cagan, R. L. (1994). Intrinsic and extrinsic signals in the developing vertebrate and fly eyes: Viewing vertebrate and invertebrate eyes in the same light. *Perspect. Dev. Neurobiol.* **2**, 183–190.

Reh, T. A., and Kljavin, I. J. (1989). Age of differentiation determines rat retinal germinal cell phenotype: Induction of differentiation by dissociation. *J. Neurosci.* **9**, 4179–4189.

Reh, T. A., Nagy, T., *et al.* (1987). Retinal pigmented epithelial cells induced to transdifferentiate to neurons by laminin. *Nature* **330**(6143), 68–71.

Reh, T. A., and Nagy, T. (1989). Characterization of Rana germinal neuroepithelial cells in normal and regenerating retina. *Neurosci. Res. Suppl.* **10**, S151–S161.

Sakaguchi, D. S., Janick, L. M., *et al.* (1997). Basic fibroblast growth factor (FGF-2) induced transdifferentiation of retinal pigment epithelium: Generation of retinal neurons and glia. *Dev. Dyn.* **209**(4), 387–398.

Sakami, S., Hisatomi, O., *et al.* (2005). Downregulation of Otx2 in the dedifferentiated RPE cells of regenerating newt retina. *Brain. Res. Dev. Brain. Res.* **155**(1), 49–59.

Sidman, R. L. (1961). Histogenesis of mouse retina studied with ^{3}H-thymidine. *In* "The structure of the eye" (G. Smelser, ed.) Academic Press, New York.

Stone, L. S. (1950). Neural retina degeneration followed by regeneration from surviving retinal pigment cells in grafted adult salamander eyes. *Anat. Rec.* **106**, 89–109.

Stone, L. S., and Steinitz, H. (1957). Regeneration of neural retina and lens form retina pigment cell grafts in adult newts. *J. Exp. Zool.* **135**, 301–317.

Stroeva, O. G., and Mitashov, V. I. (1983). Retinal pigment epithelium: Proliferation and differentiation during development and regeneration. *Int. Rev. Cytol.* **83**, 221–293.

Taylor, M., and Reh, T. A. (1990). Induction of differentiation of rat retinal, germinal, neuroepithelial cells by dbcAMP. *J. Neurobiol.* **21**(3), 470–481.

Tropepe, V., Coles, B. L., Chiasson, B. J., Horsford, D. J., Elia, A. J., McInnes, R. R., and van der Kooy, D. (2000). Retinal stem cells in the adult mammalian eye. *Science* **287,** 2032–2036.

Turner, D. L., and Cepko, C. L. (1987). A common progenitor for neurons and glia persists in rat retina late in development. *Nature* **328,** 131–136.

Walcott, J. C., and Provis, J. M. (2003). Muller cells express the neuronal progenitor cell marker nestin in both differentiated and undifferentiated human foetal retina. *Clin. Exp. Ophthalmol.* **31,** 246–249.

Young, R. W. (1985). Cell proliferation during postnatal development of the retina in the mouse. *Brain Res.* **353,** 229–239.

Yurco, P., and Cameron, D. A. (2005). Responses of Muller glia to retinal injury in adult zebrafish. *Vision Res.* **45**(8), 991–1002.

Zhao, S., Thornquist, S. C., *et al.* (1995). *In vitro* transdifferentiation of embryonic rat retinal pigment epithelium to neural retina. *Brain Res.* **677**(2), 300–310.

Zhao, S., Chen, Q., *et al.* (2002). BMP signaling is required for development of the ciliary body. *Development* **129**(19), 4435–4442.

Zuber, M. E., Gestri, G., Viczian, A. S., Barsacchi, G., and Harris, W. A. (2003). Specification of the vertebrate eye by a network of eye field transcription factors. *Development* **130**(21), 5155–5167.

CHAPTER 4

Dental Pulp Stem Cells

He Liu,★ Stan Gronthos,† and Songtao Shi‡

★Peking University School of Stomatology
 Beijing, China

†Mesenchymal Stem Cell Group
 Division of Haematology, Institute of Medical
 and Veterinary Science, Adelaide, South Australia, Australia

‡Salivary Gland Disease Center and the Molecular Laboratory for Gene Therapy
 Capital Medical University School of Stomatology
 Beijing, China

Abstract
1. Introduction
2. Identification of Dental Pulp Stem Cells (DPSCs)
3. Isolation of DPSCs
    3.1. Protocol for DPSC isolation
4. Differentiation of DPSCs
    4.1. Protocol for DPSC transplantation
References

Abstract

Postnatal stem cells have been isolated from a variety of tissues. These stem cells are thought to possess great therapeutic potential for repairing damaged and/or defective tissues. Clinically, hematopoietic stem cells have been successfully used for decades in the treatment of various diseases and disorders. However, the therapeutic potential of other postnatal stem cell populations has yet to be realized, due to the lack of a detailed understanding of their stem cell characteristics at the cellular and molecular levels. Furthermore, there is limited knowledge of their therapeutic value at the preclinical level. Therefore, it is necessary to develop optimal strategies and approaches to overcome the substantial challenges currently faced by researchers examining the clinical efficacy of different postnatal stem cell populations. In this review, we introduce methodologies for isolating postnatal stem cells from human dental pulp and discuss their potential role in tissue regeneration.

1. Introduction

A tooth can be grossly described as having two basic components, the crown and the root. The crown is gradually exposed in the oral cavity following the tooth eruption process, while the root is permanently embedded in the alveolar sockets and may only be exposed to the oral cavity under pathological conditions, such as periodontal tissue resorption caused by disease. A mature tooth is comprised of the hard tissues enamel, dentin, and cementum, as well as a soft tissue core known as dental pulp (Fig. 4.1). Enamel, the hardest mineralized tissue in the human body, covers the tooth crown, while cementum, a thin layer of orofacial bone-like tissue, covers the root surface (Fig. 4.1). Underlying the enamel and cementum is the dentin, which is a specialized and vital mineralized matrix that harbors odonto-blastic processes and nerve fibers and is directly linked to the pulp core by odontoblasts. Furthermore, blood vessels and nerve bundles enter the pulp through the apical foramen to provide nutrition and sensation for responding to external stimuli. The dental pulp contains connective tissue, mesenchymal cells, neural fibers, blood vessels, and lymphatics. The tooth is surrounded by condensed connective tissue called the periodontal ligament, which contains Sharpey's fibers embedded in the cementum on one side and the alveolar bone on the other, to attach and suspend the tooth inside the alveolar socket (Fig. 4.1). Alveolar bone is in the base of the alveolar socket and maintains teeth *in situ* and supports them for masticatory function.

Enamel is formed by epithelial cell-derived ameloblasts in a process involving reciprocal induction with odontoblasts (Slavkin and Diekwisch, 1997; Thesleff and Aberg, 1999). After the completion of crown formation, ameloblasts undergo

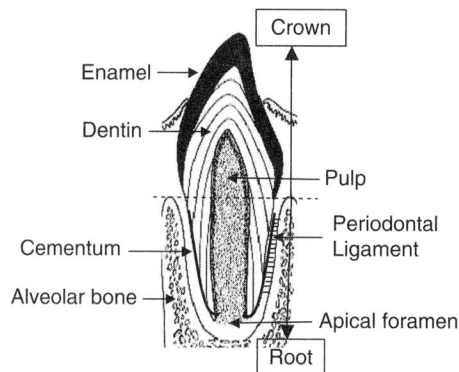

Figure 4.1 Diagram of tooth structure.

programmed cell death and totally lose the potential to repair enamel *in vivo*. In contrast, following mutual induction with ameloblasts, odontoblasts form primary dentin and subsequently line the inner surface of newly formed dentin to maintain a long cytoplasmic process inside the dentinal tubules (Mjor *et al.*, 2001; Ruch, 1998; Thesleff, 2003). Although odontoblasts are postmitotic cells that are not able to divide and repair damaged dentin, the progenitor or stem cells of odontoblasts are capable of migrating into the dentin surface and differentiating into odonto-blasts to form reparative or tertiary dentin (About *et al.*, 2001; Batouli *et al.*, 2003; Murray *et al.*, 2000, 2001; Ruch, 1998). Unlike the primary dentin, this reparative dentin is poorly organized, with irregular dentinal tubules embedded in the dentin matrix, offering a protective barrier to the dental pulp and facilitating maintenance of the integrity and vitality of dental pulp tissue.

Dentin is composed of collagens and noncollagenous proteins (NCPs). Collagen accounts for most of the matrix and is primarily type I collagen. It is believed that collagen provides a scaffold for mineralization, while the NCPs initiate and regu-late the process of mineralization (Bulter and Ritchie, 1995; Butler *et al.*, 1997). Dentin phosphoprotein (DPP) and dentin sialoprotein (DSP) are two important NCPs involved in dentin formation and mineralization (D'Souza *et al.*, 1997; Feng *et al.*, 1998; MacDougall *et al.*, 1997). In fact, they are specific cleavage products of a single gene named dentin sialophosphoprotein (DSPP). Initially, DPP and DSP were considered to be dentin-specific proteins (Ritchie *et al.*, 1994), but later it was found that other tissues such as bone, periodontium, and kidney tissue also express DSP, but at a much lower level in comparison with the expression in dentin and odontoblasts (Baba *et al.*, 2004; Ogbureke and Fisher, 2005; Qin *et al.*, 2003). Previously, DSP has been used as a marker to differentiate odontoblasts from their stem progenitors and osteoblastic cells, because the latter two cell populations express undetectable levels of DSP based on immunohistochemical staining (Gronthos *et al.*, 2002).

2. Identification of Dental Pulp Stem Cells (DPSCs)

Tooth regeneration is one of the ultimate goals of restoring the loss of natural teeth. Recent studies have indicated that cell-based strategies show promising potential for regenerating the whole tooth structure in rodents (Chai and Slavkin, 2003; Duailibi *et al.*, 2004; Ohazama *et al.*, 2004; Yen and Sharpe, 2006). Moreover, stem cell-based regeneration of human tooth structures has been achieved in immunocompromised mouse models (Gronthos *et al.*, 2000, 2002; Seo *et al.*, 2004). So far, several human tooth-associated stem cell popula-tions, including DPSCs, periodontal ligament stem cells (PDLSCs), and stem cells from human exfoliated deciduous teeth (SHED), have been isolated from dental pulp and periodontal ligament tissues (Gronthos *et al.*, 2000; Miura *et al.*, 2004; Seo *et al.*, 2004). More importantly, DPSCs and PDLSCs are capable of forming a dentin/pulp complex and cementum/periodontal ligament, respectively, when transplanted subcutaneously into immunocompromised mice (Gronthos *et al.*, 2000; Seo *et al.*, 2004), demonstrating their potential for regenerating human dental tissues *in vivo*. In addition, SHED are able to form significant amounts of bone *in vivo*, providing an alternative population of postnatal stem cells for alveolar and orofacial bone regeneration (Miura *et al.*, 2004). However, to achieve the goal of functional biotooth regeneration, a more thorough study of the intricate processes of tooth growth and development is necessary, in order to fully understand the processes of mutual induction between odontoblasts and ameloblasts and the complicated multiple structures of tooth composition (Slavkin and Diekwisch, 1997; Thesleff and Aberg, 1999).

Over the years, researchers have recognized that dental pulp contains *ex vivo* expandable cells called dental pulp cells. These cells express osteogenic markers, such as alkaline phosphatase, type I collagen, bone sialoprotein, osteocalcin, osteopontin, TGFβ, and BMPs (Chen *et al.*, 2005; Kuo *et al.*, 1992; Nakashima *et al.*, 1994; Pavasant *et al.*, 2003; Shiba *et al.*, 1998). They also respond to the induction of BMP, FGF-2, MEPE, and TGFβ by undergoing osteogenic differentiation (Alliot-Licht *et al.*, 2005; Dobie *et al.*, 2002; Iohara *et al.*, 2004; Liu *et al.*, 2005; Nakao *et al.*, 2004; Saito *et al.*, 2004; Unda *et al.*, 2000). Collectively, this experimental evidence suggests that dental pulp cells might be similar to osteoblast-like cells in terms of expressing bone markers and forming mineralized nodules when cultured under osteo-inductive conditions. Recently, it was discovered that dental pulp cells might represent a distinct population of cells because of their unique capability of forming specific crystalline structures in mineralized nodules, similar to physiological dentin but different from bone structures (About *et al.*, 2000a). In addition, dental pulp cells express high levels of DSP, which is usually expressed at a very low level in osteogenic cells (Batouli *et al.*, 2003; Gronthos *et al.*, 2002). Based on microarray analysis, multiple genes are differentially expressed between dental pulp cells and osteogenic cells (Shi *et al.*, 2001). For instance, cyclin-dependent kinase 6 and D-cyclins are highly expressed in dental pulp cells; these molecules are able to promote cell proliferation. In contrast, *IGFBP7*, an inhibitor gene of cell growth, is highly expressed in osteogenic cells (Shi *et al.*, 2001). This evidence may account for the fact that dental pulp cells show a high proliferation capacity compared to osteogenic cells (Gronthos *et al.*, 2000). Given what we know about the unique characteristics of dental pulp cells and their capacity for forming reparative dentin *in vivo*, it is reasonable to presume that dental pulp may contain stem progenitors. Moreover, it may be possible to isolate these stem cells using strategies analogous to those for retrieving mesenchymal stem cells from bone marrow aspirates (Gronthos *et al.*, 2000; Shi and Gronthos, 2003).

Human DPSCs were initially identified based on their traits of forming single colonies in culture, self renewal *in vivo*, and multidifferentiation *in vitro* (Gronthos *et al.*, 2000). Recently, one of the most important advances in DPSC research was the revelation of their stem cell niche in the perivascular region (Shi and Gronthos, 2003; Shi *et al.*, 2005). Extensive immunophenotyping of *ex vivo* expanded DPSCs demonstrated their expression of various markers associated with endothelial and/ or smooth muscle cells, such as STRO-1, VCAM-1, MUC-18, and smooth muscle α-actin (Gronthos *et al.*, 2000). In addition, smooth muscle α-actin-positive cells have also been detected close to mineralized deposits in human dental pulp cultures (Alliot-Licht *et al.*, 2001). Further characterization of DPSCs using current molecular technology will hopefully provide novel markers that will be useful in their identification *in situ*, and isolation and purification for *ex vivo* expansion.

3. Isolation of DPSCs

Dental pulp tissues can be easily collected from clinically extracted human teeth and the individual cells can be enzymatically released from the pulp tissue (Gronthos *et al.*, 2000). After plating at a low density, DPSCs are able to attach to the culture dish and form a single colony cluster after 10–14 days of culture (Gronthos *et al.*, 2000, 2002). The colonies contain fibroblast-like cells similar to colony-forming unit-fibroblasts (CFU-F) formed by bone marrow mesenchymal stem cells (BMMSCs) (Castro-Malaspina *et al.*, 1980; Friedenstein, 1976). At plating densities of $0.1–2.5 \times 10^4$ cells, the frequency of colony-forming cells derived from dental pulp tissue is 22–70 colonies/10^4 cells (Gronthos *et al.*, 2000). Single-colony-derived cells can be selected with a cloning ring and subjected to secondary *ex vivo* culture to expand cell numbers for further experimental requirements. Interestingly, individual colonies demonstrated marked differences in their proliferation rates according to a BrdU uptake assay. Only 20% of colonies are able to proliferate beyond 20 population doublings under regular culture conditions (Gronthos *et al.*, 2002). This result implies that the progeny of this small percentage of highly proliferative colonies will finally dominate the cell population in multicolony-derived cells. These data, combined with immunohistochemistry studies showing the uneven expression of markers between subsets of cells, support the concept that DPSCs are a heterogeneous population of postnatal stem cells akin to BMMSCs.

STRO-1 has been shown to be an early marker of different mesenchymal stem cell populations (Gronthos *et al.*, 1994, 2003). Currently, based on the high expression of STRO-1 in combination with expression of CD146 (MUC-18) and VCAM-1, selected BMMSCs showed characteristics of stem cells with high purity (Gronthos *et al.*, 2003; Shi and Gronthos, 2003). Because of the limited cell numbers in dental pulp, several pulp tissues obtained from three or four different third molars have to be pooled. Subsequently, immunomagnetic bead selection is carried out to assess whether STRO-1 or CD146 antigens can be used to select for the highly purified DPSC populations. Results showed that the majority of colony-forming cells (82%) were represented in the minor STRO-1$^+$ cell fraction, sixfold greater than unfractionated pulp cells, while a high proportion (96%) of CFU-F cells were present in the CD146$^+$ population, sevenfold greater than unfractionated pulp cells (Shi and Gronthos, 2003). Purified DPSC populations were subsequently expanded *in vitro* and then transplanted into immunocompromised mice and showed a similar capacity for regenerating the dentin-pulp complex as the multicolony-derived DPSCs.

3.1. Protocol for DPSC isolation

1. Collect extracted normal human teeth and place into a sterile container containing saline. Third molars are the most common resource for dental stem-cell isolation in dental clinics because of the widespread extraction of wisdom teeth. As the third molar is the last tooth developed in humans, it is usually at an earlier developmental stage and is capable of providing an optimal amount of dental pulp tissue for DPSC isolation.

2. Scrape the gingival and periodontal tissue, if there is any, from the tooth surface. Clean the tooth surfaces using iodine and 70% ethanol to prevent contamination from oral bacteria. Wash five times with $1\times$ PBS to remove iodine and ethanol.

3. Use sterilized high-speed dental fissure burs to cut around the cementum–enamel junction to separate the crown and root and expose the pulp tissue in the pulp chamber. Alternatively, the tooth can be wrapped in sterile gauze and squeezed in a vice until cracked to expose the pulp tissue.

4. Pick up the pulp tissue with sterile tweezers and mince it into tiny pieces using a scalple blade in moist conditions (specify media type). Then digest the minced pulp tissue in a solution containing 3 mg/ml collagenase type I and 4 mg/ml dispase for 30–45 min at 37 °C to totally digest the pulp tissue. It is important to always keep the pulp pellet moist during the DPSC isolation procedures.

5. After digestion, add five volumes of culture medium (αMEM?) containing 10% serum. Following centrifugation at 1200 rpm for 10 min, suspend the cell pellet in regular culture medium and pass the cells through a 70-μm strainer to obtain single-cell suspensions.

6. Seed the cells into culture plates, dishes, or flasks with α-modified Eagle's medium supplemented with 10% FCS, 100 μM L-ascorbic acid 2-phosphate, 2 mM L-glutamine, 100 U/ml penicillin, and 100 mg/ml streptomycin, then incubate the cells at 37 °C with 5% CO_2.

7. Cells grow slowly in the initial stage, with attachment to the culture dish. Single colonies can be identified after 10–14 days of culture if DPSCs are plated at a low density.

8. To select purified DPSCs, several pulp tissues obtained from three to four third molars should be pooled to obtain a sufficient number of cells for magnetic bead sorting. Following collagenase/dispase digestion, the pulp cells should be washed twice and then incubated with 0.5 ml of either STRO-1 (mouse IgM anti-BMSSC), 3G5 (mouse IgM anti-pericyte), or CC9 (mouse IgG2a anti-CD146) (Shi and Gronthos, 2002) for 1 h on ice. Supernatants are used neat, while purified

antibodies are used at 20 μg/ml. After washing twice with PBS/1% bovine serum albumin, the cells are incubated with either sheep antimouse IgG-conjugated or rat antimouse IgM-conjugated magnetic Dynabeads (Dynal, add details of company) for 1 h on a rotary mixer at 4 °C. Separation of the bead positive cells is performed using an MPC-1 magnetic particle concentrator (Dynal). Repeat magnetic wash step three times to reduce contamination of bead negative cells. After immuno-magnetic selection, single-cell suspensions of bead positive dental pulp cells are cultured as described above.

4. Differentiation of DPSCs

One of the most important characteristics of stem cells is their capacity to differentiate into multiple cell lineages. To prove that DPSCs satisfy this stem cell property, a variety of inductive culture conditions can be used to assess whether DPSCs can differentiate into multiple types of cells *in vitro*.

Previous studies have shown that cells derived from dental pulp are capable of forming mineralized nodules *in vitro* in the presence of inductive media containing ascorbic acid, dexamethasone, and an excess of phosphate (Hanks *et al.*, 1998; Tsukamoto *et al.*, 1992; Unda *et al.*, 2000; Yokose *et al.*, 2000). The use of methodology such as infrared microspectroscopic examination and X-ray diffrac-tion electron microscopy has confirmed the dentin-like nature of the crystalline structures that comprise the mineralized nodules *in vitro*, which are distinct from the crystal structures of mineralized enamel and bone *in vivo* (About *et al.*, 2000a). Furthermore, we demonstrated that *ex vivo* expanded human DPSCs are capable not only of forming mineralized nodules in osteo-inductive culture conditions, but also of generating a dentin/pulp-like complex *in vivo* in conjunction with hydroxy-apatite/tricalcium phosphate (HA/TCP) as a carrier vehicle (Fig. 4.2) (Gronthos *et al.*, 2000, 2002). These data clearly indicated the tissue regeneration capacity of DPSCs.

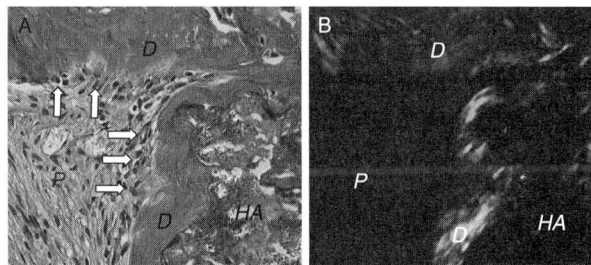

Figure 4.2 After 8 weeks post transplantation, DPSCs are capable of differentiating into odonto-blasts (arrows) that are responsible of dentin formation (*D*) on the surface of the hydroxyapatite tricalcium (*HA*) carrier. A pulp-like (*P*) tissue are generated with newly formed dentin as shown by H and E staining (A) and polarizing light microscopy (B).

Recent studies have explored whether DPSCs possess the ability to self-renew. To answer this question, we harvested primary DPSC implants at 2 months post-transplantation and liberated the cells by enzymatic digestion for subsequent expansion *in vitro*. Donor human cells were isolated from the cultures by FACS using a human β_1-integrin specific monoclonal antibody, then retransplanted into immunodeficient mice for 2 months. Recovered secondary transplants yielded the same dentin/pulp-like structures as observed in the primary transplants (Gronthos *et al.*, 2002). In addition, human DSP protein was found to be present in the dentin matrix by immunohistochemical staining, and *in situ* hybridization studies confirmed the human origin of the odontoblast/pulp cells contained within the secondary DPSC transplants (Gronthos *et al.*, 2002).

To determine whether DPSCs represent multipotent stem cells, we cultured the DPSCs in various inductive media previously shown to promote the differentiation of adipocytes. Our data suggest that DPSCs are capable of forming adipocytes when cultured with a potent cocktail of adipogenic inductive agents (0.5 mM methylisobutylxanthine, 0.5 μM hydrocortisone, 60 μM indomethacin) (Gimble *et al.*, 1995), showing the presence of Oil red O-positive fat-containing adipocytes in DPSC culture after several weeks of induction (Gronthos *et al.*, 2002). This was also correlated with an upregulation of the early adipogenic master regulatory gene, *PPARγ2*, and the mature adipocyte marker, lipoprotein lipase, using RT-PCR (Gronthos *et al.*, 2002). These observations highlight the flexibility of the DPSC population to develop into functional stromal cell types not normally associated with dental pulp tissue.

During development, odontoblasts are presumed to originate from craniofacial neural crest cells (Chai *et al.*, 2000). Recent investigations have explored the possibility that DPSCs have the potential to differentiate into neural-like cells (Gronthos *et al.*, 2002; Miura *et al.*, 2003; Nosrat *et al.*, 2004). *Ex vivo* expanded DPSCs were shown to constitutively express nestin, an early marker of neural precursor cells, and glial fibrillary acid protein (GFAP), an antigen characteristic of glial cells (Fig. 4.3). In accord with these findings, other investigators have identified the same markers in dental pulp tissue *in situ* (About *et al.*, 2000b; Hainfellner *et al.*, 2001). When DPSCs were cultured under defined neural-inductive conditions, there was enhanced expression of both nestin and GFAP. Morphological assessment of induced DPSCs identified long cytoplasmic processes protruding from rounded cell bodies, in contrast to their usual bipolar fibroblastic-like appearance. Moreover, dental pulp cells under neural-inductive culture conditions were found to express the neuron-specific marker, neuronal nuclei (NeuN), by immunohistochemical staining (Gronthos *et al.*, 2002; Miura *et al.*, 2004). These studies provide the first experimental evidence that adult human DPSCs may possess the potential to differentiate into neural-like cells, with the expression of nestin, GFAP, and NeuN *in vitro*.

To determine the capacity of *ex vivo* expanded DPSCs to generate a functional dentin/pulp-like tissue *in vivo*, we used an established transplantation system previously optimized for the formation of ectopic bone by cultured BMMSCs

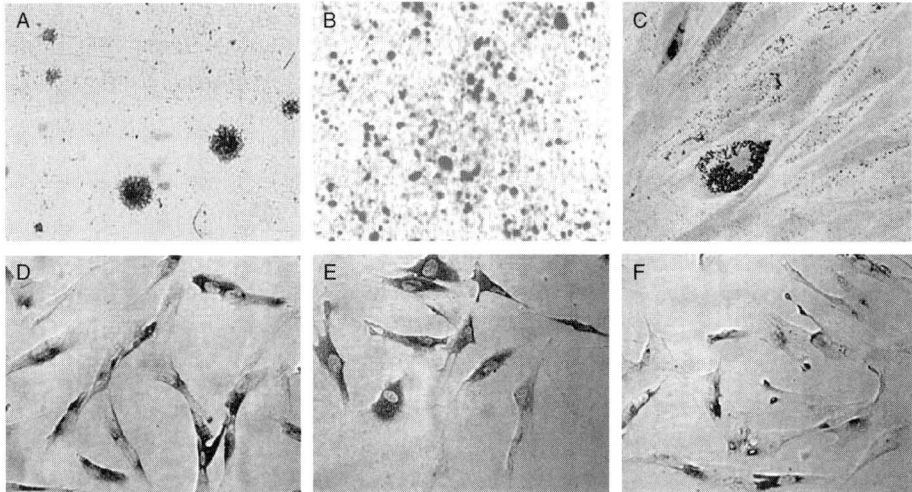

Figure 4.3 Multipotent differentiation of DPSC. DPSCs recovered from human dental pulp were capable of forming heterogenous single colony clusters after being plated at low density and cultured with regular culture medium for 10 days (A). DPSCs were cultured with L-ascorbate-2-phosphate, dexamethasone, and inorganic phosphate for 4 weeks. Alizarin red staining showed mineralized nodule formation (B). DPSCs were able to form Oil red O positive lipid clusters following 5 weeks of induction in the presence of 0.5 mM isobutylmethylxanthine, 0.5 μM hydrocortisone, and 60 μM indomethacin (C). Immunocytochemical staining depicts cultured DPSCs expressing Nestin (D), GFAP (E), and Neurofilament M (F), with culture medium containing Neurobasal A (Gibco-BRL), B27 supplement (Gibco-BRL), 1% penicillin, EGF 20ng/ml (BD Bioscience), and FGF 40 ng/ml (BD Bioscience).

(Krebsbach *et al.*, 1997; Kuznetsov *et al.*, 1997). Previously, it was demonstrated that rodent or bovine developing dental papilla tissues are capable of forming ectopic dentin *in vivo* (Holtgrave and Donath, 1995; Ishizeki *et al.*, 1990; Lyaruu *et al.*, 1999; Prime and Reade, 1980). However, similar studies using human intact developing dental papilla or adult dental pulp tissue failed to generate a mineralized dentin matrix and/or odontoblast-like cells following transplantation into immunocompromised mice (Gimble *et al.*, 1995; Prime *et al.*, 1982). It is likely that human mesenchymal cells require a suitable conductive carrier such as HA/TCP particles to initiate the mineralization process (Holtgrave and Donath, 1995; Krebsbach *et al.*, 1997). HA/TCP and other biomaterials have also been clinically used, with partial success, to stimulate a pulpal proliferation response to aid in the repair of damaged dentin (Kaigler and Mooney, 2001; Levin, 1998).

Typical DPSC transplants developed areas of vascularized pulp tissue surrounded by a well-defined layer of odontoblast-like cells aligned around mineralized dentin, with their processes extending into tubular structures. The odontoblast-like cells and fibrous pulp tissue in the transplants were shown to have originated from the donor material by their reactivity to the human-specific alu cDNA probe (Gronthos *et al.*, 2000). In addition, orientation of the collagen fibers within the dentin was characteristic of ordered primary dentin, perpendicular to the odontoblast layer. Backscatter EM analysis demonstrated that the dentin-like material formed in the transplants had a globular appearance consistent with the structure of dentin *in situ*. Moreover, the presence of human DSP detected in the transplants confirmed the ability of DPSCs to regenerate a human dentin/pulp microenvironment *in vivo* (Gronthos *et al.*, 2002).

4.1. Protocol for DPSC transplantation

1. Adherent cultures are washed once with serum free PBS and the cells liberated by enzymatic digestion by the addition of 2 ml of 0.5% Trypsin/EDTA solution per T75 flask for 5–10 min at 37 °C.

2. Single cell suspensions of DPSCs are washed in growth medium, by centrifugation and then resuspend them in 1 ml of culture medium.

3. Previous studies have shown that human dental pulp cells need a carrier to induce ectopic dentin formation. Hydroxyapatite/tricalcium phosphate (HA/TCP) particles have been identified as a suitable carrier for human DPSCs. *Ex vivo* expanded DPSCs (approximately $2.0–4.0\times10^6$) are mixed with 40 mg of HA/TCP ceramic particles and then incubated at 37 °C for 90 min under rotation at 25 rpm. After centrifugation at 1200 rpm for 15 min, the supernatant is discarded and DPSC-HA/TCP particles are suitable for transplantation.

4. Immunocompromised bg/nu/xid mice (8–12 weeks old) are used for the transplantation. The mice are first anesthetized by intraperitoneal injection with 2.5% tribromoethanol at 0.018 ml/g body weight. Approximately 1-cm-long mid-longitudinal skin incisions are then prepared on the dorsal surface of recipient mice and blunt dissection is used to produce subcutaneous pockets for transplantation of DPSC-HA/TCP particles.

5. Animal wound clips are then used to seal the wound.

6. The transplants are recovered at 8–12 weeks posttransplantation, fixed with 4% formalin, decalcified with buffered 10% EDTA (pH 8.0), and then embedded in paraffin. Sections (5 μm) are deparaffinized for histological and immunological analysis by hematoxylin/eosin and immunohistochemical staining.

7. Human-specific alu and murine-specific pf1 sequences labeled with digoxigenin are used as probes for *in situ* hybridization as previously described (PNAS). The primers used were as follows. Human alu: sense, 5'-TGGCTCACGCCTG-TAATCC-3' (base numbers 90–108); antisense, 5'-TTTTTTGAGACG-GAGTCTCGC-3' (base numbers 344–364, GenBank accession number AC004024). Murine pf1: sense, 5'-CCGGGCAGTGGTGGCGCATGCCTT-TAAATCCC-3' (base numbers 170–201); antisense, 5'-GTTTGGTTTTTGAG-CAGGGTTCTCTGTGTAGC-3' (base numbers 275–306, ank accession number X78319). The probes are prepared by PCR containing 1×PCR buffer (Perkin Elmer, Foster City, CA), 0.1 mM dATP, 0.1 mM dCTP, 0.1 mM dGTP, 0.065

mM dTTP, 0.035 mM digoxigenin-11-dUTP, with 10 pmol of specific primers, and 100 ng of human genomic DNA as templates. Unstained sections are deparaffinized and hybridized with the digoxigenin-labeled alu probe using the mRNAlocator-Hyb Kit (catalog no. #1800; Ambion, Inc., Austin, Texas). After hybridization, the presence of alu or pf1 in tissue sections are detected by immunoreactivity with an anti-digoxigenin alkaline phosphatase-conjugated Fab fragment (Boehringer Mannheim, Indianapolis, Indiana).

References

About, I., Bottero, M. J., de Denato, P., Camps, J., Franquin, J. C., and Mitsiadis, T. A. (2000). Human dentin production *in vitro. Exp. Cell Res.* **258**(1), 33–41.

About, I., Laurent-Maquin, D., Lendahl, U., and Mitsiadis, T. A. (2000). Nestin expression in embryonic and adult human teeth under normal and pathological conditions. *Am. J. Pathol.* **157** (1), 287–295.

About, I., Murray, P. E., Franquin, J. C., Remusat, M., and Smith, A. J. (2001). The effect of cavity restoration variables on odontoblast cell numbers and dental repair. *J. Dent.* **29**(2), 109–117.

Alliot-Licht, B., Bluteau, G., Magne, D., Lopez-Cazaux, S., Lieubeau, B., Daculsi, G., and Guicheux, J. (2005). Dexamethasone stimulates differentiation of odontoblast-like cells in human dental pulp cultures. *Cell Tissue Res.* **321**(3), 391–400.

Alliot-Licht, B., Hurtrel, D., and Gregoire, M. (2001). Characterization of alpha-smooth muscle actin positive cells in mineralized human dental pulp cultures. *Arch. Oral Biol.* **46**(3), 221–228.

Baba, O., Qin, C., Brunn, J. C., Jones, J. E., Wygant, J. N., Mclntyre, B. W., and Butler, W. T. (2004). Detection of dentin sialoprotein in rat periodontium. *Eur. J. Oral Sci.* **112**(2), 163–170.

Batouli, S., Miura, M., Brahim, J., Tsutsui, T. W., Fisher, L. W., Gronthos, S., Robey, P. G., and Shi, S. (2003). Comparison of stem-cell-mediated osteogenesis and dentinogenesis. *J. Dent. Res.* **82**(12), 976–981.

Bulter, W. T., and Ritchie, H. (1995). The nature and functional significance of dentin extracellular matrix proteins. *Int. J. Dev. Biol.* **39**(1), 169–179.

Butler, W. T., Ritchie, H. H., and Bronckers, A. L. (1997). Extracellular matrix proteins of dentine. *Ciba Found. Symp.* **205,** 107–115.

Castro-Malaspina, H., Gay, R. E., Resnick, G., Kapoor, N., Meyers, P., Chiarieri, D., McKenzie, S., Broxmeyer, H. E., and Moore, M. A. (1980). Characterization of human bone marrow fibroblast colony-forming cells (CFU-F) and their progeny. *Blood* **56**(2), 289–301.

Chai, Y., Jiang, X., Ito, Y., Bringas, P., Jr., Han, J., Rowitch, D. H., Soriano, P., McMahon, A. P., and Sucov, H. M. (2000). Fate of the mammalian cranial neural crest during tooth and mandibular morphogenesis. *Development* **127**(8), 1671–1679.

Chai, Y., and Slavkin, H. C. (2003). Prospects for tooth regeneration in the 21st century: A perspective. *Microsc. Res. Tech.* **60**(5), 469–479.

Chen, S., Santos, L., Wu, Y., Vuong, R., Gay, I., Schulze, J., Chuang, H. H., and MacDougall, M. (2005). Altered gene expression in human cleidocranial dysplasia dental pulp cells. *Arch. Oral Biol.* **50**(2), 227–236.

Dobie, K., Smith, G., Sloan, A. J., and Smith, A. J. (2002). Effects of alginate hydrogels and TGF-beta 1 on human dental pulp repair *in vitro. Connect. Tissue Res.* **43**(2–3), 387–390.

D'Souza, R. N., Cavender, A., Sunavala, G., Alvarez, J., Ohshima, T., Kulkarni, A. B., and MacDougall, M. (1997). Gene expression patterns of murine dentin matrix protein 1 (Dmp1) and dentin sialophosphoprotein (DSPP) suggest distinct developmental functions *in vivo. J. Bone Miner. Res.* **12**(12), 2040–2049.

Duailibi, M. T., Duailibi, S. E., Young, C. S., Bartlett, J. D., Vacanti, J. P., and Yelick, P. C. (2004). Bioengineered teeth from cultured rat tooth bud cells. *J. Dent. Res.* **83**(7), 523–528.

Feng, J. Q., Luan, X., Wallace, J., Jing, D., Ohshima, T., Kulkarni, A. B., D'Souza, R. N., Kozak, C. A., and MacDougall, M. (1998). Genomic organization, chromosomal mapping, and promoter analysis of the mouse dentin sialophosphoprotein (Dspp) gene, which codes for both dentin sialoprotein and dentin phosphoprotein. *J. Biol. Chem.* **273**(16), 9457–9464.

Friedenstein, A. J. (1976). Precursor cells of mechanocytes. *Int. Rev. Cytol.* **47**(5), 327–359.

Gimble, J. M., Morgan, C., Kelly, K., Wu, X., Dandapani, V., Wang, C. S., and Rosen, V. (1995). Bone morphogenetic proteins inhibit adipocyte differentiation by bone marrow stromal cells. *J. Cell. Biochem.* **58**(3), 393–402.

Gronthos, S., Brahim, J., Li, W., Fisher, L. W., Cherman, N., Boyde, A., DenBesten, P., Robey, P. G., and Shi, S. (2002). Stem cell properties of human dental pulp stem cells. *J. Dent. Res.* **81**(8), 531–535.

Gronthos, S., Graves, S. E., Ohta, S., and Simmons, P. J. (1994). The STRO-1 + fraction of adult human bone marrow contains the osteogenic precursors. *Blood* **84**(12), 4164–4173.

Gronthos, S., Mankani, M., Brahim, J., Robey, P. G., and Shi, S. (2000). Postnatal human dental pulp stem cells (DPSCs) *in vitro* and *in vivo*. *Proc. Natl. Acad. Sci. USA* **97**(25), 13625–13630.

Gronthos, S., Zannettino, A. C., Hay, S. J., Shi, S., Graves, S. E., Kortesidis, A., and Simmons, P. J. (2003). Molecular and cellular characterisation of highly purified stromal stem cells derived from human bone marrow. *J. Cell Sci.* **116,** 1827–1835.

Hainfellner, J. A., Voigtlander, T., Strobel, T., Mazal, P. R., Maddalena, A. S., Aguzzi, A., and Budka, H. (2001). Fibroblasts can express glial fibrillary acidic protein (GFAP) *in vivo*. *J. Neuropathol. Exp. Neurol.* **60**(5), 449–461.

Hanks, C. T., Sun, Z. L., Fang, D. N., Edwards, C. A., Wataha, J. C., Ritchie, H. H., and Butler, W. T. (1998). Cloned 3T6 cell line from CD-1 mouse fetal molar dental papillae. *Connect. Tissue Res.* **37**(3–4), 233–249.

Holtgrave, E. A., and Donath, K. (1995). Response of odontoblast-like cells to hydroxyapatite ceramic granules. *Biomaterials* **16**(2), 155–159.

Iohara, K., Nakashima, M., Ito, M., Ishikawa, M., Nakasima, A., and Akamine, A. (2004). Dentin regeneration by dental pulp stem cell therapy with recombinant human bone morphogenetic protein 2. *J. Dent. Res.* **83**(8), 590–595.

Ishizeki, K., Nawa, T., and Sugawara, M. (1990). Calcification capacity of dental papilla mesenchymal cells transplanted in the isogenic mouse spleen. *Anat. Rec.* **226**(3), 279–287.

Kaigler, D., and Mooney, D. (2001). Tissue engineering's impact on dentistry. *J. Dent. Educ.* **65**(5), 456–462.

Krebsbach, P. H., Kuznetsov, S. A., Satomura, K., Emmons, R. V., Rowe, D. W., and Robey, P. G. (1997). Bone formation *in vivo*: Comparison of osteogenesis by transplanted mouse and human marrow stromal fibroblasts. *Transplantation* **63**(8), 1059–1069.

Kuo, M. Y., Lan, W. H., Lin, S. K., Tsai, K. S., and Hahn, L. J. (1992). Collagen gene expression in human dental pulp cell cultures. *Arch. Oral Biol.* **37**(11), 945–952.

Kuznetsov, S. A., Krebsbach, P. H., Satomura, K., Kerr, J., Riminucci, M., Benayahu, D., and Robey, P. G. (1997). Single-colony derived strains of human marrow stromal fibroblasts form bone after transplantation *in vivo*. *J. Bone Miner. Res.* **12**(9), 1335–1347.

Levin, L. G. (1998). Pulpal regeneration. *Pract. Periodontics Aesthet. Dent.* **10**(5), 621–624.

Liu, H., Li, W., Shi, S., Habelitz, S., Gao, C., and Denbesten, P. (2005). MEPE is downregulated as dental pulp stem cells differentiate. *Arch. Oral. Biol.* **50**(11), 923–928.

Lyaruu, D. M., van Croonenburg, E. J., van Duin, M. A., Bervoets, T. J., Woltgens, J. H., and de Blieck-Hogervorst, J. M. (1999). Development of transplanted pulp tissue containing epithelial sheath into a tooth-like structure. *J. Oral. Pathol. Med.* **28**(7), 293–296.

MacDougall, M., Simmons, D., Luan, X., Nydegger, J., Feng, J., and Gu, T. T. (1997). Dentin phosphoprotein and dentin sialoprotein are cleavage products expressed from a single transcript coded by a gene on human chromosome 4. Dentin phosphoprotein DNA sequence determination. *J. Biol. Chem.* **272**(2), 835–842.

Miura, M., Gronthos, S., Zhao, M., Lu, B., Fisher, L. W., Robey, P. G., and Shi, S. (2003). SHED: Stem cells from human exfoliated deciduous teeth. *Proc. Natl. Acad. Sci. USA* **100**, 5807–5812.

Mjor, I. A., Sveen, O. B., and Heyeraas, K. J. (2001). Pulp-dentin biology in restorative dentistry. Part 1. Normal structure and physiology. *Quintessence Int.* **32**(6), 427–446.

Murray, P. E., About, I., Franquin, J. C., Remusat, M., and Smith, A. J. (2001). Restorative pulpal and repair responses. *J. Am. Dent. Assoc.* **132**(4), 482–491.

Murray, P. E., About, I., Lumley, P. J., Smith, G., Franquin, J. C., and Smith, A. J. (2000). Postoperative pulpal and repair responses. *J. Am. Dent. Assoc.* **131**(3), 321–329.

Nakao, K., Itoh, M., Tomita, Y., Tomooka, Y., and Tsuji, T. (2004). FGF-2 potently induces both proliferation and DSP expression in collagen type I gel cultures of adult incisor immature pulp cells. *Biochem. Biophys. Res. Commun.* **325**(3), 1052–1059.

Nakashima, M., Nagasawa, H., Yamada, Y., and Reddi, A. H. (1994). Regulatory role of transforming growth factor-beta, bone morphogenetic protein-2, and protein-4 on gene expression of extracellular matrix proteins and differentiation of dental pulp cells. *Dev. Biol.* **162**(1), 18–28.

Nosrat, I. V., Smith, C. A., Mullally, P., Olson, L., and Nosrat, C. A. (2004). Dental pulp cells provide neurotrophic support for dopaminergic neurons and differentiate into neurons *in vitro*; implications for tissue engineering and repair in the nervous system. *Eur. J. Neurosci.* **19**(9), 2388–2398.

Ogbureke, K. U., and Fisher, L. W. (2005). Renal expression of SIBLING proteins and their partner matrix metalloproteinases (MMPs). *Kidney Int.* **68**(1), 155–166.

Ohazama, A., Modino, S. A., Miletich, I., and Sharpe, P. T. (2004). Stem-cell-based tissue engineering of murine teeth. *J. Dent. Res.* **83**(7), 518–522.

Pavasant, P., Yongchaitrakul, T., Pattamapun, K., and Arksornnukit, M. (2003). The synergistic effect of TGF-beta and 1,25-dihydroxyvitamin D3 on SPARC synthesis and alkaline phosphatase activity in human pulp fibroblasts. *Arch. Oral Biol.* **48**(10), 717–722.

Prime, S. S., and Reade, P. C. (1980). Xenografts of recombind bovine odontogenic tissues and cultured cells to hypothymic mice. *Transplantation* **30**(2), 149–152.

Prime, S. S., Sim, F. R., and Reade, P. C. (1982). Xenografts of human ameloblastoma tissue and odontogenic mesenchyme to hypothymic mice. *Transplantation* **33**(5), 561–562.

Qin, C., Brunn, J. C., Cadena, E., Ridall, A., and Butler, W. T. (2003). Dentin sialoprotein in bone and dentin sialophosphoprotein gene expressed by osteoblasts. *Connect. Tissue Res.* **44**(Suppl. 1), 179–183.

Ritchie, H. H., Hou, H., Veis, A., and Butler, W. T. (1994). Cloning and sequence determination of rat dentin sialoprotein, a novel dentin protein. *J. Biol. Chem.* **269**(5), 3698–3702.

Ruch, J. V. (1998). Odontoblast commitment and differentiation. *Biochem. Cell Biol.* **76**(6), 923–938.

Saito, T., Ogawa, M., Hata, Y., and Bessho, K. (2004). Acceleration effect of human recombinant bone morphogenetic protein-2 on differentiation of human pulp cells into odontoblasts. *J. Endod.* **30**(4), 205–208.

Seo, B. M., Miura, M., Gronthos, S., Bartold, P. M., Batouli, S., Brahim, J., Young, M., Robey, P. G., Wang, C. Y., and Shi, S. (2004). Investigation of multipotent postnatal stem cells from human periodontal ligament. *Lancet* **364**(9429), 149–155.

Shi, S., Bartold, P. M., Miura, M., Seo, B. M., Robey, P., and Gronthos, S. (2005). The efficacy of mesenchymal stem cells to regenerate and repair dental structures. *Orthod. Craniofac. Res.* **8**, 191–199.

Shi, S., and Gronthos, S. (2003). Perivascular niche of postnatal mesenchymal stem cells in human bone marrow and dental pulp. *J. Bone Miner. Res.* **18**(4), 696–704.

Shi, S., Robey, P. G., and Gronthos, S. (2001). Comparison of gene expression profiles between human bone marrow stromal cells and human dental pulp stem cells by microarray analysis. *Bone* **29**(6), 532–539.

Shiba, H., Fujita, T., Doi, N., Nakamura, S., Nakanishi, K., Takemoto, T., Hino, T., Noshiro, M., Kawamoto, T., Kurihara, H., and Kato, Y. (1998). Differential effects of various growth factors and cytokines on the syntheses of DNA, type I collagen, laminin, fibronectin, osteonectin/secreted protein, acidic and rich in cysteine (SPARC), and alkaline phosphatase by human pulp cells in culture. *J. Cell Physiol.* **174**(2), 194–205.

Slavkin, H. C., and Diekwisch, T. G. (1997). Molecular strategies of tooth enamel formation are highly conserved during vertebrate evolution. *Ciba Found. Symp.* **205,** 73–80.

Thesleff, I. (2003). Developmental biology and building a tooth. *Quintessence Int.* **34**(8), 613–620.

Thesleff, I., and Aberg, T. (1999). Molecular regulation of tooth development. *Bone* **25**(1), 123–125.

Tsukamoto, Y., Fukutani, S., Shin-Ike, T., Kubota, T., Sato, S., Suzuki, Y., and Mori, M. (1992). Mineralized nodule formation by cultures of human dental pulp-derived fibroblasts. *Arch. Oral Biol.* **37**(12), 1045–1055.

Unda, F. J., Martin, A., Hilario, E., Begue-Kirn, C., Ruch, J. V., and Arechaga, J. (2000). Dissection of the odontoblast differentiation process *in vitro* by a combination of FGF1, FGF2, and TGFbeta1. *Dev. Dyn.* **218**(3), 480–489.

Yen, A. H., and Sharpe, P. T. (2006). Regeneration of teeth using stem cell-based tissue engineering. *Expert Opin. Biol. Ther.* **6**(1), 9–16.

Yokose, S., Kadokura, H., Tajima, Y., Fujieda, K., Katayama, I., Matsuoka, T., and Katayama, T. (2000). Establishment and characterization of a culture system for enzymatically released rat dental pulp cells. *Calcif. Tissue Int.* **66**(2), 139–144.

CHAPTER 5

Mouse Spermatogonial Stem Cells: Culture and Transplantation

Jon M. Oatley and Ralph L. Brinster

Department of Animal Biology
School of Veterinary Medicine
University of Pennsylvania
Philadelphia, Pennsylvania 19104

Abstract
1. Introduction
2. Spermatogonial Stem Cells (SSCs)
3. SSC Transplantation
    3.1. Choice of donor animals
    3.2. Preparation of recipient mice
    3.3. Microinjection of donor cell suspension
    3.4. Analysis of recipient testes for donor-derived spermatogenesis
4. SSC Culture
    4.1. Isolation of an SSC enriched population from the mouse testis
    4.2. Serum-free culture on feeder cell monolayers
    4.3. Subculturing practices that optimize expansion of SSC numbers
5. siRNA Transfection of Cultured Mouse SSCs
    5.1. Transient transfection of cultured germ cells with siRNAs
    5.2. Determining the level of gene knockdown
    5.3. Transplanting transfected cells to quantify SSC numbers
6. Implications
References

Abstract

The biological activities of spermatogonial stem cells (SSCs) are the foundation for spermatogenesis and thus sustained male fertility. Therefore, understanding the mechanisms governing their ability to both self-renew and differentiate is essential. Moreover, since SSCs are the only adult stem cell to contribute genetic information to the next generation, they are an excellent target for genetic modification. In this chapter, we discuss three important approaches to investigate SSCs and their cognate niche microenvironment in the mouse, the SSC transplantation assay, the long-term serum-free SSC culture method, and RNA-interference methodology in cultured SSCs. These techniques can be used to enhance our understating of SSC biology as well as to produce genetically modified animals.

1. Introduction

In mammals, spermatozoa are the vehicle by which a male's genetic information is passed to the next generation. Thus, the process of sperm production, termed spermatogenesis, is essential for species preservation and genetic diversity. Spermatogenesis is one of the most productive cell producing systems in adult animals, generating approximately 100 million sperm each day in adult humans (Sharpe, 1994). The entire process is comprised of three main phases. The first phase, termed the proliferative phase, consists of mitotic amplifying divisions of spermatogonia. These cells ultimately become primary spermatocytes which enter the second phase of spermatogenesis, termed the meiotic phase, in which meiosis and genetic recombination occur. Following two meiotic divisions, haploid spermatids are produced which undergo the third phase, termed spermiogenesis, consisting of a dramatic transformation from round germ cells to specialized spermatozoa. This complex process occurs within seminiferous tubules of the testis and is supported by the somatic Sertoli cells which are evenly spaced throughout the seminiferous tubules and form a complex architecture (Russell, 1993). The Sertoli cells are anchored to a basement membrane, which is generated by contributions from both Sertoli and myoid cells. Together, Sertoli cells and their intricate association with germ cells, form the seminiferous epithelium. The tight junctions formed between adjacent Sertoli cells divide the seminferous epithelium into a basal compartment, exposed to many lymph and blood-borne substances, and an adluminal compartment, to which blood-borne substances have limited direct access. The tight junction separation of these two compartments is often referred to as the blood-testis barrier. The adluminal compartment contains late meiotic stage germ cells, and post meiotic spermatids and spermatozoa and is often considered to be an immune privileged site. Similar to other adult self-renewing tissues that rely on differentiated cells to be replenished at a constant rate or rapidly following toxic injury, the testis contains an adult tissue-specific stem cell population.

2. Spermatogonial Stem Cells (SSCs)

The SSCs are the supporting foundation for continual sperm production throughout the majority of a male's lifespan. SSCs first arise in the testis from development of more undifferentiated germ cells called gonocytes, which are themselves derived from primordial germ cells or PGCs that migrate from the urogenital ridge to the embryonic gonad during prenatal development (McLaren, 2003). The developmental transition of gonocytes to SSCs occurs between postnatal days 0 and 5 in the mouse (McLean et al., 2003), and single spermatogonia (termed A_{single} or A_s) appear at about 6 days-of-age in the mouse (Bellve et al., 1977; Huckins and Clermont, 1968). This A_s spermatogonia population is generally agreed upon to contain the SSCs. Like other adult stem cells, SSCs have the ability to undergo both self-renewal and differentiation. Self-renewal is the key function needed for SSCs to maintain tissue continuity throughout the lifetime of a male. The very first step in spermatogenesis is the fate decision of an SSC to produce differentiated daughter progeny that are destined to eventually become spermatozoa.

Theoretically, 4096 spermatozoa are capable of being produced from a single SSC in the adult rat testis (Russell et al., 1990). However, the overall efficiency of spermatogenesis has been estimated to be only 10–25% (Barratt, 1995; Tegelenbosch and de Rooij, 1993) because of germ cell apoptosis, thus considerably fewer than the potential number are actually generated. Spermatogenesis is initiated when SSC division results in the production of daughter progeny that are committed to differentiation rather than self-renewal. These newly formed differentiating daughters are termed A_{paired} (A_{pr}) spermatogonia because of their connecting intercellular bridge. On the other hand, self-renewing divisions of SSCs rather than production of A_{pr} daughters generates more SSCs (A_s) and provides the basis for continued spermatogenesis. The A_{pr} spermatogonia then undergo a series of mitotic cell divisions becoming $A_{aligned}$ (A_{al}) spermatogonia; the number of these divisions varies among mammalian species. The A_{al} spermatogonia give rise to differentiating spermatogonia, termed A_1 through A_4 spermatogonia in the mouse, indicating the number of mitotic amplifying divisions they undergo. The differentiating A_4 spermatogonia are capable of further maturation into intermediate and type B spermatogonia which enter meiosis becoming primary and secondary spermatocytes, reduce their DNA content and become haploid spermatids which undergo a transformation into fertilization competent spermatozoa (Russell et al., 1990).

Collectively, the A_s (including SSCs), A_{pr}, and A_{al} germ cells are referred to as proliferating spermatogonia and all share very similar phenotypic and likely molecular characteristics (De Rooij and Russell, 2000; Russell et al., 1990). Currently, there is no phenotypic, biochemical, or molecular characteristics that distinguish these proliferating spermatogonia populations from one another. The differentiating spermatogonia (A_1–A_4) are distinguishable from undifferentiated spermatogonia based on several phenotypic characteristics including being c-kit positive (Ohta *et al.*, 2003). Research has clearly demonstrated that proliferating spermatogonia, including SSCs, are c-kit negative (Kubota *et al.*, 2003). Moreover, a reciprocal expression of c-kit and Thy1 (CD90) occurs in the spermatogonia subpopulation (Kubota *et al.*, 2003). Essentially all SSC activity is found in the Thy1$^+$ c-kit$^-$ cells indicating that Thy1 is a surface marker specifically expressed on proliferating spermatogonia. Thus, c-kit and Thy1 expression on the cell surface can be used as distinguishing markers between proliferating and differentiating spermatogonia. There are likely many other phenotypical, biochemical, and molecular differences between the different types of spermatogonia in the mammalian testis that await discovery.

Following prepubertal development of the seminiferous epithelium and the initial wave of spermatogenesis, SSC biological function is supported by a niche microenvironment. SSC and niche development during the pre-pubertal growth period is an essential process for shaping the frame-work of a normal functioning SSC-niche unit in the adult. The beginnings of SSC-niche formation occurs during early postnatal development coinciding with Sertoli cell maturation. The number of available SSC niches change according to age, increasing during the developmental period from birth to sexual maturity at which point the number becomes stable in the mouse (Shinohara *et al.*, 2001), but may increase in species where testes enlarge during adult life as in the rat (Ryu *et al.*, 2003). A normal functioning SSC-niche unit in the adult animal, in which SSC self-renewal and differentiation is supported, is essential for continual fertility and thus, preservation of a species. Also, declining fertility that is often associated with aging in males may be due to reduced support of the SSC niche by somatic cell contributions (Ryu *et al.*, 2006), which may be secondary to impaired endocrine support of Sertoli cell function.

A key contribution to the SSC niche is growth factors produced and secreted by Sertoli cells. One such factor is glial cell line-derived neurotrophic factor (GDNF), a related member of the TGFβ superfamily of growth factors produced by Sertoli cells in the mammalian testis (Meng *et al.*, 2000; Tadokoro *et al.*, 2002). When GDNF was overexpressed by both germ cells and somatic cells of the testis *in vivo*, a dramatic accumulation of proliferating spermatogonia occurred resulting in germ cell seminoma in the mouse (Meng *et al.*, 2000). Also, mice with one GDNF-null allele showed an impairment of spermatogenesis (Meng *et al.*, 2000). These studies demonstrated that GDNF has a dramatic effect on the proliferating spermatogonia population and the process of spermatogenesis. In subsequent *in vivo* experiments, forced expression of human GDNF by Sertoli cells was shown to result in amplification of SSCs in the mouse testis (Yomogida *et al.*, 2003), strongly suggesting that GDNF affects the proliferation of SSCs. Subsequently, critical experiments were done *in vitro* to establish the essential role of GDNF. A serum-free culture medium for SSCs was developed (Kubota *et al.*, 2004a), which was able to maintain SSCs *in vitro* without loss of stem cell activity for a short period of time. Subsequent improvements in the culture system allowed expansion in number of the SSC population and culture for extended periods.

These culture studies demonstrated that GDNF is essential for mouse SSC self-renewal *in vitro* (Kubota *et al.*, 2004b). Previous *in vitro* experiments aimed at culturing SSCs had also included GDNF as part of growth factor cocktails in less defined serum-containing culture conditions (Kanatsu-Shinohara *et al.*, 2003; Nagano *et al.*, 2003; Oatley *et al.*, 2004), suggesting its importance in SSC proliferation and survival. All these studies, culminating with Kubota *et al.* (2004b), have now clearly established GDNF as a central factor regulating the self-renewal fate decision of SSCs in the mouse and rat (Ryu *et al.*, 2005). Recently, studying GDNF-regulated changes in gene expression within SSC enriched populations of germ cells, the molecular mechanisms governing SSC self-renewal have begun to be identified (Oatley *et al.*, 2006, 2007). There are undoubtedly multiple other niche factors that influence SSC fate decisions of self-renewal, differentiation, and survival to be discovered.

Even though SSCs are of critical importance for male fertility, little is known about mechanisms regulating their function. They are a rare cell type in the testis, only constituting approximately 0.03% of the total testicular cell population in the adult mouse (Tegelenbosch and de Rooij, 1993). Currently, there are no known specific markers that can be used to distinguish SSCs from other types of proliferating spermatogonia. Also, the self-renewing proliferation rate of SSCs is slow, calculated to be ~6 days *in vitro* (Kubota *et al.*, 2004b). For these reasons, it has been very difficult to study accurately SSC biology. Currently, the only means to directly study SSCs is by functional transplantation (Brinster and Avarbock, 1994; Brinster and Zimmermann, 1994). Also, the ability to maintain self-renewing populations of mouse and rat SSCs *in vitro* has recently been achieved (Hamra *et al.*, 2005; Kanatsu-Shinohara *et al.*, 2003, 2005a; Kubota *et al.*, 2004a,b; Ryu *et al.*, 2005). These two techniques, *in vitro* maintenance of SSCs and transplantation, provide essential tools that can be used to make new discoveries and define mechanisms regulating SSC biological functions at the cell and molecular levels. Additionally, the use of RNA-interference (RNAi) to knockdown the expression of specific genes in cultured mouse SSCs provides a powerful tool to evaluate the importance of molecular mechanisms controlling SSC self-renewal. This chapter contains detailed protocols of how to perform the SSC transplantation assay technique and to establish long-term cultures of self-renewing mouse SSCs and conduct RNAi experiments to study genes regulating SSC self-renewal.

3. SSC Transplantation

For many years, knowledge of SSC biology was mainly theory, based on morphological observations because direct assays to identify and study them were lacking. In 1994, a transplantation technique was described (Fig.5.1), which not only validated the presence of SSCs in the mammalian testis but also provided a means to investigate directly their functional characteristics (Brinster and Avarbock, 1994; Brinster and Zimmermann, 1994). Currently, it is the only

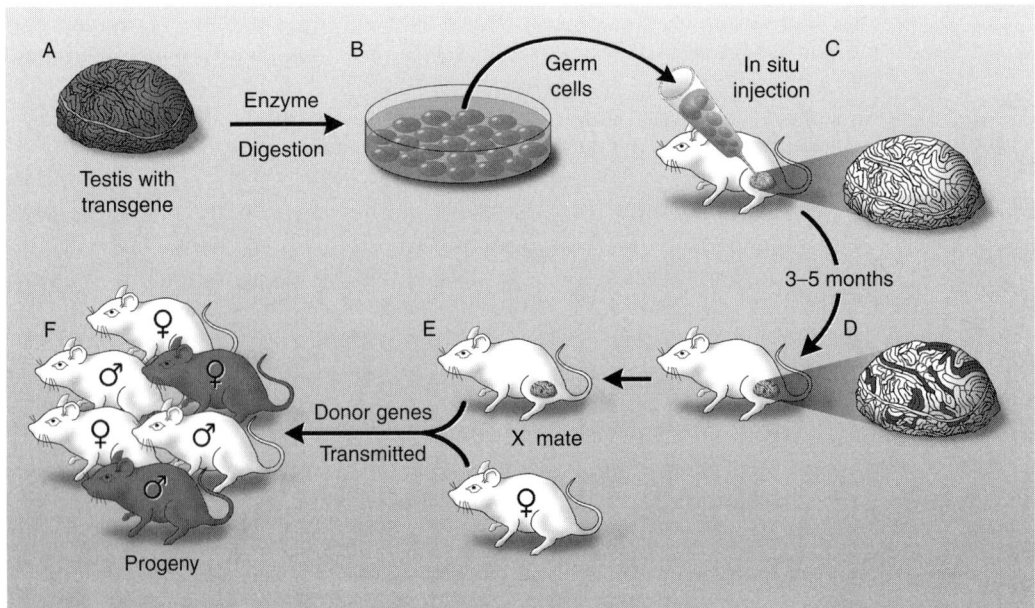

Figure 5.1 The spermatogonial stem cell transplantation technique in mice. (A) The testis from a male in which a reporter transgene is expressed in germ cells is digested to generate a single cell suspension. (B and C) The isolated cells are subsequently cultured or microinjected as a fresh cell suspension into the seminiferous tubules of an infertile germ cell-depleted recipient. (D) Colonies of donor-derived spermatogenesis can be detected in the recipient testis based on reporter transgene expression by donor germ cells. Only SSCs are capable of establishing colonies of spermatogenesis following transplantation. Each donor-derived colony of spermatogenesis in a recipient testis is generated from a single transplanted spermatogonial stem cell. (E and F) Upon mating of the recipient male to a wild-type female, offspring can be produced containing the donor haplotype and thus must have been produced by sperm generated from a donor spermatogonial stem cell-derived colony of spermatogenesis. (Reproduced with permission from Brinster (2002) © American Association for the Advancement of Science.)

means for accurately identifying SSCs, and studies reporting biological characteristics of putative SSCs should require stem cell identity to be confirmed by transplantation. The technique involves injection of a donor testicular cell suspension containing SSCs into the seminiferous tubules of a germ cell depleted or deficient recipient male. Following injection, donor SSCs migrate/relocate from the center of the seminiferous tubules and colonize stem cell niches on the basement membrane. The donor SSCs are then capable of expansion and forming colonies of donor-derived spermatogenesis. Upon breeding of recipient males, offspring with donor haplotype can be generated, and thus, must be derived from donor stem-cell derived sperm produced within the recipient testis. Each colony of donor-derived spermatogenesis within a recipient testis, in general, is clonally derived from a single transplanted SSC (Fig.5.2A) (Dobrinski *et al.*, 1999; Kanatsu-Shinohara *et al.*, 2006a; Nagano *et al.*, 1999; Zhang *et al.*, 2003). For this reason, the SSC transplantation technique can be used as an accurate and reproducible assay to study directly characteristics of SSC biology.

Using the transplantation assay to study SSC biology involves (1) choice of donor animals, (2) preparation of recipient mice, (3) microinjection of the donor cell suspension, and (4) analysis of the recipient testis for donor-derived spermatogenesis. All four of these points will be described to provide researchers with the information necessary to utilize the transplantation assay in mice.

3.1. Choice of donor animals

The choice of donor and recipient animal is essential for effective utilization of the transplantation assay. The most accurate and direct assessment of SSCs in a cell population is made by counting the number of donor cell-derived colonies of spermatogenesis and measuring the length of each colony. The following points

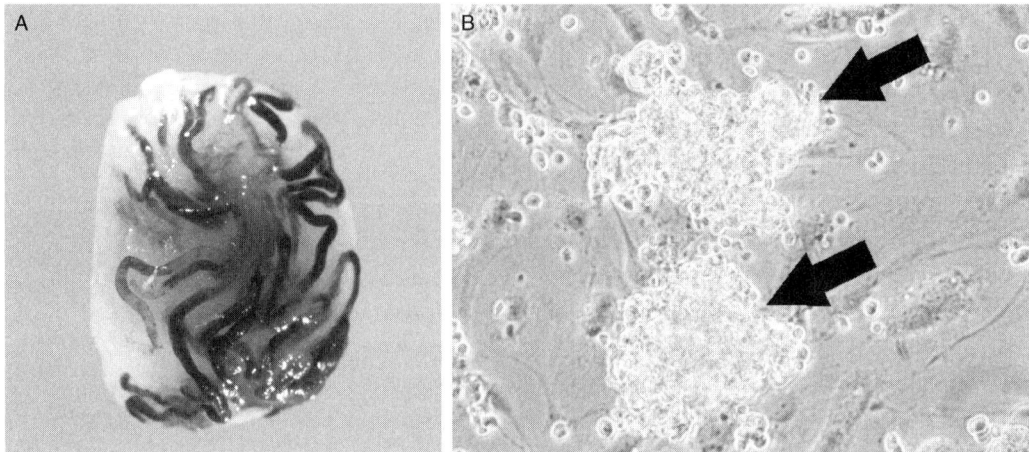

Figure 5.2 Transplantation and culture of mouse SSCs. (A) Busulfan treated recipient testis transplanted with cultured Thy1⁺ germ cells from ROSA donors. Donor-derived spermatogenesis is easily detectable by staining for *β*-galactosidase expression. Each blue colony of spermatogenesis represents colonization of a single donor spermatogonial stem cell. Assessing the number and length of blue colonies provides an unequivocal quantitative determination of spermatogonial stem cell presence, number, and biological function in the injected cell population. (B) An established spermatogonial stem cell enriched germ cell culture from 6-day-old ROSA donor mice. The germ cells grow as clumps of stem cells with tightly adhering membranes (arrows) and are loosely attached to the underlying STO feeder cell monolayer. In healthy, robust cultures of self-renewing SSCs, the germ cells form tight clumps where individual cells are difficult to distinguish.

must be taken into consideration for the transplantation technique to be utilized accurately as an assay for studying SSC biology:

• Colony number is directly correlated to the number of SSCs in a population and colony length provides a measure of an SSC's ability to proliferate and reestablish spermatogenesis (Dobrinski et al., 1999; Nagano et al., 1999). Thus, donor-derived spermatogenesis following transplantation must be capable of being unequivocally identified. The best means to achieve positive identification is use of a donor animal that expresses a marker transgene in germ cells, for example, LacZ (B6.129S7-Gtrosa26, The Jackson Laboratory; designated ROSA) or green fluorescent protein (GFP; C57BL/6-TgN ACTbEGFP 1Osb, The Jackson Laboratory) expressing mice.

• Methods for depleting endogenous germ cells in recipient testes generally are not completely effective, the possibility that some endogenous spermatogenesis will return is always present. Thus, simply identifying spermatogenesis in the testis of a recipient animal following transplantation is not an accurate means for assessing donor SSC biological function since donor-derived spermatogenesis cannot be unequivocally distinguished from reestablished endogenous spermatogenesis of the recipient, unless the donor cell expresses a transgene.

• Use of a naturally sterile recipient, such as homozygous W mice (W^v/W^v or W/W^v) (Brinster and Zimmermann, 1994; Silvers, 1979), can overcome the problem of reestablished recipient spermatogenesis, and thus can be used to confirm SSC presence in a cell population without a reporter gene. However, colony number cannot be determined; therefore, transplantation cannot be used for determining the number of SSC in the injected cell population when unmarked donor cells are used. Consequently, it is essential to utilize a donor strain in which donor-derived spermatogenesis can be identified visually by transgene expression.

3.2. Preparation of recipient mice

In the normal adult mouse testis active spermatogenesis limits, but does not completely block, the access of donor SSCs to the basal compartment of the recipient testis (Shinohara et al., 2002). In order for the injected SSCs to migrate efficiently from the lumen to the basement membrane, endogenous germ cells must be depleted. Several techniques have been used to achieve this, including irradiation (Meistrich et al., 1978; Van Beek et al., 1990; Withers et al., 1974), and chemotherapeutic drugs (Brinster and Avarbock, 1994; Bucci and Meistrich, 1987). In our experience, the most effective means for depleting endogenous germ cells to prepare an adult recipient is use of the chemotherapeutic drug busulfan (Brinster and Avarbock, 1994). These recipients consistently produce reliable and reproducible results. Neonatal W pup recipients may also be used to enhance colonization efficiency and generation of offspring; however, busulfan-treated adults are more ideal, when using transplantation as an assay to study SSCs.

● Busulfan-treated adult recipients

● Choice of recipient strain is important. Even though the adluminal compartment of the seminiferous tubules is immune-privileged, the basal compartment to which SSCs migrate and then reside is not protected by the blood-testis barrier. Also, Sertoli cells have immunological protective characteristics (Bellgrau *et al.*, 1995; Tung, 1993). For these reasons, it is essential that the recipient strain used be immunologically compatible with the transplanted donor cells. Use of nude mice as recipients is a means for avoiding immunological rejection, when suitable recipients may not be readily available for the donor cells used (e.g., transplantation of rat SSCs into mouse testes). In our experiments, an F1 hybrid cross of 129 SvCP × C57BL/6 (129 × C57) males are used as recipients. These animals are suitable as recipients for both ROSA (129 × C57 ROSA and C57 Homozygous ROSA) and GFP (C57 GFP) donor cells.

● Age of the recipient mouse influences the effectiveness of the busulfan treatment. In our experience, the optimal age is 6–8 weeks.

● Busulfan is prepared by first dissolving it in DMSO. An equal volume of warm (~40 °C) distilled water is then added slowly to avoid mixing with the DMSO to a final volume in which the concentration of busulfan will be 4 mg/ml. The DMSO and water phases are mixed just before injection to avoid busulfan precipitation. The solution is then maintained at 35–40 °C. Mice are administered an intraperitoneal injection at a dose of 60 mg/kg of body weight. This dose will eliminate nearly all endogenous spermatogenesis. Nude mice or other less hardy strains should be administered a lower dose of 45 mg/kg of body weight to avoid excessive stress to the animal.

● Following busulfan treatment, males are maintained for at least 5 weeks prior to transplantation, thus providing sufficient time for endogenous germ cells to be depleted from the seminiferous tubules.

● Pup recipients

● The normal pup testis is more receptive to colonization by donor SSCs than adult busulfan-treated males (Shinohara *et al.*, 2001). This is likely due to immature Sertoli cells lacking tight junctions, coupled with receptive niches at the time of transplantation. Donor SSCs can be injected into the seminiferous tubules of pup recipients without any prior treatment for eliminating endogenous germ cells and the stem cells will form colonies of spermatogenesis. However, the consistency and reproducibility of the transplant procedure is variable (Brinster *et al.*, 2003). Thus, the busulfan-treated adult recipient is more reliable for using transplantation as an assay system to study SSCs. Both endogenous and donor-derived spermatogenesis will occur in mice that were transplanted with donor SSCs as pups.

● W mice lack endogenous spermatogenesis due to a mutation of the c-kit receptor gene. These animals can, therefore, be used as recipients for transplantation of wild-type donor SSCs that lack a reporter gene. W mouse pups are receptive to colonization and unlike wild-type pups, endogenous spermatogenesis will not occur when the mice reach puberty. For this reason, W pups are excellent recipients for generating offspring containing the donor haplotype following transplantation. Homozygous W pup recipients are difficult to generate, microinjection of donor cells is challenging, and colonization can be variable; therefore, they are less than ideal for use in the quantitative transplantation assay.

3.3. Microinjection of donor cell suspension

There are multiple routes for introducing donor germ cells into the tubules of recipient mouse testes, including injection into the efferent ducts, the rete testis, or seminiferous tubules. The most effective injection method in mice and rats is injection into the rete testis via the efferent duct bundle. The rete testis is a central collecting network of tubuli that connects both ends of all seminiferous tubules. In the mouse, the rete lies cranial to the vascular pedicle and extends out from the vessels becoming the efferent ducts which lead into the caput epididymis. For injection, a micropipette is inserted into an efferent duct at a point preceding the rete and gently threaded a short way into the duct. The testis is relatively small in size, and the efferent ducts or rete testis are not clearly visible. Thus, microinjection of donor cell suspensions is facilitated by using a dissecting microscope. An effective system has been described by Ogawa *et al.* (1997). For the purpose of this protocol the basic needs will be presented.

- Injection pipette—The microinjection pipette is created from a 3-inch length of broscillate glass (internal diameter 0.75 mm and external diameter 1 mm). A pipette puller is used to draw out the glass tube into two pipettes, and a rotating grinding stone is then used to create a sharpened beveled point. Suitable pipettes typically have a tip with an external diameter of 40–60 μm and a beveled angel of approximately 30° (Ogawa *et al.*, 1997).

- Preparation of a donor cell suspension—A number of different techniques can be used to prepare donor cell suspensions for transplantation. We will not describe a specific method for this protocol; however, collection of an SSC enriched cell suspension by Thy1$^+$ MACS selection is described in the SSC culture protocol of this chapter. Donor cells are suspended at a concentration of $1–10° \times 10^6$ cells/ml. The choice of concentration is dependent on SSC concentration in the cell suspension. Use of a concentrated SSC testis cell suspension may result in excessive colonization, where individual colonies cannot be counted or analyzed accurately. Typically, a mouse Thy1$^+$ cell population, which is highly enriched with SSCs, is injected at a concentration of 10^6 cells/ml. Testicular cell suspensions that have not been enriched for SSCs can be injected at higher concentrations. Several different injection media have been used based on experiments performed. In most experiments, we use a Minimal Essential Media Alpha (MEMα) based serum-free media (Kubota *et al.*, 2004b) to suspend cells for injection. Prior to injection, cell suspensions should be maintained at 4 °C and can be stored up to 4 h without significant cell death.

• Microinjection—An essential piece of equipment is a dissecting microscope to visualize the testis and efferent ducts. The recipient mouse is anesthetized and placed in dorsal recumbence on the microscope stage. A midline incision is made through the abdomen; the testis is then brought through the incision, immobilized outside the body and visualized at 6–20× magnification. The micropipette is then filled with 10 μl of donor cell suspension, inserted into the efferent duct and gently threaded almost to the rete. This step is critical as the efferent ducts are translucent and hard to immobilize, but can be visualized running from the caput of the epididyimis to the rete testis. Often, a fat pad covers all or a portion of the ducts and must be gently dissected away. Care must be taken not to destroy the ducts as this will make injection nearly impossible. Positioning the needle can be conducted using a micromanipulator as described by Ogawa *et al.* (1997). Alternatively, the micropipette can be placed into the efferent ducts manually; however, this requires a steady hand. Piercing of the efferent ducts with a 30-gauge needle to create an entry point may facilitate insertion and positioning of the micropipette. Once in place, pressure is applied to infuse the donor cell suspension into the seminiferous tubules. Addition of trypan blue will facilitate visualizing advancement of the donor cell suspension. Pressure for injection can be generated by attaching silastic or polyethylene tubing between the micropipette and a mouth pipette, syringe, or pressure injector. Generally, the flow rate should be relatively slow to avoid harmful effects to the tubules. Injecting 10 μl of donor cell suspension typically fills 75–85% of the surface tubules of the mouse testis. It is important that the cell suspension is visually seen coursing through the tubules, this signifies a successful injection into the tubules. Injection into the interstitial space will result in a cloudy appearance where the tubules will be outlined by the typan blue. This scenario can be deceptive and lead one to believe a successful injection into the seminiferous tubules has been achieved. However, an injection of this type will result in poor colonization, even if some tubules were filled, because fibrosis of the testis often occurs.

3.4. Analysis of recipient testes for donor–derived spermatogenesis

Using the transplantation technique as an assay to investigate SSC biology requires that colony number and length be determined following transplantation. Generally, each colony of donor-derived spermatogenesis is clonally derived from a single SSC (Dobrinski *et al.*, 1999; Kanatsu-Shinohara *et al.*, 2006a; Nagano *et al.*, 1999; Zhang *et al.*, 2003), thus colony number is directly related to the number of SSCs in the cell suspension injected and reflects the efficiency of colonization. It is estimated that approximately 5% of the stem cells injected will produce colonies of spermatogenesis (Ogawa *et al.*, 2003; Shinohara *et al.*, 2001). Because many factors can influence this efficiency, it is critical to have adequate experimental controls and particularly to standardize recipient males for accurate results. Measuring colony length provides an assessment of the ability for a donor SSC to proliferate and reestablish spermatogenesis. In order to obtain both these measurements, colonies of donor-derived spermatogenesis must be identified following transplantation. Very useful donor cells are obtained from lacZ expressing ROSA mice; however, GFP donor cells may also be used but analysis is technically more difficult. Colonies of spermatogenesis from donor ROSA mice will stain a strong blue making identification and quantification simple using a dissecting microscope. The following protocol is for recipient testes transplanted with germ cells from donor ROSA mice.

- Recipient males are generally maintained for 8 weeks after transplantation to allow time for robust donor-derived spermatogenesis to occur.
- Testes are recovered and the overlying tunica albuginea is gently removed followed by fixation in 4% paraformaldahye at 4 °C for 2 h. A small piece of epididymis is included as a positive control for staining since it has endogenous β-galactosidase activity.
- The testes are then washed in 3 changes of LacZ rinse buffer at 4 °C for 30–60 min intervals and stained with 5-bromo-4-chloro-3-indolyly β-D-galactosidase (X-gal) overnight at 32 °C on a mechanical shaker as previously described (Brinster and Avarbock, 1994). On the next day, stained testes are postfixed in 10% neutral buffered formalin for 24 h.
- The number of blue-stained donor-derived colonies of spermatogenesis is counted using a dissecting microscope at 6–12× magnification. The tubules are gently separated to visually inspect each colony and digital pictures are captured.
- The length of each colony can be determined using a software program (e.g., NIH image 1.62).
- Histological cross sections of blue-stained colonies may be made to assess the degree of spermatogenesis. Complete spermatogenesis should be observed 8 weeks after transplantation.
- Statistical comparisons of colony numbers and lengths generated from experimental and control cell populations provide accurate assessments of treatment effects on SSC biological function.

4. SSC Culture

A key tool for studying SSC biology at the cell and molecular level is the ability to maintain an enriched population of stem cells *in vitro* for extended periods of time. Also, genetic modification and preservation of a male's genotype is facilitated by long-term maintenance *in vitro*. Early reports indicated that SSCs could be maintained *in vitro* under using STO feeder cells and a testis cell suspension (Nagano *et al.*, 1998). However, SSC expansion in number by self-renewal *in vitro* required a more specific system (Kanatsu-Shinohara *et al.*, 2003, 2005a; Kubota *et al.*, 2004a,b), as described in the following paragraphs.

An important prerequisite for establishing a long-term culture of SSCs is the purity of the starting cell population. Early experiments demonstrated that the first 2 weeks of a culture period are critical, during which SSC survival is sensitive to culture conditions (Kubota *et al.*, 2004a; Nagano *et al.*, 2003). If the concentration of the contaminating somatic cell population is too high, they will out grow the SSCs. In addition, some somatic cells have a deleterious affect on SSC self-renewal and expansion. Thus, it is imperative to begin with an enriched SSC population. There are several different methods for collecting an SSC enriched population from the testis, including differential plating (Shinohara *et al.*, 2000a), use of experimental cryptorchid donors (Shinohara *et al.*, 2000b), fluorescent activated cell sorting (FACS) (Kubota *et al.*, 2003; Shinohara *et al.*, 1999), and magnetic activated cell sorting (MACS) (Kubota *et al.*, 2004a,b). The purest population of SSCs can be obtained by FACS or MACS selection based on cell surface phenotype. In fact, FACS has been used to define the cell surface phenotype of SSCs from mice (Kubota *et al.*, 2003; Shinohara *et al.*, 1999) and rats (Ryu *et al.*, 2004). The Thy1 (CD90) cell population in the mouse testis has been shown to contain nearly all the SSCs at all developmental ages from neonate to adult (Kubota *et al.*, 2004b). Selection of cells from mouse testes based on Thy1 expression, using FACS or MACS, results in significant SSC enrichment in which ~1 in 15 cells is an SSC (Kubota *et al.*, 2003, 2004a). Even though several surface markers have been identified on mouse SSCs, this is the highest level of enrichment that has been achieved. It is likely that all proliferating spermatogonia (A_s, A_{pr}, A_{al}) have a similar surface phenotype. Thus, no currently known markers can be used to specifically identify or isolate SSCs. The markers basically enrich the testis cell population for proliferating spermatogonia.

A second critical component for establishing a long-term expanding culture of SSCs is ideal conditions which support SSC self-renewal and survival without promoting contaminating somatic cell growth. A serum-free condition was recently shown to be important for providing a proper *in vitro* environment (Kubota *et al.*, 2004a,b). Classically, serum has been utilized to promote proliferation of many different types of cells. However, the use of serum is less than ideal, since its components are undefined and vary among lots. In addition, serum can be toxic to some cell types and promotes the growth of contaminating somatic fibroblast cells that overgrow germ cells *in vitro*. Moreover, fetal bovine serum (FBS) has been shown to have a deleterious effect on both mouse and rat SSC proliferation *in vitro* (Kubota *et al.*, 2004b; Ryu *et al.*, 2005). However, the absence of FBS in the culture medium increases the reliance of the SSCs on added growth factors.

In past attempts, many growth factors have been tested for their ability to promote SSC expansion *in vitro*, including GDNF, basic fibroblast growth factor (bFGF), epidermal growth factor (EGF), transforming growth factor beta (TGFβ), insulin-like growth factor 1 (IGF1), leukemia inhibitory factor (LIF), and stem cell factor (SCF, kit ligand). A recent report has now demonstrated that GDNF is the critical growth factor for maintaining and stimulating SSC self-renewal *in vitro* (Kubota *et al.*, 2004b). Earlier *in vitro* experiments aimed at culturing SSCs had also included GDNF as part of growth factor cocktails or in less defined serum-containing media (Kanatsu-Shinohara *et al.*, 2003; Nagano *et al.*, 2003; Oatley *et al.*, 2004), providing initial suggestion of its importance in SSC proliferation and survival. However, the definitive proof of the central importance of GDNF came from Kubota *et al.* (2004a,b). In addition, self-renewal of rat SSCs *in vitro* has also been shown to be stimulated by GDNF (Hamra *et al.*, 2005; Ryu *et al.*, 2005). Because SSCs from many species, including humans, proliferate on the basement membrane of mouse seminiferous tubules (Brinster, 2002), it is likely that GDNF is a key regulator of SSC self-renewal in most mammalian species. Therefore, culture conditions described for mouse SSCs should be applicable as a starting point to develop similar systems in other species.

Using a serum-free chemically-defined medium containing GDNF, soluble GDNF family receptor α1 (GFRα1) and bFGF, Kubota *et al.* (2004a) demonstrated a dramatic expansion of SSCs *in vitro* from several different strains of mice. This study clearly established that GDNF is an essential factor that promotes self-renewal of mouse SSCs. A similar expansion of SSCs *in vitro* using a serum-free condition was reported by Kanatsu-Shinohara *et al.* (2005a,b,c) using a more complex medium containing proprietary supplements, and a cocktail of growth factors including GDNF; however, this system appears to be restricted to mice with a DBA background and does not support expansion of SSCs from other strains.

In the chemically defined system reported by Kubtoa *et al.* (2004b), an enriched self-renewing population of SSCs was maintained for up to 6 months *in vitro* on STO feeder layers. In this culture system, the SSCs are present in clumps of densely packed germ cells with tightly adhering membranes (Fig.5.2B). The germ cell clumps contain SSCs at a concentration of ~1 in 15 cells, similar to Thy1$^+$ ckit$^-$ MHC$^-$ cells recovered by FACS (Kubota *et al.*, 2003, 2004b). The remaining non SSCs are likely A$_{pr}$ and A$_{al}$ spermatogonia, which are the differentiating progeny of SSCs described previously in this chapter that are incapable of reestablishing spermatogenesis following transplantation. Based on transplantation studies, more than a 5000-fold increase in SSC number over a 10-week period was demonstrated, and a stem cell doubling time of approximately 6 days was determined. In this section, we will describe the methods necessary for establishing and maintaining enriched self-renewing populations of mouse SSCs for extended periods of time (Kubota *et al.*, 2004b). There are three main aspects (1) isolation of an SSC enriched cell population from the testis, (2) serum-free culture conditions using mitotically arrested feeder cells, and (3) subculturing practices that optimize expansion of SSC numbers. Once a culture of self-renewing SSCs is established, it can be used to investigate the mechanisms regulating SSC biological functions. However, it is essential to frequently confirm the stem cell potential of the *in vitro* manipulated cells using the transplantation assay to authenticate the biological relevance of the results from experimental manipulations. It is important to remember that these cultures are not composed of pure SSCs, and the SSC concentration within germ cell clumps can be highly variable and very low. In fact, one study estimated the SSC content in germ cell clumps to be 0.02% of the cell population (Kanatsu-Shinohara *et al.*, 2005c). Thus, studying SSCs in cultures of proliferating spermatogonia cannot rely on evaluating the morphology and number of germ cell clumps and must include functional germ cell transplantations. Without this direct examination of SSCs, the true effects of manipulating culture conditions or applying experimental treatments cannot be evaluated in regards to effects on SSC functions.

4.1. Isolation of an SSC enriched population from the mouse testis

In this protocol, we will describe isolation of an SSC enriched $Thy1^+$ cell population from the mouse testis using MACS separation (Kubota *et al.*, 2004a).

- Testis digestion

  - The donor age at which cells are collected has an effect on the overall purity of the collected cell population. In our experience, 6–8-day-old mouse pups produce the most enriched population of stem cells. In a typical preparation, 8–10 pups are used; however, a greater number can be used if the volumes are adjusted appropriately.
  - Testes are collected in sterile Hank's Balanced Salt Solution (HBSS) using aseptic technique, and the tunica albuginea is manually removed.
  - The tissue is then placed in 4.5 ml of trypsin-EDTA (0.25% trypsin and 1mM EDTA) solution and 0.5 ml of DNase I solution (7 mg/ml in HBSS) is added.
  - The tubules are gently dispersed and incubated at 37 °C for 5 min followed by gentle pipetting and addition of another 0.5 ml of DNase I solution.
  - Digestion is continued at 37 °C for another 5 min followed by addition of 1 ml of FBS and 0.5 ml of DNase I solution.
  - Pipetting is then used to ensure a single cell suspension is achieved. More DNase solution can be added if needed.
  - The cell suspension is passed through a 40 μm cell strainer, which is subsequently washed with HBSS, and the cells are pelleted by centrifugation at 600g for 7 min at 4 °C.

- Percoll selection

  - Following centrifugation, the supernatant is removed, the pellet is resuspended in 10 ml of HBSS, and the cell concentration is determined.
  - The cell suspension is then slowly layered on top of a 30% Percoll solution (Phosphate Buffered Saline (PBS) with 1% FBS, 50 U/ml Penicillin, 50 μg/ml Streptomycin, and 30% Percoll) in 15 ml conical centrifuge tubes. Five milliliter of cell suspension is layered onto 2 ml of Percoll solution per individual tube. No more than 20×10^6 total cells should be layered on top of 2 ml of Percoll. It is

beneficial to tilt the tube at an angle while pipetting in order to minimize mixing of the cell suspension with the Percoll solution.

- Tubes are then centrifuged at 600*g* for 8 min at 4 °C.

- The top solution phases (HBSS and Percoll) are removed, and the pellets are resuspended in 2 ml of PBS-S each (PBS with 1% FBS, 10 mM HEPES, 1 mM pyruvate, 1 mg/ml glucose and 50 U/ml Penicillin and 50 μg/ml Streptomycin). At this point, resuspended cell suspensions from all tubes are combined and cell concentration determined. A typical recovery after Percoll selection is 60–80% of the cells.

- The cells are again pelleted by centrifugation at 600*g* for 7 min at 4 °C.

- The supernatant is removed and the pellet resuspended in 180 μl of PBS-S.

- MACS Thy1 selection

- Thy1 antibody conjugated microbeads (anti-mouse CD90 (Thy1.2); Miltenyi Biotech Inc., Auburn, California) are added to the pelleted cells resuspended in PBS-S at a dilution of 1:10 and mixed gently followed by incubation at 4 °C for 20 min, with mixing after 10 min (see manufacturer's instructions).

- An additional volume of 2 ml of PBS-S is then added, and the cells are pelleted by centrifugation at 600*g* for 7 min at 4 °C. The pellet is then resuspended in 1 ml of PBS-S.

- The MACS columns are setup and pre-washed with 0.5 ml of PBS-S followed by addition of the Thy1 labeled cell suspension. The number of columns needed is determined based on total cell number as described by the manufacturer. Typically, only one column is needed for a preparation involving 8–10 mouse pups that are 6 days-of-age.

- After the suspension has passed through the column, it is washed 3 times with 0.5 ml of PBS-S and the Thy1$^+$ cells are eluted from the column in 1 ml of serum-free media (SFM; for formulation see Kubota *et al.*, 2004a,b).

- The cells are pelleted by centrifugation (600*g* for 7 min at 4 °C), resuspended in 1 ml of SFM, and the cell concentration determined. A typical yield should be approximately 20–25×10^4 Thy1$^+$ cells/testis. Microscopically, the cell population should be relatively homogenous, and 80–90% of the cells are large spherical germ cells and 70% are Thy1$^+$ spermatogonia (Kubota *et al.*, 2004a,b).

4.2. Serum-free culture on feeder cell monolayers

Serum-free conditions provide the most suitable environment for expansion of SSCs *in vitro* (Kubota *et al.*, 2004b). Also, use of a chemically-defined medium simplifies investigation of molecular and biochemical regulatory mechanisms of SSC biology. Development of this system clearly demonstrated that GDNF is an essential regulator of SSC self-renewal. The following protocol describes an ideal culture method for supporting mouse SSC self-renewal and expansion of their numbers.

- STO (SIM mouse embryo-derived thioguanine and oubain resistant)-feeder cell monolayer preparation

  - STO cells are grown in 10-cm dishes in Dulbecco's Modified Eagle's Medium (DMEM) growth medium (DMEM with 7% FBS, 100 mM 2-mercaptoethanol, 50 U/ml Penicillin, and 50 μg/ml Streptomycin).

  - When confluent, the cells are treated with mictomycin-C (10 μg/ml in DMEM growth medium) for 3 h at 37 °C.

  - The cells are then washed three times with 10 ml of HBSS to remove traces of mictomycin-C and collected by tryspin-EDTA (0.25% Trypsin and 1 mM EDTA) digestion at 37 °C for 5 min.

  - Growth media containing FBS is added and the cells are pipetted to generate a single cell suspension. Cells are pelleted by centrifugation (600g for 7 min at 4 °C), resuspended in freeze medium (DMEM growth medium containing 10% dimethyl sulfoxide) at a concentration of 4×10^6 cells/ml, and 1 ml is aliquoted to each cryo vial and stored at -70 °C.

  - Frozen stocks can be stored for several months without significant loss of cell viability.

  - STO monlayers for SSC culture are created using either fresh or thawed mitomycin-C treated cells plated at a concentration of 0.5×10^5 cells/cm^2 in 12-well tissue culture plates pre-coated with 0.1% gelatin. The feeders are maintained in DMEM growth media prior to adding SSCs.

  - The optimal time for plating STO monolayers is 3–4 days before addition of SSCs.

- SSC culture

  - Following MACS selection, 5–8×10^4 Thy1$^+$ cells are placed in individual wells of 12-well tissue culture plates (BD Biosciences; San Jose, California) containing STO feeder monolayers.

  - Prior to adding germ cell, STO monolayers are washed with HBSS to reduce residual FBS from the STO media.

  - It is important to avoid placing more than 10^5 cells/well in the initial culture. A high number of cells per well often results in contaminating somatic cells over-growing the SSCs, and the culture will fail.

  - The cells are cultured in SFM with the addition of 20 ng/ml recombinant human GDNF (R&D Systems; Minneapolis, Minnesota), 150 ng/ml rat GFRα1-Fc fusion protein (R&D Systems), and 1 ng/ml recombinant human bFGF (BD Biosciences).

  - The SFM components are described by Kubota *et al.* (2004a,b).

  - A major component of the medium is bovine serum albumin (BSA), consti-tuting 0.2%. The source, type, and lot of BSA used is important. We have found that ICM cat. no. 810661 lot no. 2943C and Sigma cat. no. A3803, lot nos.

064K0720, 025K1497, and 124K0729 support SSC survival and expansion. Other BSA types and lots will likely also be effective; however, each should be independently tested before use in experiments.

- Cultures are maintained at 37 °C in a humidified incubator with an atmosphere of 21% O_2, 5% CO_2, and balance N_2.
- Media is changed with the addition of the three growth factors every 2–3 days.

4.3. Subculturing practices that optimize expansion of SSC numbers

The initial 2-week period is essential in establishing a long-term culture of enriched self-renewing SSCs. During this time, SSC survival is greatly influenced by the culture conditions and contaminating somatic cells. If conditions are not ideal, contaminating somatic cells often overgrow SSCs, and a long-term culture cannot be established. Therefore, considerable attention must be paid to assessing the proliferation of the germ cells during this time. If successful, an established culture will contain a highly enriched population of self-renewing SSCs after 4 or 5 weeks. Prior to this time, especially during the first 2 weeks, cultures are not as enriched for SSCs, because contaminating cells often grow vigorously during this time. After several passages during a 3- or 4-week period, the majority of contaminating somatic cells disappear leaving an enriched SSC population. The following points should be emphasized for establishing a long-term SSC enriched culture.

- Purity of the starting cell preparation is essential to reduce the chances that somatic fibroblasts will outgrow SSCs. Thus, Thy1 selection or a similar SSC enrichment strategy is imperative for isolating a cell population intended for culture.
- SSCs grow as cell clumps that are loosely attached to the underlying feeders. Assessment of clump morphology provides a good indication of SSC support. Healthy germ cell clumps have tightly adhering membranes with a smooth and robust appearance in which individual cells are not easily discernable. Unhealthy clumps are loose, and often have a dark and grainy appearance.
- In our experience, a 12-well plate format is ideal for supporting SSC expansion. When larger wells are used cell proliferation declines and clumps break apart resulting in cell death.
- Germ cell clumps should become evident 3–4 days after initial plating. The first subculturing should be performed after 6–8 days, and the cells placed in the same size well as the original culture on a new mitomycin-C STO feeder layer.
- Subculturing typically involves trypsin-EDTA digestion (0.5 ml/well of a 12-well plate) for 5 min at 37 °C followed by addition of 50–100 μl of FBS. A single cell suspension is then created with gentle pipetting. Cells are pelleted by centrifugation at $600g$ for 7 min at 4 °C, the supernatant removed, and cells resuspended in 2 ml of SFM. The cells are again pelleted by centrifugation, resuspended in SFM with GDNF, GFRα1, and bFGF, and plated onto a fresh STO cell feeder monolayer.

- In healthy cultures, clumps will begin to reform and grow 3–4 days after subculturing.

- The second and third passages are usually conducted 6–8 days after the first subculturing and involve a split ratio of 1:1.5 or 1:2. After the first three passages, cultures can be split 1:2 to 1:4 every 7 days. The split ratio is determined subjectively based on visual assessment of the size and number of clumps.

- When using established SSC cultures for experimentation, cell concentration should be determined, and an equal number of cells should be plated per well.

- In some instances, such as contamination with a high concentration of somatic cells, it may beneficial to use gentle pipetting to remove the germ cell clumps from the feeder as described by Ryu *et al.* (2005). This technique will leave most of the contaminating cells and STO feeders in the well and allow collection of the majority of clump cells. The clumps can then be pelleted by centrifugation, digested with trypsin-EDTA, and subcultured as previously described. In our experience this sub-culturing technique is less effective than digesting the entire well when a high concentration of contaminating cells is not present.

- The gentle pipetting technique to remove germ cell clumps from feeders results in the collection of a cell population in which greater than 90% of the cells have the SSC cell surface phenotype (Oatley *et al.*, 2006). Thus, it is a good method for obtaining a very pure cell population for molecular and biochemical experimentation.

5. siRNA Transfection of Cultured Mouse SSCs

The application of RNAi technologies in mammalian cells has provided a tool for studying the role of specific genes in regulating various cell functions (Caplen *et al.*, 2001; Elabshir *et al.*, 2001; Paddison *et al.*, 2002). A highly effective RNAi-based method involves inducing transient reduction of specific gene expression with double stranded short interfering RNA (siRNA). When introduced into a cell, siRNAs specifically direct RISC-mediated destruction of homologous transcripts. siRNAs designed with modern algorithms can induce major reduction in specific gene expression with minimal off-target effects, which had plagued the use of first generation siRNAs. In cultured SSCs, siRNA technology can be used effectively to study the importance of specific genes in regulating their self-renewal, differentiation, and survival. We have used this experimental approach to show an importance of the GDNF-responsive genes Bcl6b, Etv5, and Lhx1 in SSC self-renewal *in vitro* (Oatley *et al.*, 2006, 2007). Unequivocal examination of the role that specific genes have in SSC self-renewal *in vitro* requires the use of functional germ cell transplantations in conjunction with siRNA. Cultures of proliferating spermatogonia, such as those derived from Thy1+testis cells and gonocyte cultures such as GS cells (Kanatsu-Shinohara *et al.*, 2003, 2005a) are composed of both stem cells and nonstem cell spermatogonia. While all cells in these cultures share identical phenotypic characteristics and express the same markers, only a small percentage of the cell population is capable of reestablishing spermatogenesis upon transplantation indicating that the number of true SSCs is actually low in these cultures. For example, in GDNF-dependent Thy1+germ cell cultures expression of GFRα1 is ubiquitous, but only a small percentage of the cell population is capable of reestablishing spermatogenesis, indicating that not all GFRα1+germ cells in these cultures are actual stem cells. Therefore, simply evaluating the number of germ cells expressing a specific spermatogonia marker, such as GFRα1 or Thy1, would provide an inaccurate interpretation in regards to treatment effects on SSCs. Thus, functional germ cell transplantation following siRNA treatments must be conducted to evaluate specific effects on SSCs and differentiate between those that are general in nature or affect the nonstem cells.

Studying the importance of specific genes in SSC self-renewal *in vivo* is challenging mostly due to the rarity of these cells in the total testis cell population and lack of specific markers to track their biological activity. All putative SSC markers described to date are also expressed by other spermatogonial sub-types. The majority of *in vivo* investigations into the importance of specific genes on SSC self-renewal have involved evaluating spermatogenesis in mice lacking expression of a specific gene due to experimental targeting or inactivating mutation. Progressive degeneration of spermatogenesis resulting in loss of the spermatogonial population and formation of Sertoli cell only seminfierous tubules has been regarded as hallmarks of impaired SSC self-renewal. Additionally, an inability of SSCs lacking expression of a specific gene to reestablish spermatogenesis following transplantation has been used as another measure of impaired SSC self-renewal. Neither of these interpretations is truly precise in regards to SSC self-renewal because impairment in differentiation at the level of SSCs or other proliferating spermatogonial subtypes would result in identical phenotypes. Thus, interpretation of results from these types of studies must be done with caution. Examining the self-renewal of SSCs in culture over a period of several self-renewal cycles may provide the best assessment of the importance that a specific gene has in regulating SSC self-renewal.

Conducting an siRNA experiment to evaluate the importance of a specific gene in SSC self-renewal involves (1) transient transfection of cultured germ cells with siRNAs, (2) determining the level of gene knock-down, and (3) transplanting the cultured cells to quantify SSC numbers after a select period of time usually encompassing 1–3 self-renewal cycles. These three points are described below for use with Thy1+germ cell cultures to provide researchers with information necessary to conduct siRNA experiments for examining the biological significance of specific genes in SSC self-renewal.

5.1. Transient transfection of cultured germ cells with siRNAs

The choice of siRNA can be user preferred and purchased as predesigned annealed oligonucleotides or designed to user dependent specifications. We have utilized pre-designed siRNAs from Ambion Inc. (Austin, Texas) and Dharmacon Inc.

(Chicago, Illinois) to silence the expression of several different genes in Thy1 + germ cell cultures. Several different methods exist for introducing siRNA oligonucleotides into cells. We have predominately utilized lipofection with lipofectamine 2000 reagent from Invitrogen Inc. which yields greater than 70% transfection efficiency on average.

- Preparation of Thy1+germ cells for transfection:

  - Clump-forming Thy1+germ cells are removed from STO feeders by gentle pipetting and pelleted by centrifugation at 600g for 7 min.
  - The pellet is then resuspended in 1–2 ml of 0.25% trypsin-1mMEDTA solution and incubated at 37 °C for 5 min.
  - FBS is added to stop trypsin activity and pipetting is used to generate a single cell suspension.
  - Cells are pelleted by centrifugation at 600g for 7 min., resuspended in 1–2 ml of mSFM and total cell number is determined.
  - Cells are again pelleted by centrifugation at 600g for 7 min. and resuspended in mSFM Containing recombinant GDNF (10 ng/ml) and bFGF (1 ng/ml) at a concentration to provide 5×10^4–1×10^5 cells/ml and 1 ml of cell suspension is aliquoted into individual wells containing STO feeders of a 12- or 24-well plate. It is often desirable to plate several wells to allow for multiple replications of each siRNA treatment.
  - Planning of experiments should include replicate wells of siRNAs targeting genes of interest, and control treatments, including non-targeting negative control siRNA and a mock siRNA control (e.g., transfection reagent only).

- siRNA Transfection:

  - For each reaction, transfection reagent is prepared by combining 2 μl of Lipofectamine 2000 with 100 μl of Opti-MEM media (Invitrogen).
  - siRNAs for each gene-of-interest and negative control reactions are prepared by combining 50–100 pmoles of siRNA (i.e., 1–2 μl if a 5 μM siRNA stock concentration is created) with 100 μl of Opti-MEM media.
  - If multiple transfections of each treatment is to be performed it is optimal to combine the entire volume of siRNA and Lipofectamine for each treatment together rather than to prepare individual aliquots.
  - The Lipofectamine 2000/Opti-MEM and siRNA/Opti-MEM aliquots are then combined and incubated at room temperature for 20 min.
  - The mixtures are then added drop-wise to culture wells containing the previously seeded Thy1+germ cells. Typically 203–204 μl of solution is added to each well.

5.2. Determining the level of gene knockdown

- The degree of reduced gene expression due to siRNA treatment can be assessed 18–48 after transfection.

- RNA and protein samples can be isolated from entire contents of the culture wells (STO feeders and germ cells). Alternatively, the germ cells can be separated from STO feeders by gentle pipetting and RNA/protein samples collected, but the total number of cells recovered will be significantly reduced.

- The level of reduced gene expression can then be evaluated using quanitative RT-PCR and Western blotting by comparing the expression levels of the targeted gene-of-interest in gene specific siRNA samples and control siRNA samples. In our experience, the typical range of reduced gene expression is 50–80%.

5.3. Transplanting transfected cells to quantify SSC numbers

- The self-renewal cycle of mouse SSCs *in vitro* is ~6 days (Kubota *et al.*, 2004b). Thus, determining the number of SSCs in culture wells 7 days after siRNA transfection provides an evaluation of treatment effects over a single self-renewal cycle.

- Alternatively, SSC self-renewal can be evaluated over several cycles, if cultures are extended to 14 or 21 days. In this experimental design, the germ cells should be sub-cultured onto new STO feeders and transfected with gene-specific or control siRNA every 7 days.

- To assay for SSC number in siRNA treated Thy1+germ cell cultures, the entire contents of culture wells should be collected by tryipsin-EDTA digestion for 5 min. at 37 °C. Separation of germ cell clumps from STO feeders is not desirable for these experiments because variations in germ cell recovery will occur which will confound results of the transplantation assays.

- FBS is added to stop trypsin activity and a single cell suspension is created by pipetting.

- The cells are pelleted by centrifugation at 600g for 7 min., washed in mSFM, and the number of cells is determined.

- Cells are again pelleted by centrifugation at 600g for 7 min and resuspended in mSFM at a concentration of 1×10^6 cells/ml.

- Cell suspensions for each siRNA treatment is then transplanted into 4–8 testes of busulfan-treated 129 × C57 recipient mice. Approximately, 10 μl of cell suspension is injected into each recipient testis.

- Approximately 2 months after transplantation the number of donor-derived colonies of spermatogenesis in recipient testes is determined. SSC colony number is then determined based on the number of total cells recovered following 1 or more self-renewal cycles post siRNA transfection and normalized to 10^5 cells transfected with different siRNAs on day 1 of the culture period. This value represents the relative number of SSCs present in the transplanted cell suspension and thus, culture well following 1 or more self-renewal cycles and can be used to compare SSC numbers in cultures treated with gene specific and control siRNAs.

6. Implications

Male fertility is dependent on SSCs, and they are the only stem cell type in the adult that contributes genetic information to the next generation. Thus, SSCs are essential for preservation of a species and genetic diversity. Therefore, understanding the mechanisms that regulate their biological functions is of critical importance. In addition, SSCs have enormous potential for use in generating genetically modified animals and gene therapy. The transplantation assay and *in vitro* culture system for long-term expansion described in this chapter are two techniques that greatly aid in the ability to study SSCs and take advantage of their applications. Using established cultures enriched for self-renewing SSCs in a chemically defined environment provides a means to conduct biochemical and molecular studies that are impossible to achieve *in vivo* because the SSC is extremely rare in the testis. By combining *in vitro* experiments with the transplantation assay, biological experiments can now be performed on these important stem cells.

The initial demonstration that SSCs could survive *in vitro* for long periods of time and then generate spermatogenesis following transplantation (Nagano *et al.*, 1998) laid the foundation for development of subsequent systems that support continual self-renewal of SSCs (Kanatsu-Shinohara *et al.*, 2003, 2005a; Kubota *et al.*, 2004a). These culture advances have paved the way for genetic modification of SSCs (Kanatsu-Shinohara *et al.*, 2005b; Nagano *et al.*, 2000, 2001, 2002) including gene targeting strategies (Kanatsu-Shinohara *et al.*, 2006b). Similar techniques have been used to introduce genes into rat SSCs (Hamra *et al.*, 2002; Orwig *et al.*, 2002). Moreover, sophisticated experiments regarding gene activity involved in self-renewal and fate determination of SSC can now be undertaken (Oatley *et al.*, 2006, 2007). Knowledge gained from investigating SSCs using culture and transplantation strategies described in this chapter will enhance our understanding of male fertility and general stem cell biology in mammals. Since spermatogenesis is a highly conserved process in mammals, discoveries made using mice will likely be applicable to other species including humans.

Acknowledgments

We thank Drs. H. Kubota and J.A. Schmidt for manuscript review and helpful comments. Research support has been from the National Institutes of Health and Robert J. Kleberg, Jr. and Helen C. Kleberg Foundation.

References

Barratt, C. L. R. (1995). Spermatogenesis. *In* "Gametes—The Spermatozoon" (J. G. Grudzinskas, and J. L. Yovich, Eds.), pp. 250–267. Cambridge University Press, New York.

Bellgrau, D., Gold, D., Selawry, H., Moore, J., Franzusoff, A., and Duke, R. C. (1995). A role of CD95 ligand in preventing graft rejection. *Nature* **377,** 630–632.

Bellve, A. R., Cavicchia, J. C., Millette, C. F., O'Brien, D. A., Bhatnagar, Y. M., and Dym, M. (1977). Spermatogenic cells of the prepuberal mouse, isolation and morphological characterization. *J. Cell. Biol.* **74,** 68–85.

Brinster, R. L. (2002). Germline stem cell transplantation and transgenesis. *Science* **296,** 2174–2176.

Brinster, R. L., and Avarbock, M. R. (1994). Germline transmission of donor haplotype following spermatogonial transplantation. *Proc. Natl. Acad. Sci. USA* **91,** 11303–11307.

Brinster, C. J., Ryu, B.-Y., Avarbock, M. R., Karagenc, L., Brinster, R. L., and Orwig, K. E. (2003). Restoration of fertility by germ cell transplantation requires effective recipient preparation. *Biol. Reprod.* **69,** 412–420.

Brinster, R. L., and Zimmermann, J. W. (1994). Spermatogenesis following male germ cell transplantation. *Proc. Natl. Acad. Sci. USA* **91,** 11298–11302.

Bucci, L. R., and Meistrich, M. L. (1987). Effects of busulfan on murine spermatogenesis: Cyto-toxicity, sterility, sperm abnormalities and dominant lethal mutations. *Mutat. Res.* **176,** 159–268.

Caplen, N. J., Parrish, S., Iamni, F., Fire, A., and Morgan, R. A. (2001). Specific inhibition of gene expression by small double-stranded RNAs in invertebrates and vertebrate systems. *Proc. Natl. Acad. Sci. USA* **98,** 9746–9747.

De Rooij, D. G., and Russell, L. D. (2000). All you wanted to know about spermatogonia but were afraid to ask. *J. Androl.* **21,** 776–798.

Dobrinski, I., Ogawa, T., Avarbock, M. R., and Brinster, R. L. (1999). Computer assisted image analysis to assess colonization of recipient seminiferous tubules by spermatogonial stem cells from transgenic donor mice. *Mol. Reprod. Dev.* **53,** 142–148.

Elabshir, S. M., Harboth, J., Lendeckel, W., Yalcin, A., Weber, K., and Tuschl, T. (2001). Duplexes of 21-nucleotide RNAs mediate RNA interefence in cultured mammalian cells. *Nature* **411,** 494–498.

Hamra, F. K., Chapman, K. M., Nguyen, D. M., Williams-Stephens, A. A., Hammer, R. E., and Garbers, D. L. (2005). Self renewal, expansion, and transfection of rat spermatogonial stem cells in culture. *Proc. Natl. Acad. Sci. USA* **102,** 17430–17435.

Hamra, F. K., Gatlin, J., Chapman, K. M., Grellhesl, D. M., Garcia, J. V., Hammer, R. E., and Garbers, D. L. (2002). Production of transgenic rats by lentiviral transduction of male germ-line stem cells. *Proc. Natl. Acad. Sci. USA* **99,** 14931–14936.

Huckins, C., and Clermont, Y. (1968). Evolution of gonocytes in the rat testis during late embryonic and early post-natal life. *Arch. Anat. Histol. Embryol.* **51,** 341–354.

Kanatsu-Shinohara, M., Ikawa, M., Takehashi, M., Ogonuki, N., Miki, H., Inoue, K., Kazuki, Y., Lee, J., Toyokuni, S., Oshimura, M., Ogura, A., and Shinohara, T. (2006). Production of knockout mice by random or targeted mutagenesis in spermatogonial stem cells. *Proc. Natl. Acad. Sci. USA* **103,** 8018–8023.

Kanatsu-Shinohara, M., Inoue, K., Miki, H., Ogonuki, N., Takehashi, M., Morimoto, T., Ogura, A., and Shinohara, T. (2006). Clonal origin of germ cell colonies after spermatogonial transplantation in mice. *Biol. Reprod.* **75,** 68–74.

Kanatsu-Shinohara, M., Miki, H., Inoue, K., Ogonuki, N., Toyokuni, S., Ogura, A., and Shinohara, T. (2005). Long-term culture of mouse male germline stem cells under serum- or feeder-free conditions. *Biol. Reprod.* **72,** 985–991.

Kanatsu-Shinohara, M., Ogonuki, N., Inoue, K., Miki, H., Ogura, A., Toyokuni, S., and Shinohara, T. (2003). Long-term proliferation in culture and germline transmission of mouse male germline stem cells. *Biol. Reprod.* **69,** 612–616.

Kanatsu-Shinohara, M., Ogonuki, N., Iwano, T., Lee, J., Kazuki, Y., Inoue, K., Miki, H., Takehashi, M., Toyokuni, S., Shinkai, Y., Oshimura, M., Ishino, F., *et al.* (2005). Genetic and epigenetic properties of mouse male germline stem cells during long-term culture. *Development* **132,** 4155–4163.

Kanatsu-Shinohara, M., Toyokuni, S., and Shinohara, T. (2005). Genetic selection of mouse male germline stem cells *in vitro*: Offspring from single stem cells. *Biol. Reprod.* **72,** 236–240.

Kubota, H., Avarbock, M. R., and Brinster, R. L. (2003). Spermatogonial stem cell share some, but not all, phenotypic and functional characteristics with other stem cells. *Proc. Natl. Acad. Sci. USA* **100,** 6487–6492.

Kubota, H., Avarbock, M. R., and Brinster, R. L. (2004). Culture conditions and single growth factors affect fate determination of mouse spermatogonial stem cells. *Biol. Reprod.* **71,** 722–731.

Kubota, H., Avarbock, M. R., and Brinster, R. L. (2004). Growth factors essential for self-renewal and expansion of mouse spermatogonial stem cells. *Proc. Natl. Acad. Sci. USA* **101,** 16489–16494.

McLaren, A. (2003). Primordial germ cells in the mouse. *Dev. Biol.* **262,** 1–15.

McLean, D. J., Friel, P. J., Johnston, D. S., and Griswold, M. D. (2003). Charatcterization of spermatogonial stem cell maturation and differentiation in neonatal mice. *Biol. Reprod.* **69,** 2085–2091.

Meistrich, M. L., Hunter, N. R., Suzuki, N., Trostle, P. K., and Withers, H. R. (1978). Gradual regeneration of mouse testicular stem cell after exposure to ionizing radiation. *Radiat. Res.* **74,** 349–362.

Meng, X., Lindahl, M., Hyvonen, M. E., Parvinen, M., de Rooij, D. G., Hess, M. W., Raatikainen-Ahokas, A., Sainio, K., Rauvala, H., Lakso, M., Pichel, J. G., Westphal, H., *et al.* (2000). Regulation of fate decision of undifferentiated spermatogonia by GDNF. *Science* **287,** 1489–1493.

Nagano, M., Avarbock, M. R., and Brinster, R. L. (1999). Pattern and kinetics of mouse donor spermatogonial stem cell colonization in recipient testes. *Biol. Reprod.* **60,** 1429–1436.

Nagano, M., Avarbock, M. R., Leonida, E. B., Brinster, C. J., and Brinster, R. L. (1998). Culture of mouse spermatogonial stem cell. *Tissue Cell* **30,** 389–397.

Nagano, M., Brinster, C. J., Orwig, K. E., Ryu, B.-Y., Avarbock, M. R., and Brinster, R. L. (2001). Transgenic mice produced by retroviral transduction of male germ-line stem cells. *Proc. Natl. Acad. Sci. USA* **98,** 13090–13095.

Nagano, M., Ryu, B.-Y., Brinster, C. J., Avarbock, M. R., and Brinster, R. L. (2003). Maintenance of mouse male germ line stem cell *in vitro*. *Biol. Reprod.* **68,** 2207–2214.

Nagano, M., Shinohara, T., Avarbock, M. R., and Brinster, R. L. (2000). Retrovrius-meditated gene delivery into male germ line stem cells. *FEBS Lett.* **475,** 7–10.

Nagano, M., Watson, D. J., Ryu, B.-Y., Wolfe, J. H., and Brinster, R. L. (2002). Lentiviral vector transduction of male germ line stem cells in mice. *FEBS Lett.* **524,** 111–115.

Oatley, J. M., Avarbock, M. R., and Brinster, R. L. (2007). Glial cell line-derived neurotrophic factor regulation of genes essential for self-renewal of mouse spermatogonial stem cells is dependent on Src family kinase signaling. *J. Biol. Chem.* **282,** 25842–25851.

Oatley, J. M., Avarbock, M. R., Teleranta, A. I., Fearon, D. T., and Brinster, R. L. (2006). Identifying genes important for spermatogonial stem cell self-renewal and survival. *Proc. Natl. Acad. Sci. USA* **103,** 9524–9529.

Oatley, J. M., Reeves, J. J., and McLean, D. J. (2004). Biological activity of cryporeserved bovine spermatogonial stem cell during *in vitro* culture. *Biol. Reprod.* **71,** 942–947.

Ogawa, T., Arechaga, J. M., Avarbock, M. R., and Brinster, R. L. (1997). Transplantation of testis germinal cells into mouse seminiferous tubules. *Int. J. Dev. Biol.* **41,** 111–122.

Ogawa, T., Ohmura, M., Yumura, Y., Sawada, H., and Kubota, Y. (2003). Expansion of murine spermatogonial stem cells through serial transplantation. *Biol. Reprod.* **68,** 316–322.

Ohta, H., Tohda, A., and Nishimune, Y. (2003). Proliferation and differentiation of spermatogonial stem cells in the W/W^v mutant mouse testis. *Biol. Reprod.* **69,** 1815–1821.

Orwig, K. E., Avarbock, M. R., and Brinster, R. L. (2002). Retrovirus-mediated modification of male germline stem cells in rats. *Biol. Reprod.* **67,** 874–879.

Paddison, P. J., Caudy, A., and Hannon, G. J. (2002). Stable suppression of gene expression by RNAi in mammalian cells. *Proc. Natl. Acad. Sci. USA* **99,** 1443–1448.

Russell, L. D. (1993). Form, dimension, and cytology of mammalian sertoli cells. *In* "The Sertoli Cell" (L. D. Russell, and M. D. Griswold, Eds.), pp. 1–37. Cache River, Clearwater.

Russell, L. D., Ettlin, R. A., Hikim, A. P., and Clegg, E. D. (1990). Mammalian spermatogenesis. *In* "Histological and Histopathological Evaluation of the Testis" pp. 1–40. Cache River, Clearwater.

Ryu, B.-Y., Kubota, H., Avarbock, M. R., and Brinster, R. L. (2005). Conservation of spermatogonial stem cell renewal signaling between mouse and rat. *Proc. Natl. Acad. Sci. USA* **102,** 14302–14307.

Ryu, B.-Y., Orwig, K. E., Avarbock, M. R., and Brinster, R. L. (2003). Stem cell and niche development in the postnatal rat testis. *Dev. Biol.* **263,** 253–263.

Ryu, B.-Y., Orwig, K. E., Kubota, H., Avarbock, M. R., and Brinster, R. L. (2004). Phenotypic and functional characteristics of spermatogonial stem cells in rats. *Dev. Biol.* **274,** 158–170.

Ryu, B.-Y., Orwig, K. E., Oatley, J. M., Avarbock, M. R., and Brinster, R. L. (2006). Effects of aging and niche microenvironment on spermatogonial stem cell self-renewal. *Stem Cells* **24,** 1505–1511.

Sharpe, R. (1994). Regulation of spermatogenesis. *In* "The Physiology of Reproduction" (E. Knobil, and J. D. Neill, Eds.), pp. 1363–1434. Raven, New York.

Shinohara, T., Avarbock, M. R., and Brinster, R. L. (1999). Beta1- and alpha6-integrin are surface markers on mouse spermatogonial stem cells. *Proc. Natl. Acad. Sci. USA* **96,** 5504–5509.

Shinohara, T., Avarbock, M. R., and Brinster, R. L. (2000). Functional analysis of spermatogonial stem cells in steel and cryptorchid infertile mouse models. *Dev. Biol.* **220,** 401–411.

Shinohara, T., Orwig, K. E., Avarbock, M. R., and Brinster, R. L. (2000). Spermatogonial stem cell enrichment by multiparameter selection of mouse testis cells. *Proc. Natl. Acad. Sci. USA* **97,** 8346–8351.

Shinohara, T., Orwig, K. E., Avarbock, M. R., and Brinster, R. L. (2001). Remodeling of the postnatal mouse testis is accompanied by dramatic changes in stem cell number and niche accessibility. *Proc. Natl. Acad. Sci. USA* **98,** 6186–6191.

Shinohara, T., Orwig, K. E., Avarbock, M. R., and Brinster, R. L. (2002). Germ line stem cell competition in postnatal mouse testes. *Biol. Reprod.* **66,** 1491–1497.

Silvers, W. K. (1979). "The Coat Colors of Mice" Springer-Verlag, New York.

Tadokoro, Y., Yomogida, K., Ohta, H., Tohda, A., and Nishimune, Y. (2002). Homeostatic regulation of germinal stem cell proliferation by the GDNF/FSH pathway. *Mech. Dev.* **113,** 29–39.

Tegelenbosch, R. A. J., and de Rooij, D. G. (1993). A quantitative study of spermatogonial multiplication and stem cell renewal in the C3H/101 F1 hybrid mouse. *Mutat. Res.* **290,** 193–200.

Tung, K. S. K. (1993). Regulation of testicular autoimmune disease. *In* "Cell and Molecular Biology of the Testis" (C. Desjardins, and L. L. Ewing, Eds.), pp. 474–490. Oxford University Press, New York.

Van Beek, M. E. A. B., Meistrich, M. L., and de Rooij, D. G. (1990). Probability of self-renewing divisions of spermatogonial stem cells in colonies formed after fission neuron irradiation. *Cell Tissue Kinet.* **23,** 1–16.

Withers, H. R., Hunter, N., Barkley, H. T., Jr., and Reid, B. O. (1974). Radiation survival and regeneration characterisitics of spermatogenic stem cells of mouse testis. *Radiat. Res.* **57,** 88–103.

Yomogida, K., Yagura, Y., Tadokoro, Y., and Nishimune, Y. (2003). Dramatic expansion of germinal stem cell by ectopically expressed human glial cell line-derived neurotrophic factor in mouse Sertoli cells. *Biol. Reprod.* **69,** 1303–1307.

Zhang, X., Ebata, K. T., and Nagano, M. C. (2003). Genetic analysis of the clonal origin of regenerating mouse spermatogenesis following transplantation. *Biol. Reprod.* **69,** 1872–1878.

CHAPTER 6

Stem Cells in the Adult Lung

Xiaoming Liu, Ryan R. Driskell, and John F. Engelhardt

Department of Anatomy and Cell Biology
College of Medicine
The University of Iowa
Iowa City, Iowa 52242

Abstract
1. Introduction
2. Anatomical and Cellular Diversity of the Adult Lung
3. Stem Cell Phenotypes and Niches in the Adult Lung
4. *In Vivo* Injury Models of the Lung
    4.1. Overview
    4.2. Mouse naphthalene injury and BrdU labeling protocol
    4.3. Mouse polidocanol injury and BrdU labeling protocol
5. *Ex Vivo* Epithelial Tracheal Xenograft Model to Study Stem Cells
Expansion in the Proximal Airway
    5.1. Overview
    5.2. Human bronchial xenograft protocol
6. *In Vitro* Colony-Forming Efficiency (CFE) Assay to Characterize Stem/progenitor
Cell Populations in Conducting Airway Epithelium
    6.1. Overview
    6.2. *In vitro* mouse tracheal epithelial cells CFE assay protocol
7. Models to Study Stem/Progenitor Cells of Airway SMG
    7.1. Overview
    7.2. *In vitro* invasion and tubulogenesis assay using mouse tracheal epithelial cells
References

Abstract

The lung is composed of two major anatomically distinct regions—the conducting airways and gas-exchanging airspaces. From a cell biology standpoint, the conducting airways can be further divided into two major compartments, tracheobronchial and bronchiolar airways, while the alveolar regions of the lung make up the gas-exchanging airspaces. Each of these regions consists of distinct epithelial cell types with unique cellular physiologies and stem/progenitor cell compartments. This chapter will focus on model systems to study stem/progenitor cells in the adult tracheobronchial airways, also referred to as the proximal airway of the lung. Important in such models is an appreciation for the diversity of stem cell niches in the conducting airways that provide localized environmental signals to both maintain and mobilize stem cells in the setting of airway injury and normal cellular turnover. Since cellular turnover in airways is relatively slow, methods for analysis of stem cells *in vivo* have required prior injury to the lung. In contrast, *ex vivo* and *in vitro* models for analysis of airway stem/progenitor cells have utilized genetic markers to track lineage relationships together with reconstitution systems that mimic airway biology. Over the past decades, several widely acceptable methods have been developed and used in the characterization of adult airway stem/progenitor cells. These include localization of label-retaining cells (LRCs), retroviral tagging of epithelial cells seeded into xenografts, air–liquid interface cultures to track clonal proliferative potential, and multiple transgenic mouse models. This chapter will review the biologic context and utility of these models while providing detailed methods for several of the more broadly useful models for studying adult airway stem/progenitor cell types.

1. Introduction

The adult lung is lined with numerous distinct types of epithelial cells in different anatomical regions, progressing proximally from the trachea→ bronchi→ bronchioles→alveoli. The conducting airways (including the trachea, bronchi, and bronchioles) constitute a minority of surface area in the human lung amounting to approximately 0.25 m^2, while the alveolar regions of the lung make up approximately 100 m^2. In a normal steady-state airway, epithelial cell proliferation in the adult lung is much lower compared to highly proliferative compartments in the gut, skin, and hematopoietic system. As in other adult tissues and organs, stem cells in the adult lung are a subset of undifferentiated cells with the capacity to maintain multipotency in the context of the physiologic domain in which they reside. In this context, adult lung stem cells give rise to transient amplifying (TA) progenitor cells capable of abundant self-renewal and regeneration of specific cell lineages in the lung. According to their position within the pulmonary tree, several cell types in the lung have been suggested to act as stem/progenitor cells in response to injury and serve to effect local repair (Emura, 2002; Engelhardt, 2001; Otto, 2002). Basal cells, Clara cells, and cells that reside in submucosal glands (SMGs) have been shown to function as progenitors or stem cells in the conducting airway of mice (Borges *et al.*, 1997; Borthwick *et al.*, 2001; Engelhardt *et al.*, 1995, 2001; Hong *et al.*, 2004a). While a subset of Clara cells (called variant Clara cell or Clara) residing within neuroepithlial bodies (NEB) (Hong *et al.*, 2001; Peake *et al.*, 2000;

Reynolds *et al.*, 2000a,b) or bronchoalveolar duct junctions (Giangreco *et al.*, 2002; Kim *et al.*, 2005) has been considered stem cells in bronchioles. In contrast, alveolar epithelial type II cells (AEC II) are thought to function as the progenitor of the alveolar epithelium based on their capacity to replicate and to give rise to terminally differentiated (TD) alveolar type I cells (AEC I) (Mason *et al.*, 1997). More recently, "side population" cells (SP cells) exhibiting Hoechst dye efflux properties similar to those of hematopoietic stem cells, have been isolated and identified as a putative adult stem cell population for alveolar regions of the adult lung (Giangreco *et al.*, 2004; Summer *et al.*, 2003). These SP cells appear to have both mesenchymal and epithelial potential.

Despite these advance in the study of lung stem cells, a single lung stem cell that can give rise to multiple epithelial lineages in both the proximal and distal airways of the lung has not been identified. To date, the concept of a pluripotent stem cell for all regions of the lung has not been widely accepted, based on a developing concept that local regional niches are required to control both the phenotype and expansion of stem cells across vast distances of biologically distinct airways. Thus, stem cell biology in the adult lung is most widely investigated in distinct regions of tracheobronchial, bronchiolar, and alveolar epithelium (Berns, 2005; Bishop, 2004; Emura, 1997, 2002; Engelhardt, 2001; Liu *et al.*, 2004; Mason *et al.*, 1997; Neuringer and Randell, 2004; Otto, 2002).

Most of our current understanding of stem cells in the adult lung originates from research using classic lung injury models and/or epithelial reconstitution models. These models will likely continue to be the most useful approaches to identify stem/progenitor cell populations and understand the regulation of stem/progenitor cell fates in the adult lung. However, as the molecular understanding of genes that control lung stem cell populations unravels, improved genetic approaches to study stem cell phenotypes in their niches will likely emerge in combination with more traditional approaches of analysis. This chapter will first review the diversity of cell types in the adult lung. We will then summarize the current existing knowledge and describe classic experiments used to characterize potential stem cell populations and stem cell niches in the adult lung. Lastly, we will describe several methods for characterizing stem/progenitor cells in the adult airway and provide detailed protocols for stem cell labeling, using lung injury models to mobilize airway stem cells, and the reconstitution of airway epithelia and SMGs using *in vitro* and *ex vivo* models.

2. Anatomical and Cellular Diversity of the Adult Lung

The adult lung can be functionally and structurally divided into three epithelial domains, the proximal cartilaginous airway (trachea and bronchi), distal bronchioles (bronchioles, terminal bronchioles and respiratory bronchioles), and gas exchanging airspaces (alveoli). Each of these domains has historically been classified by morphologic criteria that define the unique type of airway epithelium within each domain of the lung (Fig. 6.1) (Liu *et al.*, 2004). Consequently, defining

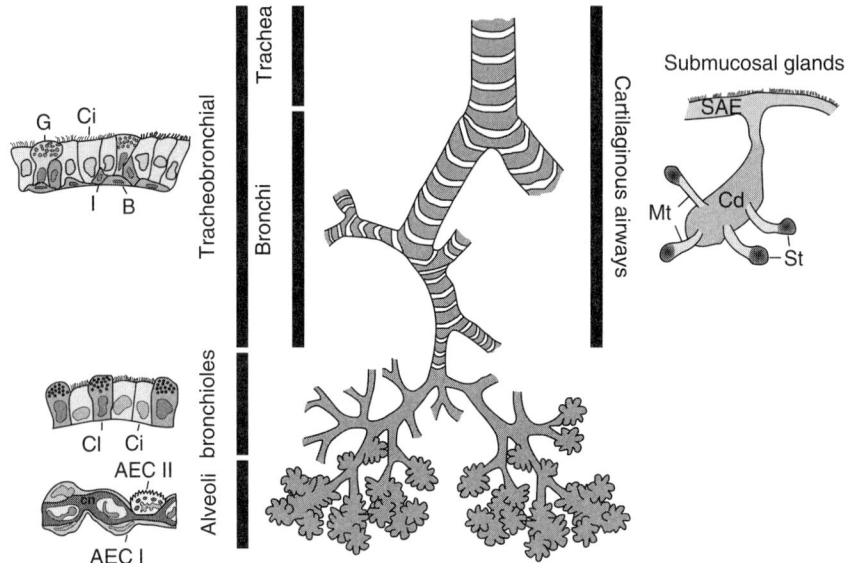

Figure 6.1 Cellular diversity of epithelial cell types in the human adult lung. When discussing progenitor/stem cells in adult airways, one must always consider the diversity of cell phenotypes that exist in spatially distinct epithelia of the lung. Three main levels of conducting airways exist in the lung, including the trachea, bronchi, and bronchioles. Predominant cells types in the human pseudostratified, columnar tracheal, and bronchial epithelia include basal (B), intermediate (I), goblet (G), and ciliated (Ci) cells; less abundant nonciliated and neuroendocrine cells are not shown. SMGs are also present only in the cartilaginous airways of the trachea and bronchi. Predominant cells types in the human columnar bronchiolar epithelia include Clara cells (C) and ciliated cells (Ci); less abundant neuroendocrine cells are not shown. Predominant cells types in the gas-exchanging alveolar airspaces include alveolar epithelial type II cells (AEC II), alveolar epithelial type I cells (AEC I), and capillary endothelial cells of the capillary networks (cn). Mt, mucous tubule; St, serous tubule; SAE, surface airway epithelium; Cd, collecting duct.

functional criteria for the diversity of epithelial cell types in the airway has been a major focus over the last decade.

In the murine proximal conducting airway and major bronchi, basal, Clara, and ciliated cells are predominant cell types, though less frequent pulmonary neuroendocrine cells (PNECs) also reside in this domain. The nonciliated, columnar, Clara cells comprise the majority of the distal bronchiolar epithelium in mice, while the alveolar epithelium is lined by squamous alveolar type I pneumocytes (AEC I) and cuboidal type II pneumocytes (AEC II). Unique to the proximal cartilaginous airways are SMGs that are contiguous with surface airway epithelium (SAE) (Fig. 6.1). Although less abundant in mice and confined to the trachea, SMGs are a major secretory structure located in the interstitium beneath the cartilaginous airway of other nonrodent species (Choi *et al.*, 2000; Liu *et al.*, 2004; Plopper *et al.*, 1986; Widdicombe *et al.*, 2001). SMGs are composed of an interconnecting network of serous and mucus tubules that secrete antibacterial factors, mucous, and fluid into the airway lumen. In contrast, the alveolar epithelium of adult lung is highly similar among mammalian species (Ten Have-Opbroek and Plopper, 1992).

Most of the current studies on stem cell biology in the adult airway have been employed using rodent animal models. To this end, it is important to note the anatomical and cellular differences between rodent and human airways. In mice, Clara cells reside throughout the tracheobronchial and bronchiolar epithelium and are the predominant secretory cell type in airways of this species. In contrast, Clara cells are limited to the bronchioles of human airways, and goblet cells are the predominant secretory cell type in the human tracheobronchial airway. Unlike humans, the predominant secretory cell type in the rat conducting airway is the serous cell. However, goblet cells in mice and rats can be induced by specific cytokine stimuli and/or injury (Liu *et al.*, 2004). A summary of cellular differences in the conducting airway epithelium and SMGs between rodents (mouse and rat) and primates (monkey and human) is listed in Table 6.1. Despite the notable differences between rodents and primates in the cell biology of the airway epithelium and the abundance of SMGs in the conducting airway, studies in mice have played significant roles in investigating stem cell biology in proximal and distal airways because of their versatile genetics.

3. Stem Cell Phenotypes and Niches in the Adult Lung

Studies in stem cell biology have demonstrated that the cell fate and the maintenance of stem-cell populations are regulated by their local anatomical and chemical microenvironment, or niche. Niches are discrete microenvironments of specialized cell types, matrix, and diffusible factors such as cytokines and growth factors, which are critical for maintaining stem cells and promoting appropriate cell fate and migration decisions (Watt and Hogan, 2000). In the adult lung, candidate stem cell populations have been identified that are restricted to the tracheal SMG ducts, NEBs of the bronchi and bronchioles, and bronchoalveolar-duct junction (BADJ) of the terminal bronchioles, suggesting distinct regional stem cell niches in the adult lung (Borthwick *et al.*, 2001; Engelhardt *et al.*, 1995, 2001; Giangreco *et al.*, 2002; Hong *et al.*, 2001, 2004a; Kim *et al.*, 2005).

Because stem cells divide very infrequently, injury to the lung has been necessary to study lung stem cell phenotypes and their niches in the airway. This slow cycling characteristic has allowed for the use of DNA labeling techniques with detectable nucleotide analogs to track stem cells *in situ*. Bromodeoxyuridine (BrdU) is one such common nucleotide analog that is classically used to track labeling-retained-cells (LRC) after a prolonged "wash out" period that dilutes the label within the more rapidly-cycling TA cells. However, it should be emphasized that nucleotide label retention is only suggestive of a stem cell phenotype. Other criteria must also be applied to make a more conclusive determination, including the ability to undergo subsequent rounds of division, and lineage fate determination of daughter cells. For example, studies in hematopoietic stem cells (HSCs) have

Table 6.1
Summary of predominant epithelial cell types and SMGs in the conducting airways between rodents and primates

Species	Basal	Intermediate	Goblet	Serous	Clara	NCC	Ciliated	SMG	References
Mouse	T,B	–	–[a]	–	T,B,Br	–	T,B,Br	T	Hansell and Moretti (1969), Pack et al. (1980), Widdicombe et al. (2001)
Rat	T,B	T,B	–[a]	T,B	Br	T,B	T,B,Br	T	Plopper et al. (1983), Souma (1987), Widdicombe et al. (2001)
Monkey	T,B	T,B	T,B	NA	NA	T,B	T,B,Br	T,B	Castleman et al. (1975), Plopper et al. (1983, 1986)
Human	T,B	T,B	T,B	–	Br	T,B	T,B,Br	T,B	Jeffery (1983), Plopper et al. (1980)

[a]Summary for rodents applies to pathogen-free animals; the abundance of goblet cells may increase in the setting of infection or cytokine simulation. NA, unknown or under investigation; T, trachea; B, bronchi; Br, bronchioles; NCC, nonciliated, nonsecretory columnar cells; SMG, submucosal glands.

challenged that the characteristic of nucleotide analog-label retention is not a specific index for HSCs (Kiel *et al.*, 2007). Recently, other thymidine analogs, including 5-chloro-2-deoxyuridine (CldU) and 5-iodo-2-deoxyuridine (IdU), have been applied to address the question of reentry of LRCs into the cell cycle (Kiel *et al.*, 2007; Teta *et al.*, 2007). These two nucleotide labels can be differentially localized with antibodies and hence, used to avoid misinterpreting long-lived TD cells that retain label as stem cells (Barker *et al.*, 2007); only label-retaining stem cells should be capable of reentry into the cell cycle and hence, labeling with a second nucleotide analog. An alternative genetic approach has also been used to localize slow-cycling stem cells in mouse hair follicles using inducible Histone H2B-GFP expression (Tumbar *et al.*, 2004). In this context, slow-cycling stem cells retain a pulse of Histone H2B-GFP label for longer periods of time. In conjunction with nucleotide label-retention, this approach could be used to track reentry of Histone H2B-GFP LRCs into the cell cycle and following their lineage fates *in vivo*. An additional advantage of the Histone H2B-GFP labeling approach is the ability to purify slow-cycling LRCs by FACS (Tumbar *et al.*, 2004).

Two methods of injury used to amplify mobilization of the stem cell compartment in the proximal airway and identify LRCs have included intratracheal instillation of polidocanol or inhalation of SO_2 (Borthwick *et al.*, 2001). Intratracheal instillation of polidocanol or inhalation of SO_2 followed by *in vivo* BrdU administration, labels airway epithelial cells throughout the entire trachea at early time points (Borthwick *et al.*, 2001). However, longer time points post labeling revealed LRC accumulation in SMG ducts of mouse trachea, suggesting that this region provides a protective niche for stem cells (Borthwick *et al.*, 2001; Engelhardt, 2001). An example of this LRC patterning following naphthalene injury in mouse trachea is shown in Fig. 6.2. The ability of glandular cell types to repopulate the SAE of the trachea was also studied by subcutaneous transplantation of epithelia-denuded tracheas into recipient mice (Borthwick *et al.*, 2001). By 28 days following transplantation, outgrowth of glandular cells led to repopulation of a ciliated SAE, suggesting that gland or gland duct cells could regenerate the tracheal epithelium.

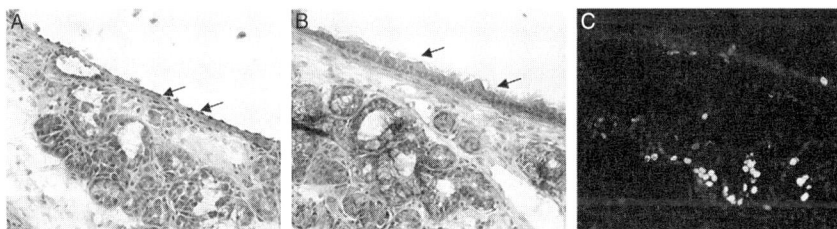

Figure 6.2 Naphthalene induced mouse tracheal epithelial injury and the subacute proliferative response. H & E stained sections of proximal trachea at 1 day (A) and 7 days (B) after intraperitoneal injection of 275 mg/kg naphthalene. Naphthalene treatment induced complete ablation of tracheal epithelium (arrows in A). By 7 days post injury, a regenerated tracheal epithelium is observed (arrows in B). (C) BrdU labeling of mice during the first week of naphthalene injury and harvested at 28 days post injury. Immunofluorescent detection of BrdU reveals high levels of incorporation in epithelia of SMGs and infrequent labeling of cells in proximal surface tracheal epithelium.

It is commonly thought that basal cells have the capacity to produce all the major cell phenotypes found in mouse trachea, including basal, ciliated, goblet, and granular secretory cells (Hong *et al.*, 2004b; Schoch *et al.*, 2004). In humans, where the proximal airway is more pseudostratafied, parabasal cells (which are located just above the basal cells, and may also be considered intermediate cells) are thought to contribute more to the proliferative cell population than basal cells based on higher-level expression of the proliferation marker MIB-1 (Ki-67) (Boers *et al.*, 1998). Hence, parabasal cells (or intermediate cells) are thought to act as a TA cell population in the proximal airway that is derived from a subset of basal stem cells. Studies using human bronchial xenografts and clonal analysis with retroviral vectors also support this concept (Engelhardt *et al.*, 1995).

Clara cells have long been thought to act as progenitor cells of the airway. This type of cell resides throughout all levels of the mouse conducting airways, but is confined to bronchiolar airways in humans. A number of studies have suggested that a subset of Clara cells act as stem cells in the bronchiolar epithelium of mice. Lung injury studies using naphthalene depletion of Clara cells revealed that a subset of variant Clara cells shows multipotent differentiation and the ability to regenerate the bronchiolar epithelium. This population of Clara cells is cytochrome P450–2F2 negative, resides in discrete pools associated with NEB and at bronchoalveolar duct junction. Further characterization of these cells has revealed that they express stem cell antigen (Sca)-1, and have the ability to efflux Hoechst dye (Reynolds *et al.*, 2000a,b).

PNECs reside within structures called NEBs at cartilage–intercartilage junctions in the proximal airway and airway junctions in the bronchioles. These cells or group of cells are thought to play a role in lung development and postnatal lung maintenance through the release of neuropeptides. Following naphthalene injury, proliferative cells that express either Clara cell secretory protein (CCSP), or the PNEC marker calcitonin gene-related peptide (CGRP), accumulate within NEBs (Hong *et al.*, 2001). To exclude the possibility that CGRP-expressing PNECs were the potential stem cell compartment in bronchioles, studies were performed using transgenic mice expressing the thymidine kinase gene under the direction of the CCSP promoter. Expression of thymidine kinase specifically in Clara cells allowed for the specific ablation of naphthalene-induced replicating CCSP-expressing cells. Cell specific ablation was facilitated by administration of the prodrug ganciclovir, which is converted by thymidine kinase to a toxic nucleotide analog that prevents DNA replication and hence, kills cells. Under these conditions that ablated all CCSP-expressing Clara cells following naphthalene injury, hyperplasia of CGRP-expressing PNECs occurred, however a ciliated epithelium could not be regenerated. Consequently, these studies demonstrated that CGRP-expressing PNECs are not a stem/progenitor cell population in the distal airway. Rather, PNECs are thought to provide a niche that regulates expansion of a CCSP-expressing stem cell population in mouse distal airways (Hong *et al.*, 2001).

More recent studies have also found that cells at the bronchioalveolar duct junctions (BADJ) represent another region found in terminal bronchioles that accumulate LRCs, called bronchioalveolar stem cells (BASCs). BASCs were previously termed DPCs (double-positive cells that express both SP-C and CCSP) (Giangreco *et al.*, 2002; Kim *et al.*, 2005). These cells appear to be resistant to bronchiolar and alveolar damage, proliferate during epithelial cell renewal *in vivo*, and are thought to function to maintain the bronchiolar Clara cell and alveolar cell populations of the distal lung. BASCs also possess characteristics of regional stem cells, such as self-renewal and multipotence in clonal assays (Kim *et al.*, 2005).

In the gas-exchanging regions of the lung, alveolar type II cells (AEC II) have been considered the stem/progenitor cell of the alveolar epithelium based on their ability to repopulate both AEC II and AEC I cells following injury (Giangreco *et al.*, 2002; Kim *et al.*, 2005; Reynolds *et al.*, 2004). Recent studies using a rat lung injury model have suggested that there may be four groups of AEC II cells, based on their expression of markers. The subpopulation of AEC II that are E-cadherin negative, proliferative, and express high levels of telomerase activity were considered stem cell candidates for alveolar epithelium (Reddy *et al.*, 2004).

In contrast to insights regarding candidate stem cells in the respiratory epithelium of adult lung, much less information is available regarding stem cells in the vascular compartment of the lung. One study demonstrated that an extremely small "spore-like" cell population in the lungs of adult sheep and rats with low oxygen demand can generate lung-like alveolar tissue *in vitro* (Vacanti *et al.*, 2001). SP cells isolated from the lung have also been shown to express stem cell markers indicative of epithelial and mesenchymal lineages. SP cells isolated from whole lung using Hoechst 33342 efflux and other stem cell markers comprise 0.03–0.07% of mouse lung cells and are Sca1 antigen positive, lin negative, heterogeneous for CD45, and express the vascular marker CD31 (Giangreco *et al.*, 2004; Summer *et al.*, 2003). Based on the expression of hematopoietic marker CD45, lung SP cells were further subdivided into hematopoietic (CD45 positive) or nonhematopoietic (CD45 negative) subpopulations. Nonhematopoietic SP cells express markers of epithelial and mesenchymal cells, share some characteristics with airway stem cells, and are currently under further investigation (Giangreco *et al.*, 2004). Consequently, there is not a single unique approach to isolate stem cell populations from the intact adult lung, however, isolation of BASCs and SP cells from the lung has been described (Giangreco *et al.*, 2004; Kim *et al.*, 2005). We have summarized in Table 6.2 the current knowledge of epithelial cell types in the conducting and respiratory airways, their specific cellular markers, their potential as stem/progenitor cells, and their potential lineage relationship.

4. *In Vivo* Injury Models of the Lung

4.1. Overview

Stem cells are slow cycling in the setting of normal tissue turnover, giving rise to TA progenitor cells that retain the capacity to replicate and impart most of the tissue renewal in the presence of injury. However, unlike stem cells that have

Table 6.2
Epithelial cell types and identified potential stem cell types in mouse adult lung

Cell type	Markers	SC potential	Daughter cells	Pollutant sensitivity	References
Basal	Cytokeratin 5, Cytokeratin 14	Yes	Basal, PNEC? Secretory, Mucous, Ciliated	NA	Hong et al. (2004a), Schoch et al. (2004)
Clara	CCSP, CyP450 2F2	Yes	Clara, PNEC? Mucous, Ciliated, AEC I, II	Naphthalene	Hong et al. (2001), Reynolds et al. (2000b)
Ciliated	Tubulin IV	NA	NA	NO_2, SO_2 Polidocanol	Borthwick et al. (2001)
Mucous	Mucin 5AC, Mucin 5B	Yes	Basal, Mucous, Ciliated	NA	Engelhardt et al. (1995), Hook et al. (1987)
AEC I	Aquaporin 5	No	NA	Bleomycin	Emura (1997)
AEC II	Lamellar body, SpA,SpB,SpC, Aquaporin 1	Yes	AEC I, AEC II, PNEC? Clara?	Bleomycin	Griffiths et al. (2005), Reddy et al. (2004)
PNEC	CGRP	No?	NA	Clara?	Hong et al. (2001), Peake et al. (2000)
SP	Sca-1, CD31 Hoechst 33342	Yes	NA	NA	Giangreco et al. (2004), Summer et al. (2003)

NA, unknown or under investigation.

Table 6.3
Some of the pollutants or regents used to generate lung injury models

Pollutants	Injury methods	Injury mechanism or targets	References
O_2, O_3, NO_2	Inhalation	Hyperoxic	Meulenbelt et al. (1992b), Smith (1985)
SO_2	Inhalation	Conducting airway	Asmundsson et al. (1973), Langley-Evans et al. (1996)
Polidocanol	Intratracheal instillation	Surface of airway and alveoli	Borthwick et al. (2001), Suzuki et al. (2000)
Naphthalene	Intraperitoneal injection	Clara cells	Reynolds et al. (2000b), West et al. (2001)
Bleomycin	Intratracheal instillation	Alveolar Type I cells, Alveolar Type II cells	Bigby et al. (1985)
Elastase	Intratracheal instillation	Alveolar walls	Dubaybo et al. (1991)
Radiation	Exposure	Exposed regions	Theise et al. (2002)

unlimited proliferative capacity, TA cells are eventually incapable of proliferation and become TD cells. Injury models have played key roles in mobilizing stem cell compartments in the lung. Different epithelial cell types in distinct areas of the lung are sensitive to various agents and have been used in lung injury models to investigate stem cells in the adult lung (Table 6.3). These include hyperoxic agents (Smith, 1985), radiation (Theise et al., 2002), SO_2 or polidocanol (Borthwick et al., 2001; Randell, 1992), bleomycin (Bigby et al., 1985; Izbicki et al., 2002), elastase (Dubaybo et al., 1991), and naphthalene (Kim et al., 2005; Van Winkle et al., 1999). Detailed methods for polidocanol and naphthalene injury will be described later, and the others are briefly introduced below.

Exposing the lung to high concentrations of oxygen (O_2) or oxidant gases, such as nitrogen dioxide (NO_2) or ozone (O_3) has been used to generate hyperoxic lung injury models. Cell death from hyperoxic injury may occur through either apoptotic or nonapoptotic pathways, possibly via free oxygen radicals. The biochemical, cellular, and morphologic characterizations of hyperoxic lung injury have been extensively studied (Smith, 1985). Mice show mild damage following 3 days of exposure to high concentrations of O_2, damage which consists of alveolar septae thickening and increases in alveolar macrophages. Extensive damage can be induced by continuous prolonged exposure to O_2, resulting in the destruction of alveolar walls, proteinaceous exudates in alveoli, and large numbers of interstitial and alveolar polymorphonuclear leukocytes (PMNs) (Smith, 1985).

Inhalation exposure of mammals to NO_2 has also been used extensively as a model of controlled acute pulmonary injury and repair. Prolonged exposure of rodents to 20–30 ppm of NO_2 results in mild emphysema and a partially reversible decrease in lung elastin and collagen content. The mechanism by which NO_2 damages AEC I is thought to be oxidation of unsaturated fatty acids of the cell membranes (Evans *et al.*, 1981). Subacute NO_2 exposure (20 ppm NO_2 for 28 days) in mice also results in swelling of the ciliary shaft, focal loss of cilia, and the formation of compound cilia in the airway epithelium. Such ciliary lesions appear to be reversible when NO_2 exposure was stopped (Ranga and Kleinerman, 1981). In addition to the mouse NO_2 injury lung models, rat (Meulenbelt *et al.*, 1992a), rabbit (Meulenbelt *et al.*, 1994), and sheep (Januszkiewicz and Mayorga, 1994) NO_2 injury lung models have also been described.

An alternative mode of lung injury includes radiation. Mice exposed to 1200 cGy of total body irradiation have lungs that appear hypocellular, with a breakdown of capillaries within alveolar septae and extravasation of erythrocytes into alveolar spaces within 3 days of exposure to radiation. Irradiation-induced alveolar epithelial damage is quickly repaired by AEC II proliferation followed by differentiation in AEC I (Theise *et al.*, 2002). The notion that bone-marrow derived stem cells have the ability to contribute to repair of alveolar epithelium was also suggested by the transplantation of retrovirally tagged GFP bone marrow cells into irradiated mice (Grove *et al.*, 2002).

Another mode of injury to alveolar regions of the lung includes bleomycin. Bleomycins are a family of compounds with antibiotic and antitumor activity that have the major side effect of pulmonary toxicity. Intratracheal instillation of bleomycin induces alveolar injury in the lung (Hay *et al.*, 1991). The cellular toxicity of bleomycin is based on its bithiazole component that partially intercalates into the DNA helix and causes oxidative damage to DNA. However, lipid peroxidation caused by bleomycin may also account for alveolar damage and subsequent pulmonary inflammation. Bleomycin lung injury is characterized by pulmonary fibrosis due to increased production of collagen and other matrix components in the lung (Bigby *et al.*, 1985; Hay *et al.*, 1991; Izbicki *et al.*, 2002). AEC I cells are particularly sensitive to bleomycin and are the first to be injured, while AEC II cells have a more variable sensitivity. Bleomycin-resistant AEC II cells undergo metaplasia in the presence of the drug (Izbicki *et al.*, 2002). More details regarding the generation of bleomycin injury for the characterization of stem cells have been described by Kim and colleagues studying the BASC niche at BADJ (Kim *et al.*, 2005).

Sulphur dioxide (SO_2), a common pollutant in the air responsible for many diseases of the respiratory system, has been used to induce injury in the proximal airways of animal models. Various species of animals have different sensitivities to SO_2—mice are most sensitive, while rats are the most tolerant. Inhalation of SO_2 has been used in pulmonary epithelial injury models of various species including mice, rats, sheep, guinea pigs and swine. SO_2 induced lung injury in the rat and sheep have also been proposed as a model for bronchitis (Borthwick *et al.*, 2001; Lamb and Reid, 1968). SO_2 is soluble in water, and therefore is easily absorbed into the wet mucous membranes of the airway causing the greatest damage to the trachea and large bronchi, while the bronchioles are invariably spared (Asmundsson *et al.*, 1973). This feature makes inhalation of SO_2 an important injury model for the proximal airway.

Another chemical used to induce injury in the proximal airway of mice includes polidocanol, a surface-active detergent clinically used to enhance absorption of small proteins/peptides. Other surface active agents, such as polyoxyethylene 9 lauryl ether (Laureth-9), sodium glycocholate, and Triton X-100 have also been shown to induce lung injury following intratracheal instillation (Borthwick *et al.*, 2001). Direct intratracheal instillation of 10 μl of 2% polidocanol in PBS to the mouse trachea causes widespread denudation of the airway epithelium by 24 h post injury (Fig. 6.3D). Using this polidocanol injury model, Borthwick and colleagues have identified the existence of stem cell niches in SMGs of the proximal mouse airway (Borthwick *et al.*, 2001). A detailed method of this approach is described in a later section.

Naphthalene, a chemical agent that is toxic to Clara cells, can induce damage to both proximal and distal airways of mice. This injury method has been used to identify two distinct stem cell niches associated with NEBs in the proximal airway and BADJ (Giangreco *et al.*, 2002; Hong *et al.*, 2001; Reynolds *et al.*, 2000a,b). Naphthalene is metabolized by the 2F2 subtype of CyP450, resulting in the production of a cytotoxic epoxide. Inhalation or intraperitoneal injection of naphthalene produces a dose-dependent cytotoxicity to Clara cells, while cells deficient in Cyp450–2F2 are spared (Buckpitt *et al.*, 1992; West *et al.*, 2001). The bronchiolar epithelium of mice is a particularly sensitive site for naphthalene toxicity, given its

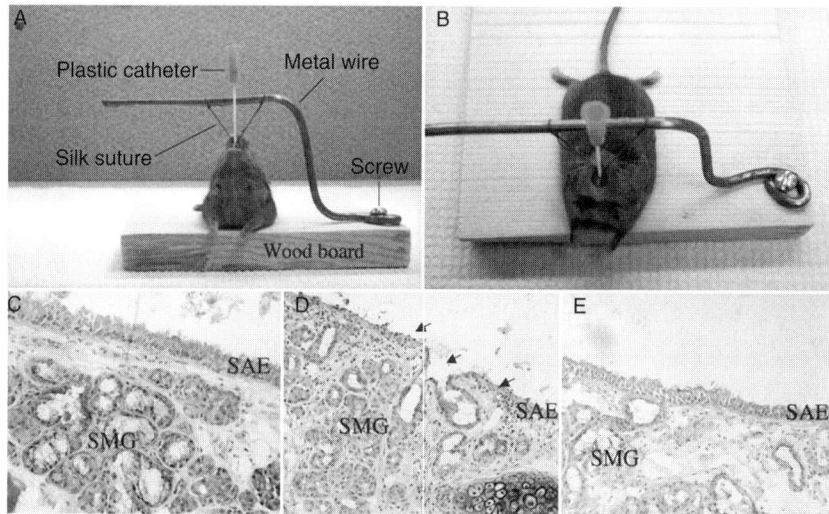

Figure 6.3 Polidocanol induced injury of the mouse tracheal epithelium. Anesthetized mice are hung on a self-made rack by placing the upper jaw incisor teeth within a suture prior to placing a plastic catheter into the trachea as shown in A (front view) and B (top view). The rack is made using a >3-mm metal wire fixed on a wood board with a screw. Following intratracheal instillation of 20 μl of 2% polidocanol, tracheas are harvested and sectioned. (C–E) H & E stained sections from (C) control uninjured trachea, (D) injured trachea at 1 day, and (E) injured trachea at 14 days. Arrows mark injury to the SAE at 1 day post polidocanol treatment in (D), as compared to injured controls (C), and regenerated epithelium (E). SAE, surface airway epithelium; SMG, submucosal gland.

high content of Clara cells (Giangreco *et al.*, 2002; Hong *et al.*, 2001). At higher doses of naphthalene (275 mg/kg), however, significant injury to the tracheobronchial epithelium can also be achieved (Fig. 6.2A). As discussed in more detail above, a subpopulation of variant CCSP expressing Clara cells deficient of CyP450–2F2 has been suggested as the principal stem/progenitor cell in the bronchiolar epithelium that facilitates repair following administration of naphthalene (Giangreco *et al.*, 2002; Hong *et al.*, 2001; Kim *et al.*, 2005; Peake *et al.*, 2000; Reynolds *et al.*, 2000a,b). Detailed methods for naphthalene-induced mouse lung injury and the identification of LRCs using BrdU-labeling are presented in the following section.

4.2. Mouse naphthalene injury and BrdU labeling protocol

1. Weigh the appropriate amount of naphthalene (Fisher Scientific, Barrington, IL. Cat # N143–500) directly into a 15 ml polypropylene conical tube and make fresh on the day of use.

2. Calculate the volume of sterile corn oil (Mazola®) needed to achieve the desired concentration of naphthalene (27.5 mg/ml) using the density of corn oil (0.9185 g/ml). Generally, the concentration of naphthalene used is 27.5 mg/ml; however, the dose can be adjusted to achieve different levels of injury to the proximal and/or distal airways. To generate a 27.5 mg/ml naphthalene stock sufficient for a 10 mouse experiment, 110 mg of naphthalene is weighed into a conical tube followed by the addition of 3.65 g of corn oil (4 ml × 0.9185 g/ml).

3. Naphthalene is difficult to dissolve. Vortex the solution at full speed for several minutes, and continue mixing on a tube rotator for at least 20 min at room temperature. Make sure that all the naphthalene has dissolved by verifying that the solution is clear of solid naphthalene.

4. Load the naphthalene solution into a 1 ml syringe using an 18-gauge needle. After loading, remove the 18-gauge needle and use a 26-gauge needle for intraperitoneal injection.

5. Weigh the mice to the nearest 0.1 gram to calculate the volume dose. It is particularly important that the dosing being exact for reproducible injury. Additionally, it is important to inject the mice before 10:00 a.m. to take advantage of the minimum glutathione levels at this time of day. Sex differences exist in the extent of lung injury at a given dose of naphthalene (female mice are less sensitive than males), so make sure all animals are the same sex (the percentage of death caused by the toxicity of naphthalene in female is lower than males).

6. Administering light anesthesia to the mice will help in obtaining an accurate weight and will also make the injection easier. Intraperitoneal injections (i.p.) are performed at 10 μl/g body weight of the freshly made naphthalene solution described above (i.e., 250 μl for a 25 g mouse given a total dose of 275 mg/kg). Naphthalene injured animals will present a ruffled appearance with crusted eyes at approximately 20–36 h post injury (To increase survival following Naphthalene

injury, 2 ml of 5% dextrose-saline twice daily can be given under the scruff of the back of the neck).

7. To identify LRCs, mice can be injected (i.p.) with a 80–100 mg/kg body weight dose of BrdU (Roche, Indianapolis, IN) PBS solution every 24 h beginning at 6 h after the naphthalene injection for 5 consecutive days. Multiple injections (at least three) are required to label a significant percentage of LRCs.

8. Maximal injury of an airway epithelium is typically observed by 2 days after naphthalene injection, and newly regenerated airway epithelium can be seen by 1 week after injury (Fig. 6.2B). To localize LRC stem cell populations, mice are sacrificed at different time points, such as 21, 42, and 90 days after injury and BrdU labeling.

9. Harvest the trachea and/or lung and remove blood by washing the organs in precooled PBS. The tissues can be either fixed in 10% formalin for paraffin embedding and sectioning or freshly frozen in OCT (Optimal Cutting Temperature) compound embedding medium (Tissue-Tek®, Sakura Finetek Inc., Torrance, California).

10. BrdU incorporation is detected by immunofluorescent or immunohistochemistry staining in 10 μm frozen sections or 6 μm paraffin sections, respectively. An immunofluorescent BrdU staining procedure for frozen sections is described below.

11. Bring frozen sections to room temperature for 10–20 min to dry. Fix the sections in 4% paraformaldehyde in PBS for 20 min before processing for epitope retrieval.

12. Use an Antigen Unmasking solution (Vector Lab Inc, Burlingame, California. Cat # H-3300) to retrieve epitopes as specified by the manufacturer.

13. Rinse sections in 3 changes of PBS (2 min for each), and then block nonspecific binding of immunoglobulin by incubating the sections in 5% rabbit serum (the same species to which the secondary antibody was produced) in PBS at room temperature for 30–60 min.

14. Directly apply a 1:100 dilution of mouse anti-BrdU antibody (Roche Diagnostic, Indianapolis, Indiana. Cat# 1-299-946) in 5% rabbit serum/PBS on sections and incubate at room temperature for 1 h.

15. Rinse sections in 3 changes of PBS (3 min for each) before incubating with a 1:200 dilution of FITC-conjugated rabbit anti-mouse IgG in 1.5% rabbit serum/ PBS (Jackson Immunoresearch, West Grove, Pennsylvania. Cat#315096045) at room temperature for 45–60 min.

16. Rinse sections in 3 changes of PBS (3 min for each) and mount sections in Vectashield® mounting medium containing DAPI (Vector Lab Inc., Burlingame, California. Cat# H-1200). Evaluate BrdU staining under a fluorescent microscope (Fig. 6.2C).

Notes: For CldU and IdU labeling and detection, detailed information is available in the following references (Kiel *et al.*, 2007; Teta *et al.*, 2007). Cautions: Terminally differentiating cells can also retain DNA labels, so alternative criteria for stem cell phenotypes must also be applied.

4.3. Mouse polidocanol injury and BrdU labeling protocol

1. Tare the balance to zero with a 15 ml polypropylene conical tube and weigh 200 mg of polidocanol (Sigma, St. Louis, Missouri. Cat # P9641) in the tube. Add sterile PBS to bring the total volume up to 10 ml, making a 2% polidocanol (weight/ volume) solution. Rotate the tube to ensure the polidocanol is completely dissolved.

2. Anesthetize the mice using 80 mg/kg ketamine and 10 mg/kg xylazine in sterile PBS. The dose of anesthesia should not be so deep as to hinder respiration, so it is advised to adjust this dose to the strain of mice being used.

3. To facilitate intubation of mice for delivery of polidocanol into the trachea, hang the mouse on a self-made wire rack with a suture loop around the incisor teeth (Fig. 6.3A and B). A small flashlight can be placed under the neck of the mouse to aid in visualization of the tracheal orifice if needed.

4. Open the mouth of the mouse and pull its tongue out down using flat jaw forceps. Remove the needle from a 0.55 mm Angiocath (BD Angiocath™, BD Medical, Sandy, Utah) and insert the plastic catheter into the trachea.

5. Use a 25-gauge needle to apply 20 μl of 2% polidocanol into the trachea through the catheter. Leave the catheter in the trachea for a minute after delivering the solution to prevent immediate expulsion into the mouth.

6. Following the instillation, intraperitoneal injection of BrdU can performed as described above.

7. Massive injury to the tracheal epithelium is seen 24 h post injury, and the epithelium is completely regenerated 1–2 weeks (Fig. 6.3D and E). LRC location can also be performed as described for the naphthalene model.

5. *Ex Vivo* Epithelial Tracheal Xenograft Model to Study Stem Cells Expansion in the Proximal Airway

5.1. Overview

The *ex vivo* epithelial xenograft model is an approach that has historically been used to study progenitor/progeny relationships in the adult airway (Duan *et al.*, 1998; Engelhardt *et al.*, 1991, 1992, 1995; Presente *et al.*, 1997; Sehgal *et al.*, 1996). This model seeds isolated airway epithelial cells onto graft tracheas that have been denuded of all endogenous airway epithelia by freeze-thawing. After the airway stem/progenitor cells are seeded, the tracheal grafts are implanted subcutaneously into immunocompromised hosts, such as nu/nu or SCID mice. A fully differentiated epithelium regenerates approximately 3–4 weeks after transplantation (Filali *et al.*, 2002).

A major advantage of this approach is that it can be applied to airway epithelia from multiple species. This model has been used study airway stem cell biology and/or lineage using purified or enriched specific epithelial cell types to reconstitute the denuded tracheal grafts. These studies have demonstrated the ability of basal and secretory cells to reconstitute a fully differentiated SAE in multiple species

(Hook *et al.*, 1987; Inayama *et al.*, 1988; Randell, 1992). Rabbit basal cells purified by centrifugal elutriation have the ability to repopulate epithelium containing basal, ciliated, and goblet cells following reconstitution of a tracheal xenograft (Inayama *et al.*, 1988). Additionally, seeding of enriched rabbit Clara cells into xenografts has demonstrated that this cell type has a limited capacity for differentiation into Clara and ciliated cells (Hook *et al.*, 1987). Although enrichment of a given cell type is greater than 90–95% in most instances, and can reach up to 98% purity following cell sorting with a combination of cellular surface markers and light scattering, one major limitation of such cell enrichment approach is the potential for small levels of contamination with unidentified airway stem cells.

To circumvent the limitations of cell enrichment approaches, retroviral-mediated gene transfer has been used to genetically tag epithelial stem/progenitor cells and follow lineage relationships using histochemical markers (Engelhardt *et al.*, 1992, 1995). Using this approach, primary human airway cells can be infected with a variety of retroviral vectors prior to reconstituting the xenograft airway epithelium (Fig. 6.4A). Histochemical staining for marker transgenes such as β-galactosidase and/or alkaline phosphatase can then be used to visualize clonal expansion of stem/progenitor cells in the reconstituted airway epithelium (Fig. 6.4B). Using two independent viral vectors with two independent marker transgenes (i.e., β-galactosidase and/or alkaline phosphatase), studies have demonstrated that transgene-expressing clones indeed arise from a single retrovirally infected progenitor/stem cell (Engelhardt *et al.*, 1995). In reconstituted xenografts with retrovirally tagged epithelial cells, the number of cells in each clone (i.e., size) can be used as an index of proliferative capacity, with the notion that clones arising from adult stem cell fractions would have a larger proliferative capacity. Phenotyping of clones (i.e., the types of transgene-expressing cells in each clone—basal, ciliated, goblet, or intermediate cells) can be used to assess the variety of progenitor cell types in the airway with either limited or pluripotent capacity for differentiation. These studies have suggested that a diverse repertoire of progenitor cells likely exists in the adult human proximal airway, with differing capacities for proliferation and differentiation. These studies demonstrated that a stem/progenitor population with the ability to differentiate into SMGs and all surface airway epithelial cell types exists in the adult human airway (Engelhardt *et al.*, 1995). This cell type is thought to be a candidate stem cell in the proximal airway.

The protocol for the generation of human bronchial xenografts described below is also suitable for studies with other species such as rat, with modifications of cell culture conditions (Engelhardt *et al.*, 1991).

5.2. Human bronchial xenograft protocol

1. Dissection of human airway tissue:

i. The bronchial/lung cassette should be chilled immediately following removal from the donor in cold physiologic saline and should remain chilled during dissection. For ideal aseptic and safety conditions, the dissection is best done in a laminar flow hood. Handling human tissue and primary airway cultures may expose the user to blood-borne pathogens, and suitable health protection measures should be followed.

ii. The proximal airways are dissected from the lung on ice and placed immediately in chilled sterile Ca^{2+}- and Mg^{2+}-free PBS containing 50 U/ml penicillin, 50 μg/ml streptomycin, 80 μg/ml tobramycin, 100 μg/ml ceftazidime, 100 μg/ml Primaxin, and 5.0 μg/ml amphotericin B. Remove the blood, mucus, and extraneous connective tissues from the airways and cut the specimen into suitably sized half-rings and place them into 50-ml polypropylene conical tubes with the above antibiotic solution. Invert the tubes for several minutes to wash the tissue, and repeat four times by moving the tissue from one conical tube to a fresh tube with new antibiotic containing cold PBS.

iii. It is important to avoid drastic warm and cold temperature changes of specimens during processing and dissection to prevent epithelial degradation. Additionally, various segments of the proximal airway (trachea and bronchi) can be separately processed since viability of the epithelial characteristics may be encountered.

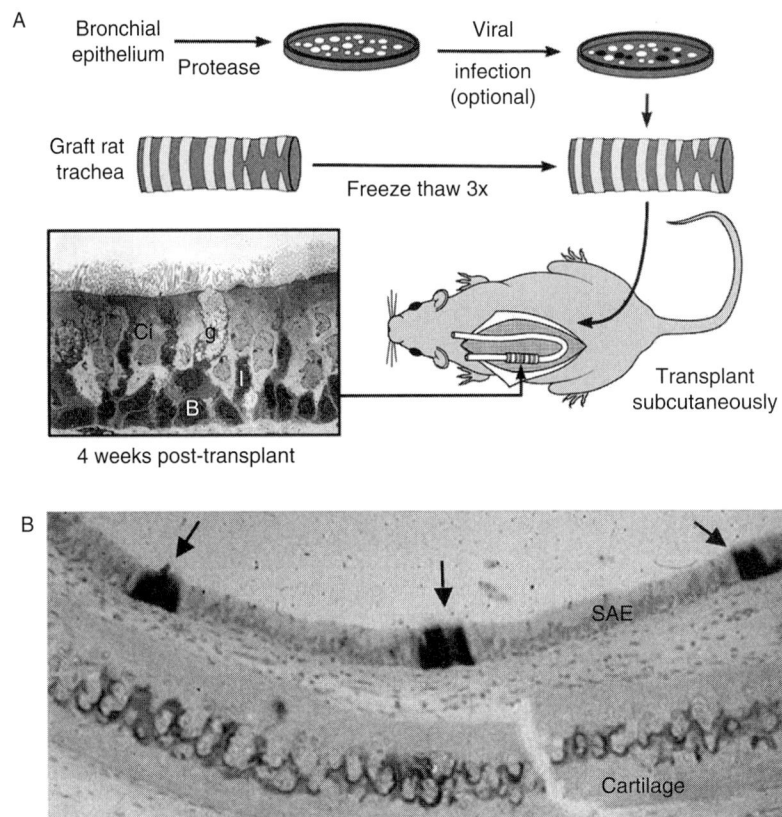

Figure 6.4   Continued

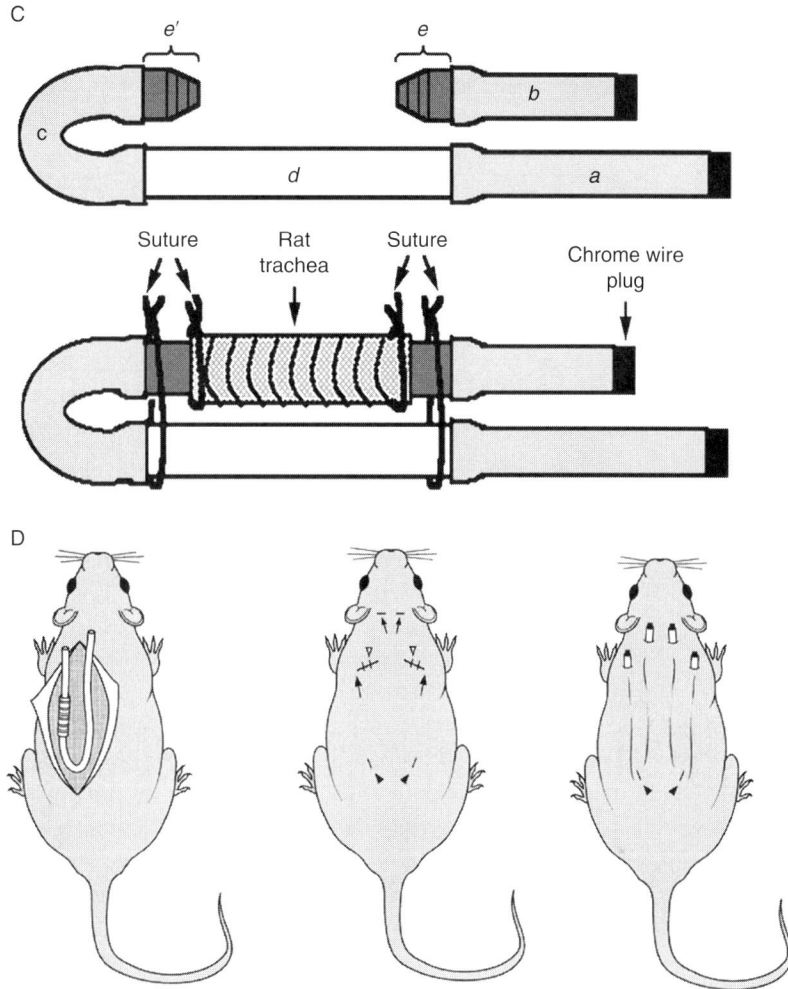

Figure 6.4 Methods for generating tracheal xenografts to study clonal expansion. (A) Schematic methods for generating proximal airway epithelial xenograft models. Primary airway epithelial cells are cultured *in vitro* and can be infected with integrating recombinant viral vectors (lentivirus or retrovirus) prior to transplantation of epithelia into denuded rat tracheas. A fully differentiated epithelium is obtained by 4 weeks post transplantation (B, basal cells; Ci, ciliated cells; I, intermediate cells; G, goblet cells). (B) CFE of stem/progenitor cells can be evaluated by detecting virally expressed transgenes in the reconstituted SAE (arrows). In this example, a recombinant retroviral vector expressing the β-galactosidase transgene was used and detected by X-gal staining. (C) Schematic view of various components of the xenograft cassette. The denuded rat trachea is connected to a sterile tubing cassette by a series of sutures as illustrated [a: 1-in. silastic tubing (Dow corning, Midland, MI; Cat# 602–175); b: 0.75-in. silastic tubing (Dow corning, Midland, MI; Cat# 602–175); c: 1.75-in. silastic tubing (Dow corning, Midland, MI; Cat# 602–175); d: 1.25-in. Teflon tubing (Thomas Scientific, Swedesboro, NJ; Cat# 9567-K10); e and e': adapter (0.8-mm borb-to-barb connector; Bio-Rad, Hercules, CA; Cat# 732–8300); f: chrome wire plug (0.0035-in. diameter Chromel A steel wire; Hoskins MFG)]. (D) Subcutaneous transplantation of the xenograft cassettes in nu/nu athymic mice. A subcutaneous view of the transplanted xenograft cassette is illustrated in the left panel. The middle panel illustrates the four incisions

iv. Following washing in antibiotic solution, airway tissue is placed in a 50 ml or 15 ml conical tube filled with *Media A* (described at the end of this section) to dissociate airway epithelial cells. It is important that the conical tubes be filled to the top with *Media A* and tightly capped to avoid changes in pH caused by alteration in CO_2 concentration over time. Airway tissues are incubated in *Media A* at 4 °C for 24–96 h. Nasal turbinates require 24–48 h for dissociation while tracheobronchial tissues require 48–72 h (depending on the desired degree of separation into single cells vs. cell clumps). However, it should be noted that dissociation times longer than 72 h will decrease cell viability, and cell clumps appear to proliferate more rapidly in culture.

2. Harvesting airway epithelial cells:

i. Add 10% FBS into the dissociation solution to inhibit the pronase while transferring the solutions and tissue to a larger or greater number of conical tubes filled approximately 1/3 full. Invert the tube(s) several times vigorously to help dissociate the cells from the airway tissue. Mild shaking is helpful to increase yields, but avoid excessive force that may damage cells (i.e., do not shake hard enough to cause FBS to foam).

ii. After shaking, allow 1 min for airway tissue to settle to the bottom of the tube (liberated airway cells will remain in suspension). Pipette the media containing the airway cells into a fresh tube on ice, and place the remaining tissue into a 100 mm tissue culture plate.

iii. Using the blunt side of a scalpel, scrape the surface of the airways to remove the remaining epithelial cells. Rinse airway samples with 10% FBS/Hams F12 media, combine the washed cells into 50 ml conical tubes, and centrifuge the tubes at $120 \times g$ for 5 min at 4 °C.

iv. Resuspend the cell pellets in *Media B* (described at the end of this section) and transfer the cell suspension to 100 mm Primaria tissue culture dishes (BD Bioscience, San Jose, California. Cat# 353803).

v. Incubate the suspension in a 5% CO_2 incubator at 37 °C for a minimum of 1–2 h to allow fibroblasts within the cell suspension to attach to the plastic surface. Airway epithelial cells will not attach rapidly to the plastic surface without collagen coating.

vi. Collect the nonattached cells by centrifugation, resuspend them in *Media C* (described at the end of this section), and count them with a hemacytometer.

(arrows) made on the back of the recipient nu/nu mouse prior to transplantation. The xenograft cassette is guided subcutaneously using forceps so that one port exits through the back of the neck and the other port exits through the main incision. Surgical staples are then used to close the main incisions (marked by open arrowheads) and one additional staple is used to anchor the xenograft tubing near the tail end of the mouse (closed arrowheads). The right panel in D illustrates the resultant xenograft cassette at 1 week post surgery when the proximal staples are removed. The staples marked by closed arrowheads in D are used to maintain the position of the cassette and prevent subcutaneous migration (it is necessary to leave these staples in for at least 2 weeks).

3. Culture of airway epithelial cells:

i. Plate $1-2 \times 10^6$ of the above airway cells per 100-mm collagen coated tissue culture dish (BD bioscience, San Jose, CA) in 10 ml airway culture *Media C*. Incubate the cells overnight in a 5% CO_2 incubator at 37 °C.

ii. On the day following plating, aspirate the culture medium and unattached cells. Wash the dish containing adherent cells with prewarmed Hams-F12 medium and refeed the cells with fresh *Media C*. Culture the cells for an additional 48 h.

iii. At 72 h post plating, feed the cells with *Media D* (containing lower levels of antibiotics and no amphotericin B). Typically, cells are ready for cryopreservation, passaging, or transplantation into xenograft models by 5 days post plating ($\sim$80% confluency). Care should be taken to not allow cells to become more than 80% confluent or they will begin to differentiate and lose their capacity for subculturing. Unpassaged fresh cultures of airway epithelia also more consistently give better reconstitution of epithelium in the xenograft model.

iv. If primary airway cells are to be expanded (typically cells can be expanded 1 time without the loss of the ability to differentiate in a xenograft model), they should be treated with 0.1% trypsin/EDTA for 1–3 min at 37 °C followed by neutralization with an equal volume of Trypsin Inhibitor Buffer (see recipe below). Cells should be closely monitored during trypisinization and harvested immediately once they are released by gentle tapping of the plate. Cell suspensions are centrifuged to remove trypsin and washed once in *Media D*, followed by plating at a 1:5 dilution.

v. For cyropreservation, epithelial cell pellets are resuspended in 4 °C *Media D* with 10% DMSO, 10% FBS and aliquoted in 2 ml cryogenic vials for slow freezing at −80 °C, O/N. Slow freezing can be performed using isopropanol-containing cryopreservation containers (Nalgene Inc.). Cells are then moved to liquid nitrogen storage. Typically, one subconfluent 100 mm plate of cells is aliquoted per vial, and when subcultured placed into five 100 mm plates.

4. Retroviral infection of primary airway epithelial cells (optional): Retroviral infection of airway epithelial cells prior to seeding into tracheal xenograft has been useful in studying airway stem/progenitor cell biology (Engelhardt *et al.*, 1995) (Fig. 6.4A and B). Briefly, freshly isolated primary airway epithelial cells on the second day post seeding (approximately 10% confluency) can be incubated in the presence of serum containing conditioned retroviral or lentiviral producer supernatant (10 ml of retroviral supernatant/100 mm plate of primary cells) for 2 h in the presence of 2 μg/ml polybrene (Sigma, St. Louis, Missouri). Following each retroviral infection, the cells are washed twice with F12 media prior to the addition of hormonally defined *Media D*. Cells can be infected up to three times on sequential days. Typically, viral titers of 1×10^6 cfu/ml are capable of transducing primary airway cells at an efficiency of 10–30% following three serial infections. Serum-free concentrated stocks of VSV-G pseudotyped retrovirus or lentivirus can also be used to achieve higher transduction efficiencies (unpublished data).

5. Preparation and construction of xenograft cassettes: A xenograft cassette is assembled by connecting tubing and adapters fitted to the length of the donor rat trachea and secured with silk sutures as illustrated in Fig. 6.4C (materials and vendors are described in the figure legend). Preassembled parts of the cassette are placed in a 100-mm tissue culture plate, sealed in a sterilization pouch, and gas sterilized prior to anchoring tracheal xenografts.

6. Preparation of denuded rat tracheas:

i. Rats are euthanized by CO_2 asphyxiation and pinned to a Styrofoam bed. Tracheas are excised from the pharynx to the carina under sterile conditions using 70% ethanol to clean the site of incision. Place the tracheas in a separate 2-ml screw cap tube and place them on ice immediately following excision.

ii. After all tracheas have been harvested, denude tracheas of all viable epithelium by freezing/thawing three times—freezing at $-80\,^\circ$C and thawing at room temperature.

iii. Following the third round of freezing/thawing, clean the tracheas of excessive fat and cut them to size (typically from the first and thirteenth cartilage ring). Rinse each tracheal lumen with 10 ml of precooled MEM.

iv. Pair tracheas of similar length in the same tube. The denuded tracheas can be stored at $-80\,^\circ$C for prolonged periods prior to transplantation.

7. Seeding tracheas with primary airway epithelial cells:

i. Airway epithelial cells cultured in 100-mm dishes (at ∼80% confluence) are harvested using trypsin and trypsin inhibitor as described earlier, resuspended in *Media D* at a concentration of $1–2 \times 10^6$ cells per 20 μl, and kept on ice.

ii. Ligate the rat tracheas to the adaptor (e) attached to tubing (b) (as shown in Fig. 6.4C) with securely tied triple knotted sutures, and loop the sutures around the tubing and trachea a total of three times.

iii. Inject 20–25 μl of media containing $1–2 \times 10^6$ cells into the open end of the rat trachea using a micropipettor under sterile conditions. Insert the pipette tip as deeply as possible into the trachea and slowly withdraw the pipette as cells are injected into the lumen of the trachea (taking care to not allow cells to leak out).

iv. Ligate the remaining open end of the rat trachea to the adapter (e') attached to tubing (c) as shown in Fig. 6.4C. Secure the length of the rat trachea by stretching it to physiologic length and clamping the tubing (b) and (a) with a hemostat. Tie the remaining two sutures as shown in Fig. 6.4C to secure the tracheas to the adapter (e') and the adapter (e') to the tubing (d).

v. Place the cell-seeded xenograft cassettes into a 100-mm tissue culture dish with 1–2 ml *Media D* overlaid on the top of the trachea to keep it in moist.

vi. Incubate the cassette in a 5% CO_2 incubator at $37\,^\circ$C for 1–2 h to equilibrate the pH prior to proceeding of transplantation.

vii. For transport, use a small, airtight, sterile container equilibrated in the incubator with the dishes of xenografts to maintain CO_2 prior to transplantation.

8. Transplanting xenografts into Nu/Nu mice:

i. Anesthetize male nu/nu athymic mice (Harlan, Indianapolis, Indiana) by intraperitoneal injection of 80 mg/kg ketamine and 10 mg/kg xylazine in sterile PBS.

ii. Place mice on a sterile drape after they are anesthetized. Clean surgical incision sites with povidone–iodine followed by 70% ethanol. Make four incisions as shown in Fig. 6.4D (center panel), two small incisions on the neck of the mouse (~0.16 cm) with just enough width to pass the tubing, and two larger incisions on the flanks of the mouse (~1.0 cm).

iii. Separate the skin from the muscle by blunt dissection and place the xenograft cassette subcutaneously by tunneling the distal end of cassette (tubing c) under the skin toward the tail of the mouse, and the proximal end of the cassette (tubing a) out of the small incision behind the neck using forceps to guide the xenograft tubing (Fig. 6.4D).

iv. Following transplantation, use 2–3 staples to close each of the largest incisions and an additional staple to anchor each xenograft to the skin at the loop of tubing (c). No stables are necessary for the exit port of tubing (a). Then transfer the mouse to a sterile cage and monitor until it is awake.

v. Irrigate xenografts weekly with 1 ml of Ham's F12 medium using a Surflo winged infusion set with an 0.75-in., 21-gauge needle, to remove excess secretions in the first 3 weeks following transplantation. Irrigate xenografts twice a week after 3 weeks post transplantation. A stratified xenograft epithelium usually is reconstituted by 4–6 weeks after transplantation (Engelhardt *et al.*, 1991, 1992, 1995).

5.2.1. Media A

Modified Eagles Medium (MEM) Ca^{2+}- and Mg^{2+}-free
100 U/ml penicillin
100 μg/ml streptomycin
80 μg/ml tobramycin
100 μg/ml ceftazidime
100 μg/ml Primaxin
5.0 μg/ml amphotericin B
100 μg/ml deoxyribronuclease (Dnase I) (Sigma, St. Louis, MO. Cat# DN 25)
1.5 mg/ml pronase (Roche, Indinapolis, IN. Cat# 1459643)
Make fresh and keep at 4 °C

5.2.2. Media B

5% FBS DMEM/Hams F12 (1:1 ratio)
1% MEM nonessential amino acids solution
100 U/ml penicillin
100 μg/ml streptomycin
80 μg/ml tobramycin

　　　　　100 μg/ml ceftazidime
　　　　　100 μg/ml Primaxin
　　　　　5.0 μg/ml amphotericin B

5.2.3. Media C

　　　　　DMEM/Hams F12 (1:1 ratio)
　　　　　15 mM HEPES,
　　　　　3.6 mM Na_2CO_3
　　　　　100 U/ml penicillin
　　　　　100 μg/ml streptomycin
　　　　　80 μg/ml tobramycin
　　　　　100 μg/ml ceftazidime
　　　　　100 μg/ml Primaxin
　　　　　0.25 μg/ml amphotericin B
Add all components of Clonetics® BEGM SingleQuots kit (Cambrex Bioscience, Walkersville, Maryland) resulting a final concentration of 10.0 μg/ml insulin, 1.0 μg/ml cholera toxin, 40 μg/ml bovine pituitary extract, 1.0 μg/ml hEGF, 1.0 μg/ml epinephrine, 20.0 μg/ml transferring, 0.0001 μg/ml retinoic acid.

5.2.4. Media D

　　　　　BGEM
　　　　　BEGM SingleQuots kit component
　　　　　50 U/ml penicillin
　　　　　50 μg/ml streptomycin
　　　　　40 μg/ml tobramycin
　　　　　50 μg/ml ceftazidime
　　　　　50 μg/ml Primaxin

5.2.5. Trypsin inhibitor buffer

　　　　　Trpsin Inhibitor Type I-S: from soybean (Sigma, St.Louis, MO. Cat. T6522)
　　　　　1 mg/ml typisin inhibitor in Ham's F12 medium
　　　　　Filter through a 0.22 μm filter, aliquot, store at $-20\,°C$.

6. *In Vitro* Colony-Forming Efficiency (CFE) Assay to Characterize Stem/progenitor Cell Populations in Conducting Airway Epithelium

6.1. Overview

Adult airway stem cells have the ability generate highly proliferative TA cells by asymmetric cell division. This feature allows stem cells to undergo multiple rounds of clonal proliferation in the setting of injury/repair that can be measured using an

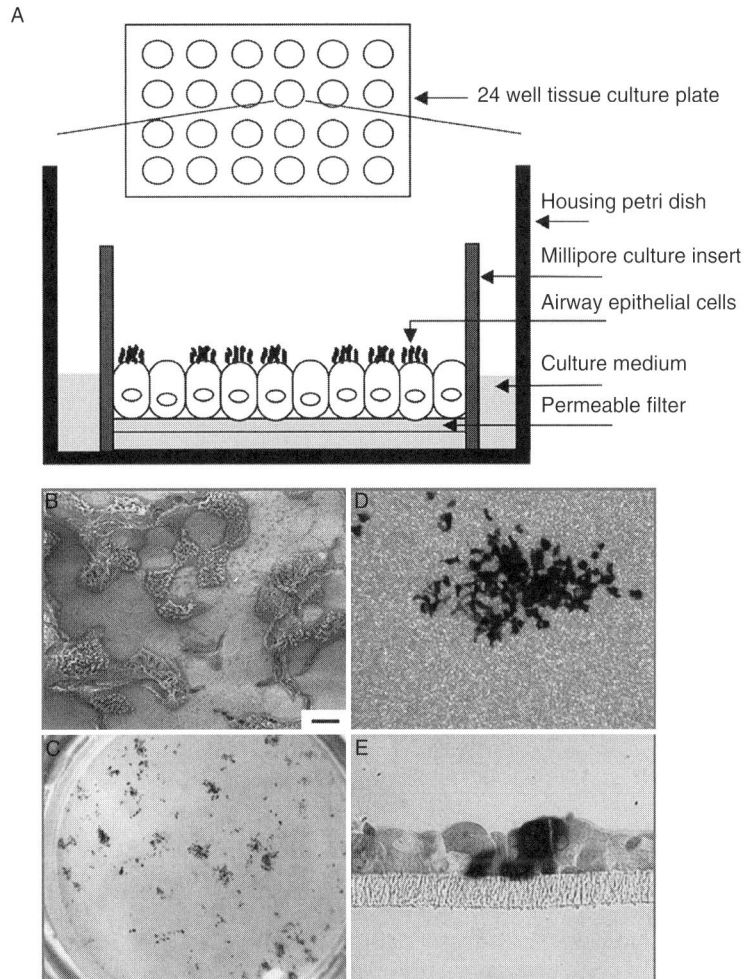

Figure 6.5 *In vitro* ALI model of CFE assays using Rosa26 mouse tracheal epithelial cells. (A) A schematic view of the ALI culture model of primary airway epithelial cells growing on collagen-coated Millipore cell culture inserts housed in a 24-well tissue culture plate. (B) Scanning electron micrograph of fully differentiated mouse airway epithelial cultures (Bar = 10 µm). (C) Enface micrograph of X-gal-stained ALI cultures seeded with a 1:50 mixture of Rosa26 to non-Rosa26 mouse tracheal epithelial cells. X-gal positive colonies of different sizes are seen. (D) Higher magnification of an individual X-gal positive colony from panel C. (E) H & E stained section of a small X-gal positive colony from panel C.

in vitro CFE assay in air–liquid interface (ALI) airway epithelial cell cultures (Randell, 1992) (Fig. 6.5). This CFE assay is similar to that used in the tracheal xenograft model with recombinant retroviral markers to assess clonal expansion of airway stem/progenitor cells (Engelhardt *et al.*, 1995; Zepeda *et al.*, 1995), but with the added advantage of being easier to carry out. Disadvantages of this *in vitro*

CFE assay include less efficient differentiation of ciliated and goblet cell types in the ALI culture as compared to the tracheal xenograft method. Hence, this *in vitro* CFE assay has been used primarily to assess the proliferative potential of stem/progenitor cell populations, and less frequently to study progenitor/progeny relationships in the airway epithelium.

Using this ALI culture system, Schoch and colleagues recently evaluated the clonal growth potential of murine tracheal epithelial cells using CFE as an index. They defined a subset of basal cells in mouse tracheal epithelium with the capacity to generate large colonies in ALI culture, suggesting they are derived from stem or TA cells (Schoch *et al.*, 2004). They diluted a single cell suspension of mouse tracheal epithelial cells derived from β-galactosidase expressing Rosa26 mice into non-Rosa26 tracheal epithelial cells and placed this mixture into the ALI culture model. After 3 weeks of ALI culture, X-gal staining revealed Rosa26 LacZ positive colonies within the polarized airway epithelium. They observed that 1.7% of the tracheal epithelial cells formed colonies of varying size, with 0.1% of the clones forming large colonies (Fig. 6.5C–E). This subset of cells with larger proliferative potential was suggested as a possible stem or early TA cell population. Previous studies from the same group have demonstrated that high keratin 5 (K5) promoter activity exists in specific niches in the mouse trachea that correspond to the location of BrdU LRCs, thought to be stem cells (Borthwick *et al.*, 2001). Transgenic mice harboring a K5 promoter-driven EGFP transgene expressed EGFP in most basal cells of the body, including a subset of tracheal basal cells apparently localized to positions similar to previously identified stem cell niches. Further studies using the same mouse model and FACS-facilitated purification of K5 promoter derived EGFP-positive tracheal epithelial cells, revealed an overall CFE 4.5-fold greater than EGFP-negative cells. Additionally, these K5 promoter derived EGFP-positive tracheal epithelial cells retained the ability to generate 12-fold larger colonies than EGFP-negative cells, consistent with the notion that K5 positive basal cells contain a proximal airway stem/progenitor cell subpopulation (Borthwick *et al.*, 2001; Schoch *et al.*, 2004). The following protocol presents the details of this CFE assay using the mouse ALI culture model and is based on the studies by Schoch and colleagues (Schoch *et al.*, 2004).

6.2. *In vitro* mouse tracheal epithelial cells CFE assay protocol

1. Isolation of mouse tracheal epithelial cells: Age 4–8 week Rosa26 mice (B6.129 S4 GT Rosa26, Jackson Lab, Bar Harbor, Maine) and non-Rosa26 mice (Harland, Indianapolis, Indiana) are euthanized by CO_2 asphyxiation and pinned to a Styrofoam bed.

   i. Excise the trachea from the pharynx to the bronchial main branches under sterile conditions and place the trachea in ice-cold Ham's F12 medium containing 100 U/ml penicillin, 100 μg/ml streptomycin, and 1.0 μg/ml Fungizone. Isolate Rosa26 and non-Rosa26 tracheal epithelial cells separately.

ii. Open the tracheas longitudinally after the muscle and fat are removed from the tracheas. Incubate the tracheas in the above Ham's F-12 pen-strep-fungizone media containing 1.5 mg/ml pronase (Roche Molecular Biochemicals, Indianapolis, Indiana) for 18–24 h at 4 °C, occasionally inverting the tube several times during dissociation (Davidson *et al.*, 2000; Liu *et al.*, 2005; You *et al.*, 2002).

iii. Add FBS to the digestion tube at a final concentration of 10%. Dissociate the tracheal epithelial cells by inverting the tube 10–20 times and transfer the cell suspension to a new tube. Wash the tracheas with Ham's F-12 pen-strep-fungizone media twice and pool the cell suspensions. Collect the cells by centrifugation at $500 \times g$ for 10 min at 4 °C.

iv. Resuspend the cell pellets in Ham's F-12 media containing 100 U/ml penicillin, 100 μg/ml streptomycin, 1.0 μg/ml Fungizone, 0.5 mg/ml crude pancreatic DNase I (Sigma-Aldrich, St. Louis, Missouri), and 10 mg/ml BSA, and incubate them on ice for 5 min. Wash the cells with Ham's F-12 pen-strep-fungizone media and centrifuge the cells at $500 \times g$ for 5 min at 4 °C.

2. The cell pellet is then resuspended in Ham's F-12 pen-strep-fungizone media with 10% FBS and cells are incubated in tissue culture plates (Primera; Becton-Dickinson Labware, Franklin Lakes, New Jersey) for 2 h in 5% CO_2 at 37 °C to adhere ?broblasts. Nonadherent epithelial cells are collected by centrifugation and resuspended in 5% FBS/BEGM medium (Cambrex, East Rutherford, New Jersey). Total cell yields are then counted using a hemacytometer. BEGM medium is made by adding reagents from one BEGM SingleQuot kit into 500 ml of 50% DMEM-50% Ham's F12 medium supplemented with 1% pen-strep, 0.25 μg/ml fungizone, 15 mM Hepes, 3.6 mM Na_2CO_3.

3. Preparation of ALI culture membranes: Primary airway epithelial cells will not attach to an ALI culture membrane supports without surface coating with collagen. Use the following protocol to prepare ALI culture membranes.

i. Mix 30 mg human placental collagen type IV (Sigma, St. Louis, Missouri, Cat# C-7521) with 50 ml deionized water and 100 μl glacial acetic acid in a 100 ml sterile beaker.

ii. Cover the holding beaker with parafilm and stir moderately at 37 °C until collagen strands are dissolved ($\sim$15–30 min).

iii. Dilute the collagen stock 1:10 with deionized sterile water and filter-sterilize the solution through a 0.22 μm membrane. The diluted sterile collagen stock (60 μg/ml) is the working solution for coating the plastic and membrane surface.

iv. Insert 12-mm diameter Millicell-PCF membrane inserts (Millipore, Bedford, Massachusetts, Cat# PIHP 01250) in a 24-well plate and apply 0.3–0.5 ml of collagen working solution on the apical surface of the inserts.

v. Collagen-coat the surface for a minimum of 18 h at room temperature. Remove the liquid collagen from the surface and air-dry the membrane surface in a laminar flow hood. Once dried, rinse both sides of the membrane support

3 times with sterile PBS or DMEM to remove all traces of free collagen. Redry the membrane inserts in a laminar flow hood and store at 4 °C (they will be stable for several months).

4. Seeding cells on the membrane:

i. Dilute an experimentally enriched subpopulation of Rosa26 tracheal epithelial cells into non-Rosa26 cells at an appropriate ratio. The ratio must be determined empirically to allow for outgrowth of isolated clones and will depend on the proliferative index of the cells being analyzed. Alternatively, one can also use retrovirally tagged cells expressing a detectable histochemical marker (i.e., EGFP, Alkaline phosphatase, or β-galactosidase) for seeding onto membranes.

ii. Seed a total of 1–2.5×10^5 cells/cm^2 in 0.2 ml volume of 5% FBS-BEGM medium on the surface of a Millicell insert (about 1–1.5×10^5 cells per 12-mm insert). The volume should be sufficient to ensure a uniform distribution of cell settling upon the membrane surface.

iii. Apply 0.3–0.5 ml 5% FBS-BEGM medium in the basal compartment of the insert to immerse the membrane without floating the insert, make sure that the membranes are level to ensure uniform cell attachment during the first 12 h.

iv. Leave the plates containing the inserts undisturbed for a minimum of 18–24 h in a 8–9% CO_2 incubator at 37 °C. The higher CO_2 level has been shown to increases successful achievement of confluence (Karp *et al.*, 2002).

5. Establishment of ALI culture:

i. The day following seeding, wash the membranes with prewarmed PBS to remove unattached cells and refeed them with fresh 5% FBS BEGM medium. The medium is changed every 2 days until the transmembrane resistance (Rt) is greater than ~ 1000 Ω/cm^2 (this usually occurs 2–5 days after seeding). The electrical resistance across the membrane is measured using the Millicell®-ERS system ohm meter (Millipore, Bedford, Massachusetts, cat. MERS 000 01).

ii. Remove the medium inside the insert (apical) and refeed the outside (basal) chamber with BEGM medium without FBS to establish an ALI. Cells are refeed with BEGM medium every 2–3 days. A polarized and differentiated airway epithelium will form ~ 10–15 days after moving the epithelium to an ALI.

6. X-gal staining and evaluation: About 3 weeks (21 days) after seeding, the mouse ALI cultures should be ready for evaluation of CFE or other assays as appropriate. Depending on the purposes of the study, the membrane with polarized mouse tracheal epithelial cells can be either freshly embedded in OCT for frozen sectioning, or directly fixed for *en face* immunostaining or X-gal staining as previously described (Schoch *et al.*, 2004). Figure 6.5C–E depicts X-gal stained clonal expansion of Rosa26 mouse tracheal epithelial progenitor cells in this model.

7. Models to Study Stem/Progenitor Cells of Airway SMG

7.1. Overview

Airway SMGs are major secretory structures that reside in the cartilaginous airways of many mammalian species (Liu *et al.*, 2004). SMGs play important roles in both innate immunity of the lung (Dajani *et al.*, 2005) and cell biology of the proximal airways (Borthwick *et al.*, 2001; Engelhardt *et al.*, 1995, 2001). As discussed in more detail in a previous section, SMGs serve as a protective niche for surface airway epithelial stem cells (Borthwick *et al.*, 2001; Engelhardt, 2001). Furthermore, the pluripotent progenitor cells that exist in the human tracheobronchial SAE (i.e., cells with the capacity to differentiate into ciliated, secretory, intermediate, and basal cells) also have a developmental capacity for SMGs (Engelhardt *et al.*, 1995). Given that adult surface airway epithelial stem/progenitor cells have the capacity to develop SMGs, the biology that controls the morphogenesis of these structures in the airway is highly relevant to defining stem cell characteristics in the airway.

The development of three-dimensional culture substrates have enabled the creation of model systems to study certain features of SMG development using a system that is far less complex than tracheal xenografts. Epithelial invasion of the extracellular matrix (ECM) is an important aspect of lung development and SMG morphogenesis. Infeld and colleagues initially developed an *in vitro* model for early airway gland development by culturing human tracheobronchial epithelial cells on a floating collagen gel substrate that contained fetal lung fibroblasts (Infeld *et al.*, 1993). Similar studies by other groups have also demonstrated the capacity of human airway epithelial cells to invade a collagen gel matrix (Emura *et al.*, 1996; Jacquot *et al.*, 1994). Together with technical advances in the isolation and culture of mouse tracheal epithelial cells, we have adapted these protocols to evaluate bud formation, tubulogenesis, and branching morphogenesis of airway epithelia using the collagen gel matrix model (Fig. 6.6). Although such models cannot reproduce the native cellular diversity founds in SMGs, they will be useful in defining fundamental epithelial/mesenchymal interactions required for airway gland morphogenesis.

7.2. *In vitro* invasion and tubulogenesis assay using mouse tracheal epithelial cells

1. Collagen gel contraction:

i. Primary mouse embryonic fibroblast cells (PMEF): PMEF are generated using a previously described general protocol (Wassarman and DePamphilis, 1993). PMEF generated from ICR or C57/BL6 mice can be used for this model with no differences in experimental outcomes. Culture PMEF using DMEM medium supplemented with 10% FBS. It is worth noting that early passages of PMEF may give better results in gel contraction than later passage cells.

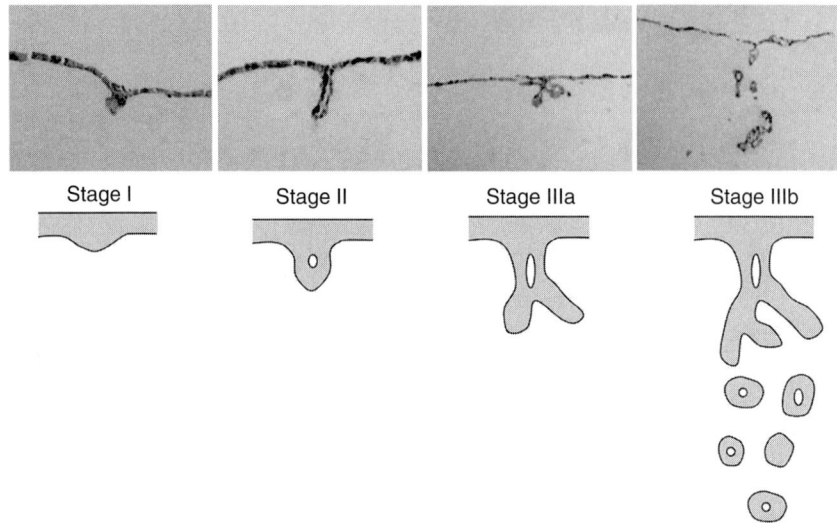

Figure 6.6 Mouse tracheal epithelial cell tubulogenesis assay. Bud formation and tubulogenesis are two early events important in SMG development and can be studied using a collagen gel matrix cultured at the ALI. The figure depicts various stages of gland-like structures formed from mouse tracheal epithelium (top panels) and schematic representations of various stages of tubulogenesis characteristic of early SMG morphogenesis (bottom panels).

However, for the most consistent results the same passages of PMEF should always be used for generating contracted collagen cells.

ii. PMEF cells are harvested by trypsinization, and the total cell number is quantified using a hemacytometer counting chamber.

iii. To generate the collagen gel matrix support on which epithelial cells are seeded, add the following reagents, in order, for each 60 mm dish at room temperature: 3.5 ml of fibroblast culture medium, 1.8 ml of FBS, 0.3 ml of 0.1 N NaOH, $3–5 \times 10^5$ fibroblast in 1.0 ml of medium, and 9.5 mg of Rat Tail Collagen Type I (BD Biosciences, Bedford, Massachusetts) in 2.7 ml H_2O to a final volume of 9.3 ml. Mix the solution gently by inverting the tube. The mixture should be light pink (if not, add 0.1 N NaOH to adjust the pH to turn the phenol red to a pink color).

iv. Apply ~9 ml of collagen gel to a 60-mm dish (or ~5 for 35-mm dish) and incubate the gel at room temperature for 10 min before moving the dish into a 5% CO_2 incubator at 37 °C for 1 h. Add a small amount of fibroblast culture medium to the gel and run a sterile pipette around the outside rim of the gel to allow contraction. Continue incubation in 5% CO_2 at 37 °C and feed twice a week with fibroblast culture medium. The gel will be fully contracted in 2–4 weeks.

2. Isolation of mouse tracheal epithelial cells: Use the same protocol described in the above section for *in vitro* CFE assays.

3. Mouse tracheal epithelial cell seeding:

i. After gel contraction, fresh isolated or cultured mouse airway cells are resuspended in 5% FBS-BEGM medium at a density of $3–5 \times 10^7$ cells/ml. BEGM medium is made by adding reagents from one BEGM SingleQuot kit to 500 ml of 50% DMEM-50% Ham's F12 medium supplemented with 1% pen-strep, 0.25 μg/ml fungizone, 15 mM Hepes, and 3.6 mM Na_2CO_3.

ii. Remove the medium from the fibroblast gel and apply 10 μl of the cell solution to the top of each gel while attempting to evenly spread the cells across the gel. Incubate the gels in 5% CO_2 at 37 °C for 30 min before adding a small amount of 5% FBS-BEGM medium around the edge of the gel. Continue incubation in 5% CO_2 at 37 °C overnight to allow adequate cell adhesion. Take care not to add much media that it covers the top of the gel and washes the cells off the surface.

iii. The following morning, add enough 5% FBS-BEGM medium to cover the top of the gel and let the gel sit for 3 days before placing the gels into 30-mm diameter Millicell-PCF membrane inserts (Millipore, Bedford, Massachusetts, Cat# PIHP 03012) in a 6-well dish to establish an Air Liquid Interface (ALI). After the ALI is established, feed the cells from the bottom of the inserts with BEGM medium (without FBS). The apical side of the gel should be exposed to air. Bud formation and tubulogenesis of airway epithelial cells can be evaluated at 1–2 weeks following the establishment of the ALI (Fig. 6.6).

Acknowledgments

We gratefully acknowledge NIDDK research funding in the area of this review (DK47967) and the editorial assistance of Ms. Leah Williams.

References

Asmundsson, T., Kilburn, K. H., and McKenzie, W. N. (1973). Injury and metaplasia of airway cells due to SO2. *Lab. Invest.* **29**, 41–53.

Barker, N., van Es, J. H., Kuipers, J., Kujala, P., van den Born, M., Cozijnsen, M., Haegebarth, A., Korving, J., Begthel, H., Peters, P. J., and Clevers, H. (2007). Identification of stem cells in small intestine and colon by marker gene Lgr5. *Nature* **449**, 1003–1007.

Berns, A. (2005). Stem cells for lung cancer? *Cell* **121**, 811–813.

Bigby, T. D., Allen, D., Leslie, C. G., Henson, P. M., and Cherniack, R. M. (1985). Bleomycin-induced lung injury in the rabbit. Analysis and correlation of bronchoalveolar lavage, morphometrics, and fibroblast stimulating activity. *Am. Rev. Respir. Dis.* **132**, 590–595.

Bishop, A. E. (2004). Pulmonary epithelial stem cells. *Cell Prolif.* **37**, 89–96.

Boers, J. E., Ambergen, A. W., and Thunnissen, F. B. (1998). Number and proliferation of basal and parabasal cells in normal human airway epithelium. *Am. J. Respir. Crit. Care Med.* **157**, 2000–2006.

Borges, M., Linnoila, R. I., van de Velde, H. J., Chen, H., Nelkin, B. D., Mabry, M., Baylin, S. B., and Ball, D. W. (1997). An achaete-scute homologue essential for neuroendocrine differentiation in the lung. *Nature* **386**, 852–855.

Borthwick, D. W., Shahbazian, M., Krantz, Q. T., Dorin, J. R., and Randell, S. H. (2001). Evidence for stem-cell niches in the tracheal epithelium. *Am. J. Respir. Cell Mol. Biol.* **24,** 662–670.

Buckpitt, A., Buonarati, M., Avey, L. B., Chang, A. M., Morin, D., and Plopper, C. G. (1992). Relationship of cytochrome P450 activity to Clara cell cytotoxicity. II. Comparison of stereoselectivity of naphthalene epoxidation in lung and nasal mucosa of mouse, hamster, rat and rhesus monkey. *J. Pharmacol. Exp. Ther.* **261,** 364–372.

Castleman, W. L., Dungworth, D. L., and Tyler, W. S. (1975). Intrapulmonary airway morphology in three species of monkeys: A correlated scanning and transmission electron microscopic study. *Am. J. Anat.* **142,** 107–121.

Choi, H. K., Finkbeiner, W. E., and Widdicombe, J. H. (2000). A comparative study of mammalian tracheal mucous glands. *J. Anat.* **197**(Pt. 3), 361–372.

Dajani, R., Zhang, Y., Taft, P. J., Travis, S. M., Starner, T. D., Olsen, A., Zabner, J., Welsh, M. J., and Engelhardt, J. F. (2005). Lysozyme secretion by submucosal glands protects the airway from bacterial infection. *Am. J. Respir. Cell Mol. Biol.* **32,** 548–552.

Davidson, D. J., Kilanowski, F. M., Randell, S. H., Sheppard, D. N., and Dorin, J. R. (2000). A primary culture model of differentiated murine tracheal epithelium. *Am. J. Physiol. Lung Cell Mol. Physiol.* **279,** L766–L778.

Duan, D., Sehgal, A., Yao, J., and Engelhardt, J. F. (1998). Lef1 transcription factor expression defines airway progenitor cell targets for *in utero* gene therapy of submucosal gland in cystic fibrosis. *Am. J. Respir. Cell Mol. Biol.* **18,** 750–758.

Dubaybo, B. A., Crowell, L. A., and Thet, L. A. (1991). Changes in tissue fibronectin in elastase induced lung injury. *Cell Biol. Int. Rep.* **15,** 675–686.

Emura, M. (1997). Stem cells of the respiratory epithelium and their *in vitro* cultivation. *In Vitro Cell. Dev. Biol. Anim.* **33,** 3–14.

Emura, M. (2002). Stem cells of the respiratory tract. *Paediatr. Respir. Rev.* **3,** 36–40.

Emura, M., Ochiai, A., and Hirohashi, S. (1996). *In vitro* reconstituted tissue as an alternative to human respiratory tract. *Toxicol. Lett.* **88,** 81–84.

Engelhardt, J. F. (2001). Stem cell niches in the mouse airway. *Am. J. Respir. Cell Mol. Biol.* **24,** 649–652.

Engelhardt, J. F., Allen, E. D., and Wilson, J. M. (1991). Reconstitution of tracheal grafts with a genetically modified epithelium. *Proc. Natl. Acad. Sci. USA* **88,** 11192–11196.

Engelhardt, J. F., Schlossberg, H., Yankaskas, J. R., and Dudus, L. (1995). Progenitor cells of the adult human airway involved in submucosal gland development. *Development* **121,** 2031–2046.

Engelhardt, J. F., Yankaskas, J. R., and Wilson, J. M. (1992). *In vivo* retroviral gene transfer into human bronchial epithelia of xenografts. *J. Clin. Invest.* **90,** 2598–2607.

Evans, M. J., Cabral-Anderson, L. J., Dekker, N. P., and Freeman, G. (1981). The effects of dietary antioxidants on NO2-induced injury to type 1 alveolar cells. *Chest* **80,** 5–8.

Filali, M., Zhang, Y., Ritchie, T. C., and Engelhardt, J. F. (2002). Xenograft model of the CF airway. *Methods Mol. Med.* **70,** 537–550.

Giangreco, A., Reynolds, S. D., and Stripp, B. R. (2002). Terminal bronchioles harbor a unique airway stem cell population that localizes to the bronchoalveolar duct junction. *Am. J. Pathol.* **161,** 173–182.

Giangreco, A., Shen, H., Reynolds, S. D., and Stripp, B. R. (2004). Molecular phenotype of airway side population cells. *Am. J. Physiol. Lung Cell. Mol. Physiol.* **286,** L624–L630.

Griffiths, M. J., Bonnet, D., and Janes, S. M. (2005). Stem cells of the alveolar epithelium. *Lancet* **366,** 249–260.

Grove, J. E., Lutzko, C., Priller, J., Henegariu, O., Theise, N. D., Kohn, D. B., and Krause, D. S. (2002). Marrow-derived cells as vehicles for delivery of gene therapy to pulmonary epithelium. *Am. J. Respir. Cell Mol. Biol.* **27,** 645–651.

Hansell, M. M., and Moretti, R. L. (1969). Ultrastructure of the mouse tracheal epithelium. *J. Morphol.* **128,** 159–169.

Hay, J., Shahzeidi, S., and Laurent, G. (1991). Mechanisms of bleomycin-induced lung damage. *Arch. Toxicol.* **65,** 81–94.

Hong, K. U., Reynolds, S. D., Giangreco, A., Hurley, C. M., and Stripp, B. R. (2001). Clara cell secretory protein-expressing cells of the airway neuroepithelial body microenvironment include a label-retaining subset and are critical for epithelial renewal after progenitor cell depletion. *Am. J. Respir. Cell Mol. Biol.* **24,** 671–681.

Hong, K. U., Reynolds, S. D., Watkins, S., Fuchs, E., and Stripp, B. R. (2004). Basal cells are a multipotent progenitor capable of renewing the bronchial epithelium. *Am. J. Pathol.* **164,** 577–588.

Hong, K. U., Reynolds, S. D., Watkins, S., Fuchs, E., and Stripp, B. R. (2004). *In vivo* differentiation potential of tracheal basal cells: Evidence for multipotent and unipotent subpopulations. *Am. J. Physiol. Lung Cell. Mol. Physiol.* **286,** L643–L649.

Hook, G. E., Brody, A. R., Cameron, G. S., Jetten, A. M., Gilmore, L. B., and Nettesheim, P. (1987). Repopulation of denuded tracheas by Clara cells isolated from the lungs of rabbits. *Exp. Lung Res.* **12,** 311–329.

Inayama, Y., Hook, G. E., Brody, A. R., Cameron, G. S., Jetten, A. M., Gilmore, L. B., Gray, T., and Nettesheim, P. (1988). The differentiation potential of tracheal basal cells. *Lab. Invest.* **58,** 706–717.

Infeld, M. D., Brennan, J. A., and Davis, P. B. (1993). Human fetal lung fibroblasts promote invasion of extracellular matrix by normal human tracheobronchial epithelial cells *in vitro*: A model of early airway gland development. *Am. J. Respir. Cell Mol. Biol.* **8,** 69–76.

Izbicki, G., Segel, M. J., Christensen, T. G., Conner, M. W., and Breuer, R. (2002). Time course of bleomycin-induced lung fibrosis. *Int. J. Exp. Pathol.* **83,** 111–119.

Jacquot, J., Spilmont, C., Burlet, H., Fuchey, C., Buisson, A. C., Tournier, J. M., Gaillard, D., and Puchelle, E. (1994). Glandular-like morphogenesis and secretory activity of human tracheal gland cells in a three-dimensional collagen gel matrix. *J. Cell Physiol.* **161,** 407–418.

Januszkiewicz, A. J., and Mayorga, M. A. (1994). Nitrogen dioxide-induced acute lung injury in sheep. *Toxicology* **89,** 279–300.

Jeffery, P. K. (1983). Morphologic features of airway surface epithelial cells and glands. *Am. Rev. Respir. Dis.* **128,** S14–S20.

Karp, P. H., Moninger, T. O., Weber, S. P., Nesselhauf, T. S., Launspach, J. L., Zabner, J., and Welsh, M. J. (2002). An *in vitro* model of differentiated human airway epithelia. Methods for establishing primary cultures. *Methods Mol. Biol.* **188,** 115–137.

Kiel, M. J., He, S., Ashkenazi, R., Gentry, S. N., Teta, M., Kushner, J. A., Jackson, T. L., and Morrison, S. J. (2007). Haematopoietic stem cells do not asymmetrically segregate chromosomes or retain BrdU. *Nature* **449,** 238–242.

Kim, C. F., Jackson, E. L., Woolfenden, A. E., Lawrence, S., Babar, I., Vogel, S., Crowley, D., Bronson, R. T., and Jacks, T. (2005). Identification of bronchioalveolar stem cells in normal lung and lung cancer. *Cell* **121,** 823–835.

Lamb, D., and Reid, L. (1968). Mitotic rates, goblet cell increase and histochemical changes in mucus in rat bronchial epithelium during exposure to sulphur dioxide. *J. Pathol. Bacteriol.* **96,** 97–111.

Langley-Evans, S. C., Phillips, G. J., and Jackson, A. A. (1996). Sulphur dioxide: A potent glutathione depleting agent. *Comp. Biochem. Physiol. C Pharmacol. Toxicol. Endocrinol.* **114,** 89–98.

Liu, X., Driskell, R. R., and Engelhardt, J. F. (2004). Airway glandular development and stem cells. *Curr. Top. Dev. Biol.* **64,** 33–56.

Liu, X., Yan, Z., Luo, M., and Engelhardt, J. F. (2005). Species-specific differences in mouse and human airway epithelial biology of rAAV transduction. *Am. J. Respir. Cell Mol. Biol.* **34,** 56–64.

Mason, R. J., Williams, M. C., Moses, H. L., Mohla, S., and Berberich, M. A. (1997). Stem cells in lung development, disease, and therapy. *Am. J. Respir. Cell Mol. Biol.* **16,** 355–363.

Meulenbelt, J., Dormans, J. A., Marra, M., Rombout, P. J., and Sangster, B. (1992). Rat model to investigate the treatment of acute nitrogen dioxide intoxication. *Hum. Exp. Toxicol.* **11,** 179–187.

Meulenbelt, J., van Bree, L., Dormans, J. A., Boink, A. B., and Sangster, B. (1992). Biochemical and histological alterations in rats after acute nitrogen dioxide intoxication. *Hum. Exp. Toxicol.* **11,** 189–200.

Meulenbelt, J., van Bree, L., Dormans, J. A., Boink, A. B., and Sangster, B. (1994). Development of a rabbit model to investigate the effects of acute nitrogen dioxide intoxication. *Hum. Exp. Toxicol.* **13,** 749–758.

Neuringer, I. P., and Randell, S. H. (2004). Stem cells and repair of lung injuries. *Respir. Res.* **5,** 6.

Otto, W. R. (2002). Lung epithelial stem cells. *J. Pathol.* **197,** 527–535.

Pack, R. J., Al-Ugaily, L. H., Morris, G., and Widdicombe, J. G. (1980). The distribution and structure of cells in the tracheal epithelium of the mouse. *Cell Tissue Res.* **208,** 65–84.

Peake, J. L., Reynolds, S. D., Stripp, B. R., Stephens, K. E., and Pinkerton, K. E. (2000). Alteration of pulmonary neuroendocrine cells during epithelial repair of naphthalene-induced airway injury. *Am. J. Pathol.* **156,** 279–286.

Plopper, C. G., Hill, L. H., and Mariassy, A. T. (1980). Ultrastructure of the nonciliated bronchiolar epithelial (Clara) cell of mammalian lung. III. A study of man with comparison of 15 mammalian species. *Exp. Lung Res.* **1,** 171–180.

Plopper, C. G., Mariassy, A. T., Wilson, D. W., Alley, J. L., Nishio, S. J., and Nettesheim, P. (1983). Comparison of nonciliated tracheal epithelial cells in six mammalian species: Ultrastructure and population densities. *Exp. Lung Res.* **5,** 281–294.

Plopper, C. G., Weir, A. J., Nishio, S. J., Cranz, D. L., and St George, J. A. (1986). Tracheal submucosal gland development in the rhesus monkey, *Macaca mulatta*: Ultrastructure and histochemistry. *Anat. Embryol. (Berl.)* **174,** 167–178.

Presente, A., Sehgal, A., Dudus, L., and Engelhardt, J. F. (1997). Differentially regulated epithelial expression of an Eph family tyrosine kinase (fHek2) during tracheal surface airway and submucosal gland development. *Am. J. Respir. Cell Mol. Biol.* **16,** 53–61.

Randell, S. H. (1992). Progenitor-progeny relationships in airway epithelium. *Chest* **101,** 11S–16S.

Ranga, V., and Kleinerman, J. (1981). A quantitative study of ciliary injury in the small airways of mice: The effects of nitrogen dioxide. *Exp. Lung Res.* **2,** 49–55.

Reddy, R., Buckley, S., Doerken, M., Barsky, L., Weinberg, K., Anderson, K. D., Warburton, D., and Driscoll, B. (2004). Isolation of a putative progenitor subpopulation of alveolar epithelial type 2 cells. *Am. J. Physiol. Lung Cell. Mol. Physiol.* **286,** L658–L667.

Reynolds, S. D., Giangreco, A., Hong, K. U., McGrath, K. E., Ortiz, L. A., and Stripp, B. R. (2004). Airway injury in lung disease pathophysiology: Selective depletion of airway stem and progenitor cell pools potentiates lung inflammation and alveolar dysfunction. *Am. J. Physiol. Lung Cell. Mol. Physiol.* **287,** L1256–L1265.

Reynolds, S. D., Giangreco, A., Power, J. H., and Stripp, B. R. (2000). Neuroepithelial bodies of pulmonary airways serve as a reservoir of progenitor cells capable of epithelial regeneration. *Am. J. Pathol.* **156,** 269–278.

Reynolds, S. D., Hong, K. U., Giangreco, A., Mango, G. W., Guron, C., Morimoto, Y., and Stripp, B. R. (2000). Conditional clara cell ablation reveals a self-renewing progenitor function of pulmonary neuroendocrine cells. *Am. J. Physiol. Lung Cell. Mol. Physiol.* **278,** L1256–L1263.

Schoch, K. G., Lori, A., Burns, K. A., Eldred, T., Olsen, J. C., and Randell, S. H. (2004). A subset of mouse tracheal epithelial basal cells generates large colonies *in vitro*. *Am. J. Physiol. Lung Cell. Mol. Physiol.* **286,** L631–L642.

Sehgal, A., Presente, A., and Engelhardt, J. F. (1996). Developmental expression patterns of CFTR in ferret tracheal surface airway and submucosal gland epithelia. *Am. J. Respir. Cell Mol. Biol.* **15,** 122–131.

Smith, L. J. (1985). Hyperoxic lung injury: Biochemical, cellular, and morphologic characterization in the mouse. *J. Lab. Clin. Med.* **106,** 269–278.

Souma, T. (1987). The distribution and surface ultrastructure of airway epithelial cells in the rat lung: A scanning electron microscopic study. *Arch. Histol. Jpn.* **50,** 419–436.

Summer, R., Kotton, D. N., Sun, X., Ma, B., Fitzsimmons, K., and Fine, A. (2003). Side population cells and Bcrp1 expression in lung. *Am. J. Physiol. Lung Cell. Mol. Physiol.* **285,** L97–L104.

Suzuki, M., Machida, M., Adachi, K., Otabe, K., Sugimoto, T., Hayashi, M., and Awazu, S. (2000). Histopathological study of the effects of a single intratracheal instillation of surface active agents on lung in rats. *J. Toxicol. Sci.* **25,** 49–55.

Ten Have-Opbroek, A. A., and Plopper, C. G. (1992). Morphogenetic and functional activity of type II cells in early fetal rhesus monkey lungs. A comparison between primates and rodents. *Anat. Rec.* **234,** 93–104.

Teta, M., Rankin, M. M., Long, S. Y., Stein, G. M., and Kushner, J. A. (2007). Growth and regeneration of adult beta cells does not involve specialized progenitors. *Dev. Cell* **12,** 817–826.

Theise, N. D., Henegariu, O., Grove, J., Jagirdar, J., Kao, P. N., Crawford, J. M., Badve, S., Saxena, R., and Krause, D. S. (2002). Radiation pneumonitis in mice: A severe injury model for pneumocyte engraftment from bone marrow. *Exp. Hematol.* **30,** 1333–1338.

Tumbar, T., Guasch, G., Greco, V., Blanpain, C., Lowry, W. E., Rendl, M., and Fuchs, E. (2004). Defining the epithelial stem cell niche in skin. *Science* **303,** 359–363.

Vacanti, M. P., Roy, A., Cortiella, J., Bonassar, L., and Vacanti, C. A. (2001). Identification and initial characterization of spore-like cells in adult mammals. *J. Cell. Biochem.* **80,** 455–460.

Van Winkle, L. S., Johnson, Z. A., Nishio, S. J., Brown, C. D., and Plopper, C. G. (1999). Early events in naphthalene-induced acute Clara cell toxicity: Comparison of membrane permeability and ultrastructure. *Am. J. Respir. Cell Mol. Biol.* **21,** 44–53.

Wassarman, P. M., and DePamphilis, M. L. (1993). Guide to Techniques in Mouse Development Academic Press, San Diego **225,** 719–732.

Watt, F. M., and Hogan, B. L. (2000). Out of Eden: Stem cells and their niches. *Science* **287,** 1427–1430.

West, J. A., Pakehham, G., Morin, D., Fleschner, C. A., Buckpitt, A. R., and Plopper, C. G. (2001). Inhaled naphthalene causes dose dependent Clara cell cytotoxicity in mice but not in rats. *Toxicol. Appl. Pharmacol.* **173,** 114–119.

Widdicombe, J. H., Chen, L. L., Sporer, H., Choi, H. K., Pecson, I. S., and Bastacky, S. J. (2001). Distribution of tracheal and laryngeal mucous glands in some rodents and the rabbit. *J. Anat.* **198,** 207–221.

You, Y., Richer, E. J., Huang, T., and Brody, S. L. (2002). Growth and differentiation of mouse tracheal epithelial cells: Selection of a proliferative population. *Am. J. Physiol. Lung Cell. Mol. Physiol.* **283,** L1315–L1321.

Zepeda, M. L., Chinoy, M. R., and Wilson, J. M. (1995). Characterization of stem cells in human airway capable of reconstituting a fully differentiated bronchial epithelium. *Somat. Cell Mol. Genet.* **21,** 61–73.

CHAPTER 7

Pancreatic Cells and Their Progenitors

Seth J. Salpeter and Yuval Dor

Department of Cellular Biochemistry and Human Genetics
The Hebrew University-Hadassah Medical School
Jerusalem 91120, Israel

Abstract
1. Introduction
2. Pancreas Development
3. Origins of Beta Cells During Postnatal Life
4. Preexisting Beta Cells
5. Ducts
6. Acini
7. Bone Marrow Cells
8. Adult Stem Cells
9. Dedifferentiation of Beta Cells
10. Summary and Perspective
11. Methods: Design of a Lineage Tracing Experiment in Mice
    References

Abstract

Both type I and type II diabetes patients would greatly benefit from transplantation of insulin-producing pancreatic beta cells, however, a severe shortage of transplantable beta cells is a major current limitation on the use of such therapy. Understanding the mechanisms by which beta cells are naturally formed is therefore a central challenge for modern pancreas biology, in the hope that insights will be applicable for regenerative of cell therapy strategies for diabetes. In particular, the cellular origins of pancreatic beta cells pose an important problem with significant basic and therapeutic implications. This chapter discusses the current controversy regarding the identity of the cells which give rise to new beta cells in the adult mammal. While numerous models suggest that beta cells can originate from adult stem cells, proposed to reside in the pancreas or in other locations, more recent work indicates that the major source for new beta cells during adult life is the proliferation of preexisting, differentiated beta cells. We present these different views, with emphasis on the methodologies employed. In particular, we focus on genetic lineage tracing using the Cre-lox system in transgenic mice, a technique considered as the gold standard for addressing *in vivo* problems of cellular origins.

1. Introduction

The mystery of pancreatic beta cell origins poses a significant scientific question on several fronts. One pressing cause for investigation lies in the necessity to improve therapy for Type I and Type II Diabetes, diseases which afflict 200 million people worldwide. In both diseases, insulin-producing beta cells, organized in clusters called the Islets of Langerhans, fail to provide enough insulin to maintain glucose homeostasis. In type I diabetes, an autoimmune response leads to the destruction of beta cells. Type II diabetes, accounting for >90% of cases of diabetes, is typically associated with peripheral insulin resistance. However, recent evidence demonstrates that in addition to insulin resistance, this disease is also associated with defects in beta cell function and with a loss of as much as 50% of beta cell mass (Butler *et al.*, 2003). Thus, the possibility of regenerating or replacing beta cells offers immense therapeutic potential in both types of diabetes. Indeed, a clinical procedure for islet transplantation, recently developed at the University of Alberta (the Edmonton protocol), offers for the first time a satisfying cure for diabetes (Shapiro *et al.*, 2000). However, a major limitation of the Edmonton protocol is the severe shortage of donor islets. As a result of this problem, significant efforts are directed towards the development of strategies for enhancement of beta cell mass, *in vivo* (regenerative therapy) or *in vitro* for transplantation (cell therapy). Such efforts will greatly benefit from a better understanding of the normal process by which new beta cells are formed, and in particular the identification of the cell type(s) capable of giving rise to new beta cells. The search for the cellular origins of beta cells, with emphasis on the methodologies employed, is the subject of this chapter.

Beyond clinical importance, understanding the origins of beta cells represents a basic challenge for developmental biology. Though long considered static, islets of Langerhans (composed mainly of beta cells and constituting the endocrine pancreas) are now appreciated as an active and developing organ that maintains the ability to respond to external stimuli (Bonner-Weir *et al.*, 1989) and continues to grow throughout their lifetime (Montanya *et al.*, 2000; Skau *et al.*, 2001). What is the basic mechanism by which this tissue is maintained? Do beta cells rely on a continuous supply of differentiated cells derived from adult stem cells, and if so, what are the characteristics of these stem cells? Or, is beta cell expansion and

maintenance based on the regulated proliferation of existing, terminally differentiated beta cells? We start by providing a brief overview of the dynamics of
pancreas and beta cell formation during embryonic development. Later sections
describe current views on the origins of new beta cells during normal adult life, as
well as following injury. Throughout the discussion, we highlight the importance
of one methodology, namely genetic lineage tracing (Fig.7.1), which has emerged
as the gold standard for addressing problems of cellular origins, in biology in
general and in the pancreas field in particular.

2. Pancreas Development

To properly understand beta cell origins in the adult pancreas, we must first
review the development of the pancreas in its early stages. The adult pancreas is
composed of two almost independent organs. The exocrine pancreas, occupying
>95% of tissue mass, is responsible for secreting digestive enzymes into the
duodenum. This is achieved by a network of converging tubes, called pancreatic
ducts, which collect the enzymes from clusters of acinar cells organized in acini.
The endocrine pancreas is organized in clusters of 100–1000 cells called the islets of
Langerhans, and is mainly responsible for the regulation of systemic glucose
homeostasis. Beta cells, composing 70% of islet cells, are responsible for accurate
sensing of blood glucose levels and appropriate secretion of insulin to the blood-
stream. Other, less abundant cell types in islets include alpha cells (producing the
hormone glucagon, negating insulin action), delta cells (producing somatostatin),
and pp cells (producing pancreatic polypeptide). There are about 1000 islets in the

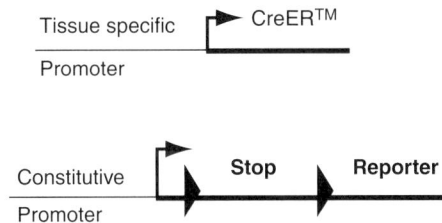

Figure 7.1 Schematic of a lineage tracing system using Cre-lox technology in transgenic mice.
Tamoxifen-dependent Cre recombinase (CreER™) is expressed from a tissue specific promoter (for
example, the insulin promoter driving expression specifically in beta cells). A second transgene expresses
a reporter gene (for example, EGFP or lacZ) from a constitutively active promoter. Expression of the
reporter is blocked by a transcriptional stop sequence flanked by loxP sites. In the absence of tamoxifen,
there is no expression of the reporter. Tamoxifen injection leads to transient activation of CreER™
(lasting ~48 h), resulting in removal of the loxP-flanked stop sequence and activation of reporter
expression. Recombined cells as well as their progeny will continue to express the reporter, allowing
for retrospective identification of their origins. Specifically, one can conclude that cells expressing the
reporter gene are the same cells or the progeny of cells that expressed CreER™ at the time of tamoxifen
injection.

adult mouse and a total of ~2,000,000 beta cells. An adult human pancreas contains about 2,000,000 islets.

We now know with certainty that all pancreatic cells derive from the embryonic primitive gut, a tissue derived from the endoderm. Careful embryological and molecular genetic studies have demonstrated that the pancreas develops from a small group of cells that bud from the gut tube, around embryonic day 9 in the mouse, that are marked by expression of the transcription factors pdx1 and ptf1a/p48. The budding epithelium then undergoes a series of differentiation and morphogenesis steps, giving rise to the endocrine, hormone producing cells, ducts, and acini with the pancreas morphogenesis completing itself around birth (Edlund, 2001, 2002; Murtaugh and Melton, 2003; Slack, 1995). Particular attention was given to the elucidation of the molecular and cellular pathway by which beta cells are generated during development, in the hope that emerging principles will be applicable for the therapeutic generation of new beta cells, *in vitro* or *in vivo*. We describe some of these studies in relative detail because they provide an example for the methodological rigor essential in this field.

As mentioned above the earliest cells of the embryonic pancreas express the transcription factors pdx1 and ptf1a/p48. Moreover, mouse knockout studies demonstrated that these genes are essential for pancreas development, including beta cell development (Ahlgren *et al.*, 1996, 1998; Jonsson *et al.*, 1994; Kawaguchi *et al.*, 2002; Offield *et al.*, 1996). At first glance, this implies that adult beta cells at one time expressed pdx1 and ptf1a/p48. However, this is not necessarily the case: beta cells could have derived from mesenchymal or ectodermal cells, and still be dependent on the noncell autonomous expression of pdx1 and ptf1a/p48. To address this possibility, it is essential to genetically mark embryonic pdx1+ and ptf1a/p48+ cells, and ask if the direct progeny of these cells give rise to beta cells. Such lineage tracing experiments are best performed using the cre-lox system in transgenic mice, in which the expression of cre recombinase in a particular cell type results in an indelible marking (usually the expression of a reporter gene) of this cell and its progeny (Branda and Dymecki, 2004; Danielian *et al.*, 1998; Hayashi and McMahon, 2002; Rossant and McMahon, 1999). Indeed, elegant studies have shown that beta cells are the progeny of cells that expressed pdx1 and ptf1a/p48 during early embryonic development (Kawaguchi *et al.*, 2002). These studies were based on the generation of transgenic mice expressing cre recombinase under control of either the pdx1 or ptf1a/p48 regulatory elements. The role of pdx1 in beta cell development posed another challenge, since adult beta cells express pdx1. Therefore, simple expression of cre from pdx1 promoter will result in the labeling of adult beta cells, regardless of their historical expression profile. To overcome this hurdle, Gu and colleagues used transgenic mice expressing a tamoxifen-inducible version of Cre recombinase under the pdx1 promoter (pdx1–CreER) (Gu *et al.*, 2002). Using such mice (together with a cre reporter strain) they were able to demonstrate that pdx1-expressing cells in the early embryonic pancreas give rise to beta cells. Along similar lines, the role of the nurogenin-3 (ngn3) transcription factor was delineated. Ngn3 is expressed transiently during embryonic development of the pancreas (in progenitor cells that do not express endocrine hormones such as insulin and glucagons), and in its absence the endocrine pancreas (islets of Langerhans) are not formed (Gradwohl *et al.*, 2000). Again, it required the indelible labeling of ngn3+ cells and their progeny, using the cre-lox system, to demonstrate that beta cells are derived from ngn3+ cells (Gu *et al.*, 2002).

Another pioneering study on the cellular origins of beta cells has focused on a particularly interesting population of cells appearing early during pancreas development, expressing both insulin and glucagon (Herrera, 2000; Herrera *et al.*, 1994). From the initial observations of such cells was concluded that insulin+ glucagon+ "double hormone positive" cells are the progenitors for both beta and alpha cells. This reasonable conclusion has important implications for the design of cell therapies in diabetes, because is suggests that efforts should be directed at increasing the numbers of insulin+ glucagon+ cells, for example, during directed differentiation of embryonic stem cells (ESC). To critically test this notion, classic experiments by Pedro Herrera have used transgenic expression of diphtheria toxin to examine the fate of insulin+ cells when glucagon+ cells are eliminated and vise versa (Herrera *et al.*, 1994). Surprisingly, specific ablation of alpha cells (using glucagon-diphtheria toxin transgenic mice) generated embryos lacking alpha cells but contain near normal levels of beta cells. These experiments indicate that beta cells are not derived from insulin+ glucagon+ cells. More recent lineage tracing experiments, this time involving only "noninvasive" tagging of insulin+ or glucagon+ cells, support the view that adult beta cells have never expressed glucagon (Herrera, 2000).

As a result of these studies and others, we now know that pdx1+ ptf1a/p48+ cells of the embryonic pancreas give rise to all pancreatic lineages in a cell-autonomous manner (Kim and MacDonald, 2002). Some progeny of these cells express transiently ngn3, during late gestation, and give rise to islet cells including beta cells. Early "double hormone positive" cells expressing both insulin and glucagon are largely considered "developmental dead ends", perhaps reflecting ancient stages in pancreas evolution but not participating in the main pathway for beta cell formation.

3. Origins of Beta Cells During Postnatal Life

After birth, the number of ngn3+ progenitor cells decline to almost zero, indicating that new beta cells formed during postnatal life must be generated in a different pathway. What are the origins of postnatal beta cells? This seemingly simple question is the subject of current heated debate, not the least because of the potentially important practical implications for the design of regenerative therapies for diabetes. For example, if beta cells derive from stem cells residing in the spleen this almost immediately suggests exciting new approaches for the cure of diabetes. Below, we discuss some notable perceptions regarding the origins of adult beta cells, with emphasis on the methodologies employed.

4. Preexisting Beta Cells

Postnatal beta cells can proliferate, even though at a low rate, as demonstrated by the incorporation of BrdU or ^{3}H-thymidine (Kassem *et al.*, 2000; Meier *et al.*, 2006; Messier and Leblond, 1960; Ritzel and Butler, 2003). Does this mean that beta cell replication is responsible for the generation of new beta cells? Not necessarily: beta cell could proliferate at a low rate, but still be derived mainly from stem cells. As described above, lineage analysis has emerged as the gold standard for proving the origins of a mature cell. Therefore, a study was undertaken by Dor *et al.* to apply inducible lineage analysis to the adult beta cell, using a new method termed "genetic pulse-chase" (Dor *et al.*, 2004). Transgenic mice were generated in which tamoxifen-dependent Cre recombinase was placed downstream of the rat insulin promoter. This transgene was crossed with a cre-reporter transgenic strain, expressing human placental alkaline phosphatase (HPAP) only after cre-mediated removal of a loxP-flanked stop sequence. Adult mice were injected with tamoxifen to label existing, differentiated beta cells ("pulse"), and the fate of these labeled beta cells was followed over time ("chase"). If beta cells derive from preexisting beta cells, the frequency of labeled beta cells should remain stable. If, however, dying beta cells are replaced by new beta cells derived from any time of stem cell (hence unlabeled), the frequency of labeled beta cells should decline with time. Strikingly, the frequency of labeled beta cells did not decline even a year after the original labeling. This led the authors to conclude that during adult life, new beta cells are largely the product of beta cell proliferation rather than stem cell differentiation. Similar experiments were performed with mice subject to partial pancreatectomy, a procedure known to result in some regeneration of beta cells. Here again, the results demonstrated that most new beta cells were the progeny of beta cells that existed prior to the operation. Importantly, this experimental system was designed to determine the major source of new beta cells during postnatal life. Stem cells could still exist and give rise to small numbers of beta cells. Such stem cells could even be recruited to action following specific injury conditions not tested so far. These are important considerations because even a small contribution of stem cells can in principle have significant biotechnological/therapeutic utility (Halban, 2004).

Other studies provide indirect support to the notion that beta cell proliferation is a dominant process in beta cell dynamics. For example, mice lacking the cell cycle components cyclinD2 or CDK4 show a progressive failure of beta cell mass (Georgia and Bhushan, 2004; Kushner *et al.*, 2005; Rane *et al.*, 1999). Since these genes are mainly expressed in differentiated beta cells, this supports the primacy of beta cell proliferation. In addition, forced expression of the cycling kinase inhibitor p27 in beta cells resulted in a progressive decline in beta cell mass, again suggesting that beta cell proliferation is essential (Uchida *et al.*, 2005).

5. Ducts

Pancreatic ducts are considered by many as a pool of adult stem cells constantly replenishing beta cells. This view is largely based on the fact that during embryonic development of the pancreas, primitive ducts give rise to ngn3+ endocrine progenitors which in turn differentiate to hormone producing cells including beta cells (Kim and MacDonald, 2002).

According to one study, taking into account of the rate of beta cell replication during the first month against the high amount of apoptosis occurring during this period, beta cell mass would drop dramatically if not for neogenesis (the formation of beta cells from stem cells) (Scaglia *et al.*, 1997). Therefore, they conclude based on their mathematical model that over 30% of new beta cells at 31 days of age were not from replication of preexisting beta cells. The authors note that the work of Dor *et al.* (2004) does not necessarily exclude the possibility of neogenesis in the one month old pancreas, as the pulse chase analysis was only undertaken starting at 6 weeks of age.

In addition to the "kinetic" argument, histological analyses of the adult pancreas, particularly in humans, show multiple islets adjacent to ducts, presumably "budding" from the ducts. While it is tempting to extrapolate a dynamic process from such static observation, caution must be taken in the interpretation. For example, what appears to be an islet budding from a duct can in fact represent a mature islet, maintained by beta cell proliferation, which happened to reside adjacent to a duct (or even be derived from this duct earlier in its life) (Bouwens *et al.*, 1994). Similar to static histological observations, the expression in islets of genes typical of ducts cannot prove lineal relationship between cell types. Finally, a number of *in vitro* studies have claimed that duct cultures can give rise to beta cells (Bonner-Weir *et al.*, 2000). However, the partial purity of these cultures and the possible "contamination" by endocrine cells prevent us from reaching a strong conclusion from such studies about the origins of beta cells (Gao *et al.*, 2005). Clearly, what is needed the most is a straight forward genetic lineage tracing experiment, in which adult duct cells are labeled and their progeny are examined. Such experiments, although underway in many laboratories, have not been published so far.

6. Acini

Similar to the hypothesis of ductal contribution to beta cells, it has been suggested that acinar cells may undergo transdifferentiation and convert into beta cells in islets. However, evidence for this model is circumstantial; *in vivo* lineage tracing experiments of the acinar pancreas have not provided so far a convincing demonstration of acinar to islet transdifferentiation. Interestingly, a recent study has used lineage tracing *in vitro* to label cultured acinar cells (by infecting cultured cells with an adenovirus expressing cre recombinase from the elastase or amylase promoter) (Minami *et al.*, 2005). If verified and extended to the *in vivo* situation, this scenario will be of great interest.

7. Bone Marrow Cells

Recent studies have raised the possibility that bone marrow may contribute to islet mass both in the normal and diseased pancreas. In one notable study, bone marrow cells from Insulin-GFP transgenic male mice were transplanted into irradiated female mice, and 2%–3% of the recipient islets were reported to contain Y chromosome positive, GFP-expressing beta cells (Ianus et al., 2003). This elegantly designed experiment ruled out the possibility of fusion between marrow cells and beta cells as an explanation. However, more recent studies failed to repeat these experiments, casting doubts on the notion that bone marrow cells can give rise to beta cells (Lechner et al., 2004; Mathews et al., 2004). Interestingly, a similar study in a setting of regeneration from a diabetogenic injury has found that bone marrow cells do contribute to the formation of new beta cells (Hess et al., 2003); however, the effect was strictly indirect: bone marrow cells were shown to differentiate into islet endothelial cells, which presumably enhanced the formation of new beta cells from other sources.

Another important study reported that in NOD mice, a model for autoimmune diabetes, neutralization of the autoimmune response resulted in dramatic regeneration of beta cells. The cellular origins of new beta cells in this case were claimed to be spleen cells that transdifferentiated to beta cells (Kodama et al., 2003). However, more recent attempts by three independent groups to repeat this study failed to detect any contribution of spleen cells to beta cells (Chong et al., 2006; Nishio et al., 2006; Suri et al., 2006). Interestingly, blockade of the autoimmune response did result in beta cell regeneration. Thus, the origins of new beta cells in this setting remain undefined, apart from the conclusion that they are not derived from marrow or spleen cells. Finally, a recent study of a human type I diabetes patient has found evidence for increased proliferation of beta cells, and not duct cells, in the early stages of the disease, strongly suggesting that beta cell regeneration in humans is initiated by beta cell proliferation rather than stem cell differentiation (Meier et al., 2006).

8. Adult Stem Cells

Several studies suggested that beta cells can derive from stem cells residing in islets. An early report suggested that after administration of the beta cell-selective toxin streptozotocin, a subpopulation of cells expressing both somatostatin, and pdx-1 started to express insulin (Guz et al., 2001). However, these studies did not demonstrate conclusively the stable formation of new beta cells. Other studies suggested that rare nestin+ cells in islets could be adult islet stem or progenitor cells (Hunziker and Stein, 2000; Lechner et al., 2002; Zulewski et al., 2001). However, expression profiling and lineage tracing experiments do not support this notion (Delacour et al., 2004; Esni et al., 2004; Means et al., 2005; Selander and Edlund, 2002; Street et al., 2004; Treutelaar et al., 2003). More recently, it has been suggested that a small number of ngn3+ embryonic endocrine progenitors remain during postnatal life, and that this population can be expanded and give rise to new beta cells under certain injury conditions (Gu *et al.*, 2002). Others have failed to detect such a contribution, leaving the issue open at this point (Lee *et al.*, 2006).

9. Dedifferentiation of Beta Cells

Perhaps most interestingly, it has been suggested recently that differentiated beta cells can undergo a process of epithelial to mesenchymal transition, giving rise to a proliferative population of insulin-negative cells capable in principle of redifferentiating into beta cells (Gershengorn *et al.*, 2004; Ouziel-Yahalom *et al.*, 2006). These studies are based on tissue culture experiments using human islets. Interestingly, *in vivo* evidence suggests that under certain conditions, beta cells can undergo significant phenotypic changes, including some that resemble epithelial to mesenchymal transition, as they proliferate (Kulkarni *et al.*, 2004). This proposal is exciting because it can potentially reconcile the view that beta cells originate from stem cells with the observation that beta cells derive from beta cells. Sophisticated lineage tracing experiments will be needed to address this possibility. Since according to this model beta cells derive from cells that express insulin, simple irreversible labeling of beta cells cannot distinguish between beta cell proliferation and beta cell dedifferentiation, proliferation, and redifferentiation. Most importantly, it has not been shown so far that beta cells can lose insulin expression and proliferate.

10. Summary and Perspective

Insulin-producing beta cells are a key target for current effort of regenerative medicine, as they could provide an effective cure for many diabetes patients. We believe that the cellular origins of beta cells must be identified for such efforts to be productive. It is clear that during embryonic development beta cells derive from progenitor cells expressing sequentially, among other genes, the transcription factors pdx1, ptf1a/p48, and ngn3. This understanding, based on robust lineage tracing experiments, provides one conceptual framework in which to direct the differentiation of stem cells, be it ESC or other types of embryonic cells, toward a beta cell fate. The cellular origins of new beta cells born postnataly are less clear. Multiple studies suggest that adult beta cells derive from stem cells residing in pancreatic ducts, acini, inside islets, or even in the bone marrow. However, these proposals are not supported so far by proper lineage tracing evidence. On the contrary, lineage analysis and other approaches have suggested that most postnatal beta cells are formed by simple duplication of preexisting, differentiated beta cells. It remains to be demonstrated using lineage tracing if, under certain injury conditions, beta cells can derive from a pool of "facultative" stem cells. One particularly attractive suggestion is that beta cells may undergo, *in vitro* and *in vivo*, a process of transient de-differentiation, perhaps via epithelial to mesenchymal transition.

Finally, all reliable evidence about the origins of beta cells has emerged from the use of transgenic mouse technology. It is acknowledged that human beta cells can in principle rely on a different mechanism for their renewal, or on a different balance between proliferation and stem cell differentiation. Performing lineage tracing experiments on humans was long considered impossible. However, ingenious approaches suggest now that it will be possible one day to determine the life history of human beta cells and their origins (Frumkin *et al.*, 2005; Spalding *et al.*, 2005).

11. Methods: Design of a Lineage Tracing Experiment in Mice

There are many approaches for lineage analysis, in mice and in other organisms, which use a wide variety of techniques. These include, among others:

• Direct labeling of particular cells with lipophilic dyes such as DiI, so that the fate of these cells and their immediate progenitors can be traced. An obvious disadvantage of this method is the twofold dilution of dye with every cell division.

• Labeling with retroviral infection, titrated such that each infection and integration event generates a unique identifiable mark. A major limitation of this method is the inability to prospectively control the identity of labeled cells.

• The injection of labeled cells into an embryo or adult animal. For example, the injection of labeled ESC into a blastocyst generates a chimeric mouse. In such an animal, the contribution of injected cells to a particular tissue can be assessed. Cells can also be injected during later stages of development, as well as in adults. A commonly used example of the latter is the transplantation of labeled bone marrow or hematopoietic stem cells to an irradiated recipient. After engraftment of the cells, their contribution to specific cell lineages, in the blood system or in solid organs, can be assessed based on marker expression. An alternative method for identification of donor cells, applicable also in humans, is the identification of Y chromosome by FISH in cases where male cells are injected into female recipients.

These methods have provided numerous important insights into problems of cellular origins and tissue dynamics. However, in recent years the Cre-lox technology has become the gold standard for lineage tracing experiments in mice, as it affords a greater control over the temporal and spatial parameters of cell labeling, including a precise selection of the labeled cell population. Here we describe some considerations when using the Cre-lox system for addressing a question about a cell of origin, similar to problems discussed in the sections dealing with pancreatic beta cells.

1. **Cre-lox system.** The basic feature of the Cre-lox system, described in detail elsewhere (Branda and Dymecki, 2004; Nagy, 2000; Rossant and McMahon, 1999), is the ability of Cre recombinase to detect DNA sequences flanked by specific 34-bp sequence called loxP. Cre activity removes the flanked sequence, and leaves a single loxP site in place. Importantly, Cre carries out this function without the need for any cofactors, allowing the use of the system in mammalian cells. This technology is mainly used for modern tissue-specific knockouts of genes in mice, but the same tools can be used to delete a transcriptional stop sequence such that a reporter gene starts to be expressed after recombination. This "indelible labeling" forms the basis for Cre-lox based lineage tracing experiments. Such experiments typically involve the use to "double transgenic" mice, containing both a Cre-expressing transgene and a Cre reporter transgene.

2. **Tissue specificity of Cre recombinase.** A major strength of the Cre-lox system is in the ability to select the cell type to be labeled based on its expression pattern, rather than its spatial location. Therefore, the system is best suitable for addressing problems of the type "do cells expressing the gene X contribute to cells of tissue Y?" This feature of the Cre-lox system relies on the use of tissue specific promoters to drive expression of Cre. As a result, the usefulness of a particular Cre-expressing transgenic mouse is determined by the specificity of the promoter driving Cre expression. The more accurate the expression of Cre (in recapitulating expression of the endogenous gene) the better. In certain applications, this is the key issue, for example, when attempting to label a rare cell type for which antibodies do not exist. In these cases, knocking-in Cre into the endogenous locus of the relevant gene is preferred. Following such a manipulation, it is guaranteed that Cre expression is subject to same regulatory mechanisms as the endogenous gene. A good example for such an approach in the case of the pancreas was provided by Kawaguchi et al., who knocked-in Cre into the ptf1a/p48 locus, and indeed were able to faithfully trace the fate of ptf1a/p48-expressing cells (Kawaguchi et al., 2002).

3. **Temporal control over Cre activity.** A major advance in the Cre-lox system was introduced with the development of Cre variants whose activity depend on an injected ligand. The most popular version is CreER™, a fusion of Cre recombinase and a mutated, tamoxifen-responsive estrogen receptor (Danielian et al., 1998). This fusion protein resides in the cytoplasm, and can access DNA only in the presence of tamoxifen which leads to a transient nuclear translocation. In transgenic mice that express CreER™, tamoxifen injection leads to a "pulse" of Cre activity, lasting about 48 h (Gu et al., 2003). The temporal control over recombination afforded by CreER™ was used in the case of the pancreas to precisely define the fate of progenitor cells in the embryonic pancreas (Gu et al., 2003). In addition, pulse-labeling of differentiated beta cells was used to demonstrate that new beta cells during adult life derive mainly from the proliferation of preexisting beta cells (Dor et al., 2004).

4. **Cre reporter.** There are numerous transgenic strains that express a reporter gene (e.g., lacZ, EGFP, luciferase, HPAP) depending on Cre activity. Reporters are designed such that only after Cre-mediated recombination will a "stop"

cassette be removed, allowing for strong expression of the reporter. Some of the most popular reporters were inserted to the constitutively active ROSA26 locus, with a "lox-stop-lox" (LSL) cassette separating the reporter from the promoter (Novak *et al.*, 2000). Other useful reporters include "Z/AP" and "Z/EG", where Cre-mediated recombination leads to removal of a lacZ gene, serving as a stop sequence, and the permanent activation of HPAP and EGFP, respectively (Lobe *et al.*, 1999; Novak *et al.*, 2000).

References

Ahlgren, U., Jonsson, J., and Edlund, H. (1996). The morphogenesis of the pancreatic mesenchyme is uncoupled from that of the pancreatic epithelium in IPF1/PDX1-deficient mice. *Development* **122,** 1409–1416.

Ahlgren, U., Jonsson, J., Jonsson, L., Simu, K., and Edlund, H. (1998). Beta-cell-specific inactivation of the mouse Ipf1/Pdx1 gene results in loss of the beta-cell phenotype and maturity onset diabetes. *Genes Dev.* **12,** 1763–1768.

Bonner-Weir, S., Deery, D., Leahy, J. L., and Weir, G. C. (1989). Compensatory growth of pancreatic beta-cells in adult rats after short-term glucose infusion. *Diabetes* **38,** 49–53.

Bonner-Weir, S., Taneja, M., Weir, G. C., Tatarkiewicz, K., Song, K. H., Sharma, A., and O'Neil, J. J. (2000). *In vitro* cultivation of human islets from expanded ductal tissue. *Proc. Natl. Acad. Sci. USA* **97,** 7999–8004.

Bouwens, L., Wang, R. N., De Blay, E., Pipeleers, D. G., and Kloppel, G. (1994). Cytokeratins as markers of ductal cell differentiation and islet neogenesis in the neonatal rat pancreas. *Diabetes* **43,** 1279–1283.

Branda, C. S., and Dymecki, S. M. (2004). Talking about a revolution: The impact of site-specific recombinases on genetic analyses in mice. *Dev. Cell* **6,** 7–28.

Butler, A. E., Janson, J., Bonner-Weir, S., Ritzel, R., Rizza, R. A., and Butler, P. C. (2003). Beta-cell deficit and increased beta-cell apoptosis in humans with type 2 diabetes. *Diabetes* **52,** 102–110.

Chong, A. S., Shen, J., Tao, J., Yin, D., Kuznetsov, A., Hara, M., and Philipson, L. H. (2006). Reversal of diabetes in non-obese diabetic mice without spleen cell-derived beta cell regeneration. *Science* **311,** 1774–1775.

Danielian, P. S., Muccino, D., Rowitch, D. H., Michael, S. K., and McMahon, A. P. (1998). Modification of gene activity in mouse embryos in utero by a tamoxifen-inducible form of Cre recombinase. *Curr. Biol.* **8,** 1323–1326.

Delacour, A., Nepote, V., Trumpp, A., and Herrera, P. L. (2004). Nestin expression in pancreatic exocrine cell lineages. *Mech. Dev.* **121,** 3–14.

Dor, Y., Brown, J., Martinez, O. I., and Melton, D. A. (2004). Adult pancreatic beta-cells are formed by self-duplication rather than stem-cell differentiation. *Nature* **429,** 41–46.

Edlund, H. (2001). Developmental biology of the pancreas. *Diabetes* **50**(Suppl. 1), S5–S9.

Edlund, H. (2002). Pancreatic organogenesis—developmental mechanisms and implications for therapy. *Nat. Rev. Genet* **3,** 524–532.

Esni, F., Stoffers, D. A., Takeuchi, T., and Leach, S. D. (2004). Origin of exocrine pancreatic cells from nestin-positive precursors in developing mouse pancreas. *Mech. Dev.* **121,** 15–25.

Frumkin, D., Wasserstrom, A., Kaplan, S., Feige, U., and Shapiro, E. (2005). Genomic variability within an organism exposes its cell lineage tree. *PLoS Comput. Biol.* **1,** e50.

Gao, R., Ustinov, J., Korsgren, O., and Otonkoski, T. (2005). *In vitro* neogenesis of human islets reflects the plasticity of differentiated human pancreatic cells. *Diabetologia.*

Georgia, S., and Bhushan, A. (2004). Beta cell replication is the primary mechanism for maintaining postnatal beta cell mass. *J. Clin. Invest.* **114,** 963–968.

Gershengorn, M. C., Hardikar, A. A., Wei, C., Geras-Raaka, E., Marcus-Samuels, B., and Raaka, B. M. (2004). Epithelial-to-mesenchymal transition generates proliferative human islet precursor cells. *Science* **306,** 2261–2264.

Gradwohl, G., Dierich, A., LeMeur, M., and Guillemot, F. (2000). neurogenin3 is required for the development of the four endocrine cell lineages of the pancreas. *Proc. Natl. Acad. Sci. USA* **97,** 1607–1611.

Gu, G., Brown, J. R., and Melton, D. A. (2003). Direct lineage tracing reveals the ontogeny of pancreatic cell fates during mouse embryogenesis. *Mech. Dev.* **120,** 35–43.

Gu, G., Dubauskaite, J., and Melton, D. A. (2002). Direct evidence for the pancreatic lineage: NGN3 + cells are islet progenitors and are distinct from duct progenitors. *Development* **129,** 2447–2457.

Guz, Y., Nasir, I., and Teitelman, G. (2001). Regeneration of pancreatic beta cells from intra-islet precursor cells in an experimental model of diabetes. *Endocrinology* **142,** 4956–4968.

Halban, P. A. (2004). Cellular sources of new pancreatic beta cells and therapeutic implications for regenerative medicine. *Nat. Cell Biol.* **6,** 1021–1025.

Hayashi, S., and McMahon, A. P. (2002). Efficient recombination in diverse tissues by a tamoxifen-inducible form of Cre: A tool for temporally regulated gene activation/inactivation in the mouse. *Dev. Biol.* **244,** 305–318.

Herrera, P. L. (2000). Adult insulin- and glucagon-producing cells differentiate from two independent cell lineages. *Development* **127,** 2317–2322.

Herrera, P. L., Huarte, J., Zufferey, R., Nichols, A., Mermillod, B., Philippe, J., Muniesa, P., Sanvito, F., Orci, L., and Vassalli, J. D. (1994). Ablation of islet endocrine cells by targeted expression of hormone-promoter-driven toxigenes. *Proc. Natl. Acad. Sci. USA* **91,** 12999–13003.

Hess, D., Li, L., Martin, M., Sakano, S., Hill, D., Strutt, B., Thyssen, S., Gray, D. A., and Bhatia, M. (2003). Bone marrow-derived stem cells initiate pancreatic regeneration. *Nat. Biotechnol.* **21,** 763–770.

Hunziker, E., and Stein, M. (2000). Nestin-expressing cells in the pancreatic islets of Langerhans. *Biochem. Biophys. Res. Commun.* **271,** 116–119.

Ianus, A., Holz, G. G., Theise, N. D., and Hussain, M. A. (2003). *In vivo* derivation of glucose-competent pancreatic endocrine cells from bone marrow without evidence of cell fusion. *J. Clin. Invest.* **111,** 843–850.

Jonsson, J., Carlsson, L., Edlund, T., and Edlund, H. (1994). Insulin-promoter-factor 1 is required for pancreas development in mice. *Nature* **371,** 606–609.

Kassem, S. A., Ariel, I., Thornton, P. S., Scheimberg, I., and Glaser, B. (2000). Beta-cell proliferation and apoptosis in the developing normal human pancreas and in hyperinsulinism of infancy. *Diabetes* **49,** 1325–1333.

Kawaguchi, Y., Cooper, B., Gannon, M., Ray, M., MacDonald, R. J., and Wright, C. V. (2002). The role of the transcriptional regulator Ptf1a in converting intestinal to pancreatic progenitors. *Nat. Genet.* **32,** 128–134.

Kim, S. K., and MacDonald, R. J. (2002). Signaling and transcriptional control of pancreatic organogenesis. *Curr. Opin. Genet. Dev.* **12,** 540–547.

Kodama, S., Kuhtreiber, W., Fujimura, S., Dale, E. A., and Faustman, D. L. (2003). Islet regeneration during the reversal of autoimmune diabetes in NOD mice. *Science* **302,** 1223–1227.

Kulkarni, R. N., Jhala, U. S., Winnay, J. N., Krajewski, S., Montminy, M., and Kahn, C. R. (2004). PDX-1 haploinsufficiency limits the compensatory islet hyperplasia that occurs in response to insulin resistance. *J. Clin. Invest.* **114,** 828–836.

Kushner, J. A., Ciemerych, M. A., Sicinska, E., Wartschow, L. M., Teta, M., Long, S. Y., Sicinski, P., and White, M. F. (2005). Cyclins D2 and D1 are essential for postnatal pancreatic beta-cell growth. *Mol. Cell Biol.* **25,** 3752–3762.

Lechner, A., Leech, C. A., Abraham, E. J., Nolan, A. L., and Habener, J. F. (2002). Nestin-positive progenitor cells derived from adult human pancreatic islets of Langerhans contain side population (SP) cells defined by expression of the ABCG2 (BCRP1) ATP-binding cassette transporter. *Biochem. Biophys. Res. Commun.* **293,** 670–674.

Lechner, A., Yang, Y. G., Blacken, R. A., Wang, L., Nolan, A. L., and Habener, J. F. (2004). No evidence for significant transdifferentiation of bone marrow into pancreatic beta-cells *in vivo*. *Diabetes* **53,** 616–623.

Lee, C. S., De Leon, D. D., Kaestner, K. H., and Stoffers, D. A. (2006). Regeneration of pancreatic islets after partial pancreatectomy in mice does not involve the reactivation of neurogenin-3. *Diabetes* **55,** 269–272.

Lobe, C. G., Koop, K. E., Kreppner, W., Lomeli, H., Gertsenstein, M., and Nagy, A. (1999). Z/AP, a double reporter for cre-mediated recombination. *Dev. Biol.* **208,** 281–292.

Mathews, V., Hanson, P. T., Ford, E., Fujita, J., Polonsky, K. S., and Graubert, T. A. (2004). Recruitment of bone marrow-derived endothelial cells to sites of pancreatic beta-cell injury. *Diabetes* **53,** 91–98.

Means, A. L., Meszoely, I. M., Suzuki, K., Miyamoto, Y., Rustgi, A. K., Coffey, R. J., Jr., Wright, C. V., Stoffers, D. A., and Leach, S. D. (2005). Pancreatic epithelial plasticity mediated by acinar cell transdifferentiation and generation of nestin-positive intermediates. *Development* **132,** 3767–3776.

Meier, J. J., Lin, J. C., Butler, A. E., Galasso, R., Martinez, D. S., and Butler, P. C. (2006). Direct evidence of attempted beta cell regeneration in an 89-year-old patient with recent-onset type 1 diabetes. *Diabetologia* **49,** 1838–1844.

Messier, B., and Leblond, C. P. (1960). Cell proliferation and migration as revealed by radioautography after injection of thymidine-H3 into male rats and mice. *Am. J. Anat.* **106,** 247–285.

Minami, K., Okuno, M., Miyawaki, K., Okumachi, A., Ishizaki, K., Oyama, K., Kawaguchi, M., Ishizuka, N., Iwanaga, T., and Seino, S. (2005). Lineage tracing and characterization of insulin-secreting cells generated from adult pancreatic acinar cells. *Proc. Natl. Acad. Sci. USA* **102,** 15116–15121.

Montanya, E., Nacher, V., Biarnes, M., and Soler, J. (2000). Linear correlation between beta-cell mass and body weight throughout the lifespan in Lewis rats: Role of beta-cell hyperplasia and hypertrophy. *Diabetes* **49,** 1341–1346.

Murtaugh, L. C., and Melton, D. A. (2003). Genes, signals, and lineages in pancreas development. *Annu. Rev. Cell Dev. Biol.* **19,** 71–89.

Nagy, A. (2000). Cre recombinase: The universal reagent for genome tailoring. *Genesis* **26,** 99–109.

Nishio, J., Gaglia, J. L., Turvey, S. E., Campbell, C., Benoist, C., and Mathis, D. (2006). Islet recovery and reversal of murine type 1 diabetes in the absence of any infused spleen cell contribution. *Science* **311,** 1775–1778.

Novak, A., Guo, C., Yang, W., Nagy, A., and Lobe, C. G. (2000). Z/EG, a double reporter mouse line that expresses enhanced green fluorescent protein upon Cre-mediated excision. *Genesis* **28,** 147–155.

Offield, M. F., Jetton, T. L., Labosky, P. A., Ray, M., Stein, R. W., Magnuson, M. A., Hogan, B. L., and Wright, C. V. (1996). PDX-1 is required for pancreatic outgrowth and differentiation of the rostral duodenum. *Development* **122,** 983–995.

Ouziel-Yahalom, L., Zalzman, M., Anker-Kitai, L., Knoller, S., Bar, Y., Glandt, M., Herold, K., and Efrat, S. (2006). Expansion and redifferentiation of adult human pancreatic islet cells. *Biochem. Biophys. Res. Commun.* **341,** 291–298.

Rane, S. G., Dubus, P., Mettus, R. V., Galbreath, E. J., Boden, G., Reddy, E. P., and Barbacid, M. (1999). Loss of Cdk4 expression causes insulin-deficient diabetes and Cdk4 activation results in beta-islet cell hyperplasia. *Nat. Genet.* **22,** 44–52.

Ritzel, R. A., and Butler, P. C. (2003). Replication increases beta-cell vulnerability to human islet amyloid polypeptide-induced apoptosis. *Diabetes* **52,** 1701–1708.

Rossant, J., and McMahon, A. (1999). "Cre"-ating mouse mutants-a meeting review on conditional mouse genetics. *Genes Dev.* **13,** 142–145.

Scaglia, L., Cahill, C. J., Finegood, D. T., and Bonner-Weir, S. (1997). Apoptosis participates in the remodeling of the endocrine pancreas in the neonatal rat. *Endocrinology* **138,** 1736–1741.

Selander, L., and Edlund, H. (2002). Nestin is expressed in mesenchymal and not epithelial cells of the developing mouse pancreas. *Mech. Dev.* **113,** 189–192.

Shapiro, A. M., Lakey, J. R., Ryan, E. A., Korbutt, G. S., Toth, E., Warnock, G. L., Kneteman, N. M., and Rajotte, R. V. (2000). Islet transplantation in seven patients with type 1 diabetes mellitus using a glucocorticoid-free immunosuppressive regimen. *N. Engl. J. Med.* **343,** 230–238.

Skau, M., Pakkenberg, B., Buschard, K., and Bock, T. (2001). Linear correlation between the total islet mass and the volume-weighted mean islet volume. *Diabetes* **50,** 1763–1770.

Slack, J. M. (1995). Developmental biology of the pancreas. *Development* **121,** 1569–1580.

Spalding, K. L., Bhardwaj, R. D., Buchholz, B. A., Druid, H., and Frisen, J. (2005). Retrospective birth dating of cells in humans. *Cell* **122,** 133–143.

Street, C. N., Lakey, J. R., Seeberger, K., Helms, L., Rajotte, R. V., Shapiro, A. M., and Korbutt, G. S. (2004). Heterogenous expression of nestin in human pancreatic tissue precludes its use as an islet precursor marker. *J. Endocrinol.* **180,** 213–225.

Suri, A., Calderon, B., Esparza, T. J., Frederick, K., Bittner, P., and Unanue, E. R. (2006). Immunological reversal of autoimmune diabetes without hematopoietic replacement of beta cells. *Science* **311,** 1778–1780.

Treutelaar, M. K., Skidmore, J. M., Dias-Leme, C. L., Hara, M., Zhang, L., Simeone, D., Martin, D. M., and Burant, C. F. (2003). Nestin-lineage cells contribute to the microvasculature but not endocrine cells of the islet. *Diabetes* **52,** 2503–2512.

Uchida, T., Nakamura, T., Hashimoto, N., Matsuda, T., Kotani, K., Sakaue, H., Kido, Y., Hayashi, Y., Nakayama, K. I., White, M. F., and Kasuga, M. (2005). Deletion of Cdkn1b ameliorates hyperglycemia by maintaining compensatory hyperinsulinemia in diabetic mice. *Nat. Med.* **11,** 175–182.

Zulewski, H., Abraham, E. J., Gerlach, M. J., Daniel, P. B., Moritz, W., Muller, B., Vallejo, M., Thomas, M. K., and Habener, J. F. (2001). Multipotential nestin-positive stem cells isolated from adult pancreatic islets differentiate *ex vivo* into pancreatic endocrine, exocrine, and hepatic phenotypes. *Diabetes* **50,** 521–533.

CHAPTER 8

Pluripotent Stem Cells from Germ Cells: Derivation and Maintenance

Candace L. Kerr, Michael J. Shamblott, and John D. Gearhart

Department of Obstetrics and Gynecology
Institute for Cell Engineering
Johns Hopkins University
Baltimore, Maryland

Abstract
1. Introduction
2. Germ Cell Development
3. EGC Derivation
   3.1. Overview
   3.2. Protocol for deriving EGCs
   3.3. Feeder layer
4. Characterization of EG Cultures
   4.1. Overview
   4.2. Protocols for immunostaining EGCs and human gonads
   4.3. Sex determination of human gonads by fluorescent *in situ* hybridization (FISH)
   4.4. Tissue nonspecific alkaline phosphatase and telomerase activity assays
   4.5. Cytogenetic analysis
5. EB Formation and Analysis
   5.1. Overview
   5.2. Protocol for embedding and immunohistochemistry
6. EB-Derived Cell Formation
   6.1. Overview
   6.2. Protocols for human EBD cells
References

Abstract

To date, stem cells have been derived from three sources of germ cells. These include embryonic germ cells (EGC), embryonal carcinoma cells (ECC), and multipotent germ-line stem cells (MGSC). EGCs are derived from primordial germ cells that arise in the late embryonic and early fetal period of development. ECCs are derived from adult testicular tumors while MGSCs have been derived by culturing spermatogonial stem cells from mouse neonates. For each of these lines, their pluripotency has been demonstrated by their contribution to all three germ layers *in vitro* and *in vivo via* chimeric animals, including germ-line transmission. These germ-line-derived stem cells have been generated in many species including human, mice, porcine, and chicken *albeit* with only slight modifications. This article describes general considerations regarding critical aspects of their derivation compared to their counterpart, embryonic stem cells (ESC). Detailed protocols for EGC derivation and maintenance from human and mouse will be discussed.

1. Introduction

Currently, three types of germ-line-derived stem cells have been identified including embryonic germ cells (EGC), embryonal carcinoma cells (ECC), and multipotent germ-line stem cells (MGSCs) (for review see Donovan and Gearhart, 2001; Smith, 2001; Turnpenny *et al.*, 2005b). Importantly, all three cell types demonstrate the properties of pluripotency including the ability of unlimited self-renewal *in vitro*, and they can give rise to derivatives of all three embryonic germ layers *in vitro* in the form of embryoid bodies (EBs) in culture and *in vivo via* chimeric contributions including in some cases, to the germ-line. ECC were the first of the three to be identified in the 1960s, originally in the mouse (Stevens, 1966) and then in human (for review see Andrews, 1998). ECCs are pluripotent cells derived from adult testicular teratomas (benign tumors) or teratocarcinomas (or malignant tumors) from which genetic, immunological, and morphological evidence suggest a primordial germ cell (PGC) origin (Stevens, 1967). However, unlike EGCs and MGSCs, ECCs are karyotypically unstable rendering their use in development studies confounded with genetic abnormalities and with very limited if any potential use in stem-cell-based therapies (Blelloch *et al.*, 2004). Nonetheless, the development and discovery of the origins of ECCs suggested that their cell type of origin, the PGCs retain an ability to become pluripotent given the appropriate cues.

PGCs are the progenitors of the germ cell lineage and are lineage restricted in that they do not survive long in culture, are unable to contribute to chimeras and do not form EBs in culture (Donovan and de Miguel, 2003). Nevertheless a decade later, culturing conditions were defined that converted mouse PGCs from nullipotent cells into pluripotent stem cells termed EGC (Matsui *et al.*, 1992; Resnick *et al.*, 1992). These conditions were then successfully recapitulated years later with the derivation of EGCs in human (Shamblott *et al.*, 1998), pig (Lee and Piedrahita, 2000; Mueller *et al.*, 1999; Piedrahita *et al.*, 1998; Shim *et al.*, 1997; Tsung *et al.*, 2003), and chicken (Park and Han, 2000). The final source of germ-line-derived stem cells has been recently produced by one group which has reported the conversion of spermatogonial stem cells isolated from mouse neonates into cells, MGCS, which under the appropriate culturing conditions display many of the pluripotent properties of ESCs. Mouse MGSCs demonstrated phenotypic characteristics of pluripotent stem cells, produced teratomas in mice, and formed germ-line chimeras when injected into blastocysts (Kanatsu-Shinohara *et al.*, 2004, 2005). Due to the paucity of information at this time regarding MGSCs and the limited therapeutic potential of ECCs, this chapter will primarily focus on EGC derivation from PGCs and their maintenance in culture. Indeed, the significance of using human EGCs (hEGCs) for cell-based therapies has been demonstrated in landmark papers where injected hEGC-derived cells have been used to restore partial motor recovery of rats with diffuse motor neuron injury (Kerr *et al.*, 2003) as well as into a mouse excitotoxic brain damage model (Mueller *et al.*, 2005) where hEGC-derived neural stem cells were able to partially restore the complement of striatum neurons in brain-damaged mice.

2. Germ Cell Development

PGCs are the progenitor cells of the germ cell lineage, which are the sole source of gametes in the adult. When strictly applied, the term PGCs refers to the diploid germ cell precursors that transiently exist in the embryo before they enter into associations with somatic cells of the gonad and become committed as germ cells. Much of the current knowledge of early mammalian germ cell development has been acquired by studies in the mouse. In the mouse, PGCs (mPGCs) are derived from a region of the epiblast that mainly gives rise to extraembryonic mesoderm. The fate of these cells into PGCs is determined by the proximity of these cells to extraembryonic ectoderm posterior during gastrulation. Here external signals are secreted which regulate PGC differentiation as demonstrated by the observation that transplantation of cells from other parts of the epiblast to this region results in the acquisition of a PGC fate (Lawson and Hage, 1994). As a result, mPGCs are first detected at ~7 days post coitum (dpc) as a small number of alkaline phosphatase (AP) positive cells near the base of the allantois. By 8.5 dpc, actively proliferating mPGCs become associated with the endoderm as it invaginates to form the hindgut and by 10.5 dpc, mPGCs are associated with dorsal mesenteries and begin infiltrating the genital ridges such that the majority of the cells (~25,000) have localized to the gonad by 12.5 dpc (Mintz and Russell, 1957; Ozdzenski, 1967; Tam and Snow, 1981). Overall, this movement of mPGCs into these areas appears to be caused by both cellular migration and association with moving tissues (Donovan *et al.*, 1986). Male and female mPGCs are indistinguishable and continue proliferation *via* mitosis until they arrive at the genital ridge at which time they are generally referred to as gonocytes. In female mice, more than 25,000 are detected by 13.5 dpc in the ovary where the association with somatic cells of the ovary induces mPGCs to enter prophase of the first meiotic division. However, in males this entry into meiosis is inhibited during fetal life by signals from the developing testis (Francavilla and Zamboni, 1985; Upadhyay and Zamboni, 1982). Instead, male PGCs once they reach the gonad continue to proliferate before undergoing mitotic arrest in G_1 until puberty.

A similar migration pattern is also seen for human PGCs (hPGCs). In humans, 50–100 PGCs first appear at $\sim$22 days. They appear in the same region as in mouse, in the endoderm of the dorsal wall of the yolk sac, near the allantois as well as in the mesenchyme of the stalk. From there, they proceed to migrate through the hindgut during the fourth week and dorsal mesentery in the fifth week to reach the genital ridge (Falin, 1969; McKay *et al.*, 1953; Witschi, 1948, 1963). By the end of the fifth week or early in the sixth week, it has been estimated that approximately 1000 hPGCs begin to actively migrate from the dorsal mesentery into the gonadal anlage (Makabe *et al.*, 1992; Motta *et al.*, 1997; Pinkerton *et al.*, 1961; Witschi, 1948). By the end of the seventh week the testis and ovary appear differentiated and germ cell proliferation is increased (Francavilla *et al.*, 1990; Rabinovici and Jaffe, 1990). At this time, in the female, premeiotic hPGCs are called oogonia which begin extensive mitotic expansion until their arrest in meiosis prophase 1 (Baker and Franchi, 1967; Makabe *et al.*, 1992; Sun and Gondos, 1984). This proliferative expansion results in $\sim$10,000 germ cells at 5–6 weeks to $\sim$600,000 by week 8 of gestation. Between 11 and 12 weeks gestation meiosis in the female begins and is concomitant with the loss of AP activity in the primary oocytes (Francavilla *et al.*, 1990; Gondos *et al.*, 1986; Ohno *et al.*, 1962; Pinkerton *et al.*, 1961; Skrzypczak *et al.*, 1981). In contrast, at 8 weeks male PGCs are now called prospermatogonia which exhibit much less mitotic expansion then in the female (at 9 weeks $\sim$30,000 have been reported) (Bendsen *et al.*, 2003; Francavilla *et al.*, 1990; Heyn *et al.*, 1998) but continue to divide until they are arrested in mitosis between 16 and 18 weeks gestation (Gondos, 1971).

3. EGC Derivation

3.1. Overview

Unlike their EGC descendents, PGCs are not pluripotent and as such do not survive past one week under standard tissue culture conditions. Early attempts to use various growth factors and feeder layers succeeded in prolonging mPGC survival, but proliferation was limited (Dolci *et al.*, 1991; Godin and Wylie, 1991; Matsui *et al.*, 1991). These factors included using a combination of mitotically inactivated mouse fibroblast feeder cells and a synergistic action of stem cell factor (SCF) and leukemia inhibitory factor (LIF). Although these conditions promoted survival and proliferation of mPGCs, it was ultimately, the addition of *β*Fgf (basic fibroblast growth factor; FGF-2) which resulted in the conversion of nullipotent mPGCs into pluripotent mEGCs (Matsui *et al.*, 1992; Resnick *et al.*, 1992). FGF-2 functions as a potent mitogen in many cell types and induces telomerase activity in neural precursor cells (De Felici *et al.*, 1993). In this case, FGF-2 along with SCF and feeder layer triggered differentiated mPGCs to proliferate from their single migratory state into multicellular colonies comprised of pluripotent, self-renewing cells similar to mouse embryonic stem cells (mESC) which were established a decade earlier (Evans and Kaufman, 1981; Martin, 1981).

Six years after the first mouse EGC cell lines were described, the first case of hEGC derivation was reported (Shamblott *et al.*, 1998). To derive hEGCs, hPGCs are isolated from the fetal gonad anywhere between 5 and 10 weeks gestation (obtained as a result of therapeutic termination pregnancies). This time period occurs during peak PGC proliferation and also encompasses the period in which the gonad undergoes sexual dimorphism into either a committed ovary or testis starting by the seventh week (Bendsen *et al.*, 2003; Heyn *et al.*, 1998; Wartenberg, 1982; Witschi, 1963). To date, four laboratories have reported successful derivation of hEGC lines (Liu *et al.*, 2004; Park *et al.*, 2004; Shamblott *et al.*, 1998; Turnpenny *et al.*, 2003).

The derivation of EGCs from human as well as from pig and chicken has been performed by employing methods from the original EGC derivation in the mouse. All laboratories relied on culturing PGCs on a feeder layer and all but chicken used one of two mouse embryonic fibroblast (MEF) lines, or **S**andoz **T**hioguanine- and **O**uabain-resistant mouse fibroblasts (STO) cells, that were either irradiated or mitomycin-C treated. For chicken EGCs, the gonadal tissue from the explants served as a feeder (Park and Han, 2000). The growth media varies depending on the species but all derivations have reportedly relied on the addition of LIF and FGF-2. Soluble stem cell factor (sSCF), also known as c-kit ligand, steel factor, or mast cell factor, was also added for mouse, pig, and chicken EGC derivation while forskolin was added for mouse and human. Forskolin is a pharmacological agent which raises intracellular cAMP levels and has been shown to stimulate mitosis in mPGCs culture (Dolci *et al.*, 1993). LIF was originally known for its inhibitory role in liver cell differentiation (Smith *et al.*, 1988) and employed for the derivation of mESC where signaling *via* the LIF receptor (LIFR), gp130 and intracellular stat3b demonstrated a significant role in maintaining mESC pluripotency (Niwa *et al.*, 1998). Interestingly, activation of this pathway does not maintain self-renewal in human ESCs but is required for hEGC culture. SCF mutations and mutations of the SCF receptor, c-kit revealed direct roles in the proliferation and maintenance of mPGCs during their migration to the gonad as well as in culture (reviewed in (Donovan and de Miguel, 2003). Although our laboratory has not seen an effect of soluble SCF in our cultures of hPGCs we have seen a positive correlation in hPGC proliferation and survival in culture with the expression of higher concentrations of transmembrane SCF present on subcloned feeder cells (Shamblott *et al.*, 2004).

As important as the culturing environment, is the stage in development at which PGCs are selected for EGC derivation. For this, a general consensus is that EGC derivation is most efficient from PGCs isolated prior to or during their migration although it is possible to acquire lines from PGCs until 12.5 dpc in the mouse (McLaren and Durcova-Hills, 2001) and up to 10 weeks gestation in human. This is significant in that a population of PGCs still exist that retain their ability of become EGCs in culture even after their gonadal niche has undergone sexual differentiation and development of the sex cords has occurred. However, it is not known whether derivation occurs here as a result of reprogramming PGCs at later stages of differentiation or whether it is a result of a few PGCs lagging in their differentiation. In either case, we have noted considerably lower efficiencies of EGC derivation from PGCs at later stages from both mouse and human.

3.2. Protocol for deriving EGCs

Although many combinations of cytokines and feeder layers have been evaluated by our laboratory, the following protocol is similar for both human and mouse ESC derivation. The few exceptions are noted in the text where they are applied.

3.2.1. Growth media components

The growth media used to derive and maintain both mouse and hEGCs is Dulbecco's modified Eagle's medium (Invitrogen) supplemented with 15% fetal bovine serum (FBS, Hyclone), 0.1-mM nonessential amino acids (Invitrogen), 0.1-mM 2-mercaptoethanol (Sigma), 2-mM glutamine (Invitrogen), 1-mM sodium pyruvate (Invitrogen), 100 U/ml penicillin (Invitrogen), and 100 μg/ml streptomycin (Invitrogen). For hEGC derivation 1000 U/ml human recombinant LIF (hrLIF, Chemicon), 1–2 ng/ml human recombinant FGF-2 (hrFGF-2, R&D systems) and 10-μM forskolin (Sigma) prepared in dimethyl sulfoxide (DMSO) were added. This media was also applied to mouse EGC derivation except that mouse 1000 U/ml LIF (ESGRO, Chemicon) was used in place of hrLIF. Another notable difference between derivations in these two species is that while sSCF can also be added to enhance mouse EGC derivation there is no effect of soluble SCF on hEGC culture. However, our laboratory has noted the requirement of transmembrane SCF on the feeder layer for hEGC derivation (Shamblott *et al.*, 2004).

3.2.2. Initial disaggregation and plating

1. For mouse EGC derivation, multiple mouse strains have been employed including 129/SV, ICR, and C57B1/6 from which mPGCs can be derived between 8 and 12.5 dpc (Durcova-Hills *et al.*, 2001; Labosky *et al.*, 1994; Matsui *et al.*, 1992;

Resnick *et al.*, 1992; Stewart *et al.*, 1994). At 8.5 dpc the lower third portion of mouse embryo starting posteriorly from the last somites are collected. At 9.5 dpc portions of the hindgut are isolated (Durcova-Hills *et al.*, 2001), while at 10.5 and 12.5 dpc genital ridges are utilized (Matsui *et al.*, 1992). The remainder of the protocol is identical to hEGC derivation with some noted exceptions.

2. For hEGC derivation, gonadal ridges and mesenteries from 5 to 10 weeks gestation are collected (as results of therapeutic termination of pregnancies using a protocol approved by the Joint Committee on Clinical Investigation of the Johns Hopkins University School of Medicine) in 1-ml ice-cold growth media.

3. The tissue is then placed in a three-well depression slide and soaked in calcium–magnesium-free Dulbecco's phosphate-buffered saline (DPBS) for 5 min then transferred to 100 μl trypsin–EDTA solution. The concentration of trypsin and EDTA is varied such that at the earliest developmental stages, a gentler 0.05% trypsin–0.5-mM EDTA is used, and at later developmental stages, a stronger 0.25% trypsin–0.5-mM EDTA solution is used.

4. The tissue is then mechanically disaggregated using fine forceps and iris scissors for 5–10 min at room temperature.

5. Next, sample is placed in a sterile 1.5 ml centrifuge tube and placed at 37 °C for 5–10 min (water bath or incubator). This disaggregation process often results in a single-cell suspension with large pieces of undigested tissue.

6. To stop the digestion, serum containing growth media is added to the tube. The digested tissue is then triturated again ~30–50 times with a 200 μl pipetman and tip.

7. 100 μl volumes of the sample is transferred to each well of a 96-well tissue culture plate previously prepared with the feeder layer (described in the later text).

8. Usually the initial plating occupies ~4–10 wells of the 96-well plate per human gonad or 0.5–1 mouse embryo per well for mEGC derivation. Plating densities are critical. Too low or too high reduces derivation efficiency.

9. The plate is incubated at 37 °C in 5% CO_2 with 95% humidity for 7 days. Approximately 90% of the growth media is removed each day, and the plate is replenished with fresh growth media containing LIF, forskolin, and FGF-2.

3.2.3. Subsequent passage of EGC cultures

In the first 7 days of derivation (passage 0), most mouse and hEGC cultures do not produce visible EGC colonies. Staining for tissue nonspecific alkaline phosphatase (TnAP) activity demonstrates the presence of solitary PGCs with either stationary or migratory morphology (Fig. 8.1A). Often, colonies of cells that do not express TnAP activity are seen (Fig. 8.1B and C), as are small clumps of tissue remaining from the initial disaggregation (Fig. 8.1D and E). After 7 days, mouse and hPGC cultures are subcultured onto a new feeder. Timing is critical here since after 7 days their mitotically inactivated feeders begin to undergo significant

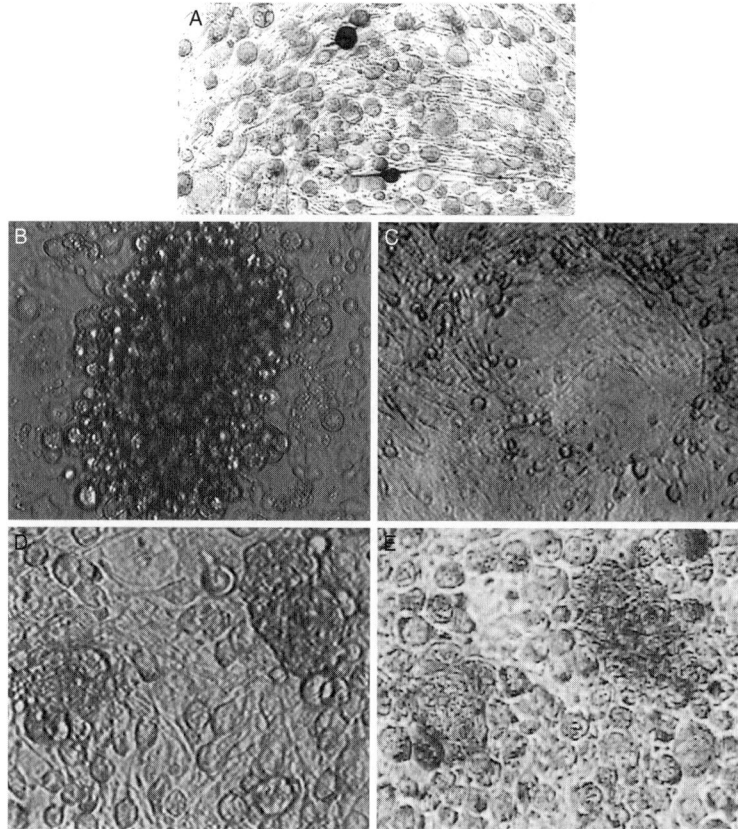

Figure 8.1 (A) TnAP staining of a single stationary (on top) and migratory (on bottom) hPGC in primary culture on a STO feeder layer. (B) Multicellular clump of gonadal tissue. (C) and (D) Flat and rounded cell colonies that do not lead to hEGC colonies and are TnAP⁻. (E) Negative staining for TnAP activity of colonies from (D).

apoptosis. Cell density from the specimens also increases at this time leading to overcrowding.

1. First media is removed and the wells are gently rinsed with calcium–magnesium-free DPBS.

2. Then, 40 μl of freshly thawed 0.05% trypsin–0.5-mM EDTA, 0.25% trypsin–0.5-mM EDTA, or a mixture of these two solutions is added to each well, and the plate is incubated on a heated platform or in a tissue culture incubator for 5 min at 37 °C. The important point at this stage is to facilitate the complete disaggregation of cells, which can be a significant challenge.

3. After incubation, a 200 μl pipetman and tip is used to scrape the bottom of the wells followed by gentle trituration ~20–30 times.

4. After the samples have been loosened, fresh growth media is added to each well, and the contents are triturated another 10–30 times. This phase is critical to successful disaggregation of STO feeder layer and EGC.

5. Each well is divided in half and placed into twice the number of wells containing freshly prepared feeders.

All subsequent passages are repeated as described above. After 2–3 weeks (during passage 1 or 2), large and recognizable EGC colonies will arise in some of the wells from both mouse and human (Fig. 8.2). At this point, wells that do not have EGC colonies are discarded. Interestingly, in our hands approximately 50% of the wells on average initially produce EGC colonies from both mouse and human. For generating mouse EGCs it is critical that once EGC colony formation occurs, FGF-2 is taken out of the media. Otherwise overgrowth of "nonEGCs" produces an adhesive monolayer of AP$^-$ cells and in subsequent passages EGC

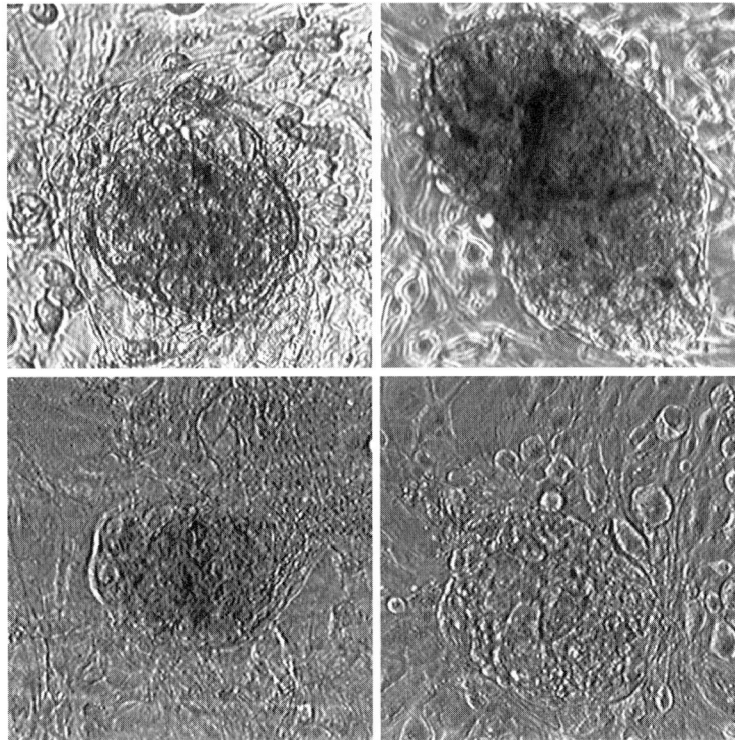

Figure 8.2 hEGC colonies from male and female gonads 6–10 weeks post conceptus on STO cell feeder layers. Colonies appear after 2–3 weeks of culture.

colonies disappear. However in humans, FGF-2 is required to sustain colonies which do not seem to produce the same overgrowth of unwanted cells as seen in the mouse cultures.

3.2.4. Troubleshooting

Several common problems occur during the passage of hEGCs that do not occur with mEGCs. One observation is that hEGC colonies do not fully disaggregate. The consequences of poor disaggregation are that the large pieces differentiate or die and fewer hEGCs are available for continued culture expansion. Although much effort has been expended to find a solution to this problem, it remains the most difficult aspect and challenging hurdle to hEGC biology. To gain some insight into this problem, a series of electron microscopic images were taken to compare the cell–cell interactions found in mouse ESC, mouse EGC, and hEGC colonies. It is evident from these images that cells within the hEGC colonies adhere more completely to each other than cells within mouse ESC and EGC colonies (Fig. 8.3). It is possible that this tight association within the colony limits the access of disaggregation reagents. At this time, neither the nature of the cell–cell interactions nor an effective solution to this problem is evident.

Because of incomplete disaggregation and other intrinsic or extrinsic signals, many hEGC colonies (10–30% per passage) differentiate to form three-dimensional structures termed EBs (Fig. 8.4A-C) or flatten into TnAP⁻ aggregates that no longer continue to proliferate (Fig. 8.4D and E). hEGC colonies that are more fully disaggregated go on to produce new colonies which under the best circumstances can routinely exceed 20 passages. Inevitably, large hEGC colonies are removed from the culture as a result of EB formation and as the cultures become sparse are discontinued for practical considerations. Efforts employing standard DMSO cryopreservation techniques have so far been unsuccessful.

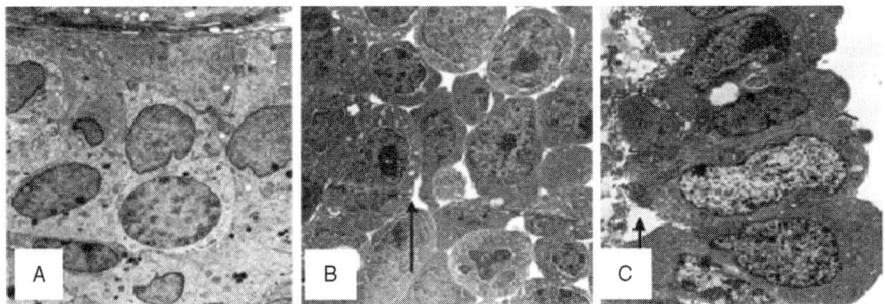

Figure 8.3   Electron micrograph of EGC versus ECC colonies. (A) hEGC colony, (B) mouse EGC colony, and (C) and mouse ESC colony. Arrows demonstrate areas of reduced adhesion which are not present in hEGC colonies contributing to the complexity of culturing hEGCs.

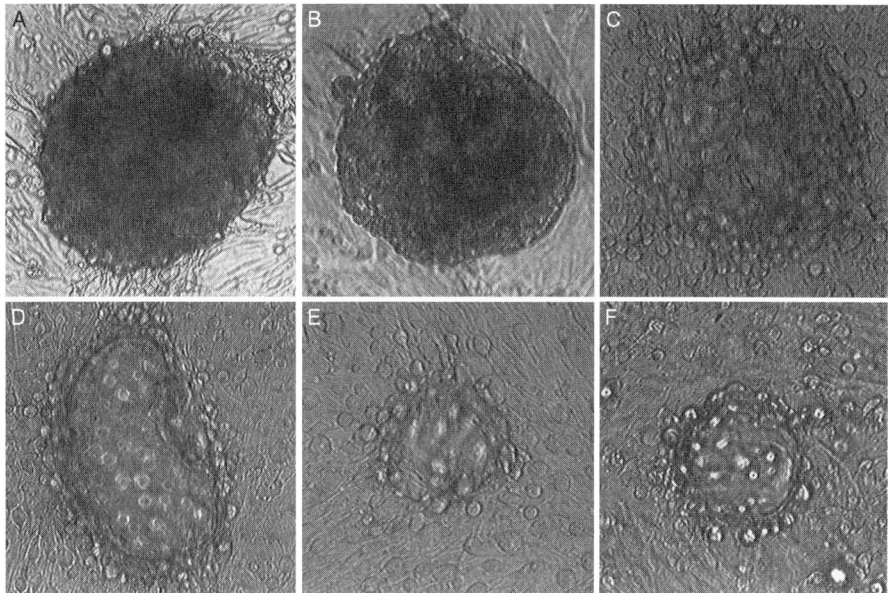

Figure 8.4 Embryoid body (EB) formation and differentiation of hEGC colonies. (A, B) hEGC colonies after incomplete disaggregation produce large three-dimensional structure called EBs on top of STO feeder layer. (C) Subsequently, EBs begin to flatten and in (D–F) eventually form cystic EBs. These colonies loose TnAP activity and do not survive after disaggregation and replating.

3.3. Feeder layer

3.3.1. Overview

EGC derivation is highly dependent on a feeder layer. Mouse and hEGCs have been derived by using primary MEF (PMEF) and STO. The majority of reports for both mEGC and hEGC derivation and those of porcine EGC derivation have utilized STO as feeders. We have also been most successful with STO cells. The factor or factors provided exclusively by this cell line are not fully understood. Although STO is a clonal cell line, individual isolates vary greatly in their ability to support hEGC derivation. This is further complicated by the known phenotypic variation of STO cells in continuous culture. Given the very limited supply of human tissue, it is prudent to screen STO cells for suitability prior to use. The most reliable screening method is to produce a number of clonal STO lines (by limiting dilution or cloning cylinder) and to evaluate them for their ability to support the derivation of mouse ESCs. However, the growth of existing mouse EGC lines is not a sufficient method, as most lines become feeder layer independent after derivation. Therefore our laboratory has screened several lines of STO for their ability to derive mouse EGCs and as expected the ability of a STO line to derive mouse EGCs directly reflected its ability to derive hEGCs. Furthermore, those lines which could not derive mouse EGCs could not derive hEGCs. Once a supportive STO fibroblast line is identified, it should be immediately cryopreserved in several low passage aliquots. One of these aliquots can then be expanded into multiple replicates and frozen for later use. After thawed, each aliquot of feeder cells is then used with limited further expansion. Continuous passage of STO fibroblasts without frequent screening should be avoided. After five passages of continuous expansion we discard the feeder and start a new culture.

3.3.2. Plating a STO feeder layer

For mitotic inactivation, the STO feeder layer can either be irradiated before plating or after plating. Most hEGC cultures are derived using the later method which requires a large γ-radiation unit.

3.3.2.1. Plating feeder cells prior to inactivation

1. STO cells are passaged for short periods (not continuously) in the EGC growth media without LIF, FGF-2, or forskolin and are disaggregated using 0.05% trypsin–EDTA solution.
2. One day prior to use, 96-well tissue culture dishes are coated with 0.1% gelatin for 30 min.
3. The gelatin is withdrawn, and 5×10^4 STO cells are plated per 96 well in EGC growth media without LIF, FGF-2, or forskolin. Similar cell densities ($\sim 1.5 \times 10^5$ cells/cm^2) can be achieved in other well configurations.
4. The cells are grown overnight and then exposed to 5000 rads (1 rad $^-$ 0.01 Gy) γ-radiation or X-ray. The cells are then returned to the tissue culture incubator until required.
5. Prior to use, the growth media is removed, 100 μl EGC growth medium with added factors is added to each well (i.e., half of the required well volume), and the dish is returned to the tissue culture incubator.

3.3.2.1. Plating feeder cells after inactivation

This method of STO cell preparation is used when a large γ-radiation unit is not available, when large amounts of cells are required, or if better control of STO cell density is desired.

1. STO cells are grown as described previously, trypsinized, counted, and resuspended in growth media without added factors.
2. The cells are placed into tubes and exposed to 5000 rads γ-radiation or X-ray.
3. Following exposure, cells are adjusted to a convenient concentration in growth media without added factors, counted, and plated into gelatinized tissue culture plates at $\sim 1.5 \times 10^5$ cells/cm^2.
4. Cells are allowed to adhere overnight.

4. Characterization of EG Cultures

4.1. Overview

EGCs are assessed on morphological, biochemical and/or immunocytological characteristics similar to ESCs. Morphologically, EGC cultures consist of tightly compacted multicellular colonies which in humans double or triple in size over 7–10 days (Shamblott *et al.*, 2004) and in mouse every 16 h (Lawson and Hage, 1994). In addition to morphological differences, EGCs as well as other germ-line-derived stem cells also express a number of pluripotent markers as well as telomerase activity. These include the stage-specific embryonic antigens (SSEA-1, SSEA-3, and SSEA-4) and tumor rejection antigen 1–60 and 1–81 (TRA-1–60, TRA-1–81). The pattern of expression is dependent in part on the species. For example, both mouse and hEGCs express SSEA-1 while hEGCs but not mEGCs also express SSEA-3 (*albeit* weakly) SSEA-4, TRA-1–60, and TRA-1–81 (Table 1). High levels of TnAP activity are associated with all EGCs such that mouse and hEGC colonies are >70–90% AP + (Fig. 8.5A and C). As colonies differentiate, loss of TnAP activity appears to occur first in cells around the periphery of the colony similar to the affect seen in ESC colony differentiation.

Nevertheless, TnAP activity and the expression of these cell surface markers are not exclusively expressed in pluripotent cells. As such the significance of the expression of these surface antigens remains unclear. While the presence of these markers is consistent in both mouse PGCs and EGCs, this does not appear to be the case for the human paradigm. Historically, the identification of hPGCs from somatic cells in the developing embryo have been largely restricted to TnAP activity and morphological characteristics including PGC rounded morphology, large eccentric nucleus, as well as the presence of glycogen particles and lipid droplets which stain with periodic acid staining (Bendsen *et al.*, 2003; McLaren,

Table 8.1
Expression profile of embryonic germ cell markers

Marker	Human EGC, ECC	Mouse EGC, ECC	Mouse MGSC	Pig EGC	Chicken EGC
AP	+	+	+	+	+
SSEA-1	+EGC	+	+	+	+
SSEA-3	+	−	ND	+	ND[a],+PGC
SSEA-4	+	−	ND	+	ND[a],+PGC
TRA-1–60	+	−	ND	ND	ND
TRA-1–81	+	−	ND	ND	ND
Oct4	+	+	+	+	ND
Nanog	+	+	+	ND	ND
hTERT	+	+	ND	ND	ND

ND[a], not determined

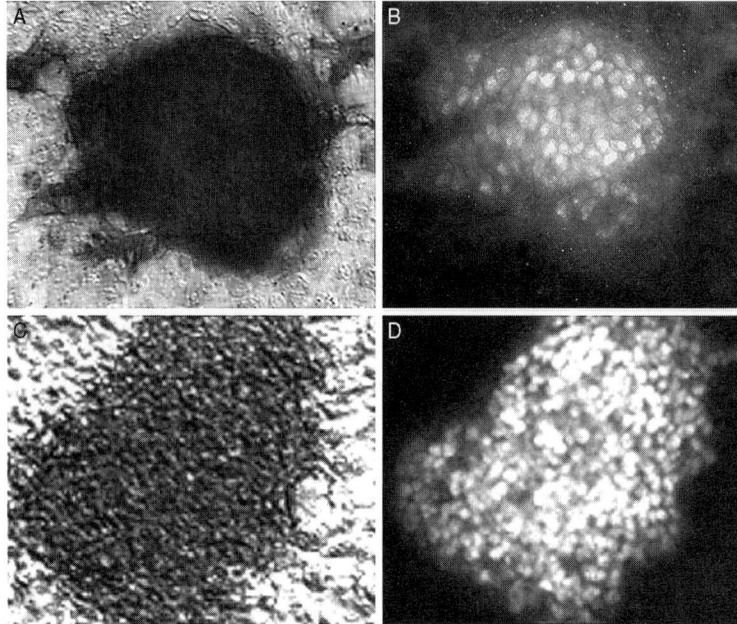

Figure 8.5 Dual labeling of EGC colonies for TnAP and Oct4 staining from human and mouse immunocytochemistry. (A) TnAP staining of a hEGC colony. (B) Indirect immunostaining of colony from (A) with Oct4 showing nuclear staining in most cells. (C) TnAP staining of a mouse EGC colony on feeder layer. (D) Oct4 staining of colony from (C). In both cases, the majority of the cells in these colonies are AP$^+$ and Oct4$^+$.

2003; Motta *et al.*, 1997; Pinkerton *et al.*, 1961). Most importantly, TnAP activity for this lineage is not indicative of pluripotency but that of the early germ cell lineage prior to meiosis. Furthermore, preliminary evidence in the immunological characterization of human gonads with stem cell markers reveal that in fact hPGCs stain SSEA-1 and SSEA-4 but do not express TRA-1–60 and TRA-1–81. Instead staining for the TRA antigens appears to be localized to the lining of the meso-nephric ducts in both male and female (data not shown). This is exemplified in Fig. 8.6 which depicts a male gonad ~7.5 week post conceptus, when PGCs are localized in the gonad and the gonad undergoes sexual dimorphism. Here a large increase in the number of PGCs staining positive for SSEA-1 and SSEA-4 in the sex cords can be seen. It is not known, however, whether the TRA antigens are a marker of the pluripotent conversion of hPGCs into hEGCs or an arbitrary artifact of their cell culture. This is particularly interesting in that the expression of the TRA antigens is also indicative of the pluripotency shared by human ECC and ESCs. Differences in marker expression between EGCs and PGCs has also been seen in the pig where porcine EGCs appear to express SSEA-1, SSEA-3, and SSEA-4 while PGCs express only SSEA-1 (Takagi *et al.*, 1997; Tsung *et al.*, 2003).

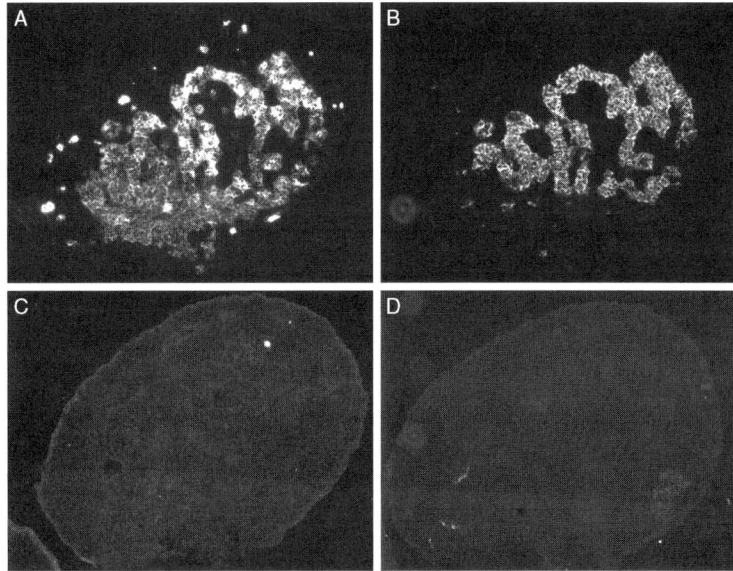

Figure 8.6 Indirect immunostaining of a middle cross-section of a male gonad at 53 days post conceptus. Cyrosections were stained with (A) Alexa^{564}-fluorescently labeled SSEA-4. Staining for SSEA4 is found in the developing rate and sex cords (B). Section from (A) dually stained with Alexa^{488}-fluorescently labeled SSEA1 localized to the developing sex cords. Adjacent sections were stained with either (C) Alexa^{564}-fluorescently labeled Tra-1–80 or with (D) Alexa^{564}-fluorescently labeled Tra-1–61.

In addition to cell surface markers, expression of certain transcription factors have also been associated with pluripotency. The most well-studied of these includes the POU domain transcription factor, Oct4, which has been associated with pluripotency in ESCs and its expression has also been detected in EGCs from human (Liu *et al.*, 2004; Park *et al.*, 2004; Turnpenny *et al.*, 2003) and mouse (Fig. 8.5B and D, respectively) (McLaren, 2003; Yeom *et al.*, 1996). To date, all of the direct evidence for the role of Oct4 in pluripotency comes from studies done in the mouse (for review see Boiani and Scholer, 2005). Here, studies have shown that during development Oct4 is originally expressed in all blastomeres and then becomes restricted to the ICM with downregulation in the trophoectoderm and the primitive endoderm (Nichols *et al.*, 1998; Niwa *et al.*, 1998; Pesce and Scholer, 2000; Scholer *et al.*, 1990). Evidence which supports a role of Oct4 in pluripotency comes from gene targeting studies in the mice which resulted in embryos devoid of a pluripotent ICM and which failed to give rise to ES colonies *in vitro* (Nichols *et al.*, 1998). Furthermore, quantitative analysis of Oct4 expression revealed that adherent levels of Oct4 expression caused differentiation in mESCs towards either extraembryonic mesoderm and endoderm lineages when increased, or trophectodermal cells when its expression was reduced (Niwa *et al.*, 1998). During gastrulation, Oct4 expression is progressively repressed in the epiblast and by 7.5 dpc is confined exclusively to newly established PGCs (Scholer *et al.*, 1990; Yeom *et al.*, 1996) and at maturity, to the developing germ cells (Pesce and Scholer, 2001; Pesce *et al.*, 1998). More recently, a new role for Oct4 in PGC survival has been reported based on a conditional gene targeting approach where restricted loss of Oct4 function in mouse PGCs resulted in significant apoptosis (Kehler *et al.*, 2004). Oct4 expression has also been detected in hPGCs (Goto *et al.*, 1999). However, the significance of this expression in hPGC development and survival is not known.

Like Oct4, nanog has also been associated with pluripotency in multiple species (Chambers *et al.*, 2003; Hatano *et al.*, 2005; Mitsui *et al.*, 2003; Yamaguchi *et al.*, 2005) including hEGCs (Turnpenny *et al.*, 2005a; Kerr and Gearhart, unpublished). Nanog is another transcription factor that is expressed specifically in mouse preimplantation embryos, mESCs, and mEGCs as well as in monkey and human ESCs (Chambers *et al.*, 2003; Hart *et al.*, 2004; Hatano *et al.*, 2005; Mitsui *et al.*, 2003). The role of *nanog* in the maintenance of pluripotency is suggested by the loss of pluripotency in nanog-deficient mESCs and in nanog-null embryos shortly after implantation (Mitsui *et al.*, 2003). In addition, nanog overexpression leads to the clonal expansion of mESC by bypassing regulation of LIF-STAT3 signaling and maintenance of Oct4 levels (Chambers *et al.*, 2003). Recent evidence also shows that Sox2 and Oct4 bind to regulatory elements of *nanog* in mEGCs suggesting nanog's potential role in the action of Oct4 and Sox2 in these cells (Kuroda *et al.*, 2005). Although nanog expression is also detected in hEGCs and in the developing human gonad (Kerr and Gearhart, unpublished), a potential role in the action of Oct4 and Sox2 still remains to be determined.

4.2. Protocols for immunostaining EGCs and human gonads

1. Gonads are cut into 5-m sections, placed on slides (ProbeOn Plus, Fisher Scientific) and immediately prepared for antibody staining.

2. Plated EGC colonies and gonad sections are first fixed in either 4% paraformaldehyde for 5 min for cell surface markers, or in 4% paraformaldehyde following a 10 min incubation with 0.2% Triton X in D-PBS; Invitrogen) to detect Oct4 (clone YL8; BD Pharmingen) and Nanog (R&D Systems) staining.

3. Cell surface glycolipid- and glycoprotein-specific monoclonal antibodies were used at 1:15–1:50 dilution. MC480 (SSEA-1), MC813–70 (SSEA-4) TRA-1–60, and TRA-1–81 antibodies were supplied by the Developmental Studies Hybridoma Bank (University of Iowa, Iowa City).

4. Antibodies are diluted in 5% goat serum in D-PBS and incubated on sections for 1 h at room temperature except Oct4 and Nanog (1:50 dilution) which were incubated overnight at 4 °C.

5. Next all antibodies are detected by using fluorescently labeled goat anti-mouse secondary antibodies (1:200 dilution; Molecular Probes) in 5% goat serum in D-PBS for 1 h at room temperature. Except for Nanog staining, rabbit anti-goat secondary antibodies (1:200 dilution; Molecular Probes) is used.

6. Sections are counterstained with DAPI (Sigma) and mounted using Anti-Fade mounting medium (Molecular Probes). Negative controls for each procedure include incubations with secondary antibodies alone and with mouse ascites fluid.

4.3. Sex determination of human gonads by fluorescent *in situ* hybridization (FISH)

The sex of each gonad was determined using CEP X SpectrumOrange/CEP Y SpectrumGreen probes (Vysis). Five-micron sections were fixed in 1 part acetic acid:3 part ethanol for 1 min, pretreated in 1 M sodium thiocyanate for 5 min at 75 °C and post fixed in 100% methanol for 1 min. Sections were then denatured in 60% formamide, 2 × SSC (sodium chloride and citric acid) buffer, pH 7.5 at 75 °C for 3 min, followed by 1 min in cold 70%, 85%, and 90% ethanol and then incubated with X, Y probe overnight at 37 °C. The following day, sections underwent three posthybridization washes in 60% formamide (Sigma), 0.3% NP-40 (Igepal, Sigma), 2 × SSC, pH 7.5. Sections were then counterstained with DAPI and mounted as described previously.

4.4. Tissue nonspecific alkaline phosphatase and telomerase activity assays

To detect TnAP activity EGC colonies are fixed on plates in 66% acetone–3% formaldehyde for 5 min and then stained with naphthol/FRV-alkaline AP substrate (Sigma) for 20 min following manufacturer's instructions. Telomerase assays are performed using a telomeric repeat amplification protocol followed by ELISA detection of amplified products (Telo TAGGG PCR ELISA PLUS, Roche).

4.5. Cytogenetic analysis

EGC colonies are incubated in growth media with 0.1 μg/ml Karyomax Colcemid (Invitrogen) for 3–4 h at 37 °C in 5% CO_2 with 95% humidity, isolated from the feeder using a 1 μl pipetman and tip and then trypsinized in 0.05%-EDTA at 37 °C. Cells are then resuspended in 2 ml of 0.75 M KCl hypotonic solution and incubated at 37 °C for 35 min. Next cells are fixed by slowly adding 10 ml of cold Carnoy's fixative 3:1 methanol/acetic acid and rinsed twice with fixative before being plated. To prepare slides ~20 μl cell suspensions are dropped onto microscope slides over a humidity chamber and then allowed to air dry. Metaphase chromosomes are stained in Giemsa staining solution (Invitrogen) and observed at × 600 magnification with oil immersion. Standard karyotyping includes approximately 20 metaphase spreads from each line to be examined for the presence of structural abnormities of chromosomes and accurate chromosomal number.

5. EB Formation and Analysis

5.1. Overview

EBs form spontaneously in hEGC cultures. Although this represents a loss of pluripotent EGCs from the culture, EBs provide evidence for the pluripotent status of the culture and provide cellular material for subsequent culture and experimentation (see the section "EBD Cells Formation"). Initially, EBs provided the only direct evidence that hEGC cultures were pluripotent, as attempts to form teratomas in mice from hEGCs have failed.

5.2. Protocol for embedding and immunohistochemistry

The cells which comprise EBs can most reliably be identified by immunohistochemistry using paraffin and staining sections with a variety of well-characterized antibodies. This process avoids the significant problem of antibody trapping which occurs on large three-dimensional structures when direct staining is attempted. Specifically, paraffin is preferred over cryosections of EBs as cryopreservation results in very poorly defined cellular morphology.

First, human EBs are collected from cultures and placed into a small drop of molten 2% low melting point agarose (FMC), prepared in DPBS, and cooled to 42 °C. Solidified agarose-containing EBs are then fixed in 3% paraformaldehyde in DPBS and embedded in paraffin. Individual 6-μm sections are placed on microscope slides (ProbeOn Plus, Fisher Scientific) and antigen retrieval is performed. Antibodies used on paraffin sections include HHF35 (muscle specific actin, Dako), M 760 (desmin, Dako), CD34 (Immunotech), Z311, (S-100, Dako), sm311 (pan-neurofilament, Sternberger Monoclonals), A008 (α-1-fetoprotein), CKERAE1/AE3 (pancytokeratin, Boehringer Mannheim), OV-TL 12/30 (cytokeratin 7, Dako), and Ks20.8 (cytokeratin 20, Dako). These antibodies are used in particular to demonstrate that when hEGC cells differentiate, they form EBs comprised of endodermal, ectodermal, and mesodermal derivatives (Liu *et al.*, 2004; Park *et al.*, 2004; Shamblott *et al.*, 1998; Turnpenny *et al.*, 2003). Primary antibodies are detected by using biotinylated anti-rabbit or anti-mouse secondary antibody, streptavidin-conjugated horseradish peroxidase, and DAB chromogen (Ventana-BioTek Solutions). After which, slides can also be counter-stained with hematoxylin if desired.

6. EB-Derived Cell Formation

6.1. Overview

hEGC-derived EBs can be used to produce multipotent populations capable of long-term and robust proliferation. These cells are referred to as EB-derived (EBD) cells. The method used to isolate cell populations from EBs is conceptually similar to microbiological selective media experiments. EBs are disaggregated and plated into several different cellular growth environments. These environments consist of various growth media and matrix supports. Although many combinations have been evaluated by our laboratory (Shamblott *et al.*, 2001), most of our EBD cell cultures have been derived from one of six environments generated by combinations of one each of two growth media and three plating surfaces. The growth media include an RPMI 1640 media supplemented with either 15% FBS or a low 5% FBS in addition to FGF-2, epidermal growth factor (EGF), insulin-like growth factor 1 (IGF1), and vascular endothelial growth factor (VEGF). The plating surfaces include bovine type I collagen, human extracellular matrix extract, and tissue culture-treated plastic alone. These are not intended to be highly selective environments; instead, they favor several basic themes: cells thriving in high serum and elevated glucose (10 mM) conditions versus cells proliferating in low glucose (5 mM) under the control of four mitogens. In general, the type I collagen and human extracellular matrices combined with the low serum media provide the most rapid and extensive cell proliferation. EBD cell lines and cultures are routinely maintained in the environment in which they were derived. To distinguish EBD cells from a culture which has developed into a lineage-restricted cell line, it is important to establish an extensive expression profile. This profile should include genetic screening using multiple markers for each of the five cell lineages (neuronal, glial, muscle, hematopoietic–vascular endothelia, and endoderm) combined with supportive immunocytochemical staining.

Using this approach, data have shown that rapidly proliferating EBD cell cultures simultaneously express a wide array of mRNA and protein markers normally associated with distinct developmental lineages (Shamblott *et al.*, 2001). This is not a surprising property considering that EBD cells are, at least during the derivation stage, a mixed-cell population. More remarkable is the finding human EBD cell lines isolated by dilution cloning also exhibit a broad multilineage gene expression profile and that this profile remains stable throughout the life span of the culture. This normally exceeds 70 population doublings but is not unlimited since human EBD cells are not immortal.

Interestingly, EBD cells can be genetically manipulated using lipofection and electroporation as well as with retroviral, adenoviral, and lentiviral vectors. Using these techniques EBD lines have been generated that constitutively and tissue-specifically express enhanced GFP and contain many different genetic selection vectors. The proliferation and expression characteristics of EBD cells suggest they may be useful in the study of human cell differentiation and as a resource for cellular transplantation therapies. One important property in this regard is that no tumor of human origin has arisen in any animal receiving EBD cells from our laboratory, including hundreds of mice, rats, and African green monkeys which have received EBD transplants in a variety of anatomical locations, often consisting of more than one million cells injected (Frimberger *et al.*, 2005; Kerr *et al.*, 2003; Mueller *et al.*, 2005). This is in contrast to the infrequent yet significant number of teratocarcinomas that have arisen following transplantation of cells produced through neural and hematopoietic differentiation of mouse ESCs.

6.2. Protocols for human EBD cells

6.2.1. EBD derivation

EBs are harvested in groups of 10 or more and are dissociated by digestion in 1 mg/ml collagenase/dispase (Roche) for 30 min–1 h at 37 °C. Cells are then spun at $200 \times g$ for 5 min and resuspended in either RPMI or EGM2MV growth media. RPMI growth media includes RPMI 1640 (LTI), 15% FCS, 0.1-mM nonessential amino acids, 2-mM L-glutamine, 100 U/ml penicillin, and 100 μg/ml streptomycin. EGM2MV media (Clonetics) includes 5% FCS, hydrocortisone, FGF-2, hVEGF, R^3-IGF1, ascorbic acid, hEGF, heparin, gentamycin, and amphotericin B. Tissue culture dishes are coated with either bovine collagen I (Collaborative Biomedical, 10 μg/cm2), human extracellular matrix (Collaborative Biomedical, 5 μg/cm^2), or tissue culture plastic alone. EBD cells are cultured at 37 °C, 5% CO_2, 95% humidity and are routinely passaged 1:10–1:40 using 0.025% trypsin–0.01% EDTA for 5 min at 37 °C. Low serum cultures are treated with 1 mg/ml trypsin inhibitor and then spun down and resuspended in growth media.

6.2.2. Immunocytochemistry

Approximately 1×10^5 cells are plated in each well of an 8-well glass bottom chamber slide. Cells are fixed in either 4% paraformaldehyde in DPBS or a 1:1 mixture of methanol–acetone for 10 min. Cells are permeabilized in 0.1% Triton X-100, followed by one DPBS rinse for 10 min, then blocked in either Powerblock (BioGenex), 5% goat serum, or 1–5% goat serum supplemented with 0.5% bovine serum albumin for 10–60 min. Primary antibodies and dilutions are as follows: neurofilament 68 kDA (Roche 1:4), neuron-specific enolase (BD Pharmingen, 1:100), tau (BD Pharmingen, 5 μg/ml), vimentin (Roche, 1:10), nestin (NIH, 1:250), galactocerebroside (Sigma, 1:500), O4 (Roche 10 μg/ml), and SMI32 (Sternberger monoclonal, 1:5000). Detection was carried out by secondary antibodies conjugated to biotin, streptavidin-conjugated horseradish peroxidase, and 3-amino-9-ethylcarbazole chromogen (BioGenex).

6.2.3. mRNA expression profiles

RNA can be prepared using the Qiagen RNeasy kit *via* manufacturers instructions from cells prepared on either 60 or 100 mm tissue culture plates. RNA preparations are digested with RNAse-free DNAse (Roche) 30 min at 37 °C, and then inactivated at 75 °C for 5 min. Synthesis of cDNA is performed on 5 μg RNA by using oligo (dT) primers and a standard MMLV (Invitrogen) reaction carried out at 42 °C. Thirty cycles of PCR are carried out in the presence of 1.5 mM $MgCl_2$ with an annealing temperature of 55 °C and incubation times of 30 s. PCRs are resolved on a 1.8% agarose gel. The efficacy of each PCR is established using human RNA tissue controls (Clontech). All amplimers are validated by Southern blot analysis by using oligonucleotide probes end-labeled with ^{32}P-ATP, hybridized in 6 × SSC, 5 × Denhardt Solution, 0.1% SDS, 0.05% sodium pyrophosphate, and 1000 μg/ml of sheared and denatured salmon sperm DNA at 45 °C. cDNA synthesis and genomic DNA contamination are monitored by primers specific to human phosphoglycerate kinase-1, which give products of ~250 base pairs (bp) and ~500 bp when amplifying cDNA and genomic DNA, respectively.

6.2.4. Cryopreservation of mouse EGC and human EBD cells

Although cryopreservation of mouse EGCs can be successful using standard protocols, hEGCs are more fragile in this process. Instead, cryopreservation of hEGC-derived EBD cells have been performed. In mouse EGC and human EBD cultures, chose cultures which are ~three-fourth—confluent in either 60 or 100 mm tissue culture dishes. To disaggregate the colonies, first wash with DPBS and add 1 ml 0.05%-Trypsin-EDTA and place in incubator for 5 min at 37 °C. After cells detach from dish, gently triturate up and down ~3–5 times with 1 ml pipet and add a 2–10-fold excess of freezing media (50% FCS/10% DMSO/40% DMEM) to stop the reaction. Spin cells at 200 × g for 5 min and aspirate off supernatant. Flick pellet so cells will not clump, add 1 ml of freezing media (1 Petri/1 cryotube vial), and freeze in cryotubes using controlled-rate freezing vessels. Store vials in liquid nitrogen and thaw using standard procedures.

Acknowledgment

We like to thank Joyce Axelman for cell culture assistance and Christine M. Hill for immunological staining and FISH of the human gonads.

References

Andrews, P. W. (1998). Teratocarcinomas and human embryology: Pluripotent human EC cell lines. Review article. *Apmis* **106,** 158–167; discussion 167–168.

Baker, T. G., and Franchi, L. L. (1967). The fine structure of oogonia and oocytes in human ovaries. *J. Cell Sci.* **2,** 213–224.

Bendsen, E., Byskov, A. G., Laursen, S. B., Larsen, H. P., Andersen, C. Y., and Westergaard, L. G. (2003). Number of germ cells and somatic cells in human fetal testes during the first weeks after sex differentiation. *Hum. Reprod.* **18,** 13–18.

Blelloch, R. H., Hochedlinger, K., Yamada, Y., Brennan, C., Kim, M., Mintz, B., Chin, L., and Jaenisch, R. (2004). Nuclear cloning of embryonal carcinoma cells. *Proc. Natl. Acad. Sci. USA* **101,** 13985–13990.

Boiani, M., and Scholer, H. R. (2005). Regulatory networks in embryo-derived pluripotent stem cells. *Nat. Rev. Mol. Cell Biol.* **6,** 872–884.

Chambers, I., Colby, D., Robertson, M., Nichols, J., Lee, S., Tweedie, S., and Smith, A. (2003). Functional expression cloning of Nanog, a pluripotency sustaining factor in embryonic stem cells. *Cell* **113,** 643–655.

De Felici, M., Dolci, S., and Pesce, M. (1993). Proliferation of mouse primordial germ cells *in vitro*: A key role for cAMP. *Dev. Biol.* **157,** 277–280.

Dolci, S., Pesce, M., and De Felici, M. (1993). Combined action of stem cell factor, leukemia inhibitory factor, and cAMP on *in vitro* proliferation of mouse primordial germ cells. *Mol. Reprod. Dev.* **35,** 134–139.

Dolci, S., Williams, D. E., Ernst, M. K., Resnick, J. L., Brannan, C. I., Lock, L. F., Lyman, S. D., Boswell, H. S., and Donovan, P. J. (1991). Requirement for mast cell growth factor for primordial germ cell survival in culture. *Nature* **352,** 809–811.

Donovan, P. J., and de Miguel, M. P. (2003). Turning germ cells into stem cells. *Curr. Opin. Genet. Dev.* **13,** 463–471.

Donovan, P. J., and Gearhart, J. (2001). The end of the beginning for pluripotent stem cells. *Nature* **414,** 92–97.

Donovan, P. J., Stott, D., Cairns, L. A., Heasman, J., and Wylie, C. C. (1986). Migratory and postmigratory mouse primordial germ cells behave differently in culture. *Cell* **44,** 831–838.

Durcova-Hills, G., Ainscough, J., and McLaren, A. (2001). Pluripotential stem cells derived from migrating primordial germ cells. *Differentiation* **68,** 220–226.

Evans, M. J., and Kaufman, M. H. (1981). Establishment in culture of pluripotential cells from mouse embryos. *Nature* **292,** 154–156.

Falin, L. I. (1969). The development of genital glands and the origin of germ cells in human embryogenesis. *Acta. Anat. (Basel)* **72,** 195–232.

Francavilla, S., Cordeschi, G., Properzi, G., Concordia, N., Cappa, F., and Pozzi, V. (1990). Ultrastructure of fetal human gonad before sexual differentiation and during early testicular and ovarian development. *J. Submicrosc. Cytol. Pathol.* **22,** 389–400.

Francavilla, S., and Zamboni, L. (1985). Differentiation of mouse ectopic germinal cells in intra- and perigonadal locations. *J. Exp. Zool.* **233,** 101–109.

Frimberger, D., Morales, N., Shamblott, M., Gearhart, J. D., Gearhart, J. P., and Lakshmanan, Y. (2005). Human embryoid body-derived stem cells in bladder regeneration using rodent model. *Urology* **65,** 827–832.

Godin, I., and Wylie, C. C. (1991). TGF beta 1 inhibits proliferation and has a chemotropic effect on mouse primordial germ cells in culture. *Development* **113,** 1451–1457.

Gondos, B., Westergaard, L., and Byskov, A. G. (1986). Initiation of oogenesis in the human fetal ovary: Ultrastructural and squash preparation study. *Am. J. Obstet. Gynecol.* **155,** 189–195.

Goto, T., Adjaye, J., Rodeck, C. H., and Monk, M. (1999). Identification of genes expressed in human primordial germ cells at the time of entry of the female germ line into meiosis. *Mol. Hum. Reprod.* **5,** 851–860.

Hart, A. H., Hartley, L., Ibrahim, M., and Robb, L. (2004). Identification, cloning and expression analysis of the pluripotency promoting Nanog genes in mouse and human. *Dev. Dyn.* **230,** 187–198.

Hatano, S. Y., Tada, M., Kimura, H., Yamaguchi, S., Kono, T., Nakano, T., Suemori, H., Nakatsuji, N., and Tada, T. (2005). Pluripotential competence of cells associated with Nanog activity. *Mech. Dev.* **122,** 67–79.

Heyn, R., Makabe, S., and Motta, P. M. (1998). Ultrastructural dynamics of human testicular cords from 6 to 16 weeks of embryonic development. Study by transmission and high resolution scanning electron microscopy. *Ital. J. Anat. Embryol.* **103,** 17–29.

Kanatsu-Shinohara, M., Inoue, K., Lee, J., Yoshimoto, M., Ogonuki, N., Miki, H., Baba, S., Kato, T., Kazuki, Y., Toyokuni, S., Toyoshima, M., Niwa, O., *et al.* (2004). Generation of pluripotent stem cells from neonatal mouse testis. *Cell* **119,** 1001–1012.

Kanatsu-Shinohara, M., Miki, H., Inoue, K., Ogonuki, N., Toyokuni, S., Ogura, A., and Shinohara, T. (2005). Long-term culture of mouse male germline stem cells under serum- or feeder-free conditions. *Biol. Reprod.* **72,** 985–991.

Kehler, J., Tolkunova, E., Koschorz, B., Pesce, M., Gentile, L., Boiani, M., Lomeli, H., Nagy, A., McLaughlin, K. J., Scholer, H. R., and Tomilin, A. (2004). Oct4 is required for primordial germ cell survival. *EMBO Rep.* **5,** 1078–1083.

Kerr, D. A., Llado, J., Shamblott, M. J., Maragakis, N. J., Irani, D. N., Crawford, T. O., Krishnan, C., Dike, S., Gearhart, J. D., and Rothstein, J. D. (2003). Human embryonic germ cell derivatives facilitate motor recovery of rats with diffuse motor neuron injury. *J. Neurosci.* **23,** 5131–5140.

Kuroda, T., Tada, M., Kubota, H., Kimura, H., Hatano, S. Y., Suemori, H., Nakatsuji, N., and Tada, T. (2005). Octamer and Sox elements are required for transcriptional *cis* regulation of Nanog gene expression. *Mol. Cell Biol.* **25,** 2475–2485.

Labosky, P. A., Barlow, D. P., and Hogan, B. L. (1994). Embryonic germ cell lines and their derivation from mouse primordial germ cells. *Ciba. Found. Symp.* **182,** 157–168; discussion 168–178.

Lawson, K. A., and Hage, W. J. (1994). Clonal analysis of the origin of primordial germ cells in the mouse. *Ciba. Found. Symp.* **182,** 68–84; discussion 84–91.

Lee, C. K., and Piedrahita, J. A. (2000). Effects of growth factors and feeder cells on porcine primordial germ cells *in vitro. Cloning* **2,** 197–205.

Liu, S., Liu, H., Pan, Y., Tang, S., Xiong, J., Hui, N., Wang, S., Qi, Z., and Li, L. (2004). Human embryonic germ cells isolation from early stages of post-implantation embryos. *Cell Tissue Res.* **318,** 525–531.

Makabe, S., Naguro, T., and Motta, P. M. (1992). A new approach to the study of ovarian follicles by scanning electron microscopy and ODO maceration. *Arch. Histol. Cytol.* **55**(Suppl.), 183–190.

Martin, G. R. (1981). Isolation of a pluripotent cell line from early mouse embryos cultured in media conditioned by teratocarcinoma stem cells. *Proc. Natl. Acad. Sci. USA* **78,** 7634–7638.

Matsui, Y., Toksoz, D., Nishikawa, S., Nishikawa, S., Williams, D., Zsebo, K., and Hogan, B. L. (1991). Effect of Steel factor and leukaemia inhibitory factor on murine primordial germ cells in culture. *Nature* **353,** 750–752.

Matsui, Y., Zsebo, K., and Hogan, B. L. (1992). Derivation of pluripotential embryonic stem cells from murine primordial germ cells in culture. *Cell* **70,** 841–847.

McKay, D. G., Hertig, A. T., Adams, E. C., and Danziger, S. (1953). Histochemical observations on the germ cells of human embryos. *Anat. Rec.* **117,** 201–219.

McLaren, A. (2003). Primordial germ cells in the mouse. *Dev. Biol.* **262,** 1–15.

McLaren, A., and Durcova-Hills, G. (2001). Germ cells and pluripotent stem cells in the mouse. *Reprod. Fertil. Dev.* **13,** 661–664.

Mintz, B., and Russell, E. S. (1957). Gene-induced embryological modifications of primordial germ cells in the mouse. *J. Exp. Zool.* **134,** 207–237.

Mitsui, K., Tokuzawa, Y., Itoh, H., Segawa, K., Murakami, M., Takahashi, K., Maruyama, M., Maeda, M., and Yamanaka, S. (2003). The homeoprotein Nanog is required for maintenance of pluripotency in mouse epiblast and ES cells. *Cell* **113,** 631–642.

Motta, P. M., Makabe, S., and Nottola, S. A. (1997). The ultrastructure of human reproduction. I. The natural history of the female germ cell: Origin, migration and differentiation inside the developing ovary. *Hum. Reprod. Update* **3,** 281–295.

Mueller, S., Prelle, K., Rieger, N., Petznek, H., Lassnig, C., Luksch, U., Aigner, B., Baetscher, M., Wolf, E., Mueller, M., and Brem, G. (1999). Chimeric pigs following blastocyst injection of transgenic porcine primordial germ cells. *Mol. Reprod. Dev.* **54,** 244–254.

Mueller, D., Shamblott, M. J., Fox, H. E., Gearhart, J. D., and Martin, L. J. (2005). Transplanted human embryonic germ cell-derived neural stem cells replace neurons and oligodendrocytes in the forebrain of neonatal mice with excitotoxic brain damage. *J. Neurosci. Res.* **82,** 592–608.

Nichols, J., Zevnik, B., Anastassiadis, K., Niwa, H., Klewe-Nebenius, D., Chambers, I., Scholer, H., and Smith, A. (1998). Formation of pluripotent stem cells in the mammalian embryo depends on the POU transcription factor Oct4. *Cell* **95,** 379–391.

Niwa, H., Burdon, T., Chambers, I., and Smith, A. (1998). Self-renewal of pluripotent embryonic stem cells is mediated *via* activation of STAT3. *Genes Dev.* **12,** 2048–2060.

Ohno, S., Klinger, H. P., and Atkin, N. B. (1962). Human oogenesis. *Cytogenetics* **1,** 42–51.

Ozdzenski, W. (1967). Observations on the origin of primordial germ cells in the mouse. *Zool. Pol.* **17,** 367–379.

Park, T. S., and Han, J. Y. (2000). Derivation and characterization of pluripotent embryonic germ cells in chicken. *Mol. Reprod. Dev.* **56,** 475–482.

Park, J. H., Kim, S. J., Lee, J. B., Song, J. M., Kim, C. G., Roh, S., II, and Yoon, H. S. (2004). Establishment of a human embryonic germ cell line and comparison with mouse and human embryonic stem cells. *Mol. Cells* **17,** 309–315.

Pesce, M., and Scholer, H. R. (2000). Oct-4: Control of totipotency and germline determination. *Mol. Reprod. Dev.* **55,** 452–457.

Pesce, M., and Scholer, H. R. (2001). Oct-4: Gatekeeper in the beginnings of mammalian development. *Stem Cells* **19,** 271–278.

Pesce, M., Wang, X., Wolgemuth, D. J., and Scholer, H. (1998). Differential expression of the Oct-4 transcription factor during mouse germ cell differentiation. *Mech. Dev.* **71,** 89–98.

Piedrahita, J. A., Moore, K., Oetama, B., Lee, C. K., Scales, N., Ramsoondar, J., Bazer, F. W., and Ott, T. (1998). Generation of transgenic porcine chimeras using primordial germ cell-derived colonies. *Biol. Reprod.* **58,** 1321–1329.

Pinkerton, J. H., Mc, K. D., Adams, E. C., and Hertig, A. T. (1961). Development of the human ovary—a study using histochemical techniques. *Obstet. Gynecol.* **18,** 152–181.

Rabinovici, J., and Jaffe, R. B. (1990). Development and regulation of growth and differentiated function in human and subhuman primate fetal gonads. *Endocr. Rev.* **11,** 532–557.

Resnick, J. L., Bixler, L. S., Cheng, L., and Donovan, P. J. (1992). Long-term proliferation of mouse primordial germ cells in culture. *Nature* **359,** 550–551.

Scholer, H. R., Ruppert, S., Suzuki, N., Chowdhury, K., and Gruss, P. (1990). New type of POU domain in germ line-specific protein Oct-4. *Nature* **344,** 435–439.

Shamblott, M. J., Axelman, J., Littlefield, J. W., Blumenthal, P. D., Huggins, G. R., Cui, Y., Cheng, L., and Gearhart, J. D. (2001). Human embryonic germ cell derivatives express a broad range of developmentally distinct markers and proliferate extensively *in vitro*. *Proc. Natl. Acad. Sci. USA* **98,** 113–118.

Shamblott, M. J., Axelman, J., Wang, S., Bugg, E. M., Littlefield, J. W., Donovan, P. J., Blumenthal, P. D., Huggins, G. R., and Gearhart, J. D. (1998). Derivation of pluripotent stem cells from cultured human primordial germ cells. *Proc. Natl. Acad. Sci. USA* **95,** 13726–13731.

Shamblott, M. J., Kerr, C. L., Axelman, J., Littlefield, J. L. W., Clark, G. O., Patterson, E. S., Addis, R. C., Kraszewski, J. L., Ket, K. C., and Gearhart, J. D. (2004). Derivation and differentiation of human embryonic germ cells. *In* "Handbook of Stem Cells" (R. Lanza, B. Hogan, D. Melton, R. Pedersen, J. Thomson, and M. West, Eds.), Vol. 1, pp. 459–469. Elsevier, New York.

Shim, H., Gutierrez-Adan, A., Chen, L. R., BonDurant, R. H., Behboodi, E., and Anderson, G. B. (1997). Isolation of pluripotent stem cells from cultured porcine primordial germ cells. *Biol. Reprod.* **57,** 1089–1095.

Skrzypczak, J., Pisarski, T., Biczysko, W., and Kedzia, H. (1981). Evaluation of germ cells development in gonads of human fetuses and newborns. *Folia. Histochem. Cytochem. (Krakow)* **19,** 17–24.

Smith, A. G. (2001). Embryo-derived stem cells: Of mice and men. *Annu. Rev. Cell Dev. Biol.* **17,** 435–462.

Smith, A. G., Heath, J. K., Donaldson, D. D., Wong, G. G., Moreau, J., Stahl, M., and Rogers, D. (1988). Inhibition of pluripotential embryonic stem cell differentiation by purified polypeptides. *Nature* **336**, 688–690.

Stevens, L. C. (1966). Development of resistance to teratocarcinogenesis by primordial germ cells in mice. *J. Natl. Cancer Inst.* **37**, 859–867.

Stevens, L. C. (1967). Origin of testicular teratomas from primordial germ cells in mice. *J. Natl. Cancer Inst.* **38**, 549–552.

Stewart, C. L., Gadi, I., and Bhatt, H. (1994). Stem cells from primordial germ cells can reenter the germ line. *Dev. Biol.* **161**, 626–628.

Sun, E. L., and Gondos, B. (1984). Squash preparation studies of germ cells in human fetal testes. *J. Androl.* **5**, 334–338.

Takagi, Y., Talbot, N. C., Rexroad, C. E., Jr., and Pursel, V. G. (1997). Identification of pig primordial germ cells by immunocytochemistry and lectin binding. *Mol. Reprod. Dev.* **46**, 567–580.

Tam, P. P., and Snow, M. H. (1981). Proliferation and migration of primordial germ cells during compensatory growth in mouse embryos. *J. Embryol. Exp. Morphol.* **64**, 133–147.

Tsung, H. C., Du, Z. W., Rui, R., Li, X. L., Bao, L. P., Wu, J., Bao, S. M., and Yao, Z. (2003). The culture and establishment of embryonic germ (EG) cell lines from Chinese mini swine. *Cell Res.* **13**, 195–202.

Turnpenny, L., Brickwood, S., Spalluto, C. M., Piper, K., Cameron, I. T., Wilson, D. I., and Hanley, N. A. (2003). Derivation of human embryonic germ cells: An alternative source of pluripotent stem cells. *Stem Cells* **21**, 598–609.

Turnpenny, L., Cameron, I. T., Spalluto, C. M., Hanley, K. P., Wilson, D. I., and Hanley, N. A. (2005). Human embryonic germ cells for future neuronal replacement therapy. *Brain Res. Bull.* **68**, 76–82.

Turnpenny, L., Spalluto, C. M., Perrett, R. M., O'Shea, M., Piper Hanley, K., Cameron, I. T., Wilson, D. I., and Hanley, N. A. (2005). Evaluating human embryonic germ cells: Concord and conflict as pluripotent stem cells. *Stem Cells* **24**, 212–220.

Upadhyay, S., and Zamboni, L. (1982). Ectopic germ cells: Natural model for the study of germ cell sexual differentiation. *Proc. Natl. Acad. Sci. USA* **79**, 6584–6588.

Wartenberg, H. (1982). Development of the early human ovary and role of the mesonephros in the differentiation of the cortex. *Anat. Embryol. (Berl.)* **165**, 253–280.

Witschi, E. (1948). Migration of the germ cells of human embryos from the yolk sac to the primitive gonadal folds. *In* "Contributions in Embryology" Vol. 32, pp. 67–80. Carnegie Institute, Washington.

Witschi, E. (1963). Embryology of the ovary. *In* "The Ovary" (H. G. Grady and D. E. Smith, Eds.) Williams and Wilkins, Baltimore.

Yamaguchi, S., Kimura, H., Tada, M., Nakatsuji, N., and Tada, T. (2005). Nanog expression in mouse germ cell development. *Gene Expr. Patterns* **5**, 639–646.

Yeom, Y. I., Fuhrmann, G., Ovitt, C. E., Brehm, A., Ohbo, K., Gross, M., Hubner, K., and Scholer, H. R. (1996). Germline regulatory element of Oct-4 specific for the totipotent cycle of embryonal cells. *Development* **122**, 881–894.

CHAPTER 9

Pluripotent Stem Cells from Amniotic Fluid and Placenta

Dawn M. Delo, Paolo DeCoppi, Anna Milanesi, Minhaj Siddiqui, and Anthony Atala

Wake Forest Institute for Regenerative Medicine
Wake Forest University School of Medicine
Medical Center Boulevard
Winston-Salem, North Carolina 27157–1094

Abstract
1. Introduction
2. Amniotic Fluid and Placenta in Developmental Biology
3. Amniotic and Placental Cells for Therapy-A New Connection
4. Isolation and Characterization of Progenitor Cells
5. Differentiation of Amniotic and Placental-Derived Progenitor Cells
   5.1. Adipocytes
   5.2. Osteocytes
   5.3. Endothelial cells
   5.4. Hepatocytes
   5.5. Myocytes
   5.6. Neuronal
6. Conclusion
References

Abstract

Human amniotic fluid has been utilized in prenatal diagnosis for over 70 years. It has proven to be a safe, reliable, and simple screening tool for a wide variety of developmental and genetic diseases. However, there is now evidence that amniotic fluid may have more utility than only as a diagnostic tool and may be a source of a powerful therapy for a multitude of congenital and adult disorders. A subset of cells found in amniotic fluid and placenta have been isolated and found to be capable of maintaining prolonged undifferentiated proliferation as well as differentiate into multiple tissue types encompassing the three germ layers. It is possible that in the near future, we will see the development of therapies utilizing progenitor cells isolated from amniotic fluid and placenta for the treatment of newborns with congenital malformations as well as in adults using cryopreserved amniotic fluid and placental stem cells. In this chapter, we will describe a number of experiments that have been isolated and characterized pluripotent progenitor cells from amniotic fluid and placenta. We will also discuss various cell lines derived from amniotic fluid and placenta and future directions of this area of research.

1. Introduction

Amniotic fluid derived progenitor cells can be obtained from a small amount of fluid during amniocentesis, a procedure that is already often performed in many pregnancies in which the fetus has a congenital abnormality. Placental derived stem cells can be obtained from a small biopsy of the chorionic villi. Recent observations on cell cultures from these two sources provide evidence that they may represent new sources for the isolation of cells with the potential to differentiate into different cell types, suggesting a new source of the cells for research and treatment.

2. Amniotic Fluid and Placenta in Developmental Biology

Gastrulation is a major milestone in early post implantation development (Snow and Bennett, 1978). At about embryo 6.5 days old (E6.5) of development, gastrulation begins in the posterior region of the embryo. Pluripotent epiblast cells are allocated to the three primary germ layers of the embryo (ectoderm, mesoderm, and endoderm) and germ cells which are the progenitors of all fetal tissue lineages as well as the extraembryonic mesoderm of the yolk sac, amnion, and allantois (Downs and Harmann, 1997; Downs *et al.*, 2004; Gardner and Beddington, 1988; Loebel *et al.*, 2003). The latter forms the umbilical cord as well as the mesenchymal part of the labyrinthine layer in the mature chorioallantoic placenta (Downs and Harmann, 1997; Moser *et al.*, 2004; Smith *et al.*, 1994). The final positions of the fetal membranes result from the process of embryonic turning, which occurs around day 8.5 of gestation and "pulls" the amnion and yolk sac around the embryo (Kinder *et al.*, 1999; Parameswaran and Tam, 1995). The specification of tissue lineages is accomplished by the restriction of developmental potency and the activation of lineage-specific gene expression (Parameswaran and Tam, 1995; Rathjen *et al.*, 1999). This process is strongly influenced by cellular interactions and signaling (Dang *et al.*, 2002; Li *et al.*, 2004).

The amniotic sac is a tough but thin transparent pair of membranes which holds a developing embryo (and later fetus) until shortly before birth. The inner membrane, the amnion, contains the amniotic fluid and the fetus. The outer membrane, the chorion, contains the amnion and is part of the placenta (Kaviani *et al.*, 2001; Kinder *et al.*, 1999; Robinson *et al.*, 2002). Amnion is derived from ectoderm and mesoderm, which grows and begins to fill mainly with water (Robinson *et al.*, 2002). Originally it is isotonic, containing proteins, carbohydrates, lipids and phospholipids, urea, and electrolytes. Later, urine excreted by the fetus increases its volume and changes its concentration (Bartha *et al.*, 2000; Heidari *et al.*, 1996; Sakuragawa *et al.*, 1999; Srivastava *et al.*, 1996). The fetus can breathe in the water, allowing normal growth and the development of lungs and the gastrointestinal tract. The fluid is swallowed by the fetus and passes *via* the fetal blood into the maternal blood. Amniotic fluid functions to ensure symmetrical structure development and growth, cushion and protect the embryo, help maintain consistent pressure and temperature, and permit freedom of fetal movement, which is important for musculoskeletal development and blood flow (Baschat and Hecher, 2004).

A wide variety of different origins has been suggested for the mixture of cells within amniotic fluid (Medina-Gomez and del Valle, 1988). Cells of different embryonic/fetal origins of all three germ layers have been reported to exist in amniotic fluid (In 't Anker *et al.*, 2003; Prusa *et al.*, 2004). These cells are thought to be sloughed from the fetal amnion, skin, and alimentary, respiratory and urogenitory tracts. In addition, it has been reported that cell culture from amniotic fluid as well as placenta provide evidence that they may represent new stem cell sources with the potential to differentiate into different cell types (Prusa and Hengstschlager, 2002). Interestingly, it has been demonstrated that the subpopulation cells in amniotic fluid produce Oct4 mRNA which is used to maintain pluripotency (Prusa *et al.*, 2003). Although research is at its early stages, these cells can be used to find treatments or even cure for many diseases in which irreplaceable cells are damaged.

3. Amniotic and Placental Cells for Therapy-A New Connection

Ideal cells for regenerative medicine are the pluripotent stem cells, which have the capability to differentiate in stages into a huge number of different types of human cells. Amniotic fluid cells can be obtained from a small amount of fluid during amniocentesis at second-trimester, a procedure that is already often performed in many pregnancies in which the fetus has a congenital abnormality and to determine characteristics such as sex (Hoehn *et al.*, 1975). Kaviani reported that "just 2 ml of amniotic fluid" can provide up to 20,000 cells, 80% of which are viable (Kaviani *et al.*, 2001, 2003). Since many pregnant women already undergo amniocentesis to screen for fetal abnormalities, cells can be simply isolated from this fluid and saved for future use. With amniotic fluid cells, it takes 20–24 h to double the number of cells collected, which is faster than umbilical cord stem cells (28–30 h) and bone marrow stem cells (over 30 h)(Tsai *et al.*, 2004). This phenomenon is a very important feature for urgent medical conditions. In addition, while scientists

have only been able to isolate and differentiate on average just 30% of mesenchymal stem cells (MSC) extracted from a child's umbilical cord shortly after birth, the
success rate for amniotic fluid-derived MSC is close to 100% (In 't Anker *et al.*,
2003; Tsai *et al.*, 2004). Furthermore, extracting the cells from amniotic fluid
bypasses the problems associated with a technique called donor–recipient HLA
matching, which involves transplanting cells (Tsai *et al.*, 2004).

4. Isolation and Characterization of Progenitor Cells

Amniotic fluid progenitor cells are isolated by centrifugation of amniotic fluid
from amniocentesis. Chorionic villi cells are isolated from single villi under light
microscopy. Amniotic fluid cells and placental cells are allowed to proliferate *in vitro*
and are maintained in culture for 4 weeks. The culture medium consists of modified
α-Modified Earls Medium (18% Chang Medium B, 2% Chang C with 15% embryonic stem cell certified fetal bovine serum, antibiotics, and L-glutamine).

A pluripotential subpopulation of progenitor cells present in the amniotic fluid
and placenta can be isolated through positive selection for cells expressing the
membrane receptor *c-kit* (Fig. 9.1) (DeCoppi, 2001; Siddiqui and Atala A., 2004).
This receptor binds to the ligand stem cell factor. Roughly 0.8%–1.4% of cells
present in amniotic fluid and placenta have been shown to be *c-kit*[pos] in analysis by
fluorescent activated cell sorting (FACS). The progenitor cells maintain a round

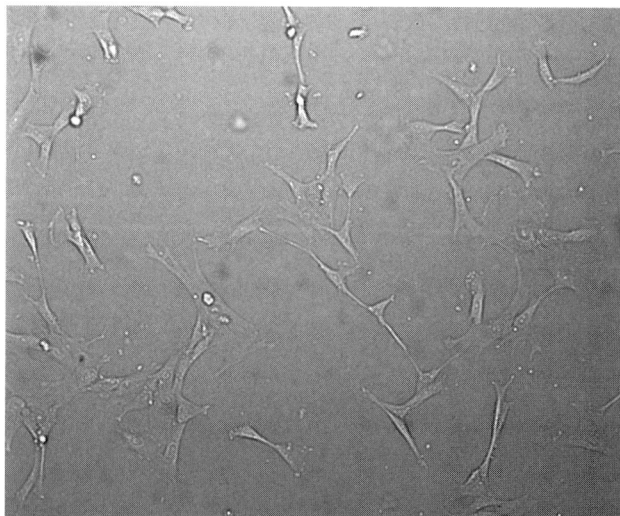

Figure 9.1 Morphology of AFSC in culture.

shape for one week post isolation when cultured in nontreated culture dishes. In this state, they demonstrate a very low proliferative capability. After the first week the cells begin to adhere to the plate and change their morphology, becoming more elongated and proliferating more rapidly, reaching 80% confluence with a need for passage every 48–72 h. No feeder layers are required either for maintenance or expansion. The progenitor cells derived show a high self-renewal capacity with >300 population doublings, far exceeding Hayflick's limit. The doubling time of the undifferentiated cells is noted to be 36 h with little variation with passages.

These cells have been shown to maintain a normal karyotype at late passages and have normal G1 and G2 cell cycle checkpoints. They demonstrate telomere length conservation while in the undifferentiated state as well as telomerase activity even in late passages (Bryan *et al.*, 1998). Analysis of surface markers shows that progenitor cells from amniotic fluid express human embryonic stage specific marker SSEA4, and the stem cell marker OCT4, and did not express SSEA1, SSEA3, CD4, CD8, CD34, CD133, C-MET, ABCG2, NCAM, BMP4, TRA1–60, and TRA1–81, to name a few. This expression profile is of interest as it demonstrates expression by the amniotic fluid derived progenitor cells of some key markers of embryonic stem cell phenotype, but not the full complement of markers expressed by embryonic stem cells. This may indicate that the amniotic cells are not quite as primitive as embryonic cells, yet maintain greater potential than most adult stem cells. Another behavior showing similarities and differences between these amniotic fluid derived cells and blastocyst derived cells are that while the amniotic fluid progenitor cells do form embryoid bodies *in vitro* which stain positive for markers of all three germ layers, these cells do not form teratomas *in vivo* when implanted in immunodeficient mice. Lastly, when expanded from a single cell, daughter cells maintained similar properties in growth and potential as the original mixed population of progenitor cells.

5. Differentiation of Amniotic and Placental–Derived Progenitor Cells

The progenitor cells derived from amniotic fluid and placenta are pluripotent and have been shown to differentiate into osteogenic, adipogenic, myogenic, neurogenic, endothelial, and hepatic phenotypes *in vitro*. Each differentiation has been performed through proof of phenotypic and biochemical changes consistent with the differentiated tissue type of interest (Fig. 9.2). We will discuss each set of differentiations separately.

5.1. Adipocytes

To promote adipogenic differentiation, the progenitor cells can be induced in dexamethasone, 3-isobutyl-1-methylxanthine, insulin, and indomethacin. The progenitor cells cultured with adipogenic supplements change their morphology from

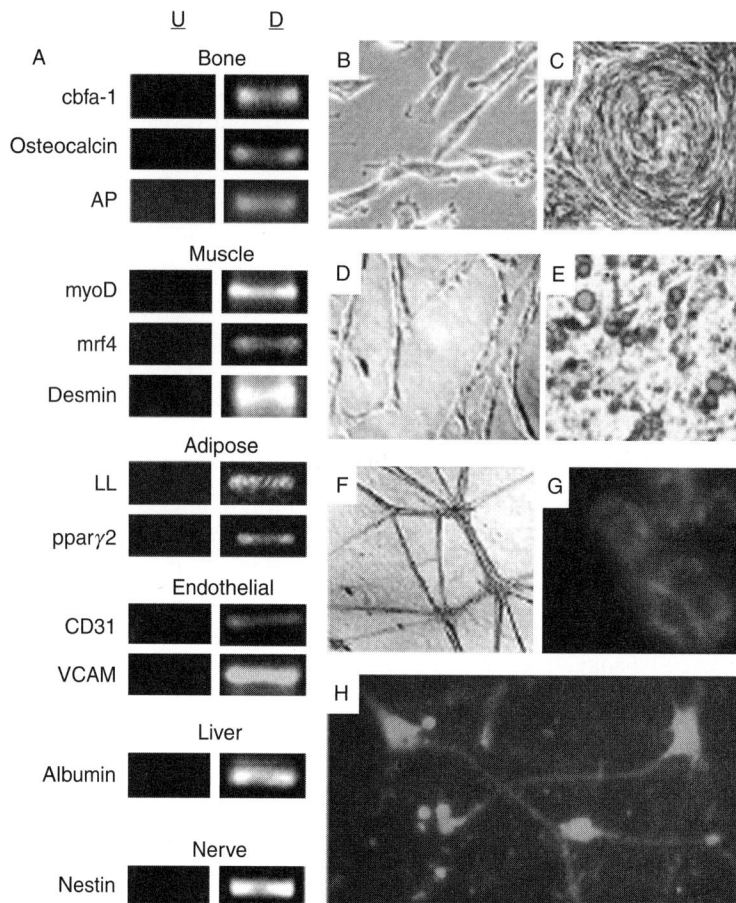

Figure 9.2 Multilineage differentiation of hAFSC. (A) RT–PCR analysis of mRNA. U: Control undifferentiated cells. D: cells maintained under conditions for differentiation to bone (8 days), muscle (8 days), adipocyte (16 days), endothelial (8 days), hepatic (45 days), and neuronal (2 days) lineages. (B) Phase contrast microscopy of control, undifferentiated cells. (b–h) Differentiated progenitor cells. (C) Bone: histochemical staining for alkaline phosphatase (D) Muscle: phase contrast showing fusion into multinucleated myotube-like cells. (E) Adipocyte: staining with Oil–Red-O (day 8) shows intracellular oil aggregation. (F) Endothelial: phase contrast microscopy of capillary-like structures. (G) Hepatic: fluorescent antibody staining (FITC–green) for albumin. (H) Neuronal: fluorescent antibody staining of nestin (day 2).

elongated to round within 8 days. This coincides with the accumulation of intracellular droplets. After 16 days in culture, more than 95% of the cells have their cytoplasm filled with lipid-rich vacuoles. Adipogenic differentiation also demonstrates the expression of peroxisome proliferation-activated receptor $\gamma2$ (ppar$\gamma2$), a transcription factor that regulates adipogenesis, and of lipoprotein lipase through RT–PCR analysis (Cremer *et al.*, 1981; Medina-Gomez and del Valle, 1988).

Expression of these genes is noted in the progenitor cells under adipogenic conditions but not in undifferentiated cells.

5.2. Osteocytes

Osteogenic differentiation was induced in the progenitor cells with use of dexamethasone, beta-glycerophosphate, and ascorbic acid-2-phosphate (Jaiswal *et al.*, 1997). The progenitor cells maintained in this medium demonstrated phenotypic changes within 4 days with a loss of spindle shape phenotype and development of an osteoblast-like appearance with fingerlike excavations into the cytoplasm. At 16 days, the cells aggregated, showing typical lamellar bone-like structures. In terms of functionality, these differentiated cells demonstrate a major feature of osteoblasts which is to precipitate calcium. Differentiated osteoblasts from the progenitor cells are able to produce alkaline phosphatase (AP) and to deposit calcium, consistent with bone differentiation. The undifferentiated progenitor cells lacked this ability. The progenitor cells in osteogenic medium express specific genes implicated in mammalian bone development [AP, core binding factor A1 (cbfa1), and osteocalcin] in a pattern consistent with the physiological analog. The progenitor cells grown in osteogenic medium show an activation of the AP gene at each time point. Expression of cbfa1, a transcription factor specifically expressed in osteoblasts and hypertrophic chondrocytes that regulates gene expression of structural proteins of the bone extracellular matrix, is highest in cells grown in osteogenic inducing medium at day 8 and decreases slightly at days 16, 24, and 32. Osteocalcin is expressed only in the progenitor cells in osteogenic conditions at 8 days (Karsenty, 2000; Komori *et al.*, 1997).

5.3. Endothelial cells

The amniotic fluid progenitor cells can be induced to form endothelial cells by culture in endothelial basal medium on gelatin coated dishes. Full differentiation is affected with 1 month in culture; however, phenotypic changes are noticed within 1 week of initiation of the protocol. Human-specific endothelial cell surface marker (P1H12), factor VIII (FVIII), and KDR are specific for differentiated endothelial cells. The differentiated cells stain positively for FVIII, KDR, and P1H12. The progenitor cells do not stain for endothelial specific markers. The amniotic fluid progenitor derived endothelial cells, once differentiated, are able to grow in culture and form capillary-like structures *in vitro*. These cells also express platelet endothelial cell adhesion molecule 1 (PECAM-1 or CD31) and vascular cell adhesion molecule (VCAM) which are not detected in the progenitor cells on RT–PCR analysis.

5.4. Hepatocytes

For hepatic differentiation, the progenitor cells are seeded on matrigel or collagen coated dishes at different stages and cultured in the presence of hepatocyte growth factor, insulin, oncostatin M, dexamethasone, fibroblast growth factor 4, and monothioglycerol for 45 days (Dunn *et al.*, 1989; Schwartz *et al.*, 2002). After 7 days of the differentiation process, cells exhibit morphological changes from an elongated to a cobblestone appearance. The cells show positive staining for albumin at day 45 post differentiation and also express the transcription factor HNF4α, the c-met receptor, the MDR membrane transporter, albumin, and α-fetoprotein. RT–PCR analysis further supports albumin production. The maximum rate of urea production for hepatic differentiation induced cells is upregulated to 1.21×10^3 ng urea/h/cell from 5.0×10^1 ng urea/h/cell for the control progenitor cell populations (Hamazaki *et al.*, 2001).

5.5. Myocytes

Myogenic differentiation is induced in the amniotic fluid derived progenitor cells by culture in media containing horse serum and chick embryo extract on a thin gel coat of matrigel (Rosenblatt *et al.*, 1995). To initiate differentiation, presence of 5-azacytidine in the media for 24 h is necessary. Phenotypically the cells can be noted to organize themselves into bundles which fuse to form multinucleated cells. These cells express sarcomeric tropomyosin and desmin, both of which are not expressed in the original progenitor population.

The development profile of cells differentiating into myogenic lineages interestingly mirrors a characteristic pattern of gene expression reflecting that seen with embryonic muscle development (Bailey *et al.*, 2001; Rohwedel *et al.*, 1994). With this protocol, *Myf6* is expressed at day 8 and suppressed at day 16. *MyoD* expression is detectable at 8 days and suppressed at 16 days in the progenitor cells. Desmin expression is induced at 8 days and increases by 16 days in the progenitor cells cultured in myogenic medium (Hinterberger *et al.*, 1991; Patapoutian *et al.*, 1995).

5.6. Neuronal

For neurogenic induction, the amniotic progenitor cells are induced in DMSO, butylated hydroxyanisole, and neuronal growth factor (Black and Woodbury, 2001; Woodbury *et al.*, 2000). The progenitor cells cultured in neurogenic conditions change their morphology within the first 24 h. Two different cell populations are apparent: morphologically large flat cells and small bipolar cells. The bipolar cell cytoplasm retracts towards the nucleus, forming contracted multipolar structures. Over the subsequent hours, the cells display primary and secondary branches, and cone-like terminal expansions. The induced progenitor cells show a characteristic sequence of expression of neural-specific proteins. At an early stage the intermediate filament protein, nestin, which is specifically expressed in neuroepithelial stem cells, is highly expressed. The expressions of βIII-tubulin and glial fibrillary acidic protein (GFAP), markers of neuron and glial differentiation, respectively, increases over time and seems to reach a plateau at about 6 days (Guan *et al.*, 2001). The progenitor cells cultured under neurogenic conditions show the presence of the neurotransmitter glutamic acid in the collected medium. Glutamic acid is usually secreted in culture by fully differentiated neurons (Carpenter *et al.*, 2001).

6. Conclusion

The pluripotent progenitor cells isolated from amniotic fluid and placenta present a very exciting possible contribution to the field of stem cell biology and regenerative medicine. These cells are an excellent source for research and therapeutic applications. The ability to isolate the progenitor cells during gestation may also be advantageous for babies born with congenital malformations. Furthermore, the progenitor cells can be cryopreserved for future self-use. When compared with embryonic stem cells, the progenitor cells isolated from amniotic fluid have many similarities: they can differentiate into all three germ layers, they express common markers, and they preserve their telomere length. However the progenitor cells isolated from amniotic fluid have considerable advantages. They easily differentiate into specific cell lineages and they avoid the current controversies associated with the use of human embryonic stem cells. The discovery of these cells has been recent, and a considerable amount of work remains to be done on the characterization and use of these cells. In the future, cells derived from amniotic fluid and placenta may represent an attractive and abundant, noncontroversial source of cells for regenerative medicine.

References

Bailey, P., Holowacz, T., and Lassar, A. B. (2001). The origin of skeletal muscle stem cells in the embryo and the adult. *Curr. Opin. Cell Biol.* **13,** 679–689.

Bartha, J. L., Romero-Carmona, R., Comino-Delgado, R., Arce, F., and Arrabal, J. (2000). Alpha-fetoprotein and hematopoietic growth factors in amniotic fluid. *Obstet. Gynecol.* **96,** 588–592.

Baschat, A. A., and Hecher, K. (2004). Fetal growth restriction due to placental disease. *Semin. Perinatol.* **28,** 67–80.

Black, I. B., and Woodbury, D. (2001). Adult rat and human bone marrow stromal stem cells differentiate into neurons. *Blood Cells Mol. Dis.* **27,** 632–636.

Bryan, T. M., Englezou, A., Dunham, M. A., and Reddel, R. R. (1998). Telomere length dynamics in telomerase-positive immortal human cell populations. *Exp. Cell Res.* **239,** 370–378.

Carpenter, M. K., Inokuma, M. S., Denham, J., Mujtaba, T., Chiu, C. P., and Rao, M. S. (2001). Enrichment of neurons and neural precursors from human embryonic stem cells. *Exp. Neurol.* **172,** 383–397.

Cremer, M., Schachner, M., Cremer, T., Schmidt, W., and Voigtlander, T. (1981). Demonstration of astrocytes in cultured amniotic fluid cells of three cases with neural-tube defect. *Hum. Genet.* **56,** 365–370.

Dang, S. M., Kyba, M., Perlingeiro, R., Daley, G. Q., and Zandstra, P. W. (2002). Efficiency of embryoid body formation and hematopoietic development from embryonic stem cells in different culture systems. *Biotechnol. Bioeng.* **78,** 442–453.

DeCoppi, P.Human fetal stem cell isolation from amniotic fluid (2001).Proceedings of the American Acadamy of Pediatrics National Conference, pp. 210–211. San Francisco, CA.

Downs, K. M., and Harmann, C. (1997). Developmental potency of the murine allantois. *Development* **124**, 2769–2780.

Downs, K. M., Hellman, E. R., McHugh, J., Barrickman, K., and Inman, K. E. (2004). Investigation into a role for the primitive streak in development of the murine allantois. *Development* **131**, 37–55.

Dunn, J. C., Yarmush, M. L., Koebe, H. G., and Tompkins, R. G. (1989). Hepatocyte function and extracellular matrix geometry: Long-term culture in a sandwich configuration. *FASEB J.* **3**, 174–177.

Gardner, R. L., and Beddington, R. S. (1988). Multi-lineage 'stem' cells in the mammalian embryo. *J. Cell Sci. Suppl.* **10**, 11–27.

Guan, K., Chang, H., Rolletschek, A., and Wobus, A. M. (2001). Embryonic stem cell-derived neurogenesis. Retinoic acid induction and lineage selection of neuronal cells.. *Cell Tissue Res.* **305**, 171–176.

Hamazaki, T., Iiboshi, Y., Oka, M., Papst, P. J., Meacham, A. M., Zon, L. I., and Terada, N. (2001). Hepatic maturation in differentiating embryonic stem cells *in vitro*. *FEBS Lett.* **497**, 15–19.

Heidari, Z., Isobe, K., Goto, S., Nakashima, I., Kiuchi, K., and Tomoda, Y. (1996). Characterization of the growth factor activity of amniotic fluid on cells from hematopoietic and lymphoid organs of different life stages. *Microbiol. Immunol.* **40**, 583–589.

Hinterberger, T. J., Sassoon, D. A., Rhodes, S. J., and Konieczny, S. F. (1991). Expression of the muscle regulatory factor MRF4 during somite and skeletal myofiber development. *Dev. Biol.* **147**, 144–156.

Hoehn, H., Bryant, E. M., Fantel, A. G., and Martin, G. M. (1975). Cultivated cells from diagnostic amniocentesis in second trimester pregnancies. III. The fetal urine as a potential source of clonable cells. *Humangenetik* **29**, 285–290.

In 't Anker, P. S., Scherjon, S. A., Kleijburg-van der Keur, C., Noort, W. A., Claas, F. H., Willemze, R., Fibbe, W. E., and Kanhai, H. H. (2003). Amniotic fluid as a novel source of mesenchymal stem cells for therapeutic transplantation. *Blood* **102**, 1548–1549.

Jaiswal, N., Haynesworth, S. E., Caplan, A. I., and Bruder, S. P. (1997). Osteogenic differentiation of purified, culture-expanded human mesenchymal stem cells *in vitro*. *J. Cell Biochem.* **64**, 295–312.

Karsenty, G. (2000). Role of Cbfa1 in osteoblast differentiation and function. *Semin. Cell Dev. Biol.* **11**, 343–346.

Kaviani, A., Guleserian, K., Perry, T. E., Jennings, R. W., Ziegler, M. M., and Fauza, D. O. (2003). Fetal tissue engineering from amniotic fluid. *J. Am. Coll. Surg.* **196**, 592–597.

Kaviani, A., Perry, T. E., Dzakovic, A., Jennings, R. W., Ziegler, M. M., and Fauza, D. O. (2001). The amniotic fluid as a source of cells for fetal tissue engineering. *J. Pediatr. Surg.* **36**, 1662–1665.

Kinder, S. J., Tsang, T. E., Quinlan, G. A., Hadjantonakis, A. K., Nagy, A., and Tam, P. P. (1999). The orderly allocation of mesodermal cells to the extraembryonic structures and the anteroposterior axis during gastrulation of the mouse embryo. *Development* **126**, 4691–4701.

Komori, T., Yagi, H., Nomura, S., Yamaguchi, A., Sasaki, K., Deguchi, K., Shimizu, Y., Bronson, R. T., Gao, Y. H., Inada, M., Sato, M., Okamoto, R., *et al.* (1997). Targeted disruption of Cbfa1 results in a complete lack of bone formation owing to maturational arrest of osteoblasts. *Cell* **89**, 755–764.

Li, L., Arman, E., Ekblom, P., Edgar, D., Murray, P., and Lonai, P. (2004). Distinct GATA6- and laminin-dependent mechanisms regulate endodermal and ectodermal embryonic stem cell fates. *Development* **131**, 5277–5286.

Loebel, D. A., Watson, C. M., De Young, R. A., and Tam, P. P. (2003). Lineage choice and differentiation in mouse embryos and embryonic stem cells. *Dev. Biol.* **264**, 1–14.

Medina-Gomez, P., and del Valle, M. (1988). The culture of amniotic fluid cells. An analysis of the colonies, metaphase and mitotic index for the purpose of ruling out maternal cell contamination. *Ginecol. Obstet. Mex.* **56**, 122–126.

Moser, M., Li, Y., Vaupel, K., Kretzschmar, D., Kluge, R., Glynn, P., and Buettner, R. (2004). Placental failure and impaired vasculogenesis result in embryonic lethality for neuropathy target esterase-deficient mice. *Mol. Cell Biol.* **24**, 1667–1679.

Parameswaran, M., and Tam, P. P. (1995). Regionalisation of cell fate and morphogenetic movement of the mesoderm during mouse gastrulation. *Dev. Genet.* **17**, 16–28.

Patapoutian, A., Yoon, J. K., Miner, J. H., Wang, S., Stark, K., and Wold, B. (1995). Disruption of the mouse MRF4 gene identifies multiple waves of myogenesis in the myotome. *Development* **121,** 3347–3358.

Prusa, A. R., and Hengstschlager, M. (2002). Amniotic fluid cells and human stem cell research: A new connection. *Med. Sci. Monit.* **8,** RA253–RA257.

Prusa, A. R., Marton, E., Rosner, M., Bernaschek, G., and Hengstschlager, M. (2003). Oct-4-expressing cells in human amniotic fluid: A new source for stem cell research? *Hum. Reprod.* **18,** 1489–1493.

Prusa, A. R., Marton, E., Rosner, M., Bettelheim, D., Lubec, G., Pollack, A., Bernaschek, G., and Hengstschlager, M. (2004). Neurogenic cells in human amniotic fluid. *Am. J. Obstet. Gynecol.* **191,** 309–314.

Rathjen, J., Lake, J. A., Bettess, M. D., Washington, J. M., Chapman, G., and Rathjen, P. D. (1999). Formation of a primitive ectoderm like cell population, EPL cells, from ES cells in response to biologically derived factors. *J. Cell Sci.* **112**(Pt. 5), 601–612.

Robinson, W. P., McFadden, D. E., Barrett, I. J., Kuchinka, B., Penaherrera, M. S., Bruyere, H., Best, R. G., Pedreira, D. A., Langlois, S., and Kalousek, D. K. (2002). Origin of amnion and implications for evaluation of the fetal genotype in cases of mosaicism. *Prenat. Diagn.* **22,** 1076–1085.

Rohwedel, J., Maltsev, V., Bober, E., Arnold, H. H., Hescheler, J., and Wobus, A. M. (1994). Muscle cell differentiation of embryonic stem cells reflects myogenesis *in vivo*: Developmentally regulated expression of myogenic determination genes and functional expression of ionic currents. *Dev. Biol.* **164,** 87–101.

Rosenblatt, J. D., Lunt, A. I., Parry, D. J., and Partridge, T. A. (1995). Culturing satellite cells from living single muscle fiber explants. *In Vitro Cell Dev. Biol. Anim.* **31,** 773–779.

Sakuragawa, N., Elwan, M. A., Fujii, T., and Kawashima, K. (1999). Possible dynamic neurotransmitter metabolism surrounding the fetus. *J. Child Neurol.* **14,** 265–266.

Schwartz, R. E., Reyes, M., Koodie, L., Jiang, Y., Blackstad, M., Lund, T., Lenvik, T., Johnson, S., Hu, W. S., and Verfaillie, C. M. (2002). Multipotent adult progenitor cells from bone marrow differentiate into functional hepatocyte-like cells. *J. Clin. Invest.* **109,** 1291–1302.

Siddiqui, M. j., and Atala, A. (2004). "Amniotic Fluid-Derived Pluripotential Cells. Handbook of Stem Cells." Vol. 2, pp. 175–179. Academic Press.

Smith, J. L., Gesteland, K. M., and Schoenwolf, G. C. (1994). Prospective fate map of the mouse primitive streak at 7.5 days of gestation. *Dev. Dyn.* **201,** 279–289.

Snow, M. H., and Bennett, D. (1978). Gastrulation in the mouse: Assessment of cell populations in the epiblast of tw18/tw18 embryos. *J. Embryol. Exp. Morphol.* **47,** 39–52.

Srivastava, M. D., Lippes, J., and Srivastava, B. I. (1996). Cytokines of the human reproductive tract. *Am. J. Reprod. Immunol.* **36,** 157–166.

Tsai, M. S., Lee, J. L., Chang, Y. J., and Hwang, S. M. (2004). Isolation of human multipotent mesenchymal stem cells from second-trimester amniotic fluid using a novel two-stage culture protocol. *Hum. Reprod.* **19,** 1450–1456.

Woodbury, D., Schwarz, E. J., Prockop, D. J., and Black, I. B. (2000). Adult rat and human bone marrow stromal cells differentiate into neurons. *J. Neurosci. Res.* **61,** 364–370.

CHAPTER 10

Hematopoietic Stem and Progenitor Cells from Cord Blood

Hal E. Broxmeyer,[*,†,§,||] **Edward Srour,**[†] **Christie Orschell,**[†]
David A. Ingram,[‡,¶] **Scott Cooper,**[*,§,||] **P. Artur Plett,**[†]
Laura E. Mead,[‡,¶] **and Mervin C. Yoder**[‡,¶]

[*]Departments of Microbiology and Immunology
Indiana University School of Medicine
Indianapolis, Indiana 46202

[†]Department of Medicine
Indiana University School of Medicine
Indianapolis, Indiana 46202

[‡]Department of Pediatrics
Indiana University School of Medicine
Indianapolis, Indiana 46202

[§]The Walther Oncology Center
Indiana University School of Medicine
Indianapolis, Indiana 46202

[¶]The Wells Center for Pediatric Research
Indiana University School of Medicine
Indianapolis, Indiana 46202

[||]The Walther Cancer Institute
Indianapolis, Indiana 46206

Abstract
1. Introduction: Cord Blood Transplantation
   1.1. Hematopoietic stem and progenitor cells
   1.2. Endothelial progenitor cells (EPC)
2. Methodologies for Assessing Hematopoietic Progenitor and Stem Cells, as Well as EPC, Present in Human Cord Blood
   2.1. Methods for enumerating human myeloid progenitor cells in umbilical cord blood by *in vitro* colony formation

2.2. CFU-GM: Colony Forming Unit-Granulocyte/Macrophage
2.3. BFU-E/CFU-GEMM: Burst forming unit-Erythroid/Colony forming
      unit-Granulocyte/Erythroid/ Macrophage/Megakaryocyte colony
      forming assay (Methylcellulose)
2.4. Colony replating (A means to estimate the limited self-renewal capacity
      of progenitor cells)
2.5. Methods for enumerating human hematopoietic stem cells in
      umbilical cord blood
3. Uses of Above Assays
A1. Appendix A
A2. Appendix B
      References

Abstract

Cord blood has served as a source of hematopoietic stem and progenitor cells for successful repopulation of the blood cell system in patients with malignant and nonmalignant disorders. It was information on these rare immature cells in cord blood that led to the first use of cord blood for transplantation. Further information on these cells and how they can be manipulated both *in vitro* and *in vivo* will likely enhance the utility and broadness of applicability of cord blood for treatment of human disease. This chapter reviews information on the clinical and biological properties of hematopoietic stem and progenitor cells, as well as the biology of endothelial progenitor cells (EPC), and serves as a source for the methods used to detect and quantitate these important functional cells. Specifically, methods are presented for enumerating human cord blood: (1) myeloid progenitor cells, including granulocyte-macrophage (CFU-GM), erythroid (BFU-E), and multipotential (CFU-GEMM or CFU-Mix) progenitors, and their replating potential; (2) hematopoietic stem cells, as assessed *in vitro* for long term culture-Initiating cells (LTC-IC), cobblestone area forming cells (CAFC), and myeloid-lymphoid-initiating cells (ML-IC), and as assessed *in vivo* for non obese diabetic (NOD)/severe combined immunodeficient (SCID) mouse repopulating cells (SRC); and (3) high- and low-proliferative potential EPC.

1. Introduction: Cord Blood Transplantation

Umbilical cord blood, collected at the birth of a baby, is a rich source of hematopoietic stem and progenitor cells (Broxmeyer, 2004, 2005). Cord blood stem and progenitor cells have been used for more than 13,000 transplants to treat a wide range of malignant and nonmalignant disorders (Broxmeyer and Smith, 2004, 2008; and unpublished information). The first cord blood transplant utilized sibling human leukocyte antigen (HLA)-matched cells to cure the hematological manifestations of Fanconi anemia (Gluckman *et al.*, 1989). The recipient of this transplant, that took place in October 1988, is still alive and well more than 20 years after the transplant. The studies and events that led up to this transplant have been reported (Broxmeyer, 1998, 2000; Broxmeyer *et al.*, 1989b, 1990b; Gluckman and Rocha, 2005).

The initial cord blood transplants were limited to HLA-matched sibling donors (Wagner *et al.*, 1995) but the encouraging results soon led to the use of partially HLA-matched sibling and related transplants, and then to the use of unrelated allogeneic cord blood transplants that were first completely, and then subsequently partially matched for HLA (Broxmeyer and Smith, 2004; Eapen *et al.*, 2006; Gluckman *et al.*, 1997; Hwang *et al.*, 2007; Kurtzberg *et al.*, 1996; Rubinstein *et al.*, 1998; Wagner *et al.*, 1996, 2002). Most of the original cord blood transplants were done in children or in low-weight recipients because of the fear that the numbers of hematopoietic stem and progenitor cells in single collections of cord blood were limiting in number, and thus, might compromise the engraftment capability of adults and higher weight recipients. However, as transplanters have gained more experience, increasing numbers of adults have been transplanted (Brunstein *et al.*, 2007; Cornetta *et al.*, 2005; Gluckman *et al.*, 2004; Laughlin *et al.*, 2001, 2004; Long *et al.*, 2003; Ooi *et al.*, 2002; Rocha *et al.*, 2004; Rubinstein *et al.*, 1998; Sanz *et al.*, 2001a,b; Takahashi *et al.*, 2004; Tse and Laughlin, 2005).

Limiting numbers of hematopoietic stem and progenitor cells still present a logistical problem for the ultimate broadness of applicability of cord blood transplantation in adults and in children. Current practice suggests the need for greater than or equal to 2×10^7 nucleated cells/Kg recipient body weight for successful transplantation, although there have been successful outcomes with fewer than this number of cells (Gluckman *et al.*, 2004).

Currently, there are three avenues of research that clinical and basic science investigators are pursuing in order to deal with the limited numbers of nucleated, as well as stem and progenitor, cells in single cord blood collections. A most recent clinical effort reflects the use of multiple cord blood units for transplantation in single recipients (Ballen *et al.*, 2007; Barker *et al.*, 2005). The results here are encouraging but only one cord blood unit eventually "wins-out" in the competition, and it is not clear yet what factors define the "winning" unit. While double cord blood transplants in adults are very encouraging, and have led to greatly increased numbers of cord blood transplants being performed in adults, there have not yet been controlled trials that definitively substantiate that multiple cord bloods are more efficacious than a single unit, and this remains to be performed.

Another clinical effort has focused on the use of cells that have first been cultured *ex vivo* (out of the body) in attempts to expand the numbers of hematopoietic stem and progenitor cells beyond that collected from the single units before infusions of these cells in recipients (Jaroscak *et al.*, 2003; Shpall *et al.*, 2002). These results have not been very encouraging so far, most likely because either the hematopoietic stem cells that would allow for long-term stable and multilineage engraftment have not been expanded, or the engrafting capabilities of these cells have been compromised by the *in vitro* culture conditions.

A third effort has focused on enhancing the homing characteristics of the hematopoietic stem cells (Christopherson and Broxmeyer, 2004; Christopherson *et al.*, 2004). A number of factors have been implicated in the homing of hematopoietic stem and progenitor cells to the environmental niches in the bone marrow that nurture these cells for survival, self-renewal, proliferation, and differentiation (Christopherson and Broxmeyer, 2004; Lapidot *et al.*, 2005). It is believed that not all hematopoietic stem cells home with absolute efficiency to their microenvironmental niche(s) after infusion into transplant recipients. Thus, increasing the homing efficiency of limiting numbers of available hematopoietic stem and progenitor cells may enhance engraftment and repopulation of blood cells. One of the means to enhance the homing and engraftment of hematopoietic stem and progenitor cells is to inhibit the Dipeptidylpeptidase IV activity of CD26. Such efforts have resulted in enhanced engraftment of limiting numbers of mouse bone marrow stem cells into lethally irradiated mice (Christopherson *et al.*, 2004). These results have now been confirmed by a number of other groups for mouse bone marrow cell engraftment in adult mice (Tian *et al.*, 2005) and also *in utero* (Peranteau *et al.*, 2005). Additionally, inhibition of CD26 on hematopoietic stem and other cells present in human cord blood resulted in enhanced engraftment of these stem cells in mice of the non obese diabetic or severe combined immunodeficiency (NOD/SCID) genotype (Campbell and Broxmeyer, 2007; Campbell *et al.*, 2007; Christopherson *et al.*, 2007). Inhibition of CD26 in recipient mice have also accelerated engraftment by limiting numbers of HSC (Broxmeyer *et al.*, 2007; Campbell and Broxmeyer, 2007; Kawai *et al.*, 2007). Enhanced engraftment may be due to CD26 inhibition enhancing the activity of stromal cell derived factor-1 (SDF-1/CXCL12), as well as other cytokines, that are involved in the survival, proliferation, self-renewal, and homing/migration of HSC (Broxmeyer *et al.*, 2008). It remains to be determined if inhibition of CD26 peptidase activity found on human cord blood hematopoietic stem cells, or other means to enhance homing and engraftment of these stem cells, will result in enhanced engraftment in human recipients transplanted with limiting numbers of stem cells.

1.1. Hematopoietic stem and progenitor cells

Hematopoietic stem cells are defined by their capacity to give rise to more of their own kind, a process termed self-renewal, and to differentiate down multiple blood lineages, including erythroid cells, neutrophils, monocytes, macrophages, platelets, T-lymphocytes, B-lymphocytes, other lymphocytes, natural kills cells, natural killer T cells, dendritic cells, etc. Hematopoietic progenitor cells have more limited capacity for self-renewal than stem cells, and they are more limited in their capacity to give rise to multiple blood cell types. The hierarchy of blood cell development begins with hematopoietic stem cells, which give rise to various hematopoietic progenitor cells, which then differentiate into precursor cells for the individual blood cell lineages. The regulation of the self-renewal, proliferation and differentiation of these cells are modulated by cytokines, chemokines, and other growth factors (Shaheen and Broxmeyer, 2005, 2008) as well as by local cell–cell interactions involving stromal cells.

There is more known about human hematopoietic progenitor cells than about human hematopoietic stem cells. This is in part due to the fact that the stem cells are rarer in frequency than the progenitors, and assays for progenitor cells, performed *in vitro*, are easier to do than assays for engrafting stem cells which are done *in vivo*. Hematopoietic progenitor cells in cord blood are increased in frequency compared to that in bone marrow (Broxmeyer *et al.*, 1989a,b) and are enhanced in proliferative capacity, generation of progeny, and also in replating capacity *in vitro* (Broxmeyer *et al.*, 1989a,b; Broxmeyer *et al.*, 1992; Cardoso *et al.*, 1993; Carow *et al.*, 1991, 1993; Lansdorp *et al.*, 1993; Lu *et al.*, 1993); replating capacity offers an estimate on the limited self-renewal capacity of progenitor cells (Carow *et al.*, 1991, 1993). Hematopoietic progenitor cells are $CD34^+$ and are found enriched in the $CD34^+ CD38^+$ subset of cells after cell sorting. Progenitors can be ranked as those that are immature with enhanced proliferative capacity, and those that are more mature, with decreased proliferative and more limited and restricted differentiation capability. It is believed that the immature subsets of progenitors are responsive to stimulation by combinations of growth factors (Shaheen and Broxmeyer, 2005). Thus, multipotential progenitor cells (termed Colony Forming Unit-Granulocyte, erythroid, macrophage, megakaryocyte; CFU-GEMM, or CFU-Mix) are denoted by the colonies of mixed lineage blood cells they give rise to in semisolid culture medium when the cells are stimulated *in vitro* by combinations of cytokines, such as erythropoietin (Epo), stem cell factor (SCF), granulocyte-macrophage colony stimulating factor (GM-CSF), and inter-leukin-3 (IL-3), in the absence or presence of thrombopoietin (TPO). Immature subsets of granulocyte-macrophage progenitor cells (Colony Forming Unit-Granuloctye, macrophage; CFU-GM) are detected by the colonies of granulocytes and macrophages they give rise to in the presence of either GM-CSF or IL-3 combined with potent costimulating cytokines SCF or Flt-3-ligand (FL). More mature subsets of CFU-GM, which give rise to smaller colonies *in vitro*, are detected after stimulation of the cells with either GM-CSF or IL-3. Granulocyte progenitors (Colony Forming Unit-Granulocyte; CFU-G) and macrophage pro-genitors (Colony Forming Unit-Macrophage; CFU-M) are detected after stimula-tion, respectively, by either granulocyte colony stimulating factor (G-CSF) or macrophage colony stimulating factor (M-CSF), and more immature subsets of CFU-G and CFU-M are detected by adding either SCF or FL with G-CSF or M-CSF. Erythroid progenitor cells (Burst Forming Unit-Erythroid; BFU-E) or megakaryocyte progenitor cells (Colony Forming Unit-Megakaryocyte; CFU-Meg) are detected by, respectively, stimulating these cells in semisolid culture medium with either Epo, or TPO, and addition of costimulating factors induces larger colonies that derive from more immature subsets of these progenitor cells. Examples of CFU-GEMM-, BFU-E-, and CFU-GM-derived colonies are shown in Fig. 10.1. Cord blood cells have been stored frozen in cryopreserved form for over 20 years without loss of functional hematopoietic stem and progenitor cell

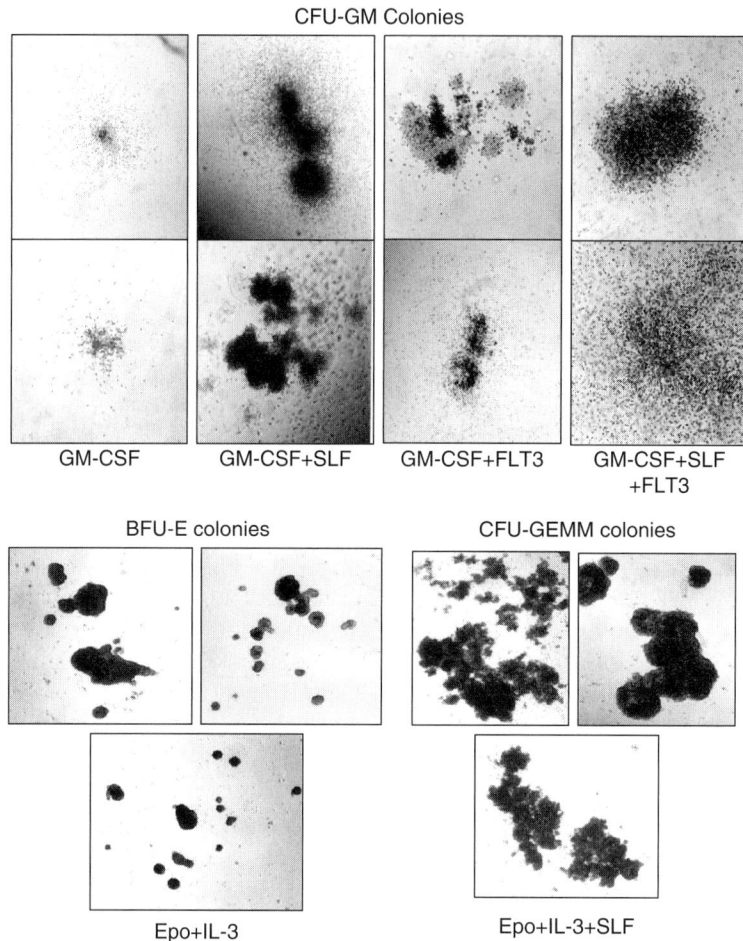

Figure 10.1 Colonies formed from cord blood CFU-GM, BFU-E, and CFU-GEMM stored frozen for 15 years, defrosted, and plated in semisolid agar culture medium for 14 days in the presence of GM-CSF alone or in combination with SLF, FL, or SLF plus FL for CFU-GM, and in semisolid methyl cellulose culture medium for 14 days, respectively, in the presence of Epo plus IL-3 and of Epo and IL-3 plus SLF for BFU-E and CFU-GEMM. These are representative of the largest colonies formed. This figure is reproduced from Fig. 10.1 of Broxmeyer *et al.* (2003) with permission from the Proceedings of the National Academies of Science, USA.

activity (Broxmeyer and Cooper, 1997; Broxmeyer *et al.*, 1992, 2003; Broxmeyer, unpublished observations, 2008).

Cord blood CD34^{+++} cells, those CD34^{+} cells expressing the highest density of CD34 antigens, are highly enriched in more immature subsets of CFU-GEMM, BFU-E, and CFU-GM (Lu *et al.*, 1993). In fact, the cloning efficiency of CD34^{+++} cord blood cells can be as high as 80% for these progenitors under optimal conditions when combinations of cytokines are added to the *in vitro* semisolid cultures. Less is known about lymphoid progenitor cells in human cord blood than of the myeloid progenitor (CFU-GEMM, BFU-E, CFU-GM) cells.

Also, less is known phenotypically and functionally about hematopoietic stem cells from humans than from mice. In mice, a phenotype of $CD34^-$ $Sca1^+$ $c\text{-}kit^+$ Lin^- defines stem cells that under optimal conditions may engraft lethally irradiated mice at levels as low as one of these phenotyped cells. The human assay for hematopoietic stem cells is not as definitive. The assay for human hematopoietic stem cells takes advantage of the engraftment of sublethally irradiated SCID mice (usually NOD/SCID mice) with human cells. Human cord blood cells have greater capacity to repopulate these conditioned NOD/SCID mice than do human bone marrow or G-CSF mobilized adult blood cells (Bock *et al.*, 1995; Bodine, 2004; Orazi *et al.*, 1994; Vormoor *et al.*, 1994). These human SCID-repopulating cells (SRC; considered to be engrafting hematopoietic stem cells) are enriched in the $CD34^+$ $CD38^-$ population of human cord blood, but the relative inadequacy of phenotypically identifying human hematopoietic stem cells is highlighted by the finding that approximately only 1/700 $CD34^+$ $CD38^-$ cord blood cells is an SRC. Nevertheless, this frequency of SRC is much greater than for SRC found in bone marrow or mobilized adult peripheral blood. Examples of flow cytometric readout of human cord blood cells infused and those that engrafted the bone marrow of sublethally irradiated NOD/SCID mice are shown in Fig. 10.2. Next generation NOD/SCID mice (NS2) (Ishikawa *et al.*, 2005; Ito *et al.*, 2002) which are IL-2 receptor gamma chain null are superior in their acceptance of human cell engraftment. The percent human cell engraftment is higher in the bone marrow, and can also be detected in blood, for over 8 months. The original NOD/SCID mice rarely detected human cell engraftment in the blood, and for only a few months in the bone marrow before the mice became ill.

1.2. Endothelial progenitor cells (EPC)

EPC have been detected, characterized, and isolated from human cord blood (Aoki *et al.*, 2004; Bompais *et al.*, 2004; Crisa *et al.*, 1999; Eggermann *et al.*, 2003; Fan *et al.*, 2003; Hildbrand *et al.*, 2004; Ingram *et al.*, 2004; Kang *et al.*, 2001; Murga *et al.*, 2004; Peichev *et al.*, 2000; Pesce *et al.*, 2003). During embryogenesis, blood vessels are formed *de novo* by the patterned assembly of angioblasts in a process termed vasculogenesis (Flamme *et al.*, 1997). Once an intact vascular system has been established, the development of new blood vessels primarily occurs via the sprouting of endothelial cells from postcapillary venules or the maturation and *de novo* growth of collateral conduits from larger diameter arteries (Simons, 2005; Skalak, 2005). These two mechanisms of new blood vessel formation are, respectively, termed angiogenesis and arteriogenesis. A population of human circulating $CD34^+$ cells was described that could differentiate *ex vivo* into cells with endothelial cell-like characteristics (Asahara *et al.*, 1997). These cells were termed "EPCs"; this study challenged traditional understanding of angiogenesis by suggesting that circulating cells in adult peripheral blood may also

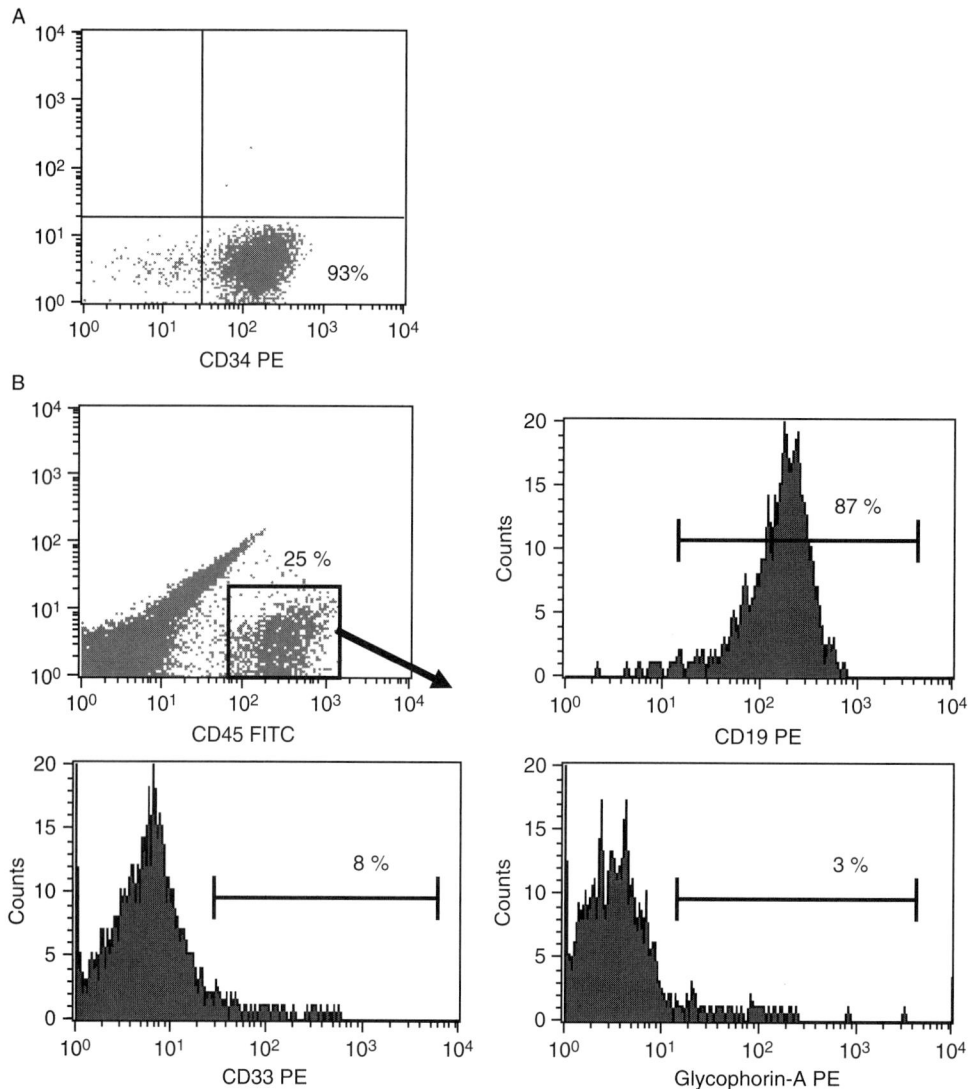

Figure 10.2 CD34+ cells were isolated utilizing anti-CD34 antibodies and magnetic beads system from Miltenyi. 1×10^5 cells to 3×10^5 cells were then infused into 300 cGy-conditioned NOD/SCID mice via tail vein injection. Mice were euthanized 8–9 weeks post transplantation, BM cells harvested and stained with anti-human CD45 Fluoroisothiocyanate(FITC) and the indicated lineage markers (CD19/lymphoid, CD33/myeloid, Glycophorin-A/Erythroid) conjugated to Phycoerythrin(PE). Flow cytometry was utilized to assess percent human chimerism and lineage marker expression on engrafted cells. **Panel A** shows a representative flow cytometric plot of purified CB CD34+ cells used in *in vitro* experiments and as grafts in NOD/SCID mice. **Panel B** shows flow cytometric plots of human cells (CD45+) in NOD/SCID mice and the gating used to obtain the expression of lineage markers on these cells.

contribute to new vessel formation. Subsequent studies showed that these cells were derived from bone marrow, circulate in peripheral blood, and home to sites of new blood vessel formation that include ischemic tissues and tumor microenvironments in a process termed "postnatal vasculogenesis" (Reviewed in: Dome *et al.*, 2008; Hristov and Weber, 2004; Iwami *et al.*, 2004; Kawamoto and Asahara, 2007; Khakoo and Finkel, 2005; Murasawa and Asahara, 2005; Rafii and Lyden, 2003; Schatteman, 2004; Urbich and Dimmeler, 2004).

Utilizing sophisticated cell marking strategies, more recent studies indicate that marrow-derived EPCs may play minimal or no role in neovascularization of tumors, vessel repair, or normal vessel growth and development (Gothert *et al.*, 2004; Purhonen, *et al.*, 2008; Stadtfeld and Graf, 2005). These conflicting reports have raised questions about the function of EPCs in vascular homeostasis and repair. The controversies surrounding these fundamental questions may, in part, originate from the heterogenous phenotypic definitions of EPCs and a lack of functional clonogenic assays to isolate and accurately describe the proliferative potential of EPCs.

Some of the coauthors of this chapter described an approach that identifies a novel hierarchy of EPCs based on their clonogenic and proliferative potential, analogous to the hematopoietic stem/progenitor cell system (Ingram *et al.*, 2004). Using this approach, they identified a previously unrecognized population of EPCs in cord blood that achieved at least 100 population doublings, could be replated into at least secondary and tertiary colonies, and expressed levels of telomerase activity that correlated with the degree of cell autonomous proliferative potential. These studies described a clonogenic method that defined a hierarchy of EPCs based on their proliferative potential, and identified a unique population of high proliferative potential-endothelial colony forming cells (HPP-ECFC) in human umbilical cord blood (Ingram *et al.*, 2004) Examples of colonies that formed *in vitro* from human cord blood high and low proliferative EPCs are shown in Fig. 10.3.

2. Methodologies for Assessing Hematopoietic Progenitor and Stem Cells, as Well as EPC, Present in Human Cord Blood

The assays described below allow detection of functional progenitor and stem cells and have been used to great advantage to determine our current knowledge of the biology of these cells in umbilical cord blood as defined above.

2.1. Methods for enumerating human myeloid progenitor cells in umbilical cord blood by *in vitro* colony formation

Human myeloid progenitor cells are not clearly definable by phenotype with the precision of mouse myeloid progenitors. Phenotype does not always recapitulate function. Therefore, it is always wise when defining a progenitor or stem cell by phenotype to confirm the functional capacity of the phenotyped cell. Functionally,

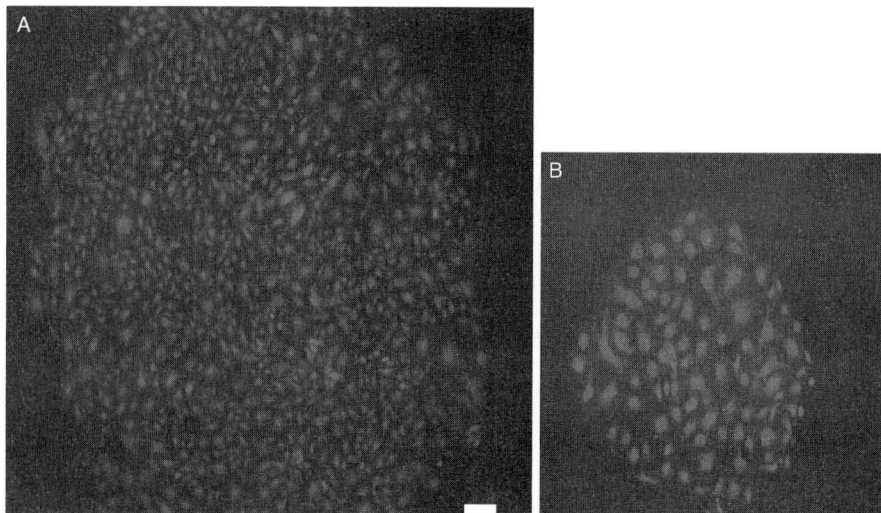

Figure 10.3 Photomicrographs of colonies formed from human cord blood ECFC. (A) Depiction of a HPP-ECFC-derived colony and (B) a low proliferative potential ECFC-derived colony. The bar in (A) is 50 μ.

myeloid progenitor cells are defined, in retrospect, at the end of proliferation and maturation by evaluating the types of cells formed within a colony at the end of a culture period and by the growth factors used for stimulation. By culturing a population of cells known to contain progenitor cells under defined conditions, one can obtain proliferation in the form of colonies ($>$40 cells per group) or clusters (3–40 cells). By utilizing a semisolid material in these cultures, such as agar, agarose, or methylcellulose, movement of the daughter cells produced by these progenitor cells, can be restricted and thereby, allow the formation of colonies. These colonies can then be counted and the number and types of progenitors present in the starting population can be determined. This assay is used primarily to quantitate the numbers of progenitors in a given sample and shed light on their proliferative potential (colony vs. cluster, as well as the number of cells within a colony).

2.2. CFU–GM: Colony Forming Unit-Granulocyte/Macrophage

Materials/Reagents (See Appendix A/B for Suppliers/Recipes)

Unless otherwise stated, all the materials should be sterile

1. Cells to be analyzed[1]
2. Agarose (0.8%) or Agar (0.6%) solution—sterile

---

[1] The cells used can be either whole, unseparated blood, or enriched populations such as low density (LD) mononuclear cells, FACS sorted, or bead sorted CD34$^+$ cells, etc.

3. 2× McCoys 5A complete medium

4. Fetal Bovine Serum (FBS)—(heat inactivated @ 37 °C for 30 min)

5. Growth factors (recombinant)

6. Petri dishes, 35 mm

7. Assortment of disposable pipettes

Procedure

All steps are performed aseptically.

1. Melt the agarose/agar in microwave and place at 45 °C until needed. This part of the culture mix is to be added last, just prior to plating.

2. For any given volume, combine the following reagents (except the agar/agarose) in the following proportions and mix well:

FBS	10%
Growth factors[2]	<1%
GM-CSF	(10–20 ng/ml)
M-CSF	(10–20 ng/ml)
SCF	(50 ng/ml)
FL	(100 ng/ml)
Cells[3]	0–5%
2× McCoys	(35–40%, qs volume)
Agar	(0.6%)50%

The agar is added last and just prior to plating. Make sure that it is not so hot that it will kill the cells.

---

[2] *The growth factors used will determine the final colony output in terms of type; For example, GM-CSF yields granulocyte-macrophage colonies, M-CSF yields macrophage colonies, etc. Single growth factors or combinations of growth factors may be used. Typically, to pick up earlier, more immature subsets of CFU-GM, we use GM-CSF (10 ng/ml) with either SCF (50 ng/ml) or FL (100 ng/ml).*

[3] *Depending on the type of cells used and final cell concentration of your cell preparation, the final cell concentration and volume of cells added to the culture mix (see below) will vary. Typically, we plate unseparated cord blood cells at 2.5–5.0×10⁴ cells/ml, low density cells at 1.25–2.5×10⁴ cells/ml, CD34⁺ cells at 50–1000 cells/ml (depending on purity). It is important that the number of cells plated yield plates that have enough colonies to score accurately. This means colonies that do not overlap with each other, or do so in minimal fashion. This may require plating cells at several different concentrations. For example, when plating CD34⁺ cells, at least three different concentrations are recommended. The volume added will be determined by the final concentration of the cell suspension after enrichment. When using unseparated blood, microliter volumes are added to the culture mix and therefore do not constitute a significant volume change with respect to the 2× McCoy's volume. However, when using low density or CD34⁺ cells, the final cell concentration may only be 10–20 times the desired culture cell concentration and thereby constituting 5–10% of the culture mix volume, thus requiring an adjustment to the 2× McCoy's volume.*

1. Add agar (~40–42 °C) and mix well.
2. Pipette 1 ml into each dish (35 mm TC dish) and swirl the plates to cover the bottom with culture. At least 3 plates per point should be plated.
3. Once all of your plates are dispensed, leave at room temp for 10 min or until all the plates have solidified.
4. Culture plates in a humidified environment @ 37 °C and 5% CO_2 for 7–14 days. We have found that culturing plated at lowered O_2 tension (e.g., 5%, rather than 20%) enhances the number of colonies detected (Broxmeyer, *et al.*, 1989a,b, 1990a,b; Lu and Broxmeyer, 1985; Smith and Broxmeyer, 1986).

2.3. BFU-E/CFU-GEMM: Burst forming unit-Erythroid/Colony forming unit-Granulocyte/ Erythroid/ Macrophage/Megakaryocyte colony forming assay (Methylcellulose)

Materials/Reagents (See Appendix A/B for Suppliers/Recipes)
Unless otherwise stated, all the materials should be sterile

1. Cells to be analyzed (same as described in footnote 1)
2. Methylcellulose, 2.1%
3. IMDM with penicillin/streptomycin
4. FBS (NOT heat inactivated)
5. Glutamine
6. 2-Mercaptoethanol (2ME), 2×10^{-5} M
7. Erythropoietin (EPO) (AmGen, Epogen)
8. Source of Growth Factors (usually recombinant)
9. Petri dishes, 35 mm
10. Assortment of disposable pipets

Procedure
All steps are performed aseptically.

1. Warm the methylcellulose in a 37 °C water bath.
2. The following quantities are for a 5 ml culture. However, by adjusting volumes proportionately, larger volumes can easily be obtained in order to test a variety of variables. Combine all the following reagents except for the methylcellulose and mix by swirling or gentle vortex.

Add methylcellulose and mix by vortexing.

3. Allow the tube to sit for several minutes to allow all the bubble to rise to the top.
4. Using a 3–5 ml syringe fitted with a 16 gauge needle, transfer 1 ml culture mix into each dish (35 mm TC dish). Swirl the plates to cover the bottom with culture mix. At least 3 plates per point should be plated. Since methylcellulose does not solidify, the entire experiment can be plated before each point is mixed.
5. Culture plates in a humidified environment at 37 °C and 5% CO_2/O_2 for 14 days.

Methylcellulose	(2.1%) 2.5 ml
FBS	1.5 ml
Glutamine	50 μl
2ME (2×10^{-5} M)	16.7 μl
EPO (200 U/ml)	25 μl
Growth Factors[4]	0.25 ml
Epo (1–2 U/ml)	
IL-3 (10 ng/ml)	
GM-CSF (10 ng/ml)	
SCF (50 ng/ml)	
TPO (10 ng/ml)	
Cells	0.25 ml
IMDM	q.s. to 5 ml

2.4. Colony replating (A means to estimate the limited self-renewal capacity of progenitor cells)

Note: This technique can only be adequately performed using primary cultures grown in methylcellulose. Agar cultures are far too solid and it is extremely difficult to remove the colonies from agar for replating into a secondary culture plate. However, to grow primary CFU-GM colonies for replating, all you need to do is replace the agar with methylcellulose and the 2× McCoy's with IMDM.

Materials/Reagents

Unless otherwise stated, all the materials should be sterile

1. Clean bench with sterile air flow
2. Inverted tissue culture microscope
3. Micropipettor (20 μl)
4. IMDM w/ pen/strep
5. 12 × 75 mm polypropylene tubes
6. All reagents as previously described for desired culture conditions

Procedure for colony harvest

All steps are performed aseptically.

1. Determine the number of colonies you wish to replate into secondary cultures. Prepare one tube for each colony by dispensing 100 μl of IMDM with 2% FBS into each tube. Recap.
2. Set up the microscope in a clean sterile air bench environment.

---

[4] *The growth factors used will determine the final colony output in terms of type and colony content. If Epo is used alone, only Erythroid colonies will be enumerated. But if a combination of growth factors are used, then multipotential progenitors will be enumerated. Typically, to pick up maximal numbers of multipotential colonies, we use IL-3, GM-CSF, SCF, Epo, with or without TPO. The addition of TPO increases this number of megakaryocytes in the mixed colonies.*

3. Place the plate containing the colonies to be harvested on the platform of the microscope and remove dish lid. The plate should be immobilized.

4. If the microscope is not equipped with a small plate holder, you may immobilize the plate by using two small loops of tape rolled sticky side out and placed on the stage. The plate can then be pushed up against the tape for the harvest.

5. Find the desired colony to be replated and using a 20 μl pipettor, gently aspirate the colony into the pipettor.

6. The best colonies for harvest will be well isolated.

7. Transfer the colony to a tube containing 100 μl IMDM medium (previously prepared) and rinse the tip with several up and down motions.

8. Once all desired colonies have been harvested, proceed to plating procedure.

Plating procedure for harvested colony culture

All steps are performed aseptically.

1. Determine the number of colonies you have harvested. Based on 1.1 ml per colony harvested mix a volume of culture medium/mix, INCLUDING methylcellulose, and set aside to allow air bubble to rise.

    For example, if 100 colonies were harvested from plates containing IL-3, Epo, and SCF as growth factors, prepare 110 ml of culture mix as described in Section 2, without the cells, containing all the growth factor as were in the primary plates and with the methylcellulose added before the cells.

2. Once the culture mixture has settled, using a 3–10 ml syringe fitted with a 16 gauge needle, dispense 1 ml into each tube containing the harvested colony and mix by vortexing.

3. Allow to settle.

4. Transfer as much of this culture mix to a new 35 mm plate and swirl to cover bottom of plate.

    Because methylcellulose is so viscous, it is virtually impossible to replate 100% of the harvested colony; some will remain behind. If a quantitative outcome is desired, you can simply plate a know percentage of the culture mix and do a corrected calculation once the secondary colonies have been scored; that is, Plate 0.9 ml of 1.1 ml cell mix (1 ml mix + 0.1 ml harvested colony). Scored colonies would then represent ~82% of total colony forming cells in harvested colony.

5. Culture plates in a humidified environment @ 37 °C and 5% CO_2/O_2 for 14 days.

2.5. Methods for enumerating human hematopoietic stem cells in umbilical cord blood

The most reliable method for assessing human hematopoietic stem cells is that involving engraftment of human cells into SCID (e.g., NOD/SCID) mice (Bodine, 2004). However, there are *in vitro* assays that have been used as surrogate assays for those stem cells detected by *in vivo* analysis. This section will first cover *in vitro* assays that are believed to detect cells within the stem cell category, although these cells may not be equivalent to the long-term marrow *in vivo* repopulating hematopoietic stem cell.

2.5.1. *In vitro* hematopoietic stem cell assays for cord blood

Although bone marrow and more recently, mobilized peripheral blood, have been reliably used as sources of hematopoietic stem cells for transplantation, umbilical cord blood is, as noted above, fast becoming recognized as a source of stem cells for transplantation of children and adults. The primary concern with cord blood is, however, the relatively lower numbers of total transplantable stem cells that can be isolated from single collections of cord blood compared to marrow or mobilized peripheral blood, and questions of whether sufficient numbers of hematopoietic stem cells are available in a typical cord blood collection for use in adults. For this reason, investigators have sought reliable *in vivo* and *in vitro* assays to estimate the stem cell content of different tissue sources.

Of the available *in vitro* assays, the long-term culture-initiating cell (LTC-IC) (Sutherland *et al.*, 1989), the extended (E) LTC-IC (Hao *et al.*, 1996), and cobblestone area-forming cell (CAFC) (Breems *et al.*, 1994; Ploemacher *et al.*, 1989) assays all measure primitive myeloid cells, while the myeloid-lymphoid initiating cell (ML-IC) assay measures cells with both myeloid and lymphoid (natural killer cell) activity (Punzel *et al.*, 1999).

In the LTC-IC assay (Sutherland *et al.*, 1989), putative stem cells are cultured in a modified Dexter style culture system (Dexter, 1993) with or without stromal cell support for 5 weeks, at which time LTC-IC will produce *de novo* colony forming cells (CFC) upon secondary culture. The CAFC assay uses a similar approach but with a visual endpoint for the presence of cobblestone areas (CA, tightly packed group of small cells imbedded in the stroma), eliminating the need for secondary culture. This assay can also be read sequentially at different time points, with presence of CAFC at later time points being indicative of more primitive HPC (Ploemacher *et al.*, 1991).

Methods for CAFC and LTC-IC:

1. A stromal feeder layer capable of supporting primitive hematopoiesis for up to 8 weeks is prepared. The use of primary stroma prepared from murine (Ploemacher *et al.*, 1989) or human bone marrow (Denning-Kendall *et al.*, 2003), as well as stromal cell lines, such as FBMD-1 cells (Kusadasi *et al.*, 2000), MS5, AFT024 (Theunissen and Verfaille, 2005), and M210B4 (Traycoff *et al.*, 1995) have been used with success as feeder layers in these assays. Selection of stromal cell lines should be approached with caution since different lines may provide variable support of CAFC and LTC-IC depending on their particular complement of stromal cell phenotypes. Stromal layers are cultured to confluency in 96-well flat bottomed plates, at which time cells are irradiated with 15–20 Gy radiation to eliminate hematopoietic cells while preserving the ability of stroma to support

hematopoiesis. An alternative method for LTC-IC eliminates the feeder layer and instead cultures test cells in suspension in IMDM, 10%FBS, L-glutamine, and cytokines, which may effectively push more committed cells to differentiate, thereby enriching the week 5 culture with primitive HPC scored as LTC-IC (Podesta *et al.*, 2001).

2. One to 6 days later, the irradiated stroma is overlaid with cells of interest in a limiting fashion, typically up to 12 successive 2-fold dilutions with 15–30 replicate wells per dilution. The first dilution should contain on the order of approximately 250 CB CD34+ cells/well. Media consist of normal cell culture media, such as α-medium+10%FBS, 10% horse serum, 0.5 mg/ml human transferrin, 10^{-5} mol/l hydrocortisone sodium succinate, and 10^{-4} mol/l β-mercaptoethanol (Ploemacher *et al.*, 1989), or Myelocult (Stem Cell Technologies, Inc.) $+ 10^{-6}$ mol/l hydrocortisone can also be used (Podesta *et al.*, 2001). IMDM +10%FBS and 1% l-glutamine, has also been used with success in the LTC-IC assay (Traycoff *et al.*, 1995)

3. Cultures are maintained by weekly half or whole media changes. For CAFC analysis, cultures are examined at predetermined time points (2, 4, 6, 8, or more weeks) for presence of cobblestone areas (CA, tightly packed group of at least 5 small cells imbedded beneath the stroma layer and visualized as "phase-dark" microscopically). The percentage of wells with at least one CA clone is used to calculate CAFC frequency using Poisson statistics. For LTC-IC analysis, the entire well is trypsinized, harvested, then cultured in standard hematopoietic progenitor cell assays in methylcellulose (e.g., as noted in part i above). For LTC-IC assays cultured in suspension without feeder layers, the supernatant is removed and a methylcellulose mixture is overlaid into the individual wells of the 96-well plate, and CFC assayed 2 weeks later (Traycoff *et al.*, 1995). The frequency of LTC-IC is also calculated using Poisson statistics.

Notes:

1. It has been reported that CB CD34^{-} lineage^{+} cells can produce CAFC-like colonies that are not analogous to LTC-IC, suggesting the need for assaying purified CD34+ cells instead of unfractionated mononuclear cells, which would contain false CAFC (Denning-Kendall *et al.*, 2003).

2. CAFC and LTC-IC are generally believed to assay similar, if not identical, primitive hematopoietic progenitor cells and thus could theoretically be used interchangeably. However, several reports show discrepancies between the frequencies of CAFC and LTC-IC (Gan *et al.*, 1997; Pettengell *et al.*, 1994; Weaver *et al.*, 1997), and investigators should validate their own results for each type of sample and feeder layer used.

3. LTC-IC frequency is reportedly lower in cord blood compared to bone marrow and mobilized peripheral blood, although each LTC-IC from cord blood produces significantly more CFC compared to bone marrow and mobilized peripheral blood, illustrating the robust proliferative potential characteristic of cord blood primitive hematopoietic progenitor cells (Podesta *et al.*, 2001;

Theunissen and Verfaille, 2005; Traycoff *et al.*, 1994). Of interest, while bone marrow LTC-IC were insensitive to temperature of incubation, cord blood produced significantly more LTC-IC at 37 °C than at 33 °C (Podesta *et al.*, 2001). In addition, cord blood LTC-IC also showed preference for type of stroma, producing 5-fold more LTC-IC when cultured over NIH3T3 cells compared to M210B4. Bone marrow produced similar numbers of LTC-IC over both types of stroma (Podesta *et al.*, 2001).

ML-IC Methods (Theunissen and Verfaillie, 2005): AFT024 feeder layers are overlaid with test cells in RPMI1640+20% FCS, 100 *μ*mol/l 2-mercaptoethanol, 100 U/ml penicillin and streptomycin, +cytokines (see Theunissen and Verfaillie, 2005 for details). In contrast to CAFC and LTC-IC assays, test cells are seeded onto feeders in the ML-IC assay in a nonlimiting manner. After 2–4 weeks, progeny of each well are divided into 4–8 new AFT024-coated wells and half assayed for LTC-IC content by overlaying with a methylcellulose mixture after 5 weeks in the presence of FL, SCF, and IL-7, and the other half for NK-IC content by staining with anti-CD56 antibody after 6–7 week culture (Miller *et al.*, 1999). ML-IC give rise to at least one LTC-IC and one NK-IC in secondary plates.

Notes: The frequency of ML-IC in cord blood was not statistically different than that of bone marrow or mobilized blood, although the generative potential of cord blood ML-IC was much higher than these other two sources of stem cells, again illustrating the HPP of cord blood primitive cord blood cells (Traycoff *et al.*, 1994).

2.5.2. *In vivo* animal models for the assay of cord blood–derived hematopoietic stem cells

As in the case of murine hematopoietic stem cells, the *in vivo* repopulating potential of these cells represents the most stringent measure of their functional capacity and their ability to sustain long-term multilineage engraftment. However, unlike the murine system, marrow repopulating studies in the human system have to rely on the use of different xenotransplantation models to adequately assess the *in vivo* function of putative HSCs (Bodine, 2004). Over the last 15–20 years, several of these transplantation models have been developed, each with its own advantages and disadvantages(Kamel-Reid and Dick, 1988; McCune *et al.*, 1988). Most, if not all, rely on the use of an immunocompromised host that offers a permissive microenvironment for the proliferation and differentiation of transplanted hematopoietic stem cells. While some of these models utilize large animals as recipients (Bodine, 2004), most rely on the use of different strains of immunodeficient mice with (Bodine, 2004) or without a human graft of either bone or other tissues that harbors and supports the expansion of human hematopoietic stem cells in the animal. The NOD/SCID mouse has become the most accepted and widely used mouse model to assess human hematopoietic function *in vivo*. Here, we restrict the description of our experimental design to that used with NOD/SCID mice.

The general concept behind the use of NOD/SCID mice for the assessment of stem cell function is that the immunodeficient status of these mice allows for the engraftment of human cells in the BM of recipient animals thus mimicking normal human hematopoiesis. However, as will be noted below in the notes section, several peculiar observations have been well documented in this system which are major deviations from normal human hematopoiesis. Although NOD/SCID mice are immunocompromised, further immunosuppression is still required prior to transplantation in order to facilitate or induce human stem cell engraftment. This immunosuppression is usually delivered in the form of sublethal dosages of total body irradiation.

Once all these conditions are met (in addition to some other important parameters discussed below in the notes section), it is almost always expected for an adequate number of human stem cells to engraft and populate the bone marrow of recipient mice with human progeny. Engrafted animals, the level of chimerism in individual recipients, and the identity of human lineages detected in these mice can be used collectively or individually to evaluate several stem cell functions including, but not limited to: (1) the potential of candidate stem cell phenotypes to engraft, (2) the least required number of putative stem cells to support hematopoiesis, (3) the multilineage differentiation potential of a particular graft, (4) the frequency of engrafting stem cells in a given graft, a measurement that is more accurately defined as the SRC, (5) the success of long-term expression of genes transduced into human hematopoietic stem cells, (6) homing of cord blood cells (progenitors and stem cells) to the bone marrow of recipient mice both short- and long-term, and (7) the ability of human putative hematopoietic stem cells to engraft sites other than classical hematopoietic tissues and organs.

Since we are primarily concerned here with the assessment of hematopoietic potential and number of hematopoietic stem cells in cord blood, procedures outlined below will be limited to the description of the first four assessments listed above. As can be appreciated from the final goal of each of these assays, a substantial degree of overlap normally takes place in both the establishment of these assays and the collection and interpretation of the data.

Materials and major equipment for transplantation

1. Cord blood cells fractionated or prepared as required for the assay.
2. Flow cytometric or magnetic cell sorters, if required, for the fractionation of cord blood cells.
3. Healthy NOD/SCID mice, preferably between the ages of 7 and 10 weeks; it is advisable to now use the next generation NOD/SCID mice (NS2) which are superior recipients for engraftment of human cells, as noted above at the end of the Introductory Section on hematopoietic stem and progenitor cells.
4. Cesium or X-ray irradiator.
5. Holding chambers for irradiating mice.
6. Holding chamber for tail vein injections.
7. Tubes, syringes, and needles (27 or 30 gauge).
8. Surgical masks, bonnets, gloves, and clean robes.

Transplantation procedure

1. All procedures involving handling and transplantation of NOD/SCID or NS2 NOD/SCID mice should be completed aseptically and all equipment used to restrain or hold these mice should be cleaned after each use with an antiseptic solution. Media in which cells are suspended, syringes, and needles should all be sterile.

2. The dose of radiation normally given to NOD/SCID or NS2 NOD/SCID mice ranges between 275 and 350 Rad given in one dose at a rate of approximately 75 Rad per minute. Higher doses of radiation are lethal to NOD/SCID mice. It is preferable that each investigator assess the level of radiation within the range given above that mice can tolerate without any morbidity and mortality. Transplantation can be completed anytime after irradiation but preferably within 12 h.

3. Working from inside a safety cabinet, mice should be removed from the cage and held in a clean chamber for tail vein injection.

4. Wipe the tail several times with an alcohol swab and locate the tail vein. If required a topical anesthetic can be applied prior to this along the tail.

5. Inject test cells intravenously in a 200 μl volume of sterile medium or phosphate buffered saline. Larger volumes can be used but these should not exceed 500 μl.

6. Return the transplanted mouse to the cage and observe for distress.

7. Wipe all the equipment and your gloves with an antiseptic solution prior to the transplantation of the next recipient.

Materials and major equipment for assessment of engraftment

1. Bone marrow cells from transplanted NOD/SCID mice collected from at least two long bones from both legs (to avoid sampling errors).

2. Anti human monoclonal antibodies recognizing at least CD45 and other lineage markers of interest as dictated by the overall goals of the study.

3. Anti murine CD45 monoclonal antibodies.

4. Flow cytometer.

Assessment of engraftment is usually performed at least 6 weeks post transplantation. Recently, it has been argued that at 6 weeks post transplantation, progeny of short-term repopulating cells predominate the bone marrow of recipient mice and therefore assessments made at this time point reflect the function of short, but not long-term repopulating cells. Since transplanted mice normally survive for several weeks after that, it is now more acceptable to analyze transplanted mice at 9–12 weeks post transplantation, in order to measure more accurately the function of long term repopulating cells.

Measurement of human cell engraftment

1. Collect the bone marrow from individual transplanted NOD/SCID mice.

2. Stain cells with appropriate monoclonal antibodies. It is preferable (if the limitations of the flow cytometer used for analysis allows) to include anti-human CD45 and anti-murine CD45 in every stain combination to allow for

accurate identification of chimeric human cells that can then be subjected to further linage analysis depending on the types of antibodies added to this combination.

3. Run samples on a flow cytometer and acquire at least 5000 events or until a minimum of 100 human CD45+ cells are collected.

4. In order to clearly recognize engrafted chimeric human cells within the analyzed BM samples, the following controls are suggested.

   a. Prepare bone marrow cells from a control, untransplanted NOD/SCID mouse.

   b. Mix some cells from "a" above with human peripheral blood cells to create a 90:10 ratio of murine to human cells.

   c. Stain both samples "a" and "b" above with anti-human CD45 and anti-murine CD45 individually and simultaneously.

   d. Use these samples to identify the exact position of human "chimeric" cells in a dotplot. Use this position to decide on the validity of chimeric human cells in test animals falling in or close to this position.

How to use data collected above for human cell engraftment to evaluate the first assay assessments listed above

There is no consensus as to what degree of human chimerism in a transplanted NOD/SCID mouse constitutes engraftment. It is, therefore, the responsibility of individual investigators to define engraftment in a consistent manner throughout the study and this level of human chimerism may be different for the NS2 NOD/SCID mice which demonstrate higher levels of human cell chimerism than the NOD/SCID mice. Upon the flow cytometric detection of human chimerism in the bone marrow of a transplanted mouse, the percentage of human CD45+ (or human cells positive for other markers) among all cells analyzed can be quantified automatically. This percentage can be adjusted to reflect that contained within the lymphocyte gate (through the use of light scatter properties of analyzed cells) or within total bone marrow cells excluding erythrocytes and debris. The latter is more accurate since focusing on only the lymphocyte population usually decreases the denominator used in the calculation of engraftment and thus artificially increases the level of chimerism.

1. Examining the ability and the minimum number of putative stem cells required for engraftment. Once chimerism is detected in transplanted mice, an average level of chimerism can be calculated from all mice receiving the same phenotype and number of putative stem cells. Assuming that chimerism is defined by a level of 0.1% human cells in the mouse bone marrow, comparisons can then be made between different phenotypes of putative stem cells to determine the levels of chimerism supported by each group of test cells. Similarly, with an escalating or dose dependent assay, the minimum number of test cells required to achieve 0.1% chimerism can be established for different test cells.

2. Multilineage engraftment can be assessed in mice harboring more than 0.5% human cells in most cases. Typically, analysis for the expression of lymphoid (B cells using CD19 or CD20), myeloid (using CD14, CD15, or CD33), and

erythroid (using Glycophorin A) cells can be done. As indicated in the notes section below, percentages of these lineages may not reflect normal hematopoiesis, however, when two groups of engrafting cells are compared, skewness towards one lineage or another can be detected and quantified.

3. In order to calculate a frequency of SRC in a given graft, groups of NOD/SCID mice are transplanted with different numbers of test cells per group. This design generates a limiting dilution analysis and allows for the detection of engrafted and nonengrafted animals (again, determined by an arbitrarily set lower limit) within each group of mice. The number of cells in a graft across the range of doses chosen for the experiment should be selected in a way that generates, in more than 2 groups, a mix of engrafted and nonengrafted mice. Once the number of positive (engrafted) mice is determined within each group, the reciprocal number of negative (nonengrafted) mice in the group can be calculated. The percentages of negative mice in each group can then be used to calculate a frequency based on poisson distribution (Taswell, 1981). Obviously, when two sets of groups of mice are used to examine two distinct stem cell candidates, different frequencies can be calculated and used to compare the relative abundance of repopulating cells in test grafts. It is recommended that at least four groups of mice be tested in a limiting dilution schema in which each group contains at least four mice. If the range of cell doses required to generate positive and negative results is not known, it is recommended to increase both the number of cell doses (to allow for a wider range of cell numbers) and the number of mice per group.

Notes

In this section, a few peculiarities of the NOD/SCID system will be listed to enable investigators to avoid common problems with the use of this model in examining cord blood hematopoietic stem cells *in vivo*.

1. NOD/SCID mice are "leaky." Therefore, some mice will develop mouse T cells that can be detected in the periphery as CD3+ cells. These T cells can interfere with and possibly eliminate human engraftment. It is, therefore, critical to examine these mice for the presence of murine T cells in the periphery and eliminate positive mice from the experimental design. Usually mice with >1% T cells are problematic.

2. Even well engrafted mice with high levels of chimerism in the bone marrow do not necessarily have detectable human cells in the periphery. Therefore, monitoring engraftment in the blood of recipient mice over time may generate negative results.

3. Human hematopoietic engraftment in NOD/SCID mice is preferentially skewed towards the B cell lineage whereby multilineage assessment reveals a large percentage of B cells. In reverse, T cell engraftment in NOD/SCID mice is almost completely absent in most cases unless specific types of NOD/SCID mice are used.

4. Human cytokines can be used to supplement transplanted mice. These cytokines were found to both protect recipient mice from the effects of radiation as well as modulate the behavior of chimeric human hematopoietic cells.

5. Use of antibiotics for a few days before and a few days after radiation was found to enhance survival and overall health of transplanted mice. (Edward F. Srour, unpublished observations).

6. In order to reduce the impact of the reticuloendothelial system of NOD/SCID mice on transplanted human graft cells, injection of up to 10×10^7 non-adherent, irradiated CD34$^-$ cells a few hours prior to transplantation or mixed with graft cells has been practiced.

7. A recently introduced technique suitable for multiple samplings of bone marrow cells from the femur can be used to temporally examine the kinetics of human engraftment. Alternatively, this approach has been used to deliver graft cells directly into the bone marrow microenvironment and avoid the need for homing of cells into the bone marrow of recipient animals.

8. When levels of engraftment are very low such that it can not be detected by flow cytometric analysis, investigators have relied upon Polymerse Chain Reaction (PCR) analysis to detect human genes in the murine bone marrow. While this method is suitable for the detection of very small, and perhaps insignificant levels of chimerism, PCR data cannot be reliably used for quantification of the level of chimerism.

9. Note that much of what we know regarding human cell engraftment of NOD/SCID mice will have to be reevaluated in the context of NS2 NOD/SCID mice.

2.5.3. Methods for enumerating EPC in umbilical cord blood

Information on the assay of high- and low-proliferative EPC (Ingram *et al.*, 2004) follows.

Materials:

Collection of umbilical cord blood (UCB)

1. Heparin sodium injection (1000 USP units/ml, Baxter; cat. no. NDC 0641–2440–41)
2. Sterile 60 cc syringe (Becton Dickenson; cat. no. 309663) fitted with a 16 gauge 1.5in. needle (Becton Dickenson; cat. no. 305198)

Preparation and culture of mononuclear cells (MNC)

1. Phophate buffered saline, pH 7.2 (PBS, Invitrogen, cat. no. 20012–027)
2. Ficoll-Paque PLUS (Ficoll, Amersham Biosciences; cat. no. 17–1440–03)
3. Sterile 20 cc syringe (Becton Dickenson; cat. no. 309661) fitted with a mixing cannula (Maersk Medical; cat. no. 500.11.012)
4. EBM-2 10:1:EBM-2 (Originally from Cambrex but now from Lonzo, Walkersville, Maryland; same cat. no. which is CC-3156) supplemented with 10% FBS (Hyclone; cat. no. SH30070.03) and 1% penicillin (10,000 U/ml)/ streptomycin (10,000 μg/ml)/ amphotericin (25 μg/ml, Invitrogen; cat. no. 15240–062)

5. 0.4% trypan blue solution (Sigma; cat. no. T8154) and a hemacytometer

6. cEGM-2:EGM-2 (Lonzo; cat. no. CC-3162) supplemented with the entire growth factor bullet kit, 10% FBS (Hyclone) and 1% penicillin (10,000 U/ml)/ streptomycin (10,000 μg/ml)/ amphotericin (25 μg/ml, Invitrogen; cat. no. 15240–062)

7. Collagen I coated 6 well plates (BD Biosciences Discovery Labware; cat. no. 356400)

Isolation and culture of clonal EPCs

1. Sterile cloning cylinders (Fisher Scientific; cat. no. 07–907–10)

2. Vacuum grease (Dow Corning; cat. no. 1658832), autoclave sterilized

3. Pasteur pipettes (Fisher Scientific; cat. no. 13–678–20C), autoclave sterilized

3. Trypsin EDTA (Invitrogen, cat. no. 25300–054)

4. Sterile forceps

Methods:

Collection of UCB

1. A 60 cc syringe is prepared by drawing up 2 ml heparin, pulling the plunger to the 60 ml mark and swirling the heparin to coat the entire inner surface of the syringe. The plunger is depressed to expel all excess air, leaving only 2 ml of heparin in the syringe.

2. UCB from the umbilical vein is collected into the prepared syringe using a 16 gauge needle. Immediately mix the blood by inverting several times to prevent clotting (Note 1).

Preparation and culture of MNCs

1. 15 ml of UCB is gently dispensed into each 50 ml conical tube and diluted with 20 ml of PBS.

2. 15 ml of Ficoll is drawn into a 20 cc syringe and a mixing cannula is attached. The end of the mixing cannula is placed at the bottom of the tube containing the diluted UCB and 15 ml of Ficoll is carefully underlayed, maintaining a clean interphase.

3. Tubes are centrifuged at 740g at room temperature for 30 min with no brake (Note 2).

4. Following centrifugation, red blood cells will form a pellet, and MNCs will form a hazy buffy coat at the interphase between the clear Ficoll layer below and the yellow serum layer above. The buffy coat MNCs are carefully removed with a tranfer pipette by placing the tip of the pipette just above the buffy coat layer and drawing up. MNCs are dispensed into a 50 ml tube containing 10 ml of EBM-2 10:1. Care should be taken to collect all buffy coat MNCs while avoiding excess collection of the Ficoll layer or the serum layer.

5. MNCs are centrifuged at 515g for 10 min at room temperature with a high brake and the supernatant is discarded. Pelleted cells are gently tapped loose and resuspended in 10 ml of EBM-2 10:1.

6. Repeat step 5 one more time.

7. The MNC suspension is mixed well by pipetting up and down several times. 30 μl of cells is removed and mixed with 30 μl of 0.4% trypan blue solution. Cells are loaded onto a hemacytometer and viable cells are counted.

8. The MNC suspension is again centrifuged at 515g for 10 min at room temperature with a high brake and the supernatant is discarded. Pelleted cells are gently tapped loose and resuspended in cEGM-2 at 1.25×10^7 cells/ml.

9. 5×10^7 MNCs (4 ml) are seeded into each well of 6 well tissue culture plates precoated with rat tail collagen I and cultured in a 37 °C, 5% CO_2 humidified incubator (Note 3).

10. After 24 h (d1), media is slowly removed from each well with a pipette and 2 ml of cEGM-2 is added to each well. The media is again removed from each well and replaced with 4 ml of cEGM-2. Culture plates are returned to a 37 °C, 5% CO_2 humidified incubator.

11. After 24 h (d2), media is slowly removed from each well with a pipette and 4 ml of cEGM-2 is added to each well. Media is changed exactly in this way each day from d3–d7, then every other day after d7 (Note 4).

12. EPC colonies appear between d4 and d7 of culture as well circumscribed areas of cobblestone-appearing cells. Individual colonies can be isolated and expanded on d7–d14. EPCs can also be allowed to grow to 80–90% confluency before subculturing to start a polyclonal EPC line.

Isolation and culture of clonal EPCs

1. EPC colonies are visualized by inverted microscopy (see for example, Fig. 10.3) and their location within the culture well outlined using a fine tipped marker.

2. Media is aspirated and the culture well is washed two times with PBS.

3. After aspirating the final wash of PBS, cloning cylinders coated on the bottom surface with a thin bead of vacuum grease are placed around each colony and pressed firmly against the plate using forceps.

4. Using a pasteur pipette, 1–2 drops of warm trypsin EDTA is added into each cloning cylinder. Plates are incubated at 37 °C for 1–5 min until the cells within the cylinder begin to ball up and detach (Note 5).

5. When all the cells within the cylinder have balled up, place the tip of a Pasteur pipette containing 200–300 μL of cEGM-2 into the center of the cylinder and pipette up and down vigorously several times. Collect the entire

    volume to a tube. Serially wash the area within the cylinder 1–3 more times with cEGM-2 until all cells are collected.

6. Each EPC colony collected can be seeded into one well of a 24-well tissue culture plate precoated with rat tail collagen I (Note 3) in a total volume of 1.5 ml of cEGM-2 and cultured in a 37 °C, 5% CO_2 humidified incubator. Media should be changed every other day.

Notes

1. If UCB cannot be processed immediately, it can be kept at room temperature with gentle rocking for up to 16 h.

2. Bringing the centrifuge up to 740g slowly over the course of 1–2 min helps result in a cleaner buffy coat.

3. 6-well tissue culture treated plates can also be coated with collagen I in the laboratory. To make the collagen coating solution, 0.575 ml of glacial acetic acid (17.4 N, Fisher; cat. no. A38–500) is diluted in 495 ml of sterile distilled water (0.02 N final concentration). This solution is sterile filtered with 0.22 μm vacuum filtration system (Millipore; cat. no. SCGPU05RE), then rat tail collagen I (BD Biosciences Discovery Labware; cat. no. 354236) is added to a final concentration of 50 μg/ml. This solution can be kept at 4 °C for 1 month. 1 ml of the collagen coating solution is placed in each well of a 6-well tissue culture treated plate (500 μl/well for 24-well plates, 4 ml/25 cm^2 flasks, 9 ml/75 cm^2 flasks) and incubated at 37 °C for at least 90 min. The collagen coating solution is removed and wells are washed two times with PBS prior to seeding of cells.

4. Removal and addition of media on d1–d7 of culture should be done very slowly (at a rate of about 1 ml per 4–5 s). After d7 media can be aspirated from the wells by vacuum. When changing the medium in the first few days, nonadherent MNCs will be removed along with the media and can be discarded.

5. Steps 3 and 4 must be done very quickly to prevent the EPC colonies from drying out.

3. Uses of Above Assays

    The assays for hematopoietic progenitor and stem cells and for EPC can be used to quantitate these cells in cord blood, as well as other human tissue sources of these cells. Perhaps, even more importantly, they have been and can continue to be used for assessment of the regulation of these cells by cytokines, chemokines, and cell–cell interactions. The more we learn of the biology of these cells and how they can be manipulated for self-renewal, survival, proliferation, differentiation, and migration/homing/mobilization, the more likely we can utilize this information for clinical efficacy and benefit.

A1. Appendix A: Suppliers

Agarose	(FMC ____)
Agar	(Difco; BactoAgar ____)
McCoy's 5A	(Sigma; M-4892)
MEM glutamine	(BioWhittaker; #BW17-605E)
MEM pen/strep	(BioWhittaker; #BW17-602E)
MEM vitamins	(BioWhittaker; #BW13-607C)
MEM essential amino acids	(Gibco; #11130-051)
MEM nonessential	(BioWhittaker; #BW13-114E)
MEM sodium pyruvate	(BioWhittaker; #BW13-115E)
L-Asparagine	(Sigma A-0884)
L-Serine	(Sigma S-4500)
FBS	(Hyclone; prescreened for colony growth)
Growth factors	(R&D Systems, Biovision, Peprotech, we have used products from all and all seem to have comparable activity)
IMDM (Iscove's Modied Dulbeccos medium)	(BioWhittaker; #BW12-722F)

A2. Appendix B: Recipes

A2.1. SAG (serine/asparagine/glutamine)

1. Dissolve 800 mg of L-asparagine and 420 mg of L-serine in 500 ml of ddH_2O.
2. Add 200 ml of L-glutamine.
3. Filter sterilize using a 0.2 μ vacuum filter and store frozen in 7.5 ml aliquots.

A2.2. McCoy's 5A complete

1. Dissolve 1 package (1 l) of powdered medium into 400 ml of tissue grade H_2O. Add Na_2HCO_2 as per manufacturer's directions for 1 l and stir until dissolved.
2. Add: Pen/Strep (10 ml), E-AA (20 ml), NE-AA (10), Vitamins (5 ml), Pyruvate (10 ml), SAG (15 ml).
3. Adjust pH to 7.0–7.2.
4. q.s. to 500 ml with tissue grade H_2O.
5. Filter sterilize using 0.2 μ vacuum filter.

A2.3. Methylcellulose (2.1%)

Methylcellulose can be rather tricky to make. If you do not wish to make your own, you may purchase the base medium or medium complete with growth factors from Stem Cell Technologies (https://www.stemcell.com).

Reagents/Supplies:

Sterile double distilled (dd)H_2O

Methylcellulose-4000 centipoise (Sigma # M0512)

IMDM Powder (Gibco #12200–036)

Penicillin–Streptomycin (BioWhittaker # BW17–602E)

Two 2-l flasks

3 in. Stir bar

Stir plate—strong

50 ml conical tubes

Procedure:

1. Make $2\times$ IMDM per vendor's instructions. Add pen/strep to 10 U/μg per ml. Filter sterilize and store at 40 °C until needed. NOTE: Must be warmed before adding to slurry below.

2. Place 21 gm of dry methylcellulose in a sterilized, 2 l flask containing a sterile 3 in. stir bar.

3. In a separate flask, bring 550 ml of sterile ddH$_2$O to a boil and continue boiling for 5 min.

4. In a hood, add the boiled H$_2$O to the dry powder. Mix thoroughly making sure that all the powder has been moistened.

5. Place flask on a stir plate and mix well until slurry has cooled to 370–400 °C. Do not let it cool too much or it will solidify.

6. In the hood, add 500 ml of warmed $2\times$ IMDM (per package instructions) to slurry and mix vigorously by hand for 1 min.

7. Recap and place on stir plate at room temperature for 1–2 h. Transfer to cold room and stir overnight. NOTE: This matrix will begin to gel as it cools down, so make sure the stir bar is big and spinning somewhat fast.

8. Aliquot into 50 ml tubes and store at -200 °C for no more than 6 months.

A2.4. 2-Mercaptoethanol (2ME; 2×10^{-5} M)

Make fresh every week

1. Place 7 μl 2ME (neet) into 10 ml IMDM medium.

2. Filter sterilize using a 0.2 μ syringe filter.

References

Aoki, M., Yasutake, M., and Murohara, T. (2004). Derivation of functional endothelial progenitor cells from human umbilical card blood mononuclear cells isolated by a novel cell filtration device. *Stem Cells* **22**, 994–1002.

Asahara, T., Murohara, T., Sullivan, A., Silver, M., van der Zee, R., Li, T., Witzenbichler, B., Schatteman, G., and Isner, J. M. (1997). Isolation of putative progenitor endothelial cells for angiogenesis. *Science* **275**, 964–967.

Ballen, K. K., Spitzer, T. R., Yeap, B. Y., McAfee, S., Dey, B. R., Attar, E., Haspel, R., Kao, G., Liney, D., Alyea, E., Lee, S., Cutler, C., *et al.* (2007). Double unrelated reduced-intensity umbilical cord blood transplantation in adults. *Biol. Blood Marrow Transplant.* **13**, 82–89.

Barker, J. N., Weisdorf, D. J., DeFor, T. E., Blazar, B. R., McGlave, P. B., Miller, J. S., Verfaillie, C. M., and Wagner, J. E. (2005). Transplantation of 2 partially HLA-matched umbilical cord blood units to enhance engraftment in adults with hematologic malignancy. *Blood* **105,** 1343–1347.

Bock, T. A., Orlic, D., Dunbar, C. E., Broxmeyer, H. E., and Bodine, D. M. (1995). Improved engraftment of human hematopoietic cells in severe combined immunodeficient (SCID) mice carrying human cytokine transgenes. *J. Exp. Med.* **182,** 2037–2043.

Bodine, D. M. (2004). Animal models for the engraftment and differentiation of human hematopoietic stem and progenitor cells. *In* "Cord Blood: Biology, Immunology, and Clinical Transplantation" (H. E. Broxmeyer, Ed.), 47–64.

Bompais, H., Chagraoui, J., Canron, X., Crisan, M., Liu, X. H., Anjo, A., Tolla-Le Port, C., Leboeuf, M., Charbord, P., Bikfalvi, A., and Uzan, G. (2004). Human endothelial cells derived from circulating progenitors display specific functional properties compared with mature vessel wall endothelial cells. *Blood* **103**(7), 2577–2584.

Breems, D. A., Blokland, E. A., Neben, S., and Ploemacher, R. E. (1994). Frequency analysis of human primitive haematopoietic stem cell subsets using a cobblestone area forming cell assay. *Leukemia* **8,** 1095–1104.

Broxmeyer, H. E. (1998). Introduction: The past, present, and future of cord blood transplantation. *In* "Cellular Characteristics of Cord Blood and Cord Blood Transplantation" (H. E. Broxmeyer, Ed.), pp. 1–9. American Association of Blood Banks Press, Bethesda, MD.

Broxmeyer, H. E. (2000). Introduction. Cord blood transplantation: Looking back and to the future. *In* "Cord Blood Characteristics: Role in Stem Cell Transplantation" (S. B. A. Cohen, E. Gluckman, P. Rubinstein, and J. A. Madrigal, Eds.), pp. 1–12: In M. Dunitz (Ed.) ** London.

Broxmeyer, H. E. (2004). Proliferation, self-renewal, and survival characteristics of cord blood hematopoietic stem and progenitor cells. *In* "Cord Blood: Biology, Immunology, Banking, and Clinical Transplantation" (H. E. Broxmeyer, Ed.), Chapter 1, pp. 1–21. American Association of Blood Banking, Bethesda, MD.

Broxmeyer, H. E. (2005). Biology of cord blood cells and future prospects for enhanced clinical benefit. *Cytotherapy* **7,** 209–218.

Broxmeyer, H. E., Campbell, T. B., Hangoc, G., Cooper, S., and Farag, S. S. (2008). Oral administration to mouse recipients of clinical grade CD26 inhibitor enhances engraftment by non-treated congenic marrow: Potential mechanisms. *Exp. Hematol.* **36** (suppl) 511.

Broxmeyer, H. E., and Cooper, S. (1997). High efficiency recovery of immature hematopoietic progenitor cells with extensive proliferative capacity from human cord blood cryopreserved for ten years. *Clin. Exp. Immunol.* **107,** 45–53.

Broxmeyer, H. E., Cooper, S., and Gabig, T. (1989a). The effects of oxidizing species derived from molecular oxygen on the proliferation *in vitro* of human granulocyte-macrophage progenitor cells. *Ann. N. Y. Acad. Sci.* **554,** 177–184.

Broxmeyer, H. E., Cooper, S., Lu, L., Miller, M. E., Langefeld, C. D., and Ralph, P. (1990a). Enhanced stimulation of human bone marrow macrophage colony formation *in vitro* by recombinant human macrophage colony stimulatng factor in agarose medium at low oxygen tension. *Blood* **76,** 323–329.

Broxmeyer, H. E., Douglas, G. W., Hangoc, G., Cooper, S., Bard, J., English, D., Arny, M., Thomas, L., and Boyse, E. A. (1989b). Human umbilical cord blood as a potential source of transplantable hematopoietic stem/progenitor cells. *Proc. Natl. Acad. Sci. USA* **86,** 3828–3832.

Broxmeyer, H. E., Gluckman, E., Auerbach, A., Douglas, G. W., Friedman, H., Cooper, S., Hangoc, G., Kurtzberg, J., Bard, J., and Boyse, E. A. (1990b). Human umbilical cord blood: A clinically useful source of transplantable hematopoietic stem/progenitor cells. *Int. J. Cell Cloning* **8,** 76–91.

Broxmeyer, H. E., Hangoc, G., Cooper, S., Campbell, T., Ito, S., and Mantel, C. (2007). AMD3100 and CD26 modulate the mobilization, engraftment, and survival of hematopoietic stem and progenitor cells mediated by the SDF-1/CXCL12-CXCR4 axis. *Ann. N. Y. Acad. Sci. USA* **1106,** 1–19.

Broxmeyer, H. E., Hangoc, G., Cooper, S., Ribeiro, R. C., Graves, V., Yoder, M., Wagner, J., Vadhan-Raj, S., Rubinstein, P., and Broun, E. R. (1992). Growth characteristics and expansion of human umbilical cord blood and estimation of its potential for transplantation of adults. *Proc. Natl. Acad. Sci. USA* **89,** 4109–4113.

Broxmeyer, H. E., and Smith, F. O. (2004). Cord blood hematopoietic cell transplantation. *In* "Thomas' Hematopoietic Cell Transplantation" (K. G. Blume, S. J. Forman, and F. R. Appelbaum, Eds.), 3rd Edn., Chapter 43, pp. 550–564. Blackwell Scientific, Cambridge, MA.

Broxmeyer, H. E., and Smith, F. O. (2008). Cordblood hematopoietic cell transplantation, (F. R. Appelbaum, S. J. Forman, S. Negrin, and K. G. Blume, Eds.), 4th Edn., Chapter 39. Blackwell Scientific, Cambridge, MA. In Press.**

Broxmeyer, H. E., Srour, E. F., Hangoc, G., Cooper, S., Anderson, J. A., and Bodine, D. (2003). High efficiency recovery of hematopoietic progenitor cells with extensive proliferative and *ex-vivo* expansion activity and of hematopoietic stem cells with NOD/SCID mouse repopulation ability from human cord blood stored frozen for 15 years. *Proc. Natl. Acad. Sci. USA* **100,** 645–650.

Brunstein, C. G., Barker, J. N., Weisdorf, D. J., DeFor, T. E., Miller, J. S., Blazar, B. R., McGlave, P. B., and Wagner, J. E. (2007). Umbilical cord blood transplantation after nonmyeloablative conditioning: Impact on transplantation outcomes in 110 adults with hematologic disease. *Blood* **110,** 3064–3070.

Campbell, T. B., and Broxmeyer, H. E. (2007). CD26 inhibition and hematopoiesis: A novel approach to enhance transplantation. (Special Issue on: Dipeptidylpeptidase IV and related molecules in health and disease). *Front. Biosci.* **13,** 1795–1805.**

Campbell, T. B., Hangoc, G., Liu, Y., Pollok, K., and Broxmeyer, H. E. (2007). Inhibition of CD26 in human cord blood CD34[+] cells enhances their engraftment of nonobese diabetic/severe combined immunodeficiency mice. *Stem Cells Dev.* **16,** 347–354.

Cardoso, A. A., Li, M. L., Batard, P., Hatzfeld, A., Brown, E. L., Levesque, J. P., Sookdeo, H., Panterne, B., Sansilvestri, P., Clark, S. C., and Hatzfeld, J. (1993). Release from quiescence of CD34[+]CD38[−] human umbilical cord blood cells reveals their potentiality to engraft adults. *Proc. Natl. Acad. Sci. USA* **90,** 8707–8711.

Carow, C. E., Hangoc, G., and Broxmeyer, H. E. (1993). Human multipotential progenitor cells (CFU-GEMM) have extensive replating capacity for secondary CFU-GEMM: An effect enhanced by cord blood plasma. *Blood* **81,** 942–949.

Carow, C., Hangoc, G., Cooper, S., Williams, D. E., and Broxmeyer, H. E. (1991). Mast cell growth factor (c-kit ligand) supports the growth of human multipotential (CFU-GEMM) progenitor cells with a high replating potential. *Blood* **78,** 2216–2221.

Christopherson, K. W., II, and Broxmeyer, H. E. (2004). Hematopoietic stem and progenitor cell homing, engraftment, and mobilization in the context of the CXCL12/SDF-1—CXCR4 axis. *In* "Cord Blood: Biology, Immunology, and Clinical Transplantation" (H. E. Broxmeyer, Ed.), pp. 65–86. American Association of Blood Banks Press, Bethesda, MD.

Christopherson, K. W., II, Hangoc, G., Mantel, C., and Broxmeyer, H. E. (2004). Modulation of hematopoietic stem cell homing and engraftment by CD26. *Science* **305,** 1000–1003.

Christophersan, K. W., II, Paganessi, L. A., Napier, S., and Porecha, N. K. (2007). CD26 inhibition on CD34[+] an lineage human umbilical cord blood donor hematopoietic stem cells/hematopoietic, progenitor cells improves long-term engraftment into NOD/SCID/Beta2[null] immunodeficient mice. *Stem Cells and Development* **16,** 355–360.

Cornetta, K., Laughlin, M., Carter, S., Wall, D., Weinthal, J., Delaney, C., Wagner, J., Sweetman, R., McCarthy, P., and Chao, N. (2005). Umbilical cord blood transplantation in adults: Results of the prospective cord blood transplantation (COBLT). *Biol. Blood Marrow Transplant.* **11,** 149–160.

Crisa, L., Cirulli, V., Smith, K. A., Ellisman, M. H., Torbett, B. E., and Salomon, D. R. (1999). Human cord blood progenitors sustain thymic t-cell development and a novel form of angiogenesis. *Blood* **94,** 3928–3940.

Denning-Kendall, P., Singha, S., Bradley, B., and Hows, J. (2003). Cobblestone area-forming cells in human cord blood are heterogeneous and differ from long-term culture-initiating cells. *Stem Cells* **21**, 694–701.

Dexter, T. M. (1993). Synergistic interactions in haematopoiesis: Biological implications and clinical use. *Eur. J. Cancer* **29A**(Suppl.), S6–S9.

Dome, B., Dobos, J., Tovari, J., Paku, S., Kovacs, G., Ostoros, G., and Timar, J. (2008). Circulating bone marrow-derived endothelial progenitor cells: Characterization, mobilization, and therapeutic considerations in malignant disease. *Cytometry A* **73**, 186–193.

Eapen, M., Rubinstein, P., Zhang, M. J., Camitta, B. M., Stevens, C., Cairo, M. S., Davies, S. M., Doyle, J. J., Kurtzberg, J., Pulsipher, M. A., Ortega, J. J., Scaradavou, A., *et al.* (2006). Comparable long-term survival after unrelated and HLA-matched sibling donor hematopoietic stem cell transplantation for acute leukemia in children younger than 18 months. *J. Clin. Oncol.* **24**, 145–151.

Eggermann, J., Kliche, S., Jarmy, G., Hoffmann, K., Mayr-Beyrle, U., Debatin, K. M., Waltenberger, J., and Beltinger, C. (2003). Endothelial progenitor cell culture and differentiation *in vitro*: A methodological comparison using human umbilical cord blood. *Cardiovasc. Res.* **58**, 478–486.

Fan, C. L., Li, Y., Gao, P. J., Liu, J. J., Zhang, X. J., and Zhu, D. L. (2003). Differentiation of endothelial progenitor cells from human umbilical cord blood CD 34+ cells *in vitro. Acta. Pharmacol. Sin.* **24**, 212–218.

Flamme, I., Frolich, T., and Risau, W. (1997). Molecular mechanisms of vasculogenesis and embryonic angiogenesis. *J. Cell. Physiol.* **173**, 206–210.

Gan, O. I., Murdoch, B., Larochelle, A., and Dick, J. E. (1997). Differential maintenance of primitive human SCID-repopulating cells, clonogenic progenitors, and long-term culture-initiating cells after incubation on human bone marrow stromal cells. *Blood* **90**, 641–650.

Gluckman, E., Broxmeyer, H. E., Auerbach, A. D., Friedman, H. S., Douglas, G. W., Devergie, A., Esperou, H., Thierry, D., Socie, G., Lehn, P., Cooper, S., English, D., *et al.* (1989). Hematopoietic reconstitution in a patient with Fanconi anemia by means of umbilical-cord blood from an HLA-identical sibling. *N. Engl. J. Med.* **321**, 1174–1178.

Gluckman, E., and Rocha, V. (2005). History of the clinical use of umbilical cord blood hematopoietic cells. *Cytotherapy* **7**, 219–227.

Gluckman, E., Rocha, V., Arcese, W., Michel, G., Sanz, G., Chan, K. W., Takahashi, T. A., Ortega, J., Filipovich, A., Locatelli, F., Asano, S., Fagioli, F., *et al.* Eurocord Group**. (2004). Factors associated with outcomes of unrelated cord blood transplant: Guidelines for donor choice. *Exp. Hematol.* **32**, 397–407.

Gluckman, E., Rocha, V., Boyer-Chammard, A., Locatelli, F., Arcese, W., Pasquini, R., Ortega, J., Souillet, G., Ferreira, E., Laporte, J. P., Fernandez, M., and Chastang, C. (1997). Outcome of cord-blood transplantation from related and unrelated donors. Eurocord transplant group and the European blood and marrow transplantation group. *N. Engl. J. Med.* **337**, 373–381.

Gothert, J. R., Gustin, S. E., van Eekelen, J. A., Schmidt, U., Hall, M. A., Jane, S. M., Green, A. R., Gottgens, B., Izon, D. J., and Begley, C. G. (2004). Genetically tagging endothelial cells *in vivo*: Bone marrow-derived cells do not contribute to tumor endothelium. *Blood* **104**, 1769–1777.

Hao, Q. L., Thiemann, F. T., Petersen, D., Smogorzewska, E. M., and Crooks, G. M. (1996). Extended long-term culture reveals a highly quiescent and primitive human hematopoietic progenitor population. *Blood* **88**, 3306–3313.

Hildbrand, P., Cirulli, V., Prinsen, R. C., Smith, K. A., Torbett, B. E., Salomon, D. R., and Crisa, L. (2004). The role of angiopoietins in the development of endothelial cells from cord blood CD34+ progenitors. *Blood* **104**, 2010–2019.

Hristov, M., and Weber, C. (2004). Endothelial progenitor cells: Characterization, pathophysiology, and possible clinical relevance. *J. Cell. Mol. Med.* **8**, 498–508.

Hwang, W. Y., Samuel, M., Tan, D., Koh, L. P., Lim, W., and Linn, Y. C. (2007). A meta-analysis of unrelated donor umbilical cord blood transplantation versus unrelated donor bone marrow transplantation in adult and pediatric patients. *Biol. Blood Marrow Transplant.* **13**, 444–453.

Ingram, D., Mead, L., Tanaka, H., Meade, V., Fenoglio, A., Mortell, K., Pollok, K., Ferkowicz, M. J., Gilley, D., and Yoder, M. C. (2004). Identification of a novel hierarchy of endothelial progenitor cells using human peripheral and umbilical cord blood. *Blood* **104,** 2752–2760.

Ishikawa, F., Yasukawa, M., Lyons, B., Yoshida, S., Miyamoto, T., Yoshimoto, G., Watanabe, T., Akashi, K., Shultz, L. D., and Harada, M. (2005). Development of functional human blood and immune systems in NOD/SCID/IL2 receptor {gamma} chain(null) mice. *Blood* **106,** 1565–1573.

Ito, M., Hiramatsu, H., Kobayashi, K., Suzue, K., Kawahata, M., Hioki, K., Ueyama, Y., Koyanagi, Y., Sugamura, K., Tsuji, K., Heike, T., and Nakahata, T. (2002). NOD/SCID/gamma (c)(null) mouse: An excellent recipient mouse model for engraftment of human cells. *Blood* **100,** 3175–3182.

Iwami, Y., Masuda, H., and Asahara, T. (2004). Endothelial progenitor cells: Past, state of the art, and future. *J. Cell. Mol. Med.* **8,** 488–497.

Jaroscak, J., Goltry, K., Smith, A., Waters-Pick, B., Martin, P. L., Driscoll, T. A., Howrey, R., Chao, N., Douville, J., Burhop, S., Fu, P., and Kurtzberg, J. (2003). Augmentation of umbilical cord blood (UCB) transplantation with *ex-vivo*-expanded UCB cells: Results of a phase I trial using the AstromReplicell system. *Blood* **101,** 5061–5067.

Kamel-Reid, S., and Dick, J. E. (1988). Engraftment of immune-deficient mice with human hematopoietic stem cells. *Science* **242,** 1706–1709.

Kang, H. J., Kim, S. C., Kim, Y. J., Kim, C. W., Kim, J. G., Ahn, H. S., Park, S. I., Jung, M. H., Choi, B. C., and Kimm, K. (2001). Short-term phytohaemagglutinin-activated mononuclear cells induce endothelial progenitor cells from cord blood CD34+ cells. *Brit. J. Haematol.* **113,** 962–969.

Kawai, T., Choi, U., Liu, P. C., Whiting-Theobald, N. L., Linton, G. F., and Malech, H. L. (2007). Diprotin A infusion with nonobese diabetic/severe combined immunodeficiency mice markedly enhances engraftment of human mobilized CD34$^+$ peripheral blood cells. *Stem Cell Dev.* **16,** 361–370.

Kawamoto, A., and Asahara, T. (2007). Role of progenitor endothelial cells in cardiovascular disease and upcoming therapies. *Catheter. Cardiovasc. Interv.* **70,** 477–484.

Khakoo, A. Y., and Finkel, T. (2005). Endothelial progenitor cells. *Annu. Rev. Med.* **56,** 79–101.

Kurtzberg, J., Laughlin, M., Graham, M. L., Smith, C., Olson, J. F., Halperin, E. C., Ciocci, G., Carrier, C., Stevens, C. E., and Rubinstein, P. (1996). Placental blood as a source of hematopoietic stem cells for transplantation into unrelated donors. *N. Engl. J. Med.* **335,** 157–166.

Kusadasi, N., van Soest, P. L., Mayen, A. E., Koevoet, J. L., and Ploemacher, R. E. (2000). Successful short-term *ex vivo* expansion of NOD/SCID repopulating ability and CAFC week 6 from umbilical cord blood. *Leukemia* **14,** 1944–1953.

Lansdorp, P. M., Dragowska, W., and Mayani, H. (1993). Ontogeny-related changes in proliferative potential of human hematopoietic cells. *J. Exp. Med.* **178,** 787–791.

Lapidot, T., Dar, A., and Kollet, O. (2005). How do stem cells find their way home? *Blood* **106,** 1901–1910.

Laughlin, M. J., Barker, J., Bambach, B., Koc, O. N., Rizzieri, D. A., Wagner, J. E., Gerson, S. L., Lazarus, H. M., Cairo, M., Stevens, C. E., Rubinstein, P., and Kurtzberg, J. (2001). Hematopoietic engraftment and survival in adult recipients of umbilical-cord blood from unrelated donors. *N. Engl. J. Med.* **344,** 1815–1822.

Laughlin, M. J., Eapen, M., Rubinstein, P., Wagner, J. E., Zhang, M. J., Champlin, R. E., Stevens, C., Barker, J. N., Gale, R. P., Lazarus, H. M., Marks, D. I., van Rood, J. J., *et al.* (2004). Outcomes after transplantation of cord blood or bone marrow from unrelated donors in adults with leukemia. *N. Eng. J. Med.* **351,** 2265–2275.

Long, G. D., Laughlin, M., Madan, B., Kurtzberg, J., Gasparetto, C., Morris, A., Rizzieri, D., Smith, C., Vredenburgh, J., Halperin, E. C., Broadwater, G., Niedzwiecki, D., *et al.* (2003). Unrelated umbilical cord blood transplantation in adult patients. *Biol. Blood Marrow Transplant.* **9,** 772–780.

Lu, L., and Broxmeyer, H. E. (1985). Comparative influences of phytohemagglutin-stimulated leukocyte conditioned medium, hemin, prostaglandin E and low oxygen tension on colony formation by erythroid progenitor cells in normal human bone marrow. *Exp. Hematol.* **13,** 989–993.

Lu, L., Xiao, M., Shen, R. N., Grigsby, S., and Broxmeyer, H. E. (1993). Enrichment, characterization and responsiveness of single primitive $CD34^{+++}$ human umbilical cord blood hematopoietic progenitor cells with high proliferative and replating potential. *Blood* **81,** 41–48.

McCune, J. M., Namikawa, R., Kaneshima, H., Shultz, L. D., Lieberman, M., and Weissman, I. L. (1988). The SCID-hu mouse: Murine model for the analysis of human hematolymphoid differentiation and function. (severe combined immunodeficiency). *Science* **241,** 1632–1639.

Miller, J. S., McCullar, V., Punzel, M., Lemischka, I. R., and Moore, K. A. (1999). Single adult human CD34(+)/Lin-/CD38(-) progenitors give rise to natural killer cells, B-lineage cells, dendritic cells, and myeloid cells. *Blood* **93,** 96–106.

Murasawa, S., and Asahara, T. (2005). Endothelial progenitor cells for vasculogenesis. *Physiology (Bethesda)* **20,** 36–42.

Murga, M., Yao, L., and Tosato, G. (2004). Derivation of endothelial cells from CD34-umbilical cord blood. *Stem Cells* **22,** 385–395.

Ooi, J., Iseki, T., Takahashi, S., Tomonari, A., Nagayama, H., Ishii, K., Ito, K., Sato, H., Takahashi, T., Shindo, M., Sekine, R., Ohno, N., *et al.* (2002). A clinical comparison of unrelated cord blood transplantation and unrelated bone marrow transplantation for adult patients with acute leukaemia in complete remission. *Br. J. Haematol.* **118,** 140–143.

Orazi, A., Braun, S. E., and Broxmeyer, H. E. (1994). Immunohistochemistry represents a useful tool to study human cell engraftment in SCID mice transplantation models. *Blood Cells* **20,** 323–330.

Peichev, M., Naiyer, A., Pereira, D., Zhu, Z., Lane, W. J., Williams, M., Oz, M. C., Hicklin, D. J., Witte, L., Moore, M. A., and Rafii, S. (2000). Expression of VEGFR-2 and AC133 by circulating human CD34+ cells identifies a population of functional endothelial precursors. *Blood* **95,** 952–958.

Peranteau, W. H., Endo, M., Adibe, O. O., Merchant, A., Zoltick, P., and Flake, A. W. (2005). CD26 inhibition enhances allogeneic donor cell homing and engraftment after *in utero* bone marrow transplantation. *Blood* **106**(Pt. I), 371a. (abstract# 1275).

Pesce, M., Orlandi, A., Iachininoto, M. G., Straino, S., Torella, A. R., Rizzuti, V., Pompilio, G., Bonanno, G., Scambia, G., and Capogrossi, M. C. (2003). Myoendothelial differentiation of human umbilical cord blood-derived stem cells in ischemic limb tissues. *Circ. Res.* **93,** e51–e62.

Pettengell, R., Luft, T., Henschler, R., Hows, J. M., Dexter, T. M., Ryder, D., and Testa, N. G. (1994). Direct comparison by limiting dilution analysis of long-term culture-initiating cells in human bone marrow, umbilical cord blood, and blood stem cells. *Blood* **84,** 3653–3659.

Ploemacher, R. E., van der Sluijs, J. P., van Beurden, C. A., Baert, M. R., and Chan, P. L. (1991). Use of limiting-dilution type long-term marrow cultures in frequency analysis of marrow-repopulating and spleen colony-forming hematopoietic stem cells in the mouse. *Blood* **78,** 2527–2533.

Ploemacher, R. E., van der Sluijs, J. P., Voerman, J. S., and Brons, N. H. (1989). An *in vitro* limiting-dilution assay of long-term repopulating hematopoietic stem cells in the mouse. *Blood* **74,** 2755–2763.

Podesta, M., Piaggio, G., Pitto, A., Zocchi, E., Soracco, M., Frassoni, F., Luchetti, S., Painelli, E., and Bacigalupo, A. (2001). Modified *in vitro* conditions for cord blood-derived long-term culture-initiaitng cells. *Exp. Hematol.* **29,** 309–314.

Punzel, M., Wissink, S. D., Miller, J. S., Moore, K. A., Lemischka, I. R., and Verfaillie, C. M. (1999). The myeloid-lymphoid initiating cell (ML-IC) assay assesses the fate of multipotent human progenitors *in vitro*. *Blood* **93,** 3750–3756.

Purhonen, S., Palm, J., Rossi, D., Kaskenpää, N., Rajantie, I., Ylä-Herttuala, S., Alitalo, K., Weissman, I. L., and Salven, P. (2008). Bone marrow-derived circulating endothelial precursors do not contribute to vascular endothelium and are not needed for tumor growth. *Proc. Natl. Acad. Sci. USA* **105,** 6620–6625.

Rafii, S., and Lyden, D. (2003). Therapeutic stem and progenitor cell transplantation for organ vascularization and regeneration. *Nat. Med.* **9,** 702–712.

Rocha, V., Labopin, M., Sanz, G., Arcese, W., Schwerdtfeger, R., Bosi, A., Jacobsen, N., Ruutu, T., de Lima, M., Finke, J., Frassoni, F., and Gluckman, E. (2004). Transplants of umbilical-cord blood or bone marrow from unrelated donors in adults with acute leukemia. *N. Eng. J. Med.* **351,** 2276–2285.

Rubinstein, P., Carrier, C., Scaradavou, A., Kurtzberg, J., Adamson, J., Migliaccio, A. R., Berkowitz, R. L., Cabbad, M., Dobrila, N. L., Taylor, P. E., Rosenfield, R. E., and Stevens, C. E. (1998). Outcomes among 562 recipients of placental-blood transplants from unrelated donors. *N. Engl. J. Med.* **339,** 1565–1577.

Sanz, G. F., Saavedra, S., Jimenez, C., Senent, L., Cervera, J., Planelles, D., Bolufer, P., Larrea, L., Martin, G., Martinez, J., Jarque, I., Moscardo, F., *et al.* (2001a). Unrelated donor cord blood transplantation in adults with chronic myelogenous leukemia: Results in nine patients from a single institution. *Bone Marrow Transplant.* **27,** 693–701.

Sanz, G. F., Saavedra, S., Planelles, D., Senent, L., Cervera, J., Barragan, E., Jimenez, C., Larrea, L., Martin, G., Martinez, J., Jarque, I., Moscardo, F., *et al.* (2001b). Standardized, unrelated donor cord blood transplantation in adults with hematologic malignancies. *Blood* **98,** 2332–2338.

Schatteman, G. C. (2004). Adult bone marrow-derived hemangioblasts, endothelial cell progenitors, and EPCs. *Curr. Top. Dev. Biol.* **64,** 141–180.

Shaheen, M., and Broxmeyer, H. E. (2005). The humoral regulation of hematopoiesis. *In* "Hematology: Basic Principles and Practice" (R. Hoffman, E. Benz, S. Shattil, B. Furie, H. Cohen, L. Silberstein, and P. McGlave, Eds.), 4th Edn., Chapter 19, pp. 233–265.

Shaheen, M., and Broxmeyer, H. E. (2008). The humoral regulation of hematopoiesis. *In* "Hematology: Basic Principles and Practice" (R. Hoffman, E. Benz, S. Shattil, B. Furie, H. Cohen, L. Silberstein, and P. McGlave, Eds.), 5th Edn., Chapter 23. Elsevier, Philadelphia, PA.

Shpall, E. J., Quinones, R., Giller, R., Zeng, C., Baron, A. E., Jones, R. B., Bearman, S. I., Nieto, Y., Freed, B., Madinger, N., Hogan, C. J., Slat-Vasquez, V., *et al.* (2002). Transplantation of *ex-vivo* expanded cord blood. *Biol. Bone Marrow Transplant.* **8,** 368–376.

Simons, M. (2005). Angiogenesis: Where do we stand now? *Circulation* **111,** 1556–1566.

Skalak, T. (2005). Angiogenesis and microvascular remodeling: A brief history and future roadmap. *Microcirculation* **12,** 47–58.

Smith, S., and Broxmeyer, H. E. (1986). The influence of oxygen tension on the long term growth *in vitro* of haematopoietic progenitor cells from human cord blood. *Brit. J. Haematol.* **63,** 29–34.

Stadtfeld, M., and Graf, T. (2005). Assessing the role of hematopoietic plasticity for endothelial and hepatocyte development by non-invasive lineage tracing. *Development* **132,** 203.

Sutherland, H. J., Eaves, C. J., Eaves, A. C., Dragowska, W., and Lansdorp, P. M. (1989). Characterization and partial purification of human marrow cells capable of initiating long-term hematopoiesis *in vitro*. *Blood* **74,** 1563–1570.

Takahashi, S., Iseki, T., Ooi, J., Tomonari, A., Takasugi, K., Shimohakamada, Y., Yamada, T., Uchimaru, K., Tojo, A., Shirafuji, N., Kodo, H., Tani, K., *et al.* (2004). Single-institute comparative analysis of unrelated bone marrow transplantation and cord blood transplantation for adult patients with hematologic malignancies. *Blood* **104,** 3813–3820.

Taswell, C. (1981). Limiting dilution assays for the determination of immunocompetent cell frequencies. I. Data analysis. *J. Immunol.* **126,** 1614–1619.

Theunissen, K., and Verfaillie, C. M. (2005). A multifactorial analysis of umbilical cord blood, adult bone marrow and mobilized peripheral blood progenitors using the improved ML-IC assay. *Exp. Hematol.* **33,** 165–172.

Tian, C., Bagley, J., Forman, D., and Iacomini, J. (2006). Inhibition of CD26 peptidase activity significantly improves engraftment of retrovirally transduced hematopoietic progenitors. *Gene Ther.* **13,** 652–658.

Traycoff, C., Abboud, M., Laver, J., Clapp, D., and Srour, E. (1994). Rapid exit from G0/G1 phases of cell cycle in response to stem cell factor confers on umbilical cord blood CD34+ cells an enhanced *ex vivo* expansion potential. *Exp. Hematol.* **22,** 1264–1272.

Traycoff, C. M., Kosak, S. T., Grigsby, S., and Srour, E. F. (1995). Evaluation of *ex vivo* expansion potential of cord blood and bone marrow hematopoietic progenitor cells using cell tracking and limiting dilution analysis. *Blood* **85,** 2059–2068.

Tse, W., and Laughlin, M. J. (2005). Cord blood transplantation in adult patients. *Cytotherapy* **7,** 228–242.

Urbich, C., and Dimmeler, S. (2004). Endothelial progenitor cells: Characterization and role in vascular biology. *Circ. Res.* **95,** 343–353.

Vormoor, J., Lapidot, T., Pflumio, F., Risdon, G., Patterson, B., Broxmeyer, H. E., and Dick, J. E. (1994). Immature human cord blood progenitors engraft and proliferate to high levels in immune-deficient SCID mice. *Blood* **83,** 2489–2497.

Wagner, J. E., Barker, J. N., DeFor, T. E., Baker, K. S., Blazar, B. R., Eide, C., Goldman, A., Kersey, J., Krivit, W., MacMillan, M. L., Orchard, P. J., Peters, C., *et al.* (2002). Transplantation of unrelated donor umbilical cord blood in 102 patients with malignant and nonmalignant diseases: Influence of CD34 cell dose and HLA disparity on treatment-related mortality and survival. *Blood* **100,** 1611–1618.

Wagner, J. E., Kernan, N. A., Steinbuch, M., Broxmeyer, H. E., and Gluckman, E. (1995). Allogeneic sibling umbilical cord blood transplantation in forty-four children with malignant and non-malignant disease. *Lancet* **346,** 214–219.

Wagner, J. E., Rosenthal, J., Sweetman, R., Shu, X. O., Davies, S. M., Ramsay, N. K., McGlave, P. B., Sender, L., and Cairo, M. S. (1996). Successful transplantation of HLA-matched and HLA-mismatched umbilical cord blood from unrelated donors: Analysis of engraftment and acute graft-versus-host disease. *Blood* **88,** 795–802.

Weaver, A., Ryder, W. D., and Testa, N. G. (1997). Measurement of long-term culture initiating cells (LTC-ICs) using limiting dilution: Comparison of endpoints and stromal support. *Exp. Hematol.* **25,** 1333–1338.

CHAPTER 11

Hematopoietic Stem Cells from Bone Marrow: Purification and Functional Analysis

K. K. Lin[*,†] and M. A. Goodell[*,†,‡]

[*]Stem Cells and Regenerative Medicine Center
Baylor College of Medicine
Houston, Texas

[†]Department of Immunology
Baylor College of Medicine
Houston, Texas

[‡]Department of Pediatrics
Baylor College of Medicine
Houston, Texas

Abstract
1. Introduction
    1.1. Cell surface markers of HSC
2. Other HSC Surface Markers: Tie-2, Endoglin, and the SLAM Family Receptors
3. Fluorescent Dye Efflux in HSCs and the SP[KLS]
4. Characteristics of HSCs from Different Purification Schemes
5. Protocol of HSC Sorting with Hoechst 33342 Staining (SP Population)
    5.1. Overview
    5.2. Hoechst staining protocol
    5.3. Protocol for antibody staining and magnetic enrichment of Hoechst-stained cells
    5.4. FACS analysis for Hoechst SP cells
References

Abstract

Hematopoietic stem cells (HSCs), primarily reside in bone marrow, are defined by their ability to maintain blood homeostasis and replenish themselves through self-renewal. Although HSC purification schemes vary from lab to lab, the resulting cell populations are similar, if not the same. This chapter will discuss different enrichment methods for HSCs and provide a detailed protocol for staining HSC with Hoechst 33342 for the side population (SP).

1. Introduction

1.1. Cell surface markers of HSC

Bone marrow hematopoietic stem cell (HSC) activity has been extensively studied since the 1950s when Ford *et al.* (1956) discovered a robust contribution of transplanted donor bone marrow in lethally irradiated recipients. HSCs were first enriched by a number of groups from murine bone marrow, most notably by Weissman and colleagues using surface marker expression and fluorescent-activated cell sorting (FACS) (Spangrude *et al.*, 1988). Thereafter, other labs have also modified or refined the approach with different combinations of surface markers. Of these surface markers, c-Kit (CD117), a tyrosine kinase receptor, Sca-1, a membrane glycoprotein, and lack of expression of hematopoietic differentiation ("lineage[negative]") markers (CD4/CD8, B220, Mac-1, Gr-1, Ter119) are the core elements to enrich for HSCs. The c-Kit$^+$Sca-1$^+$Lineage$^-$ (KSL) cells comprise ∼0.1% of the whole bone marrow, and despite being highly enriched for HSCs, it is still a very heterogeneous population that includes lineage-primed multipotent progenitors as well as short-term HSCs and long-term HSCs (Okada *et al.*, 1992). Therefore, several additional markers to exclude differentiated progenitors such as Thy1.1 (Morrison and Weissman, 1994), CD34 (Osawa *et al.*, 1996), and Flk2 (Christensen and Weissman, 2001) have been utilized to further fractionate bone marrow KSL (Table 11.1). While these approaches result in highly overlapping populations, there are also slight differences, and efforts have been made to quantify the long-term repopulating activity of these purified populations using bone marrow transplantation assays. However, because the stringency of quantification varies among different groups, depending on the precise strategy for assessment, there is not yet an absolute standard to compare the power of every enrichment scheme. Ultimately, the choice of optimal enrichment scheme will also depend on the flow cytometry equipment availability, the local expertise, and the reproducibility of the different methods in the hands of the particular investigators.

Table 11.1

Comparison of the modern canonical purification schemes

Purification strategy	Fraction of WBM (%)	Purity of functional HSC	References
c-Kit$^+$Sca-1$^+$Lin$^-$ (KSL)	0.08	+	Okada *et al.* (1992)
KSL Thy1.1$^-$ (KTSL)	0.05	++	Morrison and Weissman (1994)
KSL CD34$^-$	0.008	++	Osawa *et al.* (1996)
KSL Thy1.1^{lo}Flk2$^-$	0.01	+++	Christensen and Weissman (2001)
SP (Hoechst 33342)	0.05	++	Goodell *et al.* (1996)

The resulting purity of HSCs is based on the functional assays provided by the original articles. There is no absolute number because the calculation of enrichment power in each laboratory slightly varies.

2. Other HSC Surface Markers: Tie-2, Endoglin, and the SLAM Family Receptors

While c-Kit, Sca-1, and Lineage markers are the canonical cell surface markers used to enrich for HSCs for more than a decade, this combination of more than eight different antibodies makes HSC purification an expensive process. Furthermore, the complexity of this scheme is incompatible with standard fluorescence microscopy, if one wanted to identify HSC *in situ*. Other surface markers such as Tie-2 (Arai *et al.*, 2004) and endoglin (Chen *et al.*, 2002) have been found to be expressed in HSCs. In addition, Morrison and his colleagues recently used markers from the SLAM family, CD150, CD244, and CD48 to distinguish HSCs from differentiated hematopoietic lineages (Kiel *et al.*, 2005), identifying HSCs as CD150$^+$CD244$^-$CD48$^-$ (or CD150$^+$CD41$^-$CD48$^-$). Tie-2 and CD150 have since been used to attempt to visualize putative HSCs in their putative bone marrow niches.

3. Fluorescent Dye Efflux in HSCs and the SP^{KLS}

HSCs have been recognized to exhibit high multidrug-type efflux activity, which results in low retention of fluorescent dyes such as Hoechst 33342 and Rhodamine 123, relative to other bone marrow cells (Goodell *et al.*, 1996; Li and Johnson, 1992). This trait has been exploited in one of the most widely used HSC purification strategy, in which differential efflux of the fluorescent DNA-binding dye Hoechst 33342 results in a "side population" (SP) of low staining cells that is highly enriched for HSC. The SP (Fig. 11.1) is highly enriched for HSC functionally, and

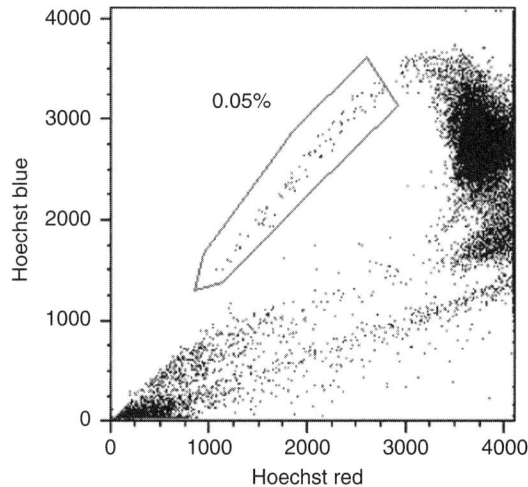

Figure 11.1 SP gate of an unenriched whole bone marrow sample. To visualize the SP population, signals are displayed in a Hoechst Blue versus Hoechst Red dot plot. The FACS PMT voltages are adjusted until the majority of the cells are at the upper right corner while the red blood cells and debris are at the lower left corner. The SP population comprises ~0.02–0.05% of whole bone marrow cells (this figure is generously provided by Stuart M. Chambers).

phenotypically overlaps with c-Kit$^+$Sca-1$^+$Lin$^-$Thy1^{lo}CD34$^-$Flk2$^-$ cells (Camargo *et al.*, 2006). While HSC purified using the SP strategy are probably one of the most pure and potent populations available, the method is sensitive to slight modifications in staining procedures, so that the resulting population is somewhat variable from lab to lab, leading to some claims that the SP is heterogeneous. Therefore, to ensure a successful purification of HSC using the SP, we recommend inclusion of cell surface markers, such as c-Kit, Lineage$^-$, and Sca-1 in the purification scheme as a purity index. The resulting population (termed SP^{KLS}, pronounced as "sparkles") includes most if not all of the murine HSCs, including long-term and short-term stem cells, and excludes other precursor cells and multipotential progenitors (Fig. 11.2).

High expression of multidrug-resistance ABC transporters such as MDR-1 and ABCG2 in HSCs has been suggested to contribute to the dye efflux ability, perhaps with a natural role of efflux of differentiation factors (Goodell *et al.*, 1996). Indeed, retroviral-mediated overexpression of human multidrug resistance-1 (MDR-1) leads to stem cell expansion (Bunting *et al.*, 1998) and an increased side population (Bunting *et al.*, 2000). However, loss of Mdr1 only impacts the efflux of Rhodamine dye, but not Hoechst, indicating that Mdr1 is not absolutely required. The expression of ABCG2 in murine bone marrow cells was found essential to detect the SP population after Hoechst staining (Zhou *et al.*, 2002). Nevertheless, the expression of ABCG2 does not always reflect the SP phenotype. A residual SP population has been found in *abcg2*-deficient murine bone marrow (Zhou *et al.*, 2002).

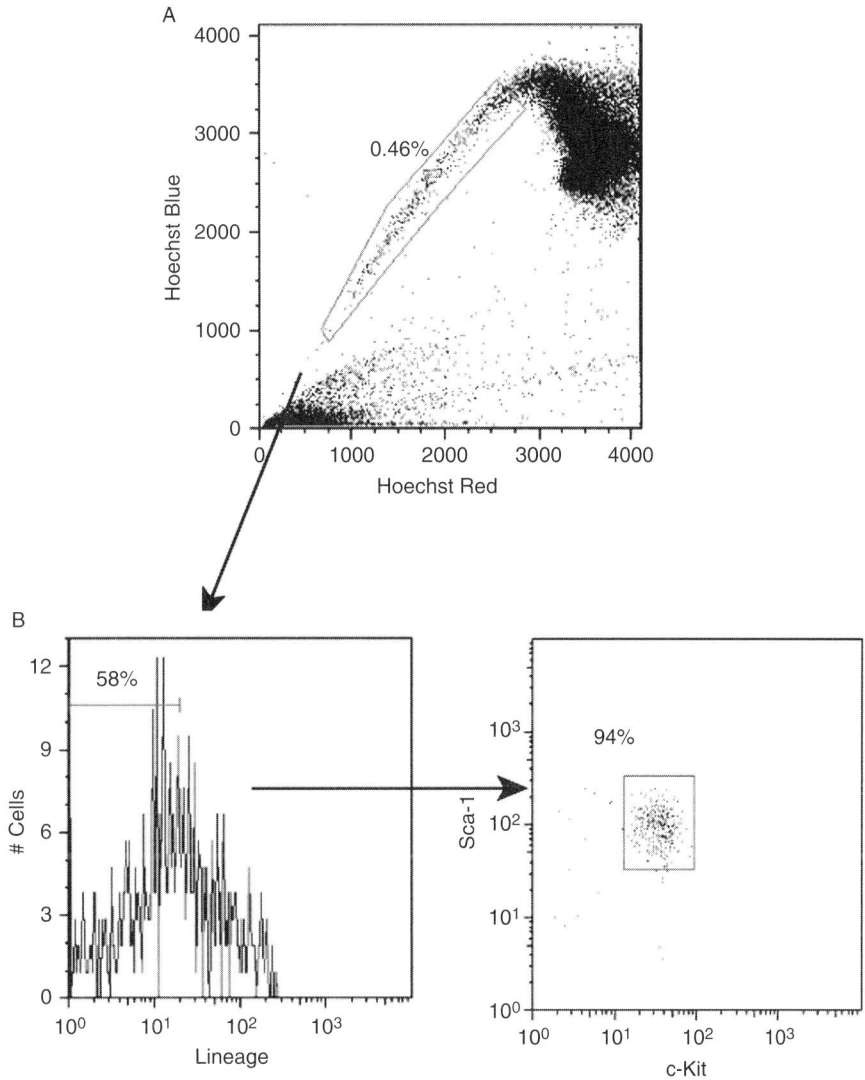

Figure 11.2 The SParKLS (SP, c-Kit$^+$, Lin$^-$, Sca-1$^+$) cells. To exclude a low level contamination of progenitor cells, we include antibodies such as c-Kit, Sca-1, and the markers of differentiated cells (Lineage) during SP analysis. The resulting population is named SParKLS. This bone marrow sample shown here was pre-enriched with Sca-1 antibody using magnetic sorting.

Although a higher expression level of ABCG2 was found in both the SP populations of murine and human, some progenitors and differentiated lineages were also found to express a detectable level of ABCG2, in the absence of a SP phenotype (Scharenberg *et al.*, 2002; Zhou *et al.*, 2001). Therefore, one cannot equate

ABCG2 expression with SP, and antibodies against ABCG2 may not detect HSC, or other stem cells, with fidelity. It is likely that multiple multidrug-resistance transporters can contribute to the SP phenotype. A microarray gene expression study of HSC revealed that a number of multidrug-resistance transporters were expressed (Venezia *et al.*, 2004).

4. Characteristics of HSCs from Different Purification Schemes

While the side population (SParKLS) cells have been demonstrated to be highly potent and homogeneous for long-term HSC (LT-HSC), and the strategy is now widely used, many labs use schemes based solely on different sets of cell surface markers. It can be difficult to compare results from different labs using different schemes, and it can also be difficult for a novice entering the field to determine which scheme to use. There is interest in the field in continuing to refine populations and in determining the most optimal HSC sorting scheme, and a few purification comparisons have been published. Overall, these papers show that most of these so-called HSC populations are highly overlapping, but there are some subtle differences. It is now well recognized that cells purified on the basis of only Lineage$^-$, Sca-1$^+$, and c-Kit$^+$ (LSK or KLS) are quite heterogeneous, including many lineage-restricted progenitors; *bona fide* stem cells are probably only around 10% of the typical KLS population. KLS subfractionated on the basis of CD34$^-$ and Flk2$^-$ are more similar to the SParKLS population when directly compared (Camargo *et al.*, 2006). Likewise, cells purified solely on the basis of the SLAM markers are still somewhat heterogeneous, but the addition of KLS criteria eliminates the remaining progenitors. Our lab has also shown that in young mice, about half of the SParKLS population would also be defined as SLAM-purified cells, but there is also a subset of cells that lack expression of CD150. Although these subpopulations have clearly different repopulating kinetics and generate a different profiles of differentiated cells, both the CD150$^+$ and CD150$^-$ SP^{KLS} cells possess long-term stem cell activity as measured by all traditional criteria (Weksberg *et al.*, 2008).

Taken together, these and other data point to a subtle heterogeneity within the entire stem cell compartment (Camargo *et al.*, 2006; Sieburg *et al.*, 2006; Smith *et al.*, 1991; Zhong *et al.*, 1996). This has been most elegantly demonstrated by recent single-cell transplantation experiments showing distinct activities of individual HSC that were all identified by the same phenotypically rigorous criteria (Dykstra *et al.*, 2007). This view represents an important shift away from a concept of a uniform HSC population which, in binary fashion, exits that state to differentiate, toward a view of a *continuum* of stem cells containing HSC with slightly different properties. Identifying markers to distinguish between those cell types is the next major challenge for the field. Likewise, understanding the relationships between the HSC types, their functional properties, and other differences such as differences in their niche will be important.

5. Protocol of HSC Sorting with Hoechst 33342 Staining (SP Population)

5.1. Overview

After staining with Hoechst 33342, HSCs appear as SP in FACS (Fig. 11.1). High dye efflux ability and the quiescent cell-cycle stage of HSCs are thought to account for the side population. Reproducible staining for SP cells is dependent on careful control of various parameters such as Hoechst dye concentration, cell concentration, and staining temperature and time. Furthermore, the protocol was originally established for murine HSC from normal C57Bl/6 bone marrow, so optimization is required when staining other tissues or species. To establish the Hoechst staining procedure *de novo*, we strongly recommend initial experiments be performed with murine bone marrows exactly as we describe. When proper Hoechst staining is performed, the SP population comprises ~0.02–0.05% of whole bone marrow cells, which often seems low to a beginner's eyes, but this small population is very highly purified for active LT-HSC. To increase the total yield, we often employ a pre-enrichment scheme based on magnetic enrichment of progenitors using a canonical surface marker (e.g., Sca-1 or c-Kit). Thus, an enrichment protocol, which gives rise to an at least 10-fold enrichment for bone marrow SP (Fig. 11.2), is also provided. Using these protocols in combination can result in roughly 50,000 purified HSC when starting from 10 mice, in a ~8–10 h procedure.

5.2. Hoechst staining protocol

1. Prewarm DMEM+ medium in a 37 °C circulating water bath. Ensure the temperature is precisely at 37 °C.

2. Using mice ~8 weeks of age, prepare bone marrow from femurs and tibias, resuspend in iced HBSS+ buffer.

3. Count nucleated cells. To count nucleated cells and avoid the similarly sized red blood cells (RBCs), one can lyse RBC in an aliquot of the bone marrow removed for counting. Mix a 5 μl bone marrow aliquot with 95 μl of RBC lysis buffer (D-5001, Gentra), vortex thoroughly and take 10 μl to count under hemacytometer. Do NOT perform RBC lysis to the whole bone marrow suspension. We typically find an average of 40–70 million nucleated cells per C57Bl/6 mouse (2 femurs and 2 tibias). Cell number is one critical parameter to a successful Hoechst staining, so count cells carefully.

4. Spin down bone marrow cells. Resuspend cells at 10^6 per ml in prewarmed DMEM+. To avoid retention of cells in tubes, polypropylene tubes must be used when Hoechst staining is performed. We find staining in 250 ml polypropylene tubes (Cat No. 430776, Corning) the most convenient when staining large volumes.

5. Add Hoechst to a final concentration of 5 μg/ml. We suggest using a 200× (1 mg/ml) stock, which is made in water by dissolving an entire bottle of Hoechst powder (B2261, Sigma), aliquoting, and storing the concentrated Hoechst solution in −20 °C freezer.

6. Incubate cells in a 37 °C water bath and periodically mix the tubes during the exactly 90-min incubation. Time and temperature are crucial to Hoechst staining, so the DMEM should be prewarmed and the tubes fully immersed in the water bath.

7. After the Hoechst stain, cells *must* be maintained at 4 °C to prevent further Hoechst dye efflux. No ficoll or other extended higher temperature procedures should be performed after the Hoechst stain, as this will result in dye efflux from other bone marrow cells that are NOT HSC, ultimately reducing the purity of the SP. The cells should be spun down in a 4 °C centrifuge and resuspend in ice-cold HBSS+.

8. Antibody staining can now be performed on ice. To ensure good HSC purification, antibodies against Sca-1, c-Kit, and one or more lineage markers are added at concentrations determined by standard antibody titration procedures, or as recommended by the manufacturer (e.g., Becton Dickinson/Pharmingen). All staining and centrifugation should be performed at 4 °C.

9. At the point when cells are ready for FACS analysis, resuspend cells in cold HBSS+ with 2 μg/ml propidium iodide (PI) to distinguish dead cells. Although staining with PI is not absolutely required to see SP, it is strongly recommended. Hoechst is toxic to some bone marrow cells, and PI allows the dead cells to be distinguished, simplifying the Hoechst emission plot.

10. Reagents:

a. DMEM+

DMEM with high glucose (Cat No. 11965-092, Gibco Invitrogen)

Penicillin/streptomycin (Cat No. 15140-122, Gibco Invitrogen)

10 mM HEPES (Cat No. 15630-080, Gibco Invitrogen)

2% fetal bovine serum/fetal calf serum

b. HBSS+

Hank's balanced salt solution (Cat No. 14170-112, Gibco Invitrogen)

10 mM HEPES

2% fetal bovine serum/fetal calf serum

c. Hoechst 33342 powder (Cat No. B2261, Sigma)

200 × stock (1 mg/ml) is made first by dissolving powder in distilled water, and filter sterilizing. From what we have obtained from sigma, a bottle of powder is good for ∼500 ml of Hoechst stock solution. It is strongly recommended that a large batch of Hoechst from a WHOLE bottle be made at once, and

frozen in small (~1 ml) aliquots. Thawed Hoechst powder may be less reliable refreezing, possibly due to acquisition of water.

d. Propidium iodide (Cat No. P-4170, Sigma)

Working stock is $100 \times$ (200 μg/ml) in PBS, and is covered with aluminum foil in 4 °C fridge. Final concentration of PI in resuspension HBSS+ should be 2 μg/ml.

5.3. Protocol for antibody staining and magnetic enrichment of Hoechst-stained cells

1. After Hoechst staining, resuspend cells at 1×10^8 cells/ml in iced HBSS+. For Sca-1 enrichment, incubate biotinylated Sca-1 antibody with cells at 1/100 dilution, on ice for 10 min.

2. Wash out unbound antibody with 10-fold volume of HBSS+, spin down in a cold centrifuge.

3. Label cells with magnetic beads. We obtain the antibiotin magnetic microbeads from Miltenyi Biotech (Cat No. 130-090-485). Incubate cells with 20% volume of microbeads at 4 °C for 15 min. We have found the binding efficiency of microbeads to antibody-labeled cells very low when incubating on ice. Incubate in a 4 °C fridge instead.

4. Wash the cells with a 10-fold volume of HBSS+, spin down in a cold centrifuge.

5. Resuspend cells at 2×10^8 cells/ml in iced HBSS+, and cells are ready to be loaded into auto-MACs column.

5.4. FACS analysis for Hoechst SP cells

1. Excitation of Hoechst 33342

To view the SP population, an ultraviolet laser is needed to excite the Hoechst 33342 dye and PI. A violet laser has also been used with good results (Simpson *et al.*, 2006). The Hoechst dye is excited at 350 nm. Additional lasers such as a 488 laser for FITC and PE can be used to excite additional fluorochromes.

2. Detection of Hoechst emission

The emission of Hoechst dye is measured with two detectors: Hoechst Blue and Hoechst Red. Hoechst Blue is measured with a 450/20 filter while the red is measured with a 675LP filter. To separate the emission wavelength, a dichroic mirror is used (we use a 610 DMSP). Fluorescence from PI is also measured with the 675LP channel, but the positive signal is much brighter than the

Hoechst red signal. The simultaneous excitation and emission of PI off the 488 lasers can be used to identify the dead cells and distinguish from red-fluorescing live Hoechst-stained cells. Although other filter sets similar to the ones we describe may work sufficiently, these are optimal.

In addition, a high power UV laser (50–100 mW) gives the best resolution of the SP. Less power may be sufficient, but the population may not be as. It is important to confirm the SP population with a transporter inhibitor, Verapamil, or antibody costaining, particularly for initial SP identification.

3. FACS analysis

The Hoechst fluorescence is displayed on Hoechst BLUE versus RED plot, with BLUE on the vertical axis and RED on the horizontal axis, both in *linear mode*. The voltage is adjusted so that RBCs appear in the lower left corner, while the dead cells that are stained with PI positive line up at the far right vertical line. The majority of the bone marrow cells can be seen in the center, or in the upper right quarter (Fig. 11.1). It is also possible to detect a major G0–G1 population with S–G2M cells going toward the upper right corner.

To be able to see the SP cells, draw a sample gate to exclude the RBCs and dead cells. With an unenriched bone marrow sample, 50,000–100,000 events within the sample gate are needed to identify the SP population. The SP region should be drawn as Fig. 11.1. The prevalence of SP is around 0.01–0.05% of an unenriched whole bone marrow sample in the mouse. A population of SP cells in normal mouse bone marrow significantly higher in proportion than this invariably indicates poor Hoechst staining, and contamination of the SP cells with non-SP. This is readily visualized by the presence of cells in the SP that do not express the canonical HSC surface markers. If the staining is done properly, between 65% and 95% of the SP cells will have canonical HSC surface markers, and these are selected for purification.

4. Tips for operating the FACS

Because Hoechst emission is displayed under the linear mode, good C.V.s are critical for SP identification. To maintain good C.V.s, it is important to perform alignment with particles that display a tight distribution in linear mode (DNA check beads, Coulter). In addition, a low sample differential pressure is important. The maximal differential pressure for samples should not be faster than the one for calibrating with alignment particle. The bone marrow cells should be run at high concentration so as to maintain a low sample pressure.

5. Confirmation of a good SP staining:

a. Include a Verapamil-treated control to confirm the SP population in one aliquot. Verapamil (Cat No. V-4629, Sigma) at the final concentration of 50 μM can be added throughout the entire Hoechst staining procedure. With Verapamil treatment, SP population will be blocked.

b. Costain with antibodies. The mouse bone marrow SP population is highly enriched with HSCs. With a good Hoechst staining, 60–80% percent of SP cells are Sca-1$^+$ and Lineage$^-$ cells (Fig. 11.2).

References

Arai, F., Hirao, A., Ohmura, M., Sato, H., Matsuoka, S., Takubo, K., Ito, K., Koh, G. Y., and Suda, T. (2004). Tie2/angiopoietin-1 signaling regulates hematopoietic stem cell quiescence in the bone marrow niche. *Cell* **118,** 149–161.

Bunting, K. D., Galipeau, J., Topham, D., Benaim, E., and Sorrentino, B. P. (1998). Transduction of murine bone marrow cells with an MDR1 vector enables *ex vivo* stem cell expansion, but these expanded grafts cause a myeloproliferative syndrome in transplanted mice. *Blood* **92,** 2269–2279.

Bunting, K. D., Zhou, S., Lu, T., and Sorrentino, B. P. (2000). Enforced P-glycoprotein pump function in murine bone marrow cells results in expansion of side population stem cells *in vitro* and repopulating cells *in vivo*. *Blood* **96,** 902–909.

Camargo, F. D., Chambers, S. M., Drew, E., McNagny, K. M., and Goodell, M. A. (2006). Hematopoietic stem cells do not engraft with absolute efficiencies. *Blood* **107,** 501–507.

Chen, C. Z., Li, M., de Graaf, D., Monti, S., Gottgens, B., Sanchez, M. J., Lander, E. S., Golub, T. R., Green, A. R., and Lodish, H. F. (2002). Identification of endoglin as a functional marker that defines long-term repopulating hematopoietic stem cells. *Proc. Natl Acad. Sci. USA* **99,** 15468–15473.

Christensen, J. L., and Weissman, I. L. (2001). Flk-2 is a marker in hematopoietic stem cell differentiation: A simple method to isolate long-term stem cells. *Proc. Natl Acad. Sci. USA* **98,** 14541–14546.

Dykstra, B., Kent, D., Bowie, M., McCaffrey, L., Hamilton, M., Lyons, K., Lee, S. J., Brinkman, R., and Eaves, C. (2007). Long-term propagation of distinct hematopoietic differentiation programs *in vivo*. *Cell Stem Cell* **1,** 218–229.

Ford, C. E., Hamerton, J. L., Barnes, D. W., and Loutit, J. F. (1956). Cytological identification of radiation-chimaeras. *Nature* **177,** 452–454.

Goodell, M. A., Brose, K., Paradis, G., Conner, A. S., and Mulligan, R. C. (1996). Isolation and functional properties of murine hematopoietic stem cells that are replicating *in vivo*. *J. Exp. Med.* **183,** 1797–1806.

Kiel, M. J., Yilmaz, O. H., Iwashita, T., Yilmaz, O. H., Terhorst, C., and Morrison, S. J. (2005). SLAM family receptors distinguish hematopoietic stem and progenitor cells and reveal endothelial niches for stem cells. *Cell* **121,** 1109–1121.

Li, C. L., and Johnson, G. R. (1992). Rhodamine123 reveals heterogeneity within murine Lin−, Sca-1+ hemopoietic stem cells. *J. Exp. Med.* **175,** 1443–1447.

Morrison, S. J., and Weissman, I. L. (1994). The long-term repopulating subset of hematopoietic stem cells is deterministic and isolatable by phenotype. *Immunity* **1,** 661–673.

Okada, S., Nakauchi, H., Nagayoshi, K., Nishikawa, S., Miura, Y., and Suda, T. (1992). *In vivo* and *in vitro* stem cell function of c-kit- and Sca-1-positive murine hematopoietic cells. *Blood* **80,** 3044–3050.

Osawa, M., Hanada, K., Hamada, H., and Nakauchi, H. (1996). Long-term lymphohematopoietic reconstitution by a single CD34-low/negative hematopoietic stem cell. *Science* **273,** 242–245.

Scharenberg, C. W., Harkey, M. A., and Torok-Storb, B. (2002). The ABCG2 transporter is an efficient Hoechst 33342 efflux pump and is preferentially expressed by immature human hematopoietic progenitors. *Blood* **99,** 507–512.

Sieburg, H. B., Cho, R. H., Dykstra, B., Uchida, N., Eaves, C. J., and Muller-Sieburg, C. E. (2006). The hematopoietic stem compartment consists of a limited number of discrete stem cell subsets. *Blood* **107,** 2311–2316.

Simpson, C., Pearce, D. J., Bonnet, D., and Davies, D. (2006). Out of the blue: A comparison of Hoechst side population (SP) analysis of murine bone marrow using 325, 363 and 407 nm excitation sources. *J. Immunol. Methods* **310,** 171–181.

Smith, L. G., Weissman, I. L., and Heimfeld, S. (1991). Clonal analysis of hematopoietic stem-cell differentiation *in vivo*. *Proc. Natl Acad. Sci. USA* **88,** 2788–2792.

Spangrude, G. J., Heimfeld, S., and Weissman, I. L. (1988). Purification and characterization of mouse hematopoietic stem cells. *Science* **241,** 58–62.

Venezia, T. A., Merchant, A. A., Ramos, C. A., Whitehouse, N. L., Young, A. S., Shaw, C. A., and Goodell, M. A. (2004). Molecular signatures of proliferation and quiescence in hematopoietic stem cells. *PLoS Biol.* **2,** e301.

Weksberg, D. C., Chambers, S. M., Boles, N. C., and Goodell, M. A. (2008). CD150-side population cells represent a functionally distinct population of long-term hematopoietic stem cells. *Blood* **111,** 2444–2451.

Zhong, R. K., Astle, C. M., and Harrison, D. E. (1996). Distinct developmental patterns of short-term and long-term functioning lymphoid and myeloid precursors defined by competitive limiting dilution analysis *in vivo*. *J. Immunol.* **157,** 138–145.

Zhou, S., Schuetz, J. D., Bunting, K. D., Colapietro, A. M., Sampath, J., Morris, J. J., Lagutina, I., Grosveld, G. C., Osawa, M., Nakauchi, H., and Sorrentino, B. P. (2001). The ABC transporter Bcrp1/ABCG2 is expressed in a wide variety of stem cells and is a molecular determinant of the side-population phenotype. *Nat. Med.* **7,** 1028–1034.

Zhou, S., Morris, J. J., Barnes, Y., Lan, L., Schuetz, J. D., and Sorrentino, B. P. (2002). Bcrp1 gene expression is required for normal numbers of side population stem cells in mice, and confers relative protection to mitoxantrone in hematopoietic cells *in vivo*. *Proc. Natl Acad. Sci. USA* **99,** 12339–12344.

CHAPTER 12

Microarray Analysis of Stem Cells and Their Differentiation

Howard Y. Chang,[*] **James A. Thomson,**[†] **and Xin Chen**[‡]

[*]Program in Epithelial Biology
Stanford University School of Medicine
Stanford, California

[†]Wisconsin National Primate Research Center
Genome Center of Wisconsin and
Department of Anatomy, Medical School
University of Wisconsin, Madison

[‡]Department of Biopharmaceutical Sciences
University of California
San Francisco

Abstract
1. Introduction
2. Overview of Microarray Technology
    2.1. cDNA microarrays
    2.2. Oligonucleotide arrays
3. Experimental Design
    3.1. Total RNA isolation and clean up
    3.2. RNA amplification
    3.3. Microarray hybridization
    3.4. Array based comparative genomic hybridization (array CGH)
    3.5. Chromatin immunoprecipitation followed by microarray
        analysis (ChIP–chip)
    3.6. MicroRNA arrays
    3.7. Data analysis
4. Confirmation Studies
5. Examples of Microarray Experiments for Stem Cell Biology and Differentiation

6. Identification of "Stemness"
7. Differentiation
8. Stem Cell Niches
9. Future Directions
   References

Abstract

Microarrays have revolutionized molecular biology and enabled biologists to perform global analysis on the expression of tens of thousands of genes simultaneously. They have been widely used in gene discovery, biomarker determination, disease classification, and studies of gene regulation. Microarrays have been applied in stem cell research to identify major features or expression signatures that define stem cells and characterize their differentiation programs toward specific lineages. Here, we provide a review of the microarray technology, including the introduction of array platforms, experimental designs, RNA isolation and amplification, array hybridization, and data analysis. We also detail examples that apply microarray technology to address several of the main questions in stem cell biology.

1. Introduction

For the past several decades, biologists have only been able to analyze one or a few genes at a time. However, the advent of complete genomic sequences of more than 800 organisms (including the human and mouse genomes), and the development of microarray technology have revolutionized molecular biology. Microarrays enable biologists to perform global analysis on the expression of tens of thousands of genes simultaneously, and they have been widely used in gene discovery, biomarker determination, disease classification, and studies of gene regulation (Brown and Botstein, 1999; Butte, 2002; Chung *et al.*, 2002; Gerhold *et al.*, 2002). Expression profiling using microarrays is generally considered "discovery research," although it can also be a powerful approach to test defined hypotheses. One advantage of microarray experiments is that at the outset, microarray experiments need not be hypothesis driven. Instead, it allows biologists a means to gather gene expression data on an unbiased basis, and can help to identify genes that may be further tested as the targets in hypothesis driven studies.

2. Overview of Microarray Technology

There are two major microarray platforms that have been widely used: cDNA microarrays and oligonucleotide microarrays.

2.1. cDNA microarrays

The principle of cDNA microarray is illustrated in Fig. 12.1. In brief, cDNA clones, which generally range from several hundred base pairs to several kilobases, are printed on a glass surface, either by mechanical or ink jet microspotting. Sample RNA and a reference RNA are differentially labeled with fluorescent Cy5 or Cy3 dyes, respectively, using reverse transcriptase. The subsequent cDNA are hybridized to the arrays overnight. The slides are washed and scanned with a fluorescence laser scanner. The relative abundance of the transcripts in the samples can be determined by the red/green ratio on each spotted array element.

One of the limitations of cDNA microarray has been that it required relatively large amounts of total RNA (≥ 10 μg) for hybridization. However, significant progress has recently been made for linear amplification of RNA, generally based on Eberwine's protocol (Wang *et al.*, 2000). In this case, RNA is converted into cDNA with oligo dT primers that contain T7 RNA polymerase promoter sequence at its 5' end. The cDNA can be subsequently used as the template for T7 RNA polymerase to transcribe into anti-sense RNA. The linear amplification protocol can produce 10^6-fold of amplification. Therefore, only very small amounts of samples are required in modern microarray experiments.

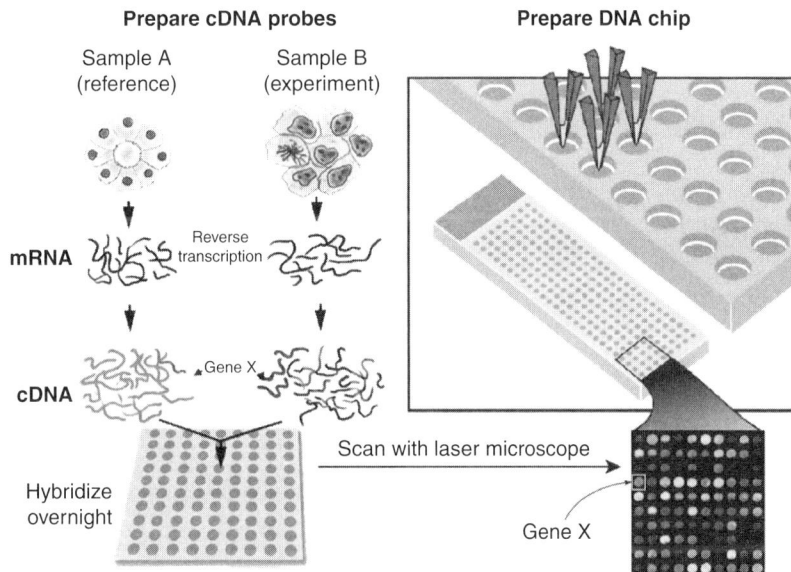

Figure 12.1 Principle of cDNA microarrays. PCR products are printed onto glass slides to produce high density cDNA microarrays. RNA is extracted from experimental samples and reference samples, and differentially labeled with Cy5 and Cy3, respectively, by reverse transcriptase. The subsequent cDNA probes are mixed and hybridized to cDNA microarray overnight. The slides are washed and scanned with fluorescence laser scanner. The relative red/green ratio of gene X indicates the relative abundance of gene X in experimental samples versus reference.

There are several advantages in using cDNA microarray. The two color competitive hybridization can reliably measure the difference between two samples since variations in spot size or amount of cDNA probe on the array will not affect the signal ratio. cDNA microarrays are relatively easy to produce. In fact, the arrayer can be easily built and allow the microarrays to be manufactured in university research labs. Also, cDNA microarrays are in general less expensive than oligonucleotide arrays and are affordable to most research biologists.

There are also some disadvantages with this system. One is that the production of cDNA microarray requires the collection of a large set of sequenced clones. The clones, however, may be misidentified or contaminated. Secondly, genes with high sequence similarity may hybridize to the same clone and generate cross hybridization. To avoid this problem, clones with 3′ end untranslated regions, which in general, are much more divergent compared with the coding sequences, should be used in producing the microarrays.

2.2. Oligonucleotide arrays

The most widely used oligonucleotide arrays are "GeneChips" produced by Affymetrix, which uses photolithography-directed synthesis of oligonucleotides on glass slides. Affymetrix GeneChip measures the absolute levels for each transcript in the sample. The principle of the Affymetrix GeneChip is shown in Fig. 12.2. In general, for each transcript, ~20 distinct and minimal overlapped 25-mer oligonucleotides are selected and synthesized on the array. For each oligonucleotide, there is also a paired mismatch control oligonucleotide, which differs from the perfect match probe by one nucleotide in the central position. Comparison of the hybridization signals from perfect match oligonucleotide with the paired mismatch oligonucleotide will allow automatic subtraction of background.

Since the sequences of oligonucleotide arrays are determined by sequencing information from the database and synthesized *de novo* for the arrays, there is no need to validate cDNA clones or PCR products that are used to print cDNA microarrays. The use of multiple oligonucleotides for each transcript also allows for the detection of splice variants and helps to distinguish genes with high sequence similarities.

The major disadvantage for oligonucleotide arrays is that they can only be produced by commercial manufacturers, and these prefabricated GeneChips are still very expensive. Besides GeneChips from Affymetrix, several other companies produce oligo arrays: Most of them use long oligos and two color comparisons that are more similar to cDNA arrays. These companies include Agilent Technologies (Palo Alto, California), NimbleGen Systems (Madison, Wisconsin), and Combi-Matrix Corporation (Mukilteo, Washington).

Systematic comparison of different microarray platforms showed that standardized protocols for microarray hybridization and data processing, rather than

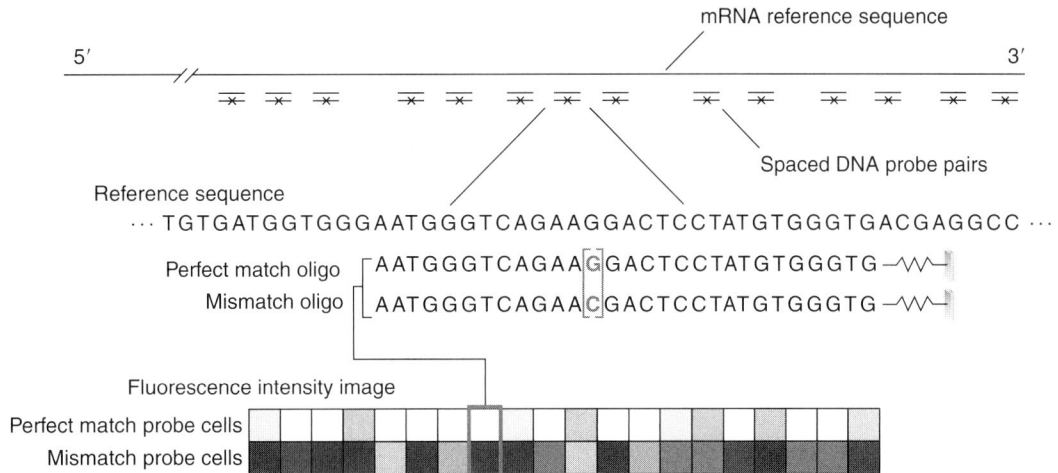

Figure 12.2 Principle of Affymetrix Genechip. For each gene, ~20 distinct and minimally overlapped 25-mer oligonucleotides are selected and synthesized on the array. For each oligonucleotide, there is also a paired mismatch control oligonucleotide probe, which differs from the perfect match probe by one nucleotide in the central position. Comparison of the hybridization signals from perfect match oligonucleotide with the paired mismatch oligonucleotide allows automatic subtraction of background. Figure is kindly provided by Affymetrix, Santa Clara, California.

array platform, appeared to be more important in achieving reproducibility among different investigators (Irizarry *et al.*, 2005; Larkin *et al.*, 2005).

Next, we will describe basic procedures and protocols for microarray experiments, including experimental design, RNA isolation and amplification, array labeling and hybridization, data analysis, and gene validation. Our discussion will focus on cDNA microarrays, but most of the general principles can be applied to both cDNA microarrays and oligonucleotide arrays.

3. Experimental Design

cDNA microarrays use two color competitive hybridization, therefore, the design of experiments, that is, which samples are labeled for each color, is very important in the subsequent data interpretation. There are two commonly used experimental designs, and in this paper, we refer to them as "type I" or "type II" designs.

In the type I experimental design, the sample RNA (for example, drug treated cells, diseased tissues) is labeled with one dye, and another sample RNA (for example, mock treated cells, normal tissues) is labeled with another dye. The two sample cDNA probes are mixed and cohybridize to microarray. The data analysis for the type I design can be quite straightforward. The relative red/green ratio represents the relative upregulation or downregulation of each gene. This design is most suited for experiments with a few perturbations, for example, to identify genes' responses to certain stimulus, or genes affected by a genetic mutation. However, it is difficult to apply this pair-wise "type I" design to a complicated system. For example, to identify genes differentially expressed in a disease and normal tissues, it may not be appropriate to compare diseased tissues versus corresponding normal tissues from the sample patients. Gene expression patterns in normal and diseased tissues are affected by many factors, including patient variation (ethnicity, sex, age, genetic background), sample variation (proximity to disease, anatomic location, and developmental range), as well as heterogeneity of cell populations within the samples. In the simple pair-wise type I design, it is impossible to distinguish all these variations within both normal and diseased tissues. Another drawback of the type I experiment is that it cannot accurately measure the relative abundance of a transcript that is not expressed or expressed at very low levels in Cy3 labeling, as this would produce a green channel signal very close to the background signal.

The type II experimental design avoids most of the problems associated with the simple pair-wise type I design. In the type II design, a common reference is used in Cy3 channel labeling, and each sample is compared with the same common reference. The two most important criteria for selection of common reference are gene coverage ("light up" most of the spots on the array) and reproducibility (can be relatively easily reproduced with minimum batch to batch variation). In most cases, the reference pool RNA is derived from a mixture of tumor cell lines (Perou *et al.*, 2000), and they are commercially available from Strategene and Clontech. Cell lines from different histological origins ensure the complex of the reference. This provides relatively good coverage of the spots on the array. Cell lines can be recultured to produce more reference RNA, however, growth conditions need to be tightly controlled to reduce batch to batch variation. The greatest advantage of the type II experimental design is that it allows the cross comparison of many samples collected over long periods of time, by different persons, as well as samples from different sources.

Another situation where type II designs are widely used is in the time series experiments, that is, to study gene expression variations in response to a stimulus, for example, drug treatment, growth factor stimulation (Boldrick *et al.*, 2002; Diehn *et al.*, 2002). A simple pair-wise type I design using treated cells against untreated cells may be quite straightforward. However, some of the genes that are induced by the stimulus may not be expressed in the untreated cells. It is difficult to obtain an accurate description of the expression profile changes over time for these genes. A better design is a type II design where the cells from different time points are mixed and used as reference. All the samples, including untreated cells ($t=0$), can then be compared with this reference. This average $t=0$ measurement can then be subtracted from each subsequent time point measurement in order to depict the temporal response patterns of expression relative to $t=0$ as the baseline. Such a design ensures that all the transcripts present in the different time points are represented in the reference RNA, and therefore, accurately describes the temporal changes in response to the stimulus.

Although there are many advantages with the type II design, this design may make the experiments more complicated and sometimes inefficient. In the analysis, the baseline reference signals have to be subtracted in order to extract biologically meaningful data. This undoubtedly will add more data variations. So the proper selection of experimental design is one of the most important steps toward the success of experiments.

3.1. Total RNA isolation and clean up

Trizol extraction is the most widely used method for isolation of total RNA. Since RNA is prone to RNase degradation, it is very important to work in an RNase-free area and use DEPC-treated water. Below is the protocol for isolating RNA from tissues. Note that when isolating RNA from tissues, it is best to handle the tissues when they are frozen. Also, before or between uses of the homogenizer, rinse the probe with DEPC-water.

1. Place 100 mg tissue into 2 ml of Trizol Reagent (1 ml of Trizol per 50 mg to 100 mg of tissue), and homogenize tissue at high speed for 1 min, incubate the homogenized samples for 5 min at room temperature.
2. Add 0.2 ml of Chloroform per ml of Trizol reagent used in step 1. Shake by hand for 15 s and incubate for 2–3 min. Centrifuge for 10 min at 14,000 rpm at 4 °C.
3. Remove the upper colorless aqueous phase containing RNA to a fresh tube. Precipitate RNA with 0.5 ml isopropanol per 1 ml of Trizol. Centrifuge at 14,000 rpm for 10 min at 4 °C.
4. Carefully remove the supernatant. Wash the pellet with 75% ethanol, and centrifuge at 14,000 rpm for 5 min at 4 °C.
5. Again carefully remove all the supernatant and air dry for 5–10 min.
6. Dissolve RNA in appropriate amount of RNase-free water. In some cases, total RNA may need to be incubated at 55–60 °C for 5 min before it can be dissolved.
7. Take 1 μl of total RNA for quality and quantity measurement, and store the rest of the total RNA at −80 °C.

It is important to obtain high quality total RNA for microarray labeling or amplification. Several methods can be used to estimate the quality of RNA extracted. For example, when measuring the RNA concentration using UV spectrophotometer, A_{260}/A_{280} ratio should be between 1.9 and 2.1. Another method is to run the RNA on an agarose gel and look for two clear bands: one 18S (0.9 kb) and the other 28S (2 kb) rRNA. Alternatively, a Bioanalyzer chip (Agilent Technologies) can be used to properly assay the quality of total RNA.

If a sufficient amount (15–50 μg) of total RNA is obtained, the RNA may be used directly for labeling. However, in most cases, further purification may be required before the labeling. One method is to reprecipitate RNA. Add 1/10 volume of 3M Sodium Acetate (pH 5.2) and 2.5 volumes of 100% ethanol. Mix and incubate at −20 °C for at least 1 h. Centrifuge at 14,000 rpm for 30 min at 4 °C. Remove the supernatant, wash the pellet with 70% ethanol, air dry, and dissolve in RNase-free water. Take 1 μl of total RNA for quality and quantity measurement. Another method is to use the RNeasy Mini Kit (Qiagen) and follow the manufacturer's protocol. Up to 100 μg of total RNA can be purified at a time. If DNA-free RNA is desired, the RNase-free DNase Set (Qiagen) provides a convenient and efficient on-column digestion of DNA during RNA purification using the RNeasy kit.

3.2. RNA amplification

If a small amount (less than 10 μg) of total RNA is isolated, amplification may be necessary before labeling and microarray hybridization. Several commercially available kits, for example, MessageAmp kit from Ambion or RiboAmp amplification kit from Arctur, have been widely used. Most of these protocols are based on Eberwine's protocol for linear amplification (Van Gelder *et al.*, 1990). Below is a protocol based on the publication by Zhao *et al.* (2002):

I. *First strand cDNA synthesis:*

I.1. Mix total RNA (0.5–5 μg) with 1 μl of Eberwine primer (1 μg/μl) together with RNase-free water to total 9 μl. Incubate at 70 °C for 3 min and cool on ice for 2 min. The sequence of Eberwine primer is 5′-AAA CGA CGG CCA GTG AAT TGT AAT ACG ACT CAC TAT AGG CGC-3′.

I.2. Briefly spin the tube. Add 4 μl of 5× first strand buffer, 2 μl of 0.1M DTT, 1 μl of RNasin, 2 μl of 10 mM dNTP mix, and 2 μl of Superscript II (Invitrogen) to a total volume of 20 μl. Mix the contents, and incubate at 42 °C for 90 min.

II. *Second strand cDNA synthesis:*

II.1. Add the following contents to the first strand synthesis reaction: RNase-free water 106 μl, 10× Advantage PCR buffer 15 μl, 10 mM dNTP mix 3 μl, RNase H (2 U/μl) 1 μl, and Advantage™ Polymerase Mix (Clontech) 3 μl.

II.2. Incubate at 37 °C for 5 min, 94 °C for 2 min, 65 °C for 1 min, and 75 °C for 30 min. Stop the reaction by adding 7.5 ml of 1M NaOH with 2 mM EDTA, and incubate at 65 °C for 10 min.

III. *Ds cDNA cleanup:*

III.1. Add 150 μl of phenol:chloroform:isoamyl alcohol (25:24:1) and mix by vortexing. Spin at 14,000 rpm for 5 min.

III.2. Transfer the aqueous layer to a new tube. Add 1 μl liner acrylamide (0.1 μg/μl), 70 μl of 7.5M NH$_4$Ac, 1 ml 100% ethanol. Spin at 14,000 rpm for 20 min at room temperature.

III.3. Wash pellet with 500 μl of 75% ethanol, spin at 14,000 rpm for 5 min. Carefully remove all the supernatant and air dry for 3 min. Suspend the pellet in 16 μl of RNase-free water.

IV. *In vitro transcription:*

IV.1. Mix the cDNA from step III with 10× reaction buffer, 75 mM NTP mix and T7 enzyme mix (from T7 MEGAscript kit of Ambion).

IV.2. Incubate at 37 °C for 5–6 h.

V. *aRNA clean up:*

Qiagen RNeasy Mini Kit was used in the aRNA clean up.

• Transfer the reaction from step IV into a new tube. Add 60 μl RNase-free water and 350 μl buffer RLT (add β-mercaptoethanol to RLT before use). Mix by pipetting. Add 250 μl 100% ethanol and mix well.

V.2. Apply the sample to the RNeasy column. Spin at 14,000 rpm for 15 s at room temperature. Transfer column to a new collection tube.

V.3. Add 500 μl buffer RPE. Spin at 14,000 rpm for 15 s. Discard flow through. Add 500 μl buffer RPE, spin at 14,000 rpm for 2 min.

V.4. Transfer column to new collection tube. Add 30 μl RNase-free water to the membrane. Let stand for 1 min, and spin at 14,000 rpm for 1 min.

V.5. Measure aRNA yield using a UV spectrometer.

Amplified RNA or isolated total RNA can be used in the labeling reaction for microarray hybridization. Two methods are commonly used for labeling. One is called direct labeling, which uses Cy3 or Cy5 coupled with dUTP in the reaction. The second method is generally referred as indirect labeling, in which aminoallyl-dUTP is used in the reverse transcription reaction. Cy3 or Cy5 dye will then need to be coupled with aminoallyl-dUTP. In general, direct labeling is faster and convenient, but may require more starting materials. Indirect labeling, on the other hand, tends to be more sensitive and therefore requires less RNA for labeling. Below, we will provide general protocols for both methods.

3.2.1. Direct labeling of RNA for microarray hybridization

Before starting the labeling reaction, unlabeled dNTPs will need to be made which include 25 mM dATP, 25 mM dCTP, 25 mM dGTP, and 10 mM dTTP.

1. For total RNA, mix 50–100 μg of total RNA with 4 mg of anchored oligo-dT primer (5′-TTT TTT TTT TTT TTT TTT TTV N-3′) in a total volume of 15.4 μl. For aRNA, use 3–5 μg of aRNA and mix with 10 mg of random hexamer primer (dN6). Heat to 65 °C for 10 min and cool on ice.

2. Add to the reaction from step 1, 6 μl of 5× first-strand buffer, 3 μl of 0.1M DTT, 0.6 μl unlabeled dNTP, 3 μl of 1 mM Cy3 or Cy5-dUTP, and 2 μl of Superscript II (Invitrogen) to total reaction of 30 μl.

3. Incubate at 42 °C for 1 h.

4. Add 1 μl of Superscript II to each sample. Incubate for an additional 0.5–1 h.

5. Degrade RNA and stop the reaction by adding 15 μl of 0.1N NaOH, 2 mM EDTA, and incubate at 65–70 °C for 10 min. Neutralize by addition of 15 μl of 0.1N HCl.

6. Use the MicroconYM-30 (Millipore) to clean up. Add 380 μl of TE (10 mM Tris, 1 mM EDTA) to the Microcon column. Combine the Cy5 and Cy3 probes, and add to the column. Spin the column for 7–8 min at 14,000 rpm. Discard the flowthrough and add 450 μl of TE and spin for 7–8 min at 14,000 rpm.

7. Remove flowthrough and add 450 μl of TE, 2 μl of Cot1 human DNA (10 μg/μl), 2 μl of polyA RNA (10 μg/μl), and 2 μl tRNA (10 μg/μl). Spin 7–10 min at 14,000 rpm so that the probe on the column is less than 28 μl.

8. Invert the Microcon into a clean tube and spin at 14,000 rpm for 1 min to recover the probe.

9. Adjust the probe volume to 28 μl. Add 5.95 μl of 20× SSC and 1.05 μl of 10% SDS to total of 35 μl. Denature probe by heating for 2 min at 100 °C, and spin at 14,000 rpm for 15–20 min. The probe is now ready for hybridization.

3.2.2. Indirect labeling of RNA for microarray hybridization

Before the reaction, make a 50× dNTP solution to include 25 mM dATP, 25 mM dCTP, 25 mM dGTP, 5 mM dTTP, and 20 mM aminoallyl-dUTP. Monofunctional NHS-ester Cy3 or Cy5 dye (Amersham) is supplied as dry pellet. Before use, add 16 μl of DMSO to each dye tube.

1. For total RNA, mix 10–15 μg of total RNA with 4 μg of anchored oligo-dT primer (5′-TTT TTT TTT TTT TTT TTT TTV N-3′) in a total volume of 14.5 μl. For aRNA, use 1–1.5 μg of aRNA and mix with 1.25 μg of random hexamer primer (dN$_6$). Heat to 72 °C for 10 min and cool on ice.

2. Add to the reaction from step 1, 3 μl of 10× StrataScript Buffer, 0.6 μl AA-dNTPs, 3 μl of 0.1M DTT, 2 μl of Stratascript RT, and 5.9 μl of DEPC water to total 30 μl. Incubate at 42 °C for 2 h.

3. Add 10 μl of 1M NaOH and 10 μl of 0.5M EDTA to each tube. Incubate for 15 min at 65 °C. Neutralize the reaction by adding 15 μl of 1M HEPES (pH 7.4).

4. Use the MicroconYM-30 (Millipore) to clean up. Add 450 μl of DEPC water to the column, and add the probes. Spin at 14,000 rpm for 8 min. Discard the flowthrough. Add again 450 μl of DEPC water to the column, and spin at 14,000 rpm for 8 min. Discard the flowthrough.

5. Invert the Microcon into a clean tube and spin at 14,000 rpm for 1 min to recover the probe.

6. Speed vac the samples until dry.

7. Resuspend cDNA in 9 μl of 50 mM NaBiocarbonate (pH 9.0), and stand at room temperature for 15 min.

8. Add 1.25 μl of Cy3 or Cy5 to the appropriate reaction tube, and incubate for 1–2 h at room temperature in the dark.

9. Add 4.5 μl 4M hydroxylamine to each sample, incubate for 15 min in dark. Add 70 μl of DEPC water and 8.5 μl of Cot-1 DNA to the sample.

10. Clean up the reaction using the QiaQuick PCR-purification kit (Qiagen). Add 500 μl of Buffer PB to the reaction from step 9, mix well. Transfer to the QiaQuick column. Spin at 14,000 rpm for 1 min and discard the flowthrough. Add 700 μl Buffer PE, spin at 14,000 rpm for 1 min, and discard the flowthrough. Spin for an additional 1 min to dry the column. Transfer the column to a new collection tube. Add 30 μl EB to the center of the filter, incubate for 2 min. Spin at 14,000 rpm for 1 min to collect the labeled cDNA.

11. Mix the corresponding Cy3 and Cy5 tubes, and speed vac to dry the samples.

12. Dissolve the sample in 25.375 μl of DEPC water, 5.25 μl of 20× SSC, 0.875 μl of 1M HEPES (pH = 7), 2.625 μl of PolyA, and 0.9 μl of 10% SDS. Denature the probe by heating for 2 min at 100 °C, cool for 1 min. The probe is now ready for hybridization.

3.3. Microarray hybridization

Before the microarray slides can be used in hybridization, it generally requires postprocessing to reduce array background and denature the cDNA on the arrays. The procedure for post processing varies, depending on the glass surface. The protocol below was developed using the UltraGAPS slides (Corning).

1. Mark boundaries of the array on the back of the slide using a diamond scriber, as the array spots will become invisible after postprocessing.
2. UV crosslink the DNA to the slide with 150–300 mJ.
3. Incubate arrays in 35–45% formamide, 5× SSC, 0.1% SDS, 0.1 mg/ml BSA at 42 °C for 30–60 min.
4. Rinse arrays briefly with water, then denature DNA on the slide by transferring into a dish of boiling water for 2 min.
5. Wash briefly with 95% ethanol and dry arrays by centrifugation.

For hybridization, add the labeled probes onto microarray slides and cover with a proper cover slip. Incubate the slide chamber in a water bath for 16 h at 65 °C.

For washing microarray slides, prepare the following solution: Wash 1A: 2 × SSC and 0.03% SDS; Wash 1B: 2 × SSC; Wash 2: 1 × SSC, and Wash 3: 0.2 × SSC. Take the slides out of the slide chamber and wash for 5 min in Wash 1A solution. Then, rinse briefly in Wash 1B, which minimizes the transfer of SDS to Wash 2. Wash the slides for 5 min in Wash 2, followed by 5 min in Wash 3. Spin dry the slide by centrifugation.

The microarray slides can be scanned. Before the arrays can be loaded into the database for analysis, array images generally need to be analyzed to exclude weak or bad spots, a process which is referred as gridding. The most widely used array image analysis software include Scanalyze, Genepix, and SpotReader.

3.4. Array based comparative genomic hybridization (array CGH)

In addition to quantifying the mRNA expression level, microarrays can be used to detect DNA copy number and other genomic aberrations in a high throughput fashion, which is generally referred as array CGH. Array CGH has been widely used in cancer research to identify chromosomal aberration in cancer cells (Albertson and Pinkel, 2003; Pinkel and Albertson, 2005). Genomic DNA of stem cells can be compared to reference DNA from somatic cells (e.g., normal blood) by two-color labeling and competitive hybridization in an analogous fashion as mRNA expression arrays. This application is particularly valuable for verifying the genomic integrity of stem cells that may be used for therapeutic purposes (Maitra *et al.*, 2005). For both basic research and clinical applications, it is essential to verify that the stem cells of interest have not acquired oncogenic or other genetic abnormalities in tissue culture. The high density of probes on microarrays can enable detection of small deletions and amplifications (Ishkanian *et al.*, 2004).

Array CGH can be performed using either cDNA array or, more commonly, BAC clone based arrays (Pinkel *et al.*, 1998; Pollack *et al.*, 1999; Snijders *et al.*, 2001). Here is the protocol for BAC clone based array CGH.

I. Preparation of probe

I.1. Mix DNA (0.2–1 μg) with 10 μl of 2.5$\times$ random primer (Invitrogen Bioprime random prime kit) together with water to total 21 μl.

I.2. Briefly spin the tube. Denature the DNA mixture at 100 °C for 10 min and immediately cool on ice.

I.3. Briefly spin the tube. Add 2.5 μl of dNTP (mixture of 4 mM each of dATP, dCTP, dGTP, and dTTP in TE) mix together with Cy3 or Cy5-dCTP, and 0.5 μl of Klenow. Mix the contents and incubate at 37 °C overnight.

I.4. Place the MicroSpin G-50 column (Amersham) into Eppendorf tube. Spin at 770 rcf for 1 min to remove excess liquid. Place the columns into a new Eppendorf tube and apply 25 μl of test probe and 25 μl of reference probe to the column. Spin at 770 rcf for 2 min. Collect the flowthrough.

II. Hybridization

II.1. Add 50 μl of Human Cot-1 DNA, 10 μl of 3M sodium acetate, and 250 μl of ethanol into the probe mixture. Precipitate the probe at 14,000 rpm for 20 min at 4 °C. Carefully remove the supernatant.

II.2. Dissolve the pellet in 5 μl of water, 10 μl of 20% SDS, and 35 μl of Master Mix. Master Mix is produced by mixing 1 g of dextransulfate, 5 ml of formamide, 1 ml of 20×SSC, and 1 ml of water. Resuspend the pellet with pipette tip by stirring. Denature the hybridization solution at 73 °C for 10 min and incubate at 37 °C for 30–60 min.

II.3. Apply the hybridization mixture to the arrays. Incubate the slide at 37 °C for 48–72 h.

III. Washing and scanning

III.1. Disassemble the slide assembly and rinse the hybe solution from the slide in a room temp stream of PN (0.1M sodium phosphate with 0.1% NP-40). Wash slides in wash solution (50% formamide, 2×SSC, pH 7) at 45 °C for 15 min. Finally, wash slides in PN at room temperature for another 15 min.

III.2. Apply DAPI mix (90% glycerol, 10% 1×PBS, 1 μl DAPI (Sigma), pH 8) on top of the array. Place cover slip on top of the array and remove excess DAPI mix.

III.3. The single color intensity image for each channel (Cy3, Cy5, and DAPI) can be collected using a charge coupled device camera.

3.5. Chromatin immunoprecipitation followed by microarray analysis (ChIP–chip)

ChIP–chip employs antibodies specific to a candidate regulator for genome-scale chromatin-immunoprecipitation combined with microarrays spotted with intergenic sequences to identify their bound targets. Protein-bound DNA fragments retrieved by the antibody, typically ~500–1000 base pairs, are hybridized to microarrays to identify the retrieved sequences (reviewed by Elnitski *et al.*, 2006). Thus, ChIP–chip analysis is similar in principle to array CGH, in that they both use microarrays to quantify copy number of genomic DNA fragments. Best results for ChIP–chip are obtained with oligonucleotide arrays that tile intergenic sequences at high resolution (Johnson *et al.*, 2008), and high throughput sequencing is emerging as an alternative method for reading ChIP results, termed ChIP-seq (Barski *et al.*, 2007; Johnson *et al.*, 2007; Mikkelsen *et al.*, 2007).

ChIP–chip analysis has been useful in stem cell biology for identifying direct targets of transcription factors and chromatin remodeling complexes that are important for pluripotency and differentiation, such as those of the Polycomb Group Proteins, Oct-4, nanog, Myc, among others (Boyer *et al.*, 2005, 2006; Lee *et al.*, 2006; Loh *et al.*, 2006; Mikkelsen *et al.*, 2007). For instance, these analyses revealed that Polycomb proteins maintain embryonic stem cells in their undifferentiated state by occupying and repressing genes encoding lineage-specific developmental regulators (Boyer *et al.*, 2006; Lee *et al.*, 2006; Loh *et al.*, 2006; Mikkelsen *et al.*, 2007). Further, ChIP–chip analysis showed that Oct4, Nanog, Sox2, and other embryonic stem cell transcription factors extensively cooccupy the promoters of each other as well as downstream target genes, providing a possible rationale for the success of induced pluripotency with just four transcription factors (reviewed by Jaenisch and Young, 2008). As another example, Lan *et al.* recently discovered that the histone demethylase UTX played a key role in animal patterning by showing that UTX is excluded from the chromatin of *HOX* loci, genes specifying tissue positional identity, in embryonic stem cells, but are specifically recruited to the *HOX* in differentiated cells (Lan *et al.*, 2007). The prospect of a large number of genome-scale ChIP–chip and ChIP-seq data set from the ENCODE project promises many insights.

3.6. MicroRNA arrays

MicroRNAs (miRNAs) are single-stranded RNAs, typically ~21 nucleiotides, noncoding RNAs (Bartel, 2004; Zamore and Haley, 2005). miRNAs are found in a wide variety of organisms including plants, insects, and animals. It has been estimated that there are 250–1000 miRNAs in mammalian cells (Bentwich *et al.*, 2005; Xie *et al.*, 2005), and a miRNA database is established and maintained by Sanger Institute (http://microrna.sanger.ac.uk) (Griffiths-Jones *et al.*, 2006). miRNAs are highly conserved molecules that regulate gene expression by binding and modulating RNA translation. miRNA tends to be multitargeted, that is, each miRNA can regulate the expression of multiple genes. Studies have demonstrated that microRNAs play important roles in multiple cellular processes, including stem cell self-renewal and differentiation (Houbaviy *et al.*, 2003; Kanellopoulou *et al.*, 2005; Lakshmipathy *et al.*, 2007; Murchison *et al.*, 2005; Sinkkonen *et al.*, 2008; Tang *et al.*, 2006; Wang *et al.*, 2007b).

Microarrays have been applied to study the variation of miRNA expression in cells. Several commercial miRNA arrays are available, including Human miRNA microarray (Agilent; Wang *et al.*, 2007a), NCode (Invitrogen; Josephson *et al.*, 2007; Palmieri *et al.*, 2008), GenoExplorer (GenoSensor; Baroukh *et al.*, 2007; Weston *et al.*, 2006), mirCURY (Exiqon; Bak *et al.*, 2008; Dore *et al.*, 2008), mirVana (Ambion; Zhang *et al.*, 2006), and µParaflo chip (LC Sciences; Cummins *et al.*, 2006). The principal microarray study design and procedure are similar to RNA microarrays, while the specific methods for RNA labeling and array hybridization vary according to the array platform, and will not be detailed here. However, several challenges remain to use the miRNA microarrays. For example, since the average size of miRNA is only 21 nucleotides, the specificity of hybridization is rather challenging due to the low melting temperatures. Higher specificity may be improved, for example, by using locked nucleic acid oligos in the case of mirCURY arrays. Another major challenge is the normalization of miRNA arrays. When normalizing RNA arrays, we generally use the assumption that total RNA levels remain the same in different biological samples. However, it is likely not true that the amount of miRNA is the same from a given amount of total RNA in different samples, since miRNA only constitutes 0.01% of the total RNA. This is further complicated by the fact that only a fraction of all miRNA is known currently. To overcome this problem, several methods to normalize miRNA arrays have been developed. One of the most popular methods is to normalize based on spiked-in controls. Finally, there is the challenge of distinguishing unprocessed pri- and pre-miRNAs from active miRNAs. The processed miRNAs can be specifically recognized using "folded-back" microarray probes that can only recognize 21 nucleotide species (Wang *et al.*, 2007a). Clearly, because of these challenges of miRNA arrays, it is important to corroborate the array results with Northern blotting and/or quantitative RT–PCR in order to obtain an accurate readout of miRNA expression in each specific condition.

3.7. Data analysis

Microarray experiments produce large amounts of data, for example, 20,000 genes × 100 samples will generate 2 million data points. Data analysis may be one of the most challenging issues facing biologists in microarray experiments. Micro-array data are often noisy and not normally distributed, and usually with missing values in its matrix. During the past several years, with the combined efforts of biologists, statisticians, and computer scientists, there has been great progress in bioinformatic techniques that can be used for the analysis of genome-wide expression data (Sherlock, 2001; Slonim, 2002). However, there is no standard or one-size-fits-all solution for interpretation of microarray data. Here, we highlight several useful methods, and provide links to freely available resources (Table 12.1).

Current methods for microarray data analysis can be divided into two major categories: supervised and unsupervised methods. Supervised approaches try to identify gene expression patterns that fit a predetermined pattern. Unsupervised approaches characterize expression components without prior input or knowledge of the predetermined pattern.

3.7.1. Supervised analysis

The purpose of supervised analysis is to identify genes that are differentially expressed between groups of samples, and to find genes that can be used to accurately predict the characteristics of groups.

There are many statistical tests developed for identification of differentially expressed genes in microarray data, for example, the *t*-test for detecting significant changes between repeated measurements of a variable in two groups; and the ANOVA F statistic for detecting significant changes in multiple groups. All these tests involve two parts: calculating a test statistic and determining the false discovery rate (FDR), or the significance of the test statistic. Here are some commonly used statistical methods for two group comparisons.

Nonparametric t-test (Dudoit and Fridlyand, 2002): In the nonparametric *t*-test, the statistical significance of each gene is calculated by computing a *p* value for it without assuming specific parametric form for the distribution of the test statistics. To determine the *p* value, a permutation procedure is used in which the class labels of the samples are permuted (10,000–500,000 times), and for each permutation,

Table 12.1
Informatic resources on the Internet

Tool	Function	Web site
Cluster	Performs hierarchical clustering. Genes and samples in microarray experiments are organized by similarity.	http://rana.lbl.gov/EisenSoftware.htm
GenePattern	Analysis and data visualization software from Broad Institute at MIT. Performs sequence and microarray analysis, including GSEA test. Curated gene sets are also available for download and analysis.	http://www.broad.mit.edu/cancer/software/genepattern/
Genomica	(i) Implements Gene Module Map. (ii) Identify enriched transcription factor binding sites in the promoters of gene modules. (iii) Implements module networks.	http://genomica.weizmann.ac.il/index.html
Gene Ontology Term Finder	Gene Ontology classifies each gene in the genome using a controlled vocabulary. Tool identified enriched GO terms within groups of select genes.	http://search.cpan.org/dist/GO-TermFinder/
GeneHopping	Identifies sets of coregulated genes between organisms and provides visualization of modules	http://barkai-serv.weizmann.ac.il/Software/GeneHopping/Hopping.html
Onto-Tools	Web-based program to (i) identify GO term enrichment, (ii) map probes among different array platforms, (iii) retrieve annotations of specific genes	http://vortex.cs.wayne.edu:8080/ontoexpress/servlet/UserInfo
GenMapp	For visualizing gene expression data along pathways, creating new pathways, and identifying global biological associated with an expression dataset.	http://www.genmapp.org/

t statistics are computed for each gene. The permutation *p* value for a particular gene is the proportion of the permutations in which the permuted test statistic exceeds the observed test statistic. A *p* value cutoff can be chosen for the dataset. And the FDR = number of genes test X cutoff *p* value/gene declared to be significant.

Wilcoxon (or Mann–Whitney) Rank Sum test (Troyanskaya *et al.*, 2002): This test rank transforms the data and looks for genes with a skewed distribution of ranks. The rank transform smoothens the data by reducing the effect of outliers.

This method is proven to be superior for decidedly nonnormal data. In general, the Wilcoxon Rank Sum test has been shown to be the most conservative, with the lower FDR.

Ideal discriminator method (Park *et al.*, 2001; Troyanskaya *et al.*, 2002): This method is based on the similarity of gene expression patterns on the array to a theoretical pattern that clearly discriminates between two groups. It potentially allows more flexibility in defining more complex theoretical patterns of behavior.

Significant Analysis of Microarray (SAM) (Tusher *et al.*, 2001): SAM uses a statistic that is similar to a *t* statistic. However, it introduces a fudge factor at the denominator when calculating the *t* statistic—therefore, it underweights those genes which have relatively small magnitude of differences and small variation within groups. It also permutes the whole dataset and sets a threshold for a FDR, instead of assigning an individual *p* value to each gene.

Overall, each statistical method has its own advantages. Each biologist may need to choose proper tests for her own dataset, and in some cases, try different statistical tests. In a situation where the most reliable list of genes is desirable, the best approach may be to examine the intersection of genes identified by different statistical tests, or by the more conservative rank sum test and nonparametric *t*-test. However, if a more inclusive list of genes is desired, a higher *p* value cutoff or SAM may be more appropriate.

3.7.2. Unsupervised analysis

Users of unsupervised analysis try to find internal structure in the data set without any prior input knowledge. Here are some widely used analytical methods.

Hierarchical clustering (Eisen *et al.*, 1998): Hierarchical clustering is a simple but proven method for analyzing gene expression data by building clusters of genes with similar patterns of expression. This is done by iteratively grouping genes that are highly correlated in their expression matrix. As the result, a dendrogram is generated. Branches in the dendrogram represent the similarities among the genes; the shorter the branch, the greater similarity of the gene expression pattern. Hierarchical clustering is also popular because it helps to visualize the overall similarities of expression profiles. The number and size of expression patterns within a dataset can be quickly estimated by biologists.

Self-organizing maps (SOM) (Tamayo *et al.*, 1999; Toronen *et al.*, 1999): In SOM, each biological sample is considered as a separate partition of the space, and after partitions are defined, genes are plotted using expression matrix as coordinate. To initiate SOM, the number of partitions to use must first be defined by the users as an input parameter. A map is set with the centers of each partitions to be (known as centroids) arranged in an initially arbitrary way. As the algorithm iterates, the centroids move toward randomly chosen genes at a decreasing rate. The method continues until there is no further significant movement of these centroids. The advantage of SOM is that it can be used to partition the data with easy two-dimensional visualization of expression patterns. It also reduces computational requirements compared with other methods. The drawbacks of the method are that the number of partitions has to be user-defined, and genes can only belong to a single partition at a time.

Singular value decomposition(SVD) (Alter *et al.*, 2000; Raychaudhuri *et al.*, 2000): SVD is also known as "principle components analysis" by statisticians. SVD transforms the genome-wide expression data into diagonalized "eigengenes" and "eigenarrays" space, where the eigengenes (or eigenarrays) are the unique orthogonal superpositions of the genes (or arrays). Sorting the gene expression data according to the eigengenes and eigenarrays will reduce the features of the data to their principal components and may help to identify the main patterns within the data. Importantly, SVD is also a powerful technique to capture any patterns within the data that may be due to experimental artifacts. For example, array hybridization on different days or hybridization on different batches of arrays occasionally gives slight differences in gene expression patterns. SVD can also remove the artifacts without removing any genes or arrays from the dataset.

3.7.3. Gene module analysis

Gene module analysis is based on the simple idea that genes typically work together in groups, such as in enzymatic pathways or regulatory cascades. Thus, the unit of analysis in a microarray experiment should be groups of functionally related genes, termed modules, and one assigns more significance to a group of genes having coordinate regulation than just one member of a group being regulated in the experiment (Mootha *et al.*, 2003; Segal *et al.*, 2004). Using previous biological knowledge, gene modules can be defined by sets of genes that are members of the same biological pathway, have a shared structural motif, are expressed in a specific tissue, or are induced by a specific stimulus. A biological pathway is typically defined by two gene modules: one for the upregulated genes in the pathway and one for the downregulated genes. Gene modules can have different numbers of member genes, and each gene can belong to multiple modules. In a microarray experiment, gene module analysis searches for coordinate regulation of genes that belong to these *a priori* defined gene modules; a statistical test is performed for each module relative to all other genes on the microarray to calculate whether the degree of coordinate regulation is more than one would expect by chance.

The gene module approach has several advantages over our current gene-by-gene methods of analysis. First, coordinated small magnitude regulation of gene expression of many genes in the same pathway can be biologically more important than a large magnitude change that is discordant with other members of the pathway; however, this type of regulation is often missed by the gene-by-gene approach. Moreover, the large number of genes examined in the gene-by-gene approach necessitates significant penalties for multiple hypothesis testing; many biologically meaningful changes can be missed. Gene module analysis takes advantage of the power of groups of genes to detect those genes that have biologically significant, albeit subtle, expression changes. Secondly, because modules are defined by groups of genes known to share certain biological functions or characteristics, defining the unit of analysis by modules improve the investigator's mechanistic interpretation of the biology underlying the gene expression changes. For example, modules can consist of groups of genes previously found to be coordinately regulated in other microarray experiments. In this way, gene module analysis allows one to compare each new microarray experiment to every previously performed experiment to identify commonalities and unifying mechanisms. By analogy with sequence searches where a newly cloned gene is compared to genes in the database for blocks of sequence similarity, gene module analysis allows one to discover features of gene expression patterns that have been observed in other microarray experiments.

In one instantiation of this strategy, Mootha *et al.* pioneered a type of gene module analysis (which they termed gene set enrichment analysis or GSEA) to discover the biological pathways underlying type II diabetes mellitus (Mootha *et al.*, 2003). They compared global gene expression patterns of skeletal muscle biopsies from individuals with normal glucose tolerance, impaired glucose tolerance and type II diabetes mellitus. After rigorous statistical tests on a gene-by-gene basis (and suffering the concomitant multiple hypothesis testing penalty), they found no single gene with a significant difference in expression. However, Mootha *et al.* noticed that many genes that showed the most consistent changes encoded enzymes involved in mitochondrial oxidative phosphorylation. To test the significance of this observation, the authors implemented gene module analysis by constructing 149 modules of various metabolic pathways or coregulated genes. The authors sorted all genes on the microarray into a ranked list, from the one best able to distinguish diabetes versus normal to the least informative. They then asked if the distribution of genes on this list is surprising given the membership of genes in modules.

Specifically, the authors applied Kolmogorov–Smirnov running sum statistic: Beginning with the gene at the top of the ranked list, the running sum increases when a gene that is a member of the gene set is encountered and decreases otherwise. The maximum enrichment score is the greatest positive deviation of the running sum across all genes. To determine the statistical significance of the maximum enrichment score and validate that the results are unlikely to arise by chance alone, permutation testing is performed, comparing the maximum enrichment score using the actual data to that seen in each of 1000 permuted data sets.

GSEA revealed that a module of genes involved in oxidative phosphorylation was significantly downregulated in patients with diabetes. Each gene in the oxidative phosphorylation gene module was transcriptionally downregulated by roughly only 20%, and thus was not clearly detected at the individual gene level. Independent work by Shulman and colleagues using magnetic resonance spectroscopy confirmed that defective mitochondrial oxidative phosphorylation is strongly associated with glucose intolerance and appears to be a strong predictor for development of diabetes (Petersen *et al.*, 2003).

In addition to looking at whether each gene module is significantly regulated in the experiment, gene module analysis also examines which particular genes within a module are contributing to the regulation. This information can refine the gene module and lead to additional mechanistic insight. For instance, Mootha *et al.* noticed that not all genes in the oxidative phosphorylation modules were equally downregulated in diabetes mellitus; the subset of genes that were downregulated consisted of many known targets of the transcriptional coactivator PGC-1α in muscle cells (Mootha *et al.*, 2003). Analysis of shared promoter elements of genes that comprise the refined oxidative phosphorylation module has identified two transcription factors, estrogen receptor related α and GA-repeat binding protein, as key regulators that cooperate with PGC-1α to regulate expression of this gene module and cellular energy metabolism (Mootha *et al.*, 2004). Thus, gene module analysis has generated a model for impaired glucose tolerance and diabetes mellitus in which down regulation of PGC-1α function in skeletal muscle results in the downregulation of genes involved in oxidative phosphorylation.

Secondly, gene module analysis has also been used for exploratory discovery of the shared biological pathways that underlie different human cancers. Using a strategy termed Gene Module Map, Segal *et al.* performed a comprehensive analysis of 1975 previously published microarrays with 2849 gene modules (Segal *et al.*, 2004). These gene modules included tissue-specific genes, coregulated genes, and genes that function in the same process, act in the same pathway, or share similar subcellular localization. Each microarray experiment was also annotated using a controlled vocabulary for several hundred biological and clinical conditions that it represents, including tumor type, stage, and clinical outcomes. For each gene module and each array, they calculated the fraction of genes from that module that was induced or repressed in that array, and asked if this fraction of enrichment was surprising based on chance alone, estimated using the hypergeometric distribution. A similar algorithm was applied to the clinical annotations, and clinical annotations that were enriched for each gene module were identified.

In this fashion, the large number of microarray experiments and their associated clinical information was distilled to a core set of relationships that defined each cancer by a specific combination of gene modules, many of which provide insight into molecular mechanisms underlying cancer phenotypes. For example, poorly differentiated tumors of many histologic types were found to share an activation of the spindle checkpoint and M phase modules, which have been previously associated with chromosomal instability and aneuploidy. Many modules that were specific to particular types of cancer or even stages of the same disease were also identified, such as deactivation of a growth inhibitory module of dual specificity phosphatases in acute lymphoblastic leukemia and repression of an intermediate filament module in breast cancers (Segal *et al.*, 2004). Although it is impressive that many of the newly described relationships between gene modules and their respective cancers can be supported by the literature, the significance of the majority of gene modules in human cancers awaits experimental validation.

3.7.4. Regulatory networks

In many biological studies, we are interested in identifying causal relationships—that is, gene A is upstream of gene B and induces B. Several investigators have applied probabilistic graphical models, specifically Bayesian networks, to identify regulatory relationships from static views of global gene expression patterns (Beer and Tavazoie, 2004; Segal *et al.*, 2003). Bayesian networks are a particularly useful type of model because they organize a set of variables into a hierarchical model of conditional probabilities; the value of the daughter variables is the joint conditional probability of the parent variables. Typically, the model is used to evaluate many combinations of specific variables, and particular models that produce good fit of the data are validated by additional computational and experimental tests (Friedman, 2004).

For example, because microarray data provide a global view of mRNA abundance, the underlying regulatory network could be the set of active transcription factors or the set of promoter and enhancer elements that produced the genome-wide transcriptional pattern. Segal *et al.* approached this problem by reasoning that many transcription factors and signal transducers are under transcriptional control themselves; thus, a regulatory model may be constructed by relating the expression pattern of genes that encode transcription factors to that of all other genes (Segal *et al.*, 2003). Segal *et al.* developed a probabilistic Bayesian algorithm, termed module networks, to identify the correlations between the expression level of a manually curated set of genes encoding transcription factors and signaling proteins, termed regulators, and, separately, all other transcribed genes (Segal *et al.*, 2003). Transcribed genes were grouped into modules based on the expression changes of the regulators, and regulators were allowed to combine into hierarchical patterns that were conditional, additive, or antagonistic. Thus, unlike hierarchical clustering that only identifies genes with similar patterns of expression, regulatory programs allowed logical operations, such as AND, OR, IF, and BUT. An iterative process of regulatory tree building and gene module assignment is performed to optimize both predictions. In each iteration, the procedure learns the best regulation program for each module, and given the inferred regulation programs, reassigns each gene to the module whose program best predicts its behavior. These two steps are iterated until convergence is reached. This method has the advantage of generating testable hypotheses about gene modules and their regulatory programs in a single analysis. However, this method is limited by current biological knowledge because it relies on compiling a list of candidate regulators. In addition, although this strategy can accommodate heterologous data, such as proteomic or enzyme activity profile data, currently most high throughput data of regulators are transcriptional analysis. Thus, the predicted regulatory trees can be wrong because it fails to take into account posttranscriptional and posttranslational regulation.

To demonstrate the power of this strategy, Segal *et al.* used a set of 466 candidate regulators and a set of 173 arrays that measure responses of *Saccharomyces cerevisiae* to various stresses, which resulted in 50 modules with regulation trees. It should be noted that this type of algorithm will always produce a regulatory tree; the key assessment is the quality of the regulatory trees and gene modules that are produced. A good regulatory tree will encompass transcription factors that are known to act or interact with one another, and the gene modules will have member genes that can be shown to have shared functions. Segal *et al.* found that 31 of the 50 modules had more than 50% of its genes that had the same functional annotation, 30 of the 50 modules included genes previously known to be regulated by at least one of the module's predicted regulators, and 15 of 50 modules had a match between the *cis*-regulatory motifs in the upstream regions of the module's genes and the regulator known to bind to that motif. This is a rather remarkable feat given that the only input information was gene expression data; no biochemical, genetic, or sequence data was used to make the predictions. To further validate this strategy, Segal *et al.* chose three novel regulatory relationships predicted by the regulatory network model, mutated each regulator, and performed global gene expression analysis. In all three cases, the deletion mutants selectively affected the gene modules that they were predicted to regulate. Thus, module networks is a useful method for generating hypotheses that can accurately predict regulators, the processes that they regulate, and the conditions under which they are active.

Hypotheses of regulatory mechanisms can also be generated from shared *cis*-regulatory DNA elements in the upstream regions of the genes in each module. Several investigators have recently developed computational methods to predict global gene regulation based on analysis of mammalian enhancers (Hallikas *et al.*, 2006) and promoters (Sinha *et al.*, 2008; Warner *et al.*, 2008). For instance, Sinha *et al.* identified sets of genes that share combinations of *cis*-regulatory elements in their promoters, taking into account the frequency, spacing, and competition among different elements. These sets, termed motif modules, can then be compared against gene expression changes identified in any experiment to infer putative regulators of the observed change (Sinha *et al.*, 2008). The motif module approach was validated by correctly predicting the targets of four uncharacterized *cis*-motifs that drive different phases of cell cycle progression (Sinha *et al.*, 2008), as well by *in vivo* experiments that demonstrated a potent requirement of NF-κB in mammalian aging (Adler *et al.*, 2007). The large number of hypotheses generated from regulatory networks analysis or *cis*-regulatory DNA elements analysis can also be validated in a high throughput fashion using ChIP–chip analysis, as described above. In addition, as mentioned above, regulator networks also can be validated by expression profiling of mutants of the regulator to see if its signature recapitulates the effect on the genes in the module that it was predicted to regulate.

4. Confirmation Studies

It is very important to independently verify array observations. There are two approaches: *in silico* analysis and experimental validation.

The *in silico* method compares array results with information available in the literature or other independent array expression database. Agreement between array results from other groups, especially using different array platforms will validate the general performance of the system and provide confidence in the overall data.

Experimental validation uses an independent experimental method to assay the expression levels of genes of interest, preferably on another sample set other than the samples that have been used in the microarray analysis. The methods that have been used widely depend on the specific scientific questions. Commonly used techniques include at the mRNA level: semi-quantitative RT–PCR, real time RT–PCR (Taqman, Applied Biosystems), Northern blot, *in situ* hybridization (ISH); at protein level: Western blot, fluorescence activated cell sorting, enzyme-linked immunoabsorbent assays, immunofluorescence, and immunohistochemistry (IHC). Both ISH and IHC can be performed in a high throughput manner via tissue arrays. In addition, both methods provide additional information on the anatomic relationship and cellular origin of gene expression programs.

5. Examples of Microarray Experiments for Stem Cell Biology and Differentiation

Classically, differentiation is defined by the expression of lineage-specific markers, appearance of unique cell morphologies, or acquisition of specific biologic functions (e.g., hemoglobin synthesis in red blood cells). From a genomic perspective, differentiation can be considered as sets of gene expression programs; these gene expression programs may be self-reinforcing, sequential, or mutually exclusive depending on the specific biologic context. Thus, the biologic state of stem cells and their subsequent differentiated states are highly amenable to microarray analysis. In exploring stem cell biology, expression profiling offers a decided advantage as an experimental approach because no specific assumptions are necessary at the outset. By observing the activity of the entire genome, one can ask what major features define stem cells and characterize their differentiation programs toward specific lineages. Here, we highlight several examples in the literature that apply microarray technology to tackle several of the main questions in stem cell biology.

6. Identification of "Stemness"

Stem cells share certain biologic properties—the capacity for self renewal and multipotency; what are the molecular programs that underlie these properties? Several investigators have approached this problem by comparing the gene expression profiles of several embryonic and adult stem cells. By comparing the intersection of the relatively enriched genes in stem cells, a set of genes that are shared by several stem cells has been identified (Ivanova *et al.*, 2002; Ramalho-Santos *et al.*, 2002). It is reassuring that traditional markers of stem cells, for example, CD34 for marrow-derived hematopoietic stem cells (HSC), were also found to be enriched in the stem cell transcriptome by microarray analysis (Ramalho-Santos *et al.*, 2002). In addition, these results provide several candidate pathways that may be involved in regulating maintenance of stem cell fate (Ramalho-Santos *et al.*, 2002). However, comparison of the genes identified in the two initial studies demonstrated little overlap in genes, and additional studies following the same strategy of intersecting stem cell enriched genes from multiple tissue types failed to identify a core set of conserved genes (Evsikov and Solter, 2003; Fortunel *et al.*, 2003; Vogel, 2003). These conflicting findings raised the importance of cross-validation in microarray studies; that is demonstrating the validity of a finding in multiple sample or data sets.

The complexity and subtlety of putative "stemness" programs that may be evident in several types of stem cells may be better approached using gene module approach than analyses focused on individual genes. Recently, Wong *et al.* employed the module map approach to systematically compare gene expression programs in hundreds of profiles of embryonic stem cells, adult tissue stem cells, and also cancer stem cells (Wong *et al.*, 2008). In contrast to inconsistencies with gene-by-gene approaches, the module map identified sets of genes that were supported by experiments from independent labs and using completely different experimental approaches. For instance, a set of genes was found to distinguish embryonic stem cells from other cells by differential mRNA expression in data generated by several labs, and the same set of genes were enriched for those that are directly occupied by stem cell transcription factors Oct-4 and were dependent on Oct-4 for their expression. This study further showed that two different transcriptional programs fundamentally distinguished embryonic stem cells and adult tissue stem cells, and that cancer stem cells resembled embryonic stem cells rather than adult tissue stem cells. Guided by this information, Wong *et al.* was able to reprogram differentiated somatic human cells directly into human cancer stem cells by introduction of three genes, which reactivated the embryonic stem-like transcriptional program. This example linking embryonic to cancer stem cells demonstrates possible advantages of integrating multiple heterogeneous sources of genome-scale data by modular approaches that considers sets of genes rather than single genes.

7. Differentiation

The excitement about stem cells arises from their ability to differentiate into many lineages and cell types, but on a practical level, the pluripotency of stem cells present experimental difficulties in guiding and assessing their development into particular lineages *in vitro* and *in vivo*. Traditionally, one may rely on certain well-established markers of the cell types in question, and in some instances, one may verify the ability of one or more stem cells to repopulate a compartment by reconstitution experiments. (e.g., reconstitution of peripheral blood cells by transplantation of HSC. However, these approaches rely on a relatively small number of protein markers or are laborious and time-consuming. The specificity of lineage markers has been explored to a limited extent in many cases, and many cell types of biologic and medical interest do not have well-defined markers. For instance, CD34, the classical marker for HSC, is also present in a number of other cell types and neoplasms, (e.g., endothelial cells and the fibrohistiocytic tumor dermatofibrosarcoma protuberans). Expression profiling of stem cells and their differentiated derivatives can help to identify the direction and progress of the differentiation program and the interrelationships of the possible differentiation states to one another.

In a prime example of this strategy, Xu *et al.* observed that exposure of human embryonic stem cells to bone morphogenetic protein 4 (BMP4) induced a substantial number of trophoblast markers, including the placental hormone human chorionic gonadotropin (Xu *et al.*, 2002). Thus, BMP4 is probably a key molecular switch that guides the first differentiation event of embryonic stem cells toward this extraembryonic lineage, and BMP4 treated ES cells provide a simple system to derive human trophoblastic cells for studying maternal-fetal interactions. In another powerful use of microarray technology, cell surface markers can also be identified in a high throughput fashion by isolating messenger RNA associated with membrane-bound polysomes (Diehn *et al.*, 2000); hybridization of such selected RNAs to microarrays allows rapid identification of membrane proteins that are likely to be useful lineage markers and receptors that respond to environmental stimuli. More recently, Eggan and colleagues used gene expression profiling to demonstrate that fusion of fibroblasts with embryonic stem cells reprogrammed the fibroblast genome to express genes in a pattern very similar to ES cells on a genome-wide scale (Cowan *et al.*, 2005).

8. Stem Cell Niches

One of the central questions in stem cell biology is the molecular features of the niches that govern the behavior of fetal and adult tissue-specific stem cells. Although some of these key molecules have been identified genetically in amenable organisms (Watt and Hogan, 2000), many of the molecular details that define stem cell niches, especially in mammalian systems, are incompletely understood and may be approached by microarray analysis (Hackney *et al.*, 2002). One approach to understanding stem niches is to explore the diversity of the normal tissue compartments, especially that provided by resident stromal cells.

An illustrative demonstration of this concept is a study of global gene expression patterns of fibroblasts derived from different anatomic sites of skin. Chang *et al.* cultured fibroblasts from multiple sites of human skin, and microarray analysis of

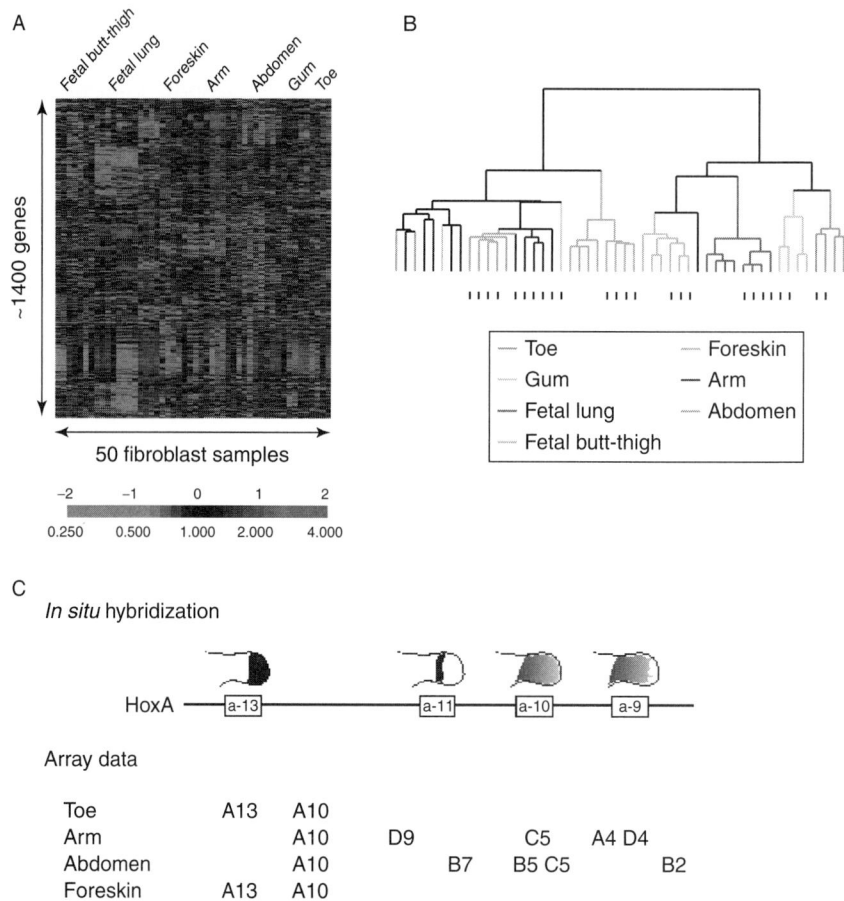

Figure 12.3 Topographic differentiation of fibroblasts identified by microarray analysis. (A) Heat map of fibroblast gene expression patterns. Fibroblasts from several anatomic sites were cultured, and their mRNAs were analyzed by cDNA microarray hybridization (Chang *et al.*, 2002). Approximately 1400 genes varied by at least threefold in 2 samples. The fibroblast samples were predominantly grouped together based on site of origin. (B) Supervised hierarchical clustering revealed the relationship of fibroblast cultures to one another. Site of origin is indicated by the color code, and high or low serum culture condition is indicated by the absence (high) or presence (low) of the black square below each branch. Because fibroblasts from the same site were grouped together irrespective of donor, passage number, or serum condition, topographic differentiation appeared to be the predominant source of gene expression variation among these cells. (C) *HOX* expression in adult fibroblasts recapitulates the embryonic Hox code. In a comparison of *HOX* expression pattern in secondary axes, schematic of expression domains of 5′ *HoxA* genes in the mouse limb bud at approximately 11.5 days post coitum is shown on top. The *HOX* genes up regulated in fibroblasts from the indicated sites are shown below. *HoxC5* is expressed in embryonic chick forelimbs, and *HoxD9* functions in proximal forelimb morphogenesis. (Discussed in detail in Chang *et al.*, 2002).

their global gene expression patterns revealed that fibroblasts from different sites have distinct gene expression programs that have the stability and diversity characteristic of differentiated cell types (Fig. 12.3) (Chang *et al.*, 2002). Some of the site-specific gene expression programs in fibroblasts included components of extracellular matrix and many cell fate signaling pathways, including members of transforming growth factor β, Wnt, receptor tyrosine kinase, and G-protein coupled receptor signaling pathways. An intriguing hint to the specification program in fibroblasts is the maintenance of key features of the embryonic Hox code (which specifies the anterior–posterior body plan) in adult fibroblasts. Thus, stromal cells such as fibroblasts are likely to encode position-specific information in a niche that specifies the developmental potential of interacting stem cell populations. In the case of fibroblasts, because their positional identities are maintained *in vitro* (as evidenced by the fidelity of the Hox expression patterns), it is likely that efforts using stem cells to develop artificial tissue and organs will benefit from incorporation of site-specific fibroblasts or their molecular signatures that recreate the stem cell niche. Identification of the specific stem cell niches—and specific cell fate determining pathways—can thus be accelerated by a comprehensive description and understanding of the stromal cell diversity using microarray analysis.

9. Future Directions

The rapid evolution of microarray technology, bioinformatics techniques, and availability of new genome sequences in model organisms will present many opportunities for harnessing genomic information in stem cell and development research. Fuzzy clustering, a method that gives proportional weight to class assignment, is a valuable technique that may help to reveal more subtle and intricate relationships among various stem cells and their differentiated progenies (Gasch and Eisen, 2002). Additional methods of selecting mRNA or DNA fragments coupled to microarray analysis provide rapid and powerful techniques to elucidate protein subcellular localization or the interaction with DNA binding proteins (Diehn *et al.*, 2000; Iyer *et al.*, 2001; Lieb *et al.*, 2001). Because of its versatility and ability for revealing unexpected features of biology, microarray analysis is likely to become one of the main work horses of the stem cell biologist.

References

Adler, A. S., Sinha, S., Kawahara, T. L., Zhang, J. Y., Segal, E., and Chang, H. Y. (2007). Motif module map reveals enforcement of aging by continual NF-kappaB activity. *Genes Dev.* **21**, 3244–3257.

Albertson, D. G., and Pinkel, D. (2003). Genomic microarrays in human genetic disease and cancer. *Hum. Mol. Genet.* **12**(Spec No 2), R145–R152.

Alter, O., Brown, P. O., and Botstein, D. (2000). Singular value decomposition for genome-wide expression data processing and modeling. *Proc. Natl. Acad. Sci. USA* **97**, 10101–10106.

Bak, M., Silahtaroglu, A., Moller, M., Christensen, M., Rath, M. F., Skryabin, B., Tommerup, N., and Kauppinen, S. (2008). MicroRNA expression in the adult mouse central nervous system. *RNA* **14,** 432–444.

Baroukh, N., Ravier, M. A., Loder, M. K., Hill, E. V., Bounacer, A., Scharfmann, R., Rutter, G. A., and Van Obberghen, E. (2007). MicroRNA-124a regulates Foxa2 expression and intracellular signaling in pancreatic beta-cell lines. *J. Biol. Chem.* **282,** 19575–19588.

Barski, A., Cuddapah, S., Cui, K., Roh, T. Y., Schones, D. E., Wang, Z., Wei, G., Chepelev, I., and Zhao, K. (2007). High-resolution profiling of histone methylations in the human genome. *Cell* **129,** 823–837.

Bartel, D. P. (2004). MicroRNAs: Genomics, biogenesis, mechanism, and function. *Cell* **116,** 281–297.

Beer, M. A., and Tavazoie, S. (2004). Predicting gene expression from sequence. *Cell* **117,** 185–198.

Bentwich, I., Avniel, A., Karov, Y., Aharonov, R., Gilad, S., Barad, O., Barzilai, A., Einat, P., Einav, U., Meiri, E., *et al.* (2005). Identification of hundreds of conserved and nonconserved human microRNAs. *Nat. Genet.* **37,** 766–770.

Boldrick, J. C., Alizadeh, A. A., Diehn, M., Dudoit, S., Liu, C. L., Belcher, C. E., Botstein, D., Staudt, L. M., Brown, P. O., and Relman, D. A. (2002). Stereotyped and specific gene expression programs in human innate immune responses to bacteria. *Proc. Natl. Acad. Sci. USA* **99,** 972–977.

Boyer, L. A., Lee, T. I., Cole, M. F., Johnstone, S. E., Levine, S. S., Zucker, J. P., Guenther, M. G., Kumar, R. M., Murray, H. L., Jenner, R. G., *et al.* (2005). Core transcriptional regulatory circuitry in human embryonic stem cells. *Cell* **122,** 947–956.

Boyer, L. A., Plath, K., Zeitlinger, J., Brambrink, T., Medeiros, L. A., Lee, T. I., Levine, S. S., Wernig, M., Tajonar, A., Ray, M. K., *et al.* (2006). Polycomb complexes repress developmental regulators in murine embryonic stem cells. *Nature* **441,** 349–353.

Brown, P. O., and Botstein, D. (1999). Exploring the new world of the genome with DNA microarrays. *Nat. Genet.* **21,** 33–37.

Butte, A. (2002). The use and analysis of microarray data. *Nat. Rev. Drug Discov.* **1,** 951–960.

Chang, H. Y., Chi, J. T., Dudoit, S., Bondre, C., van de Rijn, M., Botstein, D., and Brown, P. O. (2002). Diversity, topographic differentiation, and positional memory in human fibroblasts. *Proc. Natl. Acad. Sci. USA* **99,** 12877–12882.

Chung, C. H., Bernard, P. S., and Perou, C. M. (2002). Molecular portraits and the family tree of cancer. *Nat. Genet.* **32**(Suppl), 533–540.

Cowan, C. A., Atienza, J., Melton, D. A., and Eggan, K. (2005). Nuclear reprogramming of somatic cells after fusion with human embryonic stem cells. *Science* **309,** 1369–1373.

Cummins, J. M., He, Y., Leary, R. J., Pagliarini, R., Diaz, L. A., Jr., Sjoblom, T., Barad, O., Bentwich, Z., Szafranska, A. E., Labourier, E., *et al.* (2006). The colorectal microRNAome. *Proc. Natl. Acad. Sci. USA* **103,** 3687–3692.

Diehn, M., Alizadeh, A. A., Rando, O. J., Liu, C. L., Stankunas, K., Botstein, D., Crabtree, G. R., and Brown, P. O. (2002). Genomic expression programs and the integration of the CD28 costimulatory signal in T cell activation. *Proc. Natl. Acad. Sci. USA* **99,** 11796–11801.

Diehn, M., Eisen, M. B., Botstein, D., and Brown, P. O. (2000). Large-scale identification of secreted and membrane-associated gene products using DNA microarrays. *Nat. Genet.* **25,** 58–62.

Dore, L. C., Amigo, J. D., Dos Santos, C. O., Zhang, Z., Gai, X., Tobias, J. W., Yu, D., Klein, A. M., Dorman, C., Wu, W., *et al.* (2008). A GATA-1-regulated microRNA locus essential for erythropoiesis. *Proc. Natl. Acad. Sci. USA* **105,** 3333–3338.

Dudoit, S., and Fridlyand, J. (2002). A prediction-based resampling method for estimating the number of clusters in a dataset. *Genome Biol.* **3,** RESEARCH0036.

Eisen, M. B., Spellman, P. T., Brown, P. O., and Botstein, D. (1998). Cluster analysis and display of genome-wide expression patterns. *Proc. Natl. Acad. Sci. USA* **95,** 14863–14868.

Elnitski, L., Jin, V. X., Farnham, P. J., and Jones, S. J. (2006). Locating mammalian transcription factor binding sites: A survey of computational and experimental techniques. *Genome Res.* **16,** 1455–1464.

Evsikov, A. V., and Solter, D. (2003). Comment on " 'Stemness': Transcriptional profiling of embryonic and adult stem cells" and "a stem cell molecular signature". *Science* **302,** 393author reply 393.

Fortunel, N. O., Otu, H. H., Ng, H. H., Chen, J., Mu, X., Chevassut, T., Li, X., Joseph, M., Bailey, C., Hatzfeld, J. A., *et al.* (2003). Comment on " 'Stemness': Transcriptional profiling of embryonic and adult stem cells" and "a stem cell molecular signature". *Science* **302**, 393author reply 393.

Friedman, N. (2004). Inferring cellular networks using probabilistic graphical models. *Science* **303**, 799–805.

Gasch, A. P., and Eisen, M. B. (2002). Exploring the conditional coregulation of yeast gene expression through fuzzy k-means clustering. *Genome Biol.* **3**, RESEARCH0059.

Gerhold, D. L., Jensen, R. V., and Gullans, S. R. (2002). Better therapeutics through microarrays. *Nat. Genet.* **32**(Suppl), 547–551.

Griffiths-Jones, S., Grocock, R. J., van Dongen, S., Bateman, A., and Enright, A. J. (2006). miRBase: MicroRNA sequences, targets and gene nomenclature. *Nucleic Acids Res.* **34**, D140–D144.

Hackney, J. A., Charbord, P., Brunk, B. P., Stoeckert, C. J., Lemischka, I. R., and Moore, K. A. (2002). A molecular profile of a hematopoietic stem cell niche. *Proc. Natl. Acad. Sci. USA* **99**, 13061–13066.

Hallikas, O., Palin, K., Sinjushina, N., Rautiainen, R., Partanen, J., Ukkonen, E., and Taipale, J. (2006). Genome-wide prediction of mammalian enhancers based on analysis of transcription-factor binding affinity. *Cell* **124**, 47–59.

Houbaviy, H. B., Murray, M. F., and Sharp, P. A. (2003). Embryonic stem cell-specific MicroRNAs. *Dev. Cell* **5**, 351–358.

Irizarry, R. A., Warren, D., Spencer, F., Kim, I. F., Biswal, S., Frank, B. C., Gabrielson, E., Garcia, J. G., Geoghegan, J., Germino, G., *et al.* (2005). Multiple-laboratory comparison of microarray platforms. *Nat. Methods* **2**, 345–350.

Ishkanian, A. S., Malloff, C. A., Watson, S. K., DeLeeuw, R. J., Chi, B., Coe, B. P., Snijders, A., Albertson, D. G., Pinkel, D., Marra, M. A., *et al.* (2004). A tiling resolution DNA microarray with complete coverage of the human genome. *Nat. Genet.* **36**, 299–303.

Ivanova, N. B., Dimos, J. T., Schaniel, C., Hackney, J. A., Moore, K. A., and Lemischka, I. R. (2002). A stem cell molecular signature. *Science* **298**, 601–604.

Iyer, V. R., Horak, C. E., Scafe, C. S., Botstein, D., Snyder, M., and Brown, P. O. (2001). Genomic binding sites of the yeast cell-cycle transcription factors SBF and MBF. *Nature* **409**, 533–538.

Jaenisch, R., and Young, R. (2008). Stem cells, the molecular circuitry of pluripotency and nuclear reprogramming. *Cell* **132**, 567–582.

Johnson, D. S., Li, W., Gordon, D. B., Bhattacharjee, A., Curry, B., Ghosh, J., Brizuela, L., Carroll, J. S., Brown, M., Flicek, P., *et al.* (2008). Systematic evaluation of variability in ChIP-chip experiments using predefined DNA targets. *Genome Res.* **18**, 393–403.

Johnson, D. S., Mortazavi, A., Myers, R. M., and Wold, B. (2007). Genome-wide mapping of *in vivo* protein-DNA interactions. *Science* **316**, 1497–1502.

Josephson, R., Ording, C. J., Liu, Y., Shin, S., Lakshmipathy, U., Toumadje, A., Love, B., Chesnut, J. D., Andrews, P. W., Rao, M. S., and Auerbach, J. M. (2007). Qualification of embryonal carcinoma 2102Ep as a reference for human embryonic stem cell research. *Stem Cells* **25**, 437–446.

Kanellopoulou, C., Muljo, S. A., Kung, A. L., Ganesan, S., Drapkin, R., Jenuwein, T., Livingston, D. M., and Rajewsky, K. (2005). Dicer-deficient mouse embryonic stem cells are defective in differentiation and centromeric silencing. *Genes Dev.* **19**, 489–501.

Lakshmipathy, U., Love, B., Goff, L. A., Jornsten, R., Graichen, R., Hart, R. P., and Chesnut, J. D. (2007). MicroRNA expression pattern of undifferentiated and differentiated human embryonic stem cells. *Stem Cells Dev.* **16**, 1003–1016.

Lan, F., Bayliss, P. E., Rinn, J. L., Whetstine, J. R., Wang, J. K., Chen, S., Iwase, S., Alpatov, R., Issaeva, I., Canaani, E., *et al.* (2007). A histone H3 lysine 27 demethylase regulates animal posterior development. *Nature* **449**, 689–694.

Larkin, J. E., Frank, B. C., Gavras, H., Sultana, R., and Quackenbush, J. (2005). Independence and reproducibility across microarray platforms. *Nat. Methods* **2**, 337–344.

Lee, T. I., Jenner, R. G., Boyer, L. A., Guenther, M. G., Levine, S. S., Kumar, R. M., Chevalier, B., Johnstone, S. E., Cole, M. F., Isono, K., *et al.* (2006). Control of developmental regulators by Polycomb in human embryonic stem cells. *Cell* **125**, 301–313.

Lieb, J. D., Liu, X., Botstein, D., and Brown, P. O. (2001). Promoter-specific binding of Rap1 revealed by genome-wide maps of protein-DNA association. *Nat. Genet.* **28,** 327–334.

Loh, Y. H., Wu, Q., Chew, J. L., Vega, V. B., Zhang, W., Chen, X., Bourque, G., George, J., Leong, B., Liu, J., *et al.* (2006). The Oct4 and Nanog transcription network regulates pluripotency in mouse embryonic stem cells. *Nat. Genet.* **38,** 431–440.

Maitra, A., Arking, D. E., Shivapurkar, N., Ikeda, M., Stastny, V., Kassauei, K., Sui, G., Cutler, D. J., Liu, Y., Brimble, S. N., *et al.* (2005). Genomic alterations in cultured human embryonic stem cells. *Nat. Genet.* **37,** 1099–1103.

Mikkelsen, T. S., Ku, M., Jaffe, D. B., Issac, B., Lieberman, E., Giannoukos, G., Alvarez, P., Brockman, W., Kim, T. K., Koche, R. P., *et al.* (2007). Genome-wide maps of chromatin state in pluripotent and lineage-committed cells. *Nature* **448,** 553–560.

Mootha, V. K., Handschin, C., Arlow, D., Xie, X., St Pierre, J., Sihag, S., Yang, W., Altshuler, D., Puigserver, P., Patterson, N., *et al.* (2004). Erralpha and Gabpa/b specify PGC-1alpha-dependent oxidative phosphorylation gene expression that is altered in diabetic muscle. *Proc. Natl. Acad. Sci. USA* **101,** 6570–6575.

Mootha, V. K., Lindgren, C. M., Eriksson, K. F., Subramanian, A., Sihag, S., Lehar, J., Puigserver, P., Carlsson, E., Ridderstrale, M., Laurila, E., *et al.* (2003). PGC-1alpha-responsive genes involved in oxidative phosphorylation are coordinately downregulated in human diabetes. *Nat. Genet.* **34,** 267–273.

Murchison, E. P., Partridge, J. F., Tam, O. H., Cheloufi, S., and Hannon, G. J. (2005). Characterization of Dicer-deficient murine embryonic stem cells. *Proc. Natl. Acad. Sci. USA* **102,** 12135–12140.

Palmieri, A., Pezzetti, F., Brunelli, G., Zollino, I., Lo Muzio, L., Martinelli, M., Scapoli, L., Arlotti, M., Masiero, E., and Carinci, F. (2008). Zirconium oxide regulates RNA interfering of osteoblast-like cells. *J. Mater. Sci. Mater. Med.*

Park, P. J., Pagano, M., and Bonetti, M. (2001). A nonparametric scoring algorithm for identifying informative genes from microarray data. *Pac. Symp. Biocomput.* 52–63.

Perou, C. M., Sorlie, T., Eisen, M. B., van de Rijn, M., Jeffrey, S. S., Rees, C. A., Pollack, J. R., Ross, D. T., Johnsen, H., Akslen, L. A., *et al.* (2000). Molecular portraits of human breast tumours. *Nature* **406,** 747–752.

Petersen, K. F., Befroy, D., Dufour, S., Dziura, J., Ariyan, C., Rothman, D. L., DiPietro, L., Cline, G. W., and Shulman, G. I. (2003). Mitochondrial dysfunction in the elderly: Possible role in insulin resistance. *Science* **300,** 1140–1142.

Pinkel, D., and Albertson, D. G. (2005). Array comparative genomic hybridization and its applications in cancer. *Nat. Genet.* **37**(Suppl), S11–S17.

Pinkel, D., Segraves, R., Sudar, D., Clark, S., Poole, I., Kowbel, D., Collins, C., Kuo, W. L., Chen, C., Zhai, Y., *et al.* (1998). High resolution analysis of DNA copy number variation using comparative genomic hybridization to microarrays. *Nat. Genet.* **20,** 207–211.

Pollack, J. R., Perou, C. M., Alizadeh, A. A., Eisen, M. B., Pergamenschikov, A., Williams, C. F., Jeffrey, S. S., Botstein, D., and Brown, P. O. (1999). Genome-wide analysis of DNA copy-number changes using cDNA microarrays. *Nat. Genet.* **23,** 41–46.

Ramalho-Santos, M., Yoon, S., Matsuzaki, Y., Mulligan, R. C., and Melton, D. A. (2002). "Stemness": Transcriptional profiling of embryonic and adult stem cells. *Science* **298,** 597–600.

Raychaudhuri, S., Stuart, J. M., and Altman, R. B. (2000). Principal components analysis to summarize microarray experiments: Application to sporulation time series. *Pac. Symp. Biocomput.* 455–466.

Segal, E., Friedman, N., Koller, D., and Regev, A. (2004). A module map showing conditional activity of expression modules in cancer. *Nat. Genet.* **36,** 1090–1098.

Segal, E., Shapira, M., Regev, A., Pe'er, D., Botstein, D., Koller, D., and Friedman, N. (2003). Module networks: Identifying regulatory modules and their condition-specific regulators from gene expression data. *Nat. Genet.* **34,** 166–176.

Sherlock, G. (2001). Analysis of large-scale gene expression data. *Brief Bioinform.* **2,** 350–362.

Sinha, S., Adler, A. S., Field, Y., Chang, H. Y., and Segal, E. (2008). Systematic functional characterization of cis-regulatory motifs in human core promoters. *Genome Res.* **18,** 477–488.

Sinkkonen, L., Hugenschmidt, T., Berninger, P., Gaidatzis, D., Mohn, F., Artus-Revel, C. G., Zavolan, M., Svoboda, P., and Filipowicz, W. (2008). MicroRNAs control *de novo* DNA methylation through regulation of transcriptional repressors in mouse embryonic stem cells. *Nat. Struct. Mol. Biol.* **15**, 259–267.

Slonim, D. K. (2002). From patterns to pathways: Gene expression data analysis comes of age. *Nat. Genet.* **32**(Suppl), 502–508.

Snijders, A. M., Nowak, N., Segraves, R., Blackwood, S., Brown, N., Conroy, J., Hamilton, G., Hindle, A. K., Huey, B., Kimura, K., *et al.* (2001). Assembly of microarrays for genome-wide measurement of DNA copy number. *Nat. Genet.* **29**, 263–264.

Tamayo, P., Slonim, D., Mesirov, J., Zhu, Q., Kitareewan, S., Dmitrovsky, E., Lander, E. S., and Golub, T. R. (1999). Interpreting patterns of gene expression with self-organizing maps: Methods and application to hematopoietic differentiation. *Proc. Natl. Acad. Sci. USA* **96**, 2907–2912.

Tang, F., Hajkova, P., Barton, S. C., Lao, K., and Surani, M. A. (2006). MicroRNA expression profiling of single whole embryonic stem cells. *Nucleic Acids Res.* **34**, e9.

Toronen, P., Kolehmainen, M., Wong, G., and Castren, E. (1999). Analysis of gene expression data using self-organizing maps. *FEBS Lett.* **451**, 142–146.

Troyanskaya, O. G., Garber, M. E., Brown, P. O., Botstein, D., and Altman, R. B. (2002). Nonparametric methods for identifying differentially expressed genes in microarray data. *Bioinformatics* **18**, 1454–1461.

Tusher, V. G., Tibshirani, R., and Chu, G. (2001). Significance analysis of microarrays applied to the ionizing radiation response. *Proc. Natl. Acad. Sci. USA* **98**, 5116–5121.

Van Gelder, R. N., von Zastrow, M. E., Yool, A., Dement, W. C., Barchas, J. D., and Eberwine, J. H. (1990). Amplified RNA synthesized from limited quantities of heterogeneous cDNA. *Proc. Natl. Acad. Sci. USA* **87**, 1663–1667.

Vogel, G. (2003). Stem cells. 'Stemness' genes still elusive. *Science* **302**, 371.

Wang, E., Miller, L. D., Ohnmacht, G. A., Liu, E. T., and Marincola, F. M. (2000). High-fidelity mRNA amplification for gene profiling. *Nat. Biotechnol.* **18**, 457–459.

Wang, H., Ach, R. A., and Curry, B. (2007). Direct and sensitive miRNA profiling from low-input total RNA. *RNA* **13**, 151–159.

Wang, Y., Medvid, R., Melton, C., Jaenisch, R., and Blelloch, R. (2007). DGCR8 is essential for microRNA biogenesis and silencing of embryonic stem cell self-renewal. *Nat. Genet.* **39**, 380–385.

Warner, J. B., Philippakis, A. A., Jaeger, S. A., He, F. S., Lin, J., and Bulyk, M. L. (2008). Systematic identification of mammalian regulatory motifs' target genes and functions. *Nat. Methods* **5**, 347–353.

Watt, F. M., and Hogan, B. L. (2000). Out of Eden: Stem cells and their niches. *Science* **287**, 1427–1430.

Weston, M. D., Pierce, M. L., Rocha-Sanchez, S., Beisel, K. W., and Soukup, G. A. (2006). MicroRNA gene expression in the mouse inner ear. *Brain Res.* **1111**, 95–104.

Wong, D. J., Liu, H., Ridky, T. W., Cassarino, D., Segal, E., and Chang, H. Y. (2008). Module map of stem cell genes guides creation of epithelial cancer stem cells. *Cell Stem Cell* **2**, 333–344.

Xie, X., Lu, J., Kulbokas, E. J., Golub, T. R., Mootha, V., Lindblad-Toh, K., Lander, E. S., and Kellis, M. (2005). Systematic discovery of regulatory motifs in human promoters and 3′ UTRs by comparison of several mammals. *Nature* **434**, 338–345.

Xu, R. H., Chen, X., Li, D. S., Li, R., Addicks, G. C., Glennon, C., Zwaka, T. P., and Thomson, J. A. (2002). BMP4 initiates human embryonic stem cell differentiation to trophoblast. *Nat. Biotechnol.* **20**, 1261–1264.

Zamore, P. D., and Haley, B. (2005). Ribo-gnome: The big world of small RNAs. *Science* **309**, 1519–1524.

Zhang, L., Huang, J., Yang, N., Greshock, J., Megraw, M. S., Giannakakis, A., Liang, S., Naylor, T. L., Barchetti, A., Ward, M. R., *et al.* (2006). microRNAs exhibit high frequency genomic alterations in human cancer. *Proc. Natl. Acad. Sci. USA* **103**, 9136–9141.

Zhao, H., Hastie, T., Whitfield, M. L., Borresen-Dale, A. L., and Jeffrey, S. S. (2002). Optimization and evaluation of T7 based RNA linear amplification protocols for cDNA microarray analysis. *BMC Genomics* **3**, 31.

CHAPTER 13

Tissue Engineering Using Adult Stem Cells

Daniel Eberli and Anthony Atala

Wake Forest Institute for Regenerative Medicine
Medical Center Boulevard, Winston Salem
NC 27154-1094, USA

Abstract
1. Introduction
    1.1. Native cells and progenitor cells
2. Biomaterials
    2.1. Naturally derived materials
    2.2. Acellular tissue matrices
    2.3. Synthetic polymers
3. Angiogenic Factors
4. Adult Stem Cells for Tissue Engineering
    4.1. Engineering of bone tissue
    4.2. Engineering of cartilage tissue
    4.3. Engineering of cardiac tissue
    4.4. Engineering of neural tissue
    4.5. Fetal stem cells
5. Conclusion
    References

Abstract

Patients with a variety of diseases may be treated with transplanted tissues and organs. However, there is a shortage of donor tissues and organs, which is worsening yearly because of the aging population. Scientists in the field of tissue engineering are applying the principles of cell transplantation, material science, and bioengineering to construct biological substitutes that will restore and maintain normal function in diseased and injured tissues. The stem cell field is also advancing rapidly, opening new options for cellular therapy and tissue engineering. The use of adult stem cells for tissue engineering applications is promising. This chapter discusses applications of these new technologies for the engineering of tissues and organs. The first part provides an overview of regenerative medicine and tissue engineering techniques; the second highlights different adult stem cell populations used for tissue regeneration.

1. Introduction

Organ damage or loss can occur from congenital disorders, cancer, trauma, infection, inflammation, iatrogenic injuries, or other conditions and often necessitates reconstruction or replacement. Depending on the organ and severity of damage, autologous tissues can be used for reconstruction. However, for most tissues in the body there is not sufficient tissue and there is a degree of morbidity associated with the harvest procedure. For functional replacement, organ transplants are used for damaged tissues. However, there is a severe shortage of donor organs, which is worsening with the aging of the population. Both aforementioned approaches rarely replace the entire function of the original organ. Tissues used for reconstruction can lead to complications because of their inherent different functional parameters. The replacement of deficient tissues with functionally equivalent tissues would improve the outcome for these patients. Therefore, engineered biological substitutes that can restore and maintain normal tissue function would be useful in tissue and organ replacement applications.

Tissue engineering, one of the major components of regenerative medicine, follows the principles of cell transplantation, materials science, and engineering toward the development of biological substitutes that can restore and maintain normal function. Tissue engineering strategies generally fall into two categories: the use of acellular matrices, which depend on the body's natural ability to regenerate for proper orientation and direction of new tissue growth, and the use of matrices with cells. Acellular tissue matrices are usually prepared by manufacturing artificial scaffolds or by removing cellular components from tissues by mechanical and chemical manipulation to produce collagen-rich matrices (Chen et al., 1999; Dahms et al., 1998; Piechota et al., 1998; Yoo et al., 1998). These matrices tend to slowly degrade on implantation and are generally replaced by the extracellular matrix (ECM) proteins that are secreted by the ingrowing cells.

When native cells are used for tissue engineering, a small piece of donor tissue is dissociated into individual cells. These cells are expanded in culture, attached to a support matrix, and then reimplanted into the host after expansion. Cells can also be used for therapy through injection, either with carriers such as hydrogels or alone.

The source of donor tissue can be heterologous (different species), allogeneic (same species, different individual), or autologous (same individual). Ideally, both structural and functional tissue replacement will occur with minimal complications. The preferred cells to use are autologous cells, in which a biopsy of tissue is obtained from the host, the cells are dissociated and expanded in culture, and the expanded cells are implanted into the same host (Atala, 2001, 2005; Oberpenning *et al.*, 1999; Schultz *et al.*, 2006; Yoo *et al.*, 1999). The use of autologous cells, although it may cause an inflammatory response, avoids rejection, and thus the deleterious side effects of immunosuppressive medications can be avoided.

Most current strategies for tissue engineering depend on a sample of autologous cells from the diseased organ of the host. However, for many patients with extensive end-stage organ failure, a tissue biopsy may not yield enough normal cells for expansion and transplantation. In other instances, primary autologous human cells cannot be expanded from a particular organ, such as the pancreas. In these situations, embryonic and adult stem cells are an alternative source of cells from which the desired tissue can be derived. Embryonic stem cells can be derived from discarded human embryos or from fetal tissues. Adult stem cells can be harvested from adult tissues, including bone marrow, fat, muscle, and skin. These cells can be differentiated into the desired cell type in culture and then used for bioengineering.

To complete the list of possible cell sources for bioengineering of tissues and organs, therapeutic cloning must be mentioned. Therapeutic cloning, which has also been called *nuclear transplantation* and *nuclear transfer*, involves the introduction of a nucleus from a donor cell into an enucleated oocyte to generate an embryo with a genetic makeup identical to that of the donor. Stem cells can be derived from this source, which may have the potential to be used therapeutically.

The use of native cells and adult stem cells is ethically sound and accepted by all major religions and governments. On the other hand, the use of embryonic stem cells and therapeutical cloning are more controversial because the same methods could theoretically be used to clone human beings.

Major advances have been achieved in engineering of tissues within the past decade. Regenerative medicine may extend the treatment options for various diseases. However, like every new evolving field, regenerative medicine and tissue engineering are expensive. Several of the clinical trials involving bioengineered products have been placed on hold because of costs involved with the specific technology. With a bioengineered product, costs are usually high because of the biological nature of the therapies involved. As with any therapy, the cost that the medical health care system can allow for a specific technology is limited. Therefore, the costs of bioengineered products have to be reduced for them to have an impact clinically. This is currently being addressed for multiple tissue-engineered technologies. As the technologies advance over time and the volume of the application is considered, costs will naturally decrease.

1.1. Native cells and progenitor cells

One of the limitations of applying cell-based regenerative medicine techniques to organ replacement has been the inherent difficulty of growing specific cell types in large quantities. By studying the privileged sites for committed precursor cells in specific organs, as well as exploring the conditions that promote differentiation, one may be able to overcome the obstacles that limit cell expansion *in vitro*.

For example, in the past, urothelial cells could be grown in the institution setting, but only with limited expansion. Several protocols were developed over the past two decades that identified the undifferentiated cells and kept them undifferentiated during their growth phase (Cilento *et al.*, 1994; Puthenveettil *et al.*, 1999; Scriven *et al.*, 1997). Using these methods of cell culture, it is now possible to expand a urothelial strain from a single specimen that initially covered a surface area of 1 cm^2 to one covering a surface area of 4202 m^2 (the equivalent of one football field) within 8 weeks (Yoo *et al.*, 1998).

These studies indicated that it should be possible to collect autologous bladder cells from human patients, expand them in culture, and return them to the donor in sufficient quantities for reconstructive purposes (Cilento *et al.*, 1994; Liebert *et al.*, 1997; Nguyen *et al.*, 1999; Puthenveettil *et al.*, 1999; Strem *et al.*, 2005). Major advances have been achieved within the past decade in the possible expansion of a variety of progenitor cells and adult stem cells, with specific techniques that make the use of autologous cells possible for clinical application.

2. Biomaterials

For cell-based tissue engineering, the expanded cells are seeded onto a scaffold synthesized with the appropriate biomaterial. In tissue engineering, biomaterials replicate the biological and mechanical function of the native ECM found in tissues in the body by serving as an artificial ECM. Biomaterials provide a three-dimensional space for the cells to form into new tissues with appropriate structure and function and also can allow for the delivery of cells and appropriate bioactive factors (e.g., cell-adhesion peptides, growth factors) to desired sites in the body (Kim and Mooney, 1998). Because most mammalian cell types are anchorage dependent and will die if no cell-adhesion substrate is available, biomaterials provide a cell-adhesion substrate that can deliver cells to specific sites in the body with high loading efficiency. Biomaterials can also provide mechanical support against *in vivo* forces such that the predefined three-dimensional structure is maintained during tissue development. Furthermore, bioactive signals, such as cell-adhesion peptides and growth factors, can be loaded along with cells to help regulate cellular function.

The ideal biomaterial should be biodegradable and bioresorbable to support the replacement of normal tissue without inflammation. Incompatible materials are destined for an inflammatory or foreign-body response that eventually leads to rejection and/or necrosis. Degradation products, if produced, should be removed from the body by metabolic pathways at an adequate rate that keeps the concentration of these degradation products in the tissues at a tolerable level. The biomaterial should also provide an environment in which appropriate regulation of cell behavior (adhesion, proliferation, migration, and differentiation) can occur such that functional tissue can form. Cell behavior in the newly formed tissue has been shown to be regulated by multiple interactions of the cells with their microenvironment, including interactions with cell-adhesion ligands and with soluble growth factors (Hynes, 1992).

Because biomaterials provide temporary mechanical support while the cells undergo spatial tissue reorganization, the properly chosen biomaterial should allow the engineered tissue to maintain sufficient mechanical integrity to support itself in early development. In late development, it should have begun degradation such that it does not hinder further tissue growth (Kim and Mooney, 1998). Generally, three classes of biomaterials have been used for engineering tissues: naturally derived materials (e.g., collagen and alginate), acellular tissue matrices (e.g., bladder submucosa and small intestinal submucosa), and synthetic polymers such as polyglycolic acid (PGA), polylactic acid (PLA), and poly(lactic-coglycolic acid) (PLGA). These classes of biomaterials have been tested in respect to their biocompatibility (Pariente *et al.*, 2001, 2002). Naturally derived materials and acellular tissue matrices have the potential advantage of biological recognition. However, synthetic polymers can be produced reproducibly on a large scale with controlled properties of their strength, degradation rate, and microstructure.

2.1. Naturally derived materials

Collagen is the most abundant and ubiquitous structural protein in the body and may be readily purified from both animal and human tissues with an enzyme treatment and salt/acid extraction. Collagen implants degrade through a sequential attack by lysosomal enzymes. The *in vivo* resorption rate can be regulated by controlling the density of the implant and the extent of intermolecular crosslinking. The lower the density, the greater the interstitial space and generally the larger the pores for cell infiltration, leading to a higher rate of implant degradation. Collagen contains cell-adhesion domain sequences (e.g., RGD) that may assist in retaining the phenotype and activity of many types of cells, including fibroblasts (Silver and Pins, 1992) and chondrocytes (Sams and Nixon, 1995).

Alginate, a polysaccharide isolated from seaweed, has been used as an injectable cell delivery vehicle (Smidsrod and Skjak-Braek, 1990) and a cell immobilization matrix (Lim and Sun, 1980) because of its gentle gelling properties in the presence of divalent ions such as calcium. Alginate is relatively biocompatible and approved by the Food and Drug Administration (FDA) for human use as wound-dressing material. Alginate is a family of copolymers of D-mannuronate and L-glucuronate. The physical and mechanical properties of alginate gel are strongly correlated with the proportion and length of polyglucuronate block in the alginate chains (Smidsrod and Skjak-Braek, 1990).

2.2. Acellular tissue matrices

Acellular tissue matrices are collagen-rich matrices prepared by removing cellular components from tissues. The matrices are often prepared by mechanical and chemical manipulation of a segment of tissue (Chen *et al.*, 1999; Dahms *et al.*,

1998; Piechota *et al.*, 1998; Strem *et al.*, 2005; Yoo *et al.*, 1998). The matrices slowly degrade on implantation and are replaced and remodeled by ECM proteins synthesized and secreted by transplanted or ingrowing cells.

2.3. Synthetic polymers

Polyesters of naturally occurring α-hydroxy acids, including PGA, PLA, and PLGA, are widely used in tissue engineering. These polymers have gained FDA approval for human use in a variety of applications, including sutures (Gilding, 1981). The ester bonds in these polymers are hydrolytically labile, and these polymers degrade by nonenzymatic hydrolysis. The degradation products of PGA, PLA, and PLGA are nontoxic natural metabolites and are eventually eliminated from the body in the form of carbon dioxide and water (Gilding, 1981). The degradation rate of these polymers can be tailored from several weeks to several years by altering crystallinity, initial molecular weight, and the copolymer ratio of lactic to glycolic acid. Because these polymers are thermoplastics, they can be easily formed into a three-dimensional scaffold with a desired microstructure, gross shape, and dimension by various techniques, including molding, extrusion (Freed *et al.*, 1994), solvent casting (Mikos *et al.*, 1994), phase separation techniques, and gas-foaming techniques (Harris *et al.*, 1998). Many applications in tissue engineering require a scaffold with high porosity and ratio of surface area to volume. Other biodegradable synthetic polymers, including poly(anhydrides) and poly(orthoesters), can also be used to fabricate scaffolds for tissue engineering with controlled properties (Peppas and Langer, 1994).

3. Angiogenic Factors

The engineering of large organs will require a vascular network of arteries, veins, and capillaries to deliver nutrients to each cell. One possible method of vascularization is through the use of gene delivery of angiogenic agents such as vascular endothelial growth factor (VEGF) with the implantation of vascular endothelial cells (ECs) to enhance neovascularization of engineered tissues. Skeletal myoblasts from adult mice were cultured and transfected with an adenovirus encoding VEGF and combined with human vascular ECs (Nomi *et al.*, 2002). The mixtures of cells were injected subcutaneously in nude mice, and the engineered tissues were retrieved up to 8 weeks after implantation. The transfected cells were noted to form muscle with neovascularization by histological and immunohistochemical probing with maintenance of their muscle volume, whereas engineered muscle of nontransfected cells had a significantly smaller mass of cells with loss of muscle volume over time, less neovascularization, and no surviving ECs. These results indicate that a combination of VEGF and ECs may be useful for inducing neovascularization and volume preservation in engineered tissue.

4. Adult Stem Cells for Tissue Engineering

Investigators around the world, including our institution, have been working toward the development of several cell types, tissues, and organs for clinical application. The predominant cell type used for tissue engineering applications today is predetermined progenitor cells, which are present in almost every tissue. These cells provide replacement and repair for normal turnover or for injured tissues. Harvested through a biopsy, these cells can be expanded in culture and placed on bioscaffolds for implantation. Our institute has successfully engineered multiple tissues for organ reconstruction using tissue-specific progenitor cells, including urinary bladder (Atala, 2001), uterus (Duel et al., 1996), vagina (De Filippo et al., 2003), and penile tissue (Falke et al., 2003). The first tissue-engineered hollow organ successfully implanted into patients was the urinary bladder, a composite tissue-engineered form smooth muscle cells and urothelial cells (Atala et al., 2006).

Because of the limited availability of tissue-specific progenitor cells, there is a growing scientific interest in the potential of adult stem cells. These cells are derived from a large variety of tissues including bone marrow (Pountos and Giannoudis, 2005), fat tissue (Lin et al., 2006), muscle (Sun et al., 2005), amniotic fluid, placenta (Portmann-Lanz et al., 2006), and umbilical cord (Moise, 2005).

When compared with embryonic stem cells, adult stem cells have many similarities: they can differentiate into all three germ layers, they express common markers, and they preserve their telomere length. However, the adult stem cells demonstrate considerable advantages. They easily differentiate into specific cell lineages; they do not form teratomas if injected in vivo; they do not need any feeder layers to grow; and they do not require the sacrifice of human embryos for their isolation, thus avoiding the controversies associated with the use of human embryonic stem cells.

Unfortunately, the isolation of adult stem cells at sufficient purity and quality remains challenging. Purification is usually done by preplating techniques and through sorting for surface proteins. Adult stem cells are negative for tissue-specific markers and can be positive for embryonic cell surface antigens (Petersen et al., 1999).

Adult stem cells are unspecialized cells, which show self-renewal and can self-maintain for a long time with the potential to commit to a cell lineage with specialized functions. These cells are able to differentiate into committed cells of other tissues, a feature defined as plasticity. This would allow for engineering of composite tissues composed of multiple cell types using one single source of adult stem cells. Therefore, use of adult stem cells opens a new avenue for cellular therapy and for the engineering of tissues and organs.

The most investigated adult stem cells are mesenchymal stem cells (MSCs). This cell type holds significant promise for the engineering of musculoskeletal structures. Bone marrow stroma represents the major source of MSCs. The bone marrow is aspirated from the iliac crest or from long bones and fractionated using a Ficoll density gradient followed by centrifugation. Cells from the low-density fraction are then plated and cultured. A small percentage of these cells grow adherent fibroblastic cells in colonies. These MSCs are characterized by a high proliferation potential and the capability to differentiate into progenitor cells for distinct mesenchymal tissues (Caplan, 1991). MSCs develop into distinct terminal differentiated cells and tissues including bone, cartilage, fat, muscle, tendon, and neural tissue (Ringe *et al.*, 2002). However, even with an increasing number of cell passages, MSCs do not spontaneously differentiate (Pittenger *et al.*, 1999). Taken together MSCs seem to be an optimal cell source for the engineering of bone and cartilage.

4.1. Engineering of bone tissue

For the bridging of osseous defects, autologous bone is still one of the most effective graft materials. However, there are major disadvantages including donor site morbidity, pain during recovery, and the lack of quality and quantity of bone that can be harvested. A combination of bioactive biomaterial and bone-producing cells could eliminate this problem.

Bone is formed by osteoblasts, which arise from stem cells in a multistep development process (Caplan, 1994). This differentiation is closely guided by bioactive molecules, such as bone morphogenetic proteins and growth and differentiation factors. It involves similar transitional events as those occurring in embryogenesis (Pathi *et al.*, 1999). Although the temporal expression patterns of such molecules are not well understood, they were successfully used to enhance bone engineering. Biodegradable scaffolds impregnated with bone morphogenetic protein, BMP-2, were able to bridge critical bone damage in a rat model (Lane *et al.*, 1999). The authors conclude that the enhanced recruitment of MSCs through the chemoattractant effect of BMP-2 leads to increased bone formation. However, the regenerative capacity of a scaffold with bioactive molecules may be limited because of the decreasing number of MSCs with aging. When compared with a newborn, the number of MSCs decreases fourfold by 50 years of age and up to 20-fold for patients older than 80 (Caplan, 1994). Tissue engineering offers a solution to this problem. Adult stem cells can be isolated and expanded in culture until sufficient cell numbers are achieved. During the same period, the cells can be differentiated into the desired tissues and used for implantation. This approach was successfully applied in rats (Ohgushi *et al.*, 1990). Rat MSCs were expanded *in vitro* with a porous hydroxyapatite under osteogenic conditions. One week after implantation of these constructs, bone formation, and maturation was documented. In both control groups—whole rat bone marrow and undifferentiated MSCs—the bone formation was delayed (Ohgushi *et al.*, 1990). This study showed that tissue repair could be enhanced by expanding and differentiating the MSCs before application.

In humans, the treatment with isolated allogeneic MSCs has the potential to enhance the therapeutic effects of conventional bone marrow transplantation in patients with genetic disorders affecting mesenchymal tissues including bone, cartilage, and muscle. In a human trial, allogeneic mesenchymal cells were investigated as a therapy for osteogenesis imperfecta (Horwitz *et al.*, 2002). Each child received two infusions of allogeneic cells. Most children showed engraftment in one or more sites and had an acceleration of growth during the first 6 months after infusion. This improvement ranged from 60% to 94% of the predicted median values for age-matched and sex-matched unaffected children compared with 0–40% over the 6 months immediately preceding the infusions.

Muscle-derived stem cells might present another cell source for bone regeneration. Although muscle progenitor cells are highly heterogeneous in nature, there seems to be an adult stem cell population that is able to transdifferentiate into other tissues (Sun *et al.*, 2005). Osteogenic differentiation has been shown with cells isolated from skeletal muscle tissue (Bosch *et al.*, 2000). After exposure to BMP-2, *in vitro* cells were injected into the hind limb muscle of immune compromised mice. The injected cells seemed to actively participate in the ectopic bone formation. These results show that muscle may represent an additional source of adult stem cells for bone tissue engineering.

4.2. Engineering of cartilage tissue

Cartilage tissue is known for its slow turnover and negligible self-healing potential. Only deep osteochondral defects with damage to cartilage and bone show a small fibrous tissue repair. Tissue engineering using adult stem cells would be a promising approach to reconstruct joints after trauma or arthritis.

The reconstruction of cartilage after osteochondral defects was investigated in a rabbit model. MSCs from rabbit bone marrow were embedded in a type I collagen gel and implanted in to a large osteochondral lesion on the distal femur (Wakitani *et al.*, 1994). After 2 weeks the MSCs differentiated into chondrocytes and covered the defect zone. By 24 weeks the subchondral bone was completely repaired, and an articular cartilage layer was formed. However, the neocartilage was of low quality, with reduced thickness, minor mechanical properties, and inadequate integration to surrounding tissue. The same research group performed a study in patients with osteoarthritic knees (Wakitani *et al.*, 2002). Autologous mesenchymal cells were expanded in culture and embedded in collagen gel. The collagen–cell mix was placed into the cartilage defect and covered by autologous periosteum. To assess the tissue formation, small biopsy specimens were taken by arthroscopy. By 42 weeks, the defects were covered with soft tissue and formed a hyaline-like cartilage. Although the implantation of the collagen–MSC resulted in an improvement in tissue formation, the clinical outcome was not significantly influenced.

4.3. Engineering of cardiac tissue

Another area of great clinical and scientific interest is the use of adult stem cells for cardiac tissue engineering. Arteriosclerosis and myocardial infarction is still the number-one cause of mortality in many countries. Unlike skeletal muscle, cardiomyocytes have only limited potential to regenerate heart muscle. Necrotic cardiomyocytes in infracted tissue are replaced with fibroblasts forming a scar, which results in contractile dysfunction (Fukuda, 2001). Studies have shown that fetal rat cardiomyocytes can survive and functionally improve infarcted cardiac muscle. The use of differentiated progenitor cells is valid for a proof of principle, but this approach is not favorable for the clinical setting because of difficulties harvesting autologous heart cells. Adult stem cells with cardiomyogenic development potential may represent an alternate approach. MSCs are able to differentiate into skeletal muscle, smooth muscle, and cardiac muscle cells.

The capacity of adult stem cells from bone marrow to regenerate the heart is controversial and currently under intensive debate. Initial studies in mice showed that adult stem cells isolated from bone marrow were able to regenerate new myocardium when injected into the heart of recipient mice after myocardial infarction (Orlic et al., 2001). However, recent studies in rats using the same cell type were not able to show any significant cardiac regeneration (Balsam et al., 2004; Nygren et al., 2004).

A swine myocardial infarct model was used to evaluate the improved cardiac function after implantation of autologous MSCs in a large animal setting (Shake et al., 2002). MSCs were isolated and expanded from bone marrow aspirates. Two weeks after the myocardial infarction, 60 million MSCs were implanted into the infarct zone by direct injection. Robust engraftment of MSCs could be shown in all treated animals, and the degree of contractile dysfunction was significantly attenuated.

A German group (Strauer et al., 2001) reported the first human application of adult stem cells for the treatment of myocardial infarction. After confirmation of an occlusion of the left coronary artery causing an acute myocardial infarction, a 46-year-old man was treated by percutaneous transluminal catheter angioplasty and stent placement. Mononuclear bone marrow cells of the patient were prepared, and 6 days after infraction, 12 million cells were injected in the infarct-related artery. At 10 weeks after the stem cell transplantation, the transmural infarct area had been reduced from 24.6% to 15.7% of left ventricular circumference, whereas ejection fraction, cardiac index, and stroke volume had increased by 20–30%. These results demonstrate that selective intracoronary transplantation of human autologous adult stem cells is possible under clinical conditions. More research in a randomized setting is needed to evaluate the regeneration of the myocardial scar after stem cell therapy. To summarize, MSCs seem to be powerful candidates for bone, cartilage, and cardiac tissue engineering. The ability to form both bone and cartilage may be useful for the reconstruction of complex tissues such as joints.

4.4. Engineering of neural tissue

A similar population of adult stem cells derives from fat tissue, termed *adipose tissue-derived stem cells* (ADSCs). These cells share many of the characteristics of their counterparts in bone marrow, including high proliferative capacity and the ability for multilineage differentiation (Zuk et al., 2002). ADSC can be enzymatically digested out of adipose tissue and separated from the adipocytes by centrifugation.

Advantages of ADSCs over MSCs include their simple harvest with local anesthesia and the high density of adult stem cells in fat tissue. Although a bone marrow aspiration in healthy adults is generally limited to 40 ml, the harvest of fat tissue can easily exceed 200 ml. Therefore, the number of harvested adult stem cells is approximately 40× higher (Strem et al., 2005), a big advantage if a large number of cells is needed for tissue engineering applications. It has been demonstrated that ADSC can undergo differentiation along classical mesenchymal lineages including fat, cartilage, bone, and muscle (Strem et al., 2005). Nonmesodermal transdifferentiation has also been confirmed into endothelial cells (Planat-Benard et al., 2004) and nerve cells (Safford et al., 2004).

These findings opened up a new avenue for the cellular treatment of brain disorders. ADSC were marked with a reporter gene (LacZ) and injected into rat brains after 90 min of middle cerebral artery occlusion (Kang et al., 2003). Marked cells were seen throughout the infarct area 14 days after injection. The protein expression pattern showed signs of neural differentiation, and the ADSC-treated animals showed significant improvement in neurological tests. Similar results were shown when MSCs were injected (Zhao et al., 2002). The implanted cells underwent in vivo differentiation with expression of markers consistent with differentiation along astrocytic, oligodendrocytic, and neural lineages. The MSC-treated rats also showed significantly improved performance in a limb placement test.

4.5. Fetal stem cells

Additional promising cells for tissue engineering applications and cell therapy are fetal stem cells isolated from amniotic fluid and umbilical cord blood.

Fetal stem cells have a higher potential for expansion than cells taken from the adult individual; and for this reason they could represent a better source for therapeutic applications when large numbers of cells are needed. The ability to isolate the progenitor cells during gestation may also be advantageous for babies born with congenital malformations. Furthermore, the progenitor cells can be cryopreserved for future self-use.

Human amniotic fluid stem cells (hAFS cells) are a novel source for tissue engineering of tissues and organs. Human amniotic fluid has been used in prenatal diagnosis for more than 70 years. It has proven to be a safe, reliable, and simple screening tool for a wide variety of developmental and genetic disorders. A subset of cells found in amniotic fluid has been isolated and found to be capable of maintaining prolonged undifferentiated proliferation, as well as differentiating into multiple tissue types encompassing the three germ layers.

hAFS cells are harvested by amniocentesis at about 16 weeks of gestation. The cells are then isolated through positive selection for cells expressing the membrane receptor c-kit (Cremer et al., 1981) and can be maintained for >300 population doublings, far exceeding Hayflick's limit. The progenitor cells derived from human amniotic fluid are pluripotent and have been shown to differentiate into osteogenic, adipogenic, myogenic, neurogenic, endothelial, and hepatic phenotypes in vitro.

Recently, investigators discovered mesenchymal cells from umbilical cord blood that can be induced in culture to form a variety of tissues, including bone, cartilage, myocardial muscle, and neural tissue (Bieback *et al.*, 2004). Studies in mice were able to show significant advantages for cellular therapy in the treatment of intracranial hemorrhage (Nan *et al.*, 2005), stroke (Xiao *et al.*, 2005), and amyotrophic lateral sclerosis (Ende *et al.*, 2000).

The exact role of each cell type addressed in this chapter will be defined in future studies. The optimal combination of cell source, differentiation state, growth factors, culture conditions, biomaterials, and seeding technique will allow the engineering of tissue and organs for clinical application.

5. Conclusion

Tissue engineering efforts using adult stem cells experimentally are currently underway for virtually every type of tissue and organ within the human body. As regenerative medicine incorporates the fields of tissue engineering, cell biology, nuclear transfer, and materials science, personnel who have mastered the techniques of cell harvest, culture, expansion, transplantation, and polymer design are essential for the successful application of these technologies. Various tissues are at different stages of development, with some already being used clinically, a few in preclinical trials, and others in the discovery stage. Recent progress suggests that engineered tissues and cell-based therapies using adult stem cells may have an expanded clinical applicability in the future and may represent a viable therapeutic option for those who require tissue replacement or repair.

References

Atala, A. (2001). Bladder regeneration by tissue engineering. *BJU Int.* **88,** 765–770.
Atala, A., and Lanza, R. P. (2001). "Methods of Tissue Engineering." Academic Press, San Diego.
Atala, A., and Mooney, D. Eds. (2005). "Tissue Engineering" Birkhauser Press, Boston.
Atala, A., Bauer, S. B., Soker, S., Yoo, J. J., and Retik, A. B. (2006). Tissue-engineered autologous bladders for patients needing cystoplasty. *Lancet* **367,** 1241–1246.
Balsam, L. B., Wagers, A. J., Christensen, J. L., Kofidis, T., Weissman, I. L., and Robbins, R. C. (2004). Haematopoietic stem cells adopt mature haematopoietic fates in ischaemic myocardium. *Nature* **428,** 668–673.
Bieback, K., Kern, S., Kluter, H., and Eichler, H. (2004). Critical parameters for the isolation of mesenchymal stem cells from umbilical cord blood. *Stem Cells* **22,** 625–634.
Bosch, P., Musgrave, D. S., Lee, J. Y., Cummins, J., Shuler, T., Ghivizzani, T. C., Evans, T., Robbins, T. D., and Huard, T. D. (2000). Osteoprogenitor cells within skeletal muscle. *J. Orthop. Res.* **18,** 933–944.
Caplan, A. I. (1991). Mesenchymal stem cells. *J. Orthop. Res.* **9,** 641–650.
Caplan, A. I. (1994). The mesengenic process. *Clin. Plast. Surg.* **21,** 429–435.
Chen, F., Yoo, J. J., and Atala, A. (1999). Acellular collagen matrix as a possible "off the shelf" biomaterial for urethral repair. *Urology* **54,** 407–410.
Cilento, B. G., Freeman, M. R., Schneck, F. X., Retik, A. B., and Atala, A. (1994). Phenotypic and cytogenetic characterization of human bladder urothelia expanded *in vitro. J. Urol.* **152,** 665–670.

Cremer, M., Schachner, M., Cremer, T., Schmidt, W., and Voigtlander, T. (1981). Demonstration of astrocytes in cultured amniotic fluid cells of three cases with neural-tube defect. *Hum. Genet.* **56**, 365–370.

Dahms, S. E., Piechota, H. J., Dahiya, R., Lue, T. F., and Tanagho, E. A. (1998). Composition and biomechanical properties of the bladder acellular matrix graft: Comparative analysis in rat, pig and human. *Br. J. Urol.* **82**, 411–419.

De Filippo, R. E., Yoo, J. J., and Atala, A. (2003). Engineering of vaginal tissue *in vivo*. *Tissue Eng.* **9**, 301–306.

Duel, B. P., Hendren, W. H., Bauer, S. B., Mandell, J., Colodny, A., Peters, C. A., Atala, A., and Retik, A. B. (1996). Reconstructive options in genitourinary rhabdomyosarcoma. *J. Urol.* **156**, 1798–1804.

Ende, N., Weinstein, F., Chen, R., and Ende, M. (2000). Human umbilical cord blood effect on sod mice (amyotrophic lateral sclerosis). *Life Sci.* **67**, 53–59.

Falke, G., Yoo, J. J., Kwon, T. G., Moreland, R., and Atala, A. (2003). Formation of corporal tissue architecture *in vivo* using human cavernosal muscle and endothelial cells seeded on collagen matrices. *Tissue Eng.* **9**, 871–879.

Freed, L. E., Vunjak-Novakovic, G., and Biron, R. J. (1994). Biodegradable polymer scaffolds for tissue engineering.. *Biotechnology (NY)* **12**, 689–693.

Fukuda, K. (2001). Development of regenerative cardiomyocytes from mesenchymal stem cells for cardiovascular tissue engineering. *Artif. Organs* **25**, 187–193.

Gilding, D. K. (1981). Biodegradable polymers. *In* "Biocompatibility of Clinical Implant Materials" (D. F. Williams, Ed.) CRC Press, Boca Raton, FL.

Harris, L. D., Kim, B. S., and Mooney, D. J. (1998). Open pore biodegradable matrices formed with gas foaming. *J. Biomed. Mater. Res.* **42**, 396–402.

Horwitz, E. M., Gordon, P. L., Koo, W. K., Marx, J. C., Neel, M. D., McNall, R. Y., Muul, L., and Hofmann, T. (2002). Isolated allogeneic bone marrow-derived mesenchymal cells engraft and stimulate growth in children with osteogenesis imperfecta: Implications for cell therapy of bone. *Proc. Natl. Acad. Sci. USA* **99**, 8932–8937.

Hynes, R. O. (1992). Integrins: Versatility, modulation, and signaling in cell adhesion. *Cell* **69**, 11–25.

Kang, S. K., Lee, D. H., Bae, Y. C., Kim, H. K., Baik, S. Y., and Jung, J. S. (2003). Improvement of neurological deficits by intracerebral transplantation of human adipose tissue-derived stromal cells after cerebral ischemia in rats. *Exp. Neurol.* **183**, 355–366.

Kim, B. S., and Mooney, D. J. (1998). Development of biocompatible synthetic extracellular matrices for tissue engineering. *Trends Biotechnol.* **16**, 224–230.

Lane, J. M., Yasko, A. W., Tomin, E., Cole, B. J., Waller, S., Browne, M., Turek, T., and Gross, J. (1999). Bone marrow and recombinant human bone morphogenetic protein-2 in osseous repair. *Clin. Orthop. Relat. Res.* **361**, 216–227.

Liebert, M., Hubbel, A., Chung, M., Wedemeyer, G., Lomax, M. I., Hegeman, A., Yuan, T. Y., Brozovich, M., Wheelock, M. J., and Grossman, H. B. (1997). Expression of mal is associated with urothelial differentiation *in vitro*: Identification by differential display reverse-transcriptase polymerase chain reaction. *Differentiation* **61**, 177–185.

Lim, F., and Sun, A. M. (1980). Microencapsulated islets as bioartificial endocrine pancreas. *Science* **210**, 908–910.

Lin, Y., Chen, X., Yan, Z., Liu, L., Tang, W., Zheng, X., Li, Z., Qiao, J., Li, S., and Tian, W. (2006). Multilineage differentiation of adipose-derived stromal cells from GFP transgenic mice. *Mol. Cell. Biochem* **285**, 69–78.

Mikos, A. G., Lyman, M. D., Freed, L. E., and Langer, R. (1994). Wetting of poly(L-lactic acid) and poly(DL-lactic-co-glycolic acid) foams for tissue culture. *Biomaterials* **15**, 55–58.

Moise, K. J., Jr. (2005). Umbilical cord stem cells. *Obstet. Gynecol.* **106**, 1393–1407.

Nan, Z., Grande, A., Sanberg, C. D., Sanberg, P. R., and Low, W. C. (2005). Infusion of human umbilical cord blood ameliorates neurologic deficits in rats with hemorrhagic brain injury. *Ann. N. Y. Acad. Sci.* **1049**, 84–96.

Nguyen, H. T., Park, J. M., Peters, C. A., Adam, R. M., Orsola, A., Atala, A., and Freeman, M. R. (1999). Cell-specific activation of the HB-EGF and ErbB1 genes by stretch in primary human bladder cells. *In Vitro Cell. Dev. Biol. Anim.* **35,** 371–375.

Nomi, M., Atala, A., Coppi, P. D., and Soker, S. (2002). Principals of neovascularization for tissue engineering. *Mol. Aspects Med.* **23,** 463–483.

Nygren, J. M., Jovinge, S., Breitbach, M., Sawen, P., Roll, W., Hescheler, J., Taneera, J., Fleischmann, B. K., and Jacobsen, S. E. (2004). Bone marrow-derived hematopoietic cells generate cardiomyocytes at a low frequency through cell fusion, but not transdifferentiation. *Nat. Med.* **10,** 494–501.

Oberpenning, F., Meng, J., Yoo, J. J., and Atala, A. (1999). *De novo* reconstitution of a functional mammalian urinary bladder by tissue engineering. *Nat. Biotechnol* **17,** 149–155.

Ohgushi, H., Okumura, M., Tamai, S., Shors, E. C., and Caplan, A. I. (1990). Marrow cell induced osteogenesis in porous hydroxyapatite and tricalcium phosphate: A comparative histomorphometric study of ectopic bone formation. *J. Biomed. Mater. Res.* **24,** 1563–1570.

Orlic, D., Kajstura, J., Chimenti, S., Jakoniuk, I., Anderson, S. M., Li, B., Pickel, J., McKay, R., Nadal-Ginard, B., Bodine, D. M., Leri, A., and Anversa, P. (2001). Bone marrow cells regenerate infarcted myocardium. *Nature* **410,** 701–705.

Pariente, J. L., Kim, B. S., and Atala, A. (2001). *In vitro* biocompatibility assessment of naturally derived and synthetic biomaterials using normal human urothelial cells. *J. Biomed. Mater. Res.* **55,** 33–39.

Pariente, J. L., Kim, B. S., and Atala, A. (2002). *In vitro* biocompatibility evaluation of naturally derived and synthetic biomaterials using normal human bladder smooth muscle cells. *J. Urol.* **167,** 1867–1871.

Pathi, S., Rutenberg, J. B., Johnson, R. L., and Vortkamp, A. (1999). Interaction of Ihh and BMP/Noggin signaling during cartilage differentiation. *Dev. Biol.* **209,** 239–253.

Peppas, N. A., and Langer, R. (1994). New challenges in biomaterials. *Science* **263,** 1715–1720.

Petersen, B. E., Bowen, W. C., Patrene, K. D., Mars, W. M., Sullivan, A. K., Murase, N., Boggs, S. S., Greenberger, J. S., and Goff, J. P. (1999). Bone marrow as a potential source of hepatic oval cells. *Science* **284,** 1168–1170.

Piechota, H. J., Dahms, S. E., Nunes, L. S., Dahiya, R., Lue, T. F., and Tanagho, E. A. (1998). *In vitro* functional properties of the rat bladder regenerated by the bladder acellular matrix graft. *J. Urol.* **159,** 1717–1724.

Pittenger, M. F., Mackay, A. M., Beck, S. C., Jaiswal, R. K., Douglas, R., Mosca, J. D., Moorman, M. A., Simonetti, D. W., Craig, S., and Marshak, D. R. (1999). Multilineage potential of adult human mesenchymal stem cells. *Science* **284,** 143–147.

Planat-Benard, V., Silvestre, J. S., Cousin, B., Andre, M., Nibbelink, M., Tamarat, R., Clergue, M., Manneville, C., Saillan-Barreau, C., Duriez, M., Tedgui, A., Levy, B., *et al.* (2004). Plasticity of human adipose lineage cells toward endothelial cells: Physiological and therapeutic perspectives. *Circulation* **109,** 656–663.

Portmann-Lanz, C. B., Schoeberlein, A., Huber, A., Sager, R., Malek, A., Holzgreve, W., and Surbek, D. V. (2006). Placental mesenchymal stem cells as potential autologous graft for pre- and perinatal neuroregeneration. *Am. J. Obstet. Gynecol.* **194,** 664–673.

Pountos, I., and Giannoudis, P. V. (2005). Biology of mesenchymal stem cells. *Injury* **36**(Suppl. 3), S8–S12.

Puthenveettil, J. A., Burger, M. S., and Reznikoff, C. A. (1999). Replicative senescence in human uroepithelial cells. *Adv. Exp. Med. Biol.* **462,** 83–91.

Ringe, J., Kaps, C., Burmester, G. R., and Sittinger, M. (2002). Stem cells for regenerative medicine: Advances in the engineering of tissues and organs. *Naturwissenschaften* **89,** 338–351.

Safford, K. M., Safford, S. D., Gimble, J. M., Shetty, A. K., and Rice, H. E. (2004). Characterization of neuronal/glial differentiation of murine adipose-derived adult stromal cells. *Exp. Neurol.* **187,** 319–328.

Sams, A. E., and Nixon, A. J. (1995). Chondrocyte-laden collagen scaffolds for resurfacing extensive articular cartilage defects. *Osteoarthr. Cartil.* **3,** 47–59.

Schultz, S. S., Abraham, S., and Lucas, P. A. (2006). Stem cells isolated from adult rat muscle differentiate across all three dermal lineages. *Wound Repair Regen.* **14,** 224–231.

Scriven, S. D., Booth, C., Thomas, D. F., Trejdosiewicz, L. K., and Southgate, J. (1997). Reconstitution of human urothelium from monolayer cultures. *J. Urol.* **158,** 1147–1152.

Shake, J. G., Gruber, P. J., Baumgartner, W. A., Senechal, G., Meyers, J., Redmond, J. M., Pittenger, M. F., and Martin, B. J. (2002). Mesenchymal stem cell implantation in a swine myocardial infarct model: Engraftment and functional effects. *Ann. Thorac. Surg.* **73,** 1919–1925.

Silver, F. H., and Pins, G. (1992). Cell growth on collagen: A review of tissue engineering using scaffolds containing extracellular matrix. *J. Long Term Eff. Med. Implants* **2,** 67–80.

Smidsrod, O., and Skjak-Braek, G. (1990). Alginate as immobilization matrix for cells. *Trends Biotechnol.* **8,** 71–78.

Strauer, B. E., Brehm, M., Zeus, T., Gattermann, N., Hernandez, A., Sorg, R. V., Kogler, G., and Wernet, P. (2001). Intracoronary, human autologous stem cell transplantation for myocardial regeneration following myocardial infarction. *Dtsch. Med. Wochenschr.* **126,** 932–938.

Strem, B. M., Hicok, K. C., Zhu, M., Wulur, I., Alfonso, Z., Schreiber, R. E., Fraser, J. K., and Hedrick, M. H. (2005). Multipotential differentiation of adipose tissue-derived stem cells. *Keio J. Med.* **54,** 132–141.

Sun, J. S., Wu, S. Y., and Lin, F. H. (2005). The role of muscle-derived stem cells in bone tissue engineering. *Biomaterials* **26,** 3953–3960.

Wakitani, S., Goto, T., Pineda, S. J., Young, R. G., Mansour, J. M., Caplan, A. I., and Goldberg, V. M. (1994). Mesenchymal cell-based repair of large, full-thickness defects of articular cartilage. *J. Bone Joint Surg. Am* **76,** 579–592.

Wakitani, S., Imoto, K., Yamamoto, T., Saito, M., Murata, N., and Yoneda, M. (2002). Human autologous culture expanded bone marrow mesenchymal cell transplantation for repair of cartilage defects in osteoarthritic knees. *Osteoarthr. Cartil.* **10,** 199–206.

Xiao, J., Nan, Z., Motooka, Y., and Low, W. C. (2005). Transplantation of a novel cell line population of umbilical cord blood stem cells ameliorates neurological deficits associated with ischemic brain injury. *Stem Cells Dev.* **14,** 722–733.

Yoo, J. J., Meng, J., Oberpenning, F., and Atala, A. (1998). Bladder augmentation using allogenic bladder submucosa seeded with cells. *Urology* **51,** 221–225.

Yoo, J. J., Park, H. J., Lee, I., and Atala, A. (1999). Autologous engineered cartilage rods for penile reconstruction. *J. Urol.* **162,** 1119–1121.

Zhao, L. R., Duan, W. M., Reyes, M., Keene, C. D., Verfaillie, C. M., and Low, W. C. (2002). Human bone marrow stem cells exhibit neural phenotypes and ameliorate neurological deficits after grafting into the ischemic brain of rats. *Exp. Neurol.* **174,** 11–20.

Zuk, P. A., Zhu, M., Ashjian, P., De Ugarte, D. A., Huang, J. I., Mizuno, H., Alfonso, Z. C., Fraser, J. K., Benhaim, P., and Hedrick, M. H. (2002). Human adipose tissue is a source of multipotent stem cells. *Mol. Biol. Cell* **13,** 4279–4295.

CHAPTER 14

Tissue Engineering Using Mesenchymal Stem Cells

Jeremy J. Mao★ and Nicholas W. Marion†

★Department of Biomedical Engineering
College of Dental Medicine
Columbia University, Fu

†Foundation School of Engineering and Applied Sciences
New York

Abstract
1. MSCs: Definition and Therapeutic Promise
2. Isolation and Expansion of MSCs
   2.1. Isolation protocol
3. Multilineage Differentiation of MSCs
   3.1. Chondrogenic differentiation
   3.2. Osteogenic differentiation
   3.3. Adipogenic differentiation
4. Clinical Translation of MSC–Based Therapies
5. Conclusions
   References

Abstract

Mesenchymal stem cells (MSCs) have become one of the most studied stem cells, especially toward the healing of diseased and damaged tissues and organs. MSCs are somatic cells of rare occurrence, but can be readily isolated from multiple adult tissues. MSCs are usually quiescent, but can be activated to maintain tissue homeostasis or participate in tissue regeneration. MSCs have been differentiated into multiple cell lineages that resemble osteoblasts, chondrocytes, myoblasts, adipocytes, and other cells, and express some of the key markers typical of endothelial cells, neuron-like cells, and cardiomyocytes. MSCs have been utilized alone for cell transplantation, or seeded in biomaterial scaffolds toward the healing of tissue and organ defects. Autologous and allogeneic MSCs have been both demonstrated to participate in tissue regeneration. Since MSCs have been isolated from multiple tissues, there is a tendency to emphasize their tissue specificity. Whereas MSCs from dissimilar tissues may differ in doubling time, differentiation capacity, etc., their commonality is likely underappreciated. Where are MSCs located in various tissues? This question is being intensely investigated. There is budding evidence that MSCs in multiple tissues are located in the wall of blood vessels as pericytes. If MSCs are indeed perivascular cells across all tissues, could it be that MSCs from dissimilar tissues are more common than we previously believed? MSC's roles are likely beyond tissue homeostasis and regeneration. One of the underappreciated roles of MSCs is their role as stromal cells that support parenchyma in likely most, if not all, tissues and organs. For example, MSCs may act as precursors of cancer stromal cells. MSC-based therapies can only be realized upon our improved understanding of not only their fundamental properties such as quiescence, population doubling, and differentiation pathways, but also translational studies utilizing MSCs in diseased or damaged tissues and organs.

1. MSCs: Definition and Therapeutic Promise

"Mesenchymal stem cells" (MSCs) can be considered a name out of compromise. During embryonic development, mesenchyme or the embryonic mesoderm contains precursor cells that differentiate into virtually all connective tissue phenotypes such as bone, cartilage, bone marrow stroma, interstitial fibrous tissue, skeletal muscle, dense fibrous tissues such as tendons and ligaments, as well as adipose tissue. In vertebrates, mesenchyme is usually abundant and contains unconnected cells, in contrast to rows of tightly connected epithelial cells that derive from the ectoderm (Alberts *et al.*, 2002). Mesenchymal–epithelial interactions are critical for both appendicular skeletogenesis and craniofacial morphogenesis (Gilbert, 2000; Mao *et al.*, 2006). Upon the completion of prenatal morphogenesis, clusters of mesenchymal cells continue to reside in various tissues and are the logical sources of adult (somatic) MSC.

In the adult, the definition of a common progenitor for all connective tissues inevitably elicits controversy. Strictly speaking, mesenchyme defined as embryonic mesenchyme should not exist in the adult. Despite this textbook dilemma, it is without doubt that adult connective tissues contain precursor cells that maintain tissue homeostasis, and upon trauma or pathological conditions, participate in tissue regeneration. Whereas amicable and dispassionate debate continues regarding the appropriateness of the term "mesenchymal stem cells (MSCs)," "connective tissue progenitor cells (CTPCs)," or sibling terms such as "mesenchymal progenitor cells (MPCs)," "bone marrow progenitor cells (BMPCs)," or "bone marrow stromal cells (BMSCs)," the scientific community and health care industry demand a workable term for communication. The Editors' choice of "Mesenchymal Stem Cells" as the chapter's title is yet another indication of this need.

What we now know as MSC were first identified as colony-forming unit fibroblast-like cells in the 1970s (Friedenstein *et al.*, 1970, 1976). Numerous reports since have demonstrated that bone marrow, adipose tissue, tooth pulp, etc., contain a subset of cells that not only are capable of self-renewal for many passages, but also can differentiable into multiple end-stage cell lineages that resemble osteoblasts, adipocytes, chondrocytes, myoblasts, etc. (Alhadlaq and Mao, 2004). Recently, bone marrow-derived cells have been shown to differentiate into nonmesenchymal lineages, such as hepatic, renal, cardiac, neural cells, etc. (Alhadlaq and Mao, 2004). MSCs have been identified in an increasing number of vertebrate species, including humans (Alhadlaq and Mao, 2004).

Our understanding of MSCs has advanced tremendously due to their demonstrated and perceived therapeutic capacity (Alhadlaq and Mao, 2004; Aubin, 1998; Bianco *et al.*, 2001; Krebsbach *et al.*, 1999; Kuo and Tuan, 2003; Mao *et al.*, 2006; Pittenger *et al.*, 1999; Tuli *et al.*, 2003). Why are MSCs perceived superior to autologous tissue grafts in the regeneration of human tissue and organs? Autologous tissue grafts often represent the current clinical gold standard for the reconstruction of defects resulting from trauma, chronic diseases, congenital anomalies, and tumor resection. However, autologous grafting is based on the concept that a diseased or damaged tissue must be replaced by like tissue that is healthy. Thus, the key drawback of autologous grafting is donor site trauma and morbidity. For example, healthy cartilage must be surgically isolated to repair arthritic cartilage. A patient who receives a bone graft harvested from his/her iliac crest for facial bone reconstruction is hospitalized for extended stay because of donor site trauma and morbidity of the iliac crest, instead of facial surgery. Also, spare healthy tissue is scarce due to biological design during evolution. In contrast, MSC-based therapeutic approaches may circumvent the key deficiencies associated with autologous grafting. First, a teaspoon full of MSC-containing aspirates can be obtained from bone marrow, adipose or other sources, and expanded to heal large, clinically relevant defects. Second, MSCs can differentiate into multiple cell lineages, thus providing the possibility that a common cell source can heal many tissues, as opposed to the principle of current surgical practice to heal a defect by healthy tissue. Finally, MSCs or MSC-derived cells can be seeded in biocompatible scaffolds, which can be shaped into the anatomical structure that is to be replaced. The construct is then surgically implanted to heal the defect.

MSC-based therapies can be autologous (from self) and thus, eliminate the issues of immunorejection and pathogen transmission, or allogeneic for potentially off-the-shelf availability. Regardless of cell source (autologous vs allogeneic), MSCs will likely generate better tissue grafts than artificial materials (Barry, 2003; Mao, 2005). For example, the general lifespan of a surgically successful total joint replacement is 8–10 years, far too short for arthritic patients in their 40s, 50s, or even 60s. Recently, we and others have reported the tissue engineering of an entire articular condyle with both cartilage and bone layers from a single population of MSCs (Alhadlaq and Mao, 2004; Mao, 2005). The biological rationale for MSC-based total joint replacement is that clusters of MSCs initiate joint morphogenesis during embryonic development (Archer *et al.*, 2003; Dowthwaite *et al.*, 2003).

2. Isolation and Expansion of MSCs

When bone marrow content is isolated and cultured, a subset of fibroblast-like cells differentiate into osteoblasts (Friedenstein, 1995; Friedenstein *et al.*, 1970, 1976). Numerous reports since then have shown that these fibroblast-like cells that adhere to tissue culture polystyrene are capable of not only self-renewal, but also differentiating into multiple cell lineages in addition to osteoblasts, such as chondrocytes, myoblasts, adipocytes, etc. (Alhadlaq and Mao, 2004). The protocol by centrifugation in a density gradient to separate bone marrow-derived mononucleated cells from plasma and red blood cells is still widely used. The mononucleated cells can then be plated on tissue culture polystyrene with frequent changes of culture medium. Nonadherent cells, such as the hematopoietic lineage, are discarded upon medium change. Some of the adherent cells are MSCs (Aubin, 1998; Caplan, 1991; Marion and Mao, 2006).

The isolation of MSCs has been recently reviewed (Alhadlaq and Mao, 2004). Bone marrow extracts contain heterogeneous cell populations. MSCs represent a small fraction of total mononucleated cells within bone marrow (Barry, 2003). Further enrichment techniques have been explored, such as positive selection using cell surface markers including STRO-1 (for human MSCs), CD133 (prominin, AC133), p75LNGFR (p75, low-affinity nerve growth factor receptor), CD29, CD44, CD90, CD105, c-kit, SH2 (CD105), SH3, SH4 (CD73), CD71, CD106, CD120a, CD124, and HLA-DR (Alhadlaq and Mao, 2004; Lee *et al.*, 2004; Pittenger *et al.*, 1999). Flow cytometry is another helpful enrichment tool based on an array of cell surface markers. Negative selection is also helpful by utilizing antibody cocktails that label bone marrow-derived cells that are not MSCs (Alhadlaq and Mao, 2004; Marion and Mao, 2006). For example, CD34 can be used as a specific marker for hematopoietic cells. Once the enriched bone marrow sample is placed atop the percoll or ficoll gradient and centrifuged, the dense cells and cell-antibody units are drawn to the bottom, leaving the desired cells atop the gradient (Alhadlaq and Mao, 2004; Lennon and Caplan, 2006). The enriched layer likely will contain a high concentration of MSCs that can be plated and expanded. An example of culture-expanded human MSCs is provided in Fig. 14.1A.

MSCs can undergo a substantial, but not "unlimited" number of population doublings. Human MSCs (hMSCs) may demonstrate an initial lag phase during expansion, but this is followed by rapid proliferation with an average population doubling time of 12–24 h, and with some anticipated variation among donors and with aging (Spees *et al.*, 2004). The estimated number of hMSCs in a 2 ml bone marrow aspirate is between 12.5 and 35.5 billion (Spees *et al.*, 2004). The multipotency of hMSCs is retained up to 23 population doublings (Banfi *et al.*, 2000), whereas no visible change in morphology of MSCs takes place until after 38

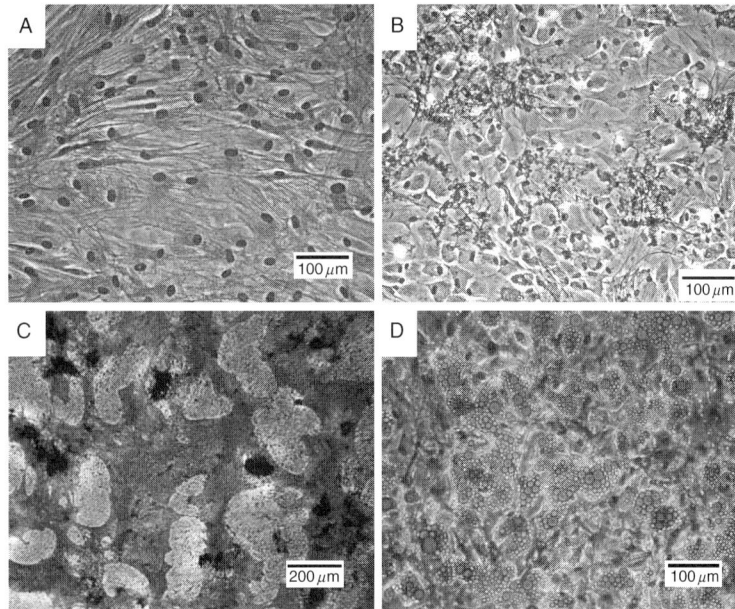

Figure 14.1 (A) Human mesenchymal stem cells (MSCs) isolated from anonymous adult human bone marrow donor following culture expansion (H&E staining). Further enrichment of MSCs can be accomplished by positive selection using cell surface markers including STRO-1, CD133 (prominin, AC133), p75LNGFR (p75, low-affinity nerve growth factor receptor), CD29, CD44, CD90, CD105, c-kit, SH2 (CD105), SH3, SH4 (CD73), CD71, CD106, CD120a, CD124, and HLA-DR or negative selection (Alhadlaq and Mao, 2004; Lee *et al.*, 2004; Pittenger *et al.*, 1999). (B) Chondrocytes derived from human MSCs showing positive staining to Alcian blue. Additional molecular and genetic markers can be used to further characterize MSC-derived chondrocytes (Alhadlaq and Mao, 2004; Lee *et al.*, 2004; Pittenger *et al.*, 1999). (C) Osteoblasts derived from human MSCs showing positive von Kossa staining for calcium deposition (black) and active alkaline phosphatase enzyme (red). Additional molecular and genetic markers can be used to further characterize MSC-derived chondrocytes (Alhadlaq and Mao, 2004; Lee *et al.*, 2004; Pittenger *et al.*, 1999). (D) Adipocytes derived from human MSCs showing positive Oil Red-O staining of intracellular lipids. Additional molecular and genetic markers can be used to further characterize MSC-derived chondrocytes (Alhadlaq and Mao, 2004; Lee *et al.*, 2004; Pittenger *et al.*, 1999).

population doublings (Bruder *et al.*, 1997). Interestingly, osteogenic differentiation of hMSCs appears to be preserved despite apparent cell senescence and slowdown in proliferation rate (Banfi *et al.*, 2000; Bruder *et al.*, 1997). Numerous studies have attempted to improve culture conditions and to increase the expandability of primary MSCs. For example, fibroblast growth factor-2 (FGF2) enhances the proliferation rate of primary MSCs without compromise of their differentiation potential (Banfi *et al.*, 2000; Bruder *et al.*, 1997; Tsutsumi *et al.*, 2001). Similar to other somatic cells, MSCs undergo telomere shortening with each cell division, and eventually results in a cessation in cell proliferation. FGF-2 delays, but not eliminates, cell senescence (Derubeis and Cancedda, 2004; Martin *et al.*, 1997).

Although telomere shortening usually leads to the cessation of cell proliferation, telomerase can repair telomeres following each cell division, thus increasing the cell's lifespan (Derubeis and Cancedda, 2004). Immortalized MSC cell lines have been developed, such as the HMPC32F (Osyczka *et al.*, 2002). HMPC32F has been shown to possess multilineage differentiation potential toward osteogenic, chondrogenic, and adipogenic lineages, while exceeding the lifespan of normal adult human MSC (Osyczka *et al.*, 2002). This MSC cell line was created by infecting primary MSCs with the human papilloma virus type-16 with E6/E7 genes within a viral vector. Immortalization was determined following multilineage differentiation for up to a year in culture and up to $\sim$20 passages. Cell lines are valuable experimental tools, but are not intended for clinical translation.

2.1. Isolation protocol

2.1.1. Human bone marrow–derived MSCs

- Ficoll-Paque—room temperature (e.g., StemCells, Inc., Vancouver, BC, Canada).
- Bone marrow sample—room temperature (e.g., AllCells, LLC., Berkeley, CA):
- 10 ml marrow + 5 ml DPBS + 125 unit/ml heparin (total volume 15 ml).
- Basal culture media (89% DMEM-low glucose, 10% fetal bovine serum (FBS), 1% antibiotics).
- RosetteSep MSC enrichment cocktail (StemCells, Inc.).
- 100 ml PBS with 2% FBS and 1 mM EDTA.
- Transfer the bone marrow sample to a 50 ml conical tube. Add 750 μl RosetteSep (50 μl per 1 ml of bone marrow, 50 μl $\times$ 15 ml = 750 μl).
- Incubate for 20 min at room temperature.
- Add 15 ml of PBS 2% FBS 1 mM EDTA solution to bone marrow. Total volume is 30 ml.
- Add 15 ml Ficoll-Paque to two new 50 ml conical tubes.
- Layer bone marrow solution gently on top of the Ficoll-Paque in each tube. Do not allow marrow to mix with the Ficoll-Paque.
- Centrifuge for 25 min at 300*g* with brake off at room temperature.
- Remove enriched cells from Ficoll-Paque interface.
- Wash enriched cells with PBS–FBS–EDTA solution in 50 ml tube and centrifuge at 1000 rpm for 10 min, brake off.
- Plate cells around 0.5–1 million total per Petri dish with basal culture media. Now referred to as primary cultures or passage 0 (P0).
- Change medium every two days. Remove nonadherent cells during medium changes. Some of the adherent colonies are of mesenchymal lineage.

3. Multilineage Differentiation of MSCs

MSCs are clearly capable of multilineage differentiation into osteoblasts, chondrocytes, myoblasts, adipocytes, etc., in *ex vivo* culture (Alhadlaq and Mao, 2004; Barry, 2003; Caplan, 1991; Caplan and Bruder, 2001; Clark *et al.*, 2008; Derubeis and Cancedda, 2004; Dezawa *et al.*, 2005; Gao and Caplan, 2003; Gregory *et al.*, 2005a,b; Indrawattana *et al.*, 2004; Pittenger *et al.*, 1999; Sekiya *et al.*, 2002). The differentiation of MSCs toward osteogenic, chondrogenic, and adipogenic lineages is reviewed below. For myogenic differentiation, the reader is referred to several comprehensive reviews (Bhagavati and Xu, 2004; Gang *et al.*, 2004; Xu *et al.*, 2004).

3.1. Chondrogenic differentiation

Chondrogenic differentiation of MSCs has tremendous significance in cartilage regeneration. Cartilage has poor capacity for self-healing, owing to the paucity of resident chondroprogenitor cells in the adult (Alberts *et al.*, 2002; Mao, 2005). Most of the sparse cells in adult articular cartilage are mature chondrocytes engaged in matrix maintenance, instead of active chondroprogenitor cells capable of proliferation and differentiation into chondrocytes (Hunziker, 2002). Recent work demonstrates that articular cartilage contains a small population of cells that possess some of the properties as progenitor cells (Dowthwaite *et al.*, 2003). However, the full capacity of these progenitor-like cells is yet to be explored. The clinical observation that injuries to articular cartilage fail to self-repair still serves as the rationale for exploring the healing capacity of MSCs in cartilage regeneration (Mao, 2005).

3.1.1. Chondrogenic stimulants

Transforming growth factors including TGF-β1, TGF-β2, or TGF-β3, frequently at a dose of 10 ng/ml, have been repeatedly demonstrated to stimulate chondrogenesis of MSCs (Barry, 2003). A combination of TGF-β3 and bone morphogenetic protein-6 (BMP-6) increases the cartilage matrix deposition (Sekiya *et al.*, 2001). Cyclical exposure to TGF-β induces a significant increase in matrix deposition, in comparison with continuous exposure of TGF-β3 alone, BMP-6 alone, or in combination (Sekiya *et al.*, 2001, 2002).

Although MSCs can differentiate into chondrocyte-like cells in 2D culture system, there is a tendency for 2D differentiated chondrocytes to dedifferentiate and/or transdifferentiate into fibroblast-like cells. Even mature chondrocytes isolated from articular cartilage have a tendency to dedifferentiate and/or transdifferentiate upon prolonged culture in 2D (Haudenschild *et al.*, 2001; Jakob *et al.*, 2001). Alternatives to 2D culture of chondrocytes include approaches such as micromass culture or pellet culture, differentiating MSCs toward chondrocytes in

3D biomaterials, or self-assembly of MSCs into 3D chondrogenic structures. By centrifuging a known quantity of MSCs in the bottom of a conical tube, for example, between 250,000 and 1.5 million cells, and using a variety of serum-free medium solutions combined with growth factor administration, hMSCs will differentiate toward the chondrogenic lineage. Chondrogenic differentiation medium is frequently high glucose DMEM, as opposed to the medium for hMSC expansion. Fetal bovine serum does not appear to be necessary for chondrogenic differentiation of MSCs. In some formulations, selected bioactive factors and chemicals are added such as L-proline at 40 μg/ml, ITS (insulin, transferring, and sodium selenite) at 1$\times$ solution, and sodium pyruvate at 100 μg/ml. Dexamethasone at 100 nM and L-ascorbic acid 2-phosphate (AsAP) at 50 μg/ml has also been incorporated (Johnstone *et al.*, 1998; Sekiya *et al.*, 2001, 2002; Yoo *et al.*, 1998).

MSCs can be differentiated into chondrocytes in 3D biocompatible scaffolds, partially to circumvent the possibility of dedifferentiation and/or trans-differentiation in extended 2D culture system. Polymeric scaffolds such as alginate, agarose, chitosan, and poly(ethylene glycol) diacrylate (PEGDA) hydrogels have been used to provide 3D environments for chondrogenic differentiation of MSCs (Alhadlaq *et al.*, 2004; Anseth *et al.*, 2002; Hung *et al.*, 2003; Kim *et al.*, 2003; Williams *et al.*, 2003; Woodfield *et al.*, 2006). The feasibility to polymerize biomaterials into complex anatomical structures makes several hydrogels well suited for cartilage tissue engineering. MSCs can be exposed to chondrogenic supplemented medium, such as TGF-β family, while encapsulated in hydrogels (Alhadlaq *et al.*, 2004; Anseth *et al.*, 2002; Kim *et al.*, 2003; Williams *et al.*, 2003). The advantage of 3D hydrogel encapsulation is to minimize the need to disrupt 2D monolayer culture or pellet culture prior to seeding cells in 3D.

Several reports have demonstrated that chondrocytes can elaborate 3D matrices and self-assemble into cartilage-like structures, sometimes when seeded on top of a biomaterial instead of within it (Klein *et al.*, 2003; Masuda *et al.*, 2003). Isolated chondrocytes are initially cultured on top of agarose or alginate gels to produce the cell-associated matrix, rather than encapsulated within gels, followed by additional 2D culture until cell–matrix structures reach a certain size. This represents a convenient variation of the 3D culture approach and may have therapeutic relevance in cartilage tissue engineering.

3.1.2. Chondrogenic differentiation protocol—Rat MSCs

- Rat chondrogenic medium: 89% DMEM-low glucose, 10% FBS, 1% antibiotic solution, supplemented with 10 ng/ml recombinant rat TGF-β1.
- Plate rat MSCs in monolayer culture with basal medium until 80% confluent.
- Remove basal medium and continue monolayer culture with rat chondrogenic medium (above) for an additional 2 weeks, and change medium biweekly.
- Monolayer cultures may be fixed for histological analysis or for quantification of biochemical markers using 1% Triton-X100.

3.1.3. Chondrogenic differentiation protocol—Human MSCs

- Human chondrogenic media: 95% DMEM-high glucose, 1% 1× ITS + 1 solution, 1% antibiotic, 100 μg/ml sodium pyruvate, 50 μg/ml AsAP, 40 μg/ml L-proline, 0.1 μM dexamethasone, 10 ng/ml recombinant human TGF-β3.
- Centrifuge approximately 2.5×10^5 hMSCs in a 15 ml conical tube at 500g for 5 min at 4 °C.
- Culture with human chondrogenic medium for at least 14 days and change medium biweekly.
- Pellets may be removed from tube by inverting and gently tapping for quantitative and histological analyses:

  a. 1% Triton-X100 may be used to disrupt cell pellets for quantitative biochemical assays such as DNA, collagen, and proteoglycans.
  b. Samples may be dehydrated and embedded in paraffin prior to sectioning and staining for histological analysis.

3.1.4. Chondrogenic differentiation markers

A number of histological dyes provide the most convenient indication of chondrogenic differentiation of MSCs. Histological dyes are reagents sensitive to the presence of proteoglycans or sulfated glycosaminoglycans. Stains for glycosaminoglycans and proteoglycans include Safranin-O/fast green and Alcain blue. These histological dyes have been conventionally used in labeling native articular cartilage and growth plate cartilage, and therefore are reliable markers of chondrogenic differentiation.

Chondrogenic differentiation is driven by a number of transcription factors such as the SOX family (Lefebvre *et al.*, 2001; Ylostalo *et al.*, 2006). *SOX9* is expressed in differentiating chondrocytes; deletions of SOX9 elicit abnormal endochondral bone formation and hypoplasia of the developing bone (Lefebvre *et al.*, 2001). *SOX5* and *SOX6* are expressed during chondrogenic differentiation. Biosynthesis of type II collagen and aggrecan are regulated by the expression of *SOX9* and *SOX5* through their activation of *COL2A1* and *aggrecan* genes (Lefebvre *et al.*, 2001; Ng *et al.*, 1997; Ylostalo *et al.*, 2006). Furthermore, collagen genes such as *COL9A1*, *COL9A2*, *COL9A3*, and *COL11A2* are expressed in response to the expression of *SOX9* (Ylostalo *et al.*, 2006). RT-PCR, Western blotting, *in situ* hybridization, and immunohistochemistry are effective approaches to identify the presence of type II collagen, type X collagen, various proteoglycans, such as aggrecan, decorin, and biglycan in engineered cartilage tissue. Quantitatively, collagen and sulfated GAG contents can be measured using commercially available reagent and ELISA kits, or biochemical assays. Genetic analysis, such as RT-PCR, aims to identify the expression of chondrogenic mRNAs, such as collagen II, collagen IX, *SOX 9, SOX5, SOX6, COL9A1, COL9A2, COL9A3,* and *COL11,* due to their presence during early chondral development. Gene arrays can provide a comprehensive portrait of not only cartilage-related genes, but also other genes that may be important in chondrogenesis. An example of chondrogenic differentiation of rat MSCs in monolayer is provided in Fig. 14.1B.

Structural analysis is necessary to determine whether tissue-engineered cartilage has micro- and ultrastructural characteristics as native cartilage. For instance, native chondrocyte matrix is characterized with pericellular matrix and interterritorial matrix (Allen and Mao, 2004; Guilak, 2000; Guilak and Mow, 2000; Poole *et al.*, 1988, 1991). Although tissue-engineered cartilage apparently has structures similar to pericellular matrix and interterritorial matrix in a number of reports, more attention has yet to be paid to structural analysis of tissue-engineered cartilage. The reader is referred to several excellent reviews of structural properties of native and engineered cartilage (Cohen *et al.*, 1998; Grodzinsky *et al.*, 2000; Hunziker, 2002; Kerin *et al.*, 2002; Nesic *et al.*, 2006; Woodfield *et al.*, 2002).

Mechanical testing is necessary to ascertain that tissue-engineered cartilage has the proper mechanical properties in addition to the "right ingredients," such as type II collagen and glycosaminoglycans, as well as having the "right" structural characteristics. The reader is referred to a number of excellent reviews on the mechanical properties of native and tissue-engineered cartilage (Hung *et al.*, 2004; Hunziker, 2002; Mow *et al.*, 1984, 1999; Troken *et al.*, 2005).

3.2. Osteogenic differentiation

Osteogenic differentiation was the first identified end-stage lineage of MSC differentiation (Friedenstein, 1995; Friedenstein *et al.*, 1970, 1976). Given that bone marrow is a rich source for MSCs, it should come as no surprise that MSCs can be readily differentiated into osteoblasts. An array of genetic and matrix markers have been utilized to verify the osteogenic differentiation of MSCs.

3.2.1. Osteogenic stimulants

Several well explored cocktails have been shown to induce MSCs to differentiate into osteoblasts. MSCs have been shown to express alkaline phosphatase following 7–14 days of exposure to 100 nM dexamethasone, 50 μg/ml AsAP, and 100 mM β-glycerophosphate (Alhadlaq and Mao, 2003, 2005; Alhadlaq *et al.*, 2004; Clark *et al.*, 2008; Marion *et al.*, 2005; Moioli *et al.*, 2008). Long-term exposure of MSCs to the formula of dexamethasone, AsAP, and β-glycerophosphate results in calcium matrix deposition and the expression of late osteogenesis markers, such as bone sialoprotein, osteocalcin, and osteonectin. Dexamethasone is a gluccocorticoid steroid capable of either stimulating or inhibiting osteogenic differentiation of MSCs depending on dosage (Bruder *et al.*, 1997). High dexamethasone dose stimulates adipogenic differentiation of MSCs, whereas lower doses stimulate osteogenic differentiation (Bruder *et al.*, 1997). The addition of AsAP further facilities osteogenic differentiation including collagen biosynthesis, in addition to its stimulatory effects on cell proliferation (Graves *et al.*, 1994a,b; Jaiswal *et al.*, 1997). A number of studies have utilized ascorbic acid, the bioactive component of AsAP as an osteogenic supplement. However, ascorbic acid is somewhat unstable at 37 °C and neutral pH, a problem not associated with AsAP (Jaiswal *et al.*, 1997). High doses of ascorbic acid can also be toxic to cells (Jaiswal *et al.*, 1997). Lastly, β-glycerophosphate is critical to stimulate calcified matrix formation, in combination with the effects of dexamethasone and AsAP (Jaiswal *et al.*, 1997). Without β-glycerophosphate, MSC-derived osteoblasts are slow to mediate a calcium–phosphate matrix (Jaiswal *et al.*, 1997).

Several members of bone morphogenetic proteins (BMPs) have also been shown to induce the osteogenic differentiation of MSCs including BMP-2, BMP-6, and BMP-9 (Dayoub *et al.*, 2003; Friedman *et al.*, 2006; Katagiri *et al.*, 1994; Li *et al.*, 2006; Long *et al.*, 1995; Rickard *et al.*, 1994; Wang *et al.*, 1990; Wozney, 1992). BMPs are usually supplemented in combination with dexamethasone to stimulate the osteogenic differentiation of MSCs (Rickard *et al.*, 1994). Osteogenic differentiation of MSCs using BMP-2 is dose dependant, with measurable effects between 25 and 100 ng/ml (Lecanda *et al.*, 1997; Rickard *et al.*, 1994).

3.2.2. Osteogenic differentiation protocol

- Osteogenic supplemented medium: 89% DMEM-low glucose, 10% FBS, 1% antibiotics, 50 μg/ml AsAP, 0.1 μM dexamethasone, and 100 mM β-glycerophosphate.
- Plate cells at $\sim$10,000 cells/cm^2 in monolayer.
- Culture 14–28 days and change medium biweekly.
- Monolayer cultures can be fixed for histological analysis or quantitative biochemical assays after 1% Triton-X100 is used to disrupt cells.
- Alkaline phosphatase activity may be detected within 2 weeks, whereas other bone markers may be detected later (Alhadlaq and Mao, 2003, 2005; Alhadlaq *et al.*, 2004; Aubin, 1998; Frank *et al.*, 2002; Malaval *et al.*, 1999; Marion *et al.*, 2005; Rodan and Noda, 1991).

3.2.3. Osteogenic differentiation markers

The osteogenic differentiation of MSCs is verified by several osteogenic matrix molecules, accumulation of mineral crystals and nodules, and ultimately, the regeneration of bone *in vivo* both ectopically, such as the dorsum of immunodeficient mice, or orthotopically, such as calvarial, axial (e.g., spinal fusion), or appendicular (e.g., segmental) defects.

Upregulation of alkaline phosphatase activity is an early indicator for the osteogenic differentiation of MSCs, and can be detected quantitatively using a commercially available kinetic kit and/or histologically using a Naphthol-based chemical stain (Aubin, 1998; Frank *et al.*, 2002; Malaval *et al.*, 1999; Rodan and Noda, 1991). Furthermore, calcium matrix synthesis is histologically verified using either von Kossa (silver nitrate) or alizarin red stains, via selective binding with calcium–phosphate matrix components (Aubin, 1998; Frank *et al.*, 2002; Malaval *et al.*, 1999; Rodan and Noda, 1991). Immunohistochemical staining for type I collagen is helpful, but nonspecific. Bone sialoprotein, osteocalcin, osteopontin, and osteonectin are late osteogenic differentiation markers and can be measured genetically using RT-PCR or proteomically using ELISA (Aubin, 1998; Clark *et al.*, 2008; Malaval *et al.*, 1999; Moioli *et al.*, 2006; Rodan and Noda, 1991). Immunohistochemistry with antibodies will localize matrix markers in relation to cells. An example of osteogenic differentiation of hMSCs is provided in Fig. 14.1C.

Tissue-engineered bone must have the appropriate structural characteristics that approximate native bone. Bone is one of the highly hierarchical structures in the body. The structure of cortical bone differs substantially from that of cancellous bone. A number of biomaterials have been utilized to simulate cortical bone and cancellous bone structures in cell-based or noncell-based approaches (Lin *et al.*, 2004; Taboas *et al.*, 2003). However, given the extent of bone modeling and remodeling, whether complete maturation of tissue-engineered bone is necessary prior to *in vivo* implantation remains an open question. The reader is referred to several excellent reviews of structural properties of native and engineered bone (El-Ghannam, 2005; Mauney *et al.*, 2005; Mistry and Mikos, 2005; Wan *et al.*, 2006). Mechanical testing of tissue-engineered bone is of paramount importance, because bone is designed to withstand mechanical stresses as its primary function. The reader is referred to a number of excellent reviews on the mechanical properties of native and tissue-engineered bone (El-Ghannam, 2005; Mauney *et al.*, 2005; Mistry and Mikos, 2005; Wan *et al.*, 2006).

3.3. Adipogenic differentiation

Adipogenic differentiation of MSCs has a number of perhaps underappreciated significance areas. First, MSC differentiation into osteoblasts and adipocytes is delicately regulated and balanced (Gregory *et al.*, 2005b). Second, our knowledge of obesity is likely improved by understanding the genetic regulation of adipogenic differentiation of MSCs. Third, adipose tissue is a key structure to restore in reconstructive and augmentative surgeries such as facial cancer reconstruction and breast cancer reconstruction. Current approaches for soft tissue reconstruction and/or augmentation suffer from shortcomings, such as donor site trauma and morbidity, suboptimal volume retention, donor site morbidity, and poor biocompatibility. One of the central issues of poor healing of adipose tissue grafts is a shortage and/or premature apoptosis of adipogenic cells. MSCs self-replenish, and as demonstrated below, can readily differentiate into adipogenic cells in 2D and 3D (Alhadlaq *et al.*, 2005; Pittenger *et al.*, 1999; Stosich and Mao, 2005, 2007; Stosich *et al.*, 2007).

3.3.1. Adipogenic stimulants

With the addition of dexamethasone (0.5 μM), 1-methyl-3-isobutylxanthine (IBMX) (0.5 μM–0.5 mM), and Indomethacin (50–100 μM), MSC in monolayer culture will undergo adipogenic differentiation (Alhadlaq and Mao, 2004; Alhadlaq *et al.*, 2005; Gregory *et al.*, 2005b; Janderova *et al.*, 2003; Lee *et al.*, 2006; Nakamura *et al.*, 2003; Pittenger *et al.*, 1999; Rosen and Spiegelman, 2000; Stosich and Mao, 2005, 2007; Stosich *et al.*, 2007; Ylostalo *et al.*, 2006). Insulin is another key ingredient, for example, in adipogenic differentiating medium (dexamethasone, IBMX, Indomethacin) for 2–5 days, and then to a maintenance supplement of insulin (Janderova *et al.*, 2003; Nakamura *et al.*, 2003). In most reports of adipogenic differentiation of MSCs, dexamethasone dose is 0.5 μM or five times higher than for osteogenic differentiation of MSCs. Adipogenic differentiation is believed to take place upon cell confluence, cell to cell contacts, a serum-free culture, or a suspension culture in methylcellulose (Rosen and Spiegelman, 2000). The growth arrest of MSC-derived chondrocytes is crucial for the subsequent activation of adipogenic differentiation processes. IBMX is a phosphodiesterase inhibitor that blocks the conversion of cAMP to 5′-AMP (Gregory *et al.*, 2005a). This causes an upregulation of protein kinase A, which results in decreased cell proliferation and upregulation of hormone-sensitive lipase (HSL). HSL has been shown to convert triacylglycerides to glycerol and free fatty acids, a known adipogenic process (Gregory *et al.*, 2005a). The activation of CCAAT/enhancer-binding proteins (C/EBP) coincides with the expression of peroxisome proliferator-activated receptor γ (PPARγ), which occurs in the presence of Indomethacin, a known ligand for PPARγ (Rosen and Spiegelman, 2000). This early transcription factor is essential for adipogenesis as it suppresses the canonical wingless (Wnt) signaling, suggesting the regulation of osteogenesis and adipogenesis by PPARγ expression in MSCs (Gregory *et al.*, 2005a). Positive Wnt signaling inhibits osteogenic differentiation but is required for adipogenic differentiation (Gregory *et al.*, 2005a). Therefore, a delicate balance exists in the regulation of adipogenic and osteogenic differentiation as PPARγ has been reported to inhibit osteogenic differentiation of progenitor cells (Cheng *et al.*, 2003; Khan and Abu-Amer, 2003). Furthermore, loss of function for PPARγ or C/EBP (C/EBPα, C/EBPβ, or C/EBPγ) results in detrimental effects for adipogenesis and reduced adipocyte proliferation as well as reduced lipid vacuole deposition (Rosen and Spiegelman, 2000).

3.3.2. Adipogenic differentiation protocol

- Human adipogenic media: 89% DMEM-low glucose, 10% FBS, 1% antibiotic, 0.5 μM dexamethasone, 0.5 μM IBMX, 50 μM Indomethacin.
- Plate cells at $\sim$20,000 cells/cm^2 or 80% confluence in monolayer.
- Continue culture in human adipogenic media for up to 28 days.

- Lipids may be visible as early as 7 days and can be viewed under phase contrast microscope.
- Monolayer cultures can be processed for histology following fixation or for quantitative biochemical analysis after 1% Triton-X100 is used to disrupt cells.

3.3.3. Adipogenic differentiation markers

One of the key transcriptional factors of adipogenic differentiation of MSCs is peroxisome proliferator-activated receptor γ 2 (PPARγ2), which can be detected by RT-PCR or gene arrays. Oil Red-O staining is a convenient and commonly performed histological stain. Upon fixing the cultures with 10% formalin, lipid vacuoles synthesized intracellularly by MSC-derived adipogenic cells bind to Oil Red-O and stain red. Hematoxylin counter staining may be used to visualize cell nuclei in blue. Free glycerol may be quantified by lysing the cells with 1% Triton X-100 and quantitatively analyzed with a glycerol kit. Glycerol-3-phosphate dehydrogenase (G-3-PDH) can be measured as one of the key enzymes in triglyceride synthesis (Pairault and Green, 1979). RT-PCR can be utilized to amplify and detect additional adipogenic gene products, such as lipoprotein lipase (LPL) and the polyclonal antibody a-P2 (Pittenger *et al.*, 1999). An example of adipogenic differentiation of hMSCs is provided in Fig. 14.1D.

Structural analysis is necessary to determine whether engineered adipose tissue has the appropriate microstructural characteristics as native adipose tissue. Adipose tissue is unique in the sense that lipid vacuoles are accumulated intracellularly. The extracellular matrix of adipose tissue consists of primarily interstitial fibrous tissue, nerve supplies, and vascular and lymphatic network.

Mechanical testing also is necessary to ascertain that tissue-engineered adipose tissue has the proper mechanical properties in addition to the "right ingredients," such as adipocytes, intracellular lipid vacuoles, and vascular and lymphatic supplies. The reader is referred to a number of excellent reviews on the mechanical properties of native and tissue-engineered adipose tissue, as well as biomaterials that have been utilized as scaffolds for adipose tissue regeneration (Beahm *et al.*, 2003; McKnight *et al.*, 2002; Patel *et al.*, 2005; Patrick, 2004; Stosich and Mao, 2007).

4. Clinical Translation of MSC-Based Therapies

MSC-based therapies are being translated toward clinical practice to heal defects resulting from trauma, chronic diseases, congenital anomalies, and tumor resection. Due to space limitation, it is impossible to outline all the ongoing effort on the clinical translation of MSC-based therapeutic approaches. Several examples are briefly introduced below.

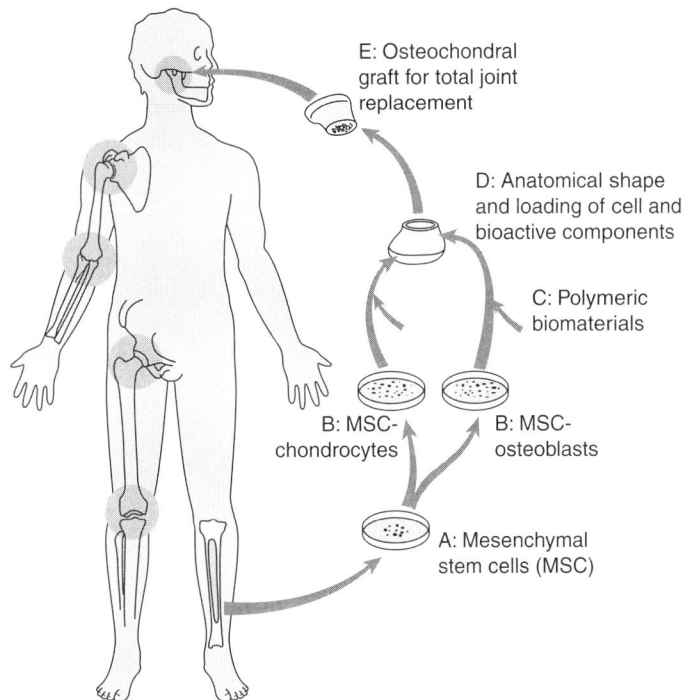

Figure 14.2 Schematic diagram of autologous, MSC-based tissue engineering therapy for total joint replacement (MSCs, mesenchymal stem cells). Progenitor cells such as MSCs are isolated from the bone marrow or other connective sources such as adipose tissue, culture-expanded and/or differentiated *ex vivo* toward chondrocytes and osteoblasts. Cells are seeded in biocompatible materials shaped into the anatomic structures of the synovial joint condyle and implanted *in vivo*. Preliminary proof of concept studies have been reported (Alhadlaq and Mao, 2003, 2005; Alhadlaq *et al.*, 2004; Mao, 2005).

Recent reports suggest the roles of MSCs in the repair of myocardial infarctions in rats and pigs upon intracardiac injection (Shake *et al.*, 2002). The precise mechanisms are unclear, although it has been suggested that MSCs may induce the homing of cardiomyocytes to the infarct site (Saito *et al.*, 2003). Also proposed is MSC differentiation into cardiomyocytes and/or paracrine effects upon intravenous MSC injection. Labeled MSCs are found in bone marrow and the site of myocardial infarction (Saito *et al.*, 2003). Several clinical trials are ongoing at universities and biotechnology companies to explore the healing effects of MSCs on myocardial infarctions (Laflamme and Murry, 2005; Pittenger and Martin, 2004).

Several experiments have demonstrated that an entire articular condyle in the same shape and dimensions of a human temporomandibular joint can be grown *in vivo* with both cartilage and bone layers from a single population of MSCs (Mao, 2005; Mao *et al.*, 2006). A visionary diagram of MSC-based therapies for total joint replacement is shown in Fig. 14.2, based on the work by us and others

(Alhadlaq and Mao, 2003, 2005; Alhadlaq *et al.*, 2004; Troken *et al.*, 2006). Although bone marrow is the most characterized source of MSCs at this time (Fig. 14.2), it is probable that MSCs needed for total joint replacement can also be isolated from adipose tissue via aspiration or lipectomy, fresh or banked human umbilical cord blood, placental tissue, or human teeth (Mao, 2005). Total joint replacement is one of the many examples of MSC-based therapies whose proof of concept has been demonstrated in recent years (Rahaman and Mao, 2005; Troken *et al.*, 2006).

An emerging concept is that MSCs have trophic effects by secreting a variety of cytokines that function via both paracrine and autocrine pathways (Caplan and Dennis, 2006). The interactions between exogenously delivered growth factors and intrinsic cytokines synthesized by MSCs are one of the most complex and meritorious approaches in cell-based therapies. Our understanding of these fundamental paracrine and autocrine pathways will undoubtedly advance the more practical approaches in MSC-based tissue engineering. The trophic effects, as proposed by Caplan and Dennis (2006), include local immune suppression, fibrosis inhibition, angiogenesis enhancement, etc. A number of translational and clinical studies in the areas of cardiac infarct, synovial joint regeneration, and stroke regeneration models may provide initial data to test the proposed trophic effects of MSCs.

5. Conclusions

MSCs are somatic stem cells that can be readily isolated from several tissues in the adult and in multiple species. MSCs are typically quiescent under physiological conditions, but can be activated to maintain tissue homeostasis and participate in tissue regeneration. It is widely believed that MSCs need to be expanded to sufficient numbers as replacement cells for cell-based therapies. Recently, there is accumulating evidence that MSCs may provide signaling cues that elaborate tissue regeneration. MSCs are able to differentiate into multiple cell lineages that resemble osteoblasts, chondrocytes, myoblasts, adipocytes, and express some of the key markers typical of endothelial cells, neuron-like cells, and cardiomyocytes. Whether a population of, presumably, MSCs utilized in a given study, are truly MSCs, is a biologically relevant question, and can be tested by arrays of cell surface and/or genetic markers. On the other hand, once a functional biological tissue is functional, it matters little whether the original population of tissue-forming cells is truly MSCs as long as they can be readily isolated from the patient. Most native tissues, and certainly all organs, are formed by heterogeneous cell populations. It appears that both stem cell biology and tissue engineering approaches are necessary to advance our understanding of how MSCs can be utilized to heal diseased and damaged tissues and organs. For example, parallel experiments can explore the healing of tissue defects from purified or cloned MSCs, as well as MSCs among heterogeneous cell populations. One of the underappreciated roles of MSCs is their role as stromal cells that support parenchyma in likely most, if not all, tissues and organs.

For example, MSCs may act as precursors of cancer stromal cells (Hall *et al.*, 2007; Spaeth *et al.*, 2008; Studeny *et al.*, 2004). MSCs are anticipated to regulate immune cells and interact with the hematopoietic lineage. Despite their first discovery in the mid-1970s, investigations of MSCs have only intensified in recent years. We submit that the true healing power of MSCs is yet to be realized.

Acknowledgments

We thank our colleagues whose work has been cited, and those whose work cannot be cited due to space limitation, for their highly meritorious work that has energized the process of composing this review. We are grateful to the remaining members of Tissue Engineering Laboratory for their dedication and hard work, in particular. Eduardo Moioli, in our laboratory, is gratefully acknowledged for his contribution of Fig. 14.1B that demonstrates chondrogenic differentiation of rat MSCs in monolayer. We thank Sarah Kennedy, Janina Acloque, and Lauren Feldman for administrative support. Generous support from the National Institutes of Health is gratefully acknowledged, through NIH grants DE15391 and EB02332 to J.J.M., for the effort spent on composing this manuscript along with some of the experimental data presented in this manuscript from our laboratory.

References

Alberts, B., Johnson, A., Lewis, J., Raff, M., Roberts, K., and Walter, P. (2002). Molecular Biology of the Cell Garland Science, New York, NY.

Alhadlaq, A., and Mao, J. J. (2003). Tissue-engineered neogenesis of human-shaped mandibular condyle from rat mesenchymal stem cells. *J. Dent. Res.* **82**, 951–956.

Alhadlaq, A., and Mao, J. J. (2004). Mesenchymal stem cells: Isolation and therapeutics. *Stem Cells Dev.* **13**, 436–448.

Alhadlaq, A., and Mao, J. J. (2005). Tissue-engineered osteochondral constructs in the shape of an articular condyle. *J. Bone Joint Surg. Am.* **87**, 936–944.

Alhadlaq, A., Elisseeff, J. H., Hong, L., Williams, C. G., Caplan, A. I., Sharma, B., Kopher, R. A., Tomkoria, S., Lennon, D. P., Lopez, A., and Mao, J. J. (2004). Adult stem cell driven genesis of human-shaped articular condyle. *Ann. Biomed. Eng.* **32**, 911–923.

Alhadlaq, A., Tang, M., and Mao, J. J. (2005). Engineered adipose tissue from human mesenchymal stem cells maintains predefined shape and dimension: Implications in soft tissue augmentation and reconstruction. *Tissue Eng.* **11**, 556–566.

Allen, D. M., and Mao, J. J. (2004). Heterogeneous nanostructural and nanoelastic properties of pericellular and interterritorial matrices of chondrocytes by atomic force microscopy. *J. Struct. Biol.* **145**, 196–204.

Anseth, K. S., Metters, A. T., Bryant, S. J., Martens, P. J., Elisseeff, J. H., and Bowman, C. N. (2002). *In situ* forming degradable networks and their application in tissue engineering and drug delivery. *J. Control. Release* **78**, 199–209.

Archer, C. W., Dowthwaite, G. P., and Francis-West, P. (2003). Development of synovial joints. *Birth Defects Res. C Embryo Today* **69**, 144–155.

Aubin, J. E. (1998). Bone stem cells. *J. Cell Biochem. Suppl.* **30–31**, 73–82.

Banfi, A., Muraglia, A., Dozin, B., Mastrogiacomo, M., Cancedda, R., and Quarto, R. (2000). Proliferation kinetics and differentiation potential of *ex vivo* expanded human bone marrow stromal cells: Implications for their use in cell therapy. *Exp. Hematol.* **28**, 707–715.

Barry, F. P. (2003). Biology and clinical applications of mesenchymal stem cells. *Birth Defects Res. C Embryo Today* **69**, 250–256.

Beahm, E. K., Walton, R. L., and Patrick, C. W., Jr. (2003). Progress in adipose tissue construct development. *Clin. Plast. Surg.* **30**, 547–558, viii.

Bhagavati, S., and Xu, W. (2004). Isolation and enrichment of skeletal muscle progenitor cells from mouse bone marrow. *Biochem. Biophys. Res. Commun.* **318,** 119–124.

Bianco, P., Riminucci, M., Gronthos, S., and Robey, P. G. (2001). Bone marrow stromal stem cells: Nature, biology, and potential applications. *Stem Cells* **19,** 180–192.

Boland, G. M., Perkins, G., Hall, D. J., and Tuan, R. S. (2004). Wnt 3a promotes proliferation and suppresses osteogenic differentiation of adult human mesenchymal stem cells. *J. Cell. Biochem.* **93,** 1210–1230.

Bruder, S. P., Jaiswal, N., and Haynesworth, S. E. (1997). Growth kinetics, self-renewal, and the osteogenic potential of purified human mesenchymal stem cells during extensive subcultivation and following cryopreservation. *J. Cell. Biochem.* **64,** 278–294.

Caplan, A. I. (1991). Mesenchymal stem cells. *J. Orthop. Res.* **9,** 641–650.

Caplan, A. I., and Dennis, J. E. (2006). Mesenchymal stem cells as trophic mediators. *J. Cell. Biochem.* **98,** 1076–1084.

Cheng, S. L., Shao, J. S., Charlton-Kachigian, N., Loewy, A. P., and Towler, D. A. (2003). MSX2 promotes osteogenesis and suppresses adipogenic differentiation of multipotent mesenchymal progenitors. *J. Biol. Chem.* **278,** 45969–45977.

Clark, P. A., Moioli, E. K., Sumner, D. R., and Mao, J. J. (2008). Porous implants as drug delivery vehicles to augment host tissue integration. *FASEB J.* **22,** 1684–1693.

Cohen, N. P., Foster, R. J., and Mow, V. C. (1998). Composition and dynamics of articular cartilage: Structure, function, and maintaining healthy state. *J. Orthop. Sports Phys. Ther.* **28,** 203–215.

Dayoub, H., Dumont, R. J., Li, J. Z., Dumont, A. S., Hankins, G. R., Kallmes, D. F., and Helm, G. A. (2003). Human mesenchymal stem cells transduced with recombinant bone morphogenetic protein-9 adenovirus promote osteogenesis in rodents. *Tissue Eng.* **9,** 347–356.

Derubeis, A. R., and Cancedda, R. (2004). Bone marrow stromal cells (BMSCs) in bone engineering: Limitations and recent advances. *Ann. Biomed. Eng.* **32,** 160–165.

Dowthwaite, G. P., Flannery, C. R., Flannelly, J., Lewthwaite, J. C., Archer, C. W., and Pitsillides, A. A. (2003). A mechanism underlying the movement requirement for synovial joint cavitation. *Matrix Biol.* **22,** 311–322.

El-Ghannam, A. (2005). Bone reconstruction: From bioceramics to tissue engineering. *Expert Rev. Med. Devices* **2,** 87–101.

Frank, O., Heim, M., Jakob, M., Barbero, A., Schafer, D., Bendik, I., Dick, W., Heberer, M., and Martin, I. (2002). Real-time quantitative RT-PCR analysis of human bone marrow stromal cells during osteogenic differentiation *in vitro. J. Cell. Biochem.* **85,** 737–746.

Friedenstein, A. J. (1995). Marrow stromal fibroblasts. *Calcif. Tissue Int.* **56**(Suppl. 1), S17.

Friedenstein, A. J., Chailakhjan, R. K., and Lalykina, K. S. (1970). The development of fibroblast colonies in monolayer cultures of guinea-pig bone marrow and spleen cells. *Cell Tissue Kinet.* **3,** 393–403.

Friedenstein, A. J., Gorskaja, J. F., and Kulagina, N. N. (1976). Fibroblast precursors in normal and irradiated mouse hematopoietic organs. *Exp. Hematol.* **4,** 267–274.

Friedman, M. S., Long, M. W., and Hankenson, K. D. (2006). Osteogenic differentiation of human mesenchymal stem cells is regulated by bone morphogenetic protein-6. *J. Cell. Biochem.* **98,** 538–554.

Frost, H. M., and Turner, C. H. (2000). Toward a mathematical description of bone biology: The principle of cellular accommodation. *Calcif. Tissue Int.* **67,** 184–187.

Gang, E. J., Jeong, J. A., Hong, S. H., Hwang, S. H., Kim, S. W., Yang, I. H., Ahn, C., Han, H., and Kim, H. (2004). Skeletal myogenic differentiation of mesenchymal stem cells isolated from human umbilical cord blood. *Stem Cells* **22,** 617–624.

Gilbert, S. F. (2000). Developmental Biology Sinauer Associates, Sunderland, MA.

Graves, S. E., Francis, M. J. O., Gundle, R., and Beresoford, J. N. (1994). Primary culture of human trabecular bone: Effects of L-ascorate-2-phosphate. *Bone* **15,** 132–133.

Graves, S. E., Gundle, R., Francis, M. J. O., and Beresoford, J. N. (1994). Ascorbate increases collagen synthesis and promote differentiation in human bone derived cell cultures. *Bone* **15,** 133.

Gregory, C. A., Gunn, W. G., Reyes, E., Smolarz, A. J., Munoz, J., Spees, J. L., and Prockop, D. J. (2005). How wnt signaling affects bone repair by mesenchymal stem cells from the bone marrow. *Ann. N. Y. Acad. Sci.* **1049**, 97–106.

Gregory, C. A., Prockop, D. J., and Spees, J. L. (2005). Non-hematopoietic bone marrow stem cells: Molecular control of expansion and differentiation. *Exp. Cell Res.* **306**, 330–335.

Grodzinsky, A. J., Levenston, M. E., Jin, M., and Frank, E. H. (2000). Cartilage tissue remodeling in response to mechanical forces. *Annu. Rev. Biomed. Eng.* **2**, 691–713.

Guilak, F. (2000). The deformation behavior and viscoelastic properties of chondrocytes in articular cartilage. *Biorheology* **37**, 27–44.

Guilak, F., and Mow, V. C. (2000). The mechanical environment of the chondrocyte: A biphasic finite element model of cell–matrix interactions in articular cartilage. *J. Biomech.* **33**, 1663–1673.

Guldberg, R. E., Oest, M., Lin, A. S., Ito, H., Chao, X., Gromov, K., Goater, J. J., Koefoed, M., Schwarz, E. M., O'Keefe, R. J., and Zhang, X. (2004). Functional integration of tissue-engineered bone constructs. *J. Musculoskelet. Neuronal. Interact.* **4**, 399–400.

Hall, B., Dembinski, J., Sasser, A. K., Studeny, M., Andreeff, M., and Marini, F. (2007). Mesenchymal stem cells in cancer: Tumor-associated fibroblasts and cell-based delivery vehicles. *Int. J. Hematol.* **86**, 8–16.

Haudenschild, D. R., McPherson, J. M., Tubo, R., and Binette, F. (2001). Differential expression of multiple genes during articular chondrocyte redifferentiation. *Anat. Rec.* **263**, 91–98.

Hu, J. C., and Athanasiou, K. A. (2006). A self-assembling process in articular cartilage tissue engineering. *Tissue Eng.* **12**, 969–979.

Hung, C. T., Lima, E. G., Mauck, R. L., Takai, E., LeRoux, M. A., Lu, H. H., Stark, R. G., Guo, X. E., and Ateshian, G. A. (2003). Anatomically shaped osteochondral constructs for articular cartilage repair. *J. Biomech.* **36**, 1853–1864.

Hung, C. T., Mauck, R. L., Wang, C. C., Lima, E. G., and Ateshian, G. A. (2004). A paradigm for functional tissue engineering of articular cartilage via applied physiologic deformational loading. *Ann. Biomed. Eng.* **32**, 35–49.

Hunziker, E. B. (2002). Articular cartilage repair: Basic science and clinical progress. A review of the current status and prospects. *Osteoarthr. Cartil.* **10**, 432–463.

Jaiswal, N., Haynesworth, S. E., Caplan, A. I., and Bruder, S. P. (1997). Osteogenic differentiation of purified, culture-expanded human mesenchymal stem cells *in vitro*. *J. Cell. Biochem.* **64**, 295–312.

Jakob, M., Démarteau, O., Schäfer, D., Hintermann, B., Dick, W., Heberer, M., and Martin, I. (2001). Specific growth factors during the expansion and redifferentiation of adult human articular chondrocytes enhance chondrogenesis and cartilaginous tissue formation *in vitro*. *J. Cell. Biochem.* **81**, 368–377.

Janderova, L., McNeil, M., Murrell, A. N., Mynatt, R. L., and Smith, S. R. (2003). Human mesenchymal stem cells as an *in vitro* model for human adipogenesis. *Obes. Res.* **11**, 65–74.

Johnstone, B., Hering, T. M., Caplan, A. I., Goldberg, V. M., and Yoo, J. U. (1998). *In vitro* chondrogenesis of bone marrow-derived mesenchymal progenitor cells. *Exp. Cell Res.* **238**, 265–272.

Katagiri, T., Yamaguchi, A., Komaki, M., Abe, E., Takahashi, N., Ikeda, T., Rosen, V., Wozney, J. M., Fujisawa-Sehara, A., and Suda, T. (1994). Bone morphogenetic protein-2 converts the differentiation pathway of C2C12 myoblasts into the osteoblast lineage. *J. Cell Biol.* **127**, 1755–1766.

Kerin, A., Patwari, P., Kuettner, K., Cole, A., and Grodzinsky, A. (2002). Molecular basis of osteoarthritis: Biomechanical aspects. *Cell Mol. Life Sci.* **59**, 27–35.

Khan, E., and Abu-Amer, Y. (2003). Activation of peroxisome proliferator-activated receptor-gamma inhibits differentiation of preosteoblasts. *J. Lab. Clin. Med.* **142**, 29–34.

Kim, T. K., Sharma, B., Williams, C. G., Ruffner, M. A., Malik, A., McFarland, E. G., and Elisseeff, J. H. (2003). Experimental model for cartilage tissue engineering to regenerate the zonal organization of articular cartilage. *Osteoarthr. Cartil.* **11**, 653–664.

Klein, T. J., Schumacher, B. L., Schmidt, T. A., Li, K. W., Voegtline, M. S., Masuda, K., Thonar, E. J., and Sah, R. L. (2003). Tissue engineering of stratified articular cartilage from chondrocyte subpopulations. *Osteoarthr. Cartil.* **11**, 595–602.

Klein-Nulend, J., Bacabac, R. G., and Mullender, M. G. (2005). Mechanobiology of bone tissue. *Pathol. Biol.* **53**, 576–580.

Krebsbach, P. H., Kuznetsov, S. A., Bianco, P., and Robey, P. G. (1999). Bone marrow stromal cells: Characterization and clinical application. *Crit. Rev. Oral Biol. Med.* **10**, 165–181.

Kuo, C. K., and Tuan, R. S. (2003). Tissue engineering with mesenchymal stem cells. *IEEE Eng. Med. Biol. Mag.* **22**, 51–56.

Laflamme, M. A., and Murry, C. E. (2005). Regenerating the heart. *Nat. Biotechnol.* **23**, 845–856.

Lecanda, F., Avioli,, L. V., and Cheng, (1997). Regulation of bone matrix protein expression and induction of differentiation of human osteoblasts and human bone marrow stromal cells by bone morphogenetic protein-2. *J. Cell. Biochem.* **67**, 386–398.

Lee, R. H., Kim, B., Choi, I., Kim, H., Choi, H. S., Suh, K., Bae, Y. C., and Jung, J. S. (2004). Characterization and expression analysis of mesenchymal stem cells from human bone marrow and adipose tissue. *Cell. Physiol. Biochem.* **14**, 311–324.

Lefebvre, V., Behringer, R. R., and de Crombrugghe, B. (2001). L-Sox5, Sox6 and Sox9 control essential steps of the chondrocyte differentiation pathway. *Osteoarthr. Cartil.* **9**(Suppl. A), S69–S75.

Lennon, D. P., and Caplan, A. I. (2006). Isolation of human marrow-derived mesenchymal stem cells. *Exp. Hematol.* **34**, 1604–1605.

Li, C., Vepari, C., Jin, H. J., Kim, H. J., and Kaplan, D. L. (2006). Electrospun silk-BMP-2 scaffolds for bone tissue engineering. *Biomaterials* **27**, 3115–3124.

Lin, C. Y., Kikuchi, N., and Hollister, S. J. (2004). A novel method for biomaterial scaffold internal architecture design to match bone elastic properties with desired porosity. *J. Biomech.* **37**, 623–636.

Long, M. W., Robinson, J. A., Ashcraft, E. A., and Mann, K. G. (1995). Regulation of human bone marrow-derived osteoprogenitor cells by osteogenic growth factors. *J. Clin. Invest.* **95**, 881–887.

Malaval, L., Liu, F., Roche, P., and Aubin, J. E. (1999). Kinetics of osteoprogenitor proliferation and osteoblast differentiation *in vitro*. *J. Cell. Biochem.* **74**, 616–627.

Mao, J. J. (2005). Stem-cell-driven regeneration of synovial joints. *Biol. Cell* **97**, 289–301.

Mao, J. J., Giannobile, W. V., Helms, J. A., Hollister, S. J., Krebsbach, P. H., Longaker, M. T., and Shi, S. (2006). Craniofacial tissue engineering. *J. Dent. Res.* **85**, 966–979.

Marion, N. W., and Mao, J. J. (2006). Mesenchymal stem cells and tissue engineering. *Methods Enzymol.* **420**, 339–361.

Marion, N. W., Liang, W., Reilly, G., Day, D. E., Rahaman, M. N., and Mao, J. J. (2005). Borate glass supports the *in vitro* osteogenic differentiation of human mesenchymal stem cells. *Mech. Adv. Mater. Struct.* **3**, 239–246.

Martin, I., Muraglia, A., Campanile, G., Cancedda, R., and Quarto, R. (1997). Fibroblast growth factor-2 supports *ex vivo* expansion and maintenance of osteogenic precursors from human bone marrow. *Endocrinology* **138**, 4456–4462.

Masuda, K., Sah, R. L., Hejna, M. J., and Thonar, E. J. (2003). A novel two-step method for the formation of tissue-engineered cartilage by mature bovine chondrocytes: The alginate-recovered-chondrocyte (ARC) method. *J. Orthop. Res.* **21**, 139–148.

Mauney, J. R., Volloch, V., and Kaplan, D. L. (2005). Role of adult mesenchymal stem cells in bone tissue engineering applications: Current status and future prospects. *Tissue Eng.* **11**, 787–802.

McKnight, A. L., Kugel, J. L., Rossman, P. J., Manduca, A., Hartmann, L. C., and Ehman, R. L. (2002). MR elastography of breast cancer: Preliminary results. *Am. J. Roentgenol.* **178**, 1411–1417.

Mistry, A. S., and Mikos, A. G. (2005). Tissue engineering strategies for bone regeneration. *Adv. Biochem. Eng. Biotechnol.* **94**, 1–22.

Moioli, E. K., Hong, L., Guardado, J., Clark, P. A., and Mao, J. J. (2006). Controlled release of TGFβ3 on early osteogenic differentiation of human mesenchymal stem cells. *Tissue Eng.* **12**, 537–546.

Moioli, E. K., Clark, P. A., Sumner, D. R., and Mao, J. J. (2008). Autologous stem cell regeneration in craniosynostosis. *Bone* **42**, 332–340.

Mow, V. C., Holmes, M. H., and Lai, W. M. (1984). Fluid transport and mechanical properties of articular cartilage: A review. *J. Biomech.* **17**, 377–394.

Mow, V. C., Wang, C. C., and Hung, C. T. (1999). The extracellular matrix, interstitial fluid and ions as a mechanical signal transducer in articular cartilage. *Osteoarthr. Cartil.* **7,** 41–58.

Nakamura, T., Shiojima, S., Hirai, Y., Iwama, T., Tsuruzoe, N., Hirasawa, A., Katsuma, S., and Tsujimoto, G. (2003). Temporal gene expression changes during adipogenesis in human mesenchymal stem cells. *Biochem. Biophys. Res. Commun.* **303,** 306–312.

Nesic, D., Whiteside, R., Brittberg, M., Wendt, D., Martin, I., and Mainil-Varlet, P. (2006). Cartilage tissue engineering for degenerative joint disease. *Adv. Drug Deliv. Rev.* **58,** 300–322.

Ng, L. J., Wheatley, S., Muscat, G. E., Conway-Campbell, J., Bowles, J., Wright, E., Bell, D. M., Tam, P. P., Cheah, K. S., and Koopman, P. (1997). SOX9 binds DNA, activates transcription, and coexpresses with type II collagen during chondrogenesis in the mouse. *Dev. Biol.* **183,** 108–121.

Osyczka, A. M., Noth, U., O'Connor, J., Caterson, E. J., Yoon, K., Danielson, K. G., and Tuan, R. S. (2002). Multilineage differentiation of adult human bone marrow progenitor cells transduced with human papilloma virus type 16 E6/E7 genes. *Calcif. Tissue Int.* **71,** 447–458.

Pairault, J., and Green, H. (1979). A study of the adipose conversion of suspended 3T3 cells by using glycerophosphate dehydrogenase as differentiation marker. *Proc. Natl. Acad. Sci. USA* **76,** 5138–5142.

Patel, P. N., Gobin, A. S., West, J. L., and Patrick, C. W., Jr. (2005). Poly(ethylene glycol) hydrogel system supports preadipocyte viability, adhesion, and proliferation. *Tissue Eng.* **11,** 1498–1505.

Patrick, C. W. (2004). Breast tissue engineering. *Annu. Rev. Biomed. Eng.* **6,** 109–130.

Pittenger, M. F., and Martin, B. J. (2004). Mesenchymal stem cells and their potential as cardiac therapeutics. *Circ. Res.* **95,** 9–20.

Pittenger, M. F., Mackay, A. M., Beck, S. C., Jaiswal, R. K., Douglas, R., Mosca, J. D., Moorman, M. A., Simonetti, D. W., Craig, S., and Marshak, D. R. (1999). Multilineage potential of adult human mesenchymal stem cells. *Science* **284,** 143–147.

Poole, C. A., Flint, M. H., and Beaumont, B. W. (1988). Chondrons extracted from canine tibial cartilage: Preliminary report on their isolation and structure. *J. Orthop. Res.* **6,** 408–419.

Poole, C. A., Glant, T. T., and Schofield, J. R. (1991). Chondrons from articular cartilage. (IV). Immunolocalization of proteoglycan epitopes in isolated canine tibial chondrons. *J. Histochem. Cytochem.* **39,** 1175–1187.

Rahaman, M. N., and Mao, J. J. (2005). Stem cell-based composite tissue constructs for regenerative medicine. *Biotechnol. Bioeng.* **91,** 261–284.

Rezwan, K., Chen, Q. Z., Blaker, J. J., and Boccaccini, A. R. (2006). Biodegradable and bioactive porous polymer/inorganic composite scaffolds for bone tissue engineering. *Biomaterials* **27,** 3413–3431.

Rickard, D. J., Sullivan, T. A., Shenker, B. J., Leboy, P. S., and Kazhdan, I. (1994). Induction of rapid osteoblast differentiation in rat bone marrow stromal cell cultures by dexamethasone and BMP-2. *Dev. Biol.* **161,** 218–228.

Rodan, G. A., and Noda, M. (1991). Gene expression in osteoblastic cells. *Crit. Rev. Eukaryot. Gene Expr.* **1,** 85–98.

Rosen, E. D., and Spiegelman, B. M. (2000). Molecular regulation of adipogenesis. *Annu. Rev. Cell Dev. Biol.* **16,** 145–171.

Saito, T., Kuang, J. Q., Lin, C. C., and Chiu, R. C. (2003). Transcoronary implantation of bone marrow stromal cells ameliorates cardiac function after myocardial infarction. *J. Thorac. Cardiovasc. Surg.* **126,** 114–123.

Sekiya, I., Colter, D. C., and Prockop, D. J. (2001). BMP-6 enhances chondrogenesis in a subpopulation of human marrow stromal cells. *Biochem. Biophys. Res. Commun.* **284,** 411–418.

Sekiya, I., Vuoristo, J. T., Larson, B. L., and Prockop, D. J. (2002). *In vitro* cartilage formation by human adult stem cells from bone marrow stroma defines the sequence of cellular and molecular events during chondrogenesis. *Proc. Natl. Acad. Sci. USA* **99,** 4397–4402.

Shake, J. G., Gruber, P. J., Baumgartner, W. A., Senechal, G., Meyers, J., Redmond, J. M., Pittenger, M. F., and Martin, B. J. (2002). Mesenchymal stem cell implantation in a swine myocardial infarct model: Engraftment and functional effects. *Ann. Thorac. Surg.* **73,** 1919–1925; discussion 1926.

Spaeth, E., Klopp, A., Dembinski, J., Andreeff, M., and Marini, F. (2008). Inflammation and tumor microenvironments: Defining the migratory itinerary of mesenchymal stem cells. *Gene Ther.* **15**, 730–738.

Spees, J. L., Gregory, C. A., Singh, H., Tucker, H. A., Peister, A., Lynch, P. J., Hsu, S. C., Smith, J., and Prockop, D. J. (2004). Internalized antigens must be removed to prepare hypoimmunogenic mesenchymal stem cells for cell and gene therapy. *Mol. Ther.* **9**, 747–756.

Stosich, M. S., and Mao, J. J. (2005). Stem cell-based soft tissue grafts for plastic and reconstructive surgeries. *Semin. Plast. Surg.* **19**, 251–260.

Stosich, M. S., and Mao, J. J. (2007). Adipose tissue engineering from human adult stem cells: Clinical implications in plastic and reconstructive surgery. *Plast. Reconstr. Surg.* **119**, 71–83.

Stosich, M. S., Bastian, B., Marion, N. W., Clark, P. A., Reilly, G., and Mao, J. J. (2007). Vascularized adipose tissue grafts from human mesenchymal stem cells with bioactive cues and microchannel conduits. *Tissue Eng.* **13**, 2881–2890.

Studeny, M., Marini, F. C., Dembinski, J. L., Zompetta, C., Cabreira-Hansen, M., Bekele, B. N., Champlin, R. E., and Andreeff, M. (2004). Mesenchymal stem cells: Potential precursors for tumor stroma and targeted-delivery vehicles for anticancer agents. *J. Natl. Cancer Inst.* **96**, 1593–1603.

Taboas, J. M., Maddox, R. D., Krebsbach, P. H., and Hollister, S. J. (2003). Indirect solid free form fabrication of local and global porous, biomimetic and composite 3D polymer-ceramic scaffolds. *Biomaterials* **24**, 181–194.

Troken, A. J., Wan, L. Q., Marion, N. W., Mao, J. J., and Mow, V. C. (2005). Properties of cartilage and meniscus. *In* The Wiley Encyclopedia of Medical Devices and Instrumentation, (J. G Webster, Ed.) Wiley, New York.

Tsutsumi, S., Shimazu, A., Miyazaki, K., Pan, H., Koike, C., Yoshida, E., Takagishi, K., and Kato, Y. (2001). Retention of multilineage differentiation potential of mesenchymal cells during proliferation in response to FGF. *Biochem. Biophys. Res. Commun.* **288**, 413–419.

Tuli, R., Seghatoleslami, M. R., Tuli, S., Wang, M. L., Hozack, W. J., Manner, P. A., Danielson, K. G., and Tuan, R. S. (2003). A simple, high-yield method for obtaining multipotential mesenchymal progenitor cells from trabecular bone. *Mol. Biotechnol.* **23**, 37–49.

Turner, C. H. (1999). Toward a mathematical description of bone biology: The principle of cellular accommodation. *Calcif. Tissue Int.* **65**, 466–471.

Wan, D. C., Nacamuli, R. P., and Longaker, M. T. (2006). Craniofacial bone tissue engineering. *Dent. Clin. North Am.* **50**, 175–190.

Wang, E. A, Rosen, V, D'Alessandro, J. S, Bauduy, M, Cordes, P, Harada, T, Israel, D. I, Hewick, R. M, Kerns, K. M, and LaPan, P. (1990). Recombinant human bone morphogenetic protein induces bone formation. *Proc. Natl. Acad. Sci. USA* **87**, 2220–2224.

Williams, C. G., Kim, T. K., Taboas, A., Malik, A., Manson, P., and Elisseeff, J. (2003). *In vitro* chondrogenesis of bone marrow-derived mesenchymal stem cells in a photopolymerizing hydrogel. *Tissue Eng.* **9**, 679–688.

Woodfield, T. B., Bezemer, J. M., Pieper, J. S., van Blitterswijk, C. A., and Riesle, J. (2002). Scaffolds for tissue engineering of cartilage. *Crit. Rev. Eukaryot. Gene Expr.* **12**, 209–236.

Wozney, J. M. (1992). The bone morphogenetic protein family and osteogenesis. *Mol. Reprod. Dev.* **32**, 160–167.

Xu, W., Zhang, X., Qian, H., Zhu, W., Sun, X., Hu, J., Zhou, H., and Chen, Y. (2004). Mesenchymal stem cells from adult human bone marrow differentiate into a cardiomyocyte phenotype *in vitro*. *Exp. Biol. Med. (Maywood)* **229**, 623–631.

Ylostalo, J., Smith, J. R., Pochampally, R. R., Matz, R., Sekiya, I., Larson, B. L., Vuoristo, J. T., and Prockop, D. J. (2006). Use of differentiating adult stem cells (MSCs) to identify new downstream target genes for transcription factors. *Stem Cells* **24**, 642–652.

Yoo, J. U., Barthel, T. S., Nishimura, K., Solchaga, L., Caplan, A. I., Goldberg, V. M., and Johnstone, B. (1998). The chondrogenic potential of human bone-marrow-derived mesenchymal progenitor cells. *J. Bone Joint Surg. Am.* **80**, 1745–1757.

SECTION II

Embryonic Stem Cells and
Their Derivatives

CHAPTER 15

Mouse Embryonic Stem Cells

Andras Nagy[*,†] and Kristina Vintersten[†]

*Department of Medical Genetics and Microbiology
University of Toronto
Toronto, Canada

†Mount Sinai Hospital
Samuel Lunenfeld Research Institute
Toronto, M5T 3H7, Canada

Abstract
1. Historical Overview
2. Factors Affecting the Efficiency of Mouse ES Cell Establishment
3. Factors Affecting the Contribution of Mouse ES Cell to Chimeric Embryos
4. Critical Events During Mouse ES Cell Establishment
    4.1. The source embryos
    4.2. Placement of the embryos into ES cell conditions and subsequent hatching
    4.3. Attachment
    4.4. Formation of the outgrowth
    4.5. The first disaggregation
    4.6. Expansion of the first ES cell-like colonies
5. Freezing of ES Cell Lines
6. Characterization
7. Protocols
    7.1. Mouse embryonic fibroblasts feeder layer preparation
    7.2. Establishment
    7.3. Method
    7.4. Maintenance
References

Abstract

Embryonic stem (ES) cells are derived from preimplantation stage mouse embryos at the time when they have reached the blastocyst stage. It is at this point that the first steps of differentiation take place during mammalian embryonic development. The individual blastomeres now start to organize themselves into three distinct locations, each encompassing a different cell type; (1) the outside epithelial cells, trophectoderm; (2) the cells at the blastocoel surface of the inner cell mass (ICM), the primitive endoderm; and (3) the inside cells of the ICM, the primitive ectoderm. ES cells originate from the third population, the primitive ectoderm, which is a transiently existing group of cells in the embryo. Primitive ectoderm cells diminish within a day as the embryo is entering into the next steps of differentiation. ES cells on the other hand, while retaining the property of their origin in terms of developmental potential, also have the ability to self renew. It is hence important to realize that ES cells do not exist *in vivo*, rather, they should be regarded simply as tissue culture artifacts. Nevertheless, these powerful cells have the potential to differentiate into all the cells of the embryo proper and postnatal animal. Furthermore, they retain the limitation of their origin through their inability to contribute to the trophectoderm lineage (the trophoblast of the placenta) and the lineages of the primitive endoderm, the visceral, and parietal endoderm. Due to these unique features, we must admit that even if we regard ES cells as products of *in vitro* culture, and should not compare them to true somatic stem cells found in the adult organism, they certainly offer us a fantastic tool for genetic, developmental, and disease studies.

1. Historical Overview

The year this book is published marks the twenty-seven anniversary of two milestone publications reporting the establishment of embryonic stem (ES) cell lines from the mouse (Evans and Kaufman, 1981; Martin, 1981). By placing the cells in specific culture conditions, the authors could block the cells in their differentiation program and induce them to self-renew. Amazingly however, once the cells were released from these conditions and placed into a differentiation-promoting environment (*in vitro* or *in vivo*), the cells proved capable of giving rise to a vast number of various cell types. The *in vivo* studies were done by injection of ES cells into a blastocyst stage host embryo. In such chimeras, ES cell derivatives could be found in all somatic cell lineages. Three years later, the ability of ES cells to contribute to the germ line was also demonstrated (Bradley *et al.*, 1984). The real breakthrough however took place two years later as two groups succeeded to pass along the genome of genetically altered ES cells through the germ line (Gossler *et al.*, 1986; Robertson *et al.*, 1986). In parallel to this development, Olivier Smithies' laboratory established the technique of homologous recombination in eukaryotic cells (Smithies *et al.*, 1985), which later became the means of targeting (mutating) any desired gene in the mouse genome. As a result, by the late 1980s, the stage was set for a revolution in mouse genetics (Koller *et al.*, 1989; Thomas and Capecchi, 1990).

During the past 15 years, nearly 6000 genes have been knocked out in the mouse genome. These mutants have revealed a tremendous amount of information on the role of various genes in normal development as well as in disease processes. Today, the technology of gene targeting has developed to the level where high throughput generation of gene knockouts has become feasible. As a result, several international consortiums have been formed to generate a bank of targeted ES cells covering the entire mouse genome (Austin *et al.*, 2004). Once this goal has been achieved, will the golden era of mouse ES cells fade? In fact, why would we need to derive further lines in addition to those that already exist and has long proven their high quality? To answer these questions, we have to look more closely into the treasure-chest, and see what other riddles these cells could help us solve.

The germ line compatibility of mouse ES cells has been utilized almost exclusively in genetic studies, while not much attention has been devoted to their somatic abilities. It has been known for long time that they are capable to support the entire embryonic (Nagy *et al.*, 1990) and adult (Nagy *et al.*, 1993) development of the mouse if some extra-embryonic lineages (trophoblast, visceral, and parietal endoderm) are provided by tetraploid embryos. One of the first—and strongest—applications of ES cell ↔ tetraploid embryo aggregation chimeras was the analysis of the peculiar *VEGF*-A knockout phenotype. This gene was identified as having a lethal heterozygous phenotype, hindering the "classical" germ line transmission-based gene targeting analysis (Carmeliet *et al.*, 1996). In order to overcome this obstacle, a homozygous *VEGF*-A null mutant ES cell line was created *in vitro* with high concentration G418 selection (Mortensen *et al.*, 1992) and then aggregated to wild-type tetraploid embryos. The embryos resulting from these experiments revealed the phenotypic consequence of *VEGF*-A deficiency, namely the lack of vessel formation (Carmeliet *et al.*, 1996). The very clear segregation of ES cell- and tetraploid embryo-derived compartments in the embryo proper and extraembryonic membranes, respectively, makes this type of chimeras extremely useful also to rescue extraembryonic phenotypes of gene deficiency (Duncan *et al.*, 1997; Guillemot *et al.*, 1994). Due to this "complementing segregation," the method earned the name: "tetraploid complementation assay" (TCA). In order to utilize the TCA, one must first establish mutant ES cell lines. The most commonly used method to generate *gain* of function mutations is by introducing a transgene into the ES cell genome. Loss-of-function can be achieved by "knock down" of a gene function though RNAi by a transgene expressing a small hairpin RNA (shRNA) (Kunath *et al.*, 2003), or by classical gene targeting. Creating a "knockout" with the latter method, however, requires the elimination of the function of both alleles of the gene in order to visualize a recessive phenotype. This is usually done by targeting the two alleles of a gene of interest in a consecutive manner. In some cases however, it may be more efficient to generate loss-of-function ES cell lines by deriving them from F2 generation embryos homozygous for the knockout allele

(Bryja *et al.*, 2006). The huge advantage of this method becomes obvious when deficiency in two or three genes is required for studies (Ding *et al.*, 2004). In all these cases, TCA can provide very fast access to the deficient phenotypes (Duncan, 2005).

2. Factors Affecting the Efficiency of Mouse ES Cell Establishment

Despite our significantly increased knowledge in ES cell biology and experience with culturing these cells, the success-rate of ES cell line establishment is still highly dependent on the genetic background of the source embryo (Gardner and Brook, 1997). The most permissive strains are the inbred 129 substrains. Consequently, ES cell lines with this genetic background have been used in the vast majority of ES cell-mediated mouse mutant generations. However, the 129 substrains come with the drawback of both minimal characterization and known anatomical (Livy and Wahlsten, 1997) and behavioral anomalies (Royce, 1972). Another relatively permissive inbred strain is the C57BL/6. This strain is strongly favored and considered to be somewhat of a "gold standard" in research using the mouse as a model. The germ line competence of the C57BL mES cell lines, however, falls behind that of 129 (Seong *et al.*, 2004) making the technology more expensive. As a result, the most common approach is to use 129 ES cells, and once mouse lines have been created, backcross these to the C57BL/6 background. The obvious drawback of this regimen lies in the extensive time required for breeding. Recent developments in establishment and maintenance conditions however, have made the C57BL/6 lines more and more feasible to work with, and this has led to the tendency of moving the ES cell-mediated mouse genetics towards the C57BL/6. Newly developed culture conditions have even affected the accessibility of so-called "nonpermissive" inbred strains, such as SVB, CBA (Roach *et al.*, 1995), and NOD (Chen *et al.*, 2005). Despite these advances, the genetic background still plays a vital role in the final potency of resulting mES cell lines. This fact is illustrated by the many attempts to derive ES cells from NOD embryos. There were no problems generating a large number of cell lines, but all of these lacked the ability to contribute to chimeras, obviously also including the germ line (Chen *et al.*, 2005). On the other extreme of this spectrum is the superior developmental potential of some F1 lines (Eggan *et al.*, 2001). ES cell-derived embryos produced with TCA using these hybrid cells not only developed to term at a very high frequency, but also survived after birth and developed into normal adults (Schwenk *et al.*, 2003). These new ES cell lines combined with TCA have tremendously improved the efficiency with which we can generate information on gene function (Nagy *et al.*, 2003).

Although mouse ES cells have been around for over 25 years, the culture conditions supporting their self-renewal and inhibiting differentiation are still not fully understood. The success rate of establishment varies from laboratory to laboratory due to different levels of expertise and different batches of undefined reagents—most notably the Fetal Bovine Serum (FBS). In an attempt to reduce these variations, efforts have been made to move toward developing defined culture conditions, for example through the use of chemically defined Serum Replacement (SR) (Invitrogen Knockout™ SR). There are signs, however, that also SR has lot-to-lot variations and may not contain all the necessary reagents for ES cell establishment. Therefore, alternation of SR and FBS was recently suggested as a solution (Bryja *et al.*, 2006). The authors reported an impressive 70% success rate of mouse ES cell establishment, compared to the about 25% that can be expected from standard derivation attempts. The ability to give rise to ES cell lines from a certain genetic background is however only a precondition—there are many more hurdles to overcome along the road to an ES cell line suitable for genetic studies.

3. Factors Affecting the Contribution of Mouse ES Cell to Chimeric Embryos

Apart from the genetic background of the ES cells, other factors also influence their capacity to contribute to the somatic tissue—and most importantly—to the germ line of chimeras. As is the case in all tissue culture, cells accumulate random genetic and epigenetic changes with time. These changes range from large chromosomal aberrations to small methylation changes in the DNA affecting critical gene expressions. If these alterations increase the proliferation rate under the given culture conditions, the population of abnormal cells could take over the culture in short time (Liu *et al.*, 1997). An ES cell culture plate always represent a competitive field favoring the "speedy." Therefore, it is essential to keep the time the cells spend in this competition as short as possible (low passage number) in order to keep the accumulation of genome damage low and so maintain high developmental potential. It is important to keep in mind that suboptimal culture conditions allow for larger a competitive field than optimal settings. Consequently, extreme care is necessary to provide the best possible physical environment both during establishment and maintenance of the cells.

Once mES cells are introduced back *in vivo*, another stage of competition is initiated; that between cells originating from the host embryo versus the ES cell. These two groups of cells are competing for colonization of the different lineages. Different genetic backgrounds have different strengths in these pairwise competitions. Taking the germ line competence as the ultimate measure, many years of world-wide experience have yielded a few, well documented successful combinations of ES cell—host embryo genetic backgrounds. 129 ES cells are generally injected into C57BL/6 blastocyst and C57BL/6 ES cells are injected into BALB/c (Kontgen *et al.*, 1993; Ledermann and Burki, 1991; Lemckert *et al.*, 1997) or coat color coisogenic (albino) C57BL/6-Tyr(c)-2J (c2J) (Schuster-Gossler *et al.*, 2001) embryos. However, in cases where 129 ES cell are aggregated with morula stage embryos, the outbred CD1 or ICR is the preferred host (Nagy *et al.*, 2003; Wood *et al.*, 1993).

4. Critical Events During Mouse ES Cell Establishment

4.1. The source embryos

Embryos for the derivation of ES cells can be obtained from either naturally mated or superovulated female mice. The former option generally yields higher quality embryos, while the latter results in higher numbers. The choice ultimately depends on the age and genetic background of the donor animals. Strains, which respond well to super-ovulation, are worth placing under this regimen, while strains that are poor responders are better naturally mated. Embryos are usually collected from the uterus of 3.5 dpc animals. At this time, they have reached the blastocyst stage and are ready for ES cell establishment without the need for further *in vitro* culture. However, all embryos from a single mouse are not always at exactly the same developmental stage at a given time. This may result in the isolation of both late morula stage as well as expanded and perhaps even a few hatched blastocysts from the same female donor. In this case, morula stage embryos should be given a few hours further culture in embryo culture media before proceeding to the next step.

Before recovery of the embryos is attempted, all necessary reagents, media, and instruments should be prepared in such a way that the procedure can be carried out speedily. The time the embryos spend outside the *in vivo* environment or the incubator greatly influence their quality. If a large cohort of embryos with the same genotype is available, it is worthwhile selecting only embryos of good quality (Fig. 15.1A). However, if embryos are isolated from mutant crosses, it is crucial to include every single embryo, since the variation of genotype may cause a slightly varied phenotype or developmental rate.

4.2. Placement of the embryos into ES cell conditions and subsequent hatching

Derivation of ES cells directly from live animals always encompasses the risk of transmitting pathogens from the animal to the tissue culture. For this reason, great care should be taken to use separate media, reagents, hoods, and incubators as far as at all possible until established cell lines have been screened and declared free of pathogens.

At the time of recovery, the embryo is still surrounded by a thick protective glycoprotein layer called the zona pellucida (ZP). *In vivo*, the ZP gradually thins, cracks, and the blastocyst hatches at around 4.5 dpc as it arrives into the uterine cavity. This hatching process occurs also *in vitro*—provided that the culture conditions are optimal (Fig. 15.1B). Once out of its protective shell, the blastocyst becomes very fragile, collapses, and immediately starts to attach to the uterine wall. *In vitro*, it is possible to lure the blastocyst to attach to the tissue culture plate or a layer of mitotically inactivated mouse embryonic fibroblasts. Needless to say, the quality of these feeder cells has to be optimal, they should be prepared fresh (no longer than 5 days prior to use) be of optimal density (see protocol 1) and low

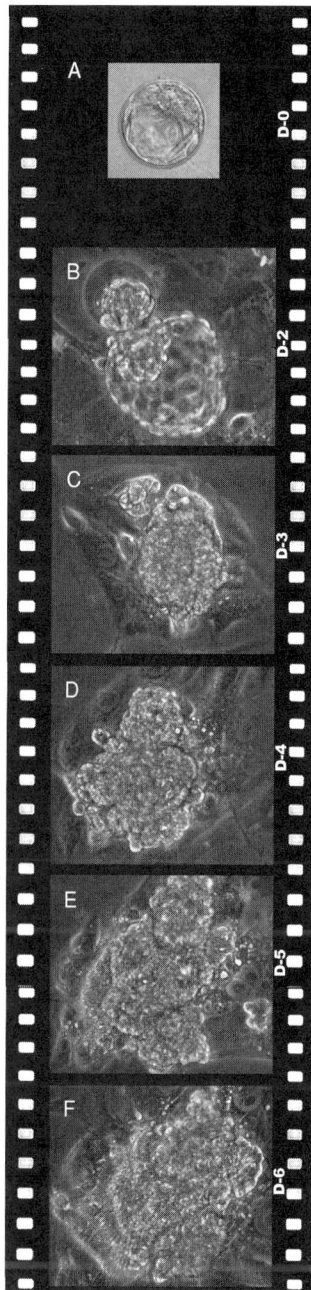

Figure 15.1 Phases of blastocyst outgrowth development during preparation for ES cell establishment. All the pictures are of the same scale. (A) High quality blastocyst ready to be plated on MEFs for ES cell derivation (day 0). (B) The embryo at the final stage of hatching (day 2). (C) Attaching embryo (day 3). The attached trophoblast cells are clearly visible under the outgrowth. (D–F) The outgrowth is increasing in size (days 4–6). Areas with ES-cell like cells become visible (E), and grow larger. This outgrowth is now ready for disaggregation (F).

passage number (no higher than p3). Also the culture media and physical environment in which the further *in vitro* culture will take place has a fundamental impact on the success-rate of derivation. Culture media and reagents should be prepared fresh and be of highest possible quality. Incubators should be checked for their temperature and CO_2 reading accuracy, and it should be ensured that the humidity remains as high as possible at all times.

4.3. Attachment

The following few days are very exciting: the blastocyst emerges from the ZP and start to attach to the feeder layer (Fig. 15.1C). The temptation is great to peek in for a quick look on this process from time to time. However, attachment is best achieved if the plates are not disturbed. As a general rule, the door of the incubator should remain shut for 48 h. Two days after plating (if the embryos were well expanded blastocysts, three days if they were smaller), the cultures should be carefully examined and the media replaced.

4.4. Formation of the outgrowth

Once the blastocyst has attached, cells will very soon start grow out on top of the feeder layer. During the coming few days, the recommendation about checking on the culture is reversed: they should be carefully investigated every day in order to determine the optimal time-point for the first disaggregation (Fig. 15.1D–F). However, it is still important to keep the time the cultures spend outside the incubator to a minimum. The media should be replaced every other day.

4.5. The first disaggregation

Timing the first disaggregation right is perhaps the most crucial determinant for the success of ES cell establishment. Done too early, the outgrowth will not contain enough cells, and the culture will die. Done too late, the outgrowth will have already started to differentiate into other, more specialized cell types, and the culture will not result in ES-like colonies. Each individual outgrowth has to be carefully assessed, and the optimal time-point determined individually. Figure 15.1 illustrates the stages the outgrowth goes through, and the optimal size/time when it should be disaggregated is illustrated in Fig. 15.1F.

The next critical step will be the dissociation procedure. The outgrowth can at this point not be compared to a simple cell colony, which after the seeding of a single-cell preparation will readily form new colonies. In fact, harsh enzymatic dissociation will inevitably result in the death of the cells. Instead, a gentle process has to be adopted where the outgrowth is divided into small cell clumps of ~5–10 cells each.

4.6. Expansion of the first ES cell–like colonies

After the first dissociation of the outgrowth, the culture usually grows slowly. Occasionally, it may appear as if all cells (except the feeders) have died. It is important to again practice patience, and leave the cultures alone until small colonies become visible. Once cell growth can be identified however, the cultures will again require daily attention. Three possible scenarios might occur: (1) growth of nonES like colonies/cell types (Fig. 15.2B), (2) growth of both ES like colonies and other cell types (i.e., Fig. 15.2C and 2A, respectively), and (3) mainly ES type

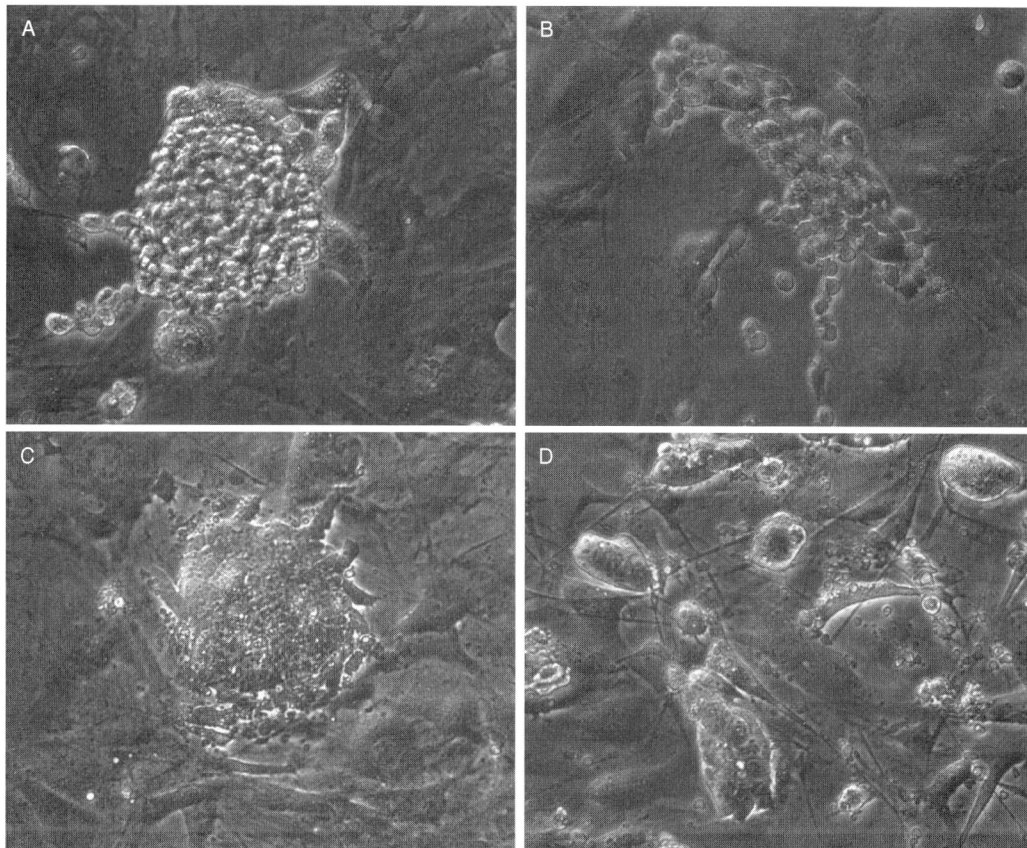

Figure 15.2 Cell colonies in the early phase of mES cell establishment. (A) Colony of mixed cell types as a result of improper disaggregation of the initial outgrowth. (B) Differentiated nonES like cell. (C) Three days after disaggregation of the outgrowth (Fig. 15.1F). ES cell like colonies can easily be recognized by the characteristic morphology. The colonies may at this stage still contain a few differentiated cells, but these usually diminish after a few passages. (D) Small colonies of pure mouse ES cells.

colonies (Fig. 15.2D). In case of no visible growth for more than 6 days, the culture can be discarded. Also cultures with only nonES-like cells can be terminated. In case both ES-like and other cell types are present, single nice colonies can be picked and transferred after gentle disaggregation to a well with fresh feeders. Once typical ES-like colonies have become visible and grown to an appropriate size (Fig. 15.2D), they can be passaged according to standard protocols for mouse ES cell culture. Each line should be expanded enough to freeze a small but safe number of vials for future characterization steps. It is important to start keeping track of the passage number right from the beginning. The most widely used method is to start counting passage no. 1 when the cells are plated for the first time onto larger surface area than what they were derived on (this usually is either two wells of a 4-well dish, or a 35 mm plate).

5. Freezing of ES Cell Lines

As mentioned in the introduction, keeping the passage number low is of vital importance for the quality of ES cell lines. Randomly acquired chromosomal or epigenetic changes that give individual cells a growth advantage will inevitably result in the accumulation of abnormal cells with increased passage numbers. For this reason, it is important to cryopreserve each line as soon as possible. Early passage vials can later be used to expand the line. Expansion should be done in a way such that vials are frozen from each consecutive passage, and a sufficiently large pool of vials is created for future use. This way, in case the cells in the final passage would have acquired suboptimal characteristics, one can fall back on the earlier passages for renewed expansion. Although time consuming, this approach is well worth all the invested efforts. Failing to establish a "ladder" of vials from earlier to later passage numbers may result in the loss of the entire line. One last word of caution: it is easy to loose the most precious early passage vials. These few valuable aliquots should not be wasted on anything else other than initial characterization and expansion.

6. Characterization

Even if all possible precautions are taken, and protocols followed to the letter, far from all resulting ES cell lines will display the desired pluripotency. Derivation also carries the innate risk of unintentional contamination with pathogens. Hence, the careful characterization of candidate lines should be given due consideration.

The first (and easiest) screening strategy for identifying potentially "good" lines is based on morphology, homogeneity of the culture, and speed of growth. These initial steps can be undertaken during the actual derivation process, already before the lines are frozen. Good morphology is depicted in Fig. 15.2D, the aim should be to achieve cultures with predominantly this kind of colonies. Optimally, established cultures should grow at a rate at which they become subconfluent in two days if passaged at a rate of 1:6 (the initial passages during establishment however should be kept at a lower expansion rate (1:2 or 1:3) until the cells have gained growth momentum. However, morphology and growth speed alone are not enough criteria for distinguishing good ES cell lines. Further characterization steps could include:

1. *Pathogen testing*: An aliquot of cells should be cultured for a minimum of three passages in media without antibiotics. Cell supernatant and/or cell suspension should then be screened for a panel of mouse pathogens. This step is not only important if future use of the ES cells is aimed at creating animals in a specific pathogen free (SPF) facility, but also as a general precaution for avoiding transfer of contamination to other cell cultures. One of the most common pathogens found in mouse ES cell cultures is Mycoplasma species.

2. *Karyotyping*: Karyotype analysis can be used to determine the sex of the line, and to detect possible chromosomal abnormalities. A complete SKY painting (Spectral Karyotyping) ultimately gives the most information, but involves considerable costs. A more economical alternative is G-banding or simple chromosome counting. In either case, a minimum of 20 metaphase spreads should be analyzed in order to correctly pinpoint the overall euploidy of the line.

3. *In vitro differentiation*: Placing ES cells in *vitro* differentiation assays can provide some information about potency. A large number of assays have been established, allowing induction of a vast number of cell types. However, all these assays are time, labor and cost intensive, and the information gain is limited to the *in vitro* potential of the cells.

4. *Teratomas* The classical method of determining the ability of ES cells to contribute to all three germ layers is by teratoma assay. This is done by injecting ES cells under the skin, kidney capsule or testicle of immunologically compatible (or compromised) mice. The tumor formation ability and composition of the teratoma gives a good indication on the developmental potency of the ES cells. However, this assay is also time and cost intensive.

5. *Chimera formation*: The most widely used and very informative method of determining the quality of ES cell lines is to introduce them back into an embryonic environment through morula aggregation or blastocyst injection (Nagy *et al.*, 2003). ES cells of high quality will contribute to all somatic tissues and the germ line of resulting chimeras.

6. *Tetraploid complementation assay (TCA)* The ultimate test of mES cell potency however can be seen when they are forced to form the entire embryo proper. This can be achieved by combining the ES cells with tetraploid host embryos, as tetraploid embryos do not contribute well to the embryo proper, but they do form normal placentas.

7. Protocols

7.1. Mouse embryonic fibroblasts feeder layer preparation

7.1.1. Materials and equipment

- Sterile horizontal flow hood
- Sterile incubator 37 °C, 5% CO_2, 100% humidity
- Centrifuge
- 70% EtOH
- Tissue culture plates (4-well, 35 mm, 60 mm, 1000 mm)
- 10 ml sterile plastic tubes
- PBS without Ca++ and Mg++
- 0.05–0.1% trypsin in saline/EDTA
- Mitomycin C (1 mg/ml Sigma M-4287)
- MEFs, early passage, frozen vial
- MEF culture media. KO-DMEM (Gibco 10829–018) supplemented with:
  - 10% FBS
  - 100 μM nonessential amino acids (100× stock, Gibco no. 11140)
  - 100 μM beta-mercaptoethanol (100× stock Sigma no. M-7522)
  - 2 mM GlutaMax™ (Invitrogen no. 35050)
  - Penicillin/streptomycin (final concentration 50 μg/ml) (100× stock Gibco no. 15070)

Note, if standard DMEM is used instead of KO-DMEM, the media should be supplemented with 1 mM Sodium Pyruvate as well (100× stock, Gibco no.11360).

7.1.2. Method

1. Thaw the vial of MEFs quickly at 37 °C. Clean the outside of the vial by wiping with 70% EtOH.
2. Add the contents of the vial to 5 ml culture medium in a 10 ml sterile plastic tube.
3. Centrifuge 3 min, at 200 g.
4. Remove the supernatant. Flick the tube to loosen the cell pellet.
5. Add the appropriate amount of culture media (depending on the cell number present in the vial) and plate the cells on tissue culture dishes.
6. Replace the media the next day, and thereafter every other day.
7. Inspect the cultures daily to determine the optimal density for inactivation. Initially the cultures will display a typical thin elongated fibroblast morphology. As the culture grows in density and space becomes sparse, the cells start

to take on a "cobble-stone" appearance. It is at this point they should be passaged. If the culture is left to grow longer, the fibroblasts will stop growing (contact inhibition), a phenomenon that should be avoided.

8. Add 10 μl Mitomycin C (1 mg/ml) per ml culture media directly to the cultures. Rock the plates gently to mix the Mitomycin C with the media. Incubate at 37 °C for 2 h.

9. Remove the media and wash the cells three times with PBS.

10. Add 0.1 ml of Trypsin per 10–15 mm diameter of plate surface.

11. Incubate for 3–5 min at 37 °C. Periodically, check the plates under a microscope and stop the Trypsin reaction when the cells start to lift off the surface.

12. Add 1 ml of culture media per 10–15 mm diameter of plate surface to stop the Trypsin reaction (the serum contained in the medium will inhibit the Trypsin immediately).

13. Resuspend the cells by pipetting up and down several times.

14. Add the cell suspension to a sterile plastic tube, and centrifuge 3 min at 200 g.

15. Remove the supernatant. Flick the tube to loosen the cell pellet.

16. Dilute the cell suspension in a small amount of culture media. Count the cell concentration and make the appropriate dilution in such a way that you plate 40,000–50,000 cells/cm^2.

17. Incubate overnight to allow the MEFs to properly adhere to the tissue culture plate.

18. Inactivated MEFs can be used as feeder layers for mES cells no layer than 5 days after plating.

7.2. Establishment

7.2.1. Material and equipment

- Sterile horizontal flow hood
- Sterile incubator 37 °C, 5% CO_2 100% humidity
- Centrifuge
- 3.5 dpc pregnant mice
- Dissecting microscope
- HEPES buffered embryo culture medium, for example M2 (Specialty Media/Chemicon MR-015-D)
- KSOM-AA (Specialty Media/Chemicon MR-121D)
- Pulled Pasteur pipettes
- Pipette P200 with sterile tips
- 5 ml syringe with 27G needle
- Tissue culture plates (4-well, 35 mm, 60 mm, 1000 mm) with mitotically inactivated MEFs

- 10 ml sterile plastic tubes
- PBS without Ca++ and Mg++
- 0.05–0.1% trypsin in saline/EDTA
- ES culture media. KO-DMEM (Gibco 10829–018) supplemented with:
  - 15% mES cell qualified FBS
  - 100 μM nonessential amino acids (100× stock, Gibco no. 11140)
  - 100 μM beta-mercaptoethanol (100× stock, $-20\,°$C, Sigma no. M-7522)
  - 2 mM L-Glutamine (100× stock, $-20\,°$C, Gibco no. 25030)
  - Penicillin/streptomycin (final concentration 50 μg/ml) (100× stock Gibco no. 15070)
  - LIF 2000 U/ml

Note, if standard DMEM is used instead of KO-DMEM, the media should be supplemented with 1 mM Sodium Pyruvate as well (100× stock, Gibco no. 11360).

7.3. Method

7.3.1. Plating

- One day prior to the experiment: remove the media in the appropriate number of 4-well tissue culture plates with MEFs (one well per embryo). Add freshly prepared ES culture media.
- Sacrifice pregnant mice at 3.5dpc in a humane way following local animal welfare practices. Dissect the uteri.
- Isolate embryos from the uterine horns by inserting a 27 G needle (with a 5 ml syringe filled with M2 medium attached) in each end of the uterus close to the ovaries and flush with ~0.5 ml medium.
- Using a finely pulled Pasteur pipette, locate the blastocysts and rinse them several times through M2 medium (for more details on this steps see Nagy *et al.*, 2003).
- Using a pulled Pasteur pipette, place one blastocyst in the centre of each well. These steps can be performed using a dissecting microscope placed in a laminar flow hood. All consecutive procedures should be carried out under strictly sterile conditions.
- Culture the blastocysts undisturbed at 37 °C, 5% CO_2 for 48 h.

7.3.2. Disagregation of the outgrowth

After 48 h of undisturbed culture, the outgrowths should be inspected daily to determine the right stage at which to perform the first disaggregation (usually the fourth–sixth day after plating). Due to variability between different embryos, it might be necessary to perform the disaggregation on different days. The inner cell mass (ICM) outgrowth ready for disaggregation should be as big as possible but not yet differentiated. The evolving morphology of outgrowths is illustrated in Fig. 15.1. During this time, the media should be replaced on the cultures every other day.

1. One day prior to the planned disaggregation: replace the media on 4-well plates with freshly inactivated MEFs using ES culture media.

2. On the day of disaggregation: remove the media in the wells with the outgrowths. Add 0.5 ml PBS per well.

3. Place 25 μl drops of Trypsin in a 96-well tissue culture plate without MEFs.

4. Using a finely pulled Pasteur pipette, gently circle the ICM clump, remove it from the surrounding trophoblast cells, and place it into the trypsin in one well of the 96-well plate. Repeat the process with up to 10–20 outgrowths (depending on experience). If a larger number is ready for disaggregation on the same day, these should be done in a separate round in order to avoid the initially picked cells to spend too long time in the enzyme.

5. Incubate at 37 °C for 3–5 min.

6. Using a P200 pipettor and yellow tips, break up the outgrowth into smaller clumps of 5–10 cells. Watch the process under a microscope as some clumps might need repeated pipeting. Take good care not to pipette too much: single cells will not survive.

7. Add 30 μl of media to each well to stop the Trypsin reaction.

8. Transfer the cell suspension into one well of the 4-well MEFs plate. Make sure that the media in the well is ES cell culture media and not MEFs media. The media should have been replaced in the wells one day in advance.

9. Change the media after overnight incubation.

7.3.3. Culture of initial colonies

A few days after disaggregation, small colonies may become visible in the cultures. However, the initial cell growth may be very slow, so that colonies may not appear for several days. During this time, the media on the cultures are best changed every other day.

• Observe the cultures every day and keep log on each well. As soon as cell growth can be seen, start changing the media every day.

• Wells which do not show sign of cell proliferation within 10 days can be discarded. Also wells in which solely cells of nonES like morphology are present can be terminated.

• In wells in which only a few mES like colonies are present among other cells of varying morphology, renewed picking can be performed: mES-like colonies are disaggregated individually as described above and placed into a new well with MEFs.

- Slow growing colonies can be Trypsinized (see protocol 3) and replated back in the same well to prevent differentiation.
- When the ES cells have reached near confluency in a well, they should be passaged. Near confluency means that the colonies cover ~75% of the surface area, but are not yet so large that they have come in contact with each other. Passaging should be done at a rate of 1:2 (into two wells of a 4-well plate) or 1:3 into a 35 mm feeder plate. This will be considered passage 1.

7.3.4. Beyond the basic derivation protocol

As with all techniques that have been utilized for a long time in many laboratories over the world, a number of alternate approaches have been developed. Depending on the individual experiment and genetic background of donor embryos, the following variations may provide useful for increasing the efficiency:

- Increasing the FBS concentration in the culture media to 25% for the initial plating.
- Using DMEM with low glucose instead of KO-DMEM for the culture media.
- Supplementing the culture media with Knockout SR (Gibco no. 10828–028) instead of FBS (Note: MEFs do not attach to the tissue culture plates when grown in SR. Culture media for feeders should always be supplemented with FBS), or using an alternating approach (Bryja *et al.*, 2006).
- Adding Nucleosides (Specialty Media mES-008D) to the culture media.
- Isolating delayed blastocysts, prevented from implantation by ovariectomy and administration of progesterone (Brook and Gardner, 1997; Nagy *et al.*, 2003).
- Removing the trophoblast cells from the blastocyst by immunosurgery (Knowles *et al.*, 1977) prior to plating.
- Using the proprietary conditioned medium ResGro (Chemicon) (Schoonjans *et al.*, 2003) instead of the standard ES cell culture media.

7.4. Maintenance

7.4.1. Method

Once an ES cell culture has been successfully initialized from an ICM outgrowth, it should be maintained at a density allowing for optimal growth. This means that the culture should be passaged every other day at a rate of ~1:6. However, this rule of thumb is only a guidance. Each individual culture plate should be inspected daily and passaged as soon it has reached subconfluency. A few additional important points to remember include:

- The media should always be kept at 4 °C but warmed to room temperature before use.

- Make sure to always create a single cell suspension during passaging. It is better to slightly over-trypsinize the cells than to leave cell clumps.
- Always keep track of the passage number of each culture dish.
- Always grow ES cells on a freshly inactivated feeder layers (ideally no older than 5 days).

7.4.2. Passaging mouse ES cell cultures

1. Aspirate the media.
2. Rinse with 1 ml of PBS per 10–15 mm diameter of plate surface (2 ml for 30 mm plates, 5 ml for 60 mm plates etc.).
3. Add 0.1 ml Trypsin per 10–15 mm diameter of plate surface.
4. Incubate for 3–5 min at 37 °C. Periodically check the plates under a microscope and stop the Trypsin reaction when the colonies start to lift off the surface.
5. Add 1 ml of culture media per 10–15 mm diameter of plate surface to stop the Trypsin reaction (the serum contained in the medium will inhibit the Trypsin immediately).
6. Resuspend the cells by pipetting up and down several times until a single cell suspension has been achieved, but not so excessively that cell damage is caused. Until experience has been gained with this step, periodically check the suspension under a microscope.
7. Add the cell suspension to a sterile plastic tube, and centrifuge 3 min at 200 g.
8. Remove the supernatant. Flick the tube to loosen the cell pellet.
9. Dilute the cell suspension in an appropriate amount of ES media and then add to a 6 times larger growing area before the passaging.

Acknowledgment

We gratefully acknowledge Marina Gertsenstein for providing her expert view on the manuscript, and for preparing the embryos photographed for the Figures.

References

Austin, C. P., Battey, J. F., Bradley, A., Bucan, M., Capecchi, M., Collins, F. S., Dove, W. F., Duyk, G., Dymecki, S., Eppig, J. T., *et al.* (2004). The knockout mouse project. *Nat. Genet.* **36,** 921–924.

Bradley, A., Evans, M., Kaufman, M. H., and Robertson, E. (1984). Formation of germ-line chimaeras from embryo-derived teratocarcinoma cell lines. *Nature* **309,** 255–256.

Brook, F. A., and Gardner, R. L. (1997). The origin and efficient derivation of embryonic stem cells in the mouse. *Proc. Natl. Acad. Sci. USA* **94,** 5709–5712.

Bryja, V., Bonilla, S., Cajanek, L., Parish, C. L., Schwartz, C. M., Luo, Y., Rao, M. S., and Arenas, E. (2006). An efficient method for the derivation of mouse embryonic stem cells. *Stem Cells* **24,** 844–849.

Carmeliet, P., Ferreira, V., Breier, G., Pollefeyt, S., Kieckens, L., Gertsenstein, M., Fahrig, M., Vandenhoeck, A., Harpal, K., Eberhardt, C., *et al.* (1996). Abnormal blood vessel development and lethality in embryos lacking a single VEGF allele. *Nature* **380,** 435–439.

Chen, J., Reifsnyder, P. C., Scheuplein, F., Schott, W. H., Mileikovsky, M., Soodeen-Karamath, S., Nagy, A., Dosch, M. H., Ellis, J., Koch-Nolte, F., *et al.* (2005). "Agouti NOD": Identification of a CBA-derived Idd locus on Chromosome 7 and its use for chimera production with NOD embryonic stem cells. *Mamm. Genome* **16,** 775–783.

Ding, H., Wu, X., Bostrom, H., Kim, I., Wong, N., Tsoi, B., O'Rourke, M., Koh, G. Y., Soriano, P., Betsholtz, C., *et al.* (2004). A specific requirement for PDGF-C in palate formation and PDGFR-alpha signaling. *Nat. Genet.* **36,** 1111–1116.

Duncan, S. A. (2005). Generation of embryos directly from embryonic stem cells by tetraploid embryo complementation reveals a role for GATA factors in organogenesis. *Biochem. Soc. Trans.* **33,** 1534–1536.

Duncan, S. A., Nagy, A., and Chan, W. (1997). Murine gastrulation requires HNF-4 regulated gene expression in the visceral endoderm: Tetraploid rescue of Hnf-4(-/-) embryos. *Development* **124,** 279–287.

Eggan, K., Akutsu, H., Loring, J., Jackson-Grusby, L., Klemm, M., Rideout, W. M., III, Yanagimachi, R., and Jaenisch, R. (2001). Hybrid vigor, fetal overgrowth, and viability of mice derived by nuclear cloning and tetraploid embryo complementation. *Proc. Natl. Acad. Sci. USA* **98,** 6209–6214.

Evans, M. J., and Kaufman, M. H. (1981). Establishment in culture of pluripotential cells from mouse embryos. *Nature* **292,** 154–156.

Gardner, R. L., and Brook, F. A. (1997). Reflections on the biology of embryonic stem (ES) cells. *Int. J. Dev. Biol.* **41,** 235–243.

Gossler, A., Doetschman, T., Korn, R., Serfling, E., and Kemler, R. (1986). Transgenesis by means of blastocyst-derived embryonic stem cell lines. *Proc. Natl. Acad. Sci. USA* **83,** 9065–9069.

Guillemot, F., Nagy, A., Auerbach, A., Rossant, J., and Joyner, A. L. (1994). Essential role of Mash-2 in extraembryonic development. *Nature* **371,** 333–336.

Knowles, B. B., Solter, D., Trinchieri, G., Maloney, K. M., Ford, S. R., and Aden, D. P. (1977). Complement-mediated antiserum cytotoxic reactions to human chromosome 7 coded antigen(s): Immunoselection of rearranged human chromosome 7 in human-mouse somatic cell hybrids. *J. Exp. Med.* **145,** 314–326.

Koller, B. H., Hagemann, L. J., Doetschman, T., Hagaman, J. R., Huang, S., Williams, P. J., First, N. L., Maeda, N., and Smithies, O. (1989). Germ-line transmission of a planned alteration made in a hypoxanthine phosphoribosyltransferase gene by homologous recombination in embryonic stem cells. *Proc. Natl. Acad. Sci. USA* **86,** 8927–8931.

Kontgen, F., Suss, G., Stewart, C., Steinmetz, M., and Bluethmann, H. (1993). Targeted disruption of the MHC class II Aa gene in C57BL/6 mice. *Int. Immunol.* **5,** 957–964.

Kunath, T., Gish, G., Lickert, H., Jones, N., Pawson, T., and Rossant, J. (2003). Transgenic RNA interference in ES cell-derived embryos recapitulates a genetic null phenotype. *Nat. Biotechnol.* **21,** 559–561.

Ledermann, B., and Burki, K. (1991). Establishment of a germ-line competent C57BL/6 embryonic stem cell line. *Exp. Cell Res.* **197,** 254–258.

Lemckert, F. A., Sedgwick, J. D., and Korner, H. (1997). Gene targeting in C57BL/6 ES cells. Successful germ line transmission using recipient BALB/c blastocysts developmentally matured *in vitro*. *Nucleic Acids Res.* **25,** 917–918.

Liu, X., Wu, H., Loring, J., Hormuzdi, S., Disteche, C. M., Bornstein, P., and Jaenisch, R. (1997). Trisomy eight in ES cells is a common potential problem in gene targeting and interferes with germ line transmission. *Dev. Dyn.* **209,** 85–91.

Livy, D. J., and Wahlsten, D. (1997). Retarded formation of the hippocampal commissure in embryos from mouse strains lacking a corpus callosum. *Hippocampus* **7,** 2–14.

Martin, G. R. (1981). Isolation of a pluripotent cell line from early mouse embryos cultured in medium conditioned by teratocarcinoma stem cells. *Proc. Natl. Acad. Sci. USA* **78,** 7634–7638.

Mortensen, R. M., Conner, D. A., Chao, S., Geisterfer-Lowrance, A. A., and Seidman, J. G. (1992). Production of homozygous mutant ES cells with a single targeting construct. *Mol. Cell. Biol.* **12,** 2391–2395.

Nagy, A., Gertsenstein, M., Vintersten, K., and Behringer, R. (2003). "Manipulating the Mouse Embryo, A Laboratory Manual", Cold Spring Harbor Press.

Nagy, A., Goczia, E., Diaz, E. M., Prideaux, V. R., Ivanyi, E., Markkula, M., and Rossant, J. (1990). Embryonic stem cells alone are able to support fetal development in the mouse. *Development* **110,** 815–821.

Nagy, A., Rossant, J., Nagy, R., Abramow-Newerly, W., and Roder, J. C. (1993). Derivation of completely cell culture-derived mice from early-passage embryonic stem cells. *Proc. Natl. Acad. Sci. USA* **90,** 8424–8428.

Roach, M. L., Stock, J. L., Byrum, R., Koller, B. H., and McNeish, J. D. (1995). A new embryonic stem cell line from DBA/1lacJ mice allows genetic modification in a murine model of human inflammation. *Exp. Cell Res.* **221,** 520–525.

Robertson, E., Bradley, A., Kuehn, M., and Evans, M. (1986). Germ-line transmission of genes introduced into cultured pluripotential cells by retroviral vector. *Nature* **323,** 445–448.

Royce, J. R. (1972). Avoidance conditioning in nine strains of inbred mice using optimal stimulus parameters. *Behav. Genet.* **2,** 107–110.

Schoonjans, L., Kreemers, V., Danloy, S., Moreadith, R. W., Laroche, Y., and Collen, D. (2003). Improved generation of germline-competent embryonic stem cell lines from inbred mouse strains. *Stem Cells* **21,** 90–97.

Schuster-Gossler, K., Lee, A. W., Lerner, C. P., Parker, H. J., Dyer, V. W., Scott, V. E., Gossler, A., and Conover, J. C. (2001). Use of coisogenic host blastocysts for efficient establishment of germline chimeras with C57BL/6J ES cell lines. *Biotechniques* **31,** 1022–10241026.

Schwenk, F., Zevnik, B., Bruning, J., Rohl, M., Willuweit, A., Rode, A., Hennek, T., Kauselmann, G., Jaenisch, R., and Kuhn, R. (2003). Hybrid embryonic stem cell-derived tetraploid mice show apparently normal morphological, physiological, and neurological characteristics. *Mol. Cell. Biol.* **23,** 3982–3989.

Seong, E., Saunders, T. L., Stewart, C. L., and Burmeister, M. (2004). To knockout in 129 or in C57BL/6: That is the question. *Trends Genet.* **20,** 59–62.

Smithies, O., Gregg, R. G., Boggs, S. S., Koralewski, M. A., and Kucherlapati, R. S. (1985). Insertion of DNA sequences into the human chromosomal beta-globin locus by homologous recombination. *Nature* **317,** 230–234.

Thomas, K. R., and Capecchi, M. R. (1990). Targeted disruption of the murine int-1 proto-oncogene resulting in severe abnormalities in midbrain and cerebellar development. *Nature* **346,** 847–850.

Wood, S. A., Allen, N. D., Rossant, J., Auerbach, A., and Nagy, A. (1993). Non-injection methods for the production of embryonic stem cell-embryo chimaeras. *Nature* **365,** 87–89.

CHAPTER 16

Human Embryo Culture

Amparo Mercader,[*] **Diana Valbuena,**[†] **and Carlos Simón**[‡,§]

[*]Instituto Universitario—Instituto Valenciano de Infertilidad
Plaza de la Policía local, 3. 46015
Valencia, Spain

[†]Centro de Investigación Príncipe Felipe
Avda. Autopista del Saler, 16-3. 46013
Valencia, Spain

[‡]Fundación IVI. Instituto Universitario—Instituto
Valenciano de Infertilidad Guadassuar
1 bajo. 46015, Valencia, Spain

[§]Centro de Investigación Príncipe Felipe
Avda. Autopista del Saler, 16-3. 46013
Valencia, Spain

Abstract
1. Introduction
2. Human Embryo Development
3. Embryo Biopsy
4. Human Embryo Culture
    4.1. General considerations
    4.2. Protocols
5. Results
References

Abstract

Human embryonic stem cells (hESC) are derived from preimplantation embryos. Approximately 60% of human embryos are blocked during *in vitro* development. Although statistics are inconclusive, experience demonstrates that hESC are more effectively derived from high-quality embryos. In this way, optimal human embryo culture conditions are a crucial aspect in any derivation laboratory. Embryos can be cultured solely with sequential media or cocultured on a monolayer of a given cell type.

This article explores general aspects of human embryonic development, the concept of sequential culture and coculture, and specific protocols and procedures in which the authors are experienced, including the results obtained in embryonic development in the *in vitro* fertilization laboratory during the last 4 years.

1. Introduction

Human embryonic stem cells (hESC) are derived from preimplantation stage embryos, a process which involves as a prerequisite the culture of the embryos to the blastocyst stage (Thomson *et al.*, 1998). hESC have also been isolated from morula stage embryos (Strelchenko *et al.*, 2004) and even from later stage blastocysts (7–8 days) (Stojkovic *et al.*, 2004). Although hESC lines have been derived from embryos of poor quality (Mitalipova *et al.*, 2003), or even from blastomeres (Chung *et al.*, 2006, 2008; Kliminskaya *et al.*, 2006) it is clear that hESC derivation is more efficient from high-quality embryos (Oh *et al.*, 2005; Simon *et al.*, 2005).

In order to optimize embryo development *in vitro,* it is essential to adopt a global approach to the embryo culture system that takes into account the media, gas phase, type of medium overlay, culture vessel, incubation chamber, ambient air quality, and the embryologists themselves. The concept of an embryo culture system highlights the interactions that exist not only between the embryo and its physical surroundings, but also with all the parameters present in a laboratory. Only by taking such a holistic approach, we can optimize embryo development *in vitro* as the previous step for optimal hESC derivation as we routinely do in *in vitro* fertilization (IVF) laboratories.

Initially, the zygote cleaves into two daughter cells, which subsequently divide to form the morula 4 days later. The transcription of the embryonic genome first occurs between the four- and eight-cell stages (Braude *et al.*, 1988), which constitutes a critical moment. Then individual blastomeres compact, and the embryo finally reaches the blastocyst stage. Approximately 50%–60% of embryos arrest during *in vitro* development.

Today, human embryology laboratories are faced not only with a multitude of embryo culture media from which to choose, but with various possibilities of how to use defined media or coculture systems.

2. Human Embryo Development

In the laboratory, embryo development from oocyte retrieval to the blastocyst stage occurs as follows:

Day 0: The human oocyte is retrieved from the follicle.

Day 1: Fertilization day. Polar bodies and pronuclei are visualized. Only fertilized eggs with two polar bodies and two pronuclei are considered to be correctly fertilized.

Day 2: First cleavage. The embryos generally have 2–4 cells. Embryos are evaluated for number of blastomeres (n), rate (%) and type (n) of fragmentation, symmetry (n), compaction (n) and multinucleation ($n \times n$), and are classified accordingly (Fig. 16.1A). Example: an embryo with 4 cells, 10% of fragmentation equally distributed throughout, with blastomeres of a similar size, without compaction and with one cell with two nuclei is classified as 4, 10, III, 2, 0, 1 × 2.

Day 3: The embryos have 6–8 cells and are evaluated as indicated earlier (Fig. 16.1B).

Day 4: Subsequent divisions form a 16–32-cell embryo: the morula stage. Individual blastomeres become indistinguishable as they come into close contact to each other. This phenomenon is named compaction. On day 4, the type of morula is classified as morula (uncompacted cells) or compacted morula (Fig. 16.1C).

Day 5: Spaces appear between the compacting cells, resulting in the formation of an external layer of cells, known as the trophoblast by trophectoderm (TE), and a group of centrally located cells, known as the inner cell mass (ICM). At this stage of development, the embryo is called a blastocyst.

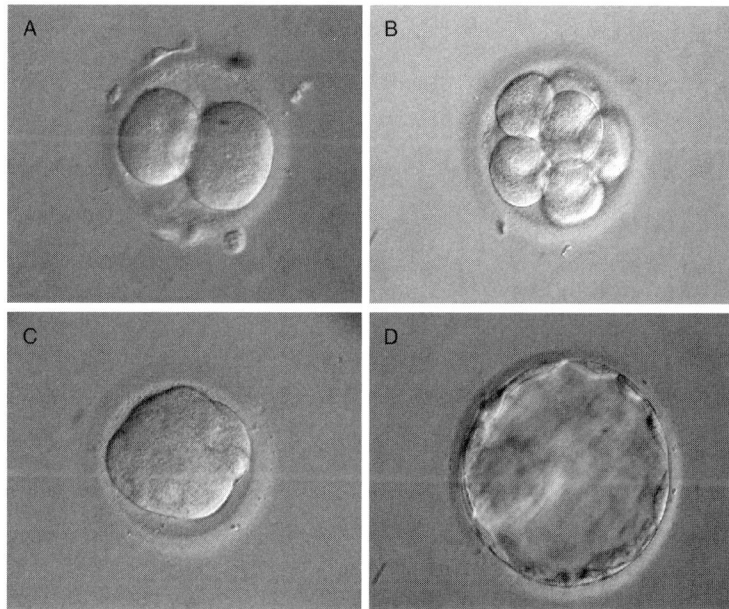

Figure 16.1 Embryo development. (A) Two-cell embryo. (B) Eight-cell embryo. (C) Compacted morula. (D) Expanded blastocyst.

Day 6: The blastocoelic cavity enlarges and causes the embryo to grow and begin to hatch out from the zona pellucida (ZP). Blastocyst expansion thies the ZP due to a series of expansions and contractions (Fig. 16.1D).

Blastocysts are classified morphologically as follows:

- Early blastocyst: when spaces appear between the compaction (Fig. 16.2A).
- Cavitated blastocyst: when the blastocoelic cavity is more than 50% of the total volume (Fig. 16.2B).
- Expanded blastocyst: when the blastocoelic cavity enlarges in size, the embryo and a monolayer, also known as the trophoectoderm (TE), and an ICM can be differentiated (Fig. 16.2C).

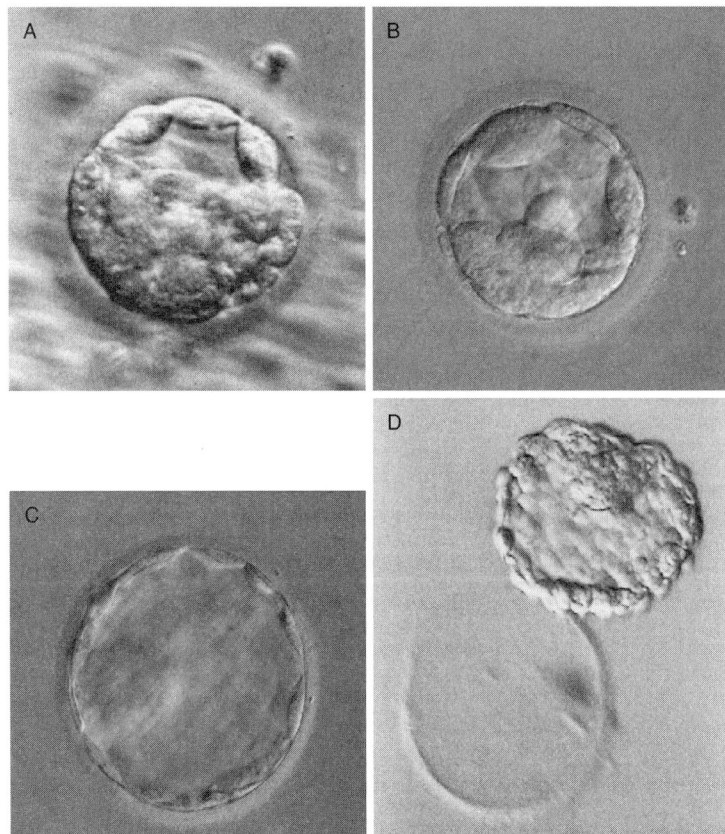

Figure 16.2   Types of blastocyst. (A) Early. (B) Cavitated. (C) Expanded. (D) Hatched.

- Hatching blastocyst: the embryo begins to hatch out of the ZP.
- Hatched blastocyst: the embryo is outside the ZP (Fig. 16.2D).

The different parts of the blastocysts—ICM and TE—can also be classified morphologically.

Inner cell mass: There are four types of ICM:

A. Dense and compact with many cells (Fig. 16.3A).
B. Several cells and not compact (Fig. 16.3B).
C. Very few cells (Fig. 16.3C).
D. Absence of a true ICM (pseudoblastocyst) (Fig. 16.3D).

Trophectoderm: There are four types:

A. Complete, with a monolayer of cells; forming a cohesive epithelium (Fig. 16.3E).
B. Incomplete, with a lineal zone (Fig. 16.3F).
C. With few large cells (Fig. 16.3G).
D. With degenerated cells (Fig. 16.3H).

Example: An expanded blastocyst with an ICM of very few cells and a trophectoderm with a lineal zone, is classified as BE (C, B).

Embryo development does not always follow an "ideal" pattern, sometimes becoming delayed or blocked due to low quality or accelerates inappropriately due to chromosomal abnormalities. Furthermore, morphology can vary considerably and is difficult to interpret at the expected developmental stage.

3. Embryo Biopsy

Embryo biopsy is routinely performed for genetic and/or chromosomal analysis of human embryos. Nowadays the field has been open to the derivation of hESC from single blastomeres (Chung *et al.*, 2006, 2008; Kliminskaya *et al.*, 2006) and therefore this technique must be included in the armamentarium of hESC derivation. Blastomere biopsy is performed on day 3 embryos with $\geq$5 nucleated blastomeres and $\leq$25% fragmentation degree. Usually one or two blastomeres are removed maintaining embryo viability.

For the biopsy, embryos are placed on a droplet containing Ca^{++} and Mg^{++} free medium (EB-10, Scandinavian IVF, Göteborg, Sweden/G-PGD, Vitrolife, Göteborg, Sweden) and laser is used for the drilling of the zona pellucida (OCTAX, Herbron, Germany) and blastomere is retrieved using a beveled micropipette that the inner diameter will range from 30–35 μm (being the best for the embryo) to 45–50 μm (being the best for the blastomere; Fig. 16.4).

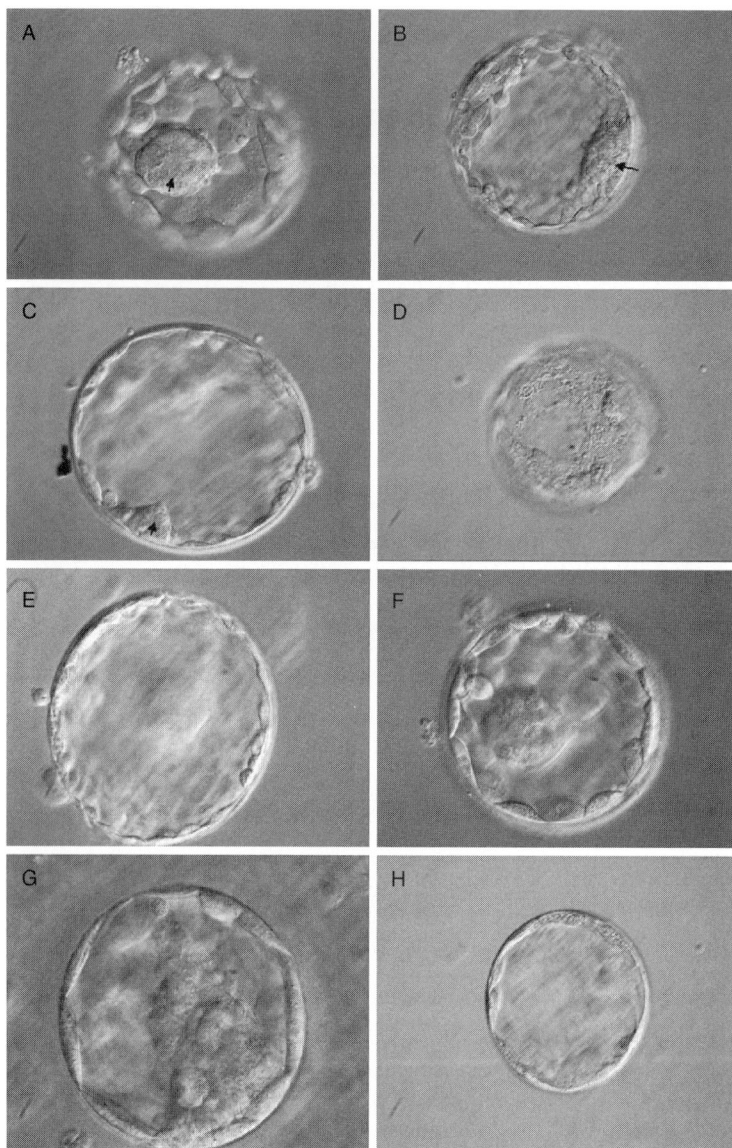

Figure 16.3 Types of inner cell mass (A–D). Types of trophoectoderm (E–H).

After the biopsy, embryos are carefully washed and cultured or cocultured on a monolayer of endometrial epithelial cells. Euploid or genetically normal embryos are transferred/freeze on day 5 at morula or blastocyst stage.

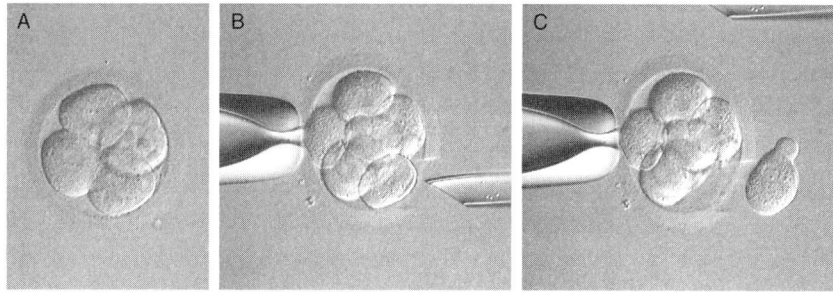

Figure 16.4 Blastomere biopsy technique.

4. Human Embryo Culture

4.1. General considerations

The dramatic changes in embryo physiology, nutrient requirements, and nutrient gradients in the female reproductive tract have led to the formulation of two culture media that are applied at different stages of human embryo development. This is the practice of sequential culture media. On the other hand, the concept of "cells helping cells," extended throughout many areas of cell biology, has prompted embryologists to coculture human embryos in the presence of other types of cells (feeder cells), resulting in the development of the coculture system (Mercader *et al.*, 2003; Simon *et al.*, 1999).

4.1.1. Sequential culture

There are several detailed treatises on the composition of embryo culture media, focusing particularly on four components: glucose, amino acids, EDTA, and macromolecules. Studies in mammals, including humans (Conaghan *et al.*, 1993; Quinn, 1995), have demonstrated the importance of relatively high concentrations of pyruvate and lactate and a relatively low level of glucose in the early stages, while the opposite metabolic conditions have been shown to be required at the blastocyst stage. Amino acids contained in culture media enhance human embryo development up until the blastocyst stage (Devreker *et al.*, 1998). In particular, there is a switch in amino acid requirements during embryo development (Lane and Gardner, 1997). The beneficial effects of divalent cation EDTA in embryo culture media have been extensively reported (Mehta and Kiessling, 1990), though said benefits are confined to the cleavage stage (Gardner and Lane, 1996; Gardner *et al.*, 2000). A commonly used protein source in human IVF and embryo culture has been patient serum, which is added to the culture medium at a concentration of 5%–20%. However, nowadays, recombinant human serum albumin (HSA) is available, eliminating the problems associated with transfusion and permitting the standardization of media formulation (Gardner *et al.*, 2000).

In general, the sequential culture is composed of two different media designed to meet metabolic requirements throughout embryo development. The first of these media is designed to support the development of the zygote to the 8-cell stage, while the second aids development from the 8-cell stage to the blastocyst stage.

4.1.2. Coculture system

Even though the formulations of embryo culture media have improved significantly over the years, and have, for the most part, a more physiologic composition, embryo development *in vitro* still lags behind that *in vivo*. For this reason, the sequential system has been opened up to include the coculture strategy.

The suggested beneficial effects of coculture include the secretion of embryotrophic factors such as nutrients and substrates, growth factors, and cytokines (for review see Bavister, 1995), and the elimination of potentially harmful substances such as heavy metals and ammonium and free radical formation, thereby detoxifying the culture medium.

Multiple cell types have been used for this purpose, including human reproductive tissues such as oviducts (Bongso *et al.*, 1989, 1992; Ouhibi *et al.*, 1989; Walker *et al.*, 1997; Weichselbaum *et al.*, 2002; Yeung *et al.*, 1996), human endometrium (Barmat *et al.*, 1998; Desai *et al.*, 1994; Jayot *et al.*, 1995; Liu *et al.*, 1999; Simón *et al.*, 1999; Spandorfer *et al.*, 2002), oviduct–endometrial sequential coculture (Bongso *et al.*, 1994), cumulus cells (Carrell *et al.*, 1999; Saito *et al.*, 1994; Quinn and Margalit, 1996), granulosa cells (Fabbri *et al.*, 2000; Plachot *et al.*, 1993), to nonhuman cells (Wiemer *et al.*, 1993), nonhuman cell lines (D'Estaing *et al.*, 2001; Hu *et al.*, 1997, 1998; Magli *et al.*, 1995; Menezo *et al.*, 1990, 1992; Sakkas *et al.*, 1994; Schillaci *et al.*, 1994; Turner and Lenton, 1996; Van Blerkom, 1993; Veiga *et al.*, 1999), and even cells from ovarian carcinoma (Ben-Chetrit *et al.*, 1996). As a consequence, the reported effects of this technology on embryonic development are cell, tissue, and species nonspecific. We have developed a coculture system using autologous human endometrial epithelial cells (Simon *et al.*, 1999), and we routinely use this system as a clinical program in our center (Mercader *et al.*, 2003).

4.2. Protocols

4.2.1. Protocol for sequential culture

1. On day 0 (oocyte retrieval day), culture dishes with human tubal fluid (HTF) (IVI, Barcelona, Spain) culture drops of 50 μl, overlaid with oil, are incubated overnight in a 5% CO_2 incubator.

2. On day 1, embryo fertilization is assessed. All correctly fertilized oocytes are rinsed with the HTF in the culture dish before being transferred to the culture drops.

3. On day 2, embryos are assessed to identify whether they have reached the cleavage stage. Culture drops containing 50 μl of COCULTURE MEDIA (CCM) medium (Vitrolife AB, Kungsbacka, Sweden) are placed in a culture dish and overlaid with oil. The culture dish is placed in the incubator overnight.

4. On day 3, embryo cleavage is assessed. Embryos are transferred from the HTF medium into the CCM medium, where they remain until derivation (day 5 or 6).

5. On day 4, the embryos are maintained in CCM. Culture drops containing 50 μl of CCM medium (Vitrolife AB, Kungsbacka, Sweden) are placed in a culture dish and overlaid with oil. The culture dish is placed in the incubator overnight.

6. On days 5 and 6, embryos are assessed according to the morphological classification previously indicated. Derivation is performed when good quality blastocysts are achieved at day 5 or 6.

4.2.2. Protocols for the coculture system with human endometrial epithelial cells (Mercader *et al.*, 2003; Simon *et al.*, 1999)

4.2.2.1. Reagents for endometrial culture

4.2.2.1.1. Collagenase type IA. The digestion is performed with 0.1% collagenase in Dulbecco's modified Eagle's medium (DMEM).

1. Add 10 ml of water for embryo transfer to 100 mg of collagenase.
2. Stock concentration is 10 mg/ml.
3. Digestion volume is 10 ml; therefore add 0.1 ml collagenase stock (10×) to 0.9 ml DMEM (10×) in order to obtain 0.1% collagenase in DMEM.
4. Store in 1 ml aliquots at $-20\,^{\circ}$C.

4.2.2.1.2. Dulbecco's Modified eagle's medium (DMEM). This is a liquid media and no supplements are added.

4.2.2.1.3. MCDB-105. This is a powdered medium. Prepare with water for embryo transfer as described in the later text. Store in the dark at 2–8 $^{\circ}$C.

1. Measure 90% of the final required volume of embryo transfer water.
2. Add the powdered medium. Stir until dissolved, but do not heat. The medium is yellow in color.
3. Rinse original package with a small amount of water to remove all traces of powder. Add to solution.
4. Adjust the pH of the medium. The pH at room temperature must be 5.1 $\pm$ 0.3. Though the final pH should be 7.4, we adjust it to 7.2 because, at 37 $^{\circ}$C, pH increases 0.1–0.3 units. Add 4–5 ml of 1 M NaOH and check pH. Add 1 M NaOH until pH = 7.2.
5. Add additional water to bring the solution to final volume.
6. Sterilize immediately by filtration using a membrane with a porosity of 0.22 μm.

4.2.2.1.4.   Fungizone.   The vial contains 250 μg/ml of Amphotericin B.

1. Rehydrate with 20 ml of embryo transfer water.

The recommended final concentration is between 0.25 and 2.5 μg/ml.

Our working dilution is 0.5 μg/ml; therefore add 2 μl/ml to the medium.

4.2.2.1.5.   Gentamicin.

The vial contains 50 mg/ml.

Our working concentration is 100 μg/ml; therefore add 2 μl/ml to the medium.

4.2.2.1.6.   Insulin.   Insulin promotes the uptake of glucose and amino acids and has a mitogenic effect. It is stable at 2–8 °C for 1 year. Soluble insulin is available in acidified water.

Our working dilution is 5 μg/ml.

1. For a vial of 100 mg, to prepare a 10 mg/ml stock solution, add 10 ml of acidified water (pH$<$ 2.0) (add 100 μl of glacial acetic acid).
2. Add 0.5 μl/ml to the medium to obtain the correct working concentration.

4.2.2.1.7.   Human serum albumin (HSA).   This serum is used to promote cell attachment. Appropriate aliquots (volume recommended is 40 ml) should be prepared using sterile containers. The serum should be stored at -20 °C.

4.2.2.2. Preparation of endometrial epithelial cell medium

We use two basic media (DMEM and MCDB-105), supplemented with 10% HSA and insulin. Additionally, antibiotics and antimycotics are added to control possible contamination.

The medium is composed of 3 DMEM: 1 MCDB-105 supplemented with 10% HSA, gentamicin (100 μg/ml), fungizone (0.5 μg/ml), and insulin (5 μg/ml).

It is sterilized by passing it through a 0.22 μm filter and is then stored in aliquots at 4 °C.

4.2.2.3. Endometrial culture

Endometrial biopsies are obtained in the luteal phase with a catheter (Genetics, Amsterdam, Belgium). Epithelial and stromal endometrial cells are isolated as follows:

1. Mince the biopsies into small pieces of less than 1 mm in length.
2. The minced biopsy pieces are placed in a conical tube with 10 ml of 0.1% collagenase type IA.
3. The biopsy is exposed to mild collagenase digestion through agitation for 1 h in a 37 °C water bath.
4. Stand the tube in a vertical position for 10 min in a horizontal laminar flow.

5. Remove the supernatant (with stromal cells) and wash by resuspending the pellet (glandular and epithelial cells) 3 times, for 5 min each time, in 3–5 ml of DMEM.

6. Finally, resuspend the pellet in 4–5 ml of 1% HSA in DMEM. Recover the mixture into a culture flask (Falcon, Beckton Dickinson, New Jersey, USA). Incubate the flask for 15 min.

7. Recover the supernatant into a fresh flask and add 3 ml of 1% HSA in DMEM. Incubate this second flask for 15 min.

8. Recover all the supernatant and place it in a tube. Check the volume.

9. Prepare 800–700 μl of endometrial epithelial cell medium and add 200–300 μl of recovered supernatant with cells into culture wells.

10. Glandular–epithelial cells are cultured for $\sim$4–6 days, until confluent (monolayer) (Fig. 16.5). The monolayer of endometrial epithelial cells is used for embryo coculture.

4.2.3. Laboratory protocol for coculture on human endometrial epithelial cells

On day 0 (oocyte retrieval day), culture dishes with IVF medium (Vitrolife AB, Kungsbacka, Sweden) are placed in the incubator. Culture drops of 50 μl are placed in a culture dish and overlaid with oil. The medium is incubated overnight in 5% CO_2.

On day 1, embryo fertilization is assessed. All correctly fertilized oocytes are rinsed in the drops in the culture dish before being transferred to the IVF culture drops. IVF and CCM media are left in the incubator (0.5 ml of each one/zygote) overnight.

On day 2, embryos are assessed to detect whether they have reached the cleavage stage. A single embryo is cocultured with 1 ml of IVF:CCM (1:1) on an endometrial epithelial monolayer in individual wells on a 24-multiwell tissue culture plate (Falcon, Becton Dickinson).

On day 3, embryo development is assessed. Embryos are transferred to CCM drops and the culture dish is placed for 10–15 min in the incubator. The embryos are then assessed. The IVF:CCM medium is removed and 1 ml of CCM medium is placed in each well.

On day 4, embryo development is not assessed.

On days 5 and 6, embryos are morphologically classified.

5. Results

Human embryonic development using the two previously described culture systems was compared in a total of 22,368 embryos from 3911 patients attending the Instituto Valenciano de Infertilidad between January 2004 and December 2007. Patients were treated for infertility and divided into two groups according

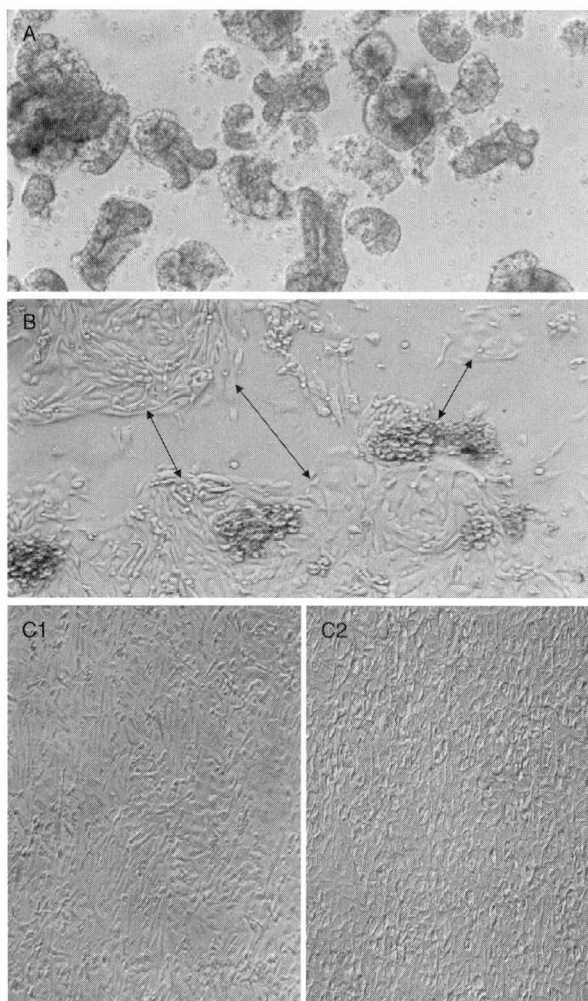

Figure 16.5 Endometrial epithelial cells culture. (A) On day 0. (B) Proliferation. (C) Monolayer. On days 4–6.

to whether they were to use their own oocytes (IVF) or receive ova from donors (ovum donation program). In addition, each group was divided into two subgroups in accordance with the culture system employed: coculture versus sequential culture. We have analyzed, at different stages of development, the blastocyst rates achieved with sequential culture versus our coculture system with endometrial epithelial cells. A Chi-square probability test was used to analyze the blastocyst rates of each group. P values <0.05 were considered to be statistically significant. Since the relevance of biopsied embryos is increasing, we have also analyzed the results from the PGD program according to same groups described earlier.

In IVF patients, statistical differences were observed in the rates of early (<0.05) and expanded blastocysts ($p < 0.01$) obtained and in the total blastocyst rates in coculture versus sequential media (56.5% vs. 47.2% ($p < 0.0001$), respectively) (Table 16.1). In ovum donation patients (Table 16.2), a statistical increase was demonstrated in early blastocysts rates ($p < 0.05$) and total blastocyst rate 66.5% (coculture) versus 51.7% (sequential) ($p < 0.0001$).

In IVF embryos from the PGD program, statistical differences were observed in the rates of all types of blastocyst obtained (except in the expanded blastocysts) and in the total blastocyst rates in coculture versus sequential media (56.0% vs. 45.9% ($p < 0.0001$), respectively) (Table 16.3). In ovum donation embryos from the

Table 16.1

Comparative embryo development in patients undergoing IVF-cycles

Type of Blastocyst	CC system	SC	P
Early (%)	104 (8.6)	251 (9.0)	<0.05
Cavitated (%)	148 (12.2)	323 (11.6)	ns
Expanded (%)	301 (24.8)	476 (17.1)	<0.01
Hatching (%)	124 (10.2)	245 (8.8)	ns
Hatched (%)	9 (0.7)	19 (0.7)	ns
TOTAL (%)	686 (56.5)	1314 (47.2)	<0.0001

Table 16.2

Comparative embryo development in patients undergoing oocyte donation-cycles

Type of Blastocyst	CC system	SC	P
Early (%)	136 (8.8)	381 (8.6)	<0.05
Cavitated (%)	239 (15.5)	473 (10.6)	ns
Expanded (%)	430 (27.9)	910 (20.4)	ns
Hatching (%)	195 (12.6)	491 (11.0)	ns
Hatched (%)	26 (1.7)	49 (1.1)	ns
TOTAL (%)	1026 (66.5)	2304 (51.7)	<0.0001

Table 16.3

Comparative embryo development in patients undergoing IVF-cycles from PGD program

Type of Blastocyst	CC system	SC	p
Early (%)	1482 (16.5)	366 (22.0)	<0.0001
Cavitated (%)	823 (9.2)	140 (8.4)	<0.0001
Expanded (%)	211 (2.3)	31 (1.9)	ns
Hatching (%)	2314 (25.8)	213 (12.8)	<0.0001
Hatched (%)	195 (2.2)	13 (0.8)	<0.01
TOTAL (%)	5025 (56.0)	763 (45.9)	<0.0001

Table 16.4

Comparative embryo development in patients undergoing oocyte donation-cycles from PGD program

Type of Blastocyst	CC system	SC	p
Early (%)	230 (15.4)	61 (24.4)	<0.0001
Cavitated (%)	156 (10.5)	32 (12.8)	<0.05
Expanded (%)	35 (2.3)	1 (0.4)	ns
Hatching (%)	573 (38.4)	47 (18.8)	<0.0001
Hatched (%)	58 (3.9)	0 (0.0)	<0.01
TOTAL (%)	1052 (70.5)	141 (56.4)	<0.0001

PGD program (Table 16.4), similar results were obtained. All blastocyst rates obtained (except expanded blastocyst) were statistically significant depending on the culture system used and the total blastocyst rate was statistically higher in coculture system (70.5%) versus sequential (56.4%) *(p* < 0.0001).

In conclusion, the experience acquired in the IVF laboratory over the last decades allows us to optimize the culture system for human embryos and, therefore, improve both implantation and derivation success rates. Many embryos cultured in these conditions can be surplus from the IVF laboratories and be used for hESC derivation after being thawed following a specific thawing protocol, after which the indicated culture systems can be applied.

References

Barmat, L. I., Liu, H. C., Spandorfer, S. D., Xu, K., Veeck, L., Damario, M. A., and Rosenwaks, Z. (1998). Human preembryo development on autologous endometrial coculture versus conventional medium. *Fertil. Steril.* **70,** 1109–1113.

Bavister, B. D. (1995). Culture of preimplantation embryos: Facts and artefacts. *Hum. Reprod. Update* **1,** 91–148.

Ben-Chetrit, A., Jurisicova, A., and Casper, R. F. (1996). Coculture with ovarian cancer cell enhances human blastocyst formation *in vitro. Fertil. Steril.* **65,** 664–666.

Bongso, A., Fong, C. Y., Ng, S. C., and Ratnam, S. (1994). Human embryonic behaviour in a sequential human oviduct-endometrial coculture system. *Fertil. Steril.* **61,** 976–978.

Bongso, A., Ng, S. C., Fong, C. Y., Anandakumar, C., Marshall, B., Edirisinghe, R., and Ratnam, S. (1992). Improved pregnancy rate after transfer of embryos grown in human fallopian tubal cell coculture. *Fertil. Steril.* **58,** 569–574.

Bongso, A., Soon-Chye, N., Sathananthan, H., Lian, N. P., Rauff, M., and Ratnam, S. (1989). Improved quality of human embryos when co-cultured with human ampullary cells. *Hum. Reprod.* **4,** 706–713.

Braude, P., Bolton, V., and Moore, S. (1988). Human gene expression first occurs between the four- and eight-cell stages of preimplantation development. *Nature* **332,** 459–461.

Carrell, D. T., Peterson, C. M., Jones, K. P., Hatasaka, H. H., Udoff, L. C., Cornwell, C. E., Thorp, C., Kuneck, P., Erickson, L., and Campbell, B. (1999). A simplified coculture system using homologous, attached cumulus tissue results in improved human embryo morphology and pregnancy rates during *in vitro* fertilization. *J. Assist. Reprod. Genet.* **16,** 344–349.

Chung, Y., Klimanskaya, I., Becker, S., Marh, J., Lu, S. J., Johnson, J., Meisner, L., and Lanza, R. (2006). Embryonic and extraembryonic stem cell lines derived from single mouse blastomeres. *Nature* **439**, 216–219.

Chung, Y., Klimanskaya, I., Becker, S., Li, T., Maserati, M., Lu, S.-J., Zdravkovic, T., Ilic, D., Genbacev, O., Fisher, S., Krtolica, A., and Lanza, R. (2008). Human embryonic stem cell lines generated without embryo destruction. *Cell Stem Cell* doi:10.1016/j.stem.2007.12.013.

Conaghan, J., Handyside, A. H., Winston, R. M., and Leese, H. J. (1993). Effects of pyruvate and glucose on the development of human preimplantation embryos *in vitro*. *J. Reprod. Fertil.* **99**, 87–95.

Desai, N. N., Kennard, E. A., Kniss, D. A., and Friedman, C. I. (1994). Novel human endometrial cell line promotes blastocyst development. *Fertil. Steril.* **61**, 760–766.

D'Estaing, S. G., Lornage, J., Hadj, S., Boulieu, D., Salle, B., and Guerin, J. F. (2001). Comparison of two blastocyst culture systems: Coculture on Vero cells and sequential media. *Fertil. Steril.* **76**, 1032–1035.

Devreker, F., Winston, R. M., and Hardy, K. (1998). Glutamine improves human preimplantation development *in vitro*. *Fertil. Steril.* **69**, 293–299.

Fabbri, R., Porcu, E., Marsella, T., Primavera, M. R., Cecconi, S., Nottola, S. A., Motta, P. M., Venturoli, S., and Flamigni, C. (2000). Human embryo development and pregnancies in a homologous granulosa cell coculture system. *J. Assist. Reprod. Genet.* **17**, 1–12.

Gardner, D. K., and Lane, M. (1996). Alleviation of the '2-cell block' and development to the blastocyst of CF1mouse embryos: Role of amino acids, EDTA and physical parameters. *Hum. Reprod.* **11**, 2703–2712.

Gardner, D. K., Lane, M. W., and Lane, M. (2000). EDTA stimulates cleavage stage bovine embryo development in culture but inhibits blastocyst development and differentiation. *Mol. Reprod. Dev.* **57**, 256–261.

Hu, Y., Maxson, W., Hoffman, D. J., Eager, S., and Dupre, J. (1997). Coculture of human embryos with buffalo rat liver cells for women with decreased prognosis in *in vitro* fertilization. *Am. J. Obstet. Gynecol.* **177**, 358–362.

Hu, Y., Maxson, W., Hoffman, D. J., Ory, S., Eager, S., Dupre, J., and Worrilow, K. (1998). Co-culture with assisted hatching of human embryos using buffalo rat liver cells. *Hum. Reprod.* **13**, 165–168.

Jayot, S., Parneix, I., Verdaguer, S., Discamps, G., Audebert, A., and Emperaire, J. C. (1995). Coculture of embryos on homologous endometrial cells in patients with repeated failures of implantation. *Fertil. Steril.* **63**, 109–114.

Klimanskaya, I., Chung, Y., Becker, S., Lu, S. J., and Lanza, R. (2006). Human embryonic stem cell lines derived from single blastomeres. *Nature* **444**, 481–485.

Lane, M., and Gardner, D. K. (1997). Differential regulation of mouse embryo development and viability by amino acids. *J. Reprod. Fertil.* **109**, 153–164.

Liu, H. C., He, Z. Y., Mele, C. A., Veeck, L., Davis, O., and Rosenwaks, Z. (1999). Human endometrial stromal cells improve embryo quality by enhancing the expression of insulin-like growth factors and their receptors in cocultured human preimplantation embryos. *Fertil. Steril.* **71**, 361–367.

Magli, M. C., Gianaroli, L., Ferraretti, A. P., Fortini, D., Fiorentino, A., and D'Errico, A. (1995). Human embryo co-culture: Results of a randomised prospective study. *Int. J. Fertil. Menopausal Stud.* **40**, 254–259.

Mehta, T. S., and Kiessling, A. A. (1990). Development potential of mouse embryos conceived *in vitro* and cultured in ethylenediaminetetraacetic acid with or without amino acids or serum. *Biol. Reprod.* **43**, 600–606.

Menezo, Y. J. R., Guerin, J. F., and Czyba, J. C. (1990). Improvement of human early embryo development *in vitro* by coculture on monolayers of Vero cells. *Biol. Reprod.* **42**, 301–306.

Menezo, Y., Hazout, A., Dumont, M., Herbaut, N., and Nicollet, B. (1992). Coculture of embryos on Vero cells and transfer of blastocysts in humans. *Hum. Reprod.* **7**, 101–106.

Mercader, A., Garcia-Velasco, J. A., Escudero, E., Remohi, J., Pellicer, A., and Simon, C. (2003). Clinical experience and perinatal outcome of blastocyst transfer after coculture of human embryos with human endometrial epithelial cells: A 5-year follow-up study. *Fertil. Steril.* **80**, 1162–1168.

Mitalipova, M., Calhoun, J., Shin, S., Wininger, D., Schulz, T., Noggle, S., Venable, A., Lyons, I., Robins, A., and Stice, S. (2003). Human embryonic stem cell lines derived from discarded embryos. *Stem Cells* **21**, 521–526.

Oh, S. K., Kim, H. S., Ahn, H. J., Seol, H. W., Kim, Y. Y., Park, Y. B., Yoon, C. J., Kim, D. W., Kim, S. H., and Moon, S. Y. (2005). Derivation and characterization of new human embryonic stem cell lines: SNUhES1, SNUhES2, and SNUhES3. *Stem Cells* **23**, 211–219.

Ouhibi, N., Menezo, Y., Benet, G., and Nicollet, B. (1989). Culture of epithelial cells derived from the oviduct of different species. *Hum. Reprod.* **4**, 229–235.

Plachot, M., Antoine, J. M., Alvarez, S., Firmin, C., Pfister, A., Mandelbaum, J., Junca, A. M., and Salat-Baroux, J. (1993). Granulosa cells improve human embryo development *in vitro*. *Hum. Reprod.* **8**, 2133–2140.

Quinn, P. (1995). Enhanced results in mouse and human embryo culture using a modified human tubal fluid medium lacking glucose and phosphate. *J. Assist. Reprod. Genet.* **12**, 97–105.

Quinn, P., and Margalit, R. (1996). Beneficial effects of coculture with cumulus cells on blastocyst formation in a prospective trial with supernumerary human embryos. *J. Assist. Reprod. Genet.* **13**, 9–14.

Saito, H., Hirayama, T., Koike, K., Saito, T., Nohara, M., and Hiroi, M. (1994). Cumulus mass maintains embryo quality. *Fertil. Steril.* **62**, 555–558.

Sakkas, D., Jaquenoud, N., Leppens, G., and Campana, A. (1994). Comparison of results after *in vitro* fertilized human embryos are cultured in routine medium and in coculture on Vero cells: A randomized study. *Fertil. Steril.* **61**, 521–525.

Schillaci, R., Ciriminna, R., and Cefalù, E. (1994). Vero cell effect on in-vitro human blastocyst development: Preliminary results. *Hum. Reprod.* **9**, 1131–1135.

Simon, C., Escobedo, C., Valbuena, D., Genbacev, O., Galan, A., Krtolica, A., Asensi, A., Sanchez, E., Esplugues, J., Fisher, S., and Pellicer, A. (2005). First derivation in Spain of human embryonic stem cell lines: Use of long-term cryopreserved embryos and animal-free conditions. *Fertil. Steril.* **83**, 246–249.

Simon, C., Mercader, A., Garcia-Velasco, J., Nikas, G., Moreno, C., Remohi, J., and Pellicer, A. (1999). Coculture of human embryos with autologous human endometrial epithelial cells in patients with implantation failure. *J. Clin. Endocrinol. Metab.* **84**, 2638–2646.

Spandorfer, S. D., Barmat, L., Navarro, J., Burmeister, L., Veeck, L., Clarke, R., Liu, H. C., and Rosenwaks, Z. (2002). Autologous endometrial coculture in patients with a previous history of poor quality embryos. *J. Assist. Reprod. Genet.* **19**, 309–312.

Stojkovic, M., Lako, M., Stojkovic, P., Stewart, R., Przyborski, S., Armstrong, L., Evans, J., Herbert, M., Hyslop, L., Ahmad, S., Murdoch, A., and Strachan, T. (2004). Derivation of human embryonic stem cells from day-8 blastocysts recovered after three-step *in vitro* culture. *Stem Cells* **22**, 790–797.

Strelchenko, N., Verlinsky, O., Kukharenko, V., and Verlinsky, Y. (2004). Morula-derived human embryonic stem cells. *Reprod. Biomed. Online* **9**, 623–629.

Thomson, J. A., Itskovitz-Eldor, J., Shapiro, S. S., Waknitz, M. A., Swiergiel, J. J., Marshall, V. S., and Jones, J. M. (1998). Embryonic stem cell lines derived from human blastocysts. *Science* **282**, 1145–1147.

Turner, K., and Lenton, E. A. (1996). The influence of Vero cell culture on human embryo development and chorionic gonadotrophin production *in vitro*. *Hum. Reprod.* **11**, 1966–1974.

Van Blerkom, J. (1993). Development of human embryos to the hatched blastocyst stage in the presence or absence of a monolayer of Vero cells. *Hum. Reprod.* **8**, 1525–1539.

Veiga, A., Torrello, M. J., Menezo, Y., Busquets, A., Sarrias, O., Coroleu, B., and Barri, P. N. (1999). Use of co-culture of human embryos on Vero cells to improve clinical implantation rate. *Hum. Reprod.* **14**, 112–120.

Walker, D. J., Vlad, M. T., and Kennedy, C. R. (1997). Establishment of human tubal epithelial cells for coculture in an IVF program. *J. Assist. Reprod. Genet.* **14**, 83–87.

Weichselbaum, A., Paltieli, Y., Philosoph, R., Rivnay, B., Coleman, R., Seibel, M. M., and Bar-Ami, S. (2002). Improved development of very-poor-quality human preembryos by coculture with human fallopian ampullary cells. *J. Assist. Reprod. Genet.* **19,** 7–13.

Wiemer, K. E., Hoffman, D. I., Maxson, W. S., Eager, S., Muhlberger, B., Fiore, I., and Cuervo, M. (1993). Embryonic morphology and rate of implantation of human embryos following co-culture on bovine oviductal ephitelial cells. *Hum. Reprod.* **8,** 97–101.

Yeung, W. S., Lau, E. Y., Chan, S. T., and Ho, P. C. (1996). Coculture with homologous oviductal cells improved the implantation of human embryos—a prospective randomized control trial. *J. Assist. Reprod.Genet.* **13,** 762–767.

CHAPTER 17

Human Embryonic Stem Cells: Derivation and Maintenance

Hidenori Akutsu,★ Chad A. Cowan,† and Douglas Melton‡

★Department of Reproductive Biology
National Research Institute for Child Health and Development,
Setagaya, Tokyo 157-8535, Japan

†MGH Center for Regenerative Medicine
Cardiovascular Research Center, and Harvard Stem Cell Institute,
Richard B. Simches Research Center,
Boston, Massachusetts

‡Department of Molecular and Cellular Biology
Howard Hughes Medical Institute, Harvard University,
Cambridge, Massachusetts

Abstract
1. Introduction
2. Derivation of hES Cell Lines
    2.1. Planning and considerations
    2.2. Preparation of MEFs
    2.3. Dissection and primary culture of MEFs
    2.4. γ-Irradiation and plating
    2.5. Preparing hES derivation media
    2.6. Isolation of inner cell mass
    2.7. Dispersion of inner cell mass
3. Maintenance of hES Cells
    3.1. Enzymatic dissociation with trypsin
    3.2. Materials
    3.3. Trypsinization
    3.4. Freezing hES cells
    3.5. Thawing hES cells
    3.6. Alternative methods
    3.7. Dissociation with collagenase or dispase
    3.8. Culture with human feeder cells
4. Conclusion
    References

359

Abstract

Human embryonic stem (hES) cells hold great promise in furthering our treatment of disease and increasing our understanding of early development. Here, we describe our protocols for the derivation and maintenance of hES cells. In addition, we briefly summarize several alternative methods for the culture of hES cells. Thus, this chapter provides a good start point for researchers interested in harnessing the potential of hES cells.

1. Introduction

In 1981, two groups succeeded in cultivating pluripotent cell lines from mouse blastocysts (Evance and Kaufman, 1981; Martin, 1981). These cell lines, termed embryonic stem (ES) cells, originate from the inner cell mass (ICM) or epiblast and could be maintained *in vitro* without an apparent loss of developmental potential. The ability of these cells to contribute to all cell lineages has been repeatedly demonstrated both *in vitro* and *in vivo* (reviewed by Wobus and Boheler, 2005). Once established, ES cells display an almost unlimited proliferative capacity while retaining their developmental potential (reviewed by Smith, 2001). The first successful derivation of human ES (hES) cell lines was reported in 1998 (Thomson *et al.*, 1998). The establishment of hES cell lines provides a unique new research tool with widespread potential clinical applications.

Under specific *in vitro* culture conditions, hES cells also proliferate indefinitely without senescence and are able to differentiate into almost all tissue specific cell lineages. These properties make hES cells an attractive candidate for cell replacement therapy and open exciting new opportunities to model human embryonic development *in vitro* (reviewed by Keller, 2005). In addition to developmental biology and cell-based therapy, the ES cell model has widespread applications in the areas of drug discovery and drug development (reviewed by Gorba and Allsopp, 2003).

Derivation of hES cell lines has not had a common uniform procedure amongst laboratories. Moreover, the culture and manipulation of hES cells differ considerably between laboratories and poses several unique challenges. To help facilitate research with hES cells we describe in detail the protocols used in our laboratory for the derivation and maintenance of hES cell lines (Cowan *et al.*, 2004; Klimanskaya and McMahon, 2004; http://mcb.harvard.edu/melton/hues). In addition, we briefly discuss alternative approaches to the maintenance of hES cells. Thus, this chapter provides a start point for researchers interested in establishing and working with hES cell lines.

2. Derivation of hES Cell Lines

Since the initial derivation of hES cell lines by Thomson *et al.* (1998), several additional hES cell lines have been established and characterized (Table 17.1 and http://stemcells.nih.gov/registry/index.asp). We previously reported the derivation and maintenance of 17 new hES cell lines that can be maintained in culture by enzymatic dissociation with trypsin (Cowan *et al.*, 2004). Our complete protocol has been previously described in detail (Klimanskaya and McMahon, 2004). The general utility and success of our approach has been validated by the transfer of this technique to several researchers and their subsequent derivation of new hES cell lines (D. Melton and Eggan, unpublished data). Here we present our most concise and current protocol for the derivation of hES cell lines.

2.1. Planning and considerations

In our experience, hES cell derivation can be rather time consuming and demanding. Until the isolated cells are frozen and thawed, they must be continually passaged and maintained. On average, one can expect 3–6 weeks of uninterrupted culture from the point of initiating an attempt to isolate hES cells from blastocyst embryos. Before deriving any new hES cell lines, we recommend that all of the reagents necessary for culture and derivation of the cells be obtained and if possible tested by routine culture of pre-existing hES cell lines. Our standard derivation protocol makes use of mouse embryonic fibroblast cells as a feeder layer, and we likewise recommend the isolation and testing of these cells before attempting to isolate new hES cell lines. Finally, our protocol is designed to derive hES cells from blastocyst stage embryos, and while we have derived several cell lines from embryos frozen at early cleavage stages, they are always first cultured until they mature into blastocysts. In the following sections, we will attempt to walk the reader through a stepwise protocol for deriving hES cell lines and when necessary we will provide specifics details as to the suppliers of certain essential reagents.

2.2. Preparation of MEFs

We use primary mouse embryo fibroblast (MEF) cells, which have been mitotically inactivated by γ-irradiation, for derivation and propagation of hES cells. MEFs are harvested from 12.5-day postcoitum (dpc) fetuses of ICR mice (Cowan *et al.*, 2004). The following reagents are required to follow our protocol for preparing MEFs:

- Sterile phosphate buffered saline (PBS), pH 7.2
- MEF media (90% Dubelco's modified Eagles media, 10% fetal bovine serum, 50 Units/ml penicillin, and 50 μg/ml streptomycin)
- 0.25% Trypsin
- 0.1% Gelatin (made by dissolving 1 g of gelatin in 1000 ml of MilliQ quality water, followed by sterile filtering)
- Freezing media (90% fetal bovine serum, 10% DMSO)
- 10 and 15 cm tissue culture dishes
- Sterile single-edged razor blade

2.3. Dissection and primary culture of MEFs

Prior to dissecting the mouse embryos several 15-cm tissue culture plates (7–8 plates per pregnant ICR female) should be coated with 0.1% gelatin. We typically cover the plates with a minimal amount of the gelatin solution (5–7 ml) and incubate them for 20 min at 37 °C with 5% CO_2. Using a microscope placed in a laminar flow hood, 12.5 dpc embryos are dissected into a 10-cm petri dish containing sterile PBS solution. The embryos are then stripped of any maternal or extraembryonic tissues and eviscerated. Eviscerated embryos are transferred to a 15-cm dish and using a sterile blade minced. 10 ml of warm 0.25% trypsin is added per 10–14 minced embryos and collected in a 50 ml conical tube. The embryos are further homogenized by tituration (pipetting up and down) until no large pieces remain. This partially dissociated mixture is then incubated at 37 °C for 1 min followed by further tituration (pipetting 5–10 more times). 40 ml prewarmed MEF media is added to the dissociated embryos and the mixture is centrifuged for 10 min at 500–600g, at room temperature. Aspirate media and then resuspended the pelleted cells with 30 ml prewarmed MEF media. Plating density is 1.5–2 embryos per 15 cm gelatin-coated plate. The final volume of media on each plate should be 20 ml. The primary MEFs are incubated at 37 °C with 5% CO_2 until confluent (typically 5–6 days). MEFs are expanded once after the initial plating (1:3–1:5 split) and then frozen (passage 1). Freeze MEFs in freezing media (90% FBS and 10% DMSO) at a rate of −1 °C per min and stored at −80 °C or in liquid nitrogen.

2.4. γ-Irradiation and plating

Thawed MEFs are only passaged once (passage 2) for expansion purposes prior to γ-irradiation. MEFs are trypsinized and resuspended in a volume of MEF media that will be accommodated by the γ-irradiator. Irradiate the MEFs for 25 min at 247.3 Rads/min for total exposure of 6182.5 Rads. After irradiation, spin cells in MEF medium for 5 min at 500–600g. To ensure a confluent monolayer, plate MEFs at a concentration of ~50,000 cells/cm^2. If there is no immediate need for mitotically inactivated MEFs, they can be frozen at a concentration of 4×10^6 to 1.2×10^7 cells/vial. MEFs feeder layers should be prepared and used within 3 days.

2.5. Preparing hES derivation media

During the isolation and early stages of ES cell cultivation hES derivation medium is used, which consisted of 75% knockout Dulbecco's modified Eagle medium (Knockout DMEM; INVITROGEN GIBCO, CAT#10829), 10% KO-serum replacement (INVITROGEN GIBCO CAT#10828-018), 10% Plasmanate (BAYER, CAT#0026-0613-20), 5% fetal bovine serum (HYCLONE CAT# SH30070.03), 2 mM Glutamax-I (INVITROGEN GIBCO, CAT#35050-061),

1% nonessential amino acids (INVITROGEN GIBCO, CAT#11140050), 50 Units/ml penicillin and 50 μg/ml streptomycin (INVITROGEN GIBCO, CAT#15070-063), 0.055 mM beta-mercaptoethanol (INVITROGEN GIBCO, CAT#21985-023), 12 ng/ml recombinant hLIF (CHEMICON INTERNATIONAL, CAT#LIF1010), and 5 ng/ml bFGF(INVITROGEN GIBCO, CAT#13256-029). The medium was sterilized by 0.22 μm filtration. Screening of FBS, Plasmanate, and serum replacement should be done and is described elsewhere (Klimanskaya and McMahon, 2004).

2.6. Isolation of inner cell mass

Fresh or frozen–thawed human embryos are cultured to the blastocyts stage in sequential media, G1.2 and G2.2 (Gardner, 1998). Recently we have derived several new hES cell lines at relatively higher efficiency from blastocysts cultured in modified KSOM media. Blastocysts are treated with acid tyrodes (Specialty Media, Cat#MR004-D) for 30–90 s to dissolve the zona pellucida. When the zona pellucida starts to dissolve, remove the embryo and wash it three times in fresh hES derivation medium. The zona-stripped embryos are then cultured in hES derivation medium at 37 °C with 5% CO_2 until immunosurgical isolation of the ICM. The process of immunosurgery includes several stages and is performed essentially as described by Solter and Knowles (1975). Initially, the embryo is incubated for approximately 30 min in rabbit antihuman RBC antibodies (resuspended as per manufacturers instructions, aliquoted and stored at −80 °C, freshly diluted 1:10 in derivation media, INTER CELL TECHNOLOGIES, CAT#0183). Penetration of the antibodies into the blastocyst is prevented because of cell–cell connections within the outer layer of the trophoblasts, leaving the ICM intact. After rinsing off any antibody residue (at least three washes with hES derivation media), the blastocyst is transferred into a guinea pig sera complement (resuspended as per manufacturers instructions, aliquoted and stored at −80 °C, freshly diluted 1:10 in derivation media, SIGMA, CAT#S-1639) diluted in hES derivation medium and incubated until cell lysis is notable, indicated by an apparent "bubbling" of the trophoblast cells (Fig. 17.1). Following the selective removal of the trophectoderm cells by gentle mouth-pipetting of the embryo in and out of a glass capillary, the intact ICM is cultured on MEF feeders plated on gelatin (SIGMA, CAT#G1890)-coated tissue culture plates at a density of ~50,000 cells/cm^2. After 2 days add a few fresh drops of hES derivation medium and then every other day change one-half the total media (e.g., for 500 μl total media then remove 250 μl of media and add 250 μl of fresh media to a final volume of 500 μl).

2.7. Dispersion of inner cell mass

Six to 10 days after the initial plating, ICM outgrowths require mechanical dissociation. Two to three pieces are cut from the initial outgrowth using a narrow glass capillary and left in the same well or moved to a new well (Fig. 17.2). When

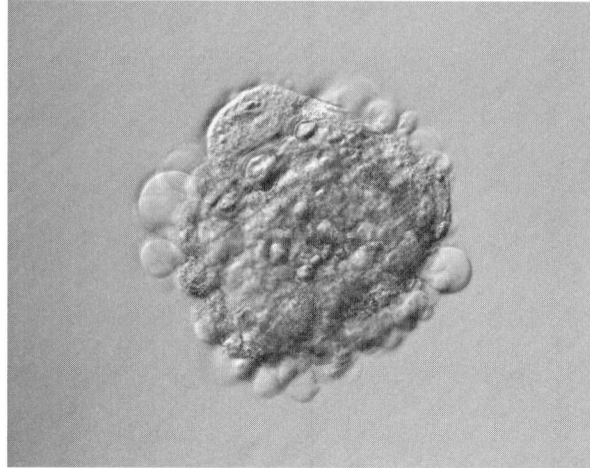

Figure 17.1 "Bubbling" of the trophoblast cells. Blastocyst after exposure to guinea pig sera complement is lysed and stop the incubation followed by removing the lyses trophoblast cells.

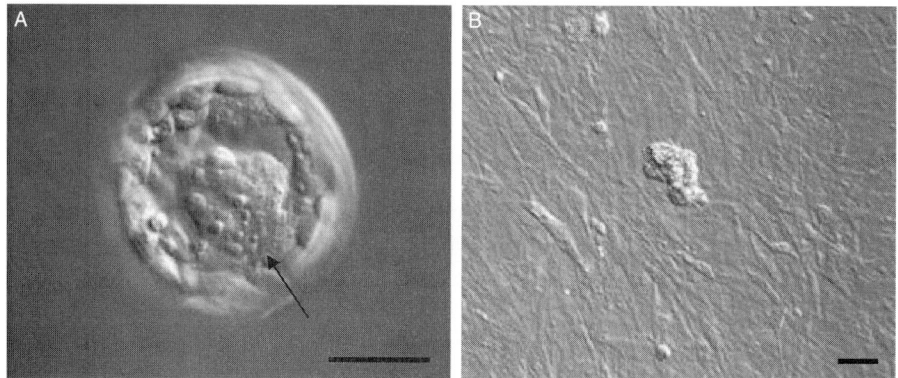

Figure 17.2 Blastocyst and ICM outgrowth. (A) Cultured blastocyst is grade 4AA. Arrow indicates ICM. (B) Isolated ICM from the blastocyst (A) is just grown ICM at day 4 after plating on mitotically inactivated MEFs. Black bar = 50 μm.

doing the initial dispersion, a part of the original colony should be left untouched as a backup, especially if the picked pieces are transferred into a new well. At this stage, it is better to concentrate on expanding the number of hES cell colonies versus freezing or proceeding to any downstream experiments. When the colonies are growing steadily, FBS is omitted from the culture media. Usually, mechanical passaging needs to be done every 5–6 days, but several larger colonies may need to be dispersed daily.

3. Maintenance of hES Cells

Variability among hES cell lines has been reported by several groups, including differences in growth characteristics, differentiation potential, karyotype, and gene expression pattern. In part, these differences might reflect the genetic heterogeneity of hES cell lines derived, as they are, from a genetically diverse, outbred population (Abeyta *et al.*, 2004; Bhattacharya *et al.*, 2004). Further confounding researchers is the fact that hES cell cultures are often heterogeneous because they contain both the undifferentiated stem cells and spontaneously arising differentiated derivatives. While there exists no single uniform protocol for the maintenance of hES cells in culture that adequately addresses all researchers concerns, we will attempt to present an overview of the techniques currently used by a number of laboratories around the world. Again, we describe in detail our method for maintaining undifferentiated hES cell growth in culture and briefly review several alternative protocols.

3.1. Enzymatic dissociation with trypsin

Human ES cell colonies are passaged by mechanical dissociation until there are sufficient colonies (50–100 average-sized colonies) or cells (usually 1×10^5 cells) to passage enzymatically. Thereafter, hES cells are propagated by enzymatic dissociation with 0.05% trypsin/EDTA (INVITROGEN GIBCO, CAT#25300-054). During the first three passages with trypsin, it is a good idea to keep a backup well of mechanically passaged cells. A mechanical backup should always be maintained until the cells are frozen. Subconfluent cultures are generally split at a 1:3 ratio (i.e., one culture well is split to three new culture wells). It is important to split colonies prior to excessive differentiation.

3.2. Materials

For the routine culture of hES cell by enzymatic dissociation with trypsin we recommend the following media and reagents:

- hES media (80% knockout DMEM, 10% KO-serum replacement, 10% Plasmanate, 2 mM Glutamax-I, 1% nonessential amino acids, 50 Units/ml penicillin and 50 μg/ml streptomycin, 0.055 mM beta-mercaptoethanol, and 5 ng/ml bFGF
- Trypsin 0.05%
- Sterile PBS, pH 7.2

3.3. Trypsinization

1. Warm hES media and trypsin in a 37 °C water bath, and keep them warm until ready for use.

2. Place MEF plate from incubator in the hood and aspirate off the medium from the well followed by 1 ml prewarmed hES media. Set the plate aside in the hood.

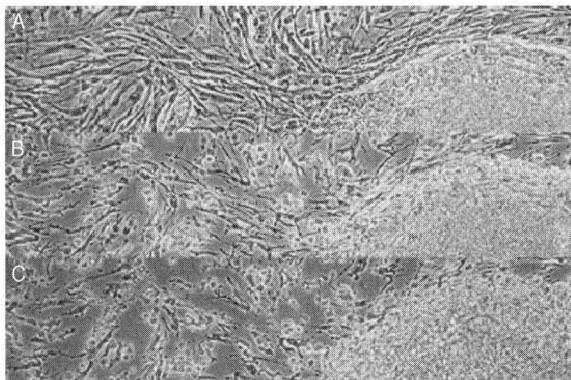

Figure 17.3 Time-lapse series of photographs showing dissociation of hES cells and MEF feeder layer with trypsin. (A) Prior to addition of trypsin. (B) Approximately 30 s after addition of trypsin. (C) Approximately 60 s after the addition of trypsin. Trypsinization should be stopped when cells appear as in (C).

3. Carefully aspirate the hES media from the culture to be split. Gently rinse the cells with a sufficient volume of PBS to completely cover the bottom of the culture dish (e.g., 5 ml for a 10-cm dish).

4. Aspirate the PBS and add a small volume of trypsin (usually 0.3 ml for a 35-mm well or 2 ml for a 10-cm dish) to the cells. Incubate in the hood at room temperature, frequently checking the cells under the microscope. The MEFs surrounding the colonies should begin to retract (Fig. 17.3). When the MEFs are sufficiently shrunk and the borders of the colonies are roughly rounded up, add 10 volumes of prewarmed hES media to the trypsinized colonies. Gently pipette up and down 5–7 times until the MEF monolayer has completely detached. Extensive pipetting should be avoided.

5. Aliquot the hES cell solution dropwise, making sure to distribute the drops evenly about the well. Without shaking the plate, carefully return to the cells to a 37 °C incubator overnight to let the colonies seed.

The time in trypsin required for the cells to detach varies depending on the hES cell density, age of MEF monolayer, etc. We recommend checking the appearance of the hES culture under the stereomicroscope and empirically determining the best incubation time for each well.

3.4. Freezing hES cells

1. Trypsinize the cells; see Section 3.3. Centrifuge the cells at 600g in 10 volumes of hES culture media.
2. Resuspended the pellet in cold freezing medium, which consists of 90% FBS and 10% DMSO.

3. Aliquot the cell suspension into prechilled freezing vials and sandwich the vials between two Styrofoam racks taping to prevent them from separating. Transfer to a −80 °C freezer overnight. Cryovials should be placed in liquid nitrogen for long-term storage.

3.5. Thawing hES cells

Ensure that the MEF plate prepared is confluent and in good condition before thawing hES cells. Prewarm hES media to 37 °C. Aliquot 10 ml hES media into a sterile and labeled 15 ml conical tube for each cell line.

All procedures should be done quickly:

1. Thaw the vial in a 37 °C water bath (Do not over-thaw, the vial should be removed from the water bath with a small ice crystal still remaining). It should be take about 45–60 s before the cells are 80% thawed.

2. Bring the tube to laminar flow hood, spray down with 70% isopropanol. Gently transfer the cells to the 10 ml of prewarmed media.

3. Centrifuge the 15 ml conical tube at 500–600g for 5 min.

4. Remove preplated MEFs from incubator to the hood. Aspirate off the MEF media and aliquot prewarmed hES media into each well of the plate, being care not to disturb the attached MEFs.

5. After the spin is complete, carefully remove the media without disturbing the pellet.

6. Gently resuspend the pellet in a small volume of prewarmed hES media.

7. Transfer the hES cell solution, in a dropwise manner, to a prepared MEF plate well which already contains hES media. Carefully return the plate to avoid swirling to a 37 °C incubator overnight to allow the hES cells to seed the MEFs.

8. The colonies usually begin to appear in 3–4 days and can be ready for splitting in 5–10 days.

3.6. Alternative methods

There exist several alternative methods for the culture of hES cells, but few have been rigorously examined over a long period of time. We will attempt to summarize some of the more common alternative methods for maintaining hES cells in culture. For detailed protocols, we advise referring to the primary literature. In addition, several alternatives such as feeder and serum free culturing of hES cells are described in subsequent chapters.

3.7. Dissociation with collagenase or dispase

Quite possibly the most widespread method for maintaining hES cells in culture depends upon their dissociation with either collagenase or dispase. For a detailed protocol, please see http://www.geron.com/PDF/scprotocols.pdf. The reported advantages of culture with these enzymes are reduced cell death and perhaps greater karyotypic stability. The disadvantages of enzymatic dissociation with collagenase or dispase include the inability to accurately assess cell number and the failure to generate definitive single cell clones.

3.8. Culture with human feeder cells

Mouse embryonic fibroblast cells have generally been used as feeder layers to support the unlimited growth of hES cells, but the use of animal feeder cells is associated with risks such as pathogen transmission and viral infection (Amit *et al.*, 2003, 2004; Richards *et al.*, 2002; Rosler *et al.*, 2004). Recently, Martin *et al.* (2005) reported that hES cells could incorporate foreign sugars into the glycoproteins on the cell surface. They also showed that an immune reaction could occur following exposure of the cells to serum from adults with high level of the antibody. These reports and other concerns have prompted many researchers to seek alternatives to mouse feeder layers.

Several groups have reported that feeder layers comprised of cells originating from human fetal and adult tissues support unlimited proliferation of hES cells without differentiation. The cell types used include human fetal skin fibroblasts, human muscle cells, adult fallopian tubal epithelial cells (Richards *et al.*, 2002), adult marrow cells (Cheng *et al.*, 2003), foreskin fibroblasts (Amit *et al.*, 2003; Hovatta *et al.*, 2003), human uterine endometrium cells, and breast parenchyma cells abortus fetus fibroblasts (Lee *et al.*, 2004). In perhaps the most comprehensive study, Richards *et al.* (2002) reported the evaluation of various human adult, fetal, and neonatal tissues as feeder cells for supporting the growth of hES cells. In addition, feeder cells derived from hES cells can be used as autogenic feeder system that efficiently supports the growth and maintenance of pluripotency of hES cells (Stojkovic *et al.*, 2005; Yoo *et al.*, 2005).

4. Conclusion

Human ES cells are viewed by many as a novel and unlimited source of cells and tissues for transplantation for the treatment of a broad spectrum of diseases (reviewed by Keller, 2005). Moreover, hES cells represent an unprecedented system suitable for the identification of new molecular targets and the development of novel drugs, which can be tested *in vitro* or used to predict or anticipate potential toxicity in humans. Finally, hES cells can yield insight into the developmental events that occur during human embryogenesis, which are for ethical reasons nearly impossible to study in the intact embryo (reviewed by Dvash and Benvenisty, 2004).

Acknowledgments

We would like to thank Jacob Zucker for providing the images used to illustrate the dissociation of hES cells by trypsin and Stephen Sullivan for suggestions and advice.

References

Abeyta, M. J., Clark, A. T., Rodriguez, R. T., Bodnar, M. S., Pera, R. A., and Firpo, M. T. (2004). Unique gene expression signatures of independently-derived human embryonic stem cell lines. *Hum. Mol. Genet.* **13,** 601–608.

Amit, M., Carpenter, M. K., Inokuma, M. S., Chiu, C. P., Harris, C. P., Waknitz, M. A., Itskovitz-Eldor, J., and Thomson, J. A. (2000). Clonally derived human embryonic stem cell lines maintain pluripotency and proliferative potential for prolonged periods of culture. *Dev. Biol.* **227,** 271–278.

Amit, M., Margulets, V., Segev, H., Shariki, K., Laevsky, I., Coleman, R., and Itskovitz-Eldor, J. (2003). Human feeder layers for human embryonic stem cells. *Biol. Reprod.* **68,** 2150–2156.

Amit, M., Shariki, C., Margulets, V., and Itskovitz-Eldor, J. (2004). Feeder layer- and serum-free culture of human embryonic stem cells. *Biol. Reprod.* **70,** 837–845.

Baharvand, H., Ashtiani, S. K., Valojerdi, M. R., Shahverdi, A., Taee, A., and Sabour, D. (2004). Establishment and *in vitro* differentiation of a new embryonic stem cell line from human blastocyst. *Differentiation* **72,** 224–229.

Bhattacharya, B., Miura, T., Brandenberger, R., Mejido, J., Luo, Y., Yang, A. X., Joshi, B. H., Ginis, I., Thies, R. S., Amit, M., Lyons, I., Condie, B. G., *et al.* (2004). Gene expression in human embryonic stem cell lines: Unique molecular signature. *Blood* **103,** 2956–2964.

Carpenter, M., Rosler, K. E., Fisk, S. G., Brandenberger, J. R., Ares, X., Miura, T., Lucero, M., and Rao, M. S. (2004). Properties of four human embryonic stem cell lines maintained in a feeder-free culture system. *Dev. Dyn.* **229,** 243–258.

Chen, H., Qian, K., Hu, J., Liu, D., Lu, W., Yang, Y., Wang, D., Yan, H., Zhang, S., and Zhu, Q. (2005). The derivation of two additional human embryonic stem cell lines from day 3 embryos with low morphological scores. *Hum. Reprod.* **20,** 2201–2206.

Cheng, L., Hammond, H., Ye, Z., Zhan, X., and Dravid, G. (2003). Human adult marrow cells support prolonged expansion of human embryonic stem cells in culture. *Stem Cells* **21,** 131–142.

Cowan, C. A., Klimanskaya, I., McMahon, J., Atienza, J., Witmyer, J., Zucker, J. P., Wang, S., Morton, C. C., McMahon, A. P., Powers, D., and Melton, D. A. (2004). Derivation of embryonic stem cell lines from human blastocysts. *N. Engl. J. Med.* **350,** 1353–1356.

Davila, J. C., Cezar, G. G., Thiede, M., Strom, S., Miki, T., and Trosko, J. (2004). Use and application of stem cells in toxicology. *Toxicol. Sci.* **79,** 214–223.

Dvash, T., and Benvenisty, N. (2004). Human embryonic stem cells as a model for early human development. *Best Pract. Res. Clin. Obstet. Gynaecol.* **18,** 929–940.

Evance, M. J., and Kaufman, M. H. (1981). Establishment in culture of pluripotential cells from mouse embryos. *Nature* **292,** 154–156.

Genbacev, O., Krtolica, A., Zdravkovic, T., Brunette, E., Powell, S., Nath, A., Caceres, E., McMaster, M., McDonagh, S., Li, Y., Mandalam, R., Lebkowski, J., *et al.* (2005). Serum-free derivation of human embryonic stem cell lines on human placental fibroblast feeders. *Fertil. Steril.* **83,** 1517–1529.

Gorba, T., and Allsopp, T. E. (2003). Pharmacological potential of embryonic stem cells. *Pharmacol. Res.* **47,** 269–278.

Heins, N., Englund, M. C. O., Sjöblom, C., Dahl, U., Tonning, A., Bergh, C., Lindahl, A., Hanson, C., and Semb, H. (2004). Derivation, characterization, and differentiation of human embryonic stem cells. *Stem Cells* **22,** 367–376.

Hochedlinger, K., and Jaenisch, R. (2003). Nuclear transplantation, embryonic stem cells, and the potential for cell therapy. *N. Engl. J. Med.* **349,** 275–286.

Hoffman, L. M., and Carpenter, M. K. (2005). Characterization and culture of human embryonic stem cells. *Nat. Biotechnol.* **23,** 699–708.

Hong-mei, P., and Gui-an, C. (2006). Serum-free medium cultivation to improve efficacy in establishment of human embryonic stem cell lines. *Hum. Reprod.* **21,** 217–222.

Hovatta, O., Mikkola, M., Gertow, K., Strömberg, A., Inzunza, J., Hreinsson, J., Rozell, B., Andäng, M., and Ährlund-Richter, L. (2003). A culture system using human foreskin fibroblasts as feeder cells allows production of human embryonic stem cells. *Hum. Reprod.* **18,** 1404–1409.

Inzunza, J., Gertow, K., Strömberg, A., Matilainen, E., Blennow, E., Skottman, H., Wolbank, S., Ährlund-Richter, L., and Hovatta, O. (2005). Derivation of human embryonic stem cell lines in serum replacement medium using postnatal human fibroblasts as feeder cells. *Stem Cells* **23,** 544–549.

Keller, G. (2005). Embryonic stem cell differentiation: Emergence of a new era in biology and medicine. *Genes Dev.* **19,** 1129–1155.

Kim, S. J., Lee, J. E., Park, J. H., Lee, J. B., Kim, J. M., Yoon, B. S., Song, J. M., Roh, S. I., Kim, C. G., and Yoon, H. S. (2005). Efficient derivation of new human embryonic stem cell lines. *Mol. Cells* **19,** 46–53.

Klimanskaya, I., and McMahon, J. (2004). Approaches for derivation and maintenance of human ES cells: Detailed procedures and alternatives. *In* "Handbook of Stem Cells. Vol. 1: Embryonic Stem Cells" (R. Lanza, J. Gearhart, and B. Hogan *et al.*, Eds.), pp. 437–449. Elsevier/Academic Press, San Diego.

Klimanskaya, I., Chung, Y., Meisner, L., Johnson, J., West, M. D., and Lanza, R. (2005). Human embryonic stem cell derived without feeder cells. *Lancet* **365,** 1636–1641.

Lee, J. B., Song, J. M., Lee, J. E., Park, J. H., Kim, S. J., Kang, S. M., Kwon, J. N., Kim, M. K., Roh, S. I., and Yoon, H. S. (2004). Available human feeder cells for the maintenance of human embryonic stem cells. *Reproduction* **128,** 727–735.

Lee, J. B., Lee, J. E., Park, J. H., Kim, S. J., Kim, M. K., Roh, S. I., and Yoon, H. S. (2005). Establishment and maintenance of human embryonic stem cell lines on human feeder cells derived from uterine endometrium under serum-free condition. *Biol. Reprod.* **72,** 42–49.

Ludwig, T. E., Levenstein, M. E., Jones, J. M., Berggren, W. T., Mitchen, E. R., Frane, J. L., Grandall, L. J., Daigh, C. A., Conard, K. R., Piekarczyk, M. S., Llanas, R. A., and Thomson, J. A. (2006). Derivation of human embryonic stem cells in defined conditions. *Nat. Biotechnol.* **24**(2), 185–187.

Martin, G. R. (1981). Isolation of a pluripotent cell line from early mouse embryos cultured in medium conditioned by teratocarcinoma stem cells. *Proc. Natl Acad. Sci. USA* **78,** 7634–7638.

Martin, M. J., Muotri, A., Gage, F., and Varki, A. (2005). Human embryonic stem cells express an immunogenic nonhuman sialic acid. *Nat. Med.* **11,** 228–232.

Mateizel, I., De Temmerman, N., Ullmann, U., Cauffman, G., Sermon, K., Van de Velde, H., De Rycke, M., Degreef, E., Devroey, P., Liebaers, I., and Van Steirteghem, A. (2006). Derivation of human embryonic stem cell lines from embryos obtained after IVF and after PGD for monogenic disorders. *Hum. Reprod.* **21**(2), 503–511.

McNeish, J. (2004). Embryonic stem cells in drug discovery. *Nat. Rev. Drug Discov.* **3,** 70–80.

Mitalipova, M., Calhoun, J., Shin, S., Wininger, D., Schulz, T., Noggle, S., Venable, A., Lyons, I., Robins, A., and Stice, S. (2003). Human embryonic stem cell lines derived from discarded embryos. *Stem Cells* **21,** 521–526.

Oh, S. K., Kim, H. S., Ahn, H. J., Seol, H. W., Kim, Y. Y., Park, Y. B., Yoon, C. J., Kim, D. W., Kim, S. H., and Moon, S. Y. (2005). Derivation and characterization of new human embryonic stem cell lines: SNUhES1, SNUhES2, and SNUhES3. *Stem Cells* **23,** 211–219.

Park, J. H., Kim, S. J., Oh, E. J., Moon, S. Y., Roh, S. I., Kim, C. G., and Yoon, H. S. (2003). Establishment and maintenance of human embryonic stem cells on STO, a permanently growing cell line. *Biol. Reprod.* **69,** 2007–2014.

Park, S., Lee, Y. J., Lee, K. S., Shin, H. A., Cho, H. Y., Chung, K. S., Kim, E. Y., and Lim, J. H. (2004). Establishment of human embryonic stem cell lines from frozen–thawed blastocysts using STO cell feeder layers. *Hum. Reprod.* **19,** 676–684.

Pera, M., and Trounson, A. (2004). Human embryonic stem cells: Prospects for development. *Development* **131,** 5515–5525.

Reubinoff, B. E., Pera, M. F., Fong, C. Y., Trounson, A., and Bongso, A. (2000). Embryonic stem cell lines from human blastocysts: Somatic differentiation *in vitro*. *Nat. Biotechnol.* **18,** 399–404.

Richards, M., Fong, C. Y., Chan, W. K., Wong, P. C., and Bongso, A. (2002). Human feeders support prolonged undifferentiated growth of human inner cell masses and embryonic stem cells. *Nat. Biotechnol.* **20,** 933–936.

Rosler, E. S., Fisk, G. J., Ares, X., Irving, J., Miura, T., Rao, M. S., and Carpenter, M. K. (2004). Long-term culture of human embryonic stem cells in feeder-free conditions. *Dev. Dyn.* **229,** 259–274.

Seiler, A., Visan, A., Buesen, R., Genschow, E., and Spielmann, E. (2004). Improvement of an *in vitro* stem cell assay for developmental toxicity: The use of molecular endpoints in the embryonic stem cell test. *Reprod. Toxicol.* **18,** 231–240.

Simon, C., Escobedo, C., Valbuena, D., Genbacev, O., Galan, A., Krtolica, A., Asensi, A., Sanchez, E., Esplugues, J., Fisher, S., and Pellicer, A. (2005). First derivation in Spain of human embryonic stem cell lines: Use of long-term cryopreserved embryos and animal-free conditions. *Fertil. Steril.* **83,** 246–249.

Smith, A. G. (2001). Embryo-derived stem cells: Of mice and men. *Annu. Rev. Cell Dev. Biol.* **17,** 435–462.

Solter, D., and Knowles, B. B. (1975). Immunosurgery of mouse blastocyst. *Proc. Natl Acad. Sci. USA* **72,** 5099–5102.

Stojkovic, M., Lako, M., Stojkovic, P., Stewart, R., Przyborski, S., Armstrong, L., Evans, J., Herbert, M., Hyslop, L., Ahmad, S., Murdoch, A., and Strachan, T. (2004). Derivation of human embryonic stem cells from day-8 blastocysts recovered after three-step *in vitro* culture. *Stem Cells* **22,** 790–797.

Stojkovic, P., Lako, M., Stewart, R., Przyborski, S., Armstrong, L., Evans, J., Murdoch, A., Strachan, T., and Stojkovic, M. (2005). An autogeneic feeder cell system that efficiently supports growth of undifferentiated human embryonic stem cell. *Stem Cells* **23,** 306–314.

Sun, B. W., Yang, A. C., Feng, Y., Sun, Y. J., Zhu, Y. F., Zhang, Y., Jiang, H., Li, C. L., Gao, F. R., Zhang, Z. H., Wang, W. C., Kong, X. Y., *et al.* (2006). Temporal and parental-specific expression of imprinted genes in a newly derived Chinese human embryonic stem cell line and embryoid bodies. *Hum. Mol. Genet.* **15,** 65–75.

Suss-Toby, E., Gerecht-Nir, S., Amit, M., Manor, D., and Itskovitz-Eldor, J. (2004). Derivation of a diploid human embryonic stem cell line from a mononuclear zygote. *Hum. Reprod.* **19,** 670–675.

Thomson, J. A., Itskovitz-Eldor, J., Shapiro, S. S., Waknitz, M. A., Swiergiel, J. J., Marshall, V. S., and Jones, J. M. (1998). Embryonic stem cell lines derived from human blastocysts. *Science* **282,** 1145–1147.

Wang, Q., Fang, Z. F., Jin, F., Lu, Y., Gai, H., and Sheng, H. Z. (2005). Derivation and growing human embryonic stem cells on feeders derived from themselves. *Stem Cells* **23,** 1221–1227.

Wobus, A. M., and Boheler, K. R. (2005). Embryonic stem cells: Prospects for development biology and cell therapy. *Physiol. Rev.* **85,** 635–678.

Xu, C., Inokuma, M. S., Denham, J., Golds, K., Kundu, P., Gold, J. D., and Carpenter, M. K. (2001). Feeder-free growth of undifferentiated human embryonic stem cells. *Nat. Biotechnol.* **19,** 971–974.

Yoo, S. J., Yoon, B. S., Kim, J. M., Song, J. M., Roh, S. I., You, S., and Yoon, H. S. (2005). Efficient culture system for human embryonic stem cells using autologous human embryonic stem cell-derived feeder cells. *Exp. Mol. Med.* **37,** 399–407.

CHAPTER 18

Human Embryonic Stem Cells: Characterization and Evaluation

Chunhui Xu

Geron Corporation
230 Constitution Drive
Menlo Park, California

Abstract
1. Introduction
2. Characterization of Undifferentiated hESCs
    2.1. Cell morphology
    2.2. Markers for undifferentiated hESCs
    2.3. Proliferation
    2.4. Karyotype analysis
    2.5. Differentiation capacity
3. Conclusion
    References

Abstract

Human embryonic stem cells (hESCs) provide great opportunities for regenerative medicine, pharmacological, and toxicological investigation, and the study of human embryonic development. These applications require proper derivation, maintenance, and extensive characterization of undifferentiated cells prior to them being used for differentiation into cells of interest. Undifferentiated hESCs possess several unique features, including their extensive proliferation capacity in undifferentiated state, ability to maintain a normal karyotype after long-term culture, expression of markers characteristic of stem cells, high constitutive telomerase activity, and capacity to differentiate into essentially all somatic cell types. This chapter will summarize the current development in culture conditions and provide technical details for the evaluation and characterization of hESCs.

1. Introduction

Human embryonic stem cells (hESCs) were first successfully isolated from inner cell masses (ICMs) of preimplantation embryos or blastocysts (Thomson *et al.*, 1998). ICMs were obtained by removal of trophoblast, the out layer of blastocysts, through immunosurgery, and the cells were plated onto a layer of mitotically inactivated mouse embryonic fibroblast (MEF) feeders to prevent cells from spontaneously differentiating. The cells were further passaged by enzymatic dissociation and replated onto new feeders for expansion. Additional hESC lines have since been isolated by several laboratories using various methods, such as culturing cells on other feeders or in feeder-free conditions, and extracting cells from frozen embryos instead of fresh ones (Hoffman and Carpenter, 2005; Klimanskaya *et al.*, 2005).

Since the original derivation of hESCs, culture conditions for the growth of undifferentiated cells have been significantly improved. An earlier modification of the culture method is the development of a feeder-free culture system, where cells are maintained on Matrigel® or laminin in medium conditioned by MEF (MEF-CM) (Carpenter *et al.*, 2004; Lebkowski *et al.*, 2001; Rosler *et al.*, 2004; Xu *et al.*, 2001). To minimize the risk of pathogen contamination, additional efforts have been devoted to screening human feeder cells, matrices, and medium supplements for the maintenance of undifferentiated cells. This led to the identification of human cells derived from a variety of tissues (Amit *et al.*, 2003; Cheng *et al.*, 2003; Genbacev *et al.*, 2005; Hovatta *et al.*, 2003; Inzunza *et al.*, 2005; Miyamoto *et al.*, 2004; Richards *et al.*, 2002) or hESCs derivatives (Stojkovic *et al.*, 2005b; Wang *et al.*, 2005c; Xu *et al.*, 2004) for their capability to serve as feeders directly or as a source of conditioned medium. Human serum, extracellular matrices extracted from MEFs and fibronectin have been identified as alternative matrices to Matrigel® or laminin (Amit *et al.*, 2004; Klimanskaya *et al.*, 2005; Stojkovic *et al.*, 2005a). In addition, several growth factors such as bFGF (Wang *et al.*, 2005a,b; Xu *et al.*, 2005a,b), TGFβ/activin/nodal signaling pathway (Amit *et al.*, 2004; Beattie *et al.*, 2005; James *et al.*, 2005; Vallier *et al.*, 2004; Xiao *et al.*, 2006), and those involved in the activation of IGF1 (Bendall *et al.*, 2007; Wang *et al.*, 2007) and ERBB2 (Wang *et al.*, 2007) receptors have been identified for their involvement in self-renewal of undifferentiated hESCs. In addition, culture systems using defined media have been developed (Li *et al.*, 2005; Lu *et al.*, 2006; Ludwig *et al.*, 2006; Vallier *et al.*, 2005; Wang *et al.*, 2007; Yao *et al.*, 2006), some of which contain proteins/factors solely derived from recombinant sources or purified from human materials (Li *et al.*, 2005; Ludwig *et al.*, 2006). One such defined medium has been used not only in the routine culture but also the derivation of new hESCs (Ludwig *et al.*, 2006). These improvements have provided solid groundwork for the generation of undifferentiated hESCs in a reproducible fashion with well-defined qualified materials.

2. Characterization of Undifferentiated hESCs

Cardinal features of undifferentiated hESCs include the extensive proliferation capacity of the cells in undifferentiated state (self-renewal) and the ability of the cells to differentiate into multiple cell types (pluripotency) that are useful for regenerative medicine and drug discovery (Fig. 18.1). Characterization of hESCs is an essential process when deriving a new hESC line or developing a new culture procedure for undifferentiated hESCs. This process had been carried out in the experiments described earlier that aimed to optimize culture conditions. For example, it was demonstrated that even after long-term culture, cells maintained on Matrigel® or laminin in MEF-CM retained a normal karyotype and a stable proliferation rate, had high telomerase activity, and expressed markers associated with undifferentiated hESCs (Xu *et al.*, 2001). In addition, hESCs maintained in MEF-CM had been induced to differentiate into cell types representing the three germ layers both *in vitro* and *in vivo*, including neural progenitors (Carpenter *et al.*, 2001), cardiomyocytes (Xu *et al.*, 2002a), trophoblast (Xu *et al.*, 2002b), hepatocyte-like cells (Rambhatla *et al.*, 2003), oligodendrocytes (Nistor *et al.*, 2003), hematopoietic progenitors (Chadwick *et al.*, 2003a), and insulin-producing islet-like clusters (Jiang *et al.*, 2007). Characterization of hESCs cultured under other feeder-free conditions is summarized in Table 18.1.

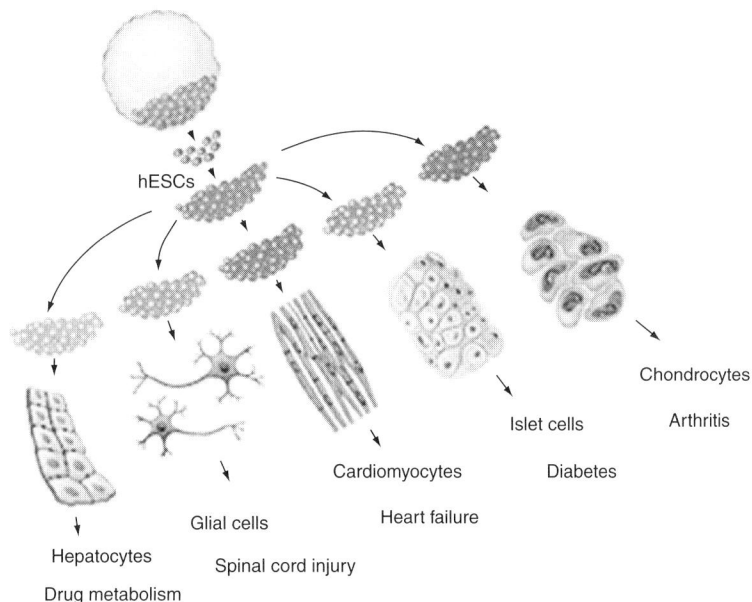

Figure 18.1 Fundamental features of undifferentiated hESCs. hESCs derived from inner cell mass have ability to proliferate in undifferentiated state (self-renewal) and to differentiate into multiple cell types (pluripotency). These derivatives include glial cells, cardiomyocytes, islet cells, chondrocytes, and hepatocytes that are valuable for treatment of diseases and drug discovery.

Table 18.1

Examples of long-term feeder-free cultures of hESCs

Cells	Matrix	Media	Characterization	References
H1, H7, H9, H14	Matrigel® or human laminin	MEF-CM using medium containing 20% SR and 4–8 ng/ml bFGF	>1 year; morphology; expression of SSEA-4, TRA-1-60, TRA-1-81, Oct3/4, hTERT, Cripto, and alkaline phosphatase; gene profiling analysis; telomerase activity; proliferation; karyotype analysis; EB formation; teratoma formation; cryopreservation; cardiomyocyte differentiation; neuron differentiation; hepatocyte-like cell differentiation; trophoblast differentiation; osteoblast differentiation; oligodendrocyte differentiation; hematopoietic cell differentiation; insulin-producing cell differentiation; and genetic modification	Carpenter et al. (2001), Chadwick et al. (2003a), Jiang et al. (2007), Laflamme et al. (2005, 2007), Lebkowski et al. (2001), Ma et al. (2003), Nistor et al. (2003), Rambhatla et al. (2003), Sottile et al. (2003), Xu et al. (2001, 2002a,b, 2006a,b), and Zwaka and Thomson (2003)
H7, H9	Matrigel®	CM from immortalized hESC-derived fibroblast-like cells using medium containing 20% SR and 4–8 ng/ml bFGF	10 passages; morphology; expression of SSEA-4, TRA-1-60, TRA-1-81, Oct3/4, hTERT, Cripto, and alkaline phosphatase; telomerase activity; karyotype analysis; EB formation; and teratoma formation	Xu et al. (2004)
I-6, I-3, H9	Human or bovine fibronectin	Medium containing 15% SR, 4 ng/ml bFGF and 0.12 ng/ml TGFb1	50 passages; morphology; expression of SSEA-4, SSEA-3, TRA-1-60, TRA-1-81, and Oct3/4; cloning efficiency; EB formation; and teratoma formation	Amit et al. (2004)
H7, H9	Matrigel®	Medium containing 20% SR and 40 ng/ml bFGF	15 passages; morphology; expression of SSEA-4, TRA-1-60, Oct3/4, hTERT, and Cripto; telomerase activity; proliferation; karyotype analysis; EB formation; and teratoma formation	Xu et al. (2005a)
H1, H9, H14	Human laminin	Medium containing 20% SR, 40 ng/ml bFGF and 500 ng/ml noggin	27 passages; morphology; expression of SSEA-4, TRA-1-60, Oct3/4, Nanog, and Zep42; proliferation; karyotype analysis; EB formation; teratoma formation; trophoblast differentiation; and cryopreservation	Xu et al. (2005b)

Cell lines	Substrate	Medium	Characterization	Reference
H1, H9	Matrigel®	Medium containing 20% SR and 24–36 ng/ml bFGF	>30 passages; morphology; expression of TRA-1-60, TRA-1-81, Oct3/4, and alkaline phosphatase; proliferation; karyotype analysis; EB formation; and hematopoietic cell differentiation	Wang et al. (2005b)
H1	Matrigel® or Human laminin	Medium containing XVIVO10, and 80 ng/ml bFGF	>20 passages; morphology; expression of SSEA-4, TRA-1-60, Oct3/4, hTERT, and Cripto; telomerase activity; proliferation; karyotype analysis; EB formation; teratoma formation; and cryopreservation	Li et al. (2005)
HSF6	Laminin	Medium containing 20% SR, 50 ng/ml activin A, 50 ng/ml KGF, and 10 mM nicotinamide	>20 passages; morphology; expression of SSEA-4, TRA-1-60, Oct3/4, hTERT, and Nanog; proliferation; karyotype analysis; EB formation; and teratoma formation	Beattie et al. (2005)
H9, HSF6	FBS	CDM (Johansson and Wiles, 1995) supplemented with 12 ng/ml bFGF and 10 ng/ml activin A or 12 ng/ml bFGF and 100 ng/ml nodal	10 passages; morphology; and expression of TRA-1-60, SSEA-3, SSEA-4, Oct3/4, and Cripto	Vallier et al. (2005)
H1, H7, H9, H14, WA15, WA16	Matrigel® or a combination of collagen IV, fibronectin, laminin, and vitronectin	Defined TeSR1 medium containing bFGF, LiCl, γ-aminobutyric acid, pipecolic acid, and TGFβ1	>7 months; morphology; expression of SSEA-4, TRA-1-60, TRA-1-81, Oct3/4, Rex1, hTERT, and Nanog; proliferation; cloning efficiency; karyotype analysis; teratoma formation; derivation of new cell lines; and sialic acid contamination	Ludwig et al. (2006)
H1	Matrigel®	Medium containing 20% SR, 5 ng/ml activin A	>20 passages; expression of Oct3/4, Nanog, SSEA-3, SSEA-4, TRA-1-60, and TRA-1-81; telomerase activity; karyotype analysis; and teratoma formation	Xiao et al. (2006)

(continues)

Table 18.1 (*continued*)

Cells	Matrix	Media	Characterization	References
H1, BG01	Matrigel® or fibronectin	Defined medium containing 4 ng/ml bFGF, 100 ng/ml Wnt3A, 160 μg/ml insulin, 88 μg/ml transferring, 2.5 ng/ml albumin, cholesterol, 100 ng/ml April or BAFF	23 passages; morphology; expression of alkaline phosphatase, Oct3/4, SSEA-3, SSEA-4, TRA-1-60, and TRA-1-81; karyotype analysis; teratoma formation; and EB formation	Lu *et al.* (2006)
H1, HSF6	Matrigel®	Defined medium containing N2 supplements, B27 supplements, and 20 ng/ml bFGF	>27 passages; morphology; expression of TRA-160, TRA-1-81, SSEA-4, Oct3/4, Nanog, and Sox2; karyotype analysis; teratoma formation; neural differentiation; endoderm differentiation; and cardiomyocyte differentiation	Yao *et al.* (2006)
H1, H9	Matrigel®	Medium containing 20% SR and 30 ng/ml IGF-II	12 passages; morphology; expression of SSEA-3; clonogenic assay; proliferation; and teratoma formation	Bendall *et al.* (2007)
BG01, BG02, CyT49	Matrigel®	Defined medium containing 10 ng/ml HRG1β, 10 ng/ml activin A, 200 ng/ml LR3-IGF1, and 8 ng/ml bFGF	Up to 9 months; morphology; expression of Oct3/4, SSEA-4, TRA-1-60, TRA-1-81; karyotype analysis; phosphorylation of IFG1R, IR, ERBB2, and FGFR2α; global transcriptional analysis; teratoma formation; EB formation; and endoderm-specific differentiation	Wang *et al.* (2007)

Abbreviations: MEF, mouse embryonic fibroblasts; CM, conditioned medium; SR, serum replacement; bFGF, basic fibroblast growth factor; TGFβ1, transforming growth factor β1; KGF, keratinocyte growth factor; April, a proliferation-inducing ligand; BAFF, B cell-activating factor belonging to TNF; HRG1β, heregulin-1β; LR3-IGF1, Long™ R^3 insulin-like growth factor 1; IGF1R, insulin-like growth factor 1 receptor; IR, insulin receptor; FGFR2α, fibroblast growth factor receptor 2α.

Following is a detailed description of the methods for evaluating hESCs, using cells maintained in the feeder-free system as examples, if not specified otherwise.

2.1. Cell morphology

Growth of undifferentiated hESCs requires specific and optimal culture conditions to prevent them from undergoing spontaneous differentiation. Nevertheless, current culture protocols with feeders or feeder-free conditions usually give rise to a mixed population of cells with various morphologies due to spontaneous differentiation. Typically, undifferentiated hESCs form compact colonies with cells tightly associated with each other through functional gap junctions (Carpenter *et al.*, 2004; Wong *et al.*, 2004; Xu *et al.*, 2004) and show a high nuclear cytoplasmic ratio with prominent nucleoli. In early passage cultures (usually <50 passages in our laboratory), some of the colonies are compact and contain cells with typical undifferentiated morphology while others are a mixture of undifferentiated cells and differentiated cell types (Fig. 18.2A–C). These differentiated cells show 3D structures or are flat and large that can be observed within or at the edge of the colonies. Most of the colonies are surrounded by a monolayer of differentiated stromal-like or fibroblast-like cells. After a longer period of maintenance (>50 passages), very few stromal-like cells remain in cultures. When reaching confluence, undifferentiated colonies fuse with each other and form a relatively uniform-looking sheet-like monolayer that is observable under a microscope at a low magnification (Fig. 18.2D). Compared with earlier passage cells, these cultures usually show a higher percentage of cells expressing markers associated with undifferentiated hESCs, suggesting selective survival of undifferentiated hESCs over long-term cultures. Since the morphology of the hESC colonies may be correlated with the degree of differentiation within the culture, it is a good practice to monitor cell morphology at each passage in conjunction with marker analysis as discussed below.

2.2. Markers for undifferentiated hESCs

Undifferentiated hESCs express several surface markers that were initially identified on human embryonic carcinoma cells using monoclonal antibodies, such as stage-specific embryonic antigen-3 (SSEA-3) and SSEA-4 for glycolipids, TRA-1-60 and TRA-1-81 for keratan sulfate-related proteoglycan antigens, and GCTM-2 for the protein core of a keratan sulfate/chondroitin sulfate pericellular matrix proteoglycan (Andrews *et al.*, 1996; Badcock *et al.*, 1999; Pera *et al.*, 1988; Reubinoff *et al.*, 2000; Thomson *et al.*, 1998; Xu *et al.*, 2001). hESCs also express a number of generic molecular markers of pluripotent stem cells that are shared with other cells, some of which appear to be critical for their self-renewal. These include Oct3/4, a POU transcription factor; hTERT, the catalytic component of telomerase; Cripto (TDGF1), a growth factor; Nanog, homeobox transcription factor; Sox2, a HMG-box transcription factor; UTF-1, undifferentiated embryonic cell

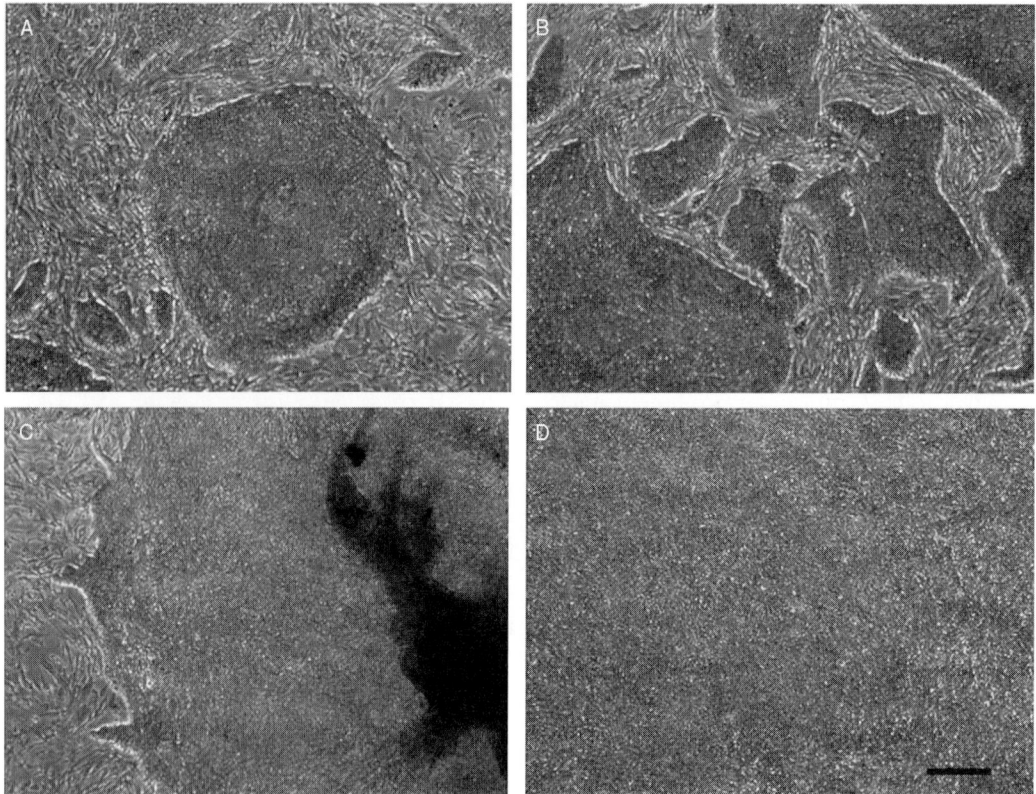

Figure 18.2 Morphology of hESCs maintained on Matrigel® in MEF-CM. (A–C) H7 passage 39 culture showed that undifferentiated colonies with different shapes were surrounded by fibroblast or stromal-like cells. Differentiated cells were also found in the center of a colony in (C). (D) H7 passage 52 culture showed relative uniform undifferentiated cell morphology. Bar = 200 μm.

transcription factor; and Rex1, zinc finger protein 42 (Abeyta *et al.*, 2004; Bhattacharya *et al.*, 2004; Boyer *et al.*, 2005; Brandenberger *et al.*, 2004a,b; Miura *et al.*, 2004; Richards *et al.*, 2004; Sperger *et al.*, 2003). A subset of hESCs also express CD133 (AC133) (Miraglia *et al.*, 1997; Uchida *et al.*, 2000; Yin *et al.*, 1997), a transmembrane glycoprotein expressed by hematopoietic stem cells, neural stem cells, and endothelial cells; CD9 (Oka *et al.*, 2002), a transmembrane protein expressed by undifferentiated mESCs; and CD90 (Thy-1), a surface marker for hematopoietic stem cells and neural cells (Carpenter *et al.*, 2004; Draper *et al.*, 2002; Kaufman *et al.*, 2001; Rosler *et al.*, 2004). In addition, undifferentiated hESC colonies display strong alkaline phosphatase activity that is also present in other pluripotent cells (Reubinoff *et al.*, 2000; Thomson *et al.*, 1998; Xu *et al.*, 2001). Together, they represent a panel of markers that are associated with pluripotent hESCs.

During differentiation, expression of undifferentiated cell markers is downregulated and, concomitantly, that of differentiated cell markers is upregulated. For example, expression of Oct3/4, Cripto, and hTERT was detected at high levels in undifferentiated hESC cultures but deceased significantly a few days after differentiation (Xu *et al.*, 2006a); SSEA-1 was undetectable in undifferentiated hESCs but its expression was elevated upon differentiation (Henderson *et al.*, 2002; Reubinoff *et al.*, 2000; Thomson *et al.*, 1998). In practice, these markers are useful for monitoring the status of undifferentiated cells in culture and for comparing culture conditions. Flow cytometry analysis, immunocytochemical analysis and/or real-time reverse transcriptase-polymerase chain reaction (RT-PCR) analysis may be used to assess expression of these markers, and alkaline phosphatase activity can be measured using a specific substrate for the enzyme. Examples of these analyses are described below.

2.2.1. Immunocytochemistry for markers of undifferentiated hESCs

Immunocytochemical analysis allows one to monitor expression of undifferentiated hESC markers in conjunction with cell morphology and provides detailed information about the location of an antigen. It is recommended to work out optimal concentrations for each specific lot of antibodies using a positive control cell line, which can be a human embryonic carcinoma cell line or an established hESC line.

Procedures for live cell staining of surface marker expression:

1. Seed cells onto chamber slides (Nunc, Rochester, NY) that have been pretreated as following. For feeder-free cultures, chamber slides are precoated with matrix such as Matrigel®, laminin, or fibronectin. For cultures with feeders, irradiated or mitomycine C-treated feeders are seeded onto the chamber slides.

2. Maintain the cells for 2–7 days before immunocytochemical analysis.

3. After the culture, remove the medium and incubate the cells with primary antibodies (SSEA-1, SSEA-4, TRA-1-60, or TRA-1-81), diluted in warm knockout DMEM (Invitrogen, Carlsbad, CA), at 37 °C for 30 min. Antibodies against SSEA-1 and SSEA-4 can be obtained from the Developmental Studies Hybridoma Bank (University of Iowa, Iowa City, IA) or Chemicon (Temecula, CA) where antibodies against TRA-1-60 and TRA-1-81 are also available.

4. After the incubation, wash the slides twice (for 5 min each) with knockout DMEM and then fix the cells with 2% paraformaldehyde in PBS for 15 min.

5. The slide is then washed with PBS, and incubated in 5% normal goat serum (NGS) in PBS at room temperature for 30 min to prevent the nonspecific binding of antibodies.

6. Incubate cells with the FITC-conjugated goat antimouse IgG (Sigma, St. Louis, MO), diluted in PBS containing 1% NGS, at room temperature for 30–60 min.

7. Wash the cells with PBS three times (for 5 min each), and then mount the slide with Vectashield® mounting media for fluorescence (Vector Laboratories, Inc., Burlingame, CA) that contains DAPI to counterstain nuclei and anti-bleaching compounds to prevent quenching fluorescence. Other DNA-binding dye, such as Hoechst, can also be used to stain nuclei. The slide is then examined under a UV microscope.

Quick tips:

For double staining, choose primary antibodies with different isotypes, for example, IgG3 type SSEA-4 antibody together with IgM type TRA-1-60 antibody. Secondary antibodies against specific isotypes of the primary antibodies conjugated with different fluorescence dyes can be used. For example, goat antimouse IgG3 conjugated with Alexa 488 and goat antimouse IgM conjugated with Alexa 594 (Invitrogen) can be used to double stain SSEA-4 and TRA-1-60.

Pitfall to avoid:

Avoid exposing the cells to the light during and after the secondary antibody incubation when performing immunocytochemical analysis or flow cytometry analysis.

Procedures for detection of intracellular antigens, such as Oct3/4 and Nanog:

1. Fix cells with 2% paraformaldehyde in PBS at room temperature for 15 min.

2. After washing with PBS, permeabilize cells in 100% ethanol for 2 min. After another washing, the slide is incubated in 5% donkey serum in PBS at room temperature for 2 h or at 4 °C overnight.

3. Incubate the slide at room temperature for 2 h with a goat antibody against Oct3/4 (Santa Cruz Biotechnology, Inc., Santa Cruz, CA), diluted with 1% donkey serum in PBS.

4. Wash the cells for three times with PBS (5–10 min for each washing).

5. Incubate the slides with FITC-conjugated donkey antigoat IgG (Jackson ImmunoResearch Laboratories, Inc., West Grove, PA), diluted in PBS containing 1% donkey serum, at room temperature for 1 h. After the incubation, wash the cells three times with PBS (5–10 min for each wash), then mount the slide with Vectashield® mounting media for microscope examination.

2.2.2. Flow cytometry analysis of surface markers

Flow cytometry analysis provides quantitative information regarding the percentage of cells expressing stem cell markers in culture and the staining intensity. A profile of multiple markers can be simultaneously obtained by this method.

Procedures:

1. After growing cells in a tissue culture plate, remove culture medium, and incubate the cells with collagenase IV (200 units/ml) (Invitrogen) in knockout DMEM at 37 °C for 5 min.

2. Remove collagenase and incubate the cells with 0.5 mM EDTA in PBS at 37 °C for 5–10 min. Alternatively, cells can be washed with PBS and dissociated with 0.5% trypsin/EDTA (Invitrogen).

3. Gently pipette the cells several times, spin to collect the cells, and resuspend them in PBS containing 0.1% BSA (diluent) to $\sim 1 \times 10^7$ cells/ml. For each test, add 50 μl primary antibodies against SSEA-1, SSEA-4, TRA-1-60, or TRA-1-81 as well as isotype controls, diluted appropriately, to 50 μl of the resuspended cells, and incubate them at room temperature for 30 min.

4. Wash the cells with the diluent and resuspend them in the diluent containing rat anti-mouse kappa chain antibodies conjugated with PE (Becton Dickinson, San Jose, CA).

5. After incubation at 4 °C for 30 min, wash cells with 1 ml of the diluent and analyzed on FACScalibur™ Flow Cytometer (Becton Dickinson, San Jose, CA) using the CellQuest software. Follow this procedure when performing multicolor flow cytometry analysis, except that different primary antibodies with different isotypes (e.g., IgG3 SSEA-4 and IgM TRA-1-60) are added together and that secondary antibodies against each of the isotypes conjugated with different fluorescence dyes are used.

Quick tips:

For both the single and multiple marker analyses, the secondary antibody incubation may be skipped if a primary antibody directly conjugated with a specific dye or different antibodies directly conjugated with different dyes are available. The latter method is recommended when developing more throughput assays.

Pitfall to avoid:

As compensation controls, single marker staining for each of the antigens needs to be carried out separately as well.

2.2.3. Detection of alkaline phosphatase

Procedures:

1. Culture cells on Matrigel® or laminin-coated chamber slides overnight or longer.
2. Remove the culture medium, wash the slide with PBS, and then fix the cells with 4% paraformaldehyde for 15 min.
3. After another wash with PBS, incubate the cells with alkaline phosphatase substrate (Vector Laboratories, Inc., Burlingame, CA) at room temperature in the dark for 1 h or until colonies show pink color.
4. Rinse the slide gently once for 2–5 min in 100% ethanol before mounting and observing under a light or UV microscopy.

2.2.4. Quantitative RT-PCR analysis of Oct3/4, hTERT, Nanog, and Cripto

Quantitative RT-PCR analysis is an alternative method to determine relative gene expression, particularly when specific antibodies are lacking.

Procedures:

1. Remove culture medium from cells maintained in 6-well plates.
2. Harvest the cells by adding 350–700 μl lysis buffer to each well.
3. Isolate RNA using Qiagen RNeasy kit (Qiagen, Valencia, CA) following the tissue isolation procedure recommended for the QiaShredder (Qiagen).
4. Prior to RT-PCR analysis, treat RNA samples with DNase I (Ambion, Austin, TX) to remove contaminating genomic DNA.
5. Set up TaqMan one-step RT-PCR using the master mix (Applied Biosystems, Foster City, CA) and specific primers and probes for Oct3/4, hTERT, Nanog, or Cripto listed in Table 18.2. Each reaction mixture contains $1\times$ RT-master mix, 300 nM of each primer, and 80 nM of probe, and $\sim$50 ng of total RNA in a final volume of 50 μl. As a control, the samples are also subjected to the analysis of 18S ribosomal RNA by real-time RT-PCR using a kit from Applied Biosystems.
6. Perform real-time RT-PCR on the ABI Prism 7700 Sequence Detection System (Applied Biosystems) using following conditions: RT at 48 °C for 30 min; denaturation and AmpliTaq gold activation at 95 °C for 10 min; and amplification for 40 cycles at 95 °C for 15 s and 60 °C for 1 min.
7. Analyze the reactions by the software from ABI Prism 7700 Sequence Detection System. The relative quantitation of gene expression can be done by normalization against endogenous 18S ribosomal RNA using the $\triangle\triangle C_T$ method described in the ABI User Bulletin #2, Relative Quantitation of Gene Expression, 1997.

Quick tips:

Triplicate cultures are recommended to derive relative levels of gene expression in mean $\pm$ standard deviation.

2.3. Proliferation

Unlike somatic cells, hESCs display extensive proliferative properties without undergoing senescence. Theoretically, these cells are functionally immortal and have unrestricted lifespan. Information regarding proliferation capacity of the cells is useful for many applications. For example, in the case of cell transplantation, it allows one to estimate the length of culture time required to expand undifferentiated cells to the desired amount for differentiation into certain cell type. Several methods can be used to monitor proliferation capacity of hESCs. Analysis of DNA content by flow cytometry can provide cell cycle profile, and measurement of BrdU or 3H thymidine incorporation can give specific information regarding cells in the S phase (Beattie *et al.*, 2005). These assays are, however, not useful to determine

Table 18.2
Primers and probes for real-time RT–PCR TaqMan assays

Genes		Sequences
Undifferentiated cells		
Oct3/4	Primers and probe	Purchased from Applied Biosystems (assay number Hs03005111_g1)
hTERT	Primers and probe	Purchased from Applied Biosystems (assay number 4319447F)
Cripto	Forward	TGAGCACGATGTGCGC
	Probe	FAM-AGAGAACTGTGGGTCTGTGCCCCATG-TAM
	Reverse	TTCTTGGGCAGCCAGGTG
Nanog	Forward	GCAGAAGGCCTCAGCACCTA
	Probe	FAM-CTACCCCAGCCTTTACTCTTCCTACCACCA-TAM
	Reverse	AGTCGGGTTCACCAGGCAT
Mesoderm		
Nkx2.5	Forward	ACCCAGCCAAGGACCCTAGA
	Probe	FAM-CGAAAAGAAAGAGCTGTGC-MGB
	Reverse	CTCCACCGCCTTCTGCAG
Tbx5	Primers and probe	Purchased from Applied Biosystems (assay number Hs00361155_m1)
MEF2C	Forward	CACATCGACCTCCAAGTGCA
	Probe	FAM-AACACAGGTGGTCTGAT-MGB
	Reverse	CCAGACGTGAGGTCTCCACC
Endoderm		
HNF3β	Forward	CCGACTGGAGCAGCTACTATG
	Probe	FAM-CAGAGCCCGAGGGCTACTCCTCC-TAM
	Reverse	TACGTGTTCATGCCGTTCAT
Sox17	Forward	CAGCAGAATCCAGACCTGCA
	Probe	FAM-ACGCCGAGTTGAGCAAGATGCTGG-TAM
	Reverse	GTCAGCGCCTTCCACGACT
Ectoderm		
NeuroD1	Forward	TCACTGCTCAGGACCTACTAACA
	Probe	FAM-TACAGCGAGAGTGGGCTGATGGG-TAM
	Reverse	GAGGACCTTGGGGCTGAG
Pax6	Primers and probe	Purchased from Applied Biosystems (assay number Hs00242217_m1)
Control		
18S	Primers and probe	Purchased from Applied Biosystems (assay number 4319413E)

the exact cell expansion. A straightforward measurement of cell expansion can be performed by direct cell counting.

Procedures:

1. Cells are generally passaged by collagenase IV dissociation, which usually only gives rise to cell clumps. For cell counting purpose, cells in parallel wells are dissociated into single cells using trypsin/EDTA (Li *et al.*, 2005). Cell expansion can then be calculated based on the number of cells harvested compared to the number of cells initially inoculated.

2. Cell proliferation rate can also be assessed by counting cells in triplicate cultures every 1–2 days for 1 week to obtain the growth curve, which is then used to derive population doubling time (Xu *et al.*, 2001).

An important feature of hESCs is their capacity to expand from a single cell. Two clonal cell lines, H9.1 and H9.2, have been successfully subcloned from H9 cells (Amit *et al.*, 2000). Single cell cloning has also been achieved using H1, H13, I3, and I6 cell lines (Amit and Itskovitz-Eldor, 2002). For clonal growth, additions of several factors such as neurotrophin (Pyle *et al.*, 2006) and pleiotrophin (Soh *et al.*, 2007) may be beneficial. In practice, clonogenic aspect of hESCs can be a useful tool when characterizing the cells maintained in novel conditions (Bendall *et al.*, 2007; Wang *et al.*, 2007).

Undifferentiated hESCs have high telomerase activity, which is consistent with their prolonged proliferation capacity and expression of hTERT. Telomerase activity can be detected by telomeric repeat amplification protocol (TRAP), as described (Kim *et al.*, 1994; Weinrich *et al.*, 1997).

2.4. Karyotype analysis

Karyotype analysis of hESC cultures should be routinely performed, since abnormal karyotypes have been reported in certain cultures maintained both on feeders and in feeder-free conditions (Draper *et al.*, 2004). In our laboratory, cultures are examined for karyotype by the standard G-banding method performed by a qualified cytogenetic laboratory (Oakland Children's Hospital, Oakland, CA).

2.5. Differentiation capacity

Undifferentiated hESCs can be induced to differentiate into a broad range of cell types that may be of potential use for treatment of degenerative diseases such as heart disease, spinal cord injury, Parkinson's disease, and diabetes. So far, cardiomyocytes, neural progenitors, oligodendricytes, trophoblasts, endothelial cells, hematopoietic lineages, hepatocytes, osteoblasts, insulin-expressing cells and, other cell types have been derived from hESCs *in vitro* (e.g., see Agarwal *et al.*, 2008; Assady *et al.*, 2001; Carpenter *et al.*, 2001; Chadwick *et al.*, 2003b; He *et al.*, 2003; Jiang *et al.*, 2007; Kehat *et al.*, 2001; Kroon *et al.*, 2008; Laflamme *et al.*, 2007; Lebkowski *et al.*, 2001; Levenberg *et al.*, 2002; Mummery *et al.*, 2002; Odorico *et al.*, 2001; Rambhatla *et al.*, 2003; Schuldiner *et al.*, 2000; Sottile *et al.*, 2003; Xu *et al.*, 2002a,b, 2006a,b; Zhang *et al.*, 2001). In addition, hESCs are also capable of undergoing spontaneous differentiation *in vivo*. When injected into immunodeficient mice, undifferentiated hESCs cells undergo differentiation, leading to the generation of tumors called teratomas consisting of multiple cell types with highly organized structures. These tissues represent those derived from all three germ layers (ectoderm, mesoderm, and endoderm), such as cartilage, bone, tooth-like structure, secretary epithelium, primitive neuroepithelial tubule, skin, gastrointestinal epithelium, and respiratory epithelium.

2.5.1. *In vivo* differentiation through teratoma formation

Procedures:

1. When cells reach confluence, harvest them by incubating in collagenase IV (200 units/ml) at 37 °C for 5 min.

2. Collect the cells as clumps by gentle scraping and transfer to a tube. The cells are then washed in PBS and resuspended in PBS to 5×10^7 cells/ml.

3. Inject the cells intramuscularly into anterior tibialis (hind limbs) of SCID/beige mice ($\sim 5 \times 10^6$ cells in 100 μl per site) or other immunodeficient mice.

4. Monitor tumor formation at least once a week for a period up to one year. When a tumor is formed or reaches 50–100 mm^3 (usually after $\sim$70–90 days in our laboratory), the animal is then euthanized, and the tumor is excised and processed for histological analysis.

Pitfall to avoid:

Do not dissociate cells into single cells with trypsin or other enzymes when harvesting cells for teratoma formation.

2.5.2. *In vitro* differentiation through embryoid bodies

In vitro spontaneous differentiation can be induced by culturing hESCs in suspension to form embryoid bodies (EBs). Multiple cell types including neurons, cardiomyocytes, endoderm cells, and epithelial cells can be detected in the EB outgrowths generated using the following procedure.

Procedures:

1. After cells reach confluence, add collagenase IV (200 units/ml) dissolved in knockout DMEM into 6-well plates (1 ml/well) or T225 flasks (15 ml/flask), and incubate at 37 °C for 5 min.

2. Aspirate the collagenase and add differentiation medium consisting of 80% knockout DMEM, 20% non-heat-inactivated FBS (Sigma), 1% non-essential amino acids (Invitrogen), 1 mM L-glutamine (Invitrogen), and 0.1 mM β-ercaptoethanol (Sigma) into the culture (2 ml/well in 6-well plates, 30 ml/T225 flask).

3. Harvest the cells with a cell scraper or pipette and transferred as clumps to 6-well low attachment plates (Corning, Cat#29443-030) (1:1 split, 1-well of 6-well plates into 1 well of the low attachment plates, one T225 into three low attachment plates; the split ratio for this procedure should be adjusted so that each well receives $\sim 3 \times 10^6$ cells).

4. Add differentiation medium to each well to give a total volume of 4 ml/well. After overnight culture in suspension, cells form floating aggregates known as EBs.

5. The differentiation medium should be changed every 2–3 days. To change the medium, transfer EBs into 15 ml or 50 ml tubes and let the aggregates settle for $\sim$5 min, aspirate the supernatant, replace with fresh differentiation medium

(4 ml/well), and then transfer to low attachment 6-well plates for further culture. During the first few days, the EBs are small with irregular outlines and increase in size by day 4 in suspension.

6. The EBs can be maintained in suspension for more than 10 days. Alternatively, EBs at different stages can be transferred to adherent tissue culture plates for further induction of differentiation.

Quick tips:

Note that EB formation become less efficient when cells maintained in the feeder-free condition reach high passages (>50 passage). These cultures usually lack stromal cells and become a sheet-like monolayer containing high percentages of cells expressing undifferentiated cell markers. To improve EB formation from high passage cells, first seed cells onto matrix in CM and then switch to the differentiation medium. The cells are then grown to confluence in differentiation medium for 5–6 days before initiation of the suspension culture.

To examine the differentiated culture, immunocytochemical or RT-PCR analyses can be used to detect cell type-specific markers. Useful diagnostic markers are β-tubulin III for neurons, α-fetal protein (AFP) for endoderm cells, cardiac troponin I (cTnI) for cardiomyocytes, and pancytokeratins for epithelial cells (Xu *et al.*, 2001, 2004, 2006b).

Procedures for immunocytochemistry analysis of EB out growths:

1. Plate EBs onto gelatin-coated chamber slides for differentiation and subsequent analysis.

2. Fix and permeablize the differentiated cultures as described in the earlier section for detecting intracellular antigens.

3. After blocking with 5% NGS in PBS, the cells are incubated at room temperature for 2 h with a monoclonal antibody against β-tubulin III (Sigma), AFP (Sigma), cTnI (Spectral Diagnostic, Inc., Toronto, Ontario, Canada), or pancytokeratins (Dako Corporation, Carpinteria, CA) diluted appropriately in 1% NGS in PBS.

4. After washing, incubate the cells with FITC-conjugated goat antimouse IgG (Sigma), diluted in PBS containing 1% NGS, at room temperature for 30 min–1 h. Alternatively, goat antimouse IgG1 conjugated with Alexa 488 or goat antimouse IgG1 conjugated with Alexa 594 (Invitrogen) can be used.

5. Wash the cells three times with PBS (5–10 min/washing) and mounted the slides with Vectashield® mounting media for examination with a UV microscopy.

Alternatively, real-time RT-PCR analysis is performed instead of immunocytochemistry to detect markers associated with cell types derived from mesoderm, endoderm, and ectoderm. Table 18.2 listed primers and probes for suggested markers. It should be noted that more in depth analyses of functional phenotypes are required to confirm the differentiation of specific cell types, details of which can be found elsewhere.

Although the EB formation procedure has been commonly used to assess *in vitro* spontaneous differentiation, other protocols without the requirement of EB formation may also be used to induce hESCs into specific cell types, particularly when EB formation is inefficient or unable to give rise to particular cells of interest. For example, two hESC lines, HES-1 and HES-2, fail to form EBs using hanging drop method or simple suspension, but have been induced to generate contracting cardiomyocytes after prolonged culture on MEF feeders (Reubinoff *et al.*, 2000) or after coculture with the endoderm-like cell line END-2 derived from mouse embryonic carcinoma cells (mECs) (Mummery *et al.*, 2002, 2003). Recently, several lineage-specific differentiation protocols have also been developed to guide hESC to differentiate efficiently into particular cell types, such as cardiomyocytes (Graichen *et al.*, 2007; Laflamme *et al.*, 2007), pancreatic endoderm cells (Jiang *et al.*, 2007; Kroon *et al.*, 2008), and hepatocytes (Agarwal *et al.*, 2008).

3. Conclusion

Research on hESCs continues to advance rapidly in various fronts, including derivation of additional hESC lines and development of novel methods to expand undifferentiated cells and induce lineage-specific differentiation. Potential application of these cells demands assurance of the highest quality as expected, which can be monitored by the methods described earlier and should be carried out in a comprehensive manner. As our understanding of human stem cell biology expands, it is of no doubt that additional methods such as those at a genomic level will be developed. These and the existing methods should be helpful to providing confidence for hESCs to be used in a clinical setting.

Acknowledgments

I would like to thank Drs. Jane Lebkowski, Calvin Harley, and Joseph Gold for their critical review of this chapter and my colleagues for their invaluable contribution to our studies.

References

Abeyta, M. J., Clark, A. T., Rodriguez, R. T., Bodnar, M. S., Pera, R. A., and Firpo, M. T. (2004). Unique gene expression signatures of independently-derived human embryonic stem cell lines. *Hum. Mol. Genet.* **13**, 601–608.

Agarwal, S., Holton, K. L., and Lanza, R. (2008). Efficient differentiation of functional hepatocytes from human embryonic stem cells. *Stem Cells* **26**, 1117–1127.

Amit, M., and Itskovitz-Eldor, J. (2002). Derivation and spontaneous differentiation of human embryonic stem cells. *J. Anat.* **200**, 225–232.

Amit, M., Carpenter, M. K., Inokuma, M. S., Chiu, C. P., Harris, C. P., Waknitz, M. A., Itskovitz-Eldor, J., and Thomson, J. A. (2000). Clonally derived human embryonic stem cell lines maintain pluripotency and proliferative potential for prolonged periods of culture. *Dev. Biol.* **227**, 271–278.

Amit, M., Margulets, V., Segev, H., Shariki, K., Laevsky, I., Coleman, R., and Itskovitz-Eldor, J. (2003). Human feeder layers for human embryonic stem cells. *Biol. Reprod.* **68**, 2150–2156.

Amit, M., Shariki, C., Margulets, V., and Itskovitz-Eldor, J. (2004). Feeder layer- and serum-free culture of human embryonic stem cells. *Biol. Reprod.* **70**, 837–845.

Andrews, P. W., Casper, J., Damjanov, I., Duggan-Keen, M., Giwercman, A., Hata, J., von Keitz, A., Looijenga, L. H., Millan, J. L., Oosterhuis, J. W., *et al.* (1996). Comparative analysis of cell surface antigens expressed by cell lines derived from human germ cell tumours. *Int. J. Cancer* **66**, 806–816.

Assady, S., Maor, G., Amit, M., Itskovitz-Eldor, J., Skorecki, K. L., and Tzukerman, M. (2001). Insulin production by human embryonic stem cells. *Diabetes* **50**, 1691–1697.

Badcock, G., Pigott, C., Goepel, J., and Andrews, P. W. (1999). The human embryonal carcinoma marker antigen TRA-1-60 is a sialylated keratan sulfate proteoglycan. *Cancer Res.* **59**, 4715–4719.

Beattie, G. M., Lopez, A. D., Bucay, N., Hinton, A., Firpo, M. T., King, C. C., and Hayek, A. (2005). Activin A maintains pluripotency of human embryonic stem cells in the absence of feeder layers. *Stem Cells* **23**, 489–495.

Bendall, S. C., Stewart, M. H., Menendez, P., George, D., Vijayaragavan, K., Werbowetski-Ogilvie, T., Ramos-Mejia, V., Rouleau, A., Yang, J., Bosse, M., *et al.* (2007). IGF and FGF cooperatively establish the regulatory stem cell niche of pluripotent human cells *in vitro*. *Nature* **448**, 1015–1021.

Bhattacharya, B., Miura, T., Brandenberger, R., Mejido, J., Luo, Y., Yang, A. X., Joshi, B. H., Ginis, I., Thies, R. S., Amit, M., *et al.* (2004). Gene expression in human embryonic stem cell lines: Unique molecular signature. *Blood* **103**, 2956–2964.

Boyer, L. A., Lee, T. I., Cole, M. F., Johnstone, S. E., Levine, S. S., Zucker, J. P., Guenther, M. G., Kumar, R. M., Murray, H. L., Jenner, R. G., *et al.* (2005). Core transcriptional regulatory circuitry in human embryonic stem cells. *Cell* **122**, 947–956.

Brandenberger, R., Khrebtukova, I., Thies, R. S., Miura, T., Jingli, C., Puri, R., Vasicek, T., Lebkowski, J., and Rao, M. (2004). MPSS profiling of human embryonic stem cells. *BMC Dev. Biol.* **4**, 10.

Brandenberger, R., Wei, H., Zhang, S., Lei, S., Murage, J., Fisk, G. J., Li, Y., Xu, C., Fang, R., Guegler, K., *et al.* (2004). Transcriptome characterization elucidates signaling networks that control human ES cell growth and differentiation. *Nat. Biotechnol.* **22**, 707–716.

Carpenter, M. K., Inokuma, M. S., Denham, J., Mujtaba, T., Chiu, C. P., and Rao, M. S. (2001). Enrichment of neurons and neural precursors from human embryonic stem cells. *Exp. Neurol.* **172**, 383–397.

Carpenter, M. K., Rosler, E. S., Fisk, G. J., Brandenberger, R., Ares, X., Miura, T., Lucero, M., and Rao, M. S. (2004). Properties of four human embryonic stem cell lines maintained in a feeder-free culture system. *Dev. Dyn.* **229**, 243–258.

Chadwick, K., Wang, L., Li, L., Menendez, P., Murdoch, B., Rouleau, A., and Bhatia, M. (2003). Cytokines and BMP-4 promote hematopoietic differentiation of human embryonic stem cells. *Blood* **102**, 906–915.

Cheng, L., Hammond, H., Ye, Z., Zhan, X., and Dravid, G. (2003). Human adult marrow cells support prolonged expansion of human embryonic stem cells in culture. *Stem Cells* **21**, 131–142.

Draper, J. S., Pigott, C., Thomson, J. A., and Andrews, P. W. (2002). Surface antigens of human embryonic stem cells: Changes upon differentiation in culture. *J. Anat.* **200**, 249–258.

Draper, J. S., Smith, K., Gokhale, P., Moore, H. D., Maltby, E., Johnson, J., Meisner, L., Zwaka, T. P., Thomson, J. A., and Andrews, P. W. (2004). Recurrent gain of chromosomes 17q and 12 in cultured human embryonic stem cells. *Nat. Biotechnol.* **22**, 53–54.

Genbacev, O., Krtolica, A., Zdravkovic, T., Brunette, E., Powell, S., Nath, A., Caceres, E., McMaster, M., McDonagh, S., Li, Y., *et al.* (2005). Serum-free derivation of human embryonic stem cell lines on human placental fibroblast feeders. *Fertil. Steril.* **83**, 1517–1529.

Graichen, R., Xu, X., Braam, S. R., Balakrishnan, T., Norfiza, S., Sieh, S., Soo, S. Y., Tham, S. C., Mummery, C., Colman, A., *et al.* (2008). Enhanced cardiomyogenesis of human embryonic stem cells by a small molecular inhibitor of p38 MAPK. *Differentiation* **76**, 357–370.

He, J. Q., Ma, Y., Lee, Y., Thomson, J. A., and Kamp, T. J. (2003). Human embryonic stem cells develop into multiple types of cardiac myocytes: Action potential characterization. *Circ. Res.* **93**, 32–39.

Henderson, J. K., Draper, J. S., Baillie, H. S., Fishel, S., Thomson, J. A., Moore, H., and Andrews, P. W. (2002). Preimplantation human embryos and embryonic stem cells show comparable expression of stage-specific embryonic antigens. *Stem Cells* **20,** 329–337.

Hoffman, L. M., and Carpenter, M. K. (2005). Characterization and culture of human embryonic stem cells. *Nat. Biotechnol.* **23,** 699–708.

Hovatta, O., Mikkola, M., Gertow, K., Stromberg, A. M., Inzunza, J., Hreinsson, J., Rozell, B., Blennow, E., Andang, M., and Ahrlund-Richter, L. (2003). A culture system using human foreskin fibroblasts as feeder cells allows production of human embryonic stem cells. *Hum. Reprod.* **18,** 1404–1409.

Inzunza, J., Gertow, K., Stromberg, M. A., Matilainen, E., Blennow, E., Skottman, H., Wolbank, S., Ahrlund-Richter, L., and Hovatta, O. (2005). Derivation of human embryonic stem cell lines in serum replacement medium using postnatal human fibroblasts as feeder cells. *Stem Cells* **23,** 544–549.

James, D., Levine, A. J., Besser, D., and Hemmati-Brivanlou, A. (2005). TGF{beta}/activin/nodal signaling is necessary for the maintenance of pluripotency in human embryonic stem cells. *Development* **132,** 1273–1282.

Jiang, J., Au, M., Lu, K., Eshpeter, A., Korbutt, G., Fisk, G., and Majumdar, A. S. (2007). Generation of insulin-producing islet-like clusters from human embryonic stem cells. *Stem Cells* **25,** 1940–1953.

Johansson, B. M., and Wiles, M. V. (1995). Evidence for involvement of activin A and bone morphogenetic protein 4 in mammalian mesoderm and hematopoietic development. *Mol. Cell. Biol.* **15,** 141–151.

Kaufman, D. S., Hanson, E. T., Lewis, R. L., Auerbach, R., and Thomson, J. A. (2001). Hematopoietic colony-forming cells derived from human embryonic stem cells. *Proc. Natl Acad. Sci. USA* **98,** 10716–10721.

Kehat, I., Kenyagin-Karsenti, D., Snir, M., Segev, H., Amit, M., Gepstein, A., Livne, E., Binah, O., Itskovitz-Eldor, J., and Gepstein, L. (2001). Human embryonic stem cells can differentiate into myocytes with structural and functional properties of cardiomyocytes. *J. Clin. Invest.* **108,** 407–414.

Kim, N. Y., Piatyszek, M. A., Prowse, K. R., Harley, C. B., West, M. D., Ho, P. L., Coviello, G. M., Wright, W. E., Weinrich, S. L., and Shay, J. W. (1994). Specific association of human telomerase activity with immortal cell lines and cancer. *Science* **266,** 2011–2015.

Klimanskaya, I., Chung, Y., Meisner, L., Johnson, J., West, M. D., and Lanza, R. (2005). Human embryonic stem cells derived without feeder cells. *Lancet* **365,** 1636–1641.

Kroon, E., Martinson, L. A., Kadoya, K., Bang, A. G., Kelly, O. G., Eliazer, S., Young, H., Richardson, M., Smart, N. G., Cunningham, J., *et al.* (2008). Pancreatic endoderm derived from human embryonic stem cells generates glucose-responsive insulin-secreting cells *in vivo. Nat. Biotechnol.* **26,** 443–452.

Laflamme, M. A., Gold, J., Xu, C., Hassanipour, M., Rosler, E., Police, S., Muskheli, V., and Murry, C. E. (2005). Formation of human myocardium in the rat heart from human embryonic stem cells. *Am. J. Pathol.* **167,** 663–671.

Laflamme, M. A., Chen, K. Y., Naumova, A. V., Muskheli, V., Fugate, J. A., Dupras, S. K., Reinecke, H., Xu, C., Hassanipour, M., Police, S., *et al.* (2007). Cardiomyocytes derived from human embryonic stem cells in pro-survival factors enhance function of infarcted rat hearts. *Nat. Biotechnol.* **25,** 1015–1024.

Lebkowski, J. S., Gold, J., Xu, C., Funk, W., Chiu, C. P., and Carpenter, M. K. (2001). Human embryonic stem cells: Culture, differentiation, and genetic modification for regenerative medicine applications. *Cancer J.* **7**(Suppl. 2), S83–S93.

Levenberg, S., Golub, J. S., Amit, M., Itskovitz-Eldor, J., and Langer, R. (2002). Endothelial cells derived from human embryonic stem cells. *Proc. Natl Acad. Sci. USA* **99,** 4391–4396.

Li, Y., Powell, S., Brunette, E., Lebkowski, J., and Mandalam, R. (2005). Expansion of human embryonic stem cells in defined serum-free medium devoid of animal-derived products. *Biotechnol. Bioeng.* **91,** 688–698.

Lu, J., Hou, R., Booth, C. J., Yang, S. H., and Snyder, M. (2006). Defined culture conditions of human embryonic stem cells. *Proc. Natl Acad. Sci. USA* **103,** 5688–5693.

Ludwig, T. E., Levenstein, M. E., Jones, J. M., Berggren, W. T., Mitchen, E. R., Frane, J. L., Crandall, L. J., Daigh, C. A., Conard, K. R., Piekarczyk, M. S., *et al.* (2006). Derivation of human embryonic stem cells in defined conditions. *Nat. Biotechnol.* **24,** 185–187.

Ma, Y., Ramezani, A., Lewis, R., Hawley, R. G., and Thomson, J. A. (2003). High-level sustained transgene expression in human embryonic stem cells using lentiviral vectors. *Stem Cells* **21,** 111–117.

Miraglia, S., Godfrey, W., Yin, A. H., Atkins, K., Warnke, R., Holden, J. T., Bray, R. A., Waller, E. K., and Buck, D. W. (1997). A novel five-transmembrane hematopoietic stem cell antigen: Isolation, characterization, and molecular cloning. *Blood* **90,** 5013–5021.

Miura, T., Luo, Y., Khrebtukova, I., Brandenberger, R., Zhou, D., Thies, R. S., Vasicek, T., Young, H., Lebkowski, J., Carpenter, M. K., and Rao, M. S. (2004). Monitoring early differentiation events in human embryonic stem cells by massively parallel signature sequencing and expressed sequence tag scan. *Stem Cells Dev.* **13,** 694–715.

Miyamoto, K., Hayashi, K., Suzuki, T., Ichihara, S., Yamada, T., Kano, Y., Yamabe, T., and Ito, Y. (2004). Human placenta feeder layers support undifferentiated growth of primate embryonic stem cells. *Stem Cells* **22,** 433–440.

Mummery, C., Ward, D., van den Brink, C. E., Bird, S. D., Doevendans, P. A., Opthof, T., Brutel de la Riviere, A., Tertoolen, L., van der Heyden, M., and Pera, M. (2002). Cardiomyocyte differentiation of mouse and human embryonic stem cells. *J. Anat.* **200,** 233–242.

Mummery, C., Ward-van Oostwaard, D., Doevendans, P., Spijker, R., van den Brink, S., Hassink, R., van der Heyden, M., Opthof, T., Pera, M., de la Riviere, A. B., *et al.* (2003). Differentiation of human embryonic stem cells to cardiomyocytes: Role of coculture with visceral endoderm-like cells. *Circulation* **107,** 2733–2740.

Nistor, I. G., Totoiu, M. O., Haque, N., Carpenter, M. K., and Keirstead, H. S. (2005). Human embryonic stem cells differentiate into oligodendrocytes in high purity and myelinate after spinal cord transplantation. *Glia* **49,** 385–396.

Odorico, J. S., Kaufman, D. S., and Thomson, J. A. (2001). Multilineage differentiation from human embryonic stem cell lines. *Stem Cells* **19,** 193–204.

Oka, M., Tagoku, K., Russell, T. L., Nakano, Y., Hamazaki, T., Meyer, E. M., Yokota, T., and Terada, N. (2002). CD9 is associated with leukemia inhibitory factor-mediated maintenance of embryonic stem cells. *Mol. Biol. Cell* **13,** 1274–1281.

Pera, M. F., Blasco-Lafita, M. J., Cooper, S., Mason, M., Mills, J., and Monaghan, P. (1988). Analysis of cell-differentiation lineage in human teratomas using new monoclonal antibodies to cytostructural antigens of embryonal carcinoma cells. *Differentiation* **39,** 139–149.

Pyle, A. D., Lock, L. F., and Donovan, P. J. (2006). Neurotrophins mediate human embryonic stem cell survival. *Nat. Biotechnol.* **24,** 344–350.

Rambhatla, L., Chiu, C. P., Kundu, P., Peng, Y., and Carpenter, M. K. (2003). Generation of hepatocyte-like cells from human embryonic stem cells. *Cell Transplant.* **12,** 1–11.

Reubinoff, B. E., Pera, M. F., Fong, C. Y., Trounson, A., and Bongso, A. (2000). Embryonic stem cell lines from human blastocysts: Somatic differentiation *in vitro. Nat. Biotechnol.* **18,** 399–404.

Richards, M., Fong, C. Y., Chan, W. K., Wong, P. C., and Bongso, A. (2002). Human feeders support prolonged undifferentiated growth of human inner cell masses and embryonic stem cells. *Nat. Biotechnol.* **20,** 933–936.

Richards, M., Tan, S. P., Tan, J. H., Chan, W. K., and Bongso, A. (2004). The transcriptome profile of human embryonic stem cells as defined by SAGE. *Stem Cells* **22,** 51–64.

Rosler, E. S., Fisk, G. J., Ares, X., Irving, J., Miura, T., Rao, M. S., and Carpenter, M. K. (2004). Long-term culture of human embryonic stem cells in feeder-free conditions. *Dev. Dyn.* **229,** 259–274.

Schuldiner, M., Yanuka, O., Itskovitz-Eldor, J., Melton, D. A., and Benvenisty, N. (2000). Effects of eight growth factors on the differentiation of cells derived from human embryonic stem cells. *Proc. Natl Acad. Sci. USA* **92,** 11307–11312.

Soh, B. S., Song, C. M., Vallier, L., Li, P., Choong, C., Yeo, B. H., Lim, E. H., Pedersen, R. A., Yang, H. H., Rao, M., and Lim, B. (2007). Pleiotrophin enhances clonal growth and long-term expansion of human embryonic stem cells. *Stem Cells* **25,** 3029–3037.

Sottile, V., Thomson, A., and McWhir, J. (2003). *In vitro* osteogenic differentiation of human ES cells. *Cloning Stem Cells* **5,** 149–155.

Sperger, J. M., Chen, X., Draper, J. S., Antosiewicz, J. E., Chon, C. H., Jones, S. B., Brooks, J. D., Andrews, P. W., Brown, P. O., and Thomson, J. A. (2003). Gene expression patterns in human embryonic stem cells and human pluripotent germ cell tumors. *Proc. Natl Acad. Sci. USA* **100,** 13350–13355.

Stojkovic, P., Lako, M., Przyborski, S., Stewart, R., Armstrong, L., Evans, J., Zhang, X., and Stojkovic, M. (2005). Human-serum matrix supports undifferentiated growth of human embryonic stem cells. *Stem Cells* **23,** 895–902.

Stojkovic, P., Lako, M., Stewart, R., Przyborski, S., Armstrong, L., Evans, J., Murdoch, A., Strachan, T., and Stojkovic, M. (2005). An autogeneic feeder cell system that efficiently supports growth of undifferentiated human embryonic stem cells. *Stem Cells* **23,** 306–314.

Thomson, J. A., Itskovitz-Eldor, J., Shapiro, S. S., Waknitz, M. A., Swiergiel, J. J., Marshall, V. S., and Jones, J. M. (1998). Embryonic stem cell lines derived from human blastocysts. *Science* **282,** 1145–1147.

Uchida, N., Buck, D. W., He, D., Reitsma, M. J., Masek, M., Phan, T. V., Tsukamoto, A. S., Gage, F. H., and Weissman, I. L. (2000). Direct isolation of human central nervous system stem cells. *Proc. Natl Acad. Sci. USA* **97,** 14720–14725.

Vallier, L., Reynolds, D., and Pedersen, R. A. (2004). Nodal inhibits differentiation of human embryonic stem cells along the neuroectodermal default pathway. *Dev. Biol.* **275,** 403–421.

Vallier, L., Alexander, M., and Pedersen, R. A. (2005). Activin/nodal and FGF pathways cooperate to maintain pluripotency of human embryonic stem cells. *J. Cell Sci.* **118,** 4495–4509.

Wang, G., Zhang, H., Zhao, Y., Li, J., Cai, J., Wang, P., Meng, S., Feng, J., Miao, C., Ding, M., *et al.* (2005). Noggin and bFGF cooperate to maintain the pluripotency of human embryonic stem cells in the absence of feeder layers. *Biochem. Biophys. Res. Commun.* **330,** 934–942.

Wang, L., Li, L., Menendez, P., Cerdan, C., and Bhatia, M. (2005). Human embryonic stem cells maintained in the absence of mouse embryonic fibroblasts or conditioned media are capable of hematopoietic development. *Blood* **105,** 4598–4603.

Wang, Q., Fang, Z., Jin, F., Lu, Y., Gai, H., and Sheng, H. Z. (2005). Derivation and growing human embryonic stem cells on feeders. *Stem Cells* **23,** 1221–1227.

Wang, L., Schulz, T. C., Sherrer, E. S., Dauphin, D. S., Shin, S., Nelson, A. M., Ware, C. B., Zhan, M., Song, C. Z., Chen, X., *et al.* (2007). Self-renewal of human embryonic stem cells requires insulin-like growth factor-1 receptor and ERBB2 receptor signaling. *Blood* **110,** 4111–4119.

Weinrich, S. L., Pruzan, R., Ma, L., Ouellette, M., Tesmer, V. M., Holt, S. E., Bodnar, A. G., Lichsteiner, S., Kim, N. W., Trager, J. B., *et al.* (1997). Reconstitution of human telomerase with the template RNA component hTR and the catalytic protein subunit hTRT. *Nat. Genet.* **17,** 498–502.

Wong, R. C., Pebay, A., Nguyen, L. T., Koh, K. L., and Pera, M. F. (2004). Presence of functional gap junctions in human embryonic stem cells. *Stem Cells* **22,** 883–889.

Xiao, L., Yuan, X., and Sharkis, S. J. (2006). Activin A maintains self-renewal and regulates FGF, Wnt and BMP pathways in human embryonic stem cells. *Stem Cells* **24,** 1476–1486.

Xu, C., Inokuma, M. S., Denham, J., Golds, K., Kundu, P., Gold, J. D., and Carpenter, M. K. (2001). Feeder-free growth of undifferentiated human embryonic stem cells. *Nat. Biotechnol.* **19,** 971–974.

Xu, C., Police, S., Rao, N., and Carpenter, M. K. (2002). Characterization and enrichment of cardiomyocytes derived from human embryonic stem cells. *Circ. Res.* **91,** 501–508.

Xu, R. H., Chen, X., Li, D. S., Li, R., Addicks, G. C., Glennon, C., Zwaka, T. P., and Thomson, J. A. (2002). BMP4 initiates human embryonic stem cell differentiation to trophoblast. *Nat. Biotechnol.* **20,** 1261–1264.

Xu, C., Jiang, J., Sottile, V., McWhir, J., Lebkowski, J., and Carpenter, M. K. (2004). Immortalized fibroblast-like cells derived from human embryonic stem cells support undifferentiated cell growth. *Stem Cells* **22,** 972–980.

Xu, C., Rosler, E., Jiang, J., Lebkowski, J. S., Gold, J. D., O'Sullivan, C., Delavan-Boorsma, K., Mok, M., Bronstein, A., and Carpenter, M. K. (2005). Basic fibroblast growth factor supports undifferentiated human embryonic stem cell growth without conditioned medium. *Stem Cells* **23,** 315–323.

Xu, R.-H., Peck, R. M., Li, D. S., Feng, X., Ludwig, T., and Thomson, J. A. (2005). Basic FGF and suppression of BMP signaling sustain undifferentiated proliferation of human ES cells. *Nat. Methods* **3,** 185–190.

Xu, C., He, J. Q., Kamp, T. J., Police, S., Hao, X., O'Sullivan, C., Carpenter, M. K., Lebkowski, J., and Gold, J. D. (2006). Human embryonic stem cell-derived cardiomyocytes can be maintained in defined medium without serum. *Stem Cells Dev.* **15,** 931–941.

Xu, C., Police, S., Hassanipour, M., and Gold, J. D. (2006). Cardiac bodies: A novel culture method for enrichment of cardiomyocytes derived from human embryonic stem cells. *Stem Cells Dev.* **15,** 631–639.

Yao, S., Chen, S., Clark, J., Hao, E., Beattie, G. M., Hayek, A., and Ding, S. (2006). Long-term self-renewal and directed differentiation of human embryonic stem cells in chemically defined conditions. *Proc. Natl Acad. Sci. USA* **103,** 6907–6912.

Yin, A. H., Miraglia, S., Zanjani, E. D., Almeida-Porada, G., Ogawa, M., Leary, A. G., Olweus, J., Kearney, J., and Buck, D. W. (1997). AC133, a novel marker for human hematopoietic stem and progenitor cells. *Blood* **90,** 5002–5012.

Zhang, S. C., Wernig, M., Duncan, I. D., Brustle, O., and Thomson, J. A. (2001). *In vitro* differentiation of transplantable neural precursors from human embryonic stem cells. *Nat. Biotechnol.* **19,** 1129–1133.

Zwaka, T. P., and Thomson, J. A. (2003). Homologous recombination in human embryonic stem cells. *Nat. Biotechnol.* **21,** 319–321.

CHAPTER 19

Human Embryonic Stem Cells: Feeder-Free Culture

Michal Amit and Joseph Itskovitz-Eldor

Department of Obstetrics and Gynecology
Rambam Medical Center, Haifa
and The Bruce Rappaport Faculty of Medicine
Technion-Israel Institute of Technology
Haifa, Israel

Abstract
1. Introduction
2. Methods for Feeder Layer-Free Culture of hESCs
    2.1. Fibronectin coating of plates
    2.2. Culture medium
    2.3. Cell splitting
    2.4. Freezing cells
    2.5. Thawing cells
References

Abstract

In addition to their contribution to research fields such as early human development, self-renewal and differentiation mechanisms, human embryonic stem cells (hESCs) may serve as a tool for drug testing and for the study of cell-based therapies. Traditionally, these cells have been cultured with mouse embryonic fibroblast (MEF) feeder layers, which allow their continuous growth in an undifferentiated state. However, for all industrial or clinical purposes, hESCs should be cultured under defined conditions, preferably in a xeno-free culture system, where exposure to animal pathogens is prevented. To this end, different culture methods for hESCs, based on serum replacements and free of supportive layers were developed. This chapter discusses a simple, feeder layer-free culture system based on medium supplemented with $TGF_{\beta 1}$ and bFGF and fibronectin as matrix.

1. Introduction

Embryonic stem cells (ESCs) are pluripotent cells derived from the inner cell mass (ICM) of embryos at the blastocyst stage. The first ESC lines were derived from mouse embryos in 1981 by two separate groups (Evans and Kaufman, 1981; Martin, 1981). In their 25 years of existence, mouse ESCs were extensively used for the study of directed differentiation into specific cell types, self-maintenance processes, early developmental events, and more. Since these pioneering studies, ESC lines have been derived from three nonhuman primates (Suemori *et al.*, 2001; Thomson *et al.*, 1995, 1996) and eventually from human blastocysts (Thomson *et al.*, 1998). Accumulating knowledge shows that human embryonic stem cells (hESCs) meet most of the acceptable criteria for ESCs (Smith, 2000). To date there several established and well-characterized hESC lines in several laboratories around the world (Amit and Itskovitz-Eldor, 2004; Cowan *et al.*, 2004; Verlinsky *et al.*, 2005). The availability of these lines provides a unique new tool with widespread potential for research and for clinical applications.

The possible future use of hESCs in the clinical and industrial arena will require a reproducible, well-defined, and animal-free culture system for their routine culture. The traditional culture and derivation methods for hESCs, however, include mouse embryonic fibroblasts (MEFs) as feeder layers and medium supplemented with fetal bovine or calf serum (FBS) (Thomson *et al.*, 1998). When cultured in these conditions, hESCs grow as flat colonies with typical spaces between the cells, high nucleus-to-cytoplasm ratio and with at least two nucleoli (illustrated in Fig. 19.1A). Culture of hESCs requires meticulous care which includes daily medium change, routine passaging every four to six days, and occasionally mechanical removal of differentiated colonies from the culture. Nevertheless, hESCs can be cultured in high quantities, and frozen and thawed with reasonable survival rates (Amit and Itskovitz-Eldor, 2002; Reubinoff *et al.*, 2000, 2001; Thomson *et al.*, 1998).

The traditional culture method for hESCs exhibits several major disadvantages: (i) the use of MEFs as feeder layers and FBS may expose the cells to animal photogenes, (ii) batch-to-batch variations of the MEFs and serum in supporting undifferentiated hESC culture, and (iii) undefined culture conditions. These three drawbacks reduce the reproducibility of the method. To obtain a reproducible, well-defined, and animal-free culture system requires that substitutes for the serum and for the feeder layer be developed. In recent years, broad research into developing these substitutes has yielded four main progresses: (1) the ability to grow cells in serum-free culture conditions (Amit *et al.*, 2000), (2) feeder layer-free culture method based on the maintenance of the cells in an undifferentiated state on Matrigel matrix with 100% MEF-conditioned medium (Xu *et al.*, 2001), (3) the use of human substitutes for MEFs such as fetal fibroblasts, Fallopian tube epithelium (Richards *et al.*, 2002), or foreskin fibroblasts (Amit *et al.*, 2003; Hovatta *et al.*, 2004) as feeder layers, and (4) prolonged culture of hESCs in feeder

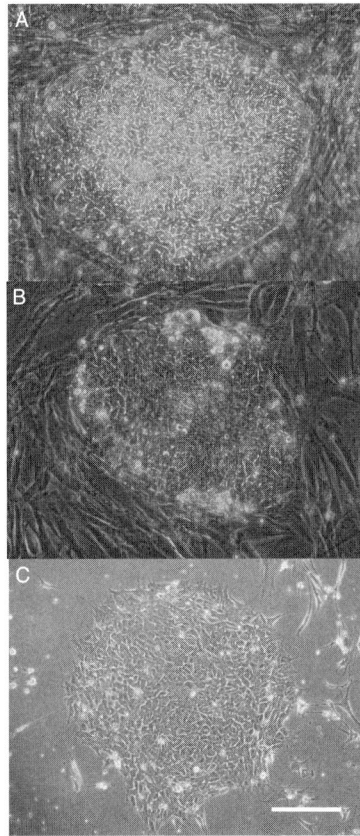

Figure 19.1 Morphology of a hESCs colony cultured in different culture conditions. (A) With MEFs, (B) with foreskin fibroblast, and (C) in animal-, serum- and feeder layer-free conditions. Bar = 100 μM.

layer-free conditions, while using selected growth factors and substitute matrices (Amit *et al.*, 2004; Ludwig *et al.*, 2006; Xu *et al.*, 2005a,b).

The simplest alternative to the culture method based on the use of MEF and FBS is the use of human supportive line and human serum or serum replacement. Among the cells proved to support continuous culture of undifferentiated hESCs one can find human fetal-derived fibroblasts (Richards *et al.*, 2002), foreskin fibroblasts (Amit *et al.*, 2003; Hovatta *et al.*, 2003), and adult marrow cells (Cheng *et al.*, 2003). Characterizations of hESCs cultured continuously with those cell lines demonstrate that the cells sustained all ESC features. Human fetal-derived fibroblasts and foreskin fibroblasts were also found to support the derivation of new hESC lines under animal-free or serum-free conditions (Hovatta *et al.*, 2003; Inzunza *et al.*, 2005; Richards *et al.*, 2002). In our experience, foreskin fibroblasts utilized as feeders enables hESCs to be cultured for over a year (124 passages, over 300 doublings) while exhibiting all stem cells features including; (i) expression of typical surface markers such as SSEA 3, SSEA4, TRA-1-60, and TRA-1-81; (ii) expression of transcription factors as Oct 4, Nanog, and Rex 1; (iii) differentiation into representative tissues of the three embryonic germ layers both in embryoid bodies (EBs) and in teratomas; (iv) high telomerase activity; and (v) maintenance of normal karyotype (Amit and Itskovitz-Eldor, unpublished data). The morphology of a hESC colony grown with foreskin fibroblast feeder layers, using an animal- and serum-free medium, is illustrated in Fig. 19.1B.

A unique source of supportive cells was offered by Xu *et al.* based on feeders derived from human EBs. These fibroblast-like cells, immortalized by vectors containing the hTERT gene, were shown to support the culture of hESCs for more than 14 passages (Xu *et al.*, 2004). The major disadvantage of this method, however, is the use of FBS for the isolation and culture of the feeder cells from EBs, as it does not provide an animal-free environment (Xu *et al.*, 2004; Yoo *et al.*, 2005).

Although human supportive cell lines were shown to maintain hESCs for prolonged cultures while preserving ESCs features, the method holds some disadvantages; first, the need for culture of the feeders themselves which will limit the large-scale culture of hESCs, and second, the culture system cannot be precisely defined due to differences between batches of feeder-layer cells and the used human serum. Ideally, the culture method for hESCs should therefore be a combination of an animal-, serum- and feeder layer-free culture system.

When developing a feeder layer-free culture method for hESCs, one should consider their differences from mouse ESCs. While mouse ESCs maintained their stem cell features and remained undifferentiated when grown on gelatine without a MEF feeder layer and medium supplemented with leukemia inhibitory factor (LIF) (Smith *et al.*, 1988; Williams *et al.*, 1988), LIF failed to support a feeder layer-free culture of hESCs (Thomson *et al.*, 1998). In fact, it had been demonstrated that the translation of proteins involved in LIF cellular pathway, such as STAT3, is weak or absent in hESCs (Daheron *et al.*, 2004; Humphrey *et al.*, 2004; Sato *et al.*, 2004). Thus, the mechanism underlining hESC self-maintenance is still unrevealed, and new candidate growth factors, cytokines and matrices should be examined.

The first supportive layer-free culture method for hESCs was reported in the study of Xu and colleagues (Xu *et al.*, 2001). The culture method relies on Matrigel, laminin, or fibronectin as matrix and 100% MEF-conditioned medium, supplemented with serum replacement (Xu *et al.*, 2001). When cultured under these conditions, hESCs can be stably maintained for over a year of continuous culture, while maintaining their ESC features. The growth and background differentiation rates were found similar to those of the traditional culture method, but several significant disadvantages still exist; (i) exposure to animal pathogens through the MEF-conditioned medium or Matrigel matrix, (ii) the culture system cannot be accurately defined due to variations between batches of MEFs used for the production of the conditioned medium and the Matrigel matrix, (iii) under these culture conditions hESCs form "auto feeders" which most probably contribute to the cells' maintenance on the one hand but results in the loss of cells to differentiation on the other, and (iv) the simultaneous culture of MEFs and hESCs limits the use of this culture system to scale-up the growth of hESCs for future clinical and industrial purposes.

Recent developments have eliminated the major disadvantage of Xu et all culture system—the use of MEFs conditioned medium. The same group proposed an improved system based on Matrigel matrix and medium supplemented with serum replacement and 40 ng/ml basic fibroblast growth factor (bFGF) (Xu et al., 2005b). The removal of the conditioned medium increased the background differentiation rate up to 28%, but this rate can be decreased to 20% by adding 75 ng/ml Flt-3 ligand to the culture medium (Xu et al., 2005b). Further improvement was achieved by a different group, through the addition of high amount of Noggin (Xu et al., 2005). The addition of inhibitor to the bone morphogenic protein (BMP) signal pathway, Noggin, allowed a further decrease in the differentiation rate to 10%, which equals that achieved by culturing the cells with Matrigel and MEF-conditioned medium (Xu et al., 2005). The same efficiency was reported when bFGF was added at a concentration of 100 ng/ml, without Noggin. Thus the importance of Noggin or other inhibitors of the BMP signal transduction pathway is still unclear.

The method described in this chapter is based on fibronectin as matrix, medium supplemented with 20% serum replacement, transforming growth factor $\beta 1$ (TGF$_{\beta 1}$) and bFGF (Amit et al., 2004). Under these culture conditions hESCs were maintained for over a year while maintaining stem cell characteristics. An example of the morphology of a colony cultured in these conditions is illustrated in Fig. 19.1C. The background differentiation, however, reached a level of 15%. The role of the components of the system is still unrevealed. TGF$_{\beta}$, a multipotent growth factor, is known to have either a positive or a negative effect on cellular proliferation, differentiation, migration, matrix deposition, or apoptosis, in the hematopoietic system, depending on the culture environment, stage of development of the cells, or in cases of response to injury or disease (Fortunel et al., 2000; Massague, 1990). The first clue as to the possible role of TGF$_{\beta 1}$ in maintaining undifferentiated hESCs came from a study by Schuldiner and colleagues, who examined the effects of eight different growth factors on hESC differentiation by evaluating cell-specific gene expression (Schuldiner et al., 2000). In that study, TGF$_{\beta 1}$ was assumed to repress cell differentiation, since it led to the production of relatively reduced cell-specific gene expression (Schuldiner et al., 2000). Additionally, TGF$_{\beta 1}$ is one of the components in Matrigel matrix, which is the most used matrix found to support hESC cell growth in feeder layer-free conditions (Xu et al., 2001).

Recently, the involvement of another member of the TGF$_{\beta}$ supper-family in the self-renewal mechanism of hESCs was put forth. Increasing evidence indicates that TGF$_{\beta 1,}$ activin, and Nodal pathway might be involved in hESC self-maintenance through transcription factor SMAD2/3 (Besser, 2004; James *et al.*, 2005; Valdimarsdottir and Mummery, 2005). SMAD2/3 signaling enhances in undifferentiated cells and reduces at early stages of differentiation, and its activation is required for the expression of markers of undifferentiated cells (Besser, 2004; James *et al.*, 2005). In contrast, SMAD 1/5 signaling, induced by BMP/GDF branch, decreases in undifferentiated cells and increases when differentiation processes begin (James *et al.*, 2005). The signal transduction pathways of $TGF_{\beta1}$/activin/Nodal and Wnt were suggested to have a positive effect through SMAD2/3 on Nanog or Oct 4 activity in the nucleus and thus take part in hESC self-renewal mechanism (Valdimarsdottir and Mummery, 2005). Indeed, the supplement of $TGF_{\beta1}$, activin, or Bio (activator of the Wnt signaling, $GSK3\beta$ inhibitor) to the culture medium contributes to the maintenance of hESCs as undifferentiated cells in supportive cell-free culture systems (Amit *et al.*, 2004; Beattie *et al.*, 2005; Sato *et al.*, 2004; Vallier *et al.*, 2005). However, none of the proposed factors, $TGF_{\beta1}$, activen, Nodal, or Wnt were proved to be directly involved in the hESC mechanism of self-maintenance, and in some cases, they failed to maintain prolonged culture of undifferentiated hESCs in a feeder layer-free environment (Dravid *et al.*, 2005). Thus, extensive research is needed to clarify the mechanism underlining hESC self-renewal, including the possibility that more than one pathway is involved, the existence of alternative pathways and the factors' synergistic effect.

Another component of the described culture system is fibronectin, which is used as a substitute matrix. Fibronectin, a basal lamina component, is frequently used to increase cell adhesion to the culture dishes. It acts through the integrin receptors, which in addition to their role as mediators of cell adhesion to extracellular matrix proteins, activate a variety of intracellular signal transduction pathways which may be involved in the cells' proliferation, apoptosis, shape, polarity, motility, gene expression profiles, and differentiation (Hynes, 2002). The fibronectin-specific integrin receptor, $\alpha_5\beta_1$, was demonstrated to be expressed in undifferentiated hESCs (Amit *et al.*, 2004). Further complementary research is required to explain the possible role of extracellular proteins such as fibronectin in the self-renewal of hESCs.

Most of the existing hESCs were derived with MEF as a feeder layer (Amit and Itskovitz-Eldor, 2004; Cowan *et al.*, 2004; Reubinoff *et al.*, 2000; Thomson *et al.*, 1998; Verlinsky *et al.*, 2005). Klimanskaya and colleagues were the first to report the isolation of hESCs in supportive cell-free culture condition (Klimanskaya *et al.*, 2005). The reported culture system includes MEF-produced matrix and medium supplemented with a high dose of bFGF (16 ng/ml) and LIF, serum replacement and plasmanate (Klimanskaya *et al.*, 2005). Six new hESCs line were successfully derived using this method, in which hESC characteristics are retained, including stable and normal karyotypes after more than 30 passages of continuous culture. Although the culture system includes some nondefined materials (MEF matrix and knockout serum replacement) it proves the feasibility of a supportive feeder layer-less derivation of hESCs. A recent publication by Ludwig and colleagues demonstrates a defined serum- and animal-free medium which not only supports a culture of undifferentiated hESCs for prolonged periods, but is also suitable for hESC isolation under feeder layer-free culture conditions (Ludwig *et al.*, 2006). Interestingly, the medium combination requires the addition of both bFGF and $TGF_{\beta 1}$ in addition to Licl, GABA, and pipecolic acid. The matrix consisted of human collagen, fibronectin and laminin. Following their culture for several months continuously, the newly derived cells sustained most hESC features. Thus, for the first time defined, animal-, serum- and feeder-free culture conditions for hESCs are presented. However, one disturbing difficulty of the new method is the karyotype stability. In the study, two new hESC lines were derived: one was reported to harbor 47,XXY after four months of continuous culture (it is unclear whether the embryo was originally defected) and the second exhibited trisomy 12 after seven months of continuous culture. It is yet to be determined whether these are exceptional events of karyotype abnormalities that occurred during the prolonged culture and which may also occur while using feeder layers, or whether the culture method does not actually sustain karyotype stability.

2. Methods for Feeder Layer-Free Culture of hESCs

2.1. Fibronectin coating of plates

The recommended concentration of fibronectin is 50 μgr/10 cm^2. Both human cellular and plasma fibronectin were tested and found suitable for hESC culture (Sigma human foreskin fibroblast cellular fibronectin F6277; Sigma human plasma fibronectin F2006; Chemicon human plasma fibronectin FC010-10) (Amit *et al.*, 2004). All plates should be covered with fibronectin 30 min before the plating of cells.

1. Dilute fibronectin to desired concentration. Recommended concentration for the stock solution is 1 mg/10 ml of sterile water (Sigma W1503).
2. Filter the fibronectin through a 22 μM filter. The stock solution can be stored up to two weeks at 4 °C.
3. Cover the dish with fibronectin solution to reach a concentration of 50 μgr/10 cm^2. Examples of recommended amounts are listed in Table 19.1.

Table 19.1
Recommended amount of fibronectin stock solution per well

Plate/dish	Volume of fibronectin stock solution per well (ml)
4 wells (2.5 cm^2)	0.3
6 wells (10 cm^2)	0.5
35 mm	0.5

4. Leave at room temperature or in incubator for at least 30 min. If desired, plates covered with fibronectin can be prepared in advance (overnight).

If water were used to dissolve the fibronectin there is no need to collect the fibronectin residues before plating the cells. It is recommended to plate the cells directly onto the fibronectin residues.

If desired, Matrigel matrix can replace the fibronectin. The recommended dilution is 1:40 (with plain medium). The matrix should be prepared according to the manufacturer's instructions.

2.2. Culture medium

The medium consists of the following materials: 85% knockout-Dulbeccos Modified Eagle Medium (ko-DMEM), supplemented with 15% knockout serum replacement, 2 mM L-glutamine, 0.1 mM β-mercaptoethanol, 1% nonessential amino acid stock, 0.12 ng/ml TGFβ1, and 4 ng/ml bFGF (all Invitrogen Corporation products, Grand Island NY, USA, but the TGFβ1 which is from R&D Systems Minneapolis MN, USA).

1. Pour all materials into a 22 μM filter unit, and filter.
2. Store at 4 °C for up to 5 days. Do not expose to light.

An example for the preparation of 500 ml of medium is described in Table 19.2. Two milliliter of medium should be added to each 10 cm^2 surface area of culture dish, and should be replaced daily by fresh medium.

2.3. Cell splitting

The most used splitting medium for hESCs is collagenase type IV. The splitting medium consists of 1 mg/ml collagenase type IV (Worthington, C.N. 17104019) diluted with DMEM (Invitrogen Corporation products, Grand Island NY, USA, C.N. 41965-039) and should be filtered through a 0.22 μM filter.

Table 19.2

Examples of recommended concentrations and amounts of culture medium ingredients

Material	Final concentrations	For 500 ml medium
Ko-DMEM	85%	414 ml
Ko-Serum replacement	15%	75 ml
Non-essential amino acid	1%	5 ml
L-Glutamine	2 mM	5 ml
β-Mercaptoethanol 50 mM	0.1 mM	1 ml
bFGF	4 ng/ml	2000 ng
TGF$_{\beta1}$	0.12 ng/ml	60 ng

1. Remove medium from well. Add 0.5 ml collagenase splitting medium and incubate for 30 min. Most colonies will float at the end of the incubation time. If needed, the incubation period can be increased up to 1 h. Do not exceed the recommended incubation time due to possible damage to the cells, which may include increasing incidence of karyotype abnormalities and decreased survival rates.
2. Add 1 ml of culture medium and gently collect cells with a 5 ml pipette.
3. Collect cell suspension and put into conical tube.
4. Centrifuge 3 min at $300 \times g$ at a recommended temperature of 4 °C.
5. Re-suspend cells in culture media and plate directly on previously fibronectin-covered plate.

Cells should be split at a recommended ratio of 1:2 very 4–5 days.

2.4. Freezing cells

The cells should be frozen at a freezing solution consisting of 20% Dimethylsolfid (DMSO, Sigma D2652), 30% FBS, serum replacement or human serum, and 50% DMEM. The recommended freezing ratio is cells from 10 cm^2 culture surface area or 2 million cells in one caryovial (volume of 250–500 μl).

1. First harvest the cells like you would when splitting; add splitting medium (collagenase) and incubate for 30 min
2. Add 1 ml culture medium, then gently scrape the cells using a 5 ml pipette and transfer cells into conical tube
3. Centrifuge 3 min at $300 \times g$ at a recommended temperature of 4 °C
4. Re-suspend cells in the culture medium. Do not fracture the cells into small clumps
5. Drop by drop add an equivalent volume of freezing medium and mix gently
6. Transfer 0.5 ml into a 1 ml cryogenic vial
7. Freeze overnight at −80 °C (the use of freezing boxes, from Nalgene C.N., increases the survival rates)
8. Transfer to liquid nitrogen on the following day

2.5. Thawing cells

The freezing process should be conducted as efficiently as possible; delays in the procedure may lead to increased cell death.

1. Remove vial from liquid nitrogen
2. Gently swirl vial in 37 °C water bath
3. When a small pellet of frozen cells remains, wash vial in 70% ethanol

4. Pipette the content of the vial up and down once to mix
5. Place contents of vial into conical tube and add, drop by drop, 2 ml of culture medium
6. Centrifuge 3 min at $300 \times g$ at a recommended temperature of 4 °C
7. Remove supernatant and re-suspend cells in 3 ml fresh culture medium
8. Place cell suspension on one well of a 6-well plate, or on a 4-well plate, previously covered with fibronectin

Acknowledgments

The authors thank Mrs. Hadas O'Neill for editing the manuscript. The research conducted by the authors was partly supported by NIH grant R24RR18405.

References

Amit, M., Carpenter, M. K., Inokuma, M. S., Chiu, C. P., Harris, C. P., Waknitz, M. A., Itskovitz-Eldor, J., and Thomson, J. A. (2000). Clonally derived human embryonic stem cell lines maintain pluripotency and proliferative potential for prolonged periods of culture. *Dev. Biol.* **227**, 271–278.

Amit, M., and Itskovitz-Eldor, J. (2002). Derivation and spontaneous differentiation of human embryonic stem cells. *J. Anat.* **200**, 225–232.

Amit, M., and Itskovitz-Eldor, J. (2004). Isolation, characterization and maintenance of primate ES cells. *In* "Handbook of Stem Cells" (R. P. Lanza, Ed.), Elsevier Science, Chapter 45, pp. 419–436. .

Amit, M., Margulets, V., Segev, H., Shariki, C., Laevsky, I., Coleman, R., and Itskovitz-Eldor, J. (2003). Human feeder layers for human embryonic stem cells. *Biol. Reprod.* **68**, 2150–2156.

Amit, M., Shariki, K., Margulets, V., and Itskovitz-Eldor, J. (2004). Feeder and serum-free culture system for human embryonic stem cells. *Biol. Reprod.* **70**, 837–845.

Beattie, G. M., Lopez, A. D., Bucay, N., Hinton, A., Firpo, M. T., King, C. C., and Hayek, A. (2005). Activin A maintains pluripotency of human embryonic stem cells in the absence of feeder layers. *Stem Cells* **23**, 489–495.

Besser, D. (2004). Expression of nodal, lefty-a, and lefty-B in undifferentiated human embryonic stem cells requires activation of Smad2/3. *J. Biol. Chem.* **279**, 45076–45084.

Cheng, L., Hammond, H., Ye, Z., Zhan, X., and Dravid, G. (2003). Human adult marrow cells support prolonged expansion of human embryonic stem cells in culture. *Stem Cells* **21**, 131–142.

Cowan, C. A., Klimanskaya, I., McMahon, J., Atienza, J., Witmyer, J., Zucker, J. P., Wang, S., Morton, C. C., McMahon, A. P., Powers, D., and Melton, D. A. (2004). Derivation of embryonic stem-cell lines from human blastocysts. *N. Engl. J. Med.* **350**, 1353–1356.

Daheron, L., Opitz, S. L., Zaehres, H., Lensch, W. M., Andrews, P. W., Itskovitz-Eldor, J., and Daley, G. Q. (2004). LIF/STAT3 signaling fails to maintain self-renewal of human embryonic stem cells. *Stem Cells* **22**, 770–778.

Dravid, G., Ye, Z., Hammond, H., Chen, G., Pyle, A., Donovan, P., Yu, X., and Cheng, L. (2005). Defining the role of Wnt/beta-catenin signaling in the survival, proliferation, and self-renewal of human embryonic stem cells. *Stem Cells* **23**, 1489–1501.

Evans, M. J., and Kaufman, M. H. (1981). Establishment in culture of pluripotential cells from mouse embryos. *Nature* **292**, 154–156.

Fortunel, N. O., Hatzfeld, A., and Hatzfeld, J. A. (2000). Transforming growth factor-*β*: Pleiotropic role in the regulation of hematopoiesis. *Blood* **96**, 2022–2036.

Hovatta, O., Mikkola, M., Gertow, K., Stromberg, A. M., Inzunza, J., Hreinsson, J., Rozell, B., Blennow, E., Andang, M., and Ahrlund-Richter, L. (2003). A culture system using human foreskin fibroblasts as feeder cells allows production of human embryonic stem cells. *Hum. Reprod.* **18,** 1404–1409.

Humphrey, R. K., Beattie, G. M., Lopez, A. D., Bucay, N., King, C. C., Firpo, M. T., Rose-John, S., and Hayek, A. (2004). Maintenance of pluripotency in human embryonic stem cells is STAT3 independent. *Stem Cells* **22,** 522–530.

Hynes, R. O. (2002). Integrins: Bidirectional, allosteric signaling machines. *Cell* **110,** 673–687.

Inzunza, J., Gertow, K., Stromberg, M. A., Matilainen, E., Blennow, E., Skottman, H., Wolbank, S., Ahrlund-Richter, L., and Hovatta, O. (2005). Derivation of human embryonic stem cell lines in serum replacement medium using postnatal human fibroblasts as feeder cells. *Stem Cells* **23,** 544–549.

James, D., Levine, A. J., Besser, D., and Hemmati-Brivanlou, A. (2005). TGFbeta/activin/nodal signaling is necessary for the maintenance of pluripotency in human embryonic stem cells. *Development* **132,** 1273–1282.

Klimanskaya, I., Chung, Y., Meisner, L., Johnson, J., West, M. D., and Lanza, R. (2005). Human embryonic stem cells derived without feeder cells. *Lancet* **365,** 1636–1641.

Ludwig, T. E., Levenstein, M. E., Jones, J. M., Berggren, W. T., Mitchen, E. R., Frane, J. L., Crandall, L. J., Daigh, C. A., Conard, K. R., Piekarczyk, M. S., Llanas, R. A., and Thomson, J. A. (2006). Derivation of human embryonic stem cells in defined conditions. *Nat. Biotechnol.* **24,** 185–187.

Martin, G. R. (1981). Isolation of a pluripotent cell line from early mouse embryos cultured in medium conditioned by teratocarcinoma stem cells. *Proc. Natl. Acad. Sci. USA* **78,** 7634–7638.

Massague, J. (1990). The transforming growth factor-beta family. *Annu. Rev. Cell Biol.* **6,** 597–641.

Reubinoff, B. E., Pera, M. F., Fong, C., Trounson, A., and Bongso, A. (2000). Embryonic stem cell lines from human blastocysts: Somatic differentiation *in vitro*. *Nat. Biotechnol.* **18,** 399–404.

Reubinoff, B. E., Pera, M. F., Vajta, G., and Trounson, A. O. (2001). Effective cryopreservation of human embryonic stem cells by the open pulled straw vitrification method. *Hum. Reprod.* **16,** 2187–2194.

Richards, M., Fong, C. Y., Chan, W. K., Wong, P. C., and Bongso, A. (2002). Human feeders support prolonged undifferentiated growth of human inner cell masses and embryonic stem cells. *Nat. Biotechnol.* **20,** 933–936.

Sato, N., Meijer, L., Skaltsounis, L., Greengard, P., and Brivanlou, A. H. (2004). Maintenance of pluripotency in human and mouse embryonic stem cells through activation of Wnt signaling by a pharmacological GSK-3-specific inhibitor. *Nat. Med.* **10,** 55–63.

Schuldiner, M., Yanuka, O., Itskovitz-Eldor, J., Melton, D. A., and Benvenisty, N. (2000). Effect of eight-growth factors on the differentiation of cells derived from human ES cells. *Proc. Natl. Acad. Sci. USA* **97,** 11307–11312.

Smith, A. G. (2000). Embryonic stem cells. *In* "Stem cell biology" (D. R. Marshak, R. L. Gardner, and D. Gottlieb, eds.), Cold spring harbor laboratory Press, Chapter 10, pp. 205–230. .

Smith, A. G., Heath, J. K., Donaldson, D. D., Wong, G. G., Moreau, J., Stahl, M., and Rogers, D. (1988). Inhibition of pluripotential embryonic stem cell differentiation by purified polypeptides. *Nature* **336,** 688–690.

Suemori, H., Tada, T., Torii, R., Hosoi, Y., Kobayashi, K., Imahie, H., Kondo, Y., Iritani, A., and Nakatsuji, N. (2001). Establishment of embryonic stem cell lines from cynomolgus monkey blastocysts produced by IVF or ICSI. *Dev. Dyn.* **222,** 273–279.

Thomson, J. A., Itskovitz-Eldor, J., Shapiro, S. S., Waknitz, M. A., Swiergiel, J. J., Marshall, V. S., and Jones, J. M. (1998). Embryonic stem cell lines derived from human blastocysts. *Science* **282,** 1145–1147 [erratum in *Science* (1998) 282, 1827].

Thomson, J. A., Kalishman, J., Golos, T. G., Durning, M., Harris, C. P., Becker, R. A., and Hearn, J. P. (1995). Isolation of a primate embryonic stem cell line. *Proc. Natl. Acad. Sci. USA* **92,** 7844–7848.

Thomson, J. A., Kalishman, J., Golos, T. G., Durning, M., Harris, C. P., and Hearn, J. P. (1996). Pluripotent cell lines derived from common marmoset (*Callithrix jacchus*) blastocysts. *Biol. Reprod.* **55,** 254–259.

Valdimarsdottir, G., and Mummery, C. (2005). Functions of the TGFbeta superfamily in human embryonic stem cells. *APMIS* **113,** 773–789.

Vallier, L., Alexander, M., and Pedersen, R. A. (2005). Activin/Nodal and FGF pathways cooperate to maintain pluripotency of human embryonic stem cells. *J. Cell Sci.* **118,** 4495–4509.

Verlinsky, Y., Strelchenko, N., Kukharenko, V., Rechitsky, S., Verlinsky, O., Galat, V., and Kuliev, A. (2005). Human embryonic stem cell lines with genetic disorders. *Reprod. Biomed. Online* **10,** 105–110.

Williams, R., Hilton, D., Pease, S., Wilson, T., Stewart, C., Gearing, D., Wagner, E., Metcalf, D., Nicola, N., and Gough, N. (1988). Myeloid leukemia inhibitory factor maintains the developmental potential of embryonic stem cells. *Nature* **336,** 684–687.

Xu, C., Inokuma, M. S., Denham, J., Golds, K., Kundu, P., Gold, J. D., and Carpenter, M. K. (2001). Feeder-free growth of undifferentiated human embryonic stem cells. *Nat. Biotechnol.* **19,** 971–974.

Xu, C., Jiang, J., Sottile, V., McWhir, J., Lebkowski, J., and Carpenter, M. K. (2004). Immortalized fibroblast-like cells derived from human embryonic stem cells support undifferentiated cell growth. *Stem Cells* **22,** 972–980.

Xu, R. H., Peck, R. M., Li, D. S., Feng, X., Ludwig, T., and Thomson, J. A. (2005a). Basic FGF and suppression of BMP signaling sustain undifferentiated proliferation of human ES cells. *Nat. Methods* **2,** 185–190.

Xu, C., Rosler, E., Jiang, J., Lebkowski, J. S., Gold, J. D., O'Sullivan, C., Delavan-Boorsma, K., Mok, M., Bronstein, A., and Carpenter, M. K. (2005b). Basic fibroblast growth factor supports undifferentiated human embryonic stem cell growth without conditioned medium. *Stem Cells* **23,** 315–323.

Yoo, S. J., Yoon, B. S., Kim, J. M., Song, J. M., Roh, S., You, S., and Yoon, H. S. (2005). Efficient culture system for human embryonic stem cells using autologous human embryonic stem cell-derived feeder cells. *Exp. Mol. Med.* **37,** 399–407.

CHAPTER 20

Neural Stem Cells, Neurons, and Glia from Embryonic Stem Cells

Steven M. Pollard,★ Alex Benchoua,[†,‡] and Sally Lowell[†]

★Wellcome Trust Centre for Stem Cell Biology
University of Cambridge
Cambridge CB2 1QR, United Kingdom

[†]Centre Development in Stem Cell Biology
Institute for Stem Cell Research
School of Biological Sciences
University of Edinburgh
Edinburgh EH9 3JQ, United Kingdom

[‡]INSERM/UEVE U861861, ISTEM/AFM
5 Rue H. Desbrueres,
91030 Evry, France

Abstract
1. Introduction
2. Protocols
   2.1. Conversion of mouse ES cells to neural progenitors in adherent monolayer
   2.2. A protocol for neural induction of human ES cells in adherent monolayer
   2.3. Converting mouse pluripotent ES cells to tissue-specific NS cells
3. Summary
4. Media and Reagents
   4.1. ES cell media
   4.2. N2B27 media
   4.3. N-2 supplement
References

407

Abstract

Embryonic stem (ES) cells are a unique resource, providing in principle access to unlimited quantities of every cell type *in vitro*. They constitute an accessible system for modeling fundamental developmental processes, such as cell fate choice, commitment, and differentiation. Furthermore the pluripotency of ES cells opens up opportunities for use of human ES (hES) cells as a source of material for pharmaceutical screening and cell-based transplantation therapies. Widespread application of ES cell-based technologies in both basic biology and medicine necessitates development of robust and reliable protocols for controlling self-renewal and differentiation in the laboratory. Here we describe protocols that enable the conversion of mouse ES cells in simple adherent conditions to either terminally differentiated neurons and glia, or self-renewing but lineage restricted neural stem (NS) cell lines. We also report current status in transfer of these approaches to hES cells.

1. Introduction

Embryonic stem (ES) cells provide a valuable and convenient source of neural cells (Gottlieb and Huettner, 1999). However, despite progress in recent years, we do not as yet have full command over these cells, and it is difficult to direct differentiation of the entire population of ES cells into neural progenitors. Studying neural specification *in vitro* will help us to further improve the efficiency and predictability with which we can generate neural cells from ES cells. It could also improve our understanding of mammalian development as neural differentiation of ES proceeds through a sequence of differentiation steps that appear to closely recapitulate neural development *in vivo* (Billon *et al.*, 2002; Conti *et al.*, 2005; Ying *et al.*, 2003b). This is particularly significant if we are to understand such events in human embryogenesis, as the earliest developmental stages are not accessible. Furthermore, ES cells have several advantages over *in vivo* or primary culture systems, in particular with regard to their immortality, which provides an unlimited cellular resource for routine biochemical analysis, genetic manipulations, and small molecule screening.

Here, we describe a protocol for neural conversion of ES cells that is designed not only for high efficiency of neural conversion but also for tractability as an experimental system. We go on to describe how this neural conversion protocol can be adapted for human ES (hES) cells. These protocols provide a means to study the process of neural induction and commitment.

Once neural cells are generated, they provide a resource for investigating the subsequent self-renewal and differentiation of neural progenitors. Progress in this area has been hampered by an inability to access homogeneous populations of true neural stem (NS) cells. However, we have recently established conditions for isolating and expanding ES cell-derived NS cell lines (Conti *et al.*, 2005).

Such somatic stem cells provide a parallel system to ES cells in which to elucidate molecular details of self-renewal, and differentiation. Protocols for the derivation and differentiation of mouse NS cells are described in the later sections. A key future challenge will be to direct these NS cells to produce specific neuronal and glial subtypes.

2. Protocols

2.1. Conversion of mouse ES cells to neural progenitors in adherent monolayer

2.1.1. Overview

Several different approaches have been developed for generating neural cells from ES cells (Stavridis and Smith, 2003). Of these, the monolayer differentiation protocol described here has several key features that make it especially useful as an experimental system for investigating the mechanisms that regulate neural specification (Ying *et al.*, 2003b). In this section, we outline some of these features.

2.1.1.1. The monolayer protocol is not based on cell selection

No current protocol is completely efficient in converting all ES cells into neural progenitors. Some neural differentiation protocols partially overcome this problem by relying on preferential survival of neural cells in minimal culture media (Li *et al.*, 2001; Tropepe *et al.*, 2001). Alternatively, targeted lineage selection can be applied to eliminate non-neural cells (Li *et al.*, 1998). These selective approaches are a useful means to obtain enriched populations of neural cells, but are less useful as an experimental model for studying the mechanisms that underlie neural fate choice.

The monolayer differentiation protocol is not selective; rather it brings about neural conversion with only modest cell death. We can therefore follow the fate of effectively all cells within a population, including that minority of cells that do not enter the neural lineage. This allows us to address several important questions. What proportion of cells initially resists neural conversion? Are they simply delayed in differentiation, or do they irreversibly commit to an alternative fate? What alternative differentiated fate do they follow? What is their distribution in relation to the neural cells within the culture? By studying those cells that resist neural specification, rather than eliminating them through negative selection, we can gain important insights into the mechanisms that regulate this fate decision.

2.1.1.2. The monolayer protocol does not rely on undefined media components or heterologous cell interactions

The monolayer protocol uses only well-characterized components. This contrasts with classical neural differentiation protocols, which rely on multicellular aggregation in serum-containing medium, usually in combination with the pleiotropic inducer retinoic acid (Bain *et al.*, 1995). Other protocols depend upon coculture of ES cells with stromal cells such as the PA6 cell line (Kawasaki *et al.*, 2000). The molecular nature of the differentiation-inducing activity delivered by PA6 cells remains obscure, and this activity can vary with cell batch or passage number. Furthermore, the rate of neural differentiation is significantly delayed in ES cells cocultured with PA6 feeders in comparison with feeder-free cultures in the same culture media (H. Kawasaki *et al.*, unpublished observations), which suggests that PA6 cells may deliver a complex mixture of pro- and antidifferentiation signals.

2.1.1.3. The monolayer protocol allows direct observation and analysis of the differentiating population

Cells attached to the dish in a two-dimensional monolayer can readily be observed throughout the differentiation process. This contrasts with floating aggregate cultures where the cells are not accessible to live microscopy at single-cell resolution. The absence of stromal cells also facilitates observation and simplifies population level analyses of mRNA and protein samples, by avoiding the need to first separate the ES cell progeny from stromal cells.

2.1.2. Protocol: Conversion of mouse ES cells to neural cells in adherent monolayer

Media and reagents for all protocols described in this chapter are described in Section 4:

1. The ES cells that are to be used in this protocol should be maintained in feeder-free culture as described (Ying and Smith, 2003). If ES cells have been maintained on a feeder layer, they must be adapted to feeder-free conditions for several passages before initiating the monolayer differentiation protocol.

2. One day before initiating the protocol, trypsinize a near-confluent ES cell culture and replate at 30–40% confluence, so that ES cells will reach 70–80% confluence on the following day. Both this trypsinization step itself, and the relatively high ES cell density, help to maximize the efficiency and consistency of neural conversion by ensuring that the starting material is a substantially pure population of undifferentiated ES cells.

3. The next day, prepare dishes by coating with 0.1% gelatin for at least 1 h at room temperature. Remove gelatin and allow plates to dry. This protocol works best in either 9 cm dishes or 6-well plates.

4. Wash ES cultures with PBS and then trypsinize for 5 min.

5. Harvest cells in serum-free media without growth factors and spin for 3 min at 1000 rpm to pellet cells. Remove media and repeat. Resuspend the washed cells in N2B27 complete media. Unsupplemented Neurobasal or DMEM–F12 is sufficient for these wash steps. Quenching of trypsin with serum medium is not necessary as long as care is taken to remove all traces of supernatant. Some workers prefer to quench with serum during the first wash: in this case, care

must be taken to remove all traces using a second serum-free wash. Even very small amounts of either residual trypsin or of serum can compromise neural differentiation.

6. Count cells carefully. The efficiency of neural differentiation is highly dependent on correct plating density. If the density is too low, cell viability is compromised, whilst relatively small increases in cell density can dramatically reduce the efficiency of neural differentiation, more cells instead remaining as undifferentiated ES cells.

7. Plate cells onto the gelatin-coated dishes in complete N2B27 media. The optimal plating density is generally around 10^4 cells/cm^2, but this can differ between cell lines and should be optimized in each case. Even the same cell line can differ in its density sensitivity depending on culture conditions, so it is advisable to plate cells over a range of densities (e.g., 8×10^3 per cm^2, 10^4 per cm^2, and 1.2×10^4 per cm^2) to maximize the probability of obtaining optimal differentiation.

8. Incubate at 37 °C 5% CO_2.

9. Replace medium every 1–2 days.

10. (Optional) For optimal neuronal differentiation and survival, cells should be replated onto laminin-coated dishes from the seventh day of differentiation.

2.1.3. Monitoring neural and neuronal differentiation

The Sox1–GFP reporter cell line is convenient for monitoring and quantitating the transition from ES cell to neural precursor. This cell line, called 46C, was generated by gene targeting of E14Tg2a cells (Aubert *et al.*, 2003). The open reading frame of Sox1 is replaced with the coding sequence for eGFP linked through an internal ribosomal entry site (IRES) to a puromycin-resistant gene. Sox1 is an early marker of neuroepithelial cells throughout the extent of the developing neural tube (Pevny *et al.*, 1998). The GFP reporter faithfully recapitulates the expression of Sox1 both *in vivo* and *in vitro* definitely (Aubert *et al.*, 2003).

This protocol has also been successfully applied to many other cell lines. Where Sox1–GFP reporter cells are not used, the transition from ES cell to neural progenitor can instead be observed in live cells by a distinctive change in morphology (Fig. 20.1). The nucleus, which is large and distinct in ES cells, becomes opaque and barely visible. Cells slightly elongate and pack closely together, often into rosette structures, with more distinct cell boundaries than ES cells. Commercially available antibodies can be used to monitor the loss of the ES cell marker Oct4 (Santa Cruz; C10, used at 1:200) and acquisition of the neural progenitor marker nestin (DSHB, Iowa; Rat401, used at 1:20).

Neuronal differentiation becomes detectable from around the fifth day, and the number of neurons progressively increases over the subsequent 2–3 days. Neurons can be easily identified by their characteristic small cell bodies with long very thin

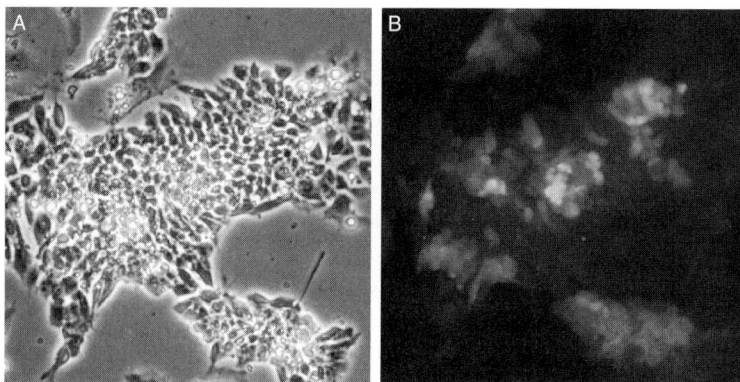

Figure 20.1 Conversion of mouse ES cells (46C) to neural progenitors in adherent monolayer. (A) Typical morphology on day 3 of differentiation. (B) Sox1–GFP expression.

projections, and by immunostaining for early neuronal markers such as beta-3-tubulin (TuJ1 antibody, Covance), or by use of a reporter cell line such as TK23 which carries an insertion of GFP into the neuronal tau-locus (Tucker *et al.*; Ying *et al.*). Proneural genes such as Mash1, Neurogenin2, and Olig2 become detectable in a subset of cells shortly before the onset of neuronal differentiation. Early markers of multipotent neural precursors such as BLBP and RC2 also become detectable at around this time. Note that these cultures contain a heterogeneous mixture of cells at different stages of differentiation, with Sox1, BLBP, and TuJ1 generally marking three distinct nonoverlapping subpopulations, and with mosaic expression of regional identity genes in contrast to the NS cells to be described later.

2.1.4. Troubleshooting

2.1.4.1. Poor plating efficiency, significant numbers of dead or floating cells 24 h after plating

Ensure that the starting population of ES cells is healthy, and that all traces of trypsin are carefully removed during the two wash steps. Coat dishes for at least 1 h with gelatin. Make sure that gelatin solution is then completely removed and that dishes are allowed to dry before plating cells.

2.1.4.2. Progressive increase in cell death during the protocol

Some cell death is normal during this protocol, but renewing the medium should reveal underlying healthy cultures throughout. However, cells sometimes begin to die in large numbers at around the third day of the protocol. The most common reasons for this are use of inappropriate basal medium or too low an initial plating density. Another common cause is use of media or N-2 that has been stored for too long. In our experience, N-2 can deteriorate after 3 weeks even when stored at −20 °C, and N2B27 should be stored at 4 °C for a maximum of 1 week. It is also important to maintain the cells throughout in a stable 37 °C 5% CO_2 environment, avoiding prolonged periods out of the incubator.

2.1.4.3. Large numbers of undifferentiated ES cells persist beyond the first 3 days

A small proportion of cells, typically 10–15% resist differentiation even after a week or more under this protocol. This is most likely due to local autocrine secretion of antidifferentiation factors. If significantly more than 15% of cells remain undifferentiated beyond the first 4–5 days, it is likely that the initial plating density was too high. The majority of residual undifferentiated ES cells will generally differentiate when replated into a fresh culture. If not, this may indicate the presence of chromosomally abnormal ES cells with impaired differentiation capacity.

2.1.4.4. Non-neural differentiation

A small subpopulation of cells, typically around 10%, differentiates into non-neural lineages under these monolayer conditions. These are readily apparent as larger flatter cells around the edge of neural colonies. High proportions of non-neural cells could indicate that the plating density is too high, or that the media is not being changed frequently enough, or that small residual traces of serum or BMP remained at the initial plating step due to incomplete washing.

It is also important to ensure that the starting population of ES cells are healthy, and do contain large numbers of differentiating cells. Trypsinization and replating of the ES cells 24 h prior to initiating the monolayer protocol can help to eliminate predifferentiated cells.

Note also that substrates such as laminin or fibronectin, which are often used to support primary cultures taken directly from neural tissue, are not suitable for this protocol because they direct ES cells to differentiate into non-neural cell types.

2.1.4.5. Variability between cell lines

Most ES cells tested undergo neural differentiation efficiently under this protocol. However, there is line-to-line variability, especially in the optimal plating density, so it is crucial to optimize this for each new cell line. Occasional cell lines may be resistant to neural differentiation using this protocol, possibly associated with genetic or epigenetic variation.

2.2. A protocol for neural induction of human ES cells in adherent monolayer

2.2.1. Overview

The protocol described earlier for neural differentiation of mouse ES cells has been successfully adapted to three different hES cell lines, Hs181, Hs238, and EDI1, with certain modifications as described later. The 181 and 237 cell hES lines were derived from supernumerary human blastocysts as described previously (Hovatta et al., 2003). The EDI1 cell line was derived following the same protocol (J. Nichols, unpublished results).

2.2.2. Protocol: Conversion of human ES cells to neural cells in adherent monolayer

1. hES cells are routinely cultivated in 6-well plates on a feeder layer of commercially available human foreskin fibroblasts (ATCC, lines HFF-1 and HFF-2), in N2B27 medium supplemented with LIF (10 ng/ml), BMP4 (3 ng/ml, R&D systems) and FGF-2 (10 ng/ml, R&D systems). Cells are passaged at a split ratio of 1:2 each week using collagenase IV (1 mg/ml).

2. When the undifferentiated ES cell culture reaches 60–70% confluence (about 100 colonies in a well with 5000/10,000 cells per colony), detach the colonies using collagenase IV and further dissociate into small clumps by gentle trituration. Feeder cells will also become detached by the collagenase treatment. These can be removed by incubating the cell mixture for 6 h in a gelatin-coated flask in fresh medium without LIF and BMP4. hES cells do not attach to gelatin whereas the feeder cells do.

3. Collect the media containing the ES cell clumps and spin down at low speed (250g) before resuspending into N2B27 medium + FGF-2 in the absence of LIF and BMP. Plate the ES cells in another 6-well plate previously coated with Matrigel growth factor-reduced matrix (BD biosciences). Exchange media fully every second day. Note that in contrast to the mouse ES cell protocol, where addition of exogenous FGF-2 is not required, FGF-2 is necessary to maintain survival of both human ES cells and neural precursors.

2.2.3. Monitoring neural induction

The first morphological changes appear after about 4 days of culture. The cells adopt an elongated morphology similar to that described earlier for mouse neural progenitors and form rosettes and neural tube-like structures (Fig. 20.2). We have analyzed expression for neural lineage markers at these stages. Pax6 is one of the

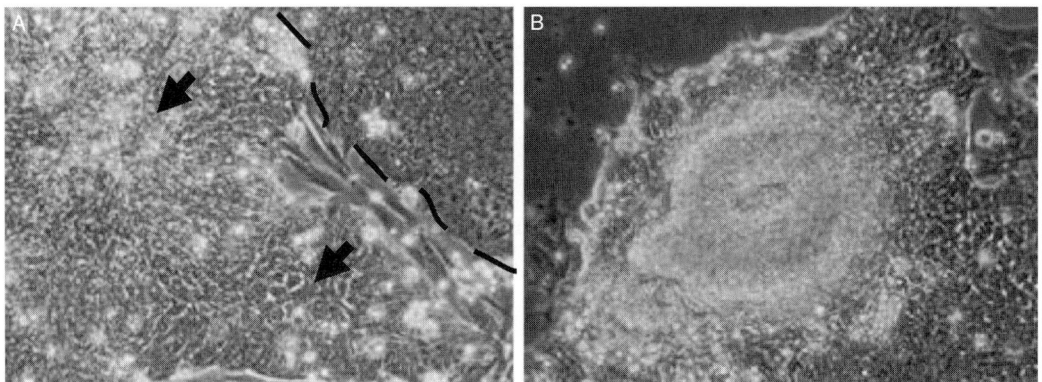

Figure 20.2 Conversion of human ES cells to neural cells in adherent monolayer. (A) Typical rosette structures at day 4 (arrows). The region above the dotted line contains cells with an undifferentiated ES-like morphology. (B) Neural tube-like structures at day 7.

earliest markers, becoming detectable after around 4 days. Sox1 becomes detectable after around 1 week. After 10 days, nestin-positive cells start to migrate radially from the rosettes.

2.2.3.1. Terminal differentiation

As for mouse cells, neuronal differentiation can be induced by passaging the neural precursors onto laminin-precoated dishes. Matrigel has an inhibitory effect on neuronal differentiation. Culture on laminin also strongly reduces attachment, survival, and proliferation of undifferentiated ES cells.

2.2.4. Troubleshooting: Human ES cells

2.2.4.1. Cell death

The most common reason for cell death during this protocol is due to dissociation of the hES cells into single cells rather than maintaining them throughout as small clumps. It is also important to store media and FGF-2 at 4 °C for no more than 1 week.

2.2.4.2. Plating density

Because hES cells are dissociated into clumps rather than into single cells, it is not possible to quantify the plating density. The optimal plating density may also differ between different cell lines. It is therefore a good idea to test different plating densities in order to get a feel for the optimal density for each cell line. As a general guideline, the plating density should be such that the hES cells cover around 50% of the surface of the culture dish on the first day of differentiation. As with mouse ES cells, if the plating density of the hES clumps is too high or if clumps are too large they tend to resist differentiation.

2.3. Converting mouse pluripotent ES cells to tissue–specific NS cells

2.3.1. Overview

ES cells divide symmetrically generating a seemingly homogeneous population of stem cells without accompanying differentiation. ES cell lines enable insightful experiments to determine stem cell self-renewal and differentiation mechanisms. Studies of neural induction, for example, have revealed similarities between those mechanisms operating *in vitro* and *in vivo* (Ying *et al.*, 2003b). We recently investigated whether transient neural precursors generated from mouse ES cells using adherent monolayer differentiation could be captured and expanded, without use of genetic immortalization strategies, by elimination of heterologous cell types and exposure to growth factors (Conti *et al.*, 2005). We found that neural precursors are readily expanded in adherent conditions using a combination of EGF and FGF-2 in serum-free medium. Importantly, these conditions do not sustain non-neural cell types or differentiating astrocytes and thus give pure NS cell populations. These cells are remarkably homogeneous and show similarities to radial glia, with all cells expressing Sox2 and BLBP (Fabp7). These Sox1$^-$/Sox2$^+$/BLBP$^+$ cultures stably retain neuronal and glial differentiation potential after prolonged culture, both *in vitro* and *in vivo* following transplantation, and can be clonally expanded as cell lines. This self-renewal capacity, immortality, and symmetrical stem cell division, is reminiscent of ES cells. We have termed these cell lines NS cells. It is now evident that BLBP-positive cells, which arise from early Sox1 neuroepithelial cells around 10.5 dpc, function as dividing precursors *in vivo* capable of generating neurons, astrocytes, and oligodendrocytes (Anthony *et al.*, 2004; Merkle *et al.*, 2004; Noctor *et al.*, 2002). Thus, conversion of ES cells to NS cells both temporally and in terms of lineage pathways reflects events in normal development, although it is not clear that self-renewal really occurs in the developing, as opposed to adult CNS (Pollard *et al.*, 2006b).

A comparison of the salient features of both ES and NS is given in Table 20.1. Potential applications for NS cells, both clinical and in basic biology, have been recently been discussed elsewhere (Pollard *et al.*, 2006a). NS cells can also be derived from fetal and adult CNS using similar protocols to that described here (Conti *et al.*, 2005; Pollard *et al.*, 2006b). Protocols for the derivation, maintenance and differentiation of NS cells from mouse ES cells are described later.

2.3.2. Protocol: NS cell derivation from ES cells

Feeder-dependent ES cells are adapted to feeder-free conditions and expanded as described earlier in this report. We have found that NS cells can be derived following initial neural induction of ES cells using either "classic" embryoid body/

Table 20.1
Similarities and differences between ES cells and NS cells

	ES cells	NS cells
Species	Rodent and primate	Rodent and primate
Source	Blastocyst	ES cells, fetal and adult CNS
Growth factor dependence	LIF plus BMP2/4 (serum-free)	EGF plus FGF-2[a] (serum-free)
Coculture with feeders	Unnecessary	Unnecessary
Expansion *in vitro*	Immortal	Immortal
Clonogenic?	Yes	Yes
Doubling time	~12 h	~25 h
Stem cell divisions	Symmetrical	Symmetrical
Karyotype	Stable diploid	Stable diploid
Niche dependence	None	None
In vivo counterpart	Similarities to inner cell mass	Similarities to radial glia
Differentiation capacity	Stable (*in vitro* and *in vivo*)	Stable (*in vitro* and *in vivo*)
Potency	Pluripotent	Multipotent
Genetic manipulation	Yes	Yes
Germline colonization	Yes	No

[a]EGF is necessary for derivation and maintenance of NS cells. Addition of FGF may only be required for NS cell derivation (Pollard *et al.*, 2006b).

retinoic acid protocols (Bain *et al.*, 1995), or adherent monolayer differentiation (Ying *et al.*, 2003b). The following protocol is based upon the latter, as this is the preferred neural induction protocol of our laboratory, for those reasons outlined earlier.

2.3.2.1. Derivation of NS cells using Sox1-lineage selection

NS cell derivation can readily be achieved using neural differentiation of 46C Sox1–GFP ES cells. Upon neural induction the endogenous Sox1 promoter is activated and cells express both GFP and puromycin resistance. Through transient exposure of differentiated cultures to puromycin one can enrich for Sox1-expressing neuroepithelial precursors cells. These can be replated and expanded adherently as NS cell lines following exposure to the growth factors EGF and FGF-2:

1. Differentiate 46C ES cells in adherent monolayer as described earlier to induce neural lineage commitment. Setup differentiations in 9-cm dishes (Iwaki). Efficient neural commitment can be estimated by monitoring GFP expression. N2B27 media should be exchanged every second day to remove debris.

2. At day 7, the majority of cells should express GFP and have characteristic neural precursor morphology, with a proportion of cells already commencing overt neuronal differentiation. To enrich for Sox1–GFP expressing cells, add 0.5 μg/ml of puromycin (Sigma) in N2B27 media, replacing media every 24 h to remove those dead/dying Sox1-negative cells.

3. Following 3 days exposure to puromycin, remove media and wash twice with $1\times$ PBS (Sigma) to remove debris. Add 1 ml of trypsin and incubate at 37 °C for 2–3 min. Flood the plate with 10 ml of N2B27 media, and immediately detach and dissociate cells by smoothly pipetting several times against the culture surface. Remove the cell suspension to a 30 ml universal tube, and centrifuge cells at 300g (1300 rpm in Eppendorf 5702) for 3 min. Resuspend the cell pellet in N2B27, plus 10 ng/ml of EGF and FGF-2 (Peprotech), *without* puromycin, and seed 2–3 $\times$ 10^6 cells into a gelatinized T25 tissue culture flask (Iwaki).

4. These GFP+ cells should readily attach and undergo expansion to near confluence within 2–3 days. Cells are typically split 1:3 to 1:4 every 2–3 days. Cells gradually extinguish expression of GFP, and acquire the NS cell phenotype within 3–5 passages. This phenotype is then stable and can be maintained indefinitely (at least 100 passages). At this point NS cells can be transferred from N2B27 media to NS cell expansion media for optimal growth.

5. Once established, NS cells can be characterized using immunocytochemistry for markers such as Rat401/nestin, vimentin, and RC2 (DSHB, Iowa) and RT-PCR for expression of genes that characterize the NS cell state, such as Sox2, Olig2, and BLBP. The cultures should be uniformly negative for astrocyte (GFAP, Sigma) and neuronal (TuJ1, Covance) markers.

2.3.2.2. Derivation of NS cells from any ES cell line

It is difficult to achieve NS cell populations from 46C ES cells using N2B27 media in the absence of Sox1-lineage selection, as contaminating ES cells and non-neural cells persist during early passages. This is perhaps not surprising as N2B27 media was optimized for both ES and NS cell survival/expansion. However, modifications to the NS cell derivation protocol will enable derivation of pure NS cells from any ES cell line. Central to this protocol is the replating of early neural precursors transiently into suspension culture in a media favoring neural lineages, thus eliminating residual non-neural cells:

1. Differentiate ES cells in adherent monolayer in 9-cm dishes to induce neural lineage commitment, as described for 46C.

2. On day 7, replate $2–5 \times 10^6$ cells (typically half the population of a 9 cm monolayer differentiation) into a *nongelatinized* T75 flask (Iwaki) in NS media (see Section 4) supplemented with EGF and FGF-2 (10 ng/ml).

3. After 24 h cultures should contain many thousands of cell aggregates in suspension culture, together with a minor population of adherent differentiated cells, and a substantial amount of cell debris. The proportion of cells in suspension varies considerably depending on the initial efficiency of neural induction, but is not a limiting factor in NS derivation. Two to three days later harvest the large aggregates by centrifugation at 700 rpm for 30 s, and resuspend in 10 ml of fresh media in a fresh gelatinized T75 flask.

4. The majority of aggregates will settle and attach over the course of 3–7 days. NS cells outgrow with characteristic morphology (Fig. 20.3). Once the flask has reached over 50% confluence, trypsinize and split cells 1:2 to 1:3 into a fresh flask. We term these cultures, passage 1 NS cells.

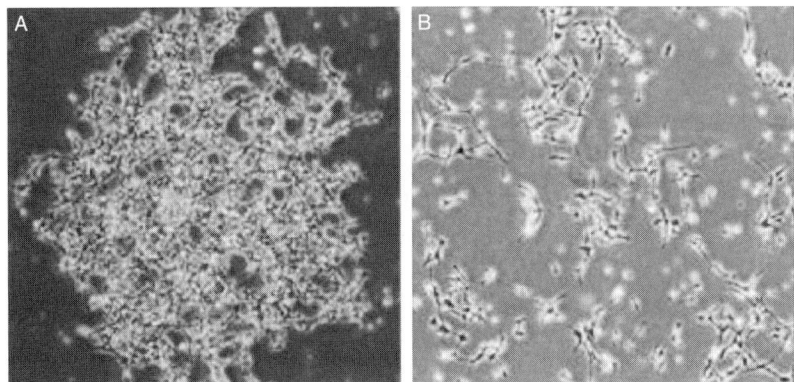

Figure 20.3 NS cell derivation from mouse ES cells IB10 (129/Ola). (A) Attachment after 4 days of a cell aggregate formed following replating of a monolayer differentiation into NS expansion media. (B) Adherent NS cell lines derived following dissociation and replating of cells from (A) (passage 2).

5. (Optional) Clonal lines can be generated once an NS cell population has been generated. Plate single cells into Terasaki microwells (Nunc) using limiting dilution. Add 10 μl of cell suspension to each well using a repeat pipette. Score wells containing a single cell after 1–2 h. Flood the plate with 5 ml of NS media to exchange media, and then aspirate excess media to leave at least 20 μl per well. Colonies appear after 7–10 days and can then be passaged into a well of a 6-well plate. Add 20 μl of trypsin to each well, incubating for 1–2 min, then pipette gently with a P20 to dissociate cells. Transfer 20 μl of cell suspension into 2 ml of NS media in a 6-well plate + EGF + FGF-2. Upon attachment after ~2 h exchange for fresh NS media.

2.3.3. Protocol: Expansion, freezing, and thawing of NS cells

We routinely expand NS cells in T75 gelatinized tissue culture flasks (Iwaki). Cultures are passaged or frozen upon reaching 70–90% confluence ($\sim$5–7 $\times$ 10^6 cells), as described later:

1. Remove media from flask and add 1 ml of trypsin solution. Note that washing of NS cells with $1\times$ PBS solution is not required at this point as cell death is minimal and media is serum-free. Place flask at 37 °C for no longer than 1–2 min and firmly tap the side of flask against the bench to dislodge cells. All cells should detach readily. Add 10 ml of NS media to the flask and pipette up and down, washing against the culture surface, several times to promote single-cell suspension.

2. Transfer cells to a universal tube and spin for 3 min at 300*g*. Aspirate the supernatant carefully ensuring all residual trypsin is removed, and resuspend the pellet into fresh NS basal media. We routinely split cells 1:2 to 1:5. Add the appropriate number of NS cells to a fresh gelatinized T75 flask, diluting with media to a final volume of 10 ml. Add EGF and FGF-2 (10 ng/ml of each) to the flask for final NS cell expansion media.

3. If NS cells are to be frozen, then following trypsinization and centrifugation resuspend the cell pellet in NS basal media + 10% DMSO (premixed). Typically, for a highly confluent T75 flask, resuspend in 2 ml and then aliquot 0.5 ml into four cryotubes. Transfer immediately to −80 °C, where NS cells can be stored and are recoverable for at least 6 months. For long-term storage frozen vials can be transferred to liquid nitrogen following at least 24 h at −80 °C.

4. To recover NS cells from frozen stocks, rapidly thaw the cryotube in a 37 °C water bath. Immediately, but gently, transfer the 0.5 ml of cell solution into 10 ml of prewarmed NS expansion media in a universal using a plugged glass pipette and mix gently. Centrifuge cells at 250*g* for 3 min. It is important that each of these steps is as quick as possible to minimize DMSO exposure.

5. Aspirate the supernatant and gently resuspend the cell pellet into 10 ml of prewarmed NS media. Transfer this cell suspension to a fresh gelatinized T25 flask and add EGF plus FGF-2 (10 ng/ml). The following morning exchange media or passage cells into a T75 flask. NS cell survival following thawing should be extremely high (>95%).

2.3.4. Differentiation of NS cells

For NS cell differentiation to astrocytes we expose cells to 1% serum. BMP4 or LIF (10 ng/ml) also induce astrocyte differentiation with high efficiency, as assessed by GFAP expression. Seed 1×10^5 NS cells into a well of a gelatinized 4-well plate (Nunc) in NS basal media supplemented with 1% fetal calf serum, but without EGF or FGF-2. More than 95% of cells will exit the cell cycle and acquire a characteristic morphology within 24–48 h. There is minimal neuronal differentiation in these conditions.

NS cells maintain a capacity to generate neurons and even after 100 passages. Neurons generated are largely GABAergic as determined by GAD67 and GABA immunoreactivity (Conti *et al.*, 2005). The protocol below results in 10–40% TuJ1 immunoreactive neurons:

1. NS cells are harvested using trypsin and 5×10^4 cells are plated onto laminin-coated 4-well plates in NS basal media + EGF + FGF-2 (10 ng/ml).
2. The following morning exchange the media fully with NS basal media supplemented with FGF-2 (5 ng/ml) + 1× B27 supplement (Gibco). Subsequently replace half-media with fresh every 3–4 days.
3. Following 1 week in these conditions exchange medium to NS basal media mixed with Neurobasal media (1:1) and supplement with 1× B27 (Gibco). Cells with neuronal morphology should emerge over the next 3–7 days. Replace half-media with fresh every 3–4 days.
4. To maintain neurons for longer periods, from day 14 of differentiation switch media to Neurobasal media supplemented with B27 + BDNF (10 ng/ml), but without N2. Such conditions should enable neuronal survival for a further 3 weeks.

2.3.4.1. Differentiation of NS cells to oligodendrocytes

NS cell lines obtained following ES cell differentiation are able to generate both astrocytes and neurons using the protocols described earlier. In a recent study, Glaser *et al.* (2007) have reported a modified differentiation protocol that enables differentiation of NS cells into functional oligodendrocytes with efficiencies around 20%. In outline, the protocol involves withdrawal of EGF and exposure of NS cells to FGF-2, platelet-derived growth factor (PDGF), and forskolin, followed by growth factor withdrawal and exposure to thyroid hormone and ascorbic acid. The oligodendrocytes generated are functional based on transplantation experiments. This protocol results in mixed cultures of neurons, astrocytes, and oligodendrocytes and clearly demonstrates that NS cells are tripotent.

2.3.5. Troubleshooting: Mouse NS cells

Unsurprisingly, we find that efficiency of initial neural lineage commitment correlates with the proportion of cells that can subsequently be expanded as NS cells. However, as the NS cells expand rapidly in low density conditions, even poor efficiency of initial neural induction provides enough founding cells to generate cell lines. Non-neural and residual ES cells are reduced/eliminated through a combination of cell death and differentiation and are not able to proliferate. A common variable is the timing of attachment of the aggregates/clusters upon replating into NS cell media. Patience is required, as apparently nonadherent cell aggregates should eventually attach. Attachment can be promoted by replating aggregates into fresh flasks if there are large amounts of debris. Once attached and outgrown these cells will remain adherent in subsequent passages.

In optimal conditions NS cultures are extremely healthy with minimal cell death or differentiation. If NS cultures start to show a general decline in viability over the course of passaging with a reduction in cell division this is likely due to one of two factors. N-2 supplement can show reduced potency if stored for long periods (>3 weeks) at $-20\ ^\circ$C, or as part of complete NS media (>3 weeks) at $4\ ^\circ$C. Over-trypsinization during passage can also result in unhealthy cultures. This can be circumvented by diluting trypsin 1:10 or through use of Accutase (Sigma) or PBS to detach/dissociate cells or reducing exposure time (<2 min). NS cells can be expanded on laminin. Here cell adhesion is strong, but so long as EGF and FGF-2 are supplied there is no cell differentiation and cells continue to self-renew. For clonal differentiations NS cells can be plated at low density (1×10^4 cells in a 9-cm plate) in NS expansion media; colonies of several hundred cells emerge after 7–10 days, and can be isolated and expanded, or induced to differentiate through growth factor removal and/or serum exposure as described earlier.

3. Summary

Neural differentiation of mouse and human ES cells can be achieved with high efficiency. Such protocols provide an accessible and genetically tractable model system with which to elucidate molecular mechanisms of lineage choice and differentiation. Initial studies have revealed a close similarity between those mechanisms responsible for ES cell differentiation *in vitro* and those operating during embryogenesis, suggesting that studies of stem cell mechanisms *in vitro* will be directly relevant to understanding embryogenesis, and vice versa. Further, understanding mechanisms of ES cell and NS cell differentiation provides a foundation for clinical applications.

4. Media and Reagents

4.1. ES cell media

500 ml	GMEM media
50 ml	Fetal calf serum
5.5 ml	MEM nonessential amino acids 100× (Gibco: final concentration 1×)
5.5 ml	Sodium pyruvate 100 mM (final concentration 1 mM)
5.5 ml	L-Glutamine 200 mM (final concentration 2 mM)
550 μl	2-Mercaptoethanol (final concentration 0.1 mM)
100 Units/ml	LIF (see below)

Alternatively, for serum-free culture, ES cells can be maintained in N2B27 media supplemented with LIF (100 Units/ml, see below) and 10 ng/ml BMP4 (3 ng/ml, R&D systems) (Ying *et al.*, 2003a).

4.2. N2B27 media

200 ml	Neurobasal media
200 ml	DMEM/F12 media
4 ml	B27 supplement (final concentration 0.5×)
2 ml	N-2 supplement (final concentration 0.5×)
400 μl	2-Mercaptoethanol (final concentration 0.1 mM)
1 ml	Glutamate (final concentration 0.2 mM)

N2B27 should be stored at 4 °C and used within 1 week.

4.3. N-2 supplement

Note that the N-2 supplement used in N2B27 is modified from the original formulation N-2, with higher insulin and addition of bovine serum albumin. This formulation increases attachment and survival of neural cells. Batches of N-2 can be stored in aliquots at −20 °C for no longer than 3 weeks.

Stock solutions are:

- Insulin 25 mg/ml (Sigma), dissolve 100 mg/4 ml 0.01 M sterile filtered HCl. Insulin should be resuspended overnight at 4 °C.
- Apotransferrin 100 mg/ml (Sigma), dissolve 500 mg/5 ml sterile filtered H_2O.
- BSA 75 mg/ml, dissolve in sterile PBS.
- Progesterone 0.6 mg/ml (Sigma), dissolve 6 mg/10 ml Ethanol, then filter sterilize.
- Putrescine 160 mg/ml (Sigma), dissolve 1.6 g/10 ml H_2O, then filter sterilize.

- Na selenite 3 mM (Sigma), dissolve 2.59 mg/5 ml H_2O, then filter sterilize.
- DMEM:F12 (L-glutamine) (Gibco).

These stocks are stored at -20 °C. We routinely use these stocks to prepare 40 ml of N-2 supplement. To 27.5 ml of DMEM:F12 add: 4 ml BSA stock, 4 ml of insulin stock (add 200 μl at a time to prevent precipitation), 4 ml apotransferrin, 40 μl sodium selenite, 400 μl putrescine, and 132 μl progesterone.

2-Mercaptoethanol. 200 μl of 2-mercaptoethanol (14.3 M) is mixed with 28.2 ml UHP water and stored at 4 °C in aliquots (final concentration 0.1 M).

Glutamate/pyruvate. 100 mM sodium pyruvate + 5.5 ml of 200 mM L-glutamine 200 mM. Store in aliquots at -20 °C.

Trypsin (1×). Add 0.186g EDTA to 500 ml UHP water and filter sterilize. Add 5 ml chick serum and 5 ml conc. Trypsin (2.5%). Store in aliquots at -20 °C.

Gelatin. One percent stock solution is prepared in UHP water, autoclaved and stored in aliquots at 4 °C. To prepare the 0.1% working solution, 1% gelatin is warmed to 37 °C until it liquefies, and then it is diluted 1:10 in sterile PBS. 0.1% gelatin can be stored for up to 2 weeks at 4 °C.

For culture and differentiation of both ES and NS cells, plates and flasks are coated with a 0.1% gelatin solution for at least 10 min at room temperature. Gelatin is aspirated prior to use. Washing with PBS is not necessary.

NS basal media. Euromed-N (formerly NS-A) is a basal media with a formulation similar to DMEM (Euroclone). We supplement Euromed-N with N-2 supplement (below) and 2 mM L-glutamine (Invitrogen). Growth factors are directly added to flasks/plates immediately prior to use to achieve the final NS cell expansion media.

EGF (Peprotech, Cat: 315-09) and *FGF-2* (Peprotech, Cat: 100-18B) are resuspended in PBS and 20 μl aliquots are stored at -20 °C. EGF and FGF-2 are added directly to flasks and plates when required. Once thawed each is used within 1 week stored at 4 °C.

LIF Recombinant. LIF is readily produced by transient transfection of Cos cells. The supernatant from these cultures is collected and the concentration of LIF assayed using CP1 indicator cells. Supernatant is then diluted in 1× PBS to give a 1000× stock concentration of 100,000 Units/ml and stored at -20 °C in 0.5 ml aliquots.

Polyornithine/laminin plates. A 0.01% solution of poly-L-ornithine (Sigma) is added to plates and flasks for at least 20 min. The solution is removed and plates/flasks are washed three times with 1× PBS. Replace PBS with a 10 μl/ml solution of laminin in PBS (Sigma) and incubate at 37 °C for at least 2 h (preferably overnight).

Acknowledgments

We thank Luciano Conti for contributions to NS protocols. Austin Smith made helpful comments on the chapter and provided support and guidance throughout these studies. This research is supported by the Biotechnology and Biological Sciences Research Council and the Medical Research Council of the United Kingdom, the Wellcome Trust, INSERM, and the Framework VI Integrated Project EuroStemCell.

References

Anthony, T. E., Klein, C., Fishell, G., and Heintz, N. (2004). Radial glia serve as neuronal progenitors in all regions of the central nervous system. *Neuron* **41,** 881–890.

Aubert, J., Stavridis, M. P., Tweedie, S., O'Reilly, M., Vierlinger, K., Li, M., Ghazal, P., Pratt, T., Mason, J. O., Roy, D., and Smith, A. (2003). Screening for mammalian neural genes *via* fluorescence-activated cell sorter purification of neural precursors from Sox1–gfp knock-in mice. *Proc. Natl Acad. Sci. USA* **100**(Suppl. 1), 11836–11841.

Bain, G., Kitchens, D., Yao, M., Huettner, J. E., and Gottlieb, D. I. (1995). Embryonic stem cells express neuronal properties *in vitro. Dev. Biol.* **168,** 342–357.

Billon, N., Jolicoeur, C., Ying, Q. L., Smith, A., and Raff, M. (2002). Normal timing of oligodendrocyte development from genetically engineered, lineage-selectable mouse ES cells. *J. Cell Sci.* **115,** 3657–3665.

Conti, L., Pollard, S. M., Gorba, T., Reitano, E., Toselli, M., Biella, G., Sun, Y., Sanzone, S., Ying, Q. L., Cattaneo, E., and Smith, A. (2005). Niche-independent symmetrical self-renewal of a mammalian tissue stem cell. *PLoS Biol.* **3,** e283.

Glaser, T., Pollard, S. M., *et al.* (2007). Tripotential differentiation of adherently expandable neural stem (NS) cells. *PLoS ONE* **2,** e298.

Gottlieb, D. I., and Huettner, J. E. (1999). An *in vitro* pathway from embryonic stem cells to neurons and glia. *Cells Tissues Organs* **165,** 165–172.

Hovatta, O., Mikkola, M., Gertow, K., Stromberg, A. M., Inzunza, J., Hreinsson, J., Rozell, B., Blennow, E., Andang, M., and Ahrlund-Richter, L. (2003). A culture system using human foreskin fibroblasts as feeder cells allows production of human embryonic stem cells. *Hum. Reprod.* **18,** 1404–1409.

Kawasaki, H., Mizuseki, K., Nishikawa, S., Kaneko, S., Kuwana, Y., Nakanishi, S., Nishikawa, S. I., and Sasai, Y. (2000). Induction of midbrain dopaminergic neurons from ES cells by stromal cell-derived inducing activity. *Neuron* **28,** 31–40.

Li, M., Price, D., and Smith, A. (2001). Lineage selection and isolation of neural precursors from embryonic stem cells. *Symp. Soc. Exp. Biol.* **53,** 29–42.

Merkle, F. T., Tramontin, A. D., Garcia-Verdugo, J. M., and Alvarez-Buylla, A. (2004). Radial glia give rise to adult neural stem cells in the subventricular zone. *Proc. Natl Acad. Sci. USA* **101,** 17528–17532.

Noctor, S. C., Flint, A. C., Weissman, T. A., Wong, W. S., Clinton, B. K., and Kriegstein, A. R. (2002). Dividing precursor cells of the embryonic cortical ventricular zone have morphological and molecular characteristics of radial glia. *J. Neurosci.* **22,** 3161–3173.

Pevny, L. H., Sockanathan, S., Placzek, M., and Lovell-Badge, R. (1998). A role for SOX1 in neural determination. *Development* **125,** 1967–1978.

Pollard, S. M., Conti, L., and Smith, A. (2006). Exploitation of adherent neural stem cells in basic and applied neurobiology. *Regen. Med.* **1,** 111–118.

Pollard, S. M., Conti, L., Sun, Y., Goffredo, D., and Smith, A. (2006). Adherent neural stem (NS) cells from foetal and adult forebrain. *Cereb. Cortex* **16,** i112–i120.

Stavridis, M. P., and Smith, A. G. (2003). Neural differentiation of mouse embryonic stem cells. *Biochem. Soc. Trans.* **31,** 45–49.

Tropepe, V., Hitoshi, S., Sirard, C., Mak, T. W., Rossant, J., and van der Kooy, D. (2001). Direct neural fate specification from embryonic stem cells: A primitive mammalian neural stem cell stage acquired through a default mechanism. *Neuron* **30,** 65–78.

Ying, Q. L., and Smith, A. G. (2003). Defined conditions for neural commitment and differentiation. *Methods Enzymol.* **365,** 327–341.

Ying, Q. L., Nichols, J., Chambers, I., and Smith, A. (2003). BMP induction of Id proteins suppresses differentiation and sustains embryonic stem cell self-renewal in collaboration with STAT3. *Cell* **115,** 281–292.

Ying, Q. L., Stavridis, M., Griffiths, D., Li, M., and Smith, A. (2003). Conversion of embryonic stem cells into neuroectodermal precursors in adherent monoculture. *Nat. Biotechnol.* **21,** 183–186.

CHAPTER 21

Hematopoietic Cell Differentiation from Embryonic Stem Cells

Malcolm A. S. Moore, Jae-Hung Shieh, and Gabsang Lee

Moore Laboratory
Cell Biology Program
Memorial Sloan-Kettering Cancer Center
New York, NY, 10021

Abstract
1. Introduction
    1.1. Murine embryonic stem cells (mESC) differentiation
2. Methods
    2.1. Maintenance of hESC
    2.2. Hematopoietic differentiation of hESC coculture with stromal cell lines
    2.3. Maintenance of stromal cell lines
    2.4. Hematopoietic differentiation of hESC in coculture with stromal cells
    2.5. Hematopoietic differentiation of hESC using stromal conditioned medium
    2.6. Embryoid body formation and hematopoietic differentiation
    2.7. Erythroid differentiation
    2.8. Megakaryocyte differentiation
    2.9. Neutrophil differentiation
    2.10. Macrophage differentiation
    2.11. Lymphoid differentiation from hESC
    2.12. *In vivo* transplantation of EB-derived CD34+ cells into immunodeficient mice
    2.13. Analysis of NOD/SCID mouse hematopoietic engraftment
    2.14. Bioluminescence Imaging (BLI)
References

Abstract

Murine embryonic stem cells (mESC) readily form embryoid bodies (EBs) that exhibit hematopoietic differentiation. Methods based on EB formation or ESC coculture with murine bone marrow stromal cell lines, have revealed pathways of both primitive and definitive hematopoietic differentiation progressing from primitive mesoderm via hemangioblasts to endothelium and hematopoietic stem and progenitor cells. The addition of specific hematopoietic growth factors and morphogens to these cultures enhances generation of neutrophils, macrophages, megakaryocyte/platelets, and hemoglobinized mature red cells. In addition, selective culture systems have been developed to support differentiation into mature T lymphocytes, NK cells, B cells, and dendritic cells. In most cases, cultures systems have been developed that support equivalent differentiation of various human ESC (hESC). The major obstacle to translation of ESC hematopoietic cultures to clinical relevance has been the general inability to produce hematopoietic stem cells (HSC) that can engraft adult, irradiated recipients. In this context, the pattern of ES hematopoietic development mirrors the yolk sac phase of hematopoiesis that precedes the appearance of engraftable HSC in the aorta-gonad-mesonephros (AGM) region. Genetic manipulation of mESC hematopoietic progeny by upregulation of HOXB4 or STAT5 has led to greatly enhanced long term or short term multilineage hematopoietic engraftment, suggesting that genetic or epigenetic manipulation of these pathways may lead to functional HSC generation from hESC.

1. Introduction

1.1. Murine embryonic stem cells (mESC) differentiation

Hematopoietic differentiation of ESC lines has been subject to a number of recent reviews covering derivations from murine, primate, and human ESC (hESC) (Bhatia, 2007; Choi *et al.*, 2005; Daley, 2003; Keller, 2005; Lengerke and Daley, 2005; Lensch and Daley, 2006; Lerou and Daley, 2005; Martin and Kaufman, 2005; Olsen *et al.*, 2006; Priddle *et al.*, 2006; Tian and Kaufman, 2005).

1.1.1. Early ESC development

When mESCs are cultured in hanging drop systems or directly in semisolid media (methyl cellulose), they proliferate and differentiate to generate colonies known as embryoid bodies (EBs) (Wiles and Keller, 1991). These EBs consist of differentiated cells from a number of lineages including those of the hematopoietic system. When EBs at 3–3.5 days of development were re-plated in methylcellulose with vascular endothelial growth factor (VEGF) and c-kit ligand/stem cell factor (SCF), colonies of blast morphology (BL-CFC) developed. Upon replating of day

6 blast colonies, colonies of primitive erythroid (EryP) cells as well as colonies of definitive erythroid (BFU-E), multilineage (CFU-Mix), and myeloid (CFU-GM) developed (Kennedy *et al.*, 1997). BL-CFC growth was greatly augmented by addition of conditioned medium from an endothelial cell line derived from EBs (Choi *et al.*, 1998). ES-derived transitional colonies express brachury, Flk1, SCL/Tal-1, Gata-1, betaH1, and beta-major-reflecting the combination of mesodermal, hematopoietic, and endothelial populations (also cardiomyocyte) (Robertson *et al.*, 2000). Replating studies demonstrated that transitional colonies contain low numbers of EryP precursors as well as a subset of precursors associated with early stage definitive hematopoiesis. BL-CFC (brachyury negative) contain higher numbers and a broader spectrum of definitive precursors than found in the transitional colonies and SCL−/− ES form transitional but not blast colonies (Robertson *et al.*, 2000). ES cultured on type IV collagen-coated dishes formed Flk1+ mesoderm. The hematopoietic developmental sequence is from proximal lateral mesoderm (E-cadherin-Flk1+ VE-cadherin−), to progenitors with hemo-angiogenic potential (Flk1+ VE-cadherin+, CD45−) then to hemopoietic progenitor (CD45+, c-Kit+), and mature blood cells (c-Kit−, CD45+, or Ter119+) (Nishikawa *et al.*, 1998). Flk1, SCL, and bFGF-mediated signaling is critical for hemangioblast development, with Activin A synergizing to increase BL-CFC (Faloon *et al.*, 2000). In mESC cultured in serum-free, chemically defined medium (CDM) supplemented with BMP-2 or BMP-4, a process resembling primitive streak formation occurred, at least at the molecular level, with the formation of mesoderm and subsequently endothelial and hematopoietic cells (Wiles and Johansson, 1997). VEGF is necessary for subsequent expansion and differentiation of hematopoietic precursors with the Smad1 and 5 and MAP kinase pathways activated by BMP-4 and VEGF, respectively (Park *et al.*, 2004). VEGF-mediated expansion of hematopoietic and endothelial cell progenitors was inhibited by TGFβ1, but was augmented by Activin A. Smad5 (−/−) EBs contained an elevated number of BL-CFCs and an increased frequency of high proliferative potential primitive hematopoietic progenitors (HPP-CFCs) (Liu *et al.*, 2003). These HPP-CFCs displayed enhanced self-renewal capacity and decreased sensitivity to TGFβ1 inhibition, suggesting a critical role of Smad5 in TGFβ1-mediated negative regulation of embryonic HPP-CFCs.

Runx1-deficient EBs form 10–15-fold fewer BL-CFC and have a complete block in definitive hematopoiesis. Runx−/− EBs and embryos generated normal numbers of EryP precursors with the latter developing from a subset of BL-CFC that can develop in a Runx1-independent fashion (Lacaud *et al.*, 2004). Runx1 heterozygosity leads to an acceleration of mesodermal commitment and specification to the BL-CFCs and to the hematopoietic lineages in EBs (Lacaud *et al.*, 2004). In contrast to normal ES cells, GATA-1 null ES cells fail to generate EryP precursors. Definitive erythroid (EryD) precursors, however, are normal in number but undergo developmental arrest and death at the proerythroblast stage (Weiss *et al.*, 1994). Flk-1−/− ESCs are capable of blast colony formation (Ema *et al.*, 2003), however, in Flk1−/− embryonic stem cell chimeras, Shalaby *et al.*

(1997) showed that Flk1 is required cell autonomously for endothelial development. Flk1 is involved in the movement of cells from the posterior primitive streak to the yolk sac and, possibly, to the intraembryonic sites of early hematopoiesis. Flk1−/− EBs showed myeloid-erythroid differentiation but ESC in OP9 stromal coculture failed to generate hematopoietic clusters even with cytokine (Hidaka *et al.*, 1999). Thus, the requirement for Flk-1 in early hematopoietic development can be abrogated by alterations in the microenvironment. This finding is consistent with a role for Flk-1 in regulating the migration of early mesodermally derived precursors into a microenvironment that is permissive for hematopoiesis (Hidaka *et al.*, 1999). An ES-derived Tie-2+, Flk1+ cell fraction is enriched for hematopoietic and endothelial progenitors but Tie2−/− ES cells had no defect in hematopoiesis (Hamaguchi *et al.*, 2006). Shp-2, a member of a small family of cytoplasmic Src homology 2 (SH2) domain-containing protein tyrosine phosphatases, is integrally necessary for bFGF-mediated hemangioblast production (Zou *et al.*, 2006). Hemangioblast formation and primitive and definitive hematopoietic progenitor formation was decreased significantly following transfection with Shp-2 siRNA.

1.1.2. EB differentiation systems

EBs can be generated by the hanging drop technique, by suspension culture or by methylcellulose culture (Dang *et al.*, 2002). For large-scale production of EBs in a controlled environment, an agarose encapsulation, stirred bioreactor system has been developed (Dang *et al.*, 2004). ES cells differentiated on porous 3D scaffold structures developed EBs similar to those in traditional two-dimensional (2D) cultures; however, unlike 2D differentiation, these EBs integrated with the scaffold and appeared embedded in a network of extracellular matrix, exhibiting enhanced progenitor (CFC) and myeloid differentiation (Liu *et al.*, 2005). Hematopoietic differentiation can proceed autonomously in developing EBs with development of erythroid precursors by day 4 and by day 6–10, 40–85% of EBs are hematopoietic, containing visible erythropoietic cells (i.e., red with hemoglobin). BMP-4 addition in the first 4 days and VEGF for a further 3 days enhanced hematopoietic development (Nakayama *et al.*, 2000). Hematopoiesis was also increased by addition of IL-11 and SCF and the kinetics of precursor development was similar to that of the yolk sac (Keller *et al.*, 1993).

βH1 globin mRNA is detectable in EBs within 5 days of differentiation, whilst β(maj)-globin RNA appears by day 6 (Wiles and Keller, 1991). Addition of erythroid stimulating factors (Epo, SCF, IGF-1, transferrin) promoted enhanced and prolonged erythropoiesis (Carotta *et al.*, 2004). BMP-4 and VEGF synergized in generation of CD45+ myelomonocytic and Ter119+ erythroid cells. The development of macrophages is significantly enhanced by the addition of IL-3 alone, or in combination with IL-1 and M-CSF or GM-CSF. When well-differentiated EBs are allowed to attach onto tissue-culture plates and grown in the presence of IL-3, a long-term output of cells of the mast cell lineage is observed (Wiles and Keller, 1991).

1.1.3. mESC–stroma coculture systems

mESC in coculture with mouse stromal cells (OP9) give rise to erythroid progenitors (EryP and EryD) sequentially, with a time course similar to that seen in murine ontogeny (Nakano *et al.*, 1996, 1997). Analysis of the role of different growth factor requirements and limiting dilution analysis of precursor frequencies indicated that most EryP and EryD probably developed from different precursors by way of distinct differentiation pathways. In OP9 cocultures, a CD41+ (dim) population was the immediate precursor of TER119+ EryP cells (Otani *et al.*, 2005). Coculture of mESC with the murine MS-5 stromal line, together with hematopoietic growth factors (KL, IL3, IL-6, IL-11, G-CSF, Epo), enhanced hematopoietic differentiation and addition of Tpo induced differentiation to megakaryocytes (Berthier *et al.*, 1997). In mESC-OP9 cocultures with Tpo, small megakaryocytes were generated that rapidly produced proplatelets by day 8 and large hyperploid megakaryocytes developed after day 12, suggesting the existence of both primitive and definitive megakaryopoiesis (Fujimoto *et al.*, 2003). From 10^4 ES cells up to 10^8 platelets could be produced. Lieber *et al.* (2004) used a 3-step culture system to generate 10^7 neutrophils from 8×10^4 ES cells. In this system, day 8 EBs were cocultured on OP9 stroma with IL-6, bFGF, Oncostatin M, SCF, IL-11, and LIF for 3 days then transferred to a medium supplemented with G-CSF, GM-CSF, and IL-6 for 4–20 days.

1.1.4. Lymphoid differentiation of ESC

B-cell development. mESC cocultured with OP9 stroma generated erythroid, myeloid and B cells (Nakano *et al.*, 1994, 1995, 1996, 1997). Isolated CD34+ cells from differentiating mESC were cultured on OP9 with IL-2 and IL-7 and developed into B220+, CD34-ve B lymphocytes and CD19+ pre-B cells (Nakayama *et al.*, 2000). OP9 stroma produces both SCF and IL-7 and addition of Flt3L led to a 10-fold increase in B cell production (CD19+, CD45R+ (B220), AA4.1+, CD24+, IgM+) with reduced erythroid and myeloid differentiation. By 4 weeks of coculture, >90% of cells were surface IgM+, IgD+ B-cells that could secrete immunoglobulin upon mitogen (LPS) stimulation (Cho *et al.*, 1999).

T-cell development. T cells can be generated from Flk-1+, CD45−, mESC-derived cells from 5–6 day OP9 cultures in a fetal thymic organ culture system or in a reaggregated thymic culture (de Pooter *et al.*, 2003). Schmitt *et al.* (2004) reported a normal program of T cell differentiation in cocultures of mESC on OP9 stroma expressing the Notch ligand Delta (DLL1). The T-cells displayed a diverse antigen receptor repertoire, and CD8+ T cells proliferated and produced interferon-γ in response to T cell receptor (TCR) stimulation. ESC-derived T cell progenitors effectively reconstituted the T cell compartment of Rag2−/− immunodeficient mice, enabling an effective response to a viral infection.

NK cell development. Culture of mESC-derived CD34+ cells on OP9 with IL2 and IL7 produced cytotoxic lymphocytes with NK markers (NKR-P1, perforin, granzyme) (de Pooter *et al.*, 2005). Lian *et al.* (2002) generated NK cells from CD34+ isolated from mEBs by coculturing on OP9 stroma with IL-6, IL-7, SCF and Flt3L, for 5 days then plating on fresh OP9 stroma with IL-2, IL-15, IL-18, and IL-12 for 7 days with a final expansion with cytokines without stroma. The ES-derived NK (ES-NK) cells expressed NK cell-associated proteins and were capable of killing certain tumor cell lines as well as MHC class I-deficient lymphoblasts. They also express CD94/NKG2 heterodimers, but not Ly49 molecules.

Dendritic cell (DC) development. mESC cocultured with OP9 and GM-CSF generated immature DCs that share many characteristics of macrophages, but upon maturation acquire the allo-stimulatory capacity and surface phenotype of classical DCs, including expression of CD11c, major histocompatibility complex (MHC) class II and costimulatory molecules CD80, and CD86 (Fairchild *et al.*, 2000; Senju *et al.*, 2003). Upon stimulation with IL-4 plus TNFα, combined with anti-CD40 monoclonal antibody or lipopolysaccharide, ES-DCs became mature DCs, characterized by a typical morphology and higher capacity to stimulate MLR and to process and present protein antigen to T cells. Immunization with ESC-derived DCs expressing a model antigen (OVA) provided protection from OVA-expressing tumor cells more potently than did immunization with OVA alone (Fukuma *et al.*, 2005; Matsuyoshi *et al.*, 2004). ESC-derived DCs may also offer prospects for reprogramming the immune system to tolerate grafted tissues. ES-derived genetically modified DC presenting myelin oligodendrocyte glycoprotein (MOG) peptide in the context of MHC class II molecules and simultaneously expressing TRAIL or Programmed Death-1 ligand (PD-L1) significantly reduced the severity of MOG peptide-induced experimental autoimmune encephalomyelitis (EAE) following pretreatment of mice (Hirata *et al.*, 2005).

1.1.5. *In vitro* engraftment of mESC-derived hematopoietic cells

There are reports that mESC-derived hematopoietic cells can produce long-term lympho-myeloid reconstitution of irradiated adult mice (Burt *et al.*, 2004; Miyagi *et al.*, 2002). Others have reported that the hematopoietic potential of ES cells *in vivo* is limited to low levels of repopulation and is restricted to the lymphoid lineage (Muller and Dzierzak, 1993). There is a limited temporal window for the derivation of multilineage repopulating hematopoietic progenitors during embryonal stem cell differentiation *in vitro*. Day 4 murine EBs can generate primitive hematopoietic progenitors but upon transplant into irradiated mice very low levels of CD45+ lymphoid and myeloid cells were detected by 12 weeks (Hole *et al.*, 1996) Upon transfer into lymphoid deficient mice, mESC-derived CD45+, B220 (CD45R)+, AA4.1+ cells generated a single transient wave of IgM+, IgD+ B cells but failed to generate T cells (Potocnik *et al.*, 1997). In contrast, transfer of the B220−, AA4.1+ fraction achieved long-term repopulation of both T and B lymphoid compartments and restored humoral and cell-mediated immune reactions in the recipients.

1.1.6. Primate ESC–derived hematopoiesis

Primate ESC lines have been shown to differentiate into multiple hematopoietic lineages. Rhesus ES cells cocultured on S17 stroma with hematopoietic growth factors and BMP-4 generated CFC and CD34+ cells that formed cobblestone areas on secondary replating (Honig *et al.*, 2004; Li *et al.*, 2001). CD34+ and CD34+, CD38− cells derived from Rhesus ES cells expressed embryonic ε and ζ as well as α, β, and γ globin genes, whereas no expression of embryonic globins could be detected in the cell preparations from BM-derived CD34+ cells (Lu *et al.*, 2002). Enhanced hemangioblast development and hematopoietic and CD34+ differentiation was reported in a rhesus EB culture system when Flt3L and SCF were supplemented with VEGF and Tpo (Wang *et al.*, 2005c). In addition, analysis of gene expression during hemangioblast development demonstrated that Tpo is capable of increasing the mRNA expression of the VEGF receptor (VEGFR) and its own receptor (c-mpl). OP-9 stromal coculture with cytokines have been used to support hematopoietic differentiation of Cynomolgus ES lines (Sasaki *et al.*, 2005; Umeda *et al.*, 2004). Primitive generation erythropoiesis was detected on day 8 of coculture without exogenous Epo, whereas definitive erythropoiesis appeared on day 16 and had an indispensable requirement for exogenous Epo (Umeda *et al.*, 2004). VEGF increased, in a dose-dependent manner, not only the number of floating hematopoietic cells, but also the number of adherent hematopoietic cell clusters containing CD34-positive immature progenitors. In colony assays, exogenous VEGF also had a dose-dependent stimulatory effect on the generation of EryP colonies. Hematopoietic cells generated in this manner have been injected intrahepatically after the first trimester in fetal sheep and microchimerism was observed up to 17 months posttransplantation with Cynmolgus cells detected in bone marrow (1–2%) and circulation (<0.1%) (Sasaki *et al.*, 2005).

1.1.7. Hematopoietic differentiation of human ESC lines

Kaufman *et al.* (2001) first reported the generation of hematopoietic cells and erythroid (BFU-E) and myeloid (CFU-GM) progenitors following 2–3 week coculture of hESC (H1 and H9) on a murine marrow stromal line (S17) or on a yolk sac endothelial line. CFCs were enriched in the CD34+ cell fraction. A greater output of CD34+ cells was seen with ESC coculture on an hTERT-immortalized human fetal liver stromal line compared to S17 (Qiu *et al.*, 2005). Human marrow stromal cells plus a low dose cocktail of hematopoietic cytokines also efficiently supported the generation of KDR+ hemangioblasts, CD34+ hematopoietic precursors, and CD45+ mature hematopoietic cells from EBs (Wang *et al.*, 2005c). The murine OP9 stromal line is superior to either MS-5 or S17 in supporting hematopoietic differentiation of hESC, with 10^7 CD34+ cells (>95% purity and 1:66 cells forming CFC) generated from a similar number of initially plated hES cells after 8–9 days of coculture (Vodyanik *et al.*, 2005, 2006). These CD34+ cells displayed the phenotype of primitive hematopoietic progenitors as defined by coexpression of CD90, CD117, CD164, and aldehyde dehydrogenase along with a lack of CD38 expression and possessing a verapamil-sensitive ability to efflux rhodamine 123. The OP9 coculture system was used to generate CD45+, CD33+, myeloperoxidase (MPO)+ myeloid precursors from an Oct4-EGFP knock-in human ES cell line, demonstrating that Oct4-EGFP expression was extinguished in these precursors (Yu *et al.*, 2006). Murine fetal liver-derived stroma has also proved efficient in supporting generation of multipotential hematopoietic progenitors from hESC (Ma *et al.*, 2007). An alternative technique for generation of hematopoietic elements involves formation of EBs from H1 or H9 ESC lines in the presence of hematopoietic growth factors (Flt3L, SCF, IL-3, IL-6, and G-CSF) (Bhatia, 2007; Chadwick *et al.*, 2003; Pick *et al.*, 2007). Hematopoietic commitment was defined by appearance of CD45+ cells in day10 EBs with an increase to day 15. Up to 90% of cells from day 22 hEBs were hematopoietic as defined by CD45 expression and CFC potential (Wang *et al.*, 2005a). EB hematopoietic development involves a temporal sequence with few or no clonogenic progenitors earlier than day 14 and none later than day 28. hEB differentiation, like that of the mouse, begins with emergence of semiadherent mesodermal-hematoendothelial (MHE) colonies that can generate endothelium and form organized, yolk sac-like structures that secondarily generate multipotent primitive hematopoietic stem and progenitor cells, erythroblasts, and CD13+, CD45+ macrophages (Zambidis *et al.*, 2005). A first wave of hematopoiesis follows MHE colony emergence and is predominated by primitive erythropoiesis characterized by a brilliant red hemoglobinization, CD71/CD325a (glycophorin A) expression, and exclusively embryonic/fetal hemoglobin expression. A second wave of definitive-type BFU-E, CFU-E, GM-CFC, and multilineage CFCs follows. These stages of hematopoiesis proceed spontaneously from hEB-derived cells without requirement for supplemental growth factors. Initiation of hematopoiesis correlated with increased levels of SCL/TAL1, GATA1, GATA2, CD34, CD31, and the homeobox gene-regulating factor CDX4 (Zambidis *et al.*, 2005). Addition of BMP-4 to the cytokine cocktail enhanced the generation of hematopoietic progenitor colonies that could undergo secondary passage. Treatment of EBs with $VEGF_{165}$ in addition to hematopoietic growth factors and BMP-4 increased the number of cells coexpressing CD34 and KDR as well as cells expressing erythroid markers (Cerdan *et al.*, 2004). Under serum-free conditions with SCF, Flt3L, Tpo, and the obligate presence of BMP-4 and VEGF, EBs generated CD45+, CD34+ hematopoietic stem/progenitor cells (Tian and Kaufman, 2005; Tian *et al.*, 2004). In a further improvement in efficiency, hESCs in serum-free medium with cytokines can be aggregated by centrifugation to foster the formation of EBs of uniform size (spin EBs) with 90% forming blood cells and CFC (Ng *et al.*, 2005). Clonal isolation of a PECAM-1 (CD31+) population coexpressing Flk1 (KDR), and VE-cadherin, and lacking CD45 (CD45-ve PFV) from day 10 EBs, has defined a human bipotent precursor (hemangioblast?) with endothelial and hematopoietic capacity (Menendez *et al.*, 2004; Wang *et al.*, 2004, 2005a,d). These cells express the GALVR-1 receptor permitting their efficient transduction with

GALV-pseudotyped retroviral vectors (Menendez *et al.*, 2004). The expression of CD31 distinguished the potential hemangioblasts of Wang *et al.* (2004) from the CD34+, KDR+, CD31-ve hemangioblast colonies (CFU-Bl) of Kennedy *et al.* (2007). Two distinct types of hemangioblasts were identified, those that give rise to EryP cells, macrophages, and endothelial cells and those that generate only the EryP population and endothelial cells (Kennedy *et al.*, 2007). Lu *et al.* (2007) also reported the isolation of CFU-BL that under limiting dilution conditions generated both adherent cells that formed vWF + capillaries and nonadhered hematopoietic cell population containing hematopoietic stem and progenitor cells.

Immune cell generation. Zhan *et al.* (2004) reported a long-term culture system in which hES-derived EBs were cultured for up to 6 weeks with a cytokine cocktail that induced hematopoietic expansion (SCF, Flt3L, Tpo, IL-3) and dendritic cell differentiation (GM-CSF, IL-4). Of the leucocytes generated ($2.3/10^6$/week from an input of 40 EBs) ~25% acquired MHC class II and costimulatory molecule (CD80 or CD86) expression. Cells expressing CD40 (a marker for antigen-presenting cells), CD83 (a dendritic cell marker), or CD14 (a macrophage and monocyte marker) were detected, confirming the findings on Wright-Giemsa staining that dendritic cells and macrophages were present. Isolated hES-derived CD34+ cells cocultured on MS-5 stroma in the presence of SCF, Flt3L, IL-7, and IL-3, differentiated into lymphoid (B and Natural Killer cells) as well as myeloid (macrophages and granulocytes) lineages (Vodyanik *et al.*, 2005). Woll *et al.* (2005) used a two-step culture method to demonstrate efficient generation of functional NK cells from hESCs. The CD56+, CD45+, hESC-derived lymphocytes express inhibitory and activating receptors typical of mature NK cells, including killer cell Ig-like receptors, natural cytotoxicity receptors, and CD16. NK cells acquire the ability to lyse human tumor cells by both direct cell-mediated cytotoxicity and antibody-dependent cellular cytotoxicity.

In vivo engraftment. CD34+ and CD34+, CD38– cells derived from hESC cocultured on OP9 have been transplanted intraperitoneally into first trimester fetal sheep and low levels of human hematopoietic engraftment detected by FACS and colony assay (Narayan *et al.*, 2006). The human cells also showed secondary passaging potential. Hemogenic precursor populations, from H1 or H9 day 10 hEBs, purified according to their expression of PECAM-1, coexpression of Flk1+ and VE-cadherin, and lack of CD45, have been transplanted by tail-vein injection into sublethally irradiated NOD/SCID mice (Wang *et al.*, 2005b). In contrast to 100% survival of recipient mice receiving a similar dose of cultured primitive somatic hematopoietic cells (cord blood CD34+), <40% of the mice transplanted with hESC-derived hematopoietic cells survived 8 week. Postinjection (24 h), numerous emboli were found lodged in small pulmonary capillaries and mice showed minimal engraftment. Up to 80% of ES-derived hematopoietic precursors underwent rapid aggregation when exposed to mouse serum for 2 h *in vitro*. In contrast, adult mouse serum did not cause aggregation of somatic hematopoietic cells. To bypass the circulatory system and complications associated with systemic delivery, Wang *et al.* (2005b) transplanted hESC-derived hematopoietic cells by means of intrabone marrow transplantation directly to the femur (i.f.). In contrast to intravenously transplanted mice, >90% of the i.f. transplanted mice survived >8 week and most demonstrated human reconstitution, indicative of human HSC function. Human engraftment in the BM of the injected femur was confirmed by Southern blot analysis for human-specific satellite sequences and human engraftment was detected in contra-lateral femurs and other bones, but at lower levels than in the injected femur. The human hematopoietic graft composition from hESC-derived HSCs was similar to that previously shown for somatic HSCs derived from cord blood and included lymphoid (CD45+/CD19+), myeloid (CD45+/CD33+), and erythroid (glycophorin A+, CD45–, human MHC-1+) hematopoietic lineages. Tian *et al.* (2006) reported that hESCs that were allowed to differentiate on S17 stromal cells for 7–24 days were capable of long-term hematopoietic engraftment when transplanted into NOD/SCID mice. Human CD45+ and CD34+ cells were identified in the mouse bone marrow more than 3 months after injection and secondary engraftment studies further confirmed long-term repopulating cells derived from hESCs. It should be noted that the level of engraftment was low.

1.1.8. Hematopoiesis and Homeobox gene expression in ESC

HOXB4 is expressed at the time of initiation of hematopoiesis in the yolk sac (McGrath and Palis, 1997) and ES cell differentiation models mimic embryonic hematopoiesis, with coexpression of HOX genes and their cofactors coinciding with the appearance of hematopoietic progenitor cells (Pineault *et al.*, 2002). Helgason *et al.* (1996) first reported that HOXB4 overexpression significantly increased the number of progenitors of mixed erythroid/myeloid lineage and definitive, but not primitive, erythroid colonies derived from EBs. Expression of HOXB4 in ES-derived primitive progenitors combined with culture on hematopoietic stroma induced a switch to the definitive HSC phenotype (Kyba *et al.*, 2002). These progenitors engrafted lethally irradiated adults and contributed to long-term, multilineage hematopoiesis in primary and secondary recipients. Initial reports employing retroviral transduction of murine bone marrow with a HOXB4 retrovirus showed no disruption in hematopoiesis but more recent data with *in vivo* transplantation of HOXB4-transduced ES-derived cells showed that while myeloid development was enforced, T and B lymphoid development was suppressed over a wide range of expression levels (Kyba *et al.*, 2003). High expression levels of HOB4 were also detrimental for erythroid development (Pilat *et al.*, 2005). Provided that HOXB4 levels are kept within a certain therapeutic window, for example, by utilizing inducible gene expression systems, ES cells carry the potential of efficient and safe somatic gene therapy.

CDX4 belongs to the caudal family of homeobox genes that have been implicated in antero-posterior patterning of the axial skeleton and are thought to regulate HOX gene expression. Davidson *et al.* (2003) reported that a CDX4 mutation in zebra fish resulted in severe anemia with complete absence of Runx1 and decreased GATA1 expression but with normal myeloid (PU.1+) and angioblast (Flk-1+) development. HOXB7 and HOXA9 mRNA almost completely rescued the mutant. Retroviral transduction of ES-derived embryoid body cells with CDX4 increased expression of HOXB4, HOXB3, HOXB8, and HOXA9 (all implicated in HSC or progenitor expansion) and significantly increased CFU-GM/CFU-MIX and EryP generation (Davidson *et al.*, 2003). Wang *et al.* (2005e) engineered mESC to express Cdx4 under a tetracyclin-inducible system and found the greatest effect on hematopoietic progenitor generation in EBs was when CDX4 was expressed on days 4–5. Ectopic CDX4 expression promoted hematopoietic mesoderm specification, increased blast colony and hematopoietic progenitor formation, and, together with HOXB4, enhanced long-term multilineage hematopoietic engraftment of lethally irradiated adult mice. The combination of ectopic HOXB4 and CDX4 expression resulted in a high degree of lymphoid engraftment and thymic repopulation with both CD4 and CD8 cells, and capacity to engraft secondary recipients. It should be noted that this long-term lymphoid engraftment was associated with transcriptional silencing of the HOXB4 provirus since the lymphoid populations lacked GFP expression (Wang *et al.*, 2005e). In human ESC studies, Bowles *et al.* (2006) developed H1 ES clones stably expressing HOXB4 and showed in an EB differentiation system that this led to greatly increased production of CD45+ cells and progenitors of all lineages.

Transplantation of *HOXB4*-transduced cord blood Lin–/CD34+ cells into NOD/SCID mice demonstrated enhanced increase in reconstitution capacity compared with vector-transduced human HSCs and a marked enhancement in the generation of primitive CD34+ cells (Schiedlmeier *et al.*, 2003). However, ectopic expression of *HOXB4* is unable to induce hematopoietic repopulating capacity from hESCs, questioning the notion, derived from murine studies, that single genes, such as *HOXB4*, represent a master gene capable of conferring engraftment potential to hESC-derived hematopoietic cells. The failure to obtain robust engraftment using hESC-derived HSCs may be due to a failure to activate a molecular program similar to somatic HSCs. Gene expression patterns of CD34+, CD38– cells derived from human ESC have been compared with those of cells isolated from adult human bone marrow (BM) using microarrays (Lu *et al.*, 2004). Flt3 gene expression was markedly decreased in cells from ESCs, whereas there was substantial Flt3 expression in cells from adult marrow. The Flt3 gene was also undetectable in Rhesus monkey ES cell-derived CD34+ and CD34+, CD38– cells (Lu *et al.*, 2002). hESC-derived hemogenic progenitors expressed higher levels of a negative hematopoietic regulator, CD164, as well as the migratory and/or adhesion proteins CKLF-1, integrin-ß3, matrix metalloproteinase 9, macrophage-inhibiting factor, and monocyte chemoattractant protein-1. These features may account for the reduced ability of hESC-derived hematopoietic cells to migrate beyond the injected site and enter the circulation (Wang *et al.*, 2005e).

2. Methods

The methodology to be outlined covers methods for optimization of hematopoietic stem/progenitor production and for specific lineage differentiation, including lymphoid. Methods are restricted to human or primate ESC systems except
where no human data are available and in such cases, methods based on murine
ESC are shown. More detailed methods for mESC hematopoietic and lymphoid
differentiation have been presented in an earlier "Methods in Enzymology"
(Fairchild *et al.*, 2003; Kitajima *et al.*, 2003; Kyba *et al.*, 2003).

2.1. Maintenance of hESC

We have undertaken our studies on hematopoietic differentiation using two
human hESC lines, H1 and Miz-4. These are maintained on mitotically inactivated
mouse embryonic fibroblast (Chemicon & Specialty Media) with Serum Replacement Media consisted of 80% Knockout Dulbecco's modified Eagle medium (KO-
DMEM, Gibco), 20% Serum Replacement (Gibco), 4 ng/mL bFGF (R&D systems, MN), 1% nonessential amino acids (Gibco), 1 mM L-glutamine (Gibco), and
0.1 mM beta-mercaptoethanol (Sigma, Canada). For propagation of undifferentiated hESC lines, hESC colonies are dissociated with 1 mg/ml Collagenase IV
(Gibco) for 5 min and split every 6–7 days. Occasionally, hESC colonies are
manually dissected during passaging. For more extensive methodology for derivation and maintenance of human ESC see Melton *et al.* (this volume).

2.2. Hematopoietic differentiation of hESC coculture with stromal cell lines

Kaufman *et al.* (2001) pioneered a stromal coculture system for generating
hematopoietic cells from hESC. As originally outlined, undifferentiated hES cells
were cultured on S17 mouse bone marrow stromal cell monolayers to derive cystic
bodies containing CD34+ hematopoietic progenitor stem cells. hES cell cultures
were treated with collagenase IV(1 mg/ml) for 10 min at 37 °C and subsequently
detached from the plate by gently scraping off the colonies. The hES cell clusters
were then transferred to irradiated (35 Gy) S17 cell layers and cultured with RPMI
differentiation medium containing 15% FBS (HyClone, Logan, Utah), 2 mM
L-glutamine, 0.1 mM β-mercaptoethanol, 1% MEM-nonessential amino acids,
and 1% penicillin/streptomycin. The medium was changed every 2–3 days during
14–17 days of culture on S17 cells. After allowing adequate time for differentiation,
hES cystic bodies were harvested and processed into a single cell suspension by
collagenase IV treatment followed by digestion with trypsin/EDTA supplemented
with 2% chick serum (Invitrogen, Carlsbad, California) for 20 min at 37 °C. Cells
were then washed twice with PBS and filtered through a 70 μM cell strainer to
obtain a single cell suspension. To assess the levels of CD34+ cells in the bulk cell
suspension, cells were labeled with PE conjugated anti-CD34 antibody (BD

Biosciences, San Jose, California) and analyzed by FACS. CD34+ cells can be isolated using a CD34 Progenitor Cell Isolation Kit (Miltenyi Biotech, Auburn, California) following the manufacturer's protocol. Subsequent adaptations of this method increased the serum concentration to 20% (Tian *et al.*, 2004) and added recombinant human cytokines and growth factors to the S17 stroma, including SCF, IL-3, IL-6, VEGF, G-CSF, Flt3L, Epo, and BMP-4) (Hematti *et al.*, 2005) or a combination of BMP-4,-2, and-7 (Honig *et al.*, 2004).

2.3. Maintenance of stromal cell lines

A number of murine and human stromal cell lines, developed for hematopoietic stem cell support, have been used to promote hESC hematopoietic differentiation.

S17. The mouse bone marrow stromal cell line S17 (Collins and Dorshkin, 1987) is maintained in α-modified Eagles minimum essential medium (α-MEM) with 2 mM L-glutamine, 1.5 g/l sodium bicarbonate, and 10% fetal bovine serum. Stromal cocultures with S17 have been established using serum-free media, either Stemline media (Sigma) or QBSF60 media (Quality Biologics, Gaithersburg, Maryland, USA) supplemented with 4 mM L-glutamine, and defined cytokines (Flt3L, SCF, Tpo; VEGF, and BMP-4) (Tian *et al.*, 2004).

AFT024. A mouse fetal liver stromal cell (Moore *et al.*, 1997) is routinely cultured in Dulbecco's modified Eagle's medium (DMEM) supplemented with 10% fetal bovine serum (FBS), 5×10^5 mol/l β-mercaptoethanol (2-ME) at 32 °C, 5% CO_2, and 100% humidity.

MS-5. The murine stromal cell line MS-5 (Itoh *et al.*, 1989) (provided by K. Mori, Kyoto University, Kyoto, Japan) is cultured in α-MEM (GIBCO-BRL Life Technologies, Grand Island, New York, USA) supplemented with 10% FBS (HyClone Laboratories, Logan, Utah, USA) and passaged weekly.

OP9. Kodama *et al.* (1994) developed a cell line (OP9) from calvaria of newborn osteopetrotic mutant op/op mice that lack M-CSF and exhibit an osteoclast formation defect. OP9-supported hematopoiesis shows a marked reduction in macrophage production and we have observed that macrophages or their products inhibit stem cell proliferation in stromal coculture systems (Feugier *et al.*, 2005). OP9 cells can be obtained from ATCC (#CRL-2749) and are maintained in α-MEM with 2 mM L-glutamine, 1.5 g/l sodium bicarbonate, 20% fetal bovine serum, and 50 μg/ml ascorbic acid (Kitajima *et al.*, 2003). OP9 cells can easily lose the ability to maintain lympho-hematopoiesis, particularly after prolonged passage or if maintained in suboptimal medium or certain lots of FCS, consequently it is important to screen a number of FCS batches to ensure optimal OP9 hematopoietic support function. While OP9 stroma is contact inhibited and confluent cultures can be used for some weeks, we observe that the unirradiated stroma may detach after two weeks under coculture conditions whereas irradiated stroma (40 Gy) can support hESC differentiation for >4 weeks. It should be noted that while OP9 can support long-term murine stem cell proliferation and differentiation, it is unable to do so with human CD34+ cells, either adult or neonatal, in the absence of additional cytokine supplementation (Feugier *et al.*, 2005). We have shown that it is one of the most effective cell lines supporting long-term human stem cell proliferation and differentiation provided that it is supplemented with either recombinant thrombopoietin, or transduced with an adenovector expressing Tpo (Feugier *et al.*, 2005; Gammaitoni *et al.*, 2004). We have shown prolonged hematopoiesis in OP9 stromal cocultures in serum-free medium (QBSF60) and have observed that the serum-free conditions inhibited overgrowth of ES-derived nonhematopoietic cells whose presence normally requires frequent repassaging onto fresh stroma.

Human bone marrow stroma. Hematopoietic differentiation of hESC has been reported using coculture with primary cultures of adult (Kim *et al.*, 2005) and fetal (Wang *et al.*, 2005c) bone marrow stroma. Ficoll (Sigma) separated marrow cells (5×10^5 cells/ml) are plated in 10 ml of IMDM plus 12.5% FCS and 12.5% horse serum and 5 μM hydrocortisone in T-25 flasks. Cultures are subject to weekly demi-depopulation with addition of fresh medium. By 2–3 weeks, a semi-confluent layer of fibroblasts with some adipocyte differentiation is apparent. The cultures at this point may be trypsinized and repassaged into fresh flasks or 6-well plates. Irradiation (15 Gy) is used to eliminate residual hematopoietic cells. At this stage, hESC can be added and differentiation monitored over 2–5 weeks. Upon continual passage of primary stroma, hematopoietic support function is progressively lost so studies should be restricted to early passage stroma.

Chorionic mesenchyme. Kim *et al.* (2006) have reported hESC differentiation on primary chorionic mesoderm. Human chorionic plate membranes are separated from placenta, a cell suspension made by enzymatic digestion and cells cultured in DMEM + 20% FCS. Upon reaching confluence, this stroma supported the hematopoietic differentiation of EBs cocultured in the presence of IMDM (Invitrogen) 12.5% HS and 12.5% FBS and L-glutamine.

2.4. Hematopoietic differentiation of hESC in coculture with stromal cells

Stromal supported hESC differentiation. OP9, MS-5, or S17 cells are plated onto gelatinized 6-well plates, 10 cm dishes, or in T-12.5 flasks with vented cap (BD Cat#:353107) in α-MEM medium supplemented with 20% heat inactivated FBS, 0.1 mM β-mercaptoethanol, 1 mM L-glutamine, 10 U/ml of penicillin and streptomycin. After formation of confluent stromal cultures on days 4–5, half of the medium is replaced and cells cultured for an additional 3–4 days. Undifferentiated hESC are harvested by treatment with 1 mg/ml collagenase IV (Invitrogen), dispersed by scraping, and added to stromal cultures at a density of 20 colonies/ 20 ml per 10 cm dish, or 4–5 colonies/4 ml per well of a 6-well plate, in α-MEM supplemented with 10% FBS (HyClone) and 100 μM MTG (Sigma, St.Louis, Missouri). When human ESCs are dispersed as individual cells or small aggregates (~50 cells), no or only a few hematopoietic cells develop in the coculture. The larger the ESC colony, the greater is the degree of generation of hematopoietic cells. This suggests that a critical ESC mass is necessary to generate the mesodermal differentiation leading to hemangioblast and hematopoietic cell differentiation. The hES cell/stromal cocultures are incubated at 37 °C in 5% CO_2 with a half-medium change on days 4, 6, and 8. When needed, single-cell suspensions are prepared by treatment of the hESC/stromal cocultures with collagenase IV (Invitrogen; 1 mg/ml in α-MEM) for 20 min at 37 °C, followed by treatment with 0.05% Trypsin-0.5 mM EDTA (Invitrogen) or Cell Dissociation Medium (Accutase, Innovative Cell Technologies) for 15 min at 37 °C. Cells are washed twice with PBS-5% FBS, filtered through a 100 μM cell strainer (BD Biosciences), counted, and used for clonogenic and flow-cytometric assays, and gene expression analysis, or for continuation of hematopoietic development, re-seeded onto fresh stroma. Hematopoietic foci are detected by 2–3 weeks, consisting of phase dark cells beneath the stroma and phase bright clusters of cells loosely attached to the stromal surface (Fig. 21.1.) The cells may be re-passaged onto OP9 stroma or switched from OP9 stroma to MS-5 stroma. The latter provides a standard for evaluation of cobblestone area formation (CAFC) or secondary colony formation (Long-Term Culture-Initiating assay-LTC-IC) at 2–5 weeks as a measure of progenitor and stem cell function (Jo *et al.*, 2000). Both assays can be undertaken under limiting-dilution condition in 96-well plates. It is useful to have control CD34+ cells from neonatal (umbilical cord blood) or adult (bone marrow or G-CSF mobilized peripheral blood) where progenitors and CAFC are to be evaluated. For progenitor (CFC) assay, 5×10^4 differentiated human ES cells or 5×10^3 ES-derived CD34$^+$ cells are plated in triplicate in 35-mm tissue culture dishes containing 1 ml assay medium consisting of IMDM, 1.2% methylcellulose

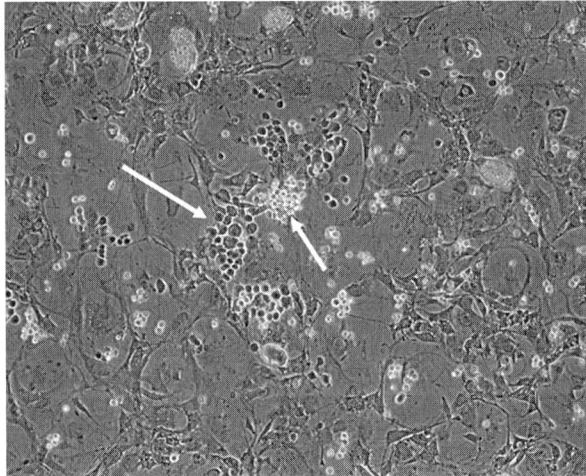

Figure 21.1 Coculture of irradiated OP9 cells and human ES cells. Two human ES colonies were cocultured with irradiated OP9 cells for 3 weeks in IMDM supplemented with 20% FCS and VEGF, SCF, Flt3L, and Tpo. Half of the medium was replaced every 4 days. Foci of loosely adherent hematopoietic cells and phase dark cobblestone area-like cells are observed (arrows).

(Fisher Scientific, Fairlawn, New Jersey, USA), 30% FBS, 5×10^{-5} M 2-mer-captoethanol, 2 mM L-glutamine (GIBCO), and 0.5 mM hemin (Sigma Chemical Co.), and supplemented with SCF, Flt3L, IL-3, G-CSF (all at 20 ng/ml), and 6 U/ml Epo (Amgen). We also routinely included one dish with no cytokine as a negative control to ensure hematopoietic colonies are specifically responding to hematopoietic cytokines. After 14 days of incubation at 37 °C in 5% CO_2 in air, CFU-GM, burst-forming unit erythroid (BFU-E), and mixed colonies (CFU-Mix) are scored (Figs. 21.2 and 21.3). Between 2 and 3 weeks of coculture on various stromal lines, 0.04–0.15% of total cells are progenitors-predominantly granulocyte-macrophage, with a lower frequency of BFU-E and mixed colonies (Table 21.1). Progenitor cells are almost exclusively present in the CD34+ fraction and the cloning efficiency of this population increases markedly from 14 days to 17–21 days of coculture on S17 (Table 21.2). FACS sorting can be used to demonstrate that CFC generated in the stromal coculture system are exclusively CD34+ and CD45+ by 2–3 weeks of culture (Fig. 21.4). The cloning efficiency of this CD34+, CD45+ population (0.68%) is considerably lower than that of neonatal or adult CD34+ populations (10–20%). OP9 cells when cocultured with hESCs support simultaneous generation of CD34+ primitive hematopoietic cells and CD73+ mesenchymal stem cells (MSCs) within the first two weeks (Trivedi and Hematti, 2007).

Stromal plus cytokine-supplemented cultures. Improved hematopoietic differentiation is achieved by coculturing hESC (H1) on irradiated (or nonirradiated) mouse stromal cell lines (S-17, MS-5, AFT-024, and OP-9) in the presence of cytokines, including *rh*VEGF (R&D System), *rh*SCF, (Kirin Brew inc), *rh* Flt-3/Flk2 ligand (Biosources), *rh*Tpo (Kirin Brewery, Inc.), *rh*EPO, (Amgen Inc.), and *rh*IL-3 (R&D system). Table 21.3 shows increasing production of CFC in OP9 coculture of hESC with addition of various cytokine combinations with maximum effect seen with a combination of SCF, Flt3L, VEGF, IL-3, and Tpo, all added at 10 ng/ml. Srivastava *et al.* (2007) reported that Tpo enhanced the generation of CD34+ cells from hESC. Further enhancement has been reported with addition of BMP-4 to a comparable cytokine cocktail (Hematti *et al.*, 2005; Honig *et al.*, 2004).

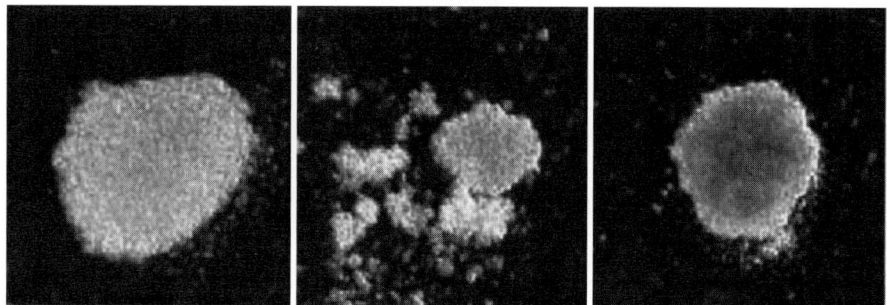

Figure 21.2 Hematopoietic colonies (erythroid, mixed, and myeloid) developing at day 14 in cytokine stimulated methyl cellulose of hESC-derived EB-CD34+ cells.

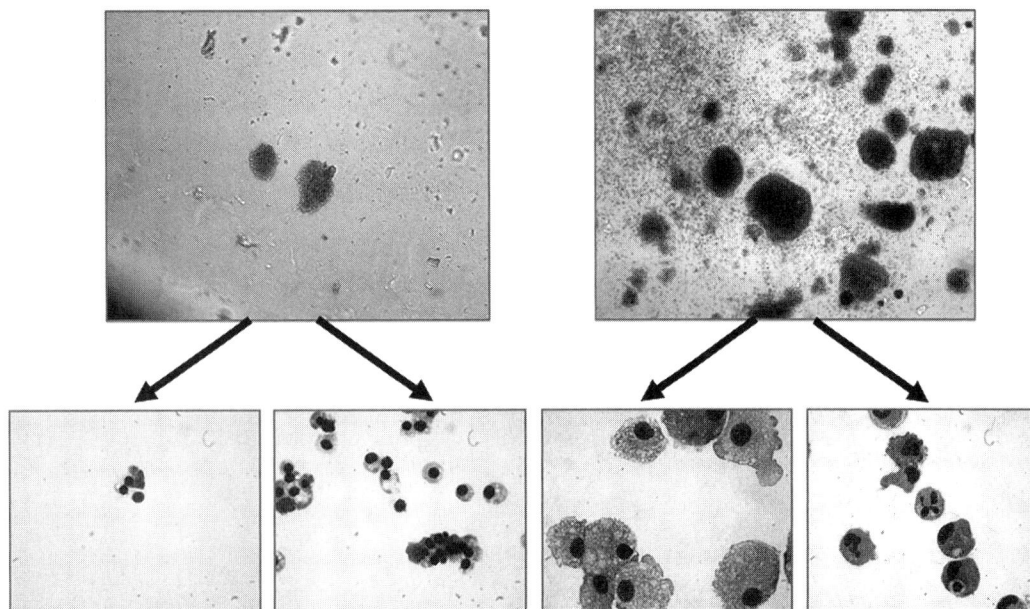

Figure 21.3 Cell morphology of CFC derived from human ESC OP9 stromal cocultures with cytokine supplementation (VEGF, SCF, Flt3L, Tpo). The hematopoietic colonies were isolated by a micropipette, washed free of methyl cellulose, cytospun on a glass slide, and stained with Giemsa. Erythroid differentiation from BFU-E (left) and macrophage and neutrophil differentiation from CFU-GM/CFU-Mix (right).

The concentration of cytokines and the frequency of cytokine replenishment are variables in a number of studies, with cytokine availability and cost as factors limiting continuous use at optimal plateau levels that can be as high as 300 ng/ml with SCF of Flt3L. We have developed a panel of replication incompetent adeno-vectors expressing SCF, Tpo, Flt3L, GM-CSF, Epo, or VEGF derived from the Ad5 E1a-, partially E1b-, partially E3-deficient vector with an expression cassette in the E1a region containing human cytokine cDNA driven by the cytomegalovirus (CMV) major immediate/early promoter/enhancer (Feugier *et al.*, 2005; Gammaitoni *et al.*, 2004). When stromal cells reach confluence, they can be transfected with 15–30 multiplicities of infection (MOI) of the adenovector in serum-free medium (*X-vivo*, Biowhittaker, Walkerville, Massachusetts) for 12 h. Following transfection, the supernatant is removed and replaced with hESC differentiation medium and hESC. Sustained levels of ~100 ng/ml of cytokine as determined by ELISA assay are produced through 4–5 weeks of culture.

2.5. Hematopoietic differentiation of hESC using stromal conditioned medium

The ability to generate hematopoietic cells from hESC in the absence of stromal-cell contamination presents certain advantages from both a practical and clinically relevant standpoint. Irradiated (40 Gy) OP9 cells are cultured with Iscove's

Table 21.1

Hematopoietic progenitor cells derived from cocultures of H1 HES cells on various stromal cell lines[a]

	Cell #	GM-CFC	BFU-E	CFU-Mix	Total CFC	% CFC
S17	13.0×10^5	1690	78	182	1970	0.15
AFT-024	33.0×10^5	660	0	0	660	0.02
MS-5	9.0×10^5	648	18	54	720	0.08
OP9	15.5×10^5	589	31	62	682	0.04

[a]Cocultures of human ES cells on various mouse stromal cells with cytokine (VEGF, SCF, Flt3L, Tpo) addition were harvested at day 18 and subjected to a clonogenic (CFC) assay.

Table 21.2

Frequency of CD34+ cells and their CFC production in hESC coculture on S17 stroma plus cytokine (VEGF, SCF, FLT3L, TPO) over 2–3 weeks

Day	CD34+ %	CFC/10^5 unseparated	CFC/10^5 CD34+ fraction	CFC/10^5 CD34− fraction
14	4.70	8 ± 2	50 ± 14	1 ± 1
17	6.02	37 ± 5	652 ± 20	11 ± 5
21	6.72	38 ± 2	638 ± 62	9 ± 2

modified Dulbecco's medium (IMDM) supplemented with 20% heat-inactivated FBS, 0.1 mM β-mercaptoethanol, 3 mM L-glutamine (Invitrogen Corp.), 5 μM hydrocortisone, 10 U/ml of penicillin (Invitrogen Corp.) and 10 μg/ml of streptomycin. After 12 h incubation at 37 °C, culture supernatants are collected and passed through Acrodisc Syringe Filters (0.2 μm, Pall Corp., Ann Arbor, Michigan). Conditioned medium (OP9-CM) can be harvested over 10 days. All the supernatants are stored at 4 °C and used within 1 week. For hematopoietic differentiation with OP9-CM, the centers of undifferentiated hESC colonies are harvested using a hand-made fine capillary. ~20 colonies are transferred into 10 cm dishes with 5 ml of OP9-CM supplemented with 20 ng/ml of rhVEGF, 20 ng/ml of rhBMP-4, 20 ng/ml of SCF, 10 ng/ml of rhFlt3L, 20 ng/ml of rhIL-3, and 10 ng/ml of rhIL-6. Medium changes are performed every 3–4 days. When needed, single-cell suspension was prepared by treatment of the hESC/OP9-CM cocultures with 0.05% Trypsin-0.5 mM EDTA (Invitrogen) or "Cell Dissociation Medium" (Accutase, Innovative Cell Technologies) for 15 min at 37 °C. Cells are washed twice with PBS + 5% FBS, filtered through a 100 μM cell strainer (BD Biosciences), counted, and used for clonogenic and flow-cytometric assays, and gene expression analysis.

2.6. Embryoid body formation and hematopoietic differentiation

Hanging Drop EB cultures. This method is based on the murine system of Kyba *et al.* (2003). Fifteen centimeter nontissue culture-treated dishes will support ~300 drops with each drop containing ~100 ES cells in 10 ul of differentiation medium.

Cell fraction	CFC/10^5 cells
Unsorted cells	68 ± 4
CD34$^+$ cells (R3)	750 ± 110
CD34$^+$ CD45$^+$ cells (R3)	680 ± 60
CD45$^-$ cells (R14)	2 ± 0

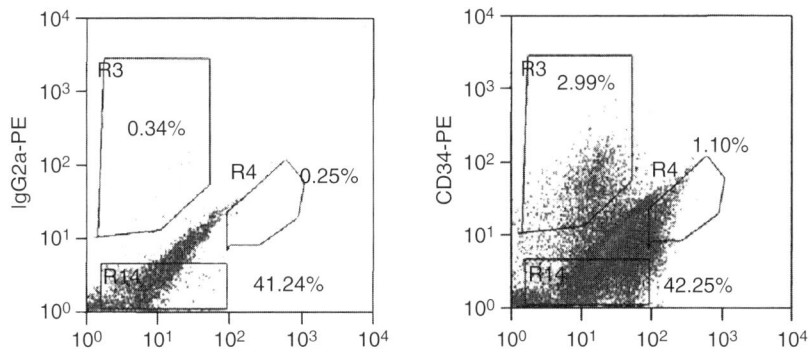

Figure 21.4 Coculture of human H1 hES cells on irradiated OP9 stroma with cytokine addition (VEGF, Flt3L, SCF, Tpo). Cells were harvested at day 18 and sorted on the basis of CD45 and CD34 expression. Sorted fractions were plated for CFC and scored after 2 weeks. Results are displayed as mean $\pm$ SD (in triplicate).

Table 21.3

Effect of cytokine combinations on induction of hESC differentiation to hematopoietic progenitors in OP9 coculture[a]

Cytokine[b]	Cell number	BFU-E (per 10^5)	GM (per 10^5)	Mix (per 10^5)	Total CFC
S/F/T	1.0×10^6	0 ± 0	16 ± 4	0 ± 0	160 ± 40
S/F/V	1.09×10^6	0 ± 0	9 ± 3	0 ± 0	99 ± 27
F/V/T	0.74×10^6	0 ± 0	44 ± 4	0 ± 0	326 ± 30
S/V/T	1.35×10^6	0 ± 0	70 ± 12	0 ± 0	945 ± 162
S/F/T/V	1.57×10^6	0 ± 0	104 ± 18	0 ± 0	1633 ± 188
S/F/T/V/3	2.14×10^6	5 ± 3	76 ± 12	2 ± 2	1766 ± 363

[a]Human ES cells (H1) were cocultured on irradiated OP9 stroma in ES induction medium and combinations of up to six different cytokine were compared. After 18 days, the cocultures were harvested and subjected to clonogenic (CFC) assays. The CFCs were scored after two weeks and data expressed as average $\pm$ SD.

[b]S, stem cell factor; F, Flt3 ligand; V, VEGF; T, Tpo; and 3, IL-3. Dose of each cytokine is 10 ng/ml.

Cells are dispensed by micropipetteing or using an 8-well multichannel pipettor. Dishes are inverted and incubated for 2 days at 37 °C in 5% CO_2. Single EBs form per drop and are collected by flushing the dish with PBS, transferring to 15 ml tubes and allowing sedimentation by gravity for 3 min. Media is aspirated and EBs suspended in 10 ml fresh differentiation medium, transferred to 10 cm bacterial-grade dishes, and cultured under slow swirling conditions on a rotating shaker

(50 rpm). Fresh Media is added every two days after removal of half the spent medium. It is important that throughout these steps, EBs are prevented from attaching to the dish. At different stages of development, EBs can be dissociated by addition of 0.5 ml of 0.25% trypsin and incubated 2 min at 37 °C, followed by addition of 5 ml IMDM + 10%FCS and passaged repeatedly through a 5 ml pipette until dissociated.

Aggregation methods for EB formation. Undifferentiated hESCs at confluence in 6-well plates are treated with collagenase IV and scraped off their Matrigel attachment and transferred to 6-well low-attachment plates to allow for EB formation by overnight incubation in differentiation medium consisting of 80% knockout (KO) DMEM (Gibco) supplemented with 20% nonheat-inactivated fetal bovine serum (FBS; HyClone, Logan, Utah), 1% nonessential amino acids, 1 mM L-glutamine, and 0.1 mM β-mercaptoethanol. After 12 h, all cells and medium are transferred to a 15 ml tube and EBs and allowed to sediment by gravity for 3 min. This step is repeated to remove cell debris and ~30 EBs are then cultured on nonadherent 10 cm tissue culture plates (Corning) with 80% KO-DMEM (Gibco), 20% FBS (Hyclone), 1% nonessential amino acids, 1 mM L-glutamine, and 0.1 mM β-mercaptoethanol (Sigma) in presence of SCF (20 ng/ml), Flt3L (20 ng/ml), Tpo (20 ng/ml), VEGF (10 ng/ml, R&D Systems), Activin A (10 ng/ml, R&D Systems) and BMP-4 (10 ng/ml, R&D Systems). Media and cytokines are replaced every 5 days. Chadwick *et al.* (2003) have used a comparable system, but have supplemented with a somewhat different, and is some cases, higher, concentration of cytokines (300 ng/ml SCF; [Amgen, Thousand Oaks, California], 300 ng/ml Flt3 lL [R&D Systems, Minneapolis, Minnesota], 10 ng/ml IL-3 [R&D Systems], 10 ng/ml IL-6 [R&D Systems], 50 ng/ml G-CSF [Amgen]), and 50 ng/ml BMP-4 [R&D Systems]).

Spin technique for EB generation. Ng *et al.* (2005) observed that only a subset of pieces of undifferentiated hESC containing 500–1000 cells, regularly formed blood cells, suggesting that a minimum number of hESC are required to generate EBs that differentiated into the mesodermal lineage. They developed a modified technique for hEB formation in which hESC are trypsinized into single cell suspension, washed in PBS and resuspended in serum-free medium consisting of a 1:1 ratio of IMDM without phenol red, and Ham's F12 (Gibco, Invitrogen Corporation), 5 mg/ml BSA (Sigma), 1:100 synthetic lipids (Sigma #11905–031), 1:100 insulin-transferrin-selenium (ITS-X, Gibco), 2 mM Glutamine, 5% protein free hybridoma mix (Gibco), and 50 μg/ml ascorbic acid. Cells (300–10,000) are seeded in 100 μl in the serum-free medium supplemented with growth factors (10 ng/ml BMP-4, 5 ng/ml hVEGF, 20 ng/ml SCF, 5 ng/ml Flt3L, 5 ng/ml IL-6, 5 ng/ml hIGF-II (R&D Systems) in each well of 96-well round-bottomed, low attachment plates (Nunc, Roskilde, Denmark), centrifuged at 1500 rpm for 4 min to aggregate the cells. After 10–12 days, EBs are transferred to 96-well flat-bottomed tissue culture plates precoated with gelatin in serum-free medium plus growth factors (VEGF, SCF, Flt3L, IL-3, Tpo, Epo) and allowed to differentiate further.

2.7. Erythroid differentiation

In the mESC system, Carotta *et al.* (2004) were able to generate $>10^{11}$ erythroid cells from an input of 20,000 CCE mESC in 10 weeks. Methylcellulose culture-generated EBs were harvested at 6–9 days and expand in Stem-Pro serum free medium (StemPro34 plus nutrient supplementation-Gibco), Epo (2 U/ml), SCF (100 ng/ml, R&D Systems, Minneapolis, Minnesota), 10^{-6} dexamethasone (Sigma), and 40 ng/ml IGF-1 (Promega, Madison Wisconsin). Cell density was maintained between 2.5 and 4×10^6 ml^{-1}. At days 1 and 3, cell aggregates and dead cells were removed by using a 70 μm cell strainer and Ficoll purification. To achieve differentiation, cells were cultured in StemPro serum free medium with 10 U Epo, insulin (10 ng/ml Actrapid HM, Novo Nordisk) and a 3×10^6 M glucocorticoid receptor antagonist ZK112.993 and 1 mg/ml human transferrin (Sigma). The cells underwent 3–4 "differentiation divisions" reduced cell size, accumulated hemoglobin, and formed enucleated erythrocytes within 72 h. Chang *et al.* (2006) has used a modification of this method for generation of human erythroid cells. CD45+ hematopoiesis peaked at late day 14 EB differentiation stage, although low levels of CD45-ve erythroid differentiation were seen earlier. By morphology, hES-derived erythroid cells were of definitive type but at both the RNA and protein level cells coexpressed high levels of embryonic and fetal (γ) globins with little or no adult (β) globin observed. This was not altered by the presence or absence of FBS, VEGF, Flt3L, or coculture on OP9 and was not culture time dependent. Thus, coexpression of both embryonic and fetal globins by EryD cells did not faithfully mimic either YS or fetal liver ontogeny. In human yolk sac, EryP cells remain mostly nucleated and synthesized mainly embryonic globins (ε, ζ, α,). Fetal cells have a macrocytic morphology and synthesize $>80\%$ adult globins ($\alpha 2\gamma 2$). Adult cells synthesize $>90\%$ adult globins ($\alpha 2\beta 2$). There are some discrepancies as regards kinetics, morphology, and globin pattern of erythroid cells generated from human ES in various published reports. Qui *et al.* (2005) have assigned ESC-derived erythroid cells exclusively to the EryP lineage while Zambidis *et al.* (2005) supported transition from primitive to definitive phenotype because of increased levels of β-globin. Two studies showed exclusive, or predominant expression of β-globin later in culture (Cerdan *et al.*, 2004; Kaufman *et al.*, 2001). It should be noted that erythroid differentiation has been reported predominantly with the H1 hES cell line (WiCell Research Inst NIH Code WA01). However, robust erythropoiesis has been reported with hES2, hES3, and hES4 lines (Ng *et al.*, 2005) and our own studies with Miz-4. It did not occur, or only at low levels in four other ES lines (hSF6-NIH Code UC06, UCSF, BG01, BG02, BG03-NIH Code BG01, BG02, BG03, BresaGen Inc, Masons, Georgia) (Chang *et al.*, 2006). Large-scale production of embryonic red blood cells from hESC cells can be obtained by coculture with a human fetal liver cell line (Olivier *et al.*, 2006). A 5000-fold increase in cell number can be obtained and the erythroid cells produced do not enucleate, are fully hemoglobinized and express a mixture of embryonic and fetal globins but no beta-globin. These primitive erythroblasts undergo a switch in hemoglobin (Hb) composition during late terminal erythroid maturation with the basophilic erythroblasts expressing predominantly Hb Gower I (zeta(2)epsilon(2)) and the orthochromatic erythroblasts hemoglobin Gower II (alpha(2)epsilon(2)). This suggests that the switch from Hb Gower I to Hb Gower II, the first hemoglobin switch in humans is a maturation switch not a lineage switch (Qiu *et al.*, 2008). Extending the coculture of the hESCs with immortalized fetal hepatocytes to 35 days yielded CD34+ cells that differentiate into more developmentally mature, fetal liver-like erythroblasts, that are smaller, express mostly fetal hemoglobin, and can enucleate (Qiu *et al.*, 2008).

2.8. Megakaryocyte differentiation

CD34+ cells derived from hEBs or ESC stromal coculture are seeded onto 24-well plates (10,000 cells/well) in QBSF-60 serum-free medium (Quality Biological, Inc, 2 ml/well), with the following human cytokines: 50 ng/ml SCF, 50 ng/ml Flt3L, 50 ng/ml Tpo for the first week. For megakaryocytic differentiation, 50 ng/ml SCF, 50 ng/ml IL-3 and 100 ng/ml Tpo are used in the second and third weeks. Only floating cells are harvested for cell phenotype analysis. The phenotype analysis by flow cytometry and cytospin are performed after the third week for CD41a+ CD45+ megakaryocytes and developing platelets.

2.9. Neutrophil differentiation

Lieber *et al.* (2004) developed a 3-step liquid culture differentiation strategy enabling reliable and abundant production of neutrophils at high purity from murine ESC. Day 8 EBs are trypsinized for 5 min at room temperature and disaggregated into a cell suspension. The cells are washed in 20 ml IMDM containing 10% FBS, centrifuged, and resuspended in secondary differentiation medium and plated onto semi-confluent OP9 cells. The secondary differentiation mix contains 10% pretested heat-inactivated FBS (Summit Biotech), 10% horse serum (Biocell Laboratories, Rancho Dominguez, California), 5% protein-free hybridoma medium (GIBCO BRL), 25 ng/ml oncostatin M (OSM), 10 ng/ml bFGF, 5 ng/ml IL-6, 5 ng/ml IL-11, and 1 ng/ml rLIF (R&D Systems, Minneapolis, Minnesota) in 74% by volume IMDM containing 100 U/ml penicillin and 100 μg/ml streptomycin and 1.5×10^{-4} M MTG. After 24 h, the adherent cells associated with the monolayers are trypsinized and replated in the same medium onto new semiconfluent OP9 monolayers along with the cells in suspension to reduce monocyte/macrophage and fibroblast-like contaminants. After 3 days in the secondary differentiation mix, cells are transferred onto a semiconfluent OP9 monolayer at a concentration of ~4×10^5 cells/ml into a tertiary neutrophil differentiation mix containing 10% platelet-depleted serum (Animal Technologies, Tyler, Texas), 2 mM L-glutamine, 88% by volume IMDM, 100 U/ml penicillin, 100 μg/ml streptomycin, 1.5×10^{-4} M MTG, 60 ng/ml G-CSF (Amgen, Thousand Oaks,

California), 3 ng/ml GM-CSF, and 5 ng/ml IL-6 (R&D Systems). After 4–20 days, the cells are harvested for assays.

2.10. Macrophage differentiation

CD34+ cells (~2.5×10^5 to 4.0×10^5 ml^{-1}) isolated from either hEBs or hESC-stromal cocultures are plated in methylcellulose culture (Methocult semisolid medium—Stem Cell Technologies, Vancouver, British Columbia), to generate myelomonocytic colonies. At day 14, colonies are harvested by the addition of 5 ml DMEM containing 10% FBS, 10 ng/ml each of GM-CSF and M-CSF. Cells (~10^6) are placed in a 35 mm well and allowed to adhere for 48 h. At 2 and 4 days postharvest, medium is replaced with fresh complete DMEM supplemented with 10 ng/ml GM-CSF and M-CSF. By 4–5 days, cells developed into mature macrophages that may be used for subsequent phenotypic and functional characterization (Anderson *et al.*, 2006). As assessed by FACS analysis, the hES-CD34 cell-derived macrophages displayed characteristic cell surface markers CD14, CD4, CCR5, CXCR4, and HLA-DR, suggesting a normal phenotype. Tests evaluating phagocytosis, upregulation of the costimulatory molecule B7.1, and cytokine secretion in response to LPS stimulation showed that these macrophages are also functionally normal.

2.11. Lymphoid differentiation from hESC

Fully grown hES colonies are cultured on irradiated OP9 stroma expressing the Delta ligand, DLL1, with α-MEM supplemented with 20% FBS (HyClone), 100 μM MTG, 5 ng/ml Flt3L for 7–8 days. Floating cells from hESC-OP9-DL1 coculture are harvested and CD34+ cell isolated using a "Direct CD34 Progenitor Cell Isolation Kit" (Miltenyi Biotech Inc) as recommended by the manufacturer. For T cell progenitor (CD5+ and CD7+) differentiation, CD34+ cells are cultured for a further 3 weeks on irradiated OP9-DLL1 with 5 ng/ml Flt3L and 5 ng/ml IL-7 (R&D Systems). For B cell (CD19+ CD45+) differentiation, CD34+ cells from differentiation EBs are cultured on MS-5 stroma for >4 weeks with α-MEM supplemented with 10% FBS (HyClone), 100 μM MTG, 10 ng/ml of each SCF and G-CSF. For Natural Killer cell (CD56+ CD45+) differentiation, CD34+ CD45+ cells from differed EBs are isolated and cultured on AFT-024 stroma with α-MEM supplemented with 10% FBS (HyClone), 100 μM MTG, 20 ng/ml SCF, 20 ng/ml IL-7, 10 ng/ml Flt3L, and 10 ng/ml IL-15 for >4 weeks. For dendritic cell differentiation, EB-derived CD34+ cells are seeded onto 24-well plates (10,000 cells/well) in QBSF-60 serum-free medium (Quality Biological, Inc., 2 ml/well), with the following human cytokines: 50 ng/ml SCF, 50 ng/ml Flt3L, 50 ng/ml Tpo for the first week and replaced with 50 ng/ml SCF, 50 ng/ml Flt3L, 50 ng/ml Tpo, 10 ng/ml GM-CSF, and 20 ng/ml TNF-α for the second and subsequent weeks. Dendritic differentiation is evaluated by FACS expression of MHC class II, CD83, CD80, CD86, and CD40 with loss of the macrophage marker CD14.

2.12. *In vivo* transplantation of EB-derived CD34+ cells into immunodeficient mice

EB-derived CD34+ cell are harvested at Day 10–12 as described above and are transplanted into sublethally irradiated 8–10 weeks old NOD/SCID β2m–/– mice using an intrafemoral bone marrow transplantation technique (IBMT) and into nonirradiated new born mice by an intrahepatic transplant method (IHT) or by facial vein injection (FVI). Cell doses ranged from 50,000–100,000 for IHT and 100,000–500,000 for IBMT. In order to enhance engraftment, human bone marrow stromal cells (HS27 and/or HS5, 250,000 cells per mouse) (Roecklein and Torok-Storb, 1995) can be coinjected with hES-derived CD34+ cells into the femur. For IMBT, 8–10 weeks old mice are exposed to 2.5 Gy Rad and on the following day, CD34+ cells (with or without human stromal cells) are suspended in PBS (20 μl) and loaded into a syringe (1/2cc 281/2 gauge, U-100 Insulin Syringe Cat #. 329461 Beckton-Dickinson). Mice are anesthetized with the mixture of ketamine (0.8 μl/g of 100 mg/ml stock solution) and of xylazine (0.2 μl/g of 20 mg/ml stock solution). The region from the groin to the knee joint is shaved with a razor, the knee flexed to 90° and a 28-gauge needle inserted into the joint surface of the tibia through the patellar tendon and into the bone marrow cavity. For IHT, on the morning of birth, pups are transplanted without irradiation. CD34+ cells are suspended in PBS (50 μl) and loaded into a 1/2cc 281/2 gauge, U-100 Insulin Syringe The pup is held so that its body is pinned between the thumb and index finger of the left hand. The middle finger is placed on the pup's abdominal part so that the body can be tilted to one side to expose the liver which is readily visualized through the skin. The needle is inserted into the liver from one side and the cells injected toward the other side. After insertion and injection, the needle is withdrawn slowly to allow minimal seepage. For FVI, on the morning of birth, pups are exposed to 0.5 Gy Rad and returned to their mother's cage. Next day (~18 h), CD34+ cells suspended with PBS (40 μl) are loaded into a syringe (30G Ultrafine Insulin Syringe Cat #. 328431, Beckton-Dickinson). The pup is held so that its body is pinned between the thumb and middle finger of the left hand. The index finger is placed on the pup's chin so that the head can be tilted to the side to expose the face and neck vessels. The needle is inserted into the facial vein and the cells injected toward the heart. In the case of IHT and FVI, transplantation is done no later than 36 h after birth. All animal experiments require approval by Institutional animal care and veterinary services.

2.13. Analysis of NOD/SCID mouse hematopoietic engraftment

To prepare mouse bone marrow (BM) for flow cytometric analysis, BM cells are washed in PBS containing 3% BSA. The presence of human cells in BM of the transplanted immunodeficient mice is determined by flow cytometry using PE or APC-conjugated antibody against human CD19, CD33, CD34, CD45, and Glycoporin A (BD Pharmingen) (Fig. 21.5). In parallel, Southern blot and PCR analyses can be performed to detect human DNA in mouse bone marrow. For

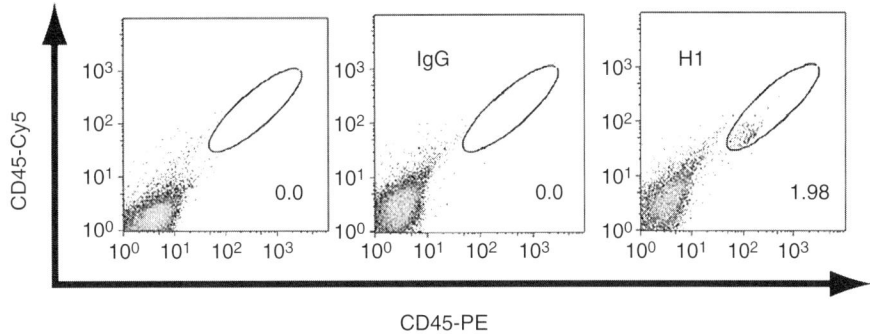

Figure 21.5 FACS analysis showing human hematopoietic (CD45+) bone marrow engraftment of NOD/SCID $\beta2-/-$ recipients 6 weeks after receiving hESC-derived hematopoietic cells by the intrahepatic route.

extraction of genomic DNA, the DNeasy kit (QIAGEN) is used according to the manufacturer's recommendations. PCR for the human chromosome 17-specific PCR is performed using A forward primer 5'-ACACTCTTTTTGCAG-GATCTA-3' and backward primer 5'-AGCAATGTGAAACTCTGGGA-3' to amplify an 1171-bp sequence (40 cycles, 94 °C for 1 min, 60 °C for 1 min, 72 °C for 1 min, followed by a final extension of 10 min at 72 °C). For Southern blot, 5 μg DNA is digested with an EcoRI restriction enzyme at 37 °C overnight, separated on an agarose gel, transferred to Hybond-N+ nylon membrane (Amersham Biosciences) and hybridized with a DIG-labeled (Roche Molecular System) human chromosome specific probe (Bhatia *et al.*, 1998).

2.14. Bioluminescence Imaging (BLI)

Stable transduction of transplantable cells with GFP/luciferase fusion gene has provided an efficient quantitative measure of cell burden and location in human tumor xenografts transplant models (Wu *et al.*, 2005). Stable transduction of hESC with this GFP/Luciferase fusion gene permits imaging of ESC-derived hematopoietic cells in NOD/SCID mice.

Production of aGFP-Luciferase-expressing Lentivirus: A GFP-luciferase fusion gene, driven by EF1 promoter is cloned into the backbone of FUGW (kindly provided by Dr. David Baltimore) after deletion a GFP gene controlled by the Ubiquitin promoter (FUEGL). 293T cells are maintained in Dulbecco's modified Eagle's medium (DMEM) supplemented with 10% fetal bovine serum (FBS), 100 units/ml penicillin, and 100 μg/ml streptomycin. The 293T cells are plated in 100-mm tissue culture dishes at least 12 h before transduction. The cell density should be 20–30% confluent when seeding and will be ~40% confluent for transfection. The culture medium is replaced with 10 ml of fresh medium 2 h before transfection. Prepare 2 ml of a calcium phosphate-DNA mixture suspension, which contains

1 ml of 2× HBS (0.05 M HEPES, 0.28 M NaCl, 1.5 mM Na_2HPO_4, pH 7.12), 150 μl of 2 M $CaCl_2$, 20 μg lentivirus vector, 10 μg pVSVG, and 15 μg p$\triangle$8.9, and distilled water (up to 2 ml), for each 100-mm plate. Allow the suspension to sit at room temperature for at least 10 min. Mix the precipitate well by vortexing and add 2 ml of calcium phosphate-DNA suspension to a 100-mm plate containing cells with drop-wise manner. Return the plates to the incubator and leave the precipitation for overnight (about 18 h). Replace fresh medium and check GFP expression the next morning. After 48 or 72 h of transfection, virus is harvested and filtered with a 0.2 μm syringe filter. Filtered medium is concentrated with 100,000 MWCO Centrifugal Filter Devices (Millipore) at 2000 RPM for 25 min and concentrated ~50 fold (we harvest ~200 μl of concentrated virus suspension from 100 ml of nonconcentrated virus-containing medium).

Transduction of Human Embryonic Stem Cells: hESC are cultured in a 24-well plate with low density (less than 10 colonies in a single well of 24-well plates) for two days. Concentrated viral supernatant (100 μl to 1 ml of Serum Replacement Media medium per well) is introduced with hESC and 4 μg/ml polybrene (Sigma) for 12 h. After 12 h, medium is replaced with fresh Serum Replacement Media and cultured for 4–5 days. Undifferentiated hESC are isolated with a 1 ml micropipette and transferred to new fresh mitotically inactivated mouse embryonic fibroblast. After 4–5 days, GFP-expressing human embryonic stem cell colonies are checked manually and reisolated for further passage. GFP-expressing colonies are maintained for more than 2 months to isolate homogenous colonies with uniform, stable GFP expression before further experiment. To confirm the luciferase activity *in vitro*, lysates of GFP-expressing colonies are analyzed by a Lumat LB9507

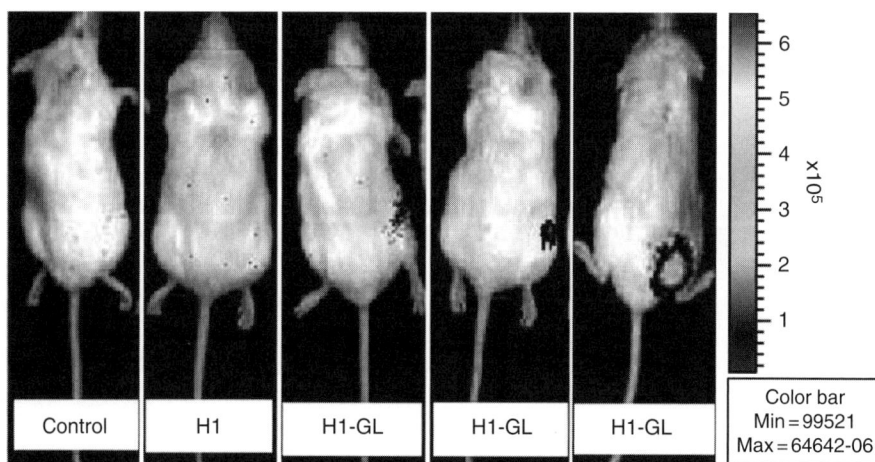

Figure 21.6 Bioluminescent images at week 5 of NOD/SCID β2−/− mice injected intrafemorally with PBS (negative control), nontransduced hESC-derived hematopoietic cells (H1), and H1-GFP/luciferase-derived hematopoietic cells (H1-GL).

Luminometer (EG &G Berthold) to measure the luciferase activity of luciferase reporter using the Dual-Luciferase reporter assay system (Promega) according to the manufacturer's recommendations.

In vivo luciferase imaging. NOD/SCID, NOD/SCID-β2M($-/-$), and NOD/SCID-γIL2R8($-/-$) mice are transplanted with GFP/Luciferase transduced ESC-derived hematopoietic cells sorted for CD34+ and GFP expression. At intervals (e.g., 3 weeks, 5 weeks) animals are subject to whole body bioimaging. Luciferin (Xenogen), the substrate for firefly luciferase, is dissolved in phosphate-buffered saline at a concentration of 15.4 mg/ml and filtered through a 0.22-μm-pore-size filter before use. Mice are injected with 200 μl of luciferin (3 mg) and immediately anesthetized in an oxygen-rich induction chamber with 2% isofluorane (Baxter Healthcare, IL). The mice are maintained for at least 10 min so that there is adequate dissemination of the injected substrate. Anesthesia is maintained during the entire imaging process by using a nose cone isofluorane–oxygen delivery device in the specimen chamber. Images are collected with 10–20 second integration times depending on the intensity of the bioluminescent signal. Data acquisition and analysis is performed by using the LivingImage (Xenogen) software with the IgorPro image analysis package (WaveMetrics) (Fig. 21.6).

References

Anderson, J. S., Bandi, S., Kaufman, D. S., and Akkina, R. (2006). Derivation of normal macrophages from human embryonic stem (hES) cells for applications in HIV gene therapy. *Retrovirology* **3**, 24–30.

Berthier, R., Prandini, M. H., Schweitzer, A., Thevenon, D., Martin-Sisteron, H., and Uzan, G. (1997). The MS-5 murine stromal cell line and hematopoietic growth factors synergize to support the megakaryocytic differentiation of embryonic stem cells. *Exp. Hematol.* **25**, 481–490.

Bhatia, M. (2005). Derivation of the hematopoietic stem cell compartment from human embryonic stem cell lines. *Ann. NY Acad. Sci.* **1044**, 24–28.

Bhatia, M. (2007). Hematopoiesis from human embryonic stem cells. *Ann. NY Acad. Sci.* **1106**, 219–222.

Bhatia, M., Bonnet, D., Murdoch, B., Gan, O. I., and Dick, J. E. (1998). A newly discovered class of human hematopoietic cells with SCID-repopulating activity. *Nat. Med.* **4**, 1038–1045.

Bowles, K. M., Vallier, L., Smith, J. R., Alexander, M. R., and Pedersen, R. A. (2006). HOXB4 overexpression promotes hematopoietic development by human embryonic stem cells. *Stem Cells* **24**, 1359–1369.

Burt, R. K., Verda, L., Kim, D. A., Oyama, Y., Luo, K., and Link, C. (2004). Embryonic stem cells as an alternate marrow donor source: Engraftment without graft-versus-host disease. *J. Exp. Med.* **199**, 895–904.

Carotta, S., Pilat, S., Mairhofer, A., Schmidt, U., Dolznig, H., Steinlein, P., and Beug, H. (2004). Directed differentiation and mass cultivation of pure erythroid progenitors from mouse embryonic stem cells. *Blood* **104**, 1873–1880.

Cerdan, C., Rouleau, A., and Bhatia, M. (2004). VEGF-A165 augments erythropoietic development from human embryonic stem cells. *Blood* **103**, 2504–2512.

Chadwick, K., Wang, L., Li, L., Menendez, P., Murdoch, B., Rouleau, A., and Bhatia, M. (2003). Cytokines and BMP-4 promote hematopoietic differentiation of human embryonic stem cells. *Blood* **102**, 906–915.

Chang, K.-H., Nelson, A. M., Cao, H., Wang, L., Nakamoto, B., Ware, C. B., and Papayannopoulou, T. (2006). Definitive-like erythroid cells derived from human embryonic stem cells co-express high levels of embryonic and fetal globins with little or no adult globin. *Blood* **108,** 1515–1523.

Choi, K., Chung, Y. S., and Zhang, W. J. (2005). Hematopoietic and endothelial development of mouse embryonic stem cells in culture. *Methods Mol. Med.* **105,** 359–368.

Choi, K., Kennedy, M., Kazarov, A., Papadimitriou, J. C., and Keller, G. (1998). A common precursor for hematopoietic and endothelial cells. *Development* **125,** 725–732.

Cho, S. K., Webber, T. D., Carlyle, J. R., Nakano, T., Lewis, S. M., and Zuniga-Pflucker, J. C. (1999). Functional characterization of B lymphocytes generated *in vitro* from embryonic stem cells. *Proc. Natl. Acad. Sci. USA* **96,** 9797–9802.

Collins, L. S., and Dorshkind, K. (1987). A stromal cell line from myeloid long-term bone marrow cultures can support myelopoiesis and B lymphopoiesis. *J. Immunol.* **138,** 1082–1087.

Daley, G. Q. (2003). From embryos to embryoid bodies, generating blood from embryonic stem cells. *Ann. NY Acad. Sci.* **996,** 122–131.

Dang, S. M., Gerecht-Nir, S., Chen, J., Itskovitz-Eldor, J., and Zandstra, P. W. (2004). Controlled, scalable embryonic stem cell differentiation culture. *Stem Cells* **22,** 275–282.

Dang, S. M., Kyba, M., Perlingeiro, R., Daley, G. Q., and Zandstra, P. W. (2002). Efficiency of embryoid body formation and hematopoietic development from embryonic stem cells in different culture systems. *Biotechnol. Bioeng.* **78,** 442–453.

Davidson, A. J., Ernst, P., Wang, Y., Dekens, M. P., Kingsley, P. D., Palis, J., Korsmeyer, S. J., Daley, G. Q., and Zon, L. I. (2003). cdx4 mutants fail to specify blood progenitors and can be rescued by multiple hox genes. *Nature* **425,** 300–306.

de Pooter, R. F., Cho, S. K., Carlyle, J. R., and Zuniga-Pflucker, J. C. (2003). *In vitro* generation of T lymphocytes from embryonic stem cell-derived prehematopoietic progenitors. *Blood* **102,** 1649–1653.

de Pooter, R. F., Cho, S. K., and Zuniga-Pflucker, J. C. (2005). *In vitro* generation of lymphocytes from embryonic stem cells. *Methods Mol. Biol.* **290,** 135–147.

Ema, M., Faloon, P., Zhang, W. J., Hirashima, M., Reid, T., Stanford, W. L., Orkin, S., Choi, K., and Rossant, J. (2003). Combinatorial effects of Flk1 and Tal1 on vascular and hematopoietic development in the mouse. *Genes Dev.* **17,** 380–393.

Fairchild, P. J., Brook, F. A., Gardner, R. L., Graca, L., Strong, V., Tone, Y., Tone, M., Nolan, K. F., and Waldmann, H. (2000). Directed differentiation of dendritic cells from mouse embryonic stem cells. *Curr. Biol.* **10,** 1515–1518.

Fairchild, P. J., Nolan, K. F., and Waldmann, H. (2003). Probing dendritic cell function by guiding the differentiation of embryonic stem cells. *Methods Enzymol.* **365,** 169–186.

Faloon, P., Arentson, E., Kazarov, A., Deng, C. X., Porcher, C., Orkin, S., and Choi, K. (2000). Basic fibroblast growth factor positively regulates hematopoietic development. *Development* **127,** 1931–1941.

Feugier, P., Li, N., Jo, D. Y., Shieh, J. H., MacKenzie, K. L., Lesesve, J. F., Latger-Cannard, V., Bensoussan, D., Crystal, R. G., Rafii, S., Stoltz, J. F., and Moore, M. A. (2005). Osteopetrotic mouse stroma with thrombopoietin, c-kit ligand, and flk-2 ligand supports long-term mobilized CD34+ hematopoiesis *in vitro*. *Stem Cells Dev.* **14,** 505–516.

Fujimoto, T. T., Kohata, S., Suzuki, H., Miyazaki, H., and Fujimura, K. (2003). Production of functional platelets by differentiated embryonic stem (ES) cells *in vitro*. *Blood* **102,** 4044–4051.

Fukuma, D., Matsuyoshi, H., Hirata, S., Kurisaki, A., Motomura, Y., Yoshitake, Y., Shinohara, M., Nishimura, Y., and Senju, S. (2005). Cancer prevention with semi-allogeneic ES cell-derived dendritic cells. *Biochem. Biophys. Res. Commun.* **335,** 5–13.

Gammaitoni, L., Weisel, K. C., Gunetti, M., Wu, K. D., Bruno, S., Pinelli, S., Bonati, A., Aglietta, M., Moore, M. A., and Piacibello, W. (2004). Elevated telomerase activity and minimal telomere loss in cord blood long-term cultures with extensive stem cell replication. *Blood* **103,** 4440–4448.

Hamaguchi, I., Morisada, T., Azuma, M., Murakami, K., Kuramitsu, M., Mizukami, T., Ohbo, K., Yamaguchi, K., Oike, Y., Dumont, D. J., and Suda, T. (2006). Loss of Tie2 receptor compromises embryonic stem cell-derived endothelial but not hematopoietic cell survival. *Blood* **107,** 1207–1213.

Helgason, C. D., Sauvageau, G., Lawrence, H. J., Largman, C., and Humphries, R. K. (1996). Overexpression of HOXB4 enhances the hematopoietic potential of embryonic stem cells differentiated *in vitro*. *Blood* **87**, 2740–2749.

Hematti, P., Obrtlikova, P., and Kaufman, D. S. (2005). Nonhuman primate embryonic stem cells as a preclinical model for hematopoietic and vascular repair. *Exp. Hematol.* **33**, 980–986.

Hidaka, M., Stanford, W. L., and Bernstein, A. (1999). Conditional requirement for the Flk-1 receptor in the *in vitro* generation of early hematopoietic cells. *Proc. Natl. Acad. Sci. USA* **96**, 7370–7375.

Hirata, S., Senju, S., Matsuyoshi, H., Fukuma, D., Uemura, Y., and Nishimura, Y. (2005). Prevention of experimental autoimmune encephalomyelitis by transfer of embryonic stem cell-derived dendritic cells expressing myelin oligodendrocyte glycoprotein peptide along with TRAIL or programmed death-1 ligand. *J. Immunol.* **174**, 1888–1897.

Hole, N., Graham, G. J., Menzel, U., and Ansell, J. D. (1996). A limited temporal window for the derivation of multilineage repopulating hematopoietic progenitors during embryonal stem cell differentiation *in vitro*. *Blood* **88**, 1266–1276.

Honig, G. R., Li, F., Lu, S. J., and Vida, L. (2004). Hematopoietic differentiation of rhesus monkey embryonic stem cells. *Blood Cells Mol. Dis.* **32**, 5–10.

Itoh, K., Tezuka, H., Sakoda, H., Konno, M., Nagata, K., Uchiyama, T., Uchino, H., and Mori, K. J. (1989). Reproducible establishment of hemopoietic supportive stromal cell lines from murine bone marrow. *Exp. Hematol.* **17**, 145–153.

Jo, D. Y., Rafii, S., Hamada, T., and Moore, M. A. (2000). Chemotaxis of primitive hematopoietic cells in response to stromal cell-derived factor-1. *J. Clin. Invest.* **105**, 101–111.

Kaufman, D. S., Hanson, E. T., Lewis, R. L., Auerbach, R., and Thomson, J. A. (2001). Hematopoietic colony-forming cells derived from human embryonic stem cells. *Proc. Natl. Acad. Sci. USA* **98**, 10716–10721.

Keller, G. (2005). Embryonic stem cell differentiation: Emergence of a new era in biology and medicine. *Genes Dev.* **19**, 1129–1155.

Keller, G., Kennedy, M., Papayannopoulou, T., and Wiles, M. V. (1993). Hematopoietic commitment during embryonic stem cell differentiation in culture. *Mol. Cell Biol.* **13**, 473–486.

Kennedy, M., D'Souza, S. L., Lynch-Kattman, M., Schwantz, S., and Keller, G. (2007). Development of the hemangioblast defines the onset of hematopoiesis in human ES cell differentiation cultures. *Blood* **109**, 2679–2687.

Kennedy, M., Firpo, M., Choi, K., Wall, C., Robertson, S., Kabrun, N., and Keller, G. (1997). A common precursor for primitive erythropoiesis and definitive haematopoiesis. *Nature* **386**, 488–493.

Kim, S. J., Kim, B. S., Ryu, S. W., Yoo, J. H., Oh, J. H., Song, C. H., Kim, S. H., Choi, D. S., Seo, J. H., Choi, C. W., Shin, S. W., Kim, Y. H., and Kim, J. S. (2005). Hematopoietic differentiation of embryoid bodies derived from the human embryonic stem cell line SNUhES3 in co-culture with human bone marrow stromal cells. *Yonsei Med. J.* **46**, 693–699.

Kim, S. J., Yoo, J. H., Kim, B. S., Oh, J. H., Song, C. H., Shin, H. J., Kim, S. H., Choi, C. W., and Kim, J. S. (2006). Mesenchymal stem cells derived from chorionic plate may promote hematopoietic differentiation of a human embryonic stem cell line, SNUhES3. *Acta Haematol.* **116**, 219–222.

Kitajima, K., Tanaka, M., Zheng, J., Sakai-Ogawa, E., and Nakano, T. (2003). *In vitro* differentiation of mouse embryonic stem cells to hematopoietic cells on an OP9 stromal cell monolayer. *Methods Enzymol.* **365**, 72–83.

Kodama, H., Nose, M., Niida, S., and Nishikawa, S. (1994). Involvement of the c-kit receptor in the adhesion of hematopoietic stem cells to stromal cells. *Exp. Hematol.* **22**, 979–984.

Kyba, M., Perlingeiro, R. C., and Daley, G. Q. (2002). HoxB4 confers definitive lymphoid-myeloid engraftment potential on embryonic stem cell and yolk sac hematopoietic progenitors. *Cell* **109**, 29–37.

Kyba, M., Perlingeiro, R. C. R., and Daley, G. Q. (2003). Development of hematopoietic repopulating cells from embryonic stem cells. *Methods Enzymol.* **365**, 114–129.

Lacaud, G., Kouskoff, V., Trumble, A., Schwantz, S., and Keller, G. (2004). Haploinsufficiency of Runx1 results in the acceleration of mesodermal development and hemangioblast specification upon *in vitro* differentiation of ES cells. *Blood* **103**, 886–889.

Lengerke, C., and Daley, G. Q. (2005). Patterning definitive hematopoietic stem cells from embryonic stem cells. *Exp. Hematol.* **33**, 971–979.

Lensch, M. W., and Daley, G. Q. (2006). Scientific and clinical opportunities for modeling blood disorders with embryonic stem cells. *Blood* **107**, 2605–2612.

Lerou, P. H., and Daley, G. Q. (2005). Therapeutic potential of embryonic stem cells. *Blood Rev.* **19**, 321–331.

Li, F., Lu, S., Vida, L., Thomson, J. A., and Honig, G. R. (2001). Bone morphogenetic protein 4 induces efficient hematopoietic differentiation of rhesus monkey embryonic stem cells *in vitro*. *Blood* **98**, 335–342.

Lian, R. H., Maeda, M., Lohwasser, S., Delcommenne, M., Nakano, T., Vance, R. E., Raulet, D. H., and Takei, F. (2002). Orderly and nonstochastic acquisition of CD94/NKG2 receptors by developing NK cells derived from embryonic stem cells *in vitro*. *J. Immunol.* **168**, 4980–4987.

Lieber, J. G., Webb, S., Suratt, B. T., Young, S. K., Johnson, G. L., Keller, G. M., and Worthen, G. S. (2004). The *in vitro* production and characterization of neutrophils from embryonic stem cells. *Blood* **103**, 852–859.

Liu, H., and Roy, K. (2005). Biomimetic three-dimensional cultures significantly increase hematopoietic differentiation efficacy of embryonic stem cells. *Tissue Eng.* **11**, 319–330.

Liu, B., Sun, Y., Jiang, F., Zhang, S., Wu, Y., Lan, Y., Yang, X., and Mao, N. (2003). Disruption of Smad5 gene leads to enhanced proliferation of high-proliferative potential precursors during embryonic hematopoiesis. *Blood* **101**, 124–133.

Lu, S. J., Feng, Q., Caballero, S., Chen, Y., Moore, M. A., Grant, M. B., and Lanza, R. (2007). Generation of functional hemangioblasts from human embryonic stem cells. *Nat. Methods* **4**, 501–509.

Lu, S. J., Li, F., Vida, L., and Honig, G. R. (2002). Comparative gene expression in hematopoietic progenitor cells derived from embryonic stem cells. *Exp. Hematol.* **30**, 58–66.

Lu, S. J., Li, F., Vida, L., and Honig, G. R. (2004). CD34+ CD38 − hematopoietic precursors derived from human embryonic stem cells exhibit an embryonic gene expression pattern. *Blood* **103**, 4134–4141.

Ma, F., Wang, D., Hanada, S., Ebihara, Y., Kawasaki, H., Zaike, Y., Heike, T., Nakahata, T., and Tsuji, K. (2007). Novel method for efficient production of multipotential hematopoietic progenitors from human embryonic stem cells. *Int. J. Hematol.* **85**, 371–379.

Martin, C. H., and Kaufman, D. S. (2005). Synergistic use of adult and embryonic stem cells to study human hematopoiesis. *Curr. Opin. Biotechnol.* **16**, 510–515.

Matsuyoshi, H., Senju, S., Hirata, S., Yoshitake, Y., Uemura, Y., and Nishimura, Y. (2004). Enhanced priming of antigen-specific CTLs *in vivo* by embryonic stem cell-derived dendritic cells expressing chemokine along with antigenic protein: Application to antitumor vaccination. *J. Immunol.* **172**, 776–786.

McGrath, K. E., and Palis, J. (1997). Expression of homeobox genes, including an insulin promoting factor, in the murine yolk sac at the time of hematopoietic initiation. *Mol. Reprod. Dev.* **48**, 145–153.

Menendez, P., Wang, L., Chadwick, K., Li, L., and Bhatia, M. (2004). Retroviral transduction of hematopoietic cells differentiated from human embryonic stem cell-derived CD45(neg)PFV hemogenic precursors. *Mol. Ther.* **10**, 1109–1120.

Miyagi, T., Takeno, M., Nagafuchi, H., Takahashi, M., and Suzuki, N. (2002). Flk1+ cells derived from mouse embryonic stem cells reconstitute hematopoiesis *in vivo* in SCID mice. *Exp. Hematol.* **30**, 1444–1453.

Moore, K. A., Ema, H., and Lemischka, I. R. (1997). *In vitro* maintenance of highly purified, transplantable hematopoietic stem cells. *Blood* **89**, 4337–4347.

Muller, A. M., and Dzierzak, E. A. (1993). ES cells have only a limited lymphopoietic potential after adoptive transfer into mouse recipients. *Development* **118**, 1343–1351.

Nakano, T. (1995). Lymphohematopoietic development from embryonic stem cells *in vitro*. *Semin. Immunol.* **7**, 197–203.

Nakano, T., Era, T., Takahashi, T., Kodama, H., and Honjo, T. (1997). Development of erythroid cells from mouse embryonic stem cells in culture: Potential use for erythroid transcription factor study. *Leukemia* **11**(Suppl. 3), 496–500.

Nakano, T., Kodama, H., and Honjo, T. (1994). Generation of lymphohematopoietic cells from embryonic stem cells in culture. *Science* **265**, 1098–1101.

Nakano, T., Kodama, H., and Honjo, T. (1996). *In vitro* development of hematopoietic system from mouse embryonic stem cells, a new approach for embryonic hematopoiesis. *Int. J. Hematol.* **65**, 1–8.

Nakayama, N., Lee, J., and Chiu, L. (2000). Vascular endothelial growth factor synergistically enhances bone morphogenetic protein-4-dependent lymphohematopoietic cell generation from embryonic stem cells *in vitro*. *Blood* **95**, 2275–2283.

Narayan, A. D., Chase, J. L., Lewis, R. L., Tian, X., Kaufman, D. S., Thomson, J. A., and Zanjani, E. D. (2006). Human embryonic stem cell-derived hematopoietic cells are capable of engrafting primary as well as secondary fetal sheep recipients. *Blood* **107**, 2180–2183.

Ng, E. S., Davis, R. P., Azzola, L., Stanley, E. G., and Elefanty, A. G. (2005). Forced aggregation of defined numbers of human embryonic stem cells into embryoid bodies fosters robust, reproducible hematopoietic differentiation. *Blood* **106**, 1601–1603.

Nishikawa, S. I., Nishikawa, S., Hirashima, M., Matsuyoshi, N., and Kodama, H. (1998). Progressive lineage analysis by cell sorting and culture identifies FLK1+ VE-cadherin+ cells at a diverging point of endothelial and hemopoietic lineages. *Development* **125**, 1747–1757.

Olivier, E. N., Qiu, C., Velho, M., Hirsch, R. E., and Bouhassira, E. E. (2006). Large-scale production of embryonic red blood cells from human embryonic stem cells. *Exp. Hematol.* **34**, 1635–1642.

Olsen, A. L., Stachura, D. L., and Weiss, M. J. (2006). Designer blood: Creating hematopoietic lineages from embryonic stem cells. *Blood* **107**, 1265–1275.

Otani, T., Inoue, T., Tsuji-Takayama, K., Ijiri, Y., Nakamura, S., Motoda, R., and Orita, K. (2005). Progenitor analysis of primitive erythropoiesis generated from *in vitro* culture of embryonic stem cells. *Exp. Hematol.* **33**, 632–640.

Park, C., Afrikanova, I., Chung, Y. S., Zhang, W. J., Arentson, E., Fong Gh, G., Rosendahl, A., and Choi, K. (2004). A hierarchical order of factors in the generation of FLK1- and SCL-expressing hematopoietic and endothelial progenitors from embryonic stem cells. *Development* **131**, 2749–2762.

Pick, M., Azzola, L., Mossman, A., Stanley, E. G., and Elefanty, A. G. (2007). Differentiation of human embryonic stem cells in serum-free medium reveals distinct roles for bone morphogenetic protein 4, vascular endothelial growth factor, stem cell factor, and fibroblast growth factor 2 in hematopoiesis. *Stem Cells* **25**, 2206–2214.

Pilat, S., Carotta, S., Schiedlmeier, B., Kamino, K., Mairhofer, A., Will, E., Modlich, U., Steinlein, P., Ostertag, W., Baum, C., Beug, H., and Klump, H. (2005). HOXB4 enforces equivalent fates of ES-cell-derived and adult hematopoietic cells. *Proc. Natl. Acad. Sci. USA* **102**, 12101–12106.

Pineault, N., Helgason, C. D., Lawrence, H. J., and Humphries, R. K. (2002). Differential expression of Hox, Meis1, and Pbx1 genes in primitive cells throughout murine hematopoietic ontogeny. *Exp. Hematol.* **30**, 49–57.

Potocnik, A. J., Kohler, H., and Eichmann, K. (1997). Hemato-lymphoid *in vivo* reconstitution potential of subpopulations derived from *in vitro* differentiated embryonic stem cells. *Proc. Natl. Acad. Sci. USA* **94**, 10295–10300.

Priddle, H., Jones, D. R., Burridge, P. W., and Patient, R. (2006). Hematopoiesis from human embryonic stem cells: Overcoming the immune barrier in stem cell therapies. *Stem Cells* **24**, 815–824.

Qiu, C., Hanson, E., Olivier, E., Inada, M., Kaufman, D. S., Gupta, S., and Bouhassira, E. E. (2005). Differentiation of human embryonic stem cells into hematopoietic cells by coculture with human fetal liver cells recapitulates the globin switch that occurs early in development. *Exp. Hematol.* **33**, 1450–1458.

Qiu, C., Olivier, E. N., Velho, M., and Bouhassira, E. E. (2008). Globin switches in yolk-sac-like primitive and fetal-like definitive red blood cells produced from human embryonic stem cells. *Blood* **111,** 2400–2408.

Robertson, S. M., Kennedy, M., Shannon, J. M., and Keller, G. (2000). A transitional stage in the commitment of mesoderm to hematopoiesis requiring the transcription factor SCL/tal-1. *Development* **127,** 2447–2459.

Roecklein, B. A., and Torok-Storb, B. (1995). Functionally distinct human marrow stromal cell lines immortalized by transduction with the human papilloma virus E6/E7 genes. *Blood* **85,** 997–1005.

Sasaki, K., Nagao, Y., Kitano, Y., Hasegawa, H., Shibata, H., Takatoku, M., Hayashi, S., Ozawa, K., and Hanazono, Y. (2005). Hematopoietic microchimerism in sheep after in utero transplantation of cultured cynomolgus embryonic stem cells. *Transplantation* **79,** 32–37.

Schiedlmeier, B., Klump, H., Will, E., Arman-Kalcek, G., Li, Z., Wang, Z., Rimek, A., Friel, J., Baum, C., and Ostertag, W. (2003). High-level ectopic HOXB4 expression confers a profound *in vivo* competitive growth advantage on human cord blood CD34+ cells, but impairs lymphomyeloid differentiation. *Blood* **101,** 1759–1768.

Schmitt, T. M., de Pooter, R. F., Gronski, M. A., Cho, S. K., Ohashi, P. S., and Zuniga-Pflucker, J. C. (2004). Induction of T cell development and establishment of T cell competence from embryonic stem cells differentiated *in vitro*. *Nat. Immunol.* **5,** 410–417.

Senju, S., Hirata, S., Matsuyoshi, H., Masuda, M., Uemura, Y., Araki, K., Yamamura, K., and Nishimura, Y. (2003). Generation and genetic modification of dendritic cells derived from mouse embryonic stem cells. *Blood* **101,** 3501–3508.

Shalaby, F., Ho, J., Stanford, W. L., Fischer, K. D., Schuh, A. C., Schwartz, L., Bernstein, A., and Rossant, J. (1997). A requirement for Flk1 in primitive and definitive hematopoiesis and vasculogenesis. *Cell* **89,** 981–990.

Srivastava, A. S., Nedelcu, E., Esmaeli-Azad, E., Mishra, R., and Carrier, E. (2007). Thrombopoietin enhances generation of CD34+ cells from human embryonic stem cells. *Stem Cells* **25,** 1456–1461.

Tian, X., and Kaufman, D. S. (2005). Hematopoietic development of human embryonic stem cells in culture. *Methods Mol. Med.* **105,** 425–436.

Tian, X., Morris, J. K., Linehan, J. L., and Kaufman, D. S. (2004). Cytokine requirements differ for stroma and embryoid body-mediated hematopoiesis from human embryonic stem cells. *Exp. Hematol.* **32,** 1000–1009.

Tian, X., Woll, P. S., Morris, J. K., Linehan, J. L., and Kaufman, D. S. (2006). Hematopoietic engraftment of human embryonic stem cell-derived cells is regulated by recipient innate immunity. *Stem Cells* **24,** 1370–1380.

Trivedi, P., and Hematti, P. (2007). Simultaneous generation of CD34+ primitive hematopoietic cells and CD73+ mesenchymal stem cells from human embryonic stem cells cocultured with murine OP9 stromal cells. *Exp. Hematol.* **35,** 146–154.

Umeda, K., Heike, T., Yoshimoto, M., Shiota, M., Suemori, H., Luo, H. Y., Chui, D. H., Torii, R., Shibuya, M., Nakatsuji, N., and Nakahata, T. (2004). Development of primitive and definitive hematopoiesis from nonhuman primate embryonic stem cells *in vitro*. *Development* **131,** 1869–1879.

Vodyanik, M. A., Bork, J. A., Thomson, J. A., and Slukvin, J. A., II (2005). Human embryonic stem cell-derived CD34+ cells: Efficient production in the coculture with OP9 stromal cells and analysis of lymphohematopoietic potential. *Blood* **105,** 617–626.

Vodyanik, M. A., Thomson, J. A., and Slukvin, J. A., II (2006). Leukosialin (CD43) defines hematopoietic progenitors in human embryonic stem cell differentiation cultures. *Blood* **108,** 2095–2105.

Wang, L., Li, L., Shojaei, F., Levac, K., Cerdan, C., Menendez, P., Martin, T., Rouleau, A., and Bhatia, M. (2004). Endothelial and hematopoietic cell fate of human embryonic stem cells originates from primitive endothelium with hemangioblastic properties. *Immunity* **21,** 31–41.

Wang, L., Menendez, P., Cerdan, C., and Bhatia, M. (2005a). Hematopoietic development from human embryonic stem cell lines. *Exp. Hematol.* **33,** 987–996.

Wang, L., Menendez, P., Shojaei, F., Li, L., Mazurier, F., Dick, J. E., Cerdan, C., Levac, K., and Bhatia, M. (2005b). Generation of hematopoietic repopulating cells from human embryonic stem cells independent of ectopic HOXB4 expression. *J. Exp. Med.* **201,** 1603–1614.

Wang, Z., Skokowa, J., Pramono, A., Ballmaier, M., and Welte, K. (2005c). Thrombopoietin regulates differentiation of rhesus monkey embryonic stem cells to hematopoietic cells. *Ann. N Y Acad. Sci.* **1044,** 29–40.

Wang, Y., Yates, F., Naveiras, O., Ernst, P., and Daley, G. Q. (2005d). Embryonic stem cell-derived hematopoietic stem cells. *Proc. Natl. Acad. Sci. USA* **102,** 19081–19086.

Wang, J., Zhao, H. P., Lin, G., Xie, C. Q., Nie, D. S., Wang, Q. R., and Lu, G. X. (2005e). *In vitro* hematopoietic differentiation of human embryonic stem cells induced by co-culture with human bone marrow stromal cells and low dose cytokines. *Cell Biol. Int.* **29,** 654–661.

Weiss, M. J., Keller, G., and Orkin, S. H. (1994). Novel insights into erythroid development revealed through *in vitro* differentiation of GATA-1 embryonic stem cells. *Genes Dev.* **8,** 1184–1197.

Wiles, M. V., and Johansson, B. M. (1997). Analysis of factors controlling primary germ layer formation and early hematopoiesis using embryonic stem cell *in vitro* differentiation. *Leukemia* **11** (Suppl. 3), 454–456.

Wiles, M. V., and Keller, G. (1991). Multiple hematopoietic lineages develop from embryonic stem (ES) cells in culture. *Development* **111,** 259–267.

Woll, P. S., Martin, C. H., Miller, J. S., and Kaufman, D. S. (2005). Human embryonic stem cell-derived NK cells acquire functional receptors and cytolytic activity. *J. Immunol.* **175,** 5095–5103.

Wu, K.-D., Cho, Y. S., Katz, J., Ponomarev, V., Chen-Kiang, S., Danishefsky, S. J., and Moore, M. A. S. (2005). Investigation of anti-tumor effects of synthetic epothilone analogs in human myeloma models *in vitro* and *in vivo*. *Proc. Natl. Acad. Sci. USA* **102,** 10640–10645.

Yu, J., Vodyanik, M. A., He, P., Slukvin, P., II, and Thomson, J. A. (2006). Human embryonic stem cells reprogram myeloid precursors following cell-cell fusion. *Stem Cells* **24,** 168–176.

Zambidis, E. T., Peault, B., Park, T. S., Bunz, F., and Civin, C. I. (2005). Hematopoietic differentiation of human embryonic stem cells progresses through sequential hematoendothelial, primitive, and definitive stages resembling human yolk sac development. *Blood* **106,** 860–870.

Zhan, X., Dravid, G., Ye, Z., Hammond, H., Shamblott, M., Gearhart, J., and Cheng, L. (2004). Functional antigen-presenting leucocytes derived from human embryonic stem cells *in vitro*. *Lancet* **364,** 163–171.

Zou, G. M., Chan, R. J., Shelley, W. C., and Yoder, M. C. (2006). Reduction of Shp-2 expression by small interfering RNA reduces murine embryonic stem cell-derived *in vitro* hematopoietic differentiation. *Stem Cells* **24,** 587–594.

CHAPTER 22

Cardiomyocyte Differentiation from Embryonic Stem Cells

X. Yang,★ X.-M. Guo,† C.-Y. Wang,† and X. Cindy Tian★

★Center for Regenerative Biology
University of Connecticut
Unit 4243, Storrs, CT 06269-4243

†Institute of Basic Medical Sciences
Tissue Engineering Research Center
Academy of Military Medical Sciences
Beijing, People's Republic of China

Abstract
1. Introduction
2. Embryoid Body Generation
   2.1. Hanging drop method
   2.2. Scalable production of EBs in a bioreactor
3. Cardiomyocyte Differentiation
4. Enrichment of Cardiomyocytes
5. Conclusion
References

Abstract

Derivation of cardiomyocytes from embryonic stem cells would be a boon for treatment of the many millions of people worldwide who suffer significant cardiac tissue damage in a myocardial infarction. Such cells could be used for transplantation, either as loose cells, as organized pieces of cardiac tissue, or even as pieces of organs. Eventual derivation of human embryonic stem cells via somatic cell nuclear cloning would provide cells that not only may replace damaged cardiac tissue, but also would replace tissue without fear that the patient's immune system will reject the implant. Embryonic stem cells can differentiate spontaneously into cardiomyocytes. *In vitro* differentiation of embryonic stem cells normally requires an initial aggregation step to form structures called embryoid bodies that differentiate into a wide variety of specialized cell types, including cardiomyocytes. Here, we discuss methods of encouraging embryoid body formation, causing pluripotent stem cells to develop into cardiomyocytes, and expanding the numbers of cardiomyocytes so that the cells may achieve functionality in transplantation, all in the mouse model system. Such methods may be adaptable and/or modifiable to produce cardiomyocytes from human embryonic stem cells.

1. Introduction

Coronary heart disease is responsible for more deaths worldwide in people aged 60 years or more than any other cause. Stroke, also a product of cardiovascular disease, ranks second, as a cause of death in this age group (MacKay and Mensah, 2004). In 2002, India, China, and the Russian Federation had the greatest numbers of deaths from coronary heart disease in the world. In the United States, cardiovascular disease is the number one cause of death in both men and women, there were more than 910,000 deaths due to cardiovascular disease in 2003 (American Heart Association, 2005). Furthermore, in 2003, nearly seven million people had in-patient cardiovascular surgery in the U.S. and more than seven million (or 3.5% of the population) had suffered a myocardial infarction at some time.

Perhaps as many as a billion or more myocytes are damaged or lost as a result of myocardial infarction (Laflamme and Murry, 2005). The end result of myocardial damage is heart failure, which, according to Laflamme and Murry (2005), results in the greatest number of hospitalizations of people over 65 years of age in the U.S. People who have suffered from myocardial infarction would be a prime group for stem cell cardiomyocyte transplantation to repair muscle damage.

The most promising cell sources for regenerative medicine are embryonic and adult stem cells. The capacity of murine embryonic stem (ES) cells to differentiate into cardiomyocytes has been demonstrated by a number of research groups (Doetschman *et al.*, 1985; Wobus *et al.*, 1991). The subsequent demonstration of spontaneous differentiation of human ES cells into the three germ layers (Schuldiner *et al.*, 2000), including the development of functional cardiomyocytes (Kehat *et al.*, 2001) has opened the avenue to a human cell source for cell-based therapies, including cardiac tissue engineering. Unlimited differentiation capacity and indefinite propagation represent the strongest advantages for use of ES cells.

Somatic cell nuclear transfer (SCNT), a technique that has been successfully used to clone cattle, rabbits, sheep, pigs, mice, and other animals, were it to work successfully in humans, would be a prime technique for production of human ES cells (hES cells). The goal of production of such nuclear transfer ES (NT-ES) cells would be to use autologous transplantation to avoid immune rejection. It is, perhaps, the strongest rationale for therapeutic cloning. To date, such cells have not been produced, but they remain a clear goal of researchers in this field.

The functional cardiomyocytes derived from ES cells can be used in cell transplantation for restoration of heart function by replacement of diseased myocardium (Reinlib and Field, 2000) or the design of artificial cardiac muscle constructs *in vitro* for later implantation *in vivo* (Zimmermann and Eschenhagen, 2003).

This chapter describes protocols for generation, differentiation, and enrichment of cardiomyocytes from ES cells.

2. Embryoid Body Generation

The most robust method for generating the most differentiated cell types is through the embryoid body (EB) system, where ES cells spontaneously differentiate as tissue-like spheroids in suspension culture. EB differentiation has been shown to recapitulate aspects of early embryogenesis, including the formation of a complex, three-dimensional architecture wherein cell–cell and cell–matrix interactions are thought to support the development of the three embryonic germ layers and their derivatives (Itskovitz-Eldor *et al.*, 2000; Keller, 1995). Presently, all human and most mouse ES cell (mESC) lines require aggregation of multiple ES cells to efficiently initiate EB formation (Dang *et al.*, 2002; Itskovitz-Eldor *et al.*, 2000). Standard methods of EB formation include hanging drop, liquid suspension and methylcellulose culture. These culture systems maintain a balance between allowing ES cell aggregation necessary for EB formation and preventing EB agglomeration for efficient cell growth and differentiation (Dang *et al.*, 2002). However, these culture systems are limited in their production capacity and are not easily amenable to process-control strategies.

In the mouse system, the ES cells spontaneously form three-dimensional aggregates and differentiate after withdrawal of leukemia inhibitory factor (LIF) and transferal onto a nonadherent surface (Maltsev *et al.*, 1994) (Fig. 22.1). These three-dimensional aggregates recapitulate early embryological development in the mouse and allow derivatives of the three germ layers to form *in vitro*. In hES cells, spontaneous differentiation toward ectoderm, endoderm, and mesoderm has also been reported when cultivated either as EBs or as a monolayer at high cell density. EB formation in ES cells is normally achieved by dissociating colonies into single cells and promoting agglomeration by seeding at high cell densities in nonadherent Petri dishes. Another way to form EBs is to suspend cells in small droplets hanging from the underside of a culture plate, often referred to as the hanging drop method. For hES cells, EB formation is promoted by detaching small clumps of hES colonies by enzymatic (collagenase/dispase) or chemical dissociation (EDTA), and keeping them in suspension in nonadherent culture dishes.

(With permission Maltsev *et al.*, 1994)

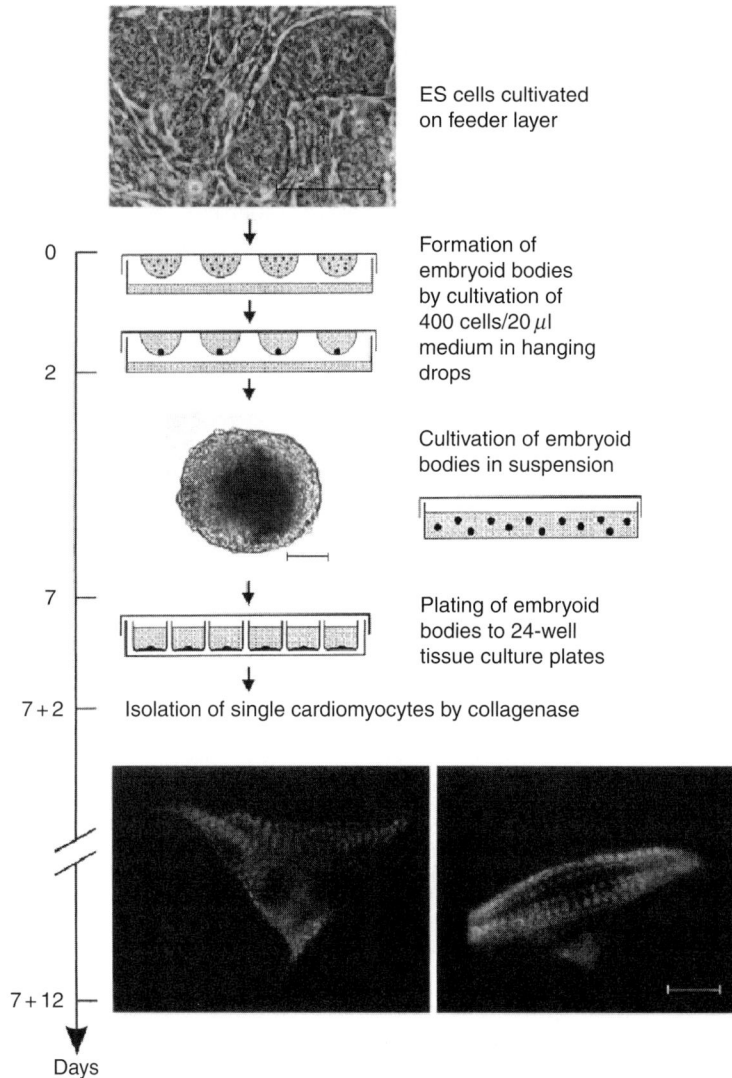

ES cells cultivated
on feeder layer

0 ⊢

Formation of
embryoid bodies
by cultivation of
400 cells/20 μl
medium in hanging
drops

2 ⊢

Cultivation of embryoid
bodies in suspension

7 ⊢

Plating of embryoid
bodies to 24-well
tissue culture plates

7 + 2 ⊢   Isolation of single cardiomyocytes by collagenase

7 + 12 ⊢

Days

Figure 22.1 Schematic presentation of the *in vitro* differentiation protocol of ES (D3) cells into cardiomyocytes. The cross-striated structure of sarcomeres of single cardiomyocytes at the terminal differentiation stage was visualized by indirect immunofluorescence with monoclonal antibodies against cardiac-specific α-cardiac myosin heavy chain and troponin TI-i (bottom, left, and right). Bar = 100 μm (top and middle); bar = 10 μm (bottom).

2.1. Hanging drop method

(Portions are adapted from the Thomson Lab with minor modifications: http://ink.primate.wisc.edu/~thomson/protocol.html#embr).

2.1.1. Mouse ES cell dissociation

1. Warm trypsin to 37 °C in a water bath.
2. Discard the culture medium, wash with 1×PBS twice.
3. Add the following amount of trypsin (0.25%:0.04% EDTA = 1:1)

   a. 0.5 ml/well of 4-well plate.
   b. 1.0 ml/well of 6-well plate.

4. Incubate at 37 °C with 5% CO_2 for about 1 min. The cells are ready when the edges of the colony are rounded up and curled away from the mouse embryonic fibroblasts (MEFs) on the plate.
5. Add 1–2 ml of fetal calf serum (FCS) to block the trypsin.
6. Scrape and wash the colonies off of the plate with a pipette.
7. Transfer the cell suspension to a 10-ml conical tube.
8. Break up the colonies by pipetting up and down against the bottom of the tube until there appears to be a fine suspension of cells, that is, without clumps of cells remaining.
9. Spin the cells at 1400 rpm for 5 min.
10. Aspirate the supernatant off and re-suspend the cells with 10 ml of differentiation medium.
11. Transfer the cell suspension into a T75 flask precoated with 0.1% gelatin and incubate at 37 °C with 5% CO_2 for an hour to allow for the adherence of fibroblasts onto the surface of the flask.
12. Collect the medium containing cells that remain unattached.
13. Spin at 1000 rpm for 5 min.
14. Aspirate off the wash medium.
15. Add another 1 ml of differentiation medium and re-suspend the cells by repetitive pipetting until there appears to be a fine suspension of cells.
16. Count the cells using a hemocytometer.

source: some of 2.1 was adopted from the Thomson Lab with minor modifications: http://ink.primate.wisc.edu/~thomson/protocol.html#embr

2.1.2. Hanging drop culture

1. Dilute the cell suspension to 400–500 cells/30 μl (1.5–2×10^4 ml^{-1}) by differentiation medium.
2. Invert a 100-mm tissue culture plate cover and put it on a "pipetting guide." The "pipetting guide" is prepared by drawing cross lines on a piece of paper with an interline distance of 8 mm; the crosses are where the cell drops are located.

3. Pipette 30 μl cell suspension onto the inner surface of the plate cover corresponding to each cross.

4. Add 10 ml PBS into the plate.

5. Gently turn the plate cover to enlarge the attaching surface for the droplets.

6. Quickly invert the plate cover and cover the plate gently.

7. Carefully transfer the plate into the incubator.

8. Culture for two or three days.

2.1.3. Suspension culture

1. Aspirate the suspended droplets and transfer them into a 100-mm bacteriological plate. Each plate receives approximately 100 droplets.

2. Add 10 ml culture medium into each plate.

3. Shake the plate gently.

4. Incubate at 37 °C with 5% CO_2 for 4–5 days.

2.1.4. Human ES cell suspension culture

1. Let human ES cells grow until the colonies are large and the cells are pretty piled up—about the time when you would normally split the colonies or even a day past that.

2. Treat cells with 0.2–0.5 mg/ml dispase. You want to use the lowest possible concentration of dispase, but it tends to vary a bit.

3. Wait until the colonies completely detach from the plate. Do not blow colonies off with a pipette. This should take about 20–30 min. If nothing is happening by that point, add more dispase.

4. Once the colonies come up, gently transfer them to a 15-ml conical tube with a 10-ml pipette. You do not want to break up the colonies.

5. The cells should sink to the bottom of the tube after a minute or two without any spinning. Aspirate off the media and wash once in hES media. If you are in a hurry and need to spin the colonies down, 1 min at 500 rpm is enough.

6. Transfer cells to a flask containing ES media without basic fibroblast growth factor (bFGF). Put all of the EBs from one 6-well plate into a T80 flask with about 25 ml media.

7. The cells will round up into actual embryoid bodies (EB) after about 12–24 h. They should then be fed every day by exchanging half the media with fresh media. The EBs should not attach; if they do, gently tap the flask to dislodge the EBs.

2.2. Scalable production of EBs in a bioreactor

All of the current protocols for hEB generation list aggregation of hESCs as a prerequisite for initiating EB formation. At later stages, however, agglomeration of the EBs may have negative effects on cell proliferation and differentiation, as was shown in the mouse system (Dang *et al.*, 2002). When formed in static cultures (mostly in flasks), agglomerated large EBs revealed extensive cell death and, eventually, large necrotic centers due to mass transport limitations. To maintain balance between these two processes and to achieve control of the extent of EB agglomeration, several methods have been developed for mouse ES cells. These methods include hanging drops and methylcellulose cultures (Dang *et al.*, 2002). Although efficient, to some extent, in preventing the agglomeration of EBs, the complex nature of these systems makes up-scaling them a rather difficult task. On the other hand, a much simpler process in spinner flasks resulted in the formation of large cell clumps within a few days, indicative of significant cell aggregations in the cultures (Wartenberg *et al.*, 2001). Attempts to increase the stirring rate to avoid agglomeration may result in massive hydrodynamic damage to the cells due to extensive mixing in the vessels (Chisti, 2001). Gerecht-Nir *et al.* introduced rotating cell culture systems (RCCS), developed by NASA, as milder bioreactors for hEB formation and differentiation (Gerecht-Nir *et al.*, 2004). In RCCS, the operating principles are: (1) whole-body rotation around a horizontal axis, which is characterized by extremely low fluid shear stress, and (2) oxygenation by active or passive diffusion to the exclusion of all but dissolved gases from the reactor chamber, yielding a vessel devoid of gas bubbles and gas/fluid interfaces (Lelkes and Unsworth, 2002). The resulting flow pattern in the RCCS is laminar with mild mixing, since the vessel rotation is slow. The settling of the cell clusters, which is associated with oscillations and tumbling, generates fluid mixing. The outcome is a very low-shear environment. Another advantage of the RCCS is that they are geometrically designed so that the membrane area to volume of medium ratio is high, thus enabling efficient gas exchange. Gerecht-Nir and associates' (Gerecht-Nit *et al.*, 2004) work indicated that cultivation of hESCs in the slow turning lateral vessel (STLV) system, a type of RCCS with oxygenator membrane in the center (Fig. 22.2), yielded nearly four-fold more hEB particles compared to static, conventional Petri dishes. These EBs were intact in shape and had less necrosis in the center (Fig. 22.3). Under the dynamic cultivation in STLV, differentiation of hEBs progressed in a normal course (Gerecht-Nir *et al.*, 2004).

1. Prepare confluent 6-well plates (60 cm^2) of undifferentiated hESCs.
2. Disperse the cells into small clumps (3–20 cells) using 0.5 mM EDTA supplemented with 1% FBS (HyClone).
3. Prepare STLV bioreactor (Synthecon, Inc., Houston, Texas) according to the instruction manual.

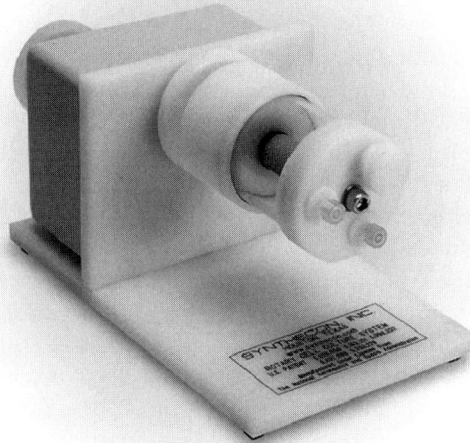

Figure 22.2 Illustration of STLV system (with permission from Synthecon, Inc.).

a. Using an Allen wrench, unscrew and remove center bolt at the top of the vessel. Gently twist outer wall while holding top end cap to disassemble vessel. Repeat procedure for rear end cap. Remove O-rings from each end cap.

b. Place all pieces of vessel in a 4-l beaker filled with a warm solution of mild detergent designed for tissue culture labware. Soak for 1 h.

c. Scrub plastic parts (except oxygenator core) with a soft bristle brush as necessary to remove any residues.

d. Very gently clean oxygenator membrane with the tip of your finger using latex laboratory gloves.

   Note: Harsh scrubbing will damage membrane material. Do not use a brush to cleanse the membrane.

e. Rinse vessel parts with continuous flow of ultra-pure water for 15–20 min.

f. Soak vessel parts in fresh, ultrapure water overnight.

g. Remove vessel parts from water and place on absorbent pads to dry.

h. Assemble vessel.

i. Fill unit with 70% ethanol and allow it to soak for 24 h.

j. Sterilize as described below.

   Autoclave method:

   i. Empty vessel of 70% ethanol.

   ii. Remove and dispose of plastic valves and their caps. Remove fill port cap and autoclave it separate from the vessel. Cover all three ports with aluminum foil.

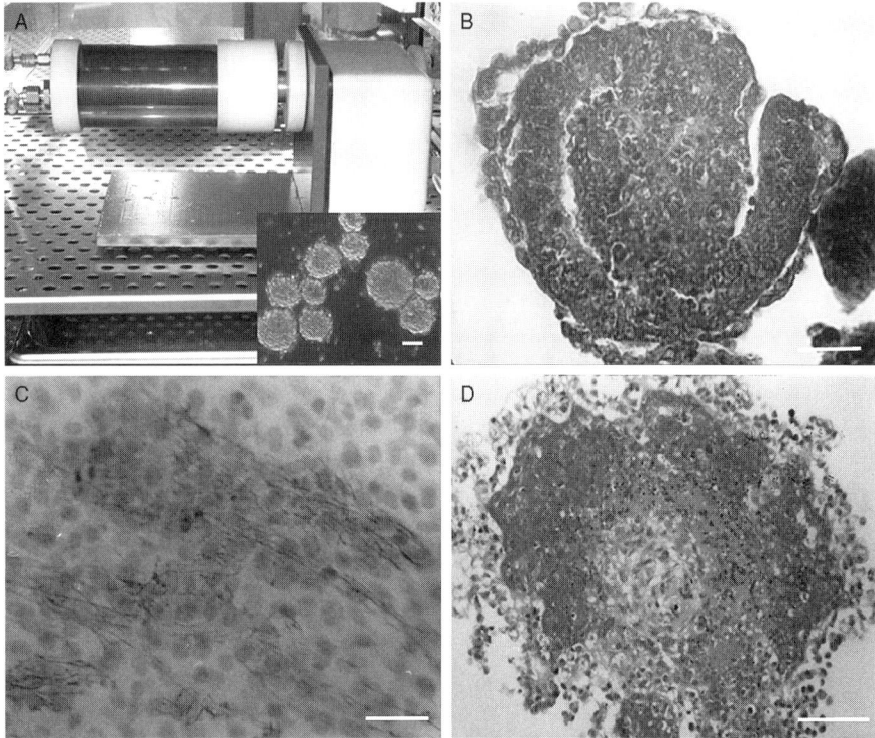

Figure 22.3 Comparison of EBs cultured in a slow-turning lateral vessel (STLV) and suspended in flasks. EBs cultured in a 250-ml STLV (A) for 5 days are intact and well differentiated, without necrosis in the center (B, H&E), these EBs demonstrated productive differentiation into cardiomyocytes (C, cardiac troponin-T (cTnT) immunohistochemistry); EBs cultured in suspension for five days are smaller in size and have significant necrosis in the middle (D, H&E). Scale bars, 50 μm (A), 100 μm (B and D), or 10 μm (Guo *et al.*, 2006).

    iii. Loosen center screw one turn.

    iv. Wrap vessel and caps and autoclave for 30 min at 105–110 °C. It is not necessary to slow-vent the autoclave.

    v. Remove from autoclave; cool to room temperature.

4. Seed hESCs into the STLV at initial cell concentrations ranging from 1×10^4 to 1×10^5 cells per 1 ml medium. The medium may be KO-DMEM, supplemented with 20% FBS, penicillin-streptomycin, 1 mM L-glutamine, 0.1 mM *h*-mercaptoethanol, and 1% nonessential amino acid stock.

Manipulation of the STLV bioreactor should follow the following instructions:

a. Transfer the vessel to a sterile hood. Remove the end caps and place them on sterile alcohol pads or sterile Petri plates.

b. Aspirate medium from the unit through 1/4-in. port.

c. Fill the vessel to 50% of total volume with growth medium minus serum. Allow space to load cells (serum addition at this time increases foaming and leads to difficulty in removing the air bubbles later.).

d. Count cells to be used or mince primary tissue (ten 1-mm pieces per 5 ml of media).

e. Dilute the cells into separate containers of media to yield desired final concentration (2–3×10^5 ml^{-1} has been used by some authors).

f. Add appropriate amount of washed, prepared microcarrier beads (5 mg/ml) to diluted cells.

g. With a 10-ml pipette, load cell/bead/medium solution through the 1/4-in. port.

h. Add the appropriate amount of serum and top off the vessel with medium.

i. Wipe the port with an alcohol pad. Replace the cap and tighten. Close the syringe port valves.

j. Fill a 10-ml sterile syringe with growth medium. Wipe one syringe port with an alcohol pad and attach syringe.

k. Wipe the other syringe port with an alcohol pad and attach an empty 3- or 5-ml sterile syringe. Open valves of both syringe ports.

l. Gently invert vessel and tap on sides to expel air bubbles from under the ports. Maneuver air bubbles under the empty syringe. With both valves open, gently press on the syringe to replace air bubbles with medium.

m. Discard the small syringe. Wipe the port with an alcohol pad and replace with the cap or another syringe.

n. Leave the large, medium-filled syringe on the unit with the valve open as the volume of the medium in the vessel may vary slightly with temperature.

o. Attach the vessel to the rotator base in a humidified CO_2 incubator. Check that the unit is level.

p. Turn power on and adjust to an initial rotation speed of 10 rpm.

Note: Cells and cell aggregates should rotate with the vessel and not settle within the vessel, nor should they collide with the cylinder wall or oxygenator core of the vessel. When the speed is properly adjusted, the cells and cell aggregates will orbit within the vessel. The rotation speed will need to be increased to compensate for the increased sedimentation rates of anchorage-dependent cells as the aggregate particles increase in size..

5. Set the bioreactor to rotate at a speed at which the suspended cell aggregates remain close to a stationary point within the reactor vessel; 10 rpm is recommended as the initial speed.

6. Change the medium every three days as described below:

   a. Turn off power and immediately remove the vessel from the base and take it to a sterile environment (biological hood).

   b. Stand the vessel vertically on its base (valves up), and let the cell/bead aggregates settle to the bottom.

   c. Close any valves that may be open.

   d. Remove and discard any syringes that may be attached. Wipe ports with sterile alcohol pads.

   e. Remove the 1/4-in. port cap and any Luer lock caps, which may be attached and placed on sterile alcohol pads.

   f. Aspirate medium through 1/4-in. port. Usually, 1/4 to 1/2 of the conditioned medium is left in the vessel. Aspirate droplets from syringe ports.

   g. Fill the vessel with medium using Luer lock syringe or a sterile pipette. Flow the medium down the wall of the vessel (Do not disturb the cell aggregate particles).

   h. Fill a 10-ml sterile syringe with growth medium. Wipe one syringe port with an alcohol pad and attach syringe.

   i. Wipe the other syringe port with an alcohol pad and attach an empty 3- or 5-ml sterile syringe.

   j. Gently invert vessel and tap on sides to expel air bubbles from under the ports. Maneuver air bubbles under the empty syringe. With both valves open, gently press on the syringe to replace air bubbles with medium.

   k. When all bubbles are removed, close the syringe valves and discard the small syringe. Wipe the port with an alcohol pad and replace the cap or leave in a small syringe for later sampling.

   l. Leave the large, medium-filled syringe on the unit with the valve open, as the volume of the medium in the vessel may vary slightly with temperature change.

   m. Attach the vessel to the rotator base and replace them in the humidified CO_2 incubator.

   n. Turn on the power and adjust the speed as necessary.

7. Take samples at different time intervals, if necessary, to monitor the EBs growing in the vessel, following the instructions described below:

   a. If a sampling syringe is not in place, stop the vessel rotation. Remove the syringe port cap and place it in a sterile Petri dish. Attach a sterile, empty 1.5- or 10-ml syringe to the valve. Both syringe port valves should be open.

   b. Turn on the power to allow the cell/bead aggregates to be evenly distributed (2 min).

    c. Push medium into the vessel with the medium-containing 10-ml syringe that is still attached from inoculation. A slight pull on the smaller, sampling syringe may also be necessary. This procedure provides a homogeneous, representative sample, however, it may take some practice, as the vessel is still rotating.

    d. When the desired sample has been drawn (usually 1–5 ml), turn off the power. Close the valve on the sampling syringe port and remove the sampling syringe.

    e. Attach another sampling syringe or replace the port cap.

    f. Turn on the power and adjust speed if necessary.

    g. If any bubbles are visible, turn off the power and utilize the bubble removal procedures provided in the prior sections.

    Note: Tissue particles too large to be drawn into a syringe can, in some cases, be removed with forceps through the fill port. Extreme care should be taken to avoid damaging the oxygenator membrane with the forceps.

3. Cardiomyocyte Differentiation

ES cells can differentiate spontaneously into cardiomyocytes. *In vitro* differentiation of ES cells normally (except for neurogenesis) requires an initial aggregation step to form structures, termed EBs, which differentiate into a wide variety of specialized cell types, including cardiomyocytes. A number of parameters specifically influence the differentiation potency of ES cells to form cardiomyocytes in culture: (1) the starting number of cells in the EB, (2) media, FBS, growth factors, and additives, (3) ES cell lines, and (4) the time of EB plating (Wobus *et al.*, 2002). Cardiomyocytes are located between an epithelial layer and a basal layer of mesenchymal cells within the developing EB (Hescheler *et al.*, 1997). Cardiomyocytes are readily identifiable, because, within 1–4 days after plating, they spontaneously contract. With continued differentiation, the number of spontaneously beating foci increases and all the EBs may contain localized beating cells. The rate of contraction within each beating area rapidly increases with differentiation, followed by a decrease in average beating rate with maturation. Depending on the number of cells in the initial aggregation step, the change in beating rate and the presence of spontaneous contractions continue from several days to more than one month. Fully differentiated cardiomyocytes often stop contracting, but can be maintained in culture for many weeks. Thus, developmental changes of cardiomyocytes may be correlated with the length of time in culture and can be readily divided into three stages of differentiation: early (pacemakerlike or primary myocardial-like cells), intermediate, and terminal (atrial-, ventricular-, nodal-, His-, and Purkinje-like cells) (Hescheler *et al.*, 1997).

Similar to mouse cells, hES cells differentiate when they are removed from feeder layers and grown in suspension. EBs of hES cells are heterogeneous, and they can express markers specific to neuronal, hematopoietic, and cardiac origin (Itskovitz-Eldor *et al.*, 2000; Schuldiner *et al.*, 2000).

Several chemicals have proven helpful for enhancement of cardiogenic differentiation of mouse or human ES cells. They are retinoic acid (Wobus *et al.*, 1997; Xu *et al.*, 2002), 0.5–1.5% DMSO (Ventura and Maioli, 2000), and 5-aza-dC (Xu *et al.*, 2002).

1. Transfer the EBs into a 100-mm tissue culture plate. For each plate, seed 100 EBs.
2. Add 15 ml induction medium to each plate. Shake or pipette gently, so as to distribute the EBs evenly.

   The induction medium varies according to different protocols. It may be differentiation medium supplemented with: ascorbic acid (Takahashi *et al.*, 2003), retinoic acid (Wobus *et al.*, 1997; Xu *et al.*, 2002), DMSO (Klug *et al.*, 1996), or 5-aza-dC (Xu *et al.*, 2002).

3. Make sure the EBs are evenly distributed across the entire plate.
4. Place plate gently into incubator. Make sure the EBs are not disturbed.
5. Let the EBs settle overnight in incubator.
6. Change medium every 3 days.
7. Observe the cells every day under the inverted optical microscope to monitor the appearance of beating areas.
8. Continue culture for approximately 7–10 days or more, as needed.

4. Enrichment of Cardiomyocytes

To use hES cell-derived cardiomyocytes in therapeutic applications, it will be beneficial to produce a population of cells highly enriched for cardiomyocytes. Xu *et al.* (2002) first demonstrated the enrichment of hES cell-derived cardiomyocytes by Percoll gradient separation and proliferation capacity of the enriched cells. These cells express appropriate cardiomyocyte-associated proteins. A subset of them appears to be proliferative, as determined by BrdU incorporation or expression of Ki-67, suggesting that these cardiomyocytes represent an early stage of cells. This strategy has been further proved efficient for enrichment of cardiomyocytes from mouse ES cells and neonatal rat ventricular cardiomyocytes (E *et al.*, 2006; Guo *et al.*, 2006) (Fig. 22.4 and Table 22.1).

1. Wash the differentiated cultures containing beating cardiomyocytes three times with PBS or a low-calcium solution (Maltsev *et al.*, 1993).

   The low-calcium solution contains 120 mM NaCl, 5.4 mM KCl, 5 mM $MgSO_4$, 5 mM Na-pyruvate, 20 mM glucose, 20 mM taurine, and 10 mM HEPES at pH 6.9.

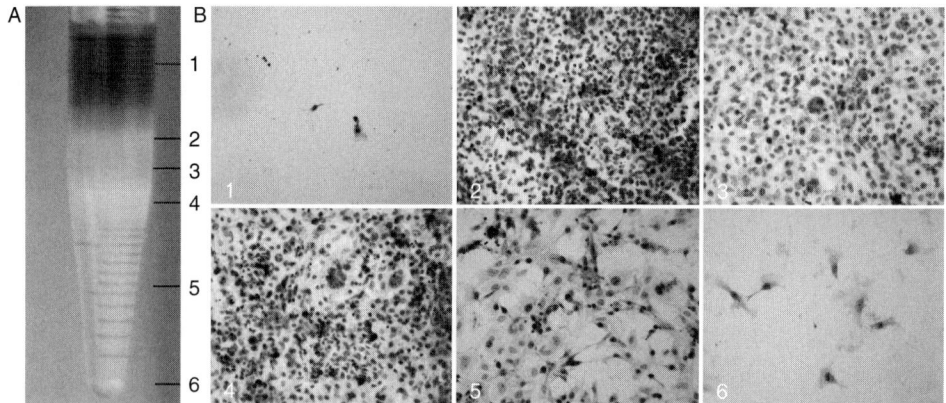

Figure 22.4 Characterization of Percoll-enriched cardiomyocytes from mESCs. (A) Six fractions of mES cell-derived cardiomyocytes after Percoll enrichment. (B) Characterization of the cells in each fraction by anticardiac troponin T (cTnT) staining after plating for 3 days in culture. cTnT positive cells were mainly in fractions 4 and 5 (E *et al.*, 2006).

Table 22.1
Enrichment of mES cell-derived cardiomyocytes by percoll gradient

Fraction	Cells collected	Beating cells[a]	% cTnT-positive cells (Day 3)
Input cells	$1–2 \times 10^8$	++	18 ± 3
II	1.77×10^7	+	3 ± 2
III	3.25×10^6	+	6 ± 3
IV	5.65×10^6	+++	40 ± 2
V	2.60×10^6	++++	88.7 ± 4

mES cell-derived cardiomyocytes differentiated for 14 days were enriched by percoll gradient separation (see methods). After separation, each layer was collected, and cells were counted and replated. Cultures were maintained for 3 days before evaluation of cTnT immunoreactivity.
[a]Amount of beating cells: ++++ > +++ > ++ >+.

2. Discard the PBS. Add an appropriate amount of 1 mg/ml collagenase B to cover the cells. The 1 mg/ml collagenase B is prepared in the low-calcium solution supplemented with 30 μM $CaCl_2$.

3. Incubate at 37 °C for 1–2 h.

4. Re-suspend the cells in a high-potassium solution. The high-potassium solution contains 85 mM KCl, 30 mM K_2HPO_4, 5 mM $MgSO_4$, 1 mM EGTA, 2 mM Na_2ATP, 5 mM Na-pyruvate, 5 mM creatine, 20 mM taurine, and 20 mM glucose at pH 7.2.

5. Incubate at 37 °C for 15 min for more complete dissociation.

6. Gently pipette to achieve a uniform cell suspension.

7. Transfer the cell suspension into a 10-ml conical tube. Spin at 1200 rpm for 5 min.

8. Aspirate the supernatant. Add 3 ml high glucose DMEM containing 20% FBS and re-suspend the cells by gentle pipetting.

9. Prepare a Percoll gradient as described below:

   a. Mix Percoll with 8.5% NaCl (9:1) to reach a physiological osmotic equilibrium.

   b. Dilute the Percoll-8.5% NaCl solution with 8.5% NaCl to a final Percoll concentration of 40.5% and 58.5%, which correspond to a physical density of 1.065 and 1.069 g/ml, respectively.

10. Add 3 ml of 58.5% Percoll to the bottom of a 10-ml conical tube, then gently add 3 ml of 40.5% Percoll on the top of the 58.5% Percoll.

11. Add 3 ml of cell suspension onto the top of the Percoll solution by using a pipette leaning against the inner wall of the tube. Be sure to do it *very gently*, so as not to disturb the Percoll layers.

12. Centrifuge at 1500 rpm for 30 min.

13. After centrifugation, 2 layers of cells will be observed: one on top of the Percoll (fraction I) and a layer of cells at the interface of the two layers of Percoll (fraction III). Cells can also be found in the 40.5% Percoll layer (fraction II) and the 58.5% Percoll layer (fraction IV).

    Generally, 20–40% of cells in fraction III and 50–70% of cells in fraction IV express cardiac-specific troponin I (cTnI), a subunit of the troponin complex that provides a calcium-sensitive molecular switch for the regulation of striated muscle contraction (Bhavsar *et al.*, 1996).

14. Carefully aspirate the cells in fraction III and cells in fraction IV.

15. Wash the cell fractions twice with PBS.

5. Conclusion

These methods have worked successfully in producing cardiomyocytes from mouse ES cells. The cardiomyocytes derived from mESCs have been successfully introduced into the process of engineering cardiac muscle (Guo *et al.* 2006). Nevertheless, in their recent review of differentiation of ES cells from mouse and human to produce cardiomyocytes, Wei *et al.* (2005) note the numerous problems in deriving human cells and differentiating them into cardiomyocytes. They point out that some studies show chromosomal abnormalities in hES cells as a result of enzymatic dissociation methods. They also indicate that differentiation of hES cells to cardiomyocytes is slower and less efficient than comparative differentiation using mouse ES cells. However, recently, Laflamme and Murry (2005) noted,

"In contrast to the limited proliferative capacity of mouse ES cell-derived cardiomyocytes, human ES cell-derived cardiomyocytes show sustained cell cycle activity both *in vitro* and after *in vivo* transplantation into the nude rat heart." Nevertheless, Wei *et al.* (2005) suggest that the numerous differences that have been noted between human and murine ES cells may be attributed to the longer gestation time allowed for heart development in the human embryo versus the mouse embryo. They outline methods of inducing cardiomyocyte development from cells derived from EBs through addition of growth factors and cytokines into the culture medium.

Mummery *et al.* (2002) used a coculture method to derive cardiomyocytes from hES cells. She and her coworkers wrote a fairly extensive review of derivation and use of cardiomyocytes (van Laake *et al.*, 2005).

These are all beyond the scope of the current chapter, which is focused on techniques. Of course, techniques may change rapidly, as new methods for growing cells are developed, and in concert with tests both in large animal models and in humans.

References

American Heart Association. (2005). "Heart Disease and Stroke Statistics—2006 Update," p. 40. American Heart Association, Dallas, Texas.

Bhavsar, P. K., Brand, N. J., Yacoub, M. H., and Barton, P. J. (1996). Isolation and characterization of the human cardiac troponin I gene (TNNI3). *Genomics* **35**, 11–23.

Chisti, Y. (2001). Hydrodynamic damage to animal cells. *Crit. Rev. Biotechnol.* **21**, 67–110.

Dang, S. M., Kyba, M., Perlingeiro, R., Daley, G. Q., and Zandstra, P. W. (2002). Efficiency of embryoid body formation and hematopoietic development from embryonic stem cells in different culture systems. *Biotechnol. Bioeng.* **78**, 442–453.

Doetschman, T. C., Eistetter, H., Katz, M., Schmidt, W., and Kemler, R. (1985). The *in vitro* development of blastocyst-derived embryonic stem cell lines: Formation of visceral yolk sac, blood islands and myocardium. *J. Embryol. Exp. Morphol.* **87**, 27–45.

E, L.-L., Zhao, Y.-S., Guo, X.-M., Wang, C.-Y., Jiang, H., Li, J., Duan, C.-M., and Song, Y. (2006). Enrichment of cardiomyocytes derived from mouse embryonic stem cells. *J. Heart Lung Transplant.* **25**, 664–674.

Gerecht-Nir, S., Cohen, S., and Itskovitz-Eldor, J. (2004). Bioreactor cultivation enhances the efficiency of human embryoid body (hEB) formation and differentiation. *Biotechnol. Bioeng.* **86**, 493–502.

Guo, X.-M., Zhao, Y.-S., Wang, C.-Y., E, L.-L., Chang, H.-X., Zhang, X.-A., Duan, C.-M., Dong, L.-Z., Jiang, H., Li, J., Song, Y., and Yang, X. (2006). Creation of engineered cardiac tissue *in vitro* from mouse embryonic stem cells. *Circulation* **113**, 2229–2237.

Hescheler, J., Fleischmann, B. K., Lentini, S., Maltsev, V. A., Rohwedel, J., Wobus, A. M., and Addicks, K. (1997). Embryonic stem cells: A model to study structural and functional properties in cardiomyogenesis. *Cardiovasc. Res.* **36**, 149–162.

Itskovitz-Eldor, J., Schuldiner, M., Karsenti, D., Eden, A., Yanuka, O., Amit, M., Soreq, H., and Benvenisty, N. (2000). Differentiation of human embryonic stem cells into embryoid bodies compromising the three embryonic germ layers. *Mol. Med.* **6**, 88–95.

Kehat, I., Kenyagin-Karsenti, D., Snir, M., Segev, H., Amit, M., Gepstein, A., Livne, E., Binah, O., Itskovitz-Eldor, J., and Gepstein, L. (2001). Human embryonic stem cells can differentiate into myocytes with structural and functional properties of cardiomyocytes. *J. Clin. Invest.* **108**, 407–414.

Keller, G. M. (1995). *In vitro* differentiation of embryonic stem cells. *Curr. Opin. Cell Biol.* **7**, 862–869.

Klug, M. G., Soonpaa, M. H., Koh, G. Y., and Field, L. J. (1996). Genetically selected cardiomyocytes from differentiating embryonic stem cells form stable intracardiac grafts. *J. Clin. Invest.* **98,** 216–224.

Laflamme, M. A., and Murry, C. E. (2005). Regenerating the heart. *Nat. Biotechnol.* **23,** 845–856.

Lelkes, P. I., and Unsworth, B. R. (2002). Neuroectodermal cell culture: Endocrine cells. *In* "Methods of Tissue Engineering" (A. Atala, and R. P. Lanza, Eds.), pp. 371–382. Academic Press, London.

MacKay, J., and Mensah, G. A. (2004). "The Atlas of Heart Disease and Stroke," pp. 48–49. World Health Organization, Zurich, Switzerland, and Centers for Disease Control and Prevention, Atlanta, Georgia.

Maltsev, V. A., Rohwedel, J., Hescheler, J., and Wobus, A. M. (1993). Embryonic stem cells differentiate *in vitro* into cardiomyocytes representing sinusnodal, atrial and ventricular cell types. *Mech. Dev.* **44,** 41–50.

Maltsev, V. A., Wobus, A. M., Rohwedel, J., Bader, M., and Hescheler, J. (1994). Cardiomyocytes differentiated *in vitro* from embryonic stem cells developmentally express cardiac-specific genes and ionic currents. *Circ. Res.* **75,** 233–244.

Mummery, C., Ward-van Oostwaard, D., Doevendans, P., Spijker, R., van den Brink, S., Hassink, R., van der Heyden, M., Opthof, T., Pera, M., de la Riviere, A. B., Passier, R., and Tertoolen, L. (2002). Differentiation of human embryonic stem cells to cardiomyocytes: Role of coculture with visceral endoderm-like cells. *Circulation* **107,** 2733–2740.

Reinlib, L., and Field, L. (2000). Cell transplantation as future therapy for cardiovascular disease?: A workshop of the National Heart, Lung, and Blood Institute. *Circulation* **101,** E182–187.

Schuldiner, M., Yanuka, O., Itskovitz-Eldor, J., Melton, D. A., and Benvenisty, N. (2000). Effects of eight growth factors on the differentiation of cells derived from human embryonic stem cells. *Proc. Natl. Acad. Sci. USA* **97,** 11307–11312.

Takahashi, T., Lord, B., Schulze, P. C., Fryer, R. M., Sarang, S. S., Gullans, S. R., and Lee, R. T. (2003). Ascorbic acid enhances differentiation of embryonic stem cells into cardiac myocytes. *Circulation* **107,** 1912–1916.

van Laake, L. W., van Hoof, D., and Mummery, C. L. (2005). Cardiomyocytes derived from stem cells. *Ann. Med.* **37,** 499–512.

Ventura, C., and Maioli, M. (2000). Opioid peptide gene expression primes cardiogenesis in embryonal pluripotent stem cells. *Circ. Res.* **87,** 189–194.

Wartenberg, M., Dönmez, F., Ling, F. C., Acker, H., Hescheler, J., and Sauer, H. (2001). Tumor-induced angiogenesis studied in confrontation cultures of multicellular tumor spheroids and embryoid bodies grown from pluripotent embryonic stem cells. *FASEB J.* **15,** 995–1005.

Wei, H., Juhasz, O., Li, J., Tarasova, Y. S., and Boheler, K. R. (2005). Embryonic stem cells and cardiomyocyte differentiation: Phenotypic and molecular analyses. *J. Cell. Mol. Med.* **9,** 804–817.

Wobus, A. M., Guan, K., Yang, H. T., and Boheler, K. R. (2002). Embryonic stem cells as a model to study cardiac, skeletal muscle, and vascular smooth muscle cell differentiation. *Methods Mol. Biol.* **185,** 127–156.

Wobus, A. M., Kaomei, G., Shan, J., Wellner, M. C., Rohwedel, J., Ji, G., Fleischmann, B., Katus, H. A., Hescheler, J., and Franz, W. M. (1997). Retinoic acid accelerates embryonic stem cell-derived cardiac differentiation and enhances development of ventricular cardiomyocytes. *J. Mol. Cell. Cardiol.* **29,** 1525–1539.

Wobus, A. M., Wallukat, G., and Hescheler, J. (1991). Pluripotent mouse embryonic stem cells are able to differentiate into cardiomyocytes expressing chronotropic responses to adrenergic and cholinergic agents and Ca2+ channel blockers. *Differentiation (Malden, MA, USA)* **48,** 173–182.

Xu, C., Police, S., Rao, N., and Carpenter, M. K. (2002). Characterization and enrichment of cardiomyocytes derived from human embryonic stem cells. *Circ. Res.* **91,** 501–508.

Zimmermann, W. H., and Eschenhagen, T. (2003). Cardiac tissue engineering for replacement therapy. *Heart Fail. Rev.* **8,** 259–269.

CHAPTER 23

Insulin–Producing Cells from Mouse Embryonic Stem Cells

Insa S. Schroeder, Sabine Sulzbacher, Thuy T. Truong, Przemyslav Blyszczuk,★ Gabriela Kania,★ and Anna M. Wobus

In Vitro Differentiation Group
Leibniz Institute of Plant Genetics and Crop Plant Research (IPK)
Gatersleben, Germany

★Present Address: University Hospital
Experimental Medical Care
CH-4031 Basel, Switzerland

Abstract
1. Introduction
2. Materials and Methods
    2.1. Culture of undifferentiated mES cells
    2.2. Generation of mES cell-derived multilineage progenitor cells
    2.3. Induction of pancreatic differentiation
    2.4. Activin A-induction of endoderm progenitor cells
    2.5. Analysis of differentiated phenotypes
3. Results
4. Summary
References

Abstract

Embryonic stem (ES) cells offer great potential for cell-replacement and tissue engineering therapies because of their almost unlimited proliferation capacity and the potential to differentiate into cellular derivatives of all three primary germ layers. Here, we describe two strategies for the *in vitro* differentiation of mouse ES cells into insulin-producing cells. Our first strategy, the "three-step protocol", includes (1) the spontaneous differentiation of ES cells via embryoid bodies (EBs), (2) the formation of progenitor cells of all three primary germ layers (multilineage progenitors) followed by (3) directed differentiation into the pancreatic lineage. However, the system does not select for nestin-expressing cells as performed in previous differentiation systems. The application of growth and extracellular matrix factors, including laminin, nicotinamide, and insulin, lead to the development of committed pancreatic progenitors, which subsequently differentiate into islet-like clusters that release insulin in response to glucose. During differentiation, transcript levels of pancreas-specific transcription factors (i.e., Pdx1, Pax4) and of genes specific for early and mature β-cells, including insulin, islet amyloid pancreatic peptide, somatostatin, and glucagon are upregulated. C-peptide/insulin-positive islet-like clusters are formed, which release insulin in response to high glucose concentrations at terminal stages. The differentiated cells reveal functional properties with respect to voltage-activated Na^+ and ATP-modulated K^+ channels and normalize blood glucose levels in Streptozotocin-treated diabetic mice. The second strategy includes the directed differentiation of ES cells into the endoderm lineage by application of activin A resulting in an increase of definitive endoderm progenitor cells.

These cells could be further induced to differentiate into the pancreatic lineage. In conclusion, we demonstrate the "proof-of-principle" for efficient differentiation of murine ES cells into insulin-producing cells, which may help in the future to establish ES cell-based therapies in diabetes mellitus.

1. Introduction

Diabetes mellitus is caused by insufficient or abolished insulin release due to autoimmune destruction or malfunction of pancreatic β-cells located in the endocrine pancreas in the so-called islets of Langerhans. The lack of insulin and the resulting inadequate control of glycemia lead to a life-threatening metabolic dysfunction that requires insulin injections to alleviate hyperglycemia. However, this does not provide dynamic control of glucose homeostasis and patients with long term diabetes suffer from complications like neuropathy, nephropathy, retinopathy, and vascular disorders. Consequently, cell replacement therapies would be beneficial to circumvent such adverse side effects. The transplantation of islets of Langerhans has been successfully established, but as the availability of human donor pancreas for islet grafting is limited, *in vitro* β-cell engineering is one promising way to overcome the limitation of donor cells.

ES cells are intensively studied as potential cellular systems to analyze lineage commitment and differentiation (Wobus and Boheler, 2005). Due to their pluripotent character they are capable to self-renew and to differentiate into practically any cell type of the endo-, ecto-, and mesodermal lineage and therefore, may serve as a promising substitute for cell therapy and organ transplantation.

For many years, ES cells of the mouse (mES cells) represent an excellent experimental system to study basic mechanisms of cell differentiation. Spontaneous differentiation of mES cells results in heterogeneous cell populations with a predominant fraction showing ectodermal characteristics, supporting the idea, that ectodermal differentiation is a default pathway and does not require complex extracellular signaling (Ying *et al.*, 2003). In contrast, the yield of endocrine pancreatic cells is relatively low (Kahan *et al.*, 2003). Therefore, the generation of sufficient amounts of insulin-producing cells requires directed differentiation either through selection/gating of pancreatic phenotypes (Leon-Quinto *et al.*, 2004; Soria *et al.*, 2000), transgenic expression of pancreatic developmental control genes (Blyszczuk *et al.*, 2003; Boretti and Gooch, 2007; Ishizaka *et al.*, 2002; Ku *et al.*, 2004, 2007; Lin *et al.*, 2007; Miyazaki *et al.*, 2004; Serafimidis *et al.*, 2008; Shiroi *et al.*, 2005; Soria *et al.*, 2000; Treff *et al.*, 2006), and/or by applying specific growth and extracellular matrix (ECM) factors (Goicoa *et al.*, 2006; Hori *et al.*, 2002; Jafary *et al.*, 2008; Ku *et al.*, 2004, 2007; Lumelsky *et al.*, 2001; Marenah *et al.*, 2006; Nakanishi *et al.*, 2007; Sipione *et al.*, 2004).

Growth and ECM factors that induce or promote pancreatic differentiation include progesterone, putrescine, insulin, transferrin, sodium selenite, fibronectin (ITSFn), nicotinamide (Vaca *et al.*, 2008), and laminin, all of which have been used in the protocol described here. Other groups suggested the use of retinoic acid (Micallef *et al.*, 2005) or the use of conditioned medium from fetal pancreatic buds containing soluble factors which promoted pancreatic differentiation of mES cells (Vaca *et al.*, 2006). Yet, Micaleff *et al.* while showing induction of Pdx1, a marker of pancreatic progenitor cells, could not show differentiation of mES cells into insulin-producing cells, and the use of conditioned medium may be questionable, as repeatable differentiation of ES cells requires a reproducible composition of the conditioned medium, which may be hard to achieve.

Several previously published protocols of pancreatic differentiation used ITSFn/FGF-2 to support proliferation of nestin-positive cells (Lumelsky *et al.*, 2001; Rajagopal *et al.*, 2003; Sipione *et al.*, 2004). However, these protocols not only promoted pancreatic, but also massive neuronal differentiation. It is well known, that during embryogenesis, neuroectodermal and pancreatic differentiation are partially regulated by the same transcription factors, such as Ngn3, Isl-1, or Pax6 in a spatially and temporally distinct manner (Gradwohl *et al.*, 2000; Habener *et al.*, 2005; Lee *et al.*, 2003; Nakagawa and O'Leary, 2001; Schwitzgebel *et al.*, 2000). Likewise, nestin is transiently expressed in ES-derived neuronal (Okabe *et al.*, 1996) and pancreatic (Blyszczuk *et al.*, 2003, 2004; Kania *et al.*, 2004) differentiation, respectively. Therefore, any induction and/or selection of cells expressing these factors leads to a parallel induction of neuronal and pancreatic differentiation, because during *in vitro* differentiation, neuroectodermal and pancreatic progenitor cells are not separated from each other. It has to be mentioned that the differentiation systems that were unsuccessful in demonstrating glucose-controlled insulin-release and other functional parameters (i.e., insulin-positive secretory granules) used the original protocol (Lumelsky *et al.*, 2001) selecting for nestin-positive cells (Hansson *et al.*, 2004; Kitano *et al.*, 2006;

Table 23.1

Comparison of protocols and parameters analyzed following pancreatic differentiation of mouse ES cells

Insulin mRNA	C-peptide / insulin coexpression	In vitro glucose response	In vitro C-peptide secretion	Rescue of diabetes in animal models	Electro-physiological studies	ELMI studies (insulin granules)	Nestin+ cell selection	Additional transgene expression	References
n.d.	n.d.	+	–	+ (but 40% of animals became hyperglycemia 12 weeks after transplantation)	n.d.	n.d.	–	gene trapping via human insulin	Soria et al. (2000)
Ins. 1: – Ins. 2: +	n.d.	+	n.d.	Survival (no sustained correction of hyperglycemia)	n.d.	n.d.	+	–	Lumelsky et al. (2001)
Ins. 1: + Ins. 2: n.d.	+	+	n.d.	+	n.d.	n.d.	+	–	Hori et al. (2002)
Ins. 1: + Ins. 2: weak	–	n.d.	n.d.	n.d.	n.d.	–	+	–	Rajagopal et al. (2003)
+ (no distinction between Ins. 1 + 2)	n.d.	+	n.d.	+	n.d.	+	with and without nestin+ cell selection	Pdx1 and Pax4	Blyszczuk et al. (2003)
Ins. 1: – Ins. 2: +	+ (C-peptide/insulin single staining)	–	n.d.	–	n.d.	n.d.	+	Pdx1	Miyazaki et al. (2004)
Ins. 1: – Ins. 2: +	–C-peptide+ and insulin+ cells, but no coexpression	–nonglucose-dependent insulin release	n.d.	–no improvement of hyperglycemia within 15–25 days	n.d.	–	+	–	Sipione et al. (2004)

Reference									
Hansson et al. (2004)	–	+ (in addition: selection for Sox2+ cells)	n.d.	n.d.	n.d.	–	+	–	–
Blyszczuk et al. (2004)	Pax4	–	n.d.	+	+	n.d.	+	+	+ (no distinction between Ins. 1 + 2*)
Shiroi et al. (2005)	Nkx2.2	–	n.d.	n.d.	n.d.	n.d.	–	n.d.	Proins. 1: + Proins. 2: + Ins. 1: + Ins. 2: n.d.
Shi et al. (2005)	–	+	n.d.	n.d.	+	n.d.	+	+	+ (no distinction between Ins. 1 + 2)
Vaca et al. (2006)	gene trapping via human insulin	–	n.d.	+	+	+	+	+	
Paek et al. (2005)	–	+	n.d.	n.d.	n.d.	n.d.	–	only when cultured w/human or bovine insulin	n.d.
Kitano et al. (2006)	–	+	n.d.	n.d.	n.d.	n.d.	n.d.	Insulin single staining: +	No induction of Ins1 or 2 (shown by qRCR)
Treff et al. (2006)	Inducible Ngn3	–	n.d.	n.d.	n.d.	n.d.	n.d.	n.d.	Ins. 1: + (Micro Array)

(continues)

Table 23.1 (*continued*)

Insulin mRNA	C-peptide / insulin coexpression	In vitro glucose response	In vitro C-peptide secretion	Rescue of diabetes in animal models	Electro-physiological studies	ELMI studies (insulin granules)	Nestin+ cell selection	Additional transgene expression	References
+ (no distinction between Ins. 1 + 2)	n.d.	+	–	n.d.	n.d.	n.d.	+ [+ GIP (LysPAL[16]) Treatment]	–	Marenah et al. (2006)
Ins. 1: + Ins. 2: +	C-peptide[+], insulin staining: n.d.	n.d.	n.d.	n.d.	n.d.	+	–	–	Nakanishi et al. (2007)
Ins. 1: weak Ins. 2: weak (data not shown)	n.d.	n.d.	n.d.	n.d.	n.d.	n.d.	–(sodium butyrate treatment)	–	Goicoa et al. (2006a)
Ins. 1: + Ins. 2: +	C-peptide: + C-peptide WB:-Insulin: weak	weak/no de novo insulin synthesis +	n.d.	n.d.	n.d.	n.d.	+	–	Mc Kiernan et al. (2007)
Ins. 1: + Ins. 2: n.d.	n.d.	+	n.d.	+	n.d.	n.d.	–	Pax4 Pax4siRNA	Lin et al. (2007)
+ (no distinction between Ins. 1 + 2)	n.d.	n.d.	n.d.	n.d.	n.d.	n.d.	–	Ngn3	Boretti and Gooch (2007)
Ins. 1: + Ins. 2: +	indirect via C-pep./Gluc. Ins./Gluc.	n.d.	n.d.	n.d.	n.d.	n.d.	–	Ngn3	Ku et al. (2004, 2007)
Ins. 1: + Ins. 2: +	Insulin: + (single staining)	n.d.	n.d.	n.d.	n.d.	n.d.	–	–	McKiernan et al. (2007b)

Ins. 1: + Ins. 2: +	Insulin: + (single staining)	n.d.	–	n.d.	n.d.	–	–	Chen et al. (2008)
Ins. 1: + Ins. 2: +	+	+	n.d.	n.d.	n.d.	–	Inducible Ngn3	Serafimidis et al. (2008)
Ins. 1: + Ins. 2: + Ins. 1: + Ins. 2: +	Insulin: + (single staining), DTZ staining	+	+	n.d.	n.d.	with and without nestin+ cell selection	–	Jafary et al. (2008)
n.d.	+	+	+	n.d.	n.d.	–(nicotinamide treatment)	gene trapping via human insulin	Vaca et al. (2008)
L^a: Ins.1: weak Ins2: + H^b: Ins.1: weak Ins2: + B^c: Ins.1: + Ins2: +	L^a: –/+ H^b: –/+ B^c: +	L^a: –/+ H^b: –/+ B^c: –/+	L^a: <2 days H^b: <2 days B^c: >2 weeks in 33% of recipients	n.d.	n.d.	L^a: + H^b: + B^c: –	–	Boyd et al. (2008)

n.d.: not done.

*Repetition of RT–PCR with primers specific for insulin 1 and 2 revealed marked induction of insulin 1, while insulin 2 was not expressed (see Fig. 23.2).

L^a: according to protocol by Lumelsky et al. (2001).

H^b: according to protocol by Hori et al. (2002).

B^c: according to protocol by Blyszczuk et al. (2004).

Mc Kiernan *et al.*, 2007a; Paek *et al.*, 2005; Rajagopal *et al.*, 2003; Sipione *et al.*, 2004; see Table 23.1).

Consequently, we avoided any selection or support of nestin-expressing cells, and found that the selection of nestin-positive cells with ITSFn and FGF-2 is neither obligatory nor profitable for successful pancreatic differentiation and does not promote the generation of specific pancreatic progenitors when applied to ES-derived cells (Blyszczuk *et al.*, 2004; Kania *et al.*, 2004). However, nestin as well as cytokeratin 19, a marker expressed in pancreatic duct epithelial cells but not in mature islets, is expressed at intermediate stages of differentiation, but down-regulated at terminal stages.

The advantage of the three-step protocol in generating glucose-responsive insulin-producing cells (according to Blyszczuk *et al.*, 2004 and Schroeder *et al.*, 2006) in comparison to the protocols established by Lumelsky *et al.* (2001) and Hori *et al.* (2002) was recently supported by Boyd *et al.* (2008).

A few years ago, Kubo *et al.* found that endoderm cells could be induced in EBs by limited exposure to serum or treatment with activin A under serum-free conditions (Kubo *et al.*, 2004). Activin A, a member of the transforming growth factor beta (TGF-β) superfamily, was also used by Shi *et al.* in combination with all-trans retinoic acid (RA) to differentiate mES cells into pancreatic β-like cells (Shi *et al.*, 2005). Still, activin A induces formation of neuronal extensions and neurofilament proteins in PC12 cells (Iwasaki *et al.*, 1996) pointing towards the involvement of this substance in neural differentiation. This is in agreement with studies of McKiernan *et al.* (2007b) who found that the combination of RA with sodium butyrate led to the formation of pancreatic endocrine insulin-expressing cells while RA/ activin A and RA/ betacellulin treatment induced neuronal- and glial-like cell types. Likewise, our own studies of pancreatic differentiation, according to the protocol of Shi *et al.*, resulted in both, the induction of pancreatic and neuronal differentiation (Rolletschek *et al.*, 2006).

Evidently, activin A treatment requires a very sophisticated application scheme for successful enrichment of endoderm progenitor cells (D'Amour *et al.*, 2005). By sequential application of endoderm- and pancreas-specific growth factors, the authors further demonstrated that human ES-derived definitive endoderm cells could be differentiated into primitive gut tube, posterior foregut, pancreatic endoderm, and endocrine progenitors and finally hormone expressing endocrine cells. However, in this study, many of the ES cell derivatives showed expression of more than one hormone (especially, insulin and glucagon coexpression) and no glucose-dependent insulin release pointing to an immature cell stage of the differentiated insulin-positive cells (D'Amour *et al.*, 2006). By improvement of the differentiation protocol and by transplantation of an immature human ES-derived pancreatic precursor cell population into animals, the same group generated glucose-responsive, insulin-secreting endocrine cells that normalized experimentally induced diabetes in SCID mice (Kroon *et al.*, 2008). These data underline that the *in vivo* maturation of appropriate pancreatic progenitor cells is still necessary to acquire functional properties of human ES-derived insulin-producing cells.

Here, we present two strategies for pancreatic differentiation of murine ES cells, the reproducible three-step differentiation system of murine ES cells via multilineage progenitors into insulin-producing cell clusters (Blyszczuk *et al.*, 2004; Schroeder *et al.*, 2006; Boyd *et al.* 2008), and the activin A-mediated induction of endoderm progenitor cells for further differentiation into the pancreatic lineage.

2. Materials and Methods

2.1. Culture of undifferentiated mES cells

Mouse R1 embryonic stem (ES) cells were cultivated on feeder layers of mouse embryonic fibroblasts [MEFs, for preparation, see (Wobus *et al.*, 2002)] on gelatin-coated (0.1%) petri dishes (Falcon Becton Dickinson, Heidelberg, Germany) in Dulbecco's modified Eagle's medium (Invitrogen, Karlsruhe, Germany) supplemented with 15% heat-inactivated fetal calf serum (FCS, selected batches, Invitrogen), L-glutamine (Invitrogen, 2 mM), β-mercaptoethanol (Serva, Heidelberg, Germany, final concentration 5×10^{-5} M), nonessential amino acids (Invitrogen, $100\times$ stock solution diluted 1:100), penicillin–streptomycin (Invitrogen, $100\times$ stock solution diluted 1:100), and 10 ng/ml recombinant human leukemia inhibitory factor prepared from LIF expression vectors (see Smith and Johnson, 1988; Wobus *et al.*, 2002) or obtained from commercial sources (Chemicon, Hampshire, United Kingdom). The use of STO feeder layers is not recommended as in our experience the supportive capacity for mES cells is dependent on specific sublines, which may not be commonly available.

To maintain their undifferentiated state, the mES cells (line R1) must be cultured at relatively high density. Good-quality, batch tested FCS is critical for long-term culture of mES cells, and for subsequent successful differentiation. As mouse ES cells divide every 12–15 h, the culture medium should be replenished daily and the cells passaged every 24–48 h onto freshly-prepared feeder layers. For passaging, ES cells must be dissociated carefully by treatment with trypsin-EDTA solution. If one or more of these requirements are not complied with, ES cells may differentiate spontaneously during culture and become unsuitable for differentiation studies.

2.2. Generation of mES cell-derived multilineage progenitor cells

Withdrawal of feeder cells and LIF leads to spontaneous differentiation of mES cells into cells of all three germ layers. This formation of the so-called multilineage progenitor cells is the basis of the ES cell differentiation protocol. Controlled production of multilineage progenitors from mES cells comprises of two steps: (i) the formation of three-dimensional aggregates or embryoid bodies (EBs) to promote differentiation into all three germ layers and (ii) the expansion and further differentiation on adhesive substrata. EB formation of mES cells may be induced either by the "hanging drop" method (Wobus *et al.*, 2002) or by "mass culture" in bacteriological-grade dishes (Doetschman *et al.*, 1985). However, the "hanging drop" method has several advantages including low variation in size of the EBs due to a defined number of mES cells in the starting aggregates as well as greater reproducibility of differentiation. Therefore, mES cells were cultivated as EBs in "hanging drops" (600 cells/ 20 μl) for 2 days, transferred into bacteriological plates (Greiner, Germany), and cultured in suspension in Iscove's modified DMEM (IMDM, Invitrogen) supplemented with 20% FCS, L-glutamine, nonessential amino acids (see above), and α-monothioglycerol (Sigma, Steinheim, Germany; final concentration 450 μM) instead of β-mercaptoethanol for another 3 days (for differentiation scheme, see Fig. 23.1). Penicillin and streptomycin may be added to the cultures (see above). EBs were seeded onto gelatin-coated (0.1%) dishes (20–30 EBs/ 60-mm) and grown in IMDM (see above). Medium was changed every second to third day until 9 days after EB plating (=stage 5 + 9 d; also differentiation of multilineage progenitors for 7 days = 5 + 7d, would be applicable dependent on culture conditions).

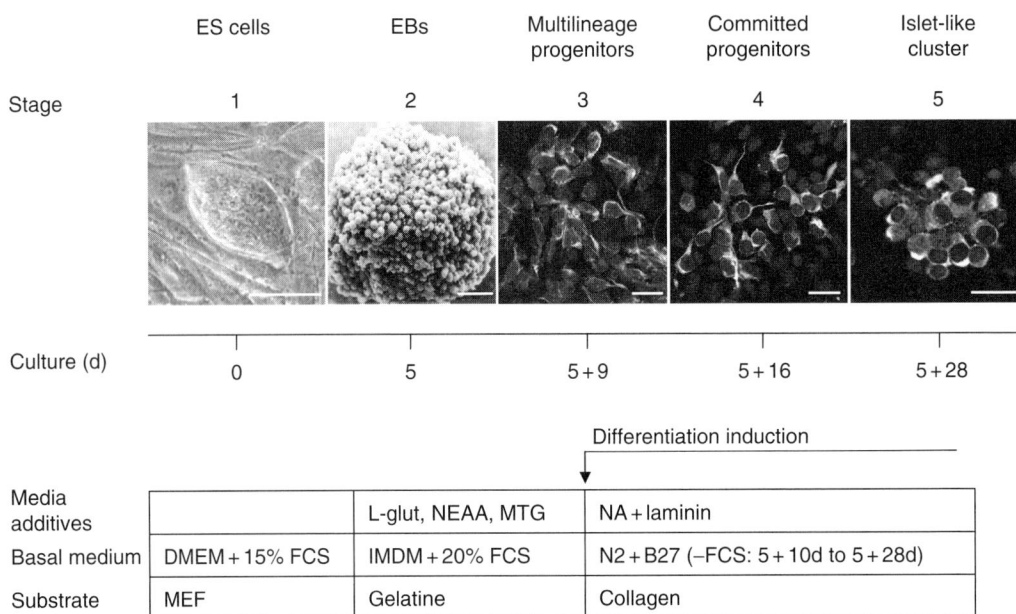

	ES cells	EBs	Multilineage progenitors	Committed progenitors	Islet-like cluster
Stage	1	2	3	4	5
Culture (d)	0	5	5 + 9	5 + 16	5 + 28

Differentiation induction

Media additives		L-glut, NEAA, MTG	NA + laminin
Basal medium	DMEM + 15% FCS	IMDM + 20% FCS	N2 + B27 (−FCS: 5 + 10d to 5 + 28d)
Substrate	MEF	Gelatine	Collagen

Figure 23.1 Schematic representation of ES cell-derived pancreatic differentiation by the three-step protocol. Mouse ES cells cultured on fibroblast feeder layers (1) are differentiated via EBs, scanning electron microscopy (2), into multilineage progenitor cells (3), after differentiation induction by growth factors into committed progenitors (4), and insulin-producing cells in islet-like clusters (5). Shown are the stages of differentiation with examples of cell morphology and media, additives and substrates used during *in vitro* differentiation. Cells at stage 5 + 9d are dissociated and replated onto collagen I-coated tissue culture plates and cultured in complex differentiation medium with differentiation factors and 10% FCS to support attachment of cells. One day later, the differentiation medium is replenished without serum and cells are differentiated for up to 28 or more days. Immunostaining shows nestin/CK19 (3), nestin/C-peptide (4) and C-peptide/insulin (5) coexpression in ES-derived cells at different stages. Bars = 20 μm (3, 4, 5), 50 μm (2), 100 μm (1). Modified according to Blyszczuk *et al.* (2004) and Wobus and Boheler (2005).

2.3. Induction of pancreatic differentiation

The differentiation of mES cells into specific cell types requires the enhancement and subsequently differentiation of lineage-committed progenitor cells via defined growth- and differentiation-inducing factors. However, parameters like the dissociation of the EBs, the choice of suitable adhesive substrata, and the cell density after replating also determine the differentiation efficiency. The way of dissociating the EB outgrowths is crucial as cell-to-cell interactions within the complex and heterogeneous structure of EBs may influence the fate of progenitor cells. Likewise, ECM factors of the adhesive substrata affect adhesion, proliferation, and migration of specific progenitor cells after replating of dissociated EBs. Finally, improper cell density may decrease differentiation efficiency: overgrowth can result in metabolic starvation, necrosis, and cell death while too low cell density can lead to reduced cell-to-cell contacts and reduced release of essential autocrine factors.

Factors that induce pancreatic differentiation include nicotinamide (Chen *et al.*, 2008; Otonkoski *et al.*, 1993; Vaca *et al.*, 2008) and laminin (Jiang *et al.*, 1999). In addition, factors required for pancreatic-cell survival, such as progesterone, putrescine, insulin, sodium selenite, and transferrin, promote also pancreatic differentiation. These factors have been used to direct the committed progenitor cells towards pancreatic insulin-producing cells. EB outgrowths generated by day 5 + 9 (see Fig. 23.1) were dissociated by 0.1% trypsin (Serva)/0.08% EDTA (Sigma) in PBS (1:1) for 1 min, collected by centrifugation and replated onto collagen-coated tissue culture plates (Nunc, Wiesbaden, Germany) in DMEM/F12 (Invitrogen) containing 20 nM progesterone, 100 μM putrescine, 1 μg/ml laminin, 10 mM nicotinamide (all from Sigma), B27 media supplement (Invitrogen), 25 μg/ml insulin, 50 μg/ml transferrin, 5 μg/ml fibronectin, and 30 nM sodium selenite (all from Sigma) supplemented with 10% FCS and penicillin–streptomycin (see above). Collagen proved to be a more desirable substrate than laminin as the latter one induced neuronal differentiation (data not shown). For immunofluorescence analysis, cells were plated onto collagen-coated cover slips, for ELISA onto 3 cm culture dishes, and for RT–PCR onto 6 cm culture dishes. One day after replating (at day 5 + 10), FCS was removed and the cells were cultivated until day 5 + 16 and 5 + 28 for further analysis (Blyszczuk *et al.*, 2004).

2.4. Activin A-induction of endoderm progenitor cells

The spontaneous differentiation of ES cells via multilineage progenitor cells generates cell types of all three primary germ layers, and therefore, the formation of endoderm cells may be inefficient. To increase the number of definitive endoderm cells, directed differentiation of murine ES cells via activin A treatment was performed. For cell differentiation in the presence of activin A all contaminating cells and proteins, such as feeder cells and sera (FCS), have to be removed, because of their interference with activin A activity. Undifferentiated murine R1 ES-cells were dissociated by trypsin: EDTA treatment and transferred to a 10 cm tissue culture plate for 1 h to remove feeder cells. The replating method allows the removal of feeder cells that will attach to the plates, whereas ES cells will remain in suspension. The supernatant containing ES cells was centrifuged and washed twice with chemically defined medium [CDM, according to (Johansson and Wiles, 1995); for detailed discussion of CDM preparation and experimentation, see (Wiles and Proetzel, 2006)]. Basal CDM is prepared from IMDM with Gluta-max-I and F12 with Glutamax-I (both Invitrogen) at a ratio of 1:1 supplemented with 450 μM MTG, 2U/ml LIF, 1% chemically defined lipid concentrate (Invitrogen), 5 mg/ml BSA (Invitrogen), 150 μg/ml transferrin, and 7 μg/ml insulin (Sigma). For the induction of endoderm differentiation, ES cells were differentiated as EBs in CDM with 100 ng/ml activin A (R&D Systems) in hanging drop (see above) or mass culture for 7 days. For mass culture, ES cells (n = 3×10^4) were cultured in 6 cm bacteriological plates to allow EB formation. For further differentiation into insulin-producing cells, additional differentiation factors have to be applied according to D'Amour *et al.* (2006) and Kroon *et al.* (2008).

2.5. Analysis of differentiated phenotypes

For proper characterization of cells in the various stages of differentiation, it is crucial to use multiple markers of pancreatic and nonpancreatic cells. It is a precondition to use several phenotypic as well as functional assays to evaluate the extent of differentiation (see currently used analytical methods in Table 23.1). As an example, immunostaining for insulin alone may lead to false positive results regarding endogenous production of this hormone as it is a compound of most treatment protocols and can be easily taken up by apoptotic cells from the culture medium (Hansson *et al.*, 2004; Rajagopal *et al.*, 2003). Therefore, C-peptide, a byproduct of insulin synthesis, is a more reliable marker of insulin production and should always be used for costaining with other pancreatic markers. Characterization should include pancreatic developmental control genes known to be specific for proper β-cell formation, like HNF3β, ngn3, Pdx1, Pax4, and Nkx6.1 (see Cerf *et al.* 2005), in addition to mature pancreatic markers such as insulin 1 and 2, Glut2, glucagon, IAPP, and pancreatic polypeptide. Moreover, it has to be taken into consideration that both endo- and ectodermal cells produce insulin: in ectodermal cells insulin acts as a growth factor and is expressed at relatively low levels, while in endoderm-derived pancreatic islets insulin is involved in hormonal regulations of glucose homeostasis. Therefore, the *in vitro* generation of ectoderm-derived insulin-producing cells can simulate pancreatic β-cell formation. However, insulin release from ectodermal cells may not be glucose-responsive. Rodents possess two insulin genes, insulin 1 and 2. The insulin 1 gene is exclusively expressed in pancreatic tissue, while insulin 2 is expressed in pancreatic islets and certain neurons (Melloul *et al.*, 2002).

In the current three-step protocol, the characterization of cells was carried out by RT–PCR, immunocytochemistry, and ELISA. This allowed the qualitative and quantitative determination of progenitor- and pancreas-specific markers at the mRNA and protein level as well as the determination of proper cell function.

2.5.1. Semiquantitative RT–PCR analysis

ES-derived cells were collected and suspended in lysis buffer composed of 4 M guanidinium thiocyanate, 25 mM sodium citrate (pH 7); 0.5% (w/v) sarcosyl, and 0.1 M β-mercaptoethanol.

Total RNA was isolated by the single step extraction method according to Chomczynski and Sacchi (1987) including a Proteinase K digest for 1 h at 56 °C. mRNA was reverse transcribed using Oligo d(T) and Revert Aid™ M-MuL-V Reverse Transcriptase (Fermentas, St. Leon-Rot, Germany). cDNAs were amplified using oligonucleotide primers complementary to transcripts of the analyzed genes (see Table 23.2) and Taq polymerase (Fermentas). The PCR reaction was electrophoretically separated on 2% (w/v) agarose gels, visualized using ethidium bromide staining, and analyzed by the TINA2.08e software (Raytest Isotopen-meßgeräte GmbH, Straubenhardt, Germany). All markers were normalized to the housekeeping gene β5-tubulin [for a detailed description of RT–PCR see (Wobus et al., 2002)].

2.5.2. Immunofluorescence analysis

For immunofluorescence, EB outgrowths of ES cells growing on cover slips were either fixed with 4% paraformaldehyde (PFA) in PBS at room temperature (RT) for 20 min or in methanol: aceton (7: 3, vol: vol) at −20 °C for 10 min, depending on the antibody used (see Table 23.3). After rinsing (3×) in PBS, bovine serum albumin (BSA, 1% in PBS) was used to inhibit unspecific labelling (30 min) at RT. Cells were incubated with the primary antibodies in specific dilutions (Table 23.3) at 37 °C for 60 min. Samples were washed (3×) in PBS and incubated with fluorescence-labelled secondary antibodies (diluted in 0.5% BSA in PBS) at 37 °C for 45 min (Table 23.4). To label the nuclei for a semiquantitative estimation of immunofluorescence signals, cells were incubated in 5 μg/ ml Hoechst 33342 in PBS at 37 °C for 10 min. After washing (3×) in PBS and (1×) in Aqua dest, the specimens were embedded in mounting medium (Vectashield, Vector Laboratories Inc., Burlingame, California, USA).

Labelled cells were analysed by the fluorescence microscope ECLIPSE TE300 (Nikon, Japan), or the confocal laser scanning microscope (CLSM) LSM-410 (Carl Zeiss, Jena, Germany) using the following excitation lines/barrier filters: 364nm/ 450–490BP (Hoechst 33342), 488nm/ 510–525BP (ALEXA), 543nm/ 570LP (Cy3).

Table 23.2
Primer sequences, annealing temperature and the length of the amplified fragment applicable for RT-PCR amplification of progenitor and pancreas-specific genes

Gene	Primer sequence (Forward/Reverse)	Annealing temperature (°C)	Product size (bp)
Sox 17	5'-CCA TAG CAG AGC TCG GGG TC-3' 5'-GTG CGG AGA CAT CAG CGG AG-3'	62	627
HNF3β (Foxa2)	5'-ACT GGA GCA GCT ACT ACG-3' 5'-CCC ACA TAG GAT GAC ATG-3'	55	169
Cytokeratin 19	5'-CTG CAG ATG ACT TCA GAA CC-3' 5'-GGC CAT GAT CTC ATA CTG AC-3'	62	300
Isl-1	5'-GTT TGT ACG GGA TCA AAT GC-3' 5'-ATG CTG CGT TTC TTG TCC TT-3'	60	503
Nestin	5'-CTA CCA GGA GCG CGT GGC-3' 5'-TCC ACA GCC AGC TGG AAC TT-3'	60	219
Ngn3 (MATH4B)	5'-TGG CGC CTC ATC CCT TGG ATG-3' 5'-AGT CAC CCA CTT CTG CTT CG-3'	60	159
Pax4	5'-ACC AGA GCT TGC ACT GGA CT-3' 5'-CCC ATT TCA GCT TCT CTT GC-3'	60	301
Pax6	5'-TCA CAG CGG AGT GAA TCA G-3' 5'-CCC AAG CAA AGA TGG AAG-3'	58	332
Pdx1 (IPF-1)	5'-CTT TCC CGT GGA TGA AAT CC-3' 5'-GTC AAG TTC AAC ATC ACT GCC-3'	60	205
Insulin 1/Preproinsulin 1	5'-TAG TGA CCA GCT ATA ATC AGA GAC-3' 5'-CGC CAA GGT CTG AAG GTC-3'	60	288 406
Insulin 2/Preproinsulin 2	5'-CCC TGC TGG CCC TGC TCT T-3' 5'-AGG TCT GAA GGT CAC CTG CT-3'	65	213 701
Glucagon	5'-CAT TCA CAG GGC ACA TTC ACC-3' 5'-CCA GCC CAA GCA ATG AAT TCC-3'	55	207
Amylase	5'-CAG GCA ATC CTG CAG GAA CAA-3' 5'-CAC TTG CGG ATA ACT GTG CCA-3'	60	484
Glut-2	5'-TTC GGC TAT GAC ATC GGT GTG-3' 5'-AGC TGA GGC CAG CAA TCT GAC-3'	60	556
IAPP	5'-TGA TAT TGC TGC CTC GGA CC-3' 5'-GGA GGA CTG GAC CAA GGT TG-3'	65	233
PP	5'-ACT AGC TCA GCA CAC AGG AT-3' 5'-AGA CAA GAG AGG CTG CAA GT-3'	60	364
Somatostatin/ Preprosomatostatin	5'-TCG CTG CTG CCT GAG GAC CT-3' 5'-GCC AAG AAG TAC TTG GCC AGT TC-3'	60	232 897
β5-Tubulin	5'-TCA CTG TGC CTG AAC TTA CC-3' 5'-GGA ACA TAG CCG TAA ACT GC-3'	60	318
TG (Brachyury)	5'-GAG AGA GAG CGA GCC TCC AAA C-3' 5'-GCT GTG ACT GCC TAC CAG AAT G-3'	59	231
Oct-4	5'-CTC GAA CCA CAT CCT TCT CT-3' 5'-GGC GTT CTC TTT GGA AAG GTG TTC-3'	61	313
Sox 7	5'-AGC GCC GGC CCC ACG AG-3' 5'-GCG CTT GCC TTG TTT CTT CCT G-3'	60	416
ZIC1	5'-AAA AGG ACA CAC ACA GGG GAG-3' 5'-GTC TCT AAA ATA GGG GGT CG-3'	65	417
GAPDH	5'-CCA TGT TTG TGA TGG GTG TGA ACC-3' 5'-TGT GAG GGA GAT GCT CAG TGT TGG-3'	65	734

Table 23.3
Selected primary antibodies to characterize progenitor and pancreatic cell types

Primary antibody	Dilution	Supplier	Fixation
rat anti-Oct3/4 IgG	1:50	R&D Systems, Germany	4% PFA
mouse anti-nestin IgG (clone rat 401)	1:3	Developmental Studies Hybridoma Bank, Indiana, USA	4% PFA
rabbit anti-Isl-1 IgG	1:50	Millipore, Germany	4% PFA
mouse anti-cytokeratin 19 IgM	1:100	Cymbus, United Kingdom	MeOH:Ac[a] 4% PFA[b]
rabbit anti-Foxa2 IgG	1:100	Millipore, Germany	4% PFA
rat anti-CXCR4 IgG	1:50	R&D Systems, Germany	4% PFA
rabbit anti-Brachyury IgG	1:50	Abcam, United Kingdom	4% PFA
mouse anti-insulin IgG (clone K36AC10)	1:100	Sigma-Aldrich, Munich, Germany	4% PFA+ 0.1% glutaraldehyde
sheep anti-C-peptide IgG	1:100	Acris, Germany	4% PFA
rabbit anti-glucagon IgG	1:40	Abcam, United Kingdom	4% PFA
rabbit anti-somatostatin IgG	1:40	Dako, Denmark	4% PFA
rabbit anti-PP IgG	1:40	Dako, Denmark	4% PFA

MeOH:Ac: methanol:acetone (7:3, vol:vol) fixation at −20 °C for 10 min.
PFA: 4% paraformaldehyde fixation at room temperature for 20 min.
[a]Filament structures.
[b]Dot-like structures.

Table 23.4
Fluorescence-labeled secondary antibodies

Secondary antibody	Dilution	Supplier
Cy[3TM]-conjugated goat anti-mouse IgG	1:600	Jackson ImmunoResearch
Cy[3TM]-conjugated goat anti-rabbit IgG	1:600	Laboratories, USA
Cy[3TM]-conjugated goat anti-mouse IgM	1:600	
ALEXA 488-conjugated rabbit anti-rat IgG	1:100	Molecular Probes, Germany
ALEXA 488-conjugated donkey anti-sheep IgG	1:100	
ALEXA 488-conjugated goat anti-mouse IgG	1:100	

Quantification of immunofluorescence signals was performed by two alternative methods depending on the cell culture status. Cells growing in monolayer may be analysed by direct determination of immunolabeled cells (percentage values), whereas, for cells growing in multilayered clusters, the "labeling index" technique is proposed.

1. Determination of percentage values of Hoechst-labeled cells:

   Cells were analysed for immunofluorescence signals and the percentage number of immunopositive cells relative to a total number of (n = 1000) Hoechst 33342-labeled cells is given.

2. Estimation of the "labeling index" (Blyszczuk *et al.*, 2003):

   For cells growing in clusters, immunofluorescence analysis was performed using the inverted fluorescence microscope ECLIPSE TE300 (Nikon, Japan) equipped with a 3CCD Color Video Camera DXC-9100P (Sony, Japan) and the LUCIA M—Version 3.52a software (LIM, Nikon). For each sample, at least 20 randomly, but representative selected pictures were analysed for the "area fraction" value, which is the ratio of the immunopositive signal area to the measured area. To discriminate the immunopositive signal from background fluorescence, the pictures were binarized with the specific threshold fluorescence values.

2.5.3. Insulin ELISA

The analysis of differentiated pancreatic endocrine cells should include the determination of insulin production as a functional assay. The intracellular insulin content can be measured by commercialised specific insulin ELISA. Additionally, the glucose responsiveness should be tested. For this purpose, insulin release in the presence of low (5.5 mM, as a control) and high (27.7 mM) glucose concentration is determined. Tolbutamide (10 μM), a sulfonylurea known to stimulate insulin secretion, together with 5.5 mM glucose can also be used. However, failure of glucose response may be dependent on insufficient maturation during differentiation. Such effects were already described during pancreatic differentiation of mouse ES cells, where insulin was secreted in response to glucose at an advanced stage of 32 days of differentiation, but not at day 28 (Blyszczuk *et al.*, 2003).

ES-derived cells differentiated into the pancreatic lineage were cultured in differentiation medium without insulin for 3 h prior to ELISA. Cells were washed in PBS (5×) and preincubated in freshly prepared KRBH (Krebs' Ringer Bicarbonate Hepes) buffer containing 118 mM sodium chloride, 4.7 mM potassium chloride, 1.1 mM potassium dihydrogen phosphate, 25 mM sodium hydrogen carbonate (all from Carl Roth GmbH & Co, Karlsruhe, Germany), 3.4 mM calcium chloride (Sigma), 2.5 mM magnesium sulphate (Merck), 10 mM Hepes, and 2 mg/ml bovine serum albumin supplemented with 2.5 mM glucose (all from Invitrogen) for 90 min at 37 °C.

To estimate glucose-induced insulin secretion, the buffer was replaced by 27.7 mM glucose and alternatively with 5.5 mM glucose and 50 μM tolbutamide dissolved in KRBH buffer for 15 min at 37 °C. The control was incubated in KRBH buffer supplemented with 5.5 mM glucose. The supernatant was collected and stored at −20 °C for determination of insulin release.

Cells were washed two times with 0.2% trypsin: 0.02% EDTA in PBS (1:1), trypsinization was stopped with 1.5 ml DMEM containing 10% FCS and cells collected by centrifugation. Proteins were extracted from the cells with 50 μl acid ethanol (1 M hydrochloric acid: absolute ethanol = 1:9), incubated at 4 °C overnight, sonicated, and stored at -20 °C for the determination of total cellular insulin and protein content, respectively.

The insulin enzyme-linked immunosorbent assay (ELISA, Mercodia AB, Sweden) was performed according to manufacturer recommendations.

The total protein content was determined by the protein Bradford assay according to manufacturer recommendations (Bio-Rad Laboratories GmbH, Munich, Germany).

Released insulin levels are presented as ratio of released insulin per 15 min and intracellular insulin content. The intracellular insulin level is given as ng insulin per mg protein (Blyszczuk *et al.*, 2003, 2004).

3. Results

By applying the three-step protocol comprising of spontaneous differentiation of mES cells via EBs and multilineage progenitors followed by directed differentiation using specific growth and ECM factors, islet-like clusters were formed (Fig. 23.1), which expressed Pdx1, Pax4, IAPP, insulin 1, glucagon, amylase, and somatostatin (Fig. 23.2A and B). Cells at the committed progenitor stage (5 + 16d) show coexpression of nestin with Isl-1 and C-peptide, and of C-peptide with CK19, respectively (Fig. 23.3A–C). Cells at the terminal differentiation stage (5 + 28d) did not coexpress C-peptide and nestin in islet-like clusters (Fig. 23.3D), but nestin-positive cells were found outside the clusters (Fig. 23.3D). CK 19 shows only a low level of coexpression with C-peptide-positive cells suggesting that the cells represent still an immature phenotype (Fig. 23.3E). C-peptide expression in insulin-producing cells (Fig. 23.3F) and glucose-dependent insulin release (Fig. 23.2C and D) present evidence that differentiated mES cells indeed produced and released insulin rather than taking it up from the medium.

For selective differentiation induction of definitive endoderm progenitors, ES cells were cultured as EBs in hanging drops or in mass culture in CDM or CDM + 100 ng/ml activin A for 7 days. Significantly smaller EBs were formed in the CDM and activin A variants in comparison to EBs grown in Iscove medium (used to generate multilineage progenitors) as shown in Fig. 23.4A. RT–PCR analysis demonstrated a significant upregulation only of the definitive endoderm-specific marker gene Sox17 by activin A treatment, whereas transcript levels of Sox7 (expressed also in extra-embryonic endoderm), the pluripotency-associated gene Oct4, neural-specific Pax6 were downregulated in activin A compared to CDM variants (Fig. 23.4B). The mRNA level of the mesodermal gene Brachyury (TG) was decreased in CDM and CDM+ activin A-treated EBs in comparison to undifferentiated ES cells. Evidently, expression of Pax6 was significantly upregulated in CDM-treated EBs suggesting differentiation into the neuroectodermal lineage in the absence of growth factors support, whereas the extra-embryonic

A

	ES	5+9d	5+16d	5+28d	Panc	Brain
Insulin 1						
Insulin 2						
Glucagon						
Amylase						
Somatostatin						
β5-tubulin						

B

Pdx1

Pax4

IAPP

β5-tubulin

5+9 5+16 5+28
Time of differentiation (d)

C

Cell line	Intracellular insulin (ng/mg protein)	Released insulin (ng/mg protein) at 5.5 mM glucose	Released insulin (ng/mg protein) at 27.7 mM glucose
wt	38.1 ± 4.7	6.1 ± 1.3	8.6 ± 1.0
Pax4⁺	168.0 ± 28.5	17.3 ± 2.9	27.0 ± 3.2

D

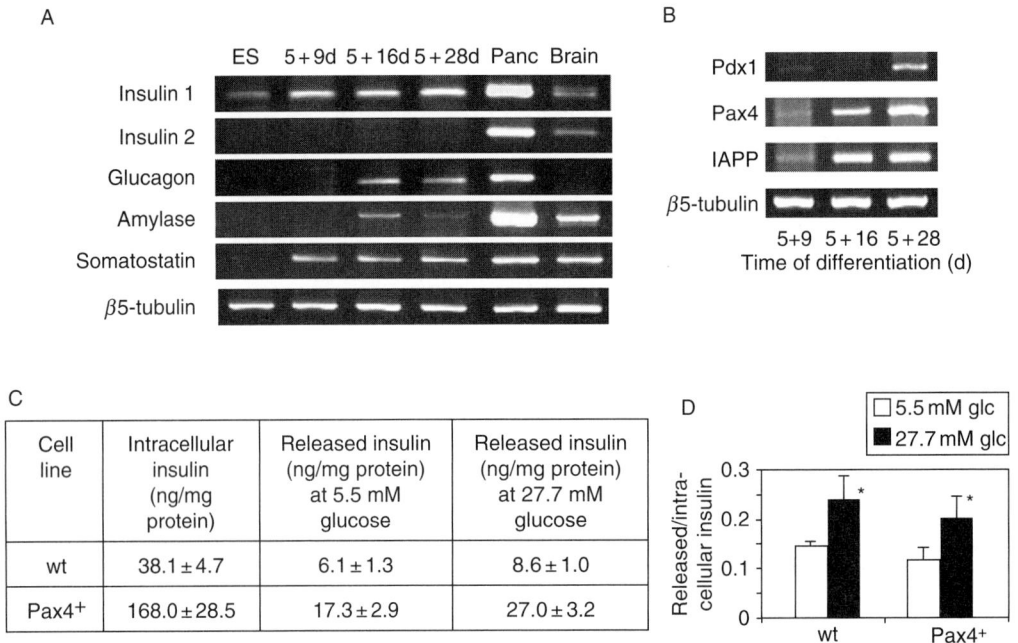

Figure 23.2 Transcript levels of pancreas-specific genes and insulin release by ELISA. (A and B) RT–PCR results of mES cells and cells at differentiation stages 5 + 9d, 5 + 16d, and 5 + 28d. Mouse pancreas and brain served as positive controls. (C and D) ELISA data of insulin levels in ES-derived cells after pancreatic differentiation. (C) Levels of intracellular and released insulin in wild type (wt) and Pax4-overexpressing cells (Pax4+). (D) Glucose-dependent insulin release shown as the ratio of secreted and intracellular insulin values. Each value represents the mean ± SEM. Statistical significance was tested by the Student t-test: *, $P < 0.05$ (B–D according to Blyszczuk *et al.*, 2004).

marker Sox7 was strongly decreased in CDM-treated EBs. Taken together, the results prove the selective differentiation of ES cells into definitive endoderm progenitor cells after treatment with high concentrations of activin A. The endoderm progenitor cell population is a suitable source for further pancreatic differentiation induction.

4. Summary

Our three-step differentiation model represents a reproducible method to generate insulin-producing cells avoiding those selection steps, which might critically affect ES cell differentiation into the pancreatic lineage. Especially, we avoid the induced propagation and specific selection of nestin-expressing cells, because it has

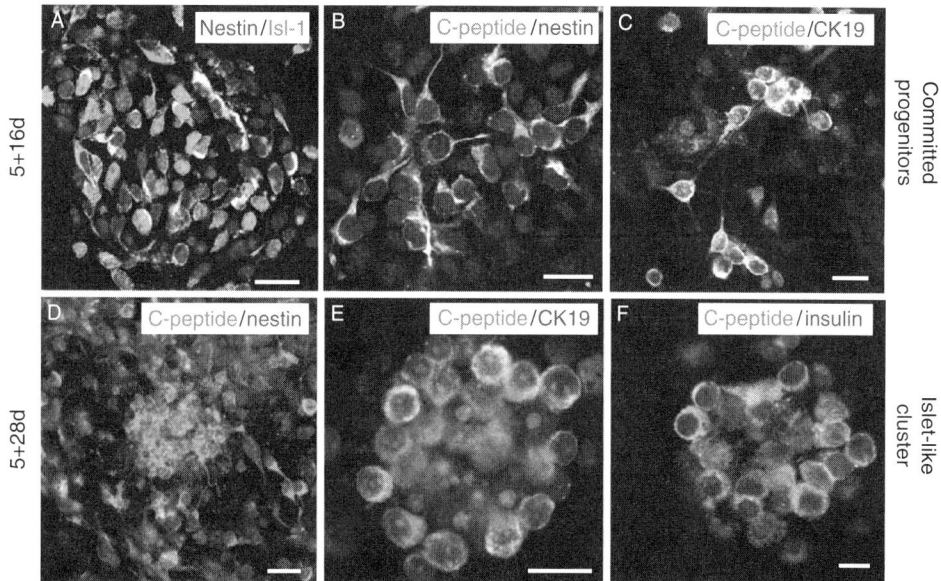

Figure 23.3 Double-immunofluorescence analysis of ES-derived cells differentiated into the pancreatic lineage. Shown are ES-derived cells at intermediate (5 + 16d, A–C) and terminal (5 + 28d, D–F) stages following the three-step pancreatic differentiation protocol. (A and B) Images show immunohisto-chemical analysis of nestin/Isl-1 and nestin/C-peptide coexpression, and (C) coexpression of C-peptide and cytokeratin 19 (CK19). (D–F) Images demonstrate the lack C-peptide/nestin coexpression in islet-like clusters, a weak coexpression of C-peptide with CK19, but C-peptide/insulin colabeling in islet-like clusters at the terminal differentiation stage, 5 + 28d. Bars = 20 μm; according to Blyszczuk *et al.* (2004).

been shown by several independent studies that this may result in apoptotic pathways, induction of neural differentiation and lack of functional pancreatic insulin-producing cells. Moreover, the application of activin A to induce definitive endo-derm progenitor cells prior to directed differentiation into the pancreatic lineage is a valid alternative to the three-step protocol and may be applied to further enhance the differentiation efficiency.

The differentiation protocols presented here allow further analysis of pancreatic differentiation factors and signaling mechanisms necessary for the generation and maturation of islet-like cells *in vitro*.

Acknowledgments

We thank Mrs. S. Sommerfeld, O. Weiss, and K. Meier for excellent technical assistance and Dr. A. Rolletschek for helpful discussion. The financial support by the German Research Foundation (DFG, WO 503/3–3) and the EU FunGenES program to A.M.W. is gratefully acknowledged.

A

B

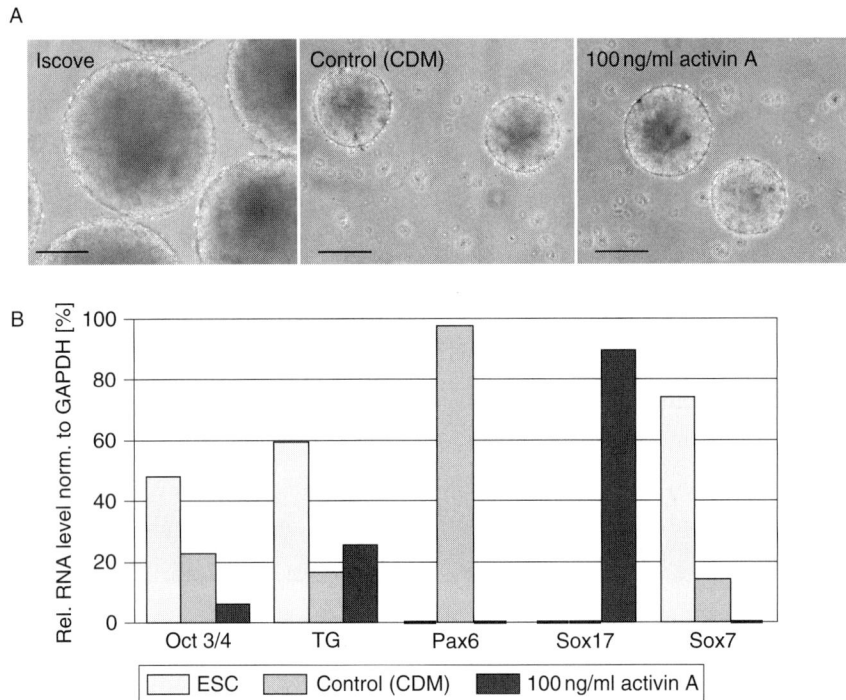

Figure 23.4 Activin A-induced differentiation into definitive endoderm progenitor cells. (A) Morphology of mouse ES-derived embryoid bodies (EBs) cultured in hanging drops for 7d in Iscove's modified Dulbecco medium (IMDM) supplemented with 20% FCS (left), CDM (control, middle) or CDM + 100 ng/ml activin A (right). Bar: 100 μm. (B) Transcript levels of pluripotency-associated (Oct3/4)-, mesoderm (TG)-, ectoderm (Pax6)-, definitive endoderm (Sox17)- and extra-embryonic endoderm (Sox7)-specific genes. Graph shows relative mRNA levels normalized to GAPDH of undifferentiated ES cells (ESC, day 0), and of EBs cultured in mass culture for 7 days in CDM (control) or in CDM + 100 ng/ml activin A.

References

Blyszczuk, P., Asbrand, C., Rozzo, A., Kania, G., St Onge, L., Rupnik, M., and Wobus, A. M. (2004). Embryonic stem cells differentiate into insulin-producing cells without selection of nestin-expressing cells. *Int. J. Dev. Biol.* **48,** 1095–1104.

Blyszczuk, P., Czyz, J., Kania, G., Wagner, M., Roll, U., St Onge, L., and Wobus, A. M. (2003). Expression of Pax4 in embryonic stem cells promotes differentiation of nestin-positive progenitor and insulin-producing cells. *Proc. Natl. Acad. Sci. USA* **100,** 998–1003.

Boretti, M. I., and Gooch, K. J. (2007). Transgene expression level and inherent differences in target gene activation determine the rate and fate of neurogenin3-mediated islet cell differentiation *in vitro.* *Tissue Eng.* **13,** 775–788.

Boyd, A. S., Wu, D. C., Higashi, Y., and Wood, K. J. (2008). A comparison of protocols used to generate insulin-producing cell clusters from mouse embryonic stem cells. *Stem Cells* **26,** 1128–1137.

Cerf, M. E., Muller, C. J., Du Toit, D. F., Louw, J., and Wolfe-Coote, S. A. (2005). Transcription factors, pancreatic development, and beta-cell maintenance. *Biochem. Biophys. Res. Commun.* **326,** 699–702.

Chen, C., Zhang, Y., Sheng, X., Huang, C., and Zang, Y. Q. (2008). Differentiation of embryonic stem cells towards pancreatic progenitor cells and their transplantation into streptozotocin-induced diabetic mice. *Cell Biol. Int.* **32,** 456–461.

Chomczynski, P., and Sacchi, N. (1987). Single-step method of RNA isolation by acid guanidinium thiocyanate-phenol-chloroform extraction. *Anal. Biochem.* **162,** 156–159.

D'Amour, K. A., Agulnick, A. D., Eliazer, S., Kelly, O. G., Kroon, E., and Baetge, E. E. (2005). Efficient differentiation of human embryonic stem cells to definitive endoderm. *Nat. Biotechnol.* **23,** 1534–1541.

D'Amour, K. A., Bang, A. G., Eliazer, S., Kelly, O. G., Agulnick, A. D., Smart, N. G., Moorman, M. A., Kroon, E., Carpenter, M. K., and Baetge, E. E. (2006). Production of pancreatic hormone-expressing endocrine cells from human embryonic stem cells. *Nat. Biotechnol.* **24,** 1392–1401.

Doetschman, T. C., Eistetter, H., Katz, M., Schmidt, W., and Kemler, R. (1985). The *in vitro* development of blastocyst-derived embryonic stem cell lines: Formation of visceral yolk sac, blood islands and myocardium. *J. Embryol. Exp. Morphol.* **87,** 27–45.

Goicoa, S., Alvarez, S., Ricordi, C., Inverardi, L., and Dominguez-Bendala, J. (2006). Sodium butyrate activates genes of early pancreatic development in embryonic stem cells. *Cloning Stem Cells* **8,** 140–149.

Gradwohl, G., Dierich, A., LeMeur, M., and Guillemot, F. (2000). Neurogenin3 is required for the development of the four endocrine cell lineages of the pancreas. *Proc. Natl. Acad. Sci. USA* **97,** 1607–1611.

Habener, J. F., Kemp, D. M., and Thomas, M. K. (2005). Minireview: Transcriptional regulation in pancreatic development. *Endocrinology* **146,** 1025–1034.

Hansson, M., Tonning, A., Frandsen, U., Petri, A., Rajagopal, J., Englund, M. C., Heller, R. S., Hakansson, J., Fleckner, J., Skold, H. N., Melton, D., Semb, H., *et al.* (2004). Artifactual insulin release from differentiated embryonic stem cells. *Diabetes* **53,** 2603–2609.

Hori, Y., Rulifson, I. C., Tsai, B. C., Heit, J. J., Cahoy, J. D., and Kim, S. K. (2002). Growth inhibitors promote differentiation of insulin-producing tissue from embryonic stem cells. *Proc. Natl. Acad. Sci. USA* **99,** 16105–16110.

Ishizaka, S., Shiroi, A., Kanda, S., Yoshikawa, M., Tsujinoue, H., Kuriyama, S., Hasuma, T., Nakatani, K., and Takahashi, K. (2002). Development of hepatocytes from ES cells after transfection with the HNF-3beta gene. *FASEB J.* **16,** 1444–1446.

Iwasaki, S., Hattori, A., Sato, M., Tsujimoto, M., and Kohno, M. (1996). Characterization of the bone morphogenetic protein-2 as a neurotrophic factor. Induction of neuronal differentiation of PC12 cells in the absence of mitogen-activated protein kinase activation. *J. Biol. Chem.* **271,** 17360–17365.

Jafary, H., Larijani, B., Farrokhi, A., Pirouz, M., Mollamohammadi, S., and Baharvand, H. (2008). Differential effect of activin on mouse embryonic stem cell differentiation in insulin-secreting cells under nestin-positive selection and spontaneous differentiation protocols. *Cell Biol. Int.* **32,** 278–286.

Jiang, F. X., Cram, D. S., DeAizpurua, H. J., and Harrison, L. C. (1999). Laminin-1 promotes differentiation of fetal mouse pancreatic beta-cells. *Diabetes* **48,** 722–730.

Johansson, B. M., and Wiles, M. V. (1995). Evidence for involvement of activin A and bone morphogenetic protein 4 in mammalian mesoderm and hematopoietic development. *Mol. Cell Biol.* **15,** 141–151.

Kahan, B. W., Jacobson, L. M., Hullett, D. A., Ochoada, J. M., Oberley, T. D., Lang, K. M., and Odorico, J. S. (2003). Pancreatic precursors and differentiated islet cell types from murine embryonic stem cells: An *in vitro* model to study islet differentiation. *Diabetes* **52,** 2016–2024.

Kania, G., Blyszczuk, P., and Wobus, A. M. (2004). The generation of insulin-producing cells from embryonic stem cells–a discussion of controversial findings. *Int. J. Dev. Biol.* **48,** 1061–1064.

Kitano, M., Kakinuma, M., Takatori, A., Negishi, T., Ishii, Y., Kyuwa, S., and Yoshikawa, Y. (2006). Gene expression profiling of mouse embryonic stem cell progeny differentiated by Lumelsky's protocol. *Cells Tissues Organs* **183,** 24–31.

Kroon, E., Martinson, L. A., Kadoya, K., Bang, A. G., Kelly, O. G., Eliazer, S., Young, H., Richardson, M., Smart, N. G., Cunningham, J., Agulnick, A. D., D'Amour, K. A., *et al.* (2008). Pancreatic endoderm derived from human embryonic stem cells generates glucose-responsive insulin-secreting cells *in vivo*. *Nat. Biotechnol.* **26,** 443–452.

Ku, H. T., Chai, J., Kim, Y. J., White, P., Purohit-Ghelani, S., Kaestner, K. H., and Bromberg, J. S. (2007). Insulin-expressing colonies developed from murine embryonic stem cell-derived progenitors. *Diabetes* **56,** 921–929.

Ku, H. T., Zhang, N., Kubo, A., O'Connor, R., Mao, M., Keller, G., and Bromberg, J. S. (2004). Committing embryonic stem cells to early endocrine pancreas *in vitro*. *Stem Cells* **22,** 1205–1217.

Kubo, A., Shinozaki, K., Shannon, J. M., Kouskoff, V., Kennedy, M., Woo, S., Fehling, H. J., and Keller, G. (2004). Development of definitive endoderm from embryonic stem cells in culture. *Development* **131,** 1651–1662.

Lee, J., Wu, Y., Qi, Y., Xue, H., Liu, Y., Scheel, D., German, M., Qiu, M., Guillemot, F., Rao, M., and Gradwohl, G. (2003). Neurogenin3 participates in gliogenesis in the developing vertebrate spinal cord. *Dev. Biol.* **253,** 84–98.

Leon-Quinto, T., Jones, J., Skoudy, A., Burcin, M., and Soria, B. (2004). *In vitro* directed differentiation of mouse embryonic stem cells into insulin-producing cells. *Diabetologia* **47,** 1442–1451.

Lin, H. T., Kao, C. L., Lee, K. H., Chang, Y. L., Chiou, S. H., Tsai, F. T., Tsai, T. H., Sheu, D. C., Ho, L. L., and Ku, H. H. (2007). Enhancement of insulin-producing cell differentiation from embryonic stem cells using pax4-nucleofection method. *World J. Gastroenterol.* **13,** 1672–1679.

Lumelsky, N., Blondel, O., Laeng, P., Velasco, I., Ravin, R., and McKay, R. (2001). Differentiation of embryonic stem cells to insulin-secreting structures similar to pancreatic islets. *Science* **292,** 1389–1394.

Marenah, L., McCluskey, J. T., Abdel-Wahab, Y. H., O'Harte, F. P., McClenaghan, N. H., and Flatt, P. R. (2006). A stable analogue of glucose-dependent insulinotropic polypeptide, GIP (lyspal16), enhances functional differentiation of mouse embryonic stem cells into cells expressing islet-specific genes and hormones. *Biol. Chem.* **387,** 941–947.

McKiernan, E., Barron, N. W., O'Sullivan, F., Barham, P., Clynes, M., and O'Driscoll, L. (2007a). Detecting *de novo* insulin synthesis in embryonic stem cell-derived populations. *Exp. Cell Res.* **313,** 1405–1414.

McKiernan, E., O'Driscoll, L., Kasper, M., Barron, N., O'Sullivan, F., and Clynes, M. (2007b). Directed differentiation of mouse embryonic stem cells into pancreatic-like or neuronal- and glial-like phenotypes. *Tissue Eng.* **13,** 2419–2430.

Melloul, D., Marshak, S., and Cerasi, E. (2002). Regulation of pdx-1 gene expression. *Diabetes* **51**(Suppl. 3), S320–S325.

Micallef, S. J., Janes, M. E., Knezevic, K., Davis, R. P., Elefanty, A. G., and Stanley, E. G. (2005). Retinoic acid induces Pdx1-positive endoderm in differentiating mouse embryonic stem cells. *Diabetes* **54,** 301–305.

Miyazaki, S., Yamato, E., and Miyazaki, J. (2004). Regulated expression of pdx-1 promotes *in vitro* differentiation of insulin-producing cells from embryonic stem cells. *Diabetes* **53,** 1030–1037.

Nakagawa, Y., and O'Leary, D. D. (2001). Combinatorial expression patterns of LIM-homeodomain and other regulatory genes parcellate developing thalamus. *J. Neurosci.* **21,** 2711–2725.

Nakanishi, M., Hamazaki, T. S., Komazaki, S., Okochi, H., and Asashima, M. (2007). Pancreatic tissue formation from murine embryonic stem cells *in vitro*. *Differentiation* **75,** 1–11.

Okabe, S., Forsberg-Nilsson, K., Spiro, A. C., Segal, M., and McKay, R. D. (1996). Development of neuronal precursor cells and functional postmitotic neurons from embryonic stem cells *in vitro*. *Mech. Dev.* **59,** 89–102.

Otonkoski, T., Beattie, G. M., Mally, M. I., Ricordi, C., and Hayek, A. (1993). Nicotinamide is a potent inducer of endocrine differentiation in cultured human fetal pancreatic cells. *J. Clin. Invest* **92**, 1459–1466.

Paek, H. J., Moise, L. J., Morgan, J. R., and Lysaght, M. J. (2005). Origin of insulin secreted from islet-like cell clusters derived from murine embryonic stem cells. *Cloning Stem Cells* **7**, 226–231.

Rajagopal, J., Anderson, W. J., Kume, S., Martinez, O. I., and Melton, D. A. (2003). Insulin staining of ES cell progeny from insulin uptake. *Science* **299**, 363.

Rolletschek, A., Kania, G., and Wobus, A. M. (2006). Generation of pancreatic insulin-producing cells from embryonic stem cells—'proof of principle', but questions still unanswered. *Diabetologia* **49**, 2541–2545.

Schroeder, I. S., Rolletschek, A., Blyszczuk, P., Kania, G., and Wobus, A. M. (2006). Differentiation of mouse embryonic stem cells to insulin-producing cells. *Nat. Protoc.* **1**, 495–507.

Schwitzgebel, V. M., Scheel, D. W., Conners, J. R., Kalamaras, J., Lee, J. E., Anderson, D. J., Sussel, L., Johnson, J. D., and German, M. S. (2000). Expression of neurogenin3 reveals an islet cell precursor population in the pancreas. *Development* **127**, 3533–3542.

Serafimidis, I., Rakatzi, I., Episkopou, V., Gouti, M., and Gavalas, A. (2008). Novel effectors of directed and Ngn3-mediated differentiation of mouse embryonic stem cells into endocrine pancreas progenitors. *Stem Cells* **26**, 3–16.

Shi, Y., Hou, L., Tang, F., Jiang, W., Wang, P., Ding, M., and Deng, H. (2005). Inducing embryonic stem cells to differentiate into pancreatic beta cells by a novel three-step approach with activin A and all-trans retinoic acid. *Stem Cells* **23**, 656–662.

Shiroi, A., Ueda, S., Ouji, Y., Saito, K., Moriya, K., Sugie, Y., Fukui, H., Ishizaka, S., and Yoshikawa, M. (2005). Differentiation of embryonic stem cells into insulin-producing cells promoted by Nkx2.2 gene transfer. *World J. Gastroenterol.* **11**, 4161–4166.

Sipione, S., Eshpeter, A., Lyon, J. G., Korbutt, G. S., and Bleackley, R. C. (2004). Insulin expressing cells from differentiated embryonic stem cells are not beta cells. *Diabetologia* **47**, 499–508.

Smith, D. B., and Johnson, K. S. (1988). Single-step purification of polypeptides expressed in Escherichia coli as fusions with glutathione S-transferase. *Gene* **67**, 31–40.

Soria, B., Roche, E., Berna, G., Leon-Quinto, T., Reig, J. A., and Martin, F. (2000). Insulin-secreting cells derived from embryonic stem cells normalize glycemia in streptozotocin-induced diabetic mice. *Diabetes* **49**, 157–162.

Treff, N. R., Vincent, R. K., Budde, M. L., Browning, V. L., Magliocca, J. F., Kapur, V., and Odorico, J. S. (2006). Differentiation of embryonic stem cells conditionally expressing neurogenin 3. *Stem Cells* **24**, 2529–2537.

Vaca, P., Berna, G., Araujo, R., Carneiro, E. M., Bedoya, F. J., Soria, B., and Martin, F. (2008). Nicotinamide induces differentiation of embryonic stem cells into insulin-secreting cells. *Exp. Cell Res.* **314**, 969–974.

Vaca, P., Martin, F., Vegara-Meseguer, J. M., Rovira, J. M., Berna, G., and Soria, B. (2006). Induction of differentiation of embryonic stem cells into insulin-secreting cells by fetal soluble factors. *Stem Cells* **24**, 258–265.

Wiles, M. V., and Proetzel, G. (2006). Controlling the differentiation of mouse ES cells *in vitro*. *In* "Embryonic Stem Cells" (E. Notarianni, and M. J. Evans, Eds.), pp. 112–129. Oxford, Oxford University Press.

Wobus, A. M., and Boheler, K. R. (2005). Embryonic stem cells: Prospects for developmental biology and cell therapy. *Physiol. Rev.* **85**, 635–678.

Wobus, A. M., Guan, K., Yang, H. T., and Boheler, K. R. (2002). Embryonic stem cells as a model to study cardiac, skeletal muscle, and vascular smooth muscle cell differentiation. *Methods Mol. Biol.* **185**, 127–156.

Ying, Q. L., Stavridis, M., Griffiths, D., Li, M., and Smith, A. (2003). Conversion of embryonic stem cells into neuroectodermal precursors in adherent monoculture. *Nat. Biotechnol.* **21**, 183–186.

CHAPTER 24

Transgene Expression and RNA Interference in Embryonic Stem Cells

Holm Zaehres and George Q. Daley

Department of Biological Chemistry and Molecular Pharmacology
Harvard Medical School; Division of Pediatric Hematology/Oncology
Children's Hospital Boston and Dana Farber Cancer Institute; Division of Hematology
Brigham and Women's Hospital; Harvard Stem Cell Institute
Boston, Massachusetts

Abstract
1. Retrovirus Expression Vectors and ESCs
2. RNA Interference and ESCs
3. siRNA Expression Vector Design
4. Retrovirus Production
5. Retroviral and Lentiviral Gene Transfer into Mouse and Human ESCs
6. Transgene and siRNA Expression in Mouse and hESCs
7. Biotechnological and Medical Applications
References

Abstract

Over the last 30 years of biomedical research, technologies for transgene expression and gene loss-of-function have been developed from animal model systems to human gene therapy, and increasingly, embryonic stem cells (ESCs) are a key target for these genetic engineering strategies. In this chapter, we describe how retrovirus/lentivirus vectors can be used to transfer and stably express genes or small interfering RNAs in mouse and human ESCs and their progeny, thereby allowing genetic gain- or loss-of-function analysis and cell lineage selection in a multitude of experimental settings to study self-renewal, directed differentiation, and reprogramming.

1. Retrovirus Expression Vectors and ESCs

The first transgenic mouse strain was developed 30 years ago when mouse embryos were exposed to infectious Moloney murine leukemia retrovirus (MoMuLV), and germ line integration and Mendelian transmission of the foreign DNA was demonstrated (Jaenisch, 1976). Recombinant DNA technology has been used to redesign retroviruses as vector systems for efficient gene transfer and stable expression of genomically integrated transgenes in a broad range of mammalian cells (Cepko *et al.*, 1984; Mann *et al.*, 1983; Shimotohno and Temin, 1981). Retroviral vector expression in mouse embryonic stem cells (mESCs) has been studied since their derivation (Evans and Kaufman, 1981) but has proven problematic, because the provirus can be transcriptionally inactivated through a process termed silencing (Cherry *et al.*, 2000; Laker *et al.*, 1998). Initial transcriptional repression of the proviruses is attributed to the effect of trans-acting repressors or the lack of transactivators to start transcription from the regulatory elements in the retroviral long terminal repeat (LTR) and leader region. DNA methylation is the main mechanism of silencing of the retroviral control elements in mESCs undergoing differentiation. This mechanism has been described in particular for murine leukemia virus (MLV)/gammaretrovirus based vectors. The mESC virus MESV (Grez *et al.*, 1990), also termed MSCV (Hawley *et al.*, 1992), allows efficient initial expression in mESCs; however this provirus is also subjected to long-term methylation-dependent silencing during embryogenesis. Additional LTR/leader modifications increased the expression levels in mESCs (Robbins *et al.*, 1997). The LTR/leader configuration of MESV with improved expression in ESCs also predicted vector design for efficient expression profiles in hematopoietic stem and progenitor cells (Baum *et al.*, 1995). MLV-derived vectors have been used for a significant number of clinical human gene therapy applications, in which transgene expression has been reported for several years. In most studies, transgene modification of cells of the blood-forming lineages allowed these cells to have competitive growth advantages, thereby enforcing *in vivo* selection of highly expressing subclones. The lack of significant provirus transcription under nonselective conditions has hampered the use of simple retroviral vectors in transgenic *in vivo* experiments. Silencing of retroviral vectors with MoMuLV–LTR driven transgenes of Oct4, Sox2, Klf4, and c-Myc has also been described in the context of generating mESC-like induced pluripotent stem cells (iPS) from mouse fibroblasts (Takahashi and Yamanaka, 2006).

The advent of more complex vectors based on lentivirus (e.g., Human Immunodeficiency Virus (HIV; Naldini *et al.*, 1996; Poznansky *et al.*, 1991) has improved the prospect of retroviral gene transfer considerably. Lentiviral vectors derived from HIV-1 with the internal phosphoglycerate kinase (PGK) promoter (Hamaguchi *et al.*, 2000), the chicken beta actin/cytomegalovirus (CMV) enhancer

(CAG) promoter (Pfeifer *et al.*, 2002), and the ubiquitin-C-promoter (Lois *et al.*, 2002) have all been used successfully to express transgenes in mESCs and *in vitro* differentiated progeny. Transfer of lentivector-transduced ESCs into blastocysts can generate chimeric mice that express the transgene in multiple tissues. Thus, germ line transmission of lentiviral vector driven transgenes has allowed efficient generation of novel transgenic mouse models.

Human embryonic stem cells (hESCs) have the ability to differentiate along all embryonic and adult developmental lineages, which makes them a valuable model to study early developmental processes and potentially a source of cells for gene and cell therapies (Thomson *et al.*, 1998). Genetic modification of hESCs is playing a key role in driving the directed differentiation and selection of particular cell lineages, thereby exploiting their potential for basic research and regenerative medicine applications. Lentivirus-mediated gene transfer and expression of the green fluorescent protein (GFP) in hESCs has been described with vectors incorporating the EF1α (elongation factor 1 alpha), the PGK, and the CMV promoter (Gropp *et al.*, 2003; Ma *et al.*, 2003; Zaehres *et al.*, 2005). Further developments include the usage of inducible lentiviral vector systems in hESCs (Zhou *et al.*, 2007).

2. RNA Interference and ESCs

RNA interference (RNAi) is an evolutionary conserved cellular pathway of sequence-specific gene silencing at the messenger RNA level. The basic mechanism behind RNAi is the cleavage of a double-stranded RNA (dsRNA) matching a specific gene sequence into short pieces called short interfering RNA (siRNA), which triggers the degradation of mRNA that matches its sequence (Fire *et al.*, 1998; McManus and Sharp, 2002). Knockout of the Dicer endonuclease that processes endogenous small RNAs, including siRNAs, results in lethality early in mouse development, with Dicer1-null embryos depleted of ESCs (Bernstein *et al.*, 2003). Transfection of chemically synthesized siRNA duplexes have been successfully used to suppress different genes in mammalian cells (Elbashir *et al.*, 2001). siRNAs can be expressed from different promoters as fold-back stem-loop structures (small hairpin RNAs (shRNAs)), that give rise to siRNAs after intracellular processing. Several groups have reported RNA polymerase III promoter based vectors for transient and stable expression of siRNAs in mammalian cells (Brummelkamp *et al.*, 2002). An increasing number of reports describe the use of shRNA expression cassettes to generate transgenic RNAi in mESCs and thereby derived mice with knockdown and knockout phenotypes (Kunath *et al.*, 2003). Several lentiviral vector systems have been developed for the stable expression of small interfering RNAs (siRNAs) in mESCs and mice (Rubinson *et al.*, 2003; Tiscornia *et al.*, 2003; Ventura *et al.*, 2004). We and others have demonstrated efficient gene knockdowns in hESCs (H9/WA09) using siRNA transfection

(Hyslop *et al.*, 2005, Vallier *et al.*, 2004) and siRNA expression delivered by retroviral and lentiviral vectors (Zaehres *et al.*, 2005). In this chapter, we introduce detailed protocols to stably express transgenes or siRNAs in mouse and human ESCs and their progeny, thereby allowing genetic gain-of-function or loss-of-function analysis and cell lineage selection.

3. siRNA Expression Vector Design

Recent chapters of *Methods in Enzymology* have described the design of siRNA and shRNA expression vector systems in detail (Li and Rossi, 2005; Sano *et al.*, 2005). Background material is available for the lentiviral *Lentilox* vector system (Rubinson *et al.*, 2003) and *Lentihair* vector system (Stewart *et al.*, 2003) that we are using for siRNA expression in mouse and human ESCs. A strategy is presented below:

- Search for sequences targeting your gene of interest using the siRNA Selection Program at the Whitehead Institute for Biomedical Research, Cambridge, Massachusetts, (http://jura.wi.mit.edu/siRNAext/) or refer to other published siRNA prediction schemes. The consensus sequence should correspond to: $AAGN_{18}TT$. Test three targets for each gene of interest.
- Design two oligonucleotides that will form the shRNA structure upon expression from the U6 promoter:

Sense oligonucleotide: 5' T-(GN_{18})-(TTCAAGAGA)-(^{81}NC)-TTTTTT
Antisense oligonucleotide: Complement of sense
Additional nucleotides can be added to create restriction sites for cloning into the lentiviral vector.

- Anneal sense and antisense oligonucleotide (60 pmol/μl) in annealing buffer (100 mM -acetate, 30 mM HEPES–KOH pH 7.4, 2 mM Mg–acetate) by slowly decreasing the temperature from 70 °C to 4 °C.
- Clone sequences into your lentiviral vector under the control of the U6 promoter.
- Verify the presence of the oligonucleotide in your vector by sequencing.

The RNAi consortium at the BROAD Institute, Cambridge, MA is creating up to 150,000 custom-designed lentiviral vectors that express short hairpin RNAs targeting 15,000 human genes and 15,000 mouse genes. This fundamental resource is made available to scientists worldwide through commercial and academic distributors. All constructs are based on the lentiviral *Lentihair* vector system (Stewart *et al.*, 2003) that we have validated for use in hESCs.

- Search for lentiviral vectors targeting your gene of interest using the RNAi Consortium shRNA Library at the BROAD Institute (http://www.broad.mit.edu/genome_bio/trc/).

4. Retrovirus Production

Retroviral and lentiviral vectors pseudotyped with the vesicular stomatitis virus surface protein (VSV-G) (Burns *et al.*, 1993) can infect all mammalian cells, and the titers produced upon ultracentrifugation are of the same order as virus stocks used in human gene therapy applications (10^8–10^9 infectious particles/ml). Therefore, refer to the local biosafety guidelines of your institution and always handle virus production under BL2/BL2+ conditions. Bleach and autoclave all used plastic materials that have been in contact with virus.

- Plate 293T cells (maintained in Dulbecco's modified Eagle's medium (DMEM) supplemented with 10% fetal bovine serum (FBS) in ten 100-mm tissue culture dishes 24 h before transfection so that they are 80% confluent for transfection.

- Cotransfect 293T cells with 3 μg retroviral/lentiviral vector, 2 μg packaging plasmid, 1 μg VSV-G expression vector (for a three plasmid system) and 18 μl *FuGENE 6* (Roche) per plate according to the suppliers conditions. *FuGENE 6* is nontoxic to the 293T so that there is no need for further medium changes.

- Collect virus supernatant from all plates 48 h or 72 h after transfection by using plastic pipettes and filter supernatant through a 0.45 μM filter.

- Transfer filtered supernatants (35 ml each) to centrifuge tubes (Beckman Polyallomer, 25 × 89 mm) and spin 23,000 rpm for 90 min at 4 °C in a Beckman SW28 swinging bucket rotor.

- Remove supernatant carefully. You normally see a pellet which is opaque at the bottom center of the tube. Add 1 ml DMEM to the tube and keep the paraffin wrapped tubes at 4 °C overnight.

- Resuspend and mix all tubes containing the same virus, aliquot into 10 freeze vials and store virus stocks at −80 °C until use. Avoid multiple freeze–thaws.

- For titer determination, plate 1 × 10^5 293T cells/well in a 6-well plate and add 1 μl, 10 μl, and 100 μl concentrated virus 6 h later; 24 h later, wash twice with PBS and supplement with fresh media. Analyze cells for GFP expression 48 h after infection by fluorescence activated cell sorter (FACS). The titer can be calculated as follows: (percentage of GFP+ cells) × 10^5× (dilution factor) and is represented as IU (infectious units)/ml concentrated vector; 100× concentrated stocks of the *Lentilox* vector have titers between 1 × 10^8 and 1 × 10^9 IU/ml. For vectors without a reporter gene, quantitative real-time (QRT) PCR can determine the number of vector DNA molecules/ml. Use 0.4% paraformaldehyde in your transduced cell solutions to inactivate remaining virus during titer determination.

- Each virus stock should be assayed for replication-competent retrovirus (RCR). Supernatant collected from infected cells (e.g., 293T) should itself be tested for presence of virus 48 h after infection (by transferring supernatant onto another dish of 293T cells). This supernatant should be virus free. The retrovirus literature describes more detailed protocols for ruling out the generation of RCR during viral passage.

5. Retroviral and Lentiviral Gene Transfer into Mouse and Human ESCs

We cultivate our hESC cultures on mouse embryo fibroblast feeder layers (MEFs) (Specialty Media) (Thomson *et al.*, 1998) or on Matrigel in basic fibroblast growth factor (bFGF)-supplemented MEF conditioned medium (Xu *et al.*, 2001). Please refer to the many described protocols for hESC culture.

• Dissociate your hESC culture by trypsinization (0.25% trypsin/5 mM ethylenediaminetetraacetic acid/phosphate buffered saline (EDTA/PBS) for 2–3 min) to a single cell suspension or alternatively disperse cells by collagenase treatment into small cell clumps.

• Plate 1×10^5 cells on a tissue culture 6-well plate pretreated with Matrigel basement membrane matrix (BD). Use MEF conditioned medium supplemented with bFGF (4 ng/ml) to keep the hESCs undifferentiated.

• Add virus supernatant 6 h later with a calculated multiplicity of infection (MOI). The MOI is defined as the number of IU used per cell to be infected. For most high-efficiency gene transfer applications into hESCs use a MOI of 50 (5×10^6 IU on 1×10^5 cells). Evaluate different transduction efficiencies by infecting with different MOIs. For your RNAi knockdown experiments, use the nontargeting vector and/ or a vector targeting GFP, luciferase, or other irrelevant gene as a control.

• Add 6 μg/ml protamine sulfate (Sigma) to your culture to enhance virus binding to the cells. Because of our highly efficient gene transfer rates, we have not tried to improve retroviral transduction further by spin infection.

• Culture cells with the virus for 24 h, wash 3 times with PBS, and then add fresh media.

• On day 3 after infection, measure for transgene activity and continue the culture on MEFs or Matrigel. To evaluate the knockdown efficiency, use QRT PCR analysis of RNA/cDNA extracts and Western blotting of protein extracts from the transduced cells.

• Optional: Determine the provirus copy number by Southern-blot or QRT–PCR analysis of genomic DNA from the transduced cells.

We also infected hESC cultures grown on MEFs during the transduction process. As expected, the gene transfer rates are reduced in comparison to transductions under feeder free conditions, because virus also transduces the feeder cells.

ESCs cultivated in suspension and in the absence of differentiation inhibitors form proliferating and differentiating multicellular clusters known as *embryoid bodies* (EBs) consisting of ectoderm-derived, endoderm-derived, and mesoderm-derived lineages. The EB culture system has proven to be very valuable to recapitulate hematopoiesis *in vitro* and to generate repopulating hematopoietic stem cells (HSCs) from ESCs in the mouse system (Kyba *et al.*, 2003). These protocols can be adapted for hESC cultures:

- To initiate EB hanging drop cultures from genetically modified hESCs, plate drops of 200 hESCs in 30 μl EB differentiation medium (hESC medium without bFGF) on 150-mm non-tissue culture-treated dishes.
- Invert the dishes and incubate for 4 days at 37 °C in 5% CO_2.
- ollect EBs in 10 ml of fresh differentiation medium and transfer to 100-mm bacterial-grade dishes. Culture under slow swirling conditions on a rotating shaker inside of an incubator. Initiate your differentiation conditions of interest.

For lentiviral gene transfer into mESCs the procedures are similar:

- Dissociate your mESC culture by trypsinization.
- Plate 1×10^5 cells on a tissue culture 6-well plate pretreated with Matrigel basement membrane matrix (BD) or gelatin under described feeder-free conditions (mESC medium supplemented with leukemia inhibitory factor (LIF) (1000 U/ml).
- Add virus supernatant 6 h later with a calculated MOI.
- Add 6 μg/ml protamine sulfate (Sigma) to your culture to enhance virus binding.
- Culture cells with the virus for 24 h, wash 3 times with PBS, and then add fresh media.
- On day 3 after infection measure for transgene activity or gene knockdown efficiency.
- To initiate EB hanging drop cultures from genetically modified mESCs plate drops of 100 mESCs in 10 μl EB differentiation medium (mESC medium lacking supplemental LIF) on 150-mm nontissue culture-treated dishes.
- Invert the dishes and incubate for 2 days at 37 °C in 5% CO_2.
- Collect EBs in 10 ml of fresh differentiation medium and transfer to 100-mm bacterial-grade dishes. Culture under slow swirling conditions on a rotating shaker inside of an incubator. Initiate your differentiation conditions of interest.

6. Transgene and siRNA Expression in Mouse and hESCs

We have compared expression from a set of VSV-G pseudotyped and concentrated retroviral and lentiviral vectors in the hESC lines H9/WA09 and HSF6/UC06 (Fig. 24.1). All vectors tested allowed efficient transgene expression in undifferentiated hESCs, and the amount of transduced cells was virus dose dependent. The percentage of enhanced GFP expression in the hESC populations correlates with the MOI for all vectors tested in both hESC lines. At a MOI of 1 we can achieve transgene expression in the range of 1%–15% of cells; at a MOI of 10 in the range of 10%–60%; and at a MOI of 50 in the range of 40%–85%, depending on the retrovirus/lentivirus used. The retroviral vector MFG drives expression from the MoMuLV–LTR and leader region, and the retroviral vector

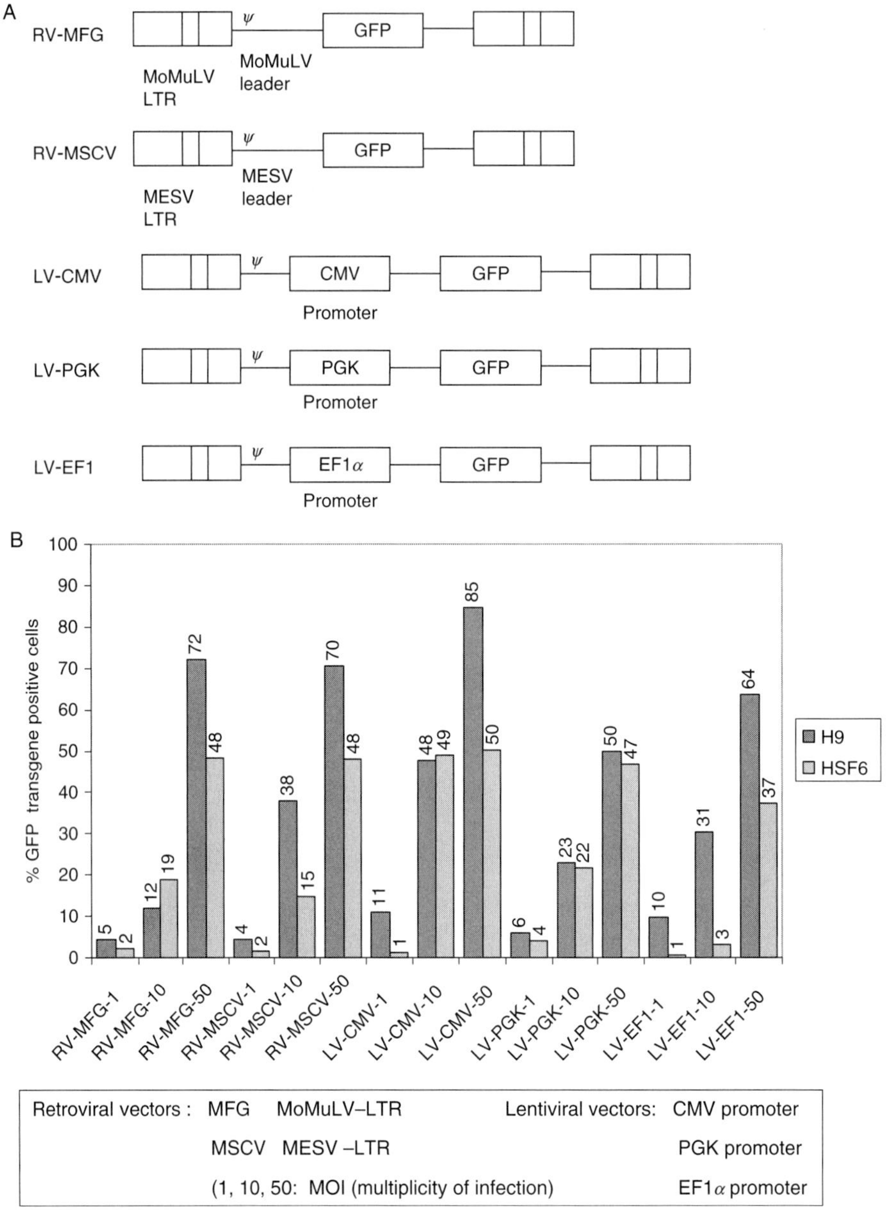

Figure 24.1 Retroviral (RV) and lentiviral (LV) vectors allow viral dose-dependent transgene expression in human embryonic stem cells. (**A**) RV and LV vectors for transgene expression in human embryonic stem cells. The RV express transgenes from the MoMuLV –LTR (long terminal repeat) or the MESV–LTR; the LV vectors express transgenes from the internal CMV (Cytomegalovirus) promoter, the PGK (phosphoglycerate kinase) promoter, or the EF1α (elongation factor 1 alpha) promoter. (**B**) Human embryonic stem cells (H9/WA09 and HSF6/UC06) were transduced with the RV and LV vectors at MOIs (multiplicities of infection) of 1, 10, and 50. Enhanced green fluorescent protein (GFP) transgene positivity was measured 72 h posttransduction by fluorescence activated cell sorter (FACS).

MSCV from the MESV–LTR and leader (Cherry *et al.*, 2000). The MESV–LTR/ leader region has point mutations that allows expression in mESCs (Grez *et al.*, 1990). MSCV, but not MFG, permits expression in mESCs before and throughout *in vitro* differentiation; however, MSCV is also subject to methylation-dependent silencing (Cherry *et al.*, 2000). In contrast to mESCs, both retroviral vector types allowed efficient transgene expression in hESCs (up to 70% GFP positive H9/ WA09 at MOI 50 [Fig. 24.1], see also (Daheron *et al.*, 2004). This is of practical importance, because many available MoMuLV-based vectors can be directly exploited to transduce hESCs in the described manner. To analyze for long-term expression during differentiation of hESCs, we injected MFG vector (MoMuLV– LTR) driven GFP transgenic H9/WA09 in NOD/SCID mice (2×10^6 cells by subcutaneous injection). Teratomas developed within 6 weeks and were found to be GFP fluorescence positive. Similarly, Koch *et al.* (2006) reported expression from a MoMuLV–LTR retroviral vector in hESC derived EBs and throughout neural differentiation. On the other side silencing of retroviral vectors with MoMuLV–LTR and MESV–LTR driven transgenes of Oct4, Sox2, Klf4, and c-Myc has been demonstrated in the context of generating hESC-like iPS from human fibroblasts (Park *et al.*, 2008, Takahashi *et al.*, 2007).

The evaluated lentiviral vectors driving transgene expression from the CMV promoter (*Lentilox* vector; Rubinson *et al.*, 2003), the PGK promoter (*RRL* vector; Zufferey *et al.*, 1998), and the EF1α promoter [*HPV* vector; Philippe Leboulch, Harvard Medical School) allowed a similar virus-dose dependent expression in hESCs, like their gammaretroviral counterparts (Fig. 24.1). In addition to the reported lentiviral vectors incorporating the EF1α promoter (Ma *et al.*, 2003) and the PGK promoter (Gropp *et al.*, 2003), these vectors are suitable for efficient transgene expression in hESC (lines H9/WA09 and HSF6/UC06). We could not detect any changes in the differentiation status of the hESCs by lentiviral transductions itself, as measured by TRA1–60 surface marker staining (Zaehres *et al.*, 2005). When inducing iPS cells from human fibroblasts, lentiviral EF1α promoter driven transgenes of Oct4, Sox2, Nanog, and Lin28 also seem to be silenced in the pluripotent state (Yu *et al.*, 2007).

Whether lentiviral vectors are less prone to silencing then their gammaretroviral counterparts when examined at the single copy provirus level still remains controversial (Ellis, 2005). Based on our data, gammaretrovirus based vectors can be used for efficient *in vitro* genetic modification of hESCs. Cellular defense mechanisms directed against foreign DNA regulatory structures (e.g., DNA methylation-dependent silencing of the LTR of murine oncoretroviruses in murine ESCs) may not be similarly effective in hESCs. Retroviral/lentiviral vector-driven transgene expression in ESCs is affected by position-effect-variegation (PEV), depending on the proviral integration side. To reduce PEV and silencing chromatin insulators and scaffold/matrix-attachment-regions can be incorporated into the vector backbones (Ma *et al.*, 2003; Rivella *et al.*, 2000; Schubeler *et al.*, 1996; Stief *et al.*, 1989). The performance of these genomic bordering modules in ESCs has yet to be evaluated in detail.

We have demonstrated high-efficiency silencing of a GFP transgene and the stem cell-specific transcription factors Oct4/POU5F1 and Nanog using lentiviral siRNA expression in hESCs (Zaehres *et al.*, 2005). As expected, gene knockdown of Oct4 and Nanog but not of GFP promoted differentiation into cells expressing trophectodermal and endodermal markers, thereby demonstrating a conserved role for these factors in hESC self-renewal. Our data demonstrated that the mouse, as well as the human, U6 promoter can be used to drive efficient expression of siRNAs/shRNAs in hESCs (Fig. 24.2A). An expression vector with the RNA polymerase III H1 promoter has proven to work as well to express siRNAs in hESCs (Vallier *et al.*, 2004). We also evaluated potential apoptotic effects of lentiviral transduction and siRNA expression on the hESC cultures; Annexin V staining was only slightly increased in the RNAi lentivector-modified cells with MOIs up to 50, suggesting that RNAi is well-tolerated.

Expression cassettes for siRNAscan also be incorporated for stable expression in ESCs by site-specific or homologous recombination. Episomal expression of siRNAs/shRNAs using adenovirus vectors might increase the efficiency of delivery but does not seem to be advantageous over direct delivery of chemically synthesized siRNAs for applications that entail transient genetic modification.

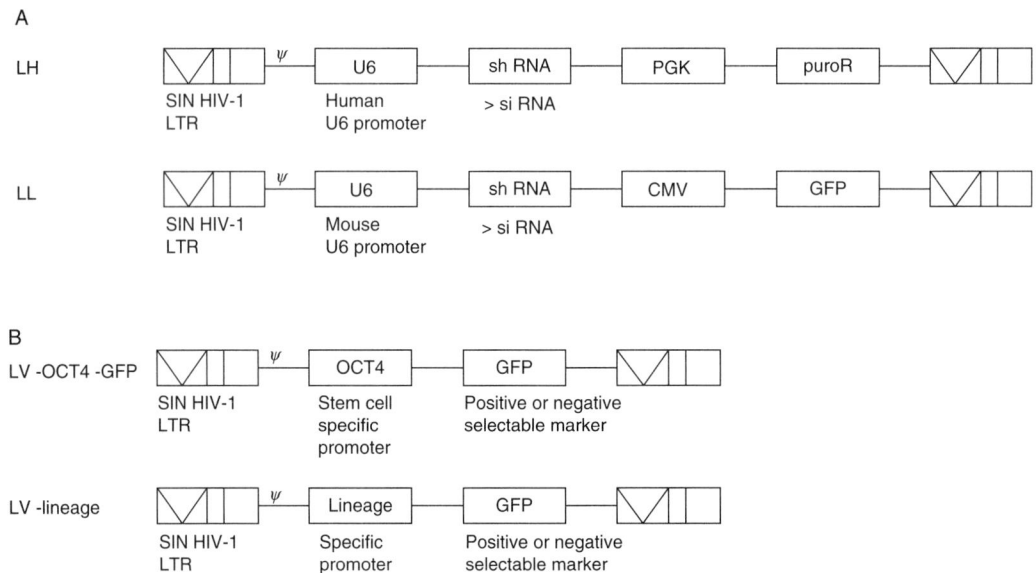

Figure 24.2 (**A**) LV vectors for expression of small interfering RNAs to mediate gene silencing in embryonic stem cells. The LV vectors express small hairpin RNAs (shRNAs) from the mouse or human U6 promoter, that are processed to small interfering RNAs (siRNAs), and further contain a selection marker to monitor vector presence. (**B**) LV vectors for lineage selection in ESCs and their progeny. LV vectors for lineage selection incorporate positive or negative selectable markers whose expression is directed by stem cell-specific (e.g., Oct4 promoter) or lineage-specific promoter/enhancer elements.

RNAi has been reported to vary in efficiency in different developmental stages derived from mESCs with the same siRNA expression cassette (Oberdoerffer *et al.*, 2005). Variable knockdown efficiency might also be observed when deriving specified lineages from hESCs. In our current experiments, U6 promoter-driven knockdowns are maintained during EB differentiation of ESCs.

To derive specified cell lineages and to produce and propagate pure populations of specific cell types from a complex culture of differentiating ESCs, lineage selection methodologies must be applied to the culture systems. Besides gene targeting by classic homologous recombination (Zwaka and Thomson, 2003) essentially all strategies employ vectors with positive or negative selectable markers whose expression is directed by stem cell or lineage specific promoter/enhancer elements (Klug *et al.*, 1996). We have developed a lentiviral vector expressing enhanced GFP under the control of 3.2 kb of the upstream promoter sequences of the human POU5F1/Oct4 gene (Fig. 24.2B). We transduced hESCs (WA09/H9) with this vector at a low MOI, isolated single cell clones using FACS, and expanded them in culture. We identified clones in which GFP expression correlated with the differentiation status of the stem cells, as measured by surface marker expression (TRA1–60). Low MOI infections seem to be necessary to conserve the lineage specificity of the promoter element in the lentiviral vector because multiple proviral integrations increase the potential for positional influences on promoter fidelity. The lentiviral vector itself has a self-inactivating (SIN) LTR configuration. It does not incorporate further promoter/enhancer elements from its HIV parent that could lead to promoter interference and derangement of lineage specificity of the internal promoter. This concept of lineage selection using lentiviral vector technology should be widely applicable to hESCs. Our strategy is as follows:

- Clone a positive or negative selectable marker under the control of a lineage-specific promoter/enhancer element in a (SIN-) lentiviral vector backbone (Fig. 24.2B).
- Transduce hESCs with lentivirus at a low MOI.
- Sort single cell clones and expand the undifferentiated stem cell population.
- Evaluate the cell population for the selectable marker upon differentiation.

7. Biotechnological and Medical Applications

Laboratories can now routinely use retroviral/lentiviral vector technology for transgene and siRNA expression in mouse and human ESCs and their differentiated progeny. We are especially interested in gain-of-function and loss-of-function analysis of transcription factors involved in self-renewal, and the developmental transitions of ESCs to HSCs, and to germ cells. Retroviral/lentiviral expression and RNAi libraries can be used to carry out genome-wide screening experiments in ESCs to identify new factors involved in self-renewal and directed differentiation.

Engineering of cardiomyocytes, blood, neurons, pancreatic cells or a multitude of other lineages from ESCs depends on defined *in vitro* differentiation protocols and a pristine starting population of undifferentiated ESCs. Transgene expression and gene knockdown technologies are crucial to the development of these *in vitro* differentiation protocols, and lineage selection methodologies will be a requirement for clinical applications. Vectors designed for transgene expression in ESCs might also be suitable for efficient expression in adult stem cells, thereby facilitating human gene therapy applications. Transgenic mice with constitutive, inducible, conditional, and cell type-specific gain-of-function or loss-of-function phenotypes have proven invaluable as advanced models for human diseases and for the evaluation of drug, protein, gene, and cell therapies. Transgenic hESCs and their *in vitro* differentiation into defined cell lineages will have a similar impact as model systems for human diseases, including the study of cancer stem cells. This immediate impact is independent of the long-term goal of providing stem cell derived allogeneic or autologous cell grafts, which may not have clinical impact for a decade or longer. Knowledge of retrovirus and lentivirus vector expression in ESCs has captured broad interest when applied to reprogram human somatic cells (Park *et al.*, 2008, Takahashi *et al.*, 2007, Yu *et al.*, 2007, Zaehres and Schöler, 2007), thereby providing potential alternative pluripotent stem cell sources. The lessons learned from transgenic ESC systems and their self-renewal properties might be further productively applied to discover and modulate endogenous regenerative cell programs in the human body and to reprogram somatic cells.

Acknowledgments

We want to thank Sheila Stewart and Philippe Leboulch for providing us with viral vectors and virus production protocols, variations of which are described herein. Joseph Itskovitz-Eldor was instrumental in launching our efforts to culture human embryonic stem cells. This work was supported by grants from the National Institutes of Health, the NIH Director's Pioneer Award of the NIH Roadmap for Medical Research, the Burroughs Wellcome Fund, and philanthropic funds from the Bekenstein Family and the Thomas Anthony Pappas Charitable Foundation.

References

Baum, C., Hegewisch-Becker, S., Eckert, H. G., Stocking, C., and Ostertag, W. (1995). Novel retroviral vectors for efficient expression of the multidrug resistance (mdr-1) gene in early hematopoietic cells. *J. Virol.* **69,** 7541–7547.

Bernstein, E., Kim, S. Y., Carmell, M. A., Murchison, E. P., Alcorn, H., Li, M. Z., Mills, A. A., Elledge, S. J., Anderson, K. V., and Hannon, G. J. (2003). Dicer is essential for mouse development. *Nat. Genet.* **35,** 215–217.

Brummelkamp, T. R., Bernards, R., and Agami, R. (2002). A system for stable expression of short interfering RNAs in mammalian cells. *Science* **296,** 550–553.

Burns, J. C., Friedmann, T., Driever, W., Burrascano, M., and Yee, J. K. (1993). Vesicular stomatitis virus G glycoprotein pseudotyped retroviral vectors: Concentration to very high titer and efficient gene transfer into mammalian and nonmammalian cells. *Proc. Natl. Acad. Sci. USA* **90,** 8033–8037.

Cepko, C. L., Roberts, B. E., and Mulligan, R. C. (1984). Construction and applications of a highly transmissible murine retrovirus shuttle vector. *Cell* **37,** 1053–1062.

Cherry, S. R., Biniszkiewicz, D., van Parijs, L., Baltimore, D., and Jaenisch, R. (2000). Retroviral expression in embryonic stem cells and hematopoietic stem cells. *Mol. Cell. Biol.* **20,** 7419–7426.

Daheron, L., Opitz, S. L., Zaehres, H., Lensch, W. M., Andrews, P. W., Itskovitz-Eldor, J., and Daley, G. Q. (2004). LIF/STAT3 signaling fails to maintain self-renewal of human embryonic stem cells. *Stem Cells* **22,** 770–778.

Elbashir, S. M., Harborth, J., Lendeckel, W., Yalcin, A., Weber, K., and Tuschl, T. (2001). Duplexes of 21-nucleotide RNAs mediate RNA interference in cultured mammalian cells. *Nature* **411,** 494–498.

Ellis, J. (2005). Silencing and variegation of gammaretrovirus and lentivirus vectors. *Hum. Gene Ther.* **16,** 1241–1246.

Evans, M. J., and Kaufman, M. H. (1981). Establishment in culture of pluripotential cells from mouse embryos. *Nature* **292,** 154–156.

Fire, A., Xu, S., Montgomery, M. K., Kostas, S. A., Driver, S. E., and Mello, C. C. (1998). Potent and specific genetic interference by double-stranded RNA in *Caenorhabditis elegans*. *Nature* **391,** 806–811.

Grez, M., Akgun, E., Hilberg, F., and Ostertag, W. (1990). Embryonic stem cell virus, a recombinant murine retrovirus with expression in embryonic stem cells. *Proc. Natl. Acad. Sci. USA* **87,** 9202–9206.

Gropp, M., Itsykson, P., Singer, O., Ben-Hur, T., Reinhartz, E., Galun, E., and Reubinoff, B. E. (2003). Stable genetic modification of human embryonic stem cells by lentiviral vectors. *Mol. Ther.* **7,** 281–287.

Hamaguchi, I., Woods, N. B., Panagopoulos, I., Andersson, E., Mikkola, H., Fahlman, C., Zufferey, R., Carlsson, L., Trono, D., and Karlsson, S. (2000). Lentivirus vector gene expression during ES cell-derived hematopoietic development *in vitro*. *J. Virol.* **74,** 10778–10784.

Hawley, R. G., Fong, A. Z., Burns, B. F., and Hawley, T. S. (1992). Transplantable myeloproliferative disease induced in mice by an interleukin 6 retrovirus. *J. Exp. Med.* **176,** 1149–1163.

Hyslop, L., Stojkovic, M., Armstrong, L., Walter, T., Stojkovic, P., Przyborski, S., Herbert, M., Murdoch, A., Strachan, T., and Lako, M. (2005). Downregulation of NANOG induces differentiation of human embryonic stem cells to extraembryonic lineages. *Stem Cells* **23,** 1035–1043.

Jaenisch, R. (1976). Germ line integration and Mendelian transmission of the exogenous Moloney leukemia virus. *Proc. Natl. Acad. Sci. USA* **73,** 1260–1264.

Klug, M. G., Soonpaa, M. H., Koh, G. Y., and Field, L. J. (1996). Genetically selected cardiomyocytes from differentiating embronic stem cells form stable intracardiac grafts. *J. Clin. Invest.* **98,** 216–224.

Koch, P., Siemen, H., Biegler, A., Itskovitz-Eldor, J., and Brustle, O. (2006). Transduction of human embryonic stem cells by ecotropic retroviral vectors. *Nucleic Acids Res.* **34,** e120.

Kunath, T., Gish, G., Lickert, H., Jones, N., Pawson, T., and Rossant, J. (2003). Transgenic RNA interference in ES cell-derived embryos recapitulates a genetic null phenotype. *Nat. Biotechnol.* **21,** 559–561.

Kyba, M., Perlingeiro, R. C., and Daley, G. Q. (2003). Development of hematopoietic repopulating cells from embryonic stem cells. *Methods Enzymol.* **365,** 114–129.

Laker, C., Meyer, J., Schopen, A., Friel, J., Heberlein, C., Ostertag, W., and Stocking, C. (1998). Host cis-mediated extinction of a retrovirus permissive for expression in embryonal stem cells during differentiation. *J. Virol.* **72,** 339–348.

Li, M. J., and Rossi, J. J. (2005). Lentiviral vector delivery of recombinant small interfering RNA expression cassettes. *Methods Enzymol.* **392,** 218–226.

Lois, C., Hong, E. J., Pease, S., Brown, E. J., and Baltimore, D. (2002). Germline transmission and tissue-specific expression of transgenes delivered by lentiviral vectors. *Science* **295,** 868–872.

Ma, Y., Ramezani, A., Lewis, R., Hawley, R. G., and Thomson, J. A. (2003). High-level sustained transgene expression in human embryonic stem cells using lentiviral vectors. *Stem Cells* **21,** 111–117.

Mann, R., Mulligan, R. C., and Baltimore, D. (1983). Construction of a retrovirus packaging mutant and its use to produce helper-free defective retrovirus. *Cell* **33**, 153–159.

McManus, M. T., and Sharp, P. A. (2002). Gene silencing in mammals by small interfering RNAs. *Nat. Rev. Genet.* **3**, 737–747.

Naldini, L., Blomer, U., Gallay, P., Ory, D., Mulligan, R., Gage, F. H., Verma, I. M., and Trono, D. (1996). *In vivo* gene delivery and stable transduction of nondividing cells by a lentiviral vector. *Science* **272**, 263–267.

Oberdoerffer, P., Kanellopoulou, C., Heissmeyer, V., Paeper, C., Borowski, C., Aifantis, I., Rao, A., and Rajewsky, K. (2005). Efficiency of RNA interference in the mouse hematopoietic system varies between cell types and developmental stages. *Mol. Cell. Biol.* **25**, 3896–3905.

Park, I. H., Zhao, R., West, J. A., Yabuuchi, A., Huo, H., Ince, T. A., Lerou, P. H., Lensch, M. W., and Daley, G. Q. (2008). Reprogramming of human somatic cells to pluripotency with defined factors. *Nature* **451**, 141–146.

Pfeifer, A., Ikawa, M., Dayn, Y., and Verma, I. M. (2002). Transgenesis by lentiviral vectors: Lack of gene silencing in mammalian embryonic stem cells and preimplantation embryos. *Proc. Natl. Acad. Sci. USA* **99**, 2140–2145.

Poznansky, M., Lever, A., Bergeron, L., Haseltine, W., and Sodroski, J. (1991). Gene transfer into human lymphocytes by a defective human immunodeficiency virus type 1 vector. *J. Virol.* **65**, 532–536.

Rivella, S., Callegari, J. A., May, C., Tan, C. W., and Sadelain, M. (2000). The cHS4 insulator increases the probability of retroviral expression at random chromosomal integration sites. *J. Virol.* **74**, 4679–4687.

Robbins, P. B., Yu, X. J., Skelton, D. M., Pepper, K. A., Wasserman, R. M., Zhu, L., and Kohn, D. B. (1997). Increased probability of expression from modified retroviral vectors in embryonal stem cells and embryonal carcinoma cells. *J. Virol.* **71**, 9466–9474.

Rubinson, D. A., Dillon, C. P., Kwiatkowski, A. V., Sievers, C., Yang, L., Kopinja, J., Zhang, M., McManus, M. T., Gertler, F. B., Scott, M. L., and Van Parijs, L. (2003). A lentivirus-based system to functionally silence genes in primary mammalian cells, stem cells and transgenic mice by RNA interference. *Nat. Genet.* **33**, 401–406.

Sano, M., Kato, Y., Akashi, H., Miyagishi, M., and Taira, K. (2005). Novel methods for expressing RNA interference in human cells. *Methods Enzymol.* **392**, 97–112.

Schubeler, D., Mielke, C., Maass, K., and Bode, J. (1996). Scaffold/matrix-attached regions act upon transcription in a context-dependent manner. *Biochemistry* **35**, 11160–11169.

Shimotohno, K., and Temin, H. M. (1981). Formation of infectious progeny virus after insertion of herpes simplex thymidine kinase gene into DNA of an avian retrovirus. *Cell* **26**, 67–77.

Stewart, S. A., Dykxhoorn, D. M., Palliser, D., Mizuno, H., Yu, E. Y., An, D. S., Sabatini, D. M., Chen, I. S., Hahn, W. C., Sharp, P. A., Weinberg, R. A., and Novina, C. D. (2003). Lentivirus-delivered stable gene silencing by RNAi in primary cells. *RNA* **9**, 493–501.

Stief, A., Winter, D. M., Stratling, W. H., and Sippel, A. E. (1989). A nuclear DNA attachment element mediates elevated and position-independent gene activity. *Nature* **341**, 343–345.

Takahashi, K., Tanabe, K., Ohnuki, M., Narita, M., Ichisaka, T., Tomoda, K., and Yamanaka, S. (2007). Induction of pluripotent stem cells from adult human fibroblasts by defined factors. *Cell* **131**, 861–872.

Takahashi, K., and Yamanaka, S. (2006). Induction of pluripotent stem cells from mouse embryonic and adult fibroblast cultures by defined factors. *Cell* **126**, 663–676.

Thomson, J. A., Itskovitz-Eldor, J., Shapiro, S. S., Waknitz, M. A., Swiergiel, J. J., Marshall, V. S., and Jones, J. M. (1998). Embryonic stem cell lines derived from human blastocysts. *Science* **282**, 1145–1147.

Tiscornia, G., Singer, O., Ikawa, M., and Verma, I. M. (2003). A general method for gene knockdown in mice by using lentiviral vectors expressing small interfering RNA. *Proc. Natl. Acad. Sci. USA* **100**, 1844–1848.

Vallier, L., Rugg-Gunn, P. J., Bouhon, I. A., Andersson, F. K., Sadler, A. J., and Pedersen, R. A. (2004). Enhancing and diminishing gene function in human embryonic stem cells. *Stem Cells* **22,** 2–11.

Ventura, A., Meissner, A., Dillon, C. P., McManus, M., Sharp, P. A., Van Parijs, L., Jaenisch, R., and Jacks, T. (2004). Cre-lox-regulated conditional RNA interference from transgenes. *Proc. Natl. Acad. Sci. USA* **101,** 10380–10385.

Xu, C., Inokuma, M. S., Denham, J., Golds, K., Kundu, P., Gold, J. D., and Carpenter, M. K. (2001). Feeder-free growth of undifferentiated human embryonic stem cells. *Nat. Biotechnol.* **19,** 971–974.

Yu, J., Vodyanik, M. A., Smuga-Otto, K., Antosiewicz-Bourget, J., Frane, J. L., Tian, S., Nie, J., Jonsdottir, G. A., Ruotti, V., Stewart, R., *et al.* (2007). Induced pluripotent stem cell lines derived from human somatic cells. *Science* **318,** 1917–1920.

Zaehres, H., Lensch, M. W., Daheron, L., Stewart, S. A., Itskovitz-Eldor, J., and Daley, G. Q. (2005). High-efficiency RNA interference in human embryonic stem cells. *Stem Cells* **23,** 299–305.

Zaehres, H., and Schöler, H. R. (2007). Induction of pluripotency: From mouse to human. *Cell* **131,** 834–835.

Zhou, B. Y., Ye, Z., Chen, G., Gao, Z. P., Zhang, Y. A., and Cheng, L. (2007). Inducible and reversible transgene expression in human stem cells after efficient and stable gene transfer. *Stem Cells* **25,** 779–789.

Zufferey, R., Dull, T., Mandel, R. J., Bukovsky, A., Quiroz, D., Naldini, L., and Trono, D. (1998). Self-inactivating lentivirus vector for safe and efficient *in vivo* gene delivery. *J. Virol.* **72,** 9873–9880.

Zwaka, T. P., and Thomson, J. A. (2003). Homologous recombination in human embryonic stem cells. *Nat. Biotechnol.* **21,** 319–321.

CHAPTER 25

Lentiviral Vector-Mediated Gene Delivery into Human Embryonic Stem Cells

Michal Gropp and Benjamin Reubinoff

The Hadassah Human Embryonic Stem Cell Research Center
The Goldyne Savad Institute of Gene Therapy and the Department of Gynecology
Hadassah University Medical Center
Jerusalem 91120, Israel

Abstract
1. Introduction
    1.1. Development of lentiviral-based vectors
    1.2. Design of HIV-1-based vectors
    1.3. Transduction of hESCs by lentiviral-based vectors
    1.4. Potential applications of gene delivery into hESCs by lentiviral-based vectors
2. Design of HIV-1 Based Vectors for Transduction of hESCs
    2.1. Choice of HIV-1 based vector
    2.2. Choice of internal promoter
    2.3. Virus pseudotyping
3. Generation of Recombinant Viral Particles
    3.1. Plasmid constructs
    3.2. Production of viral particles
    3.3. FuGene-6/TransIT-LT1 transfection protocol
    3.4. Collection of recombinant viral particles
    3.5. Concentration of viral particles
4. Transduction of hESCs
    4.1. Transduction of hESCs cultured in serum-containing medium and passaged as cell clusters

    4.2. Transduction of hESCs cultured in serum-free medium on feeders and
         passaged as single cells
    4.3. Transduction of hESCs cultured in serum-free medium without feeders,
         and passaged enzymatically as small clusters
 5. Measurement of Transduction Efficiency
 6. Enrichment for Transduced hESCs Expressing High Levels of the Transgene
    6.1. Mechanical enrichment of transduced hESC colonies expressing high levels
         of a reporter gene
    6.2. Isolation of transduced hESC clones expressing high levels of a reporter
         gene by limiting dilution
    6.3. Enrichment of transduced hESCs expressing high levels of a reporter gene
         by FACS sorting
    6.4. Enrichment of transduced hESC cells expressing high levels of the transgene
         using antibiotic selection
 7. Determination of Viral Titer
    References

Abstract

Human embryonic stem cells (hESCs) are pluripotent cells derived from the inner cell mass of preimplantation embryos. These cells can be cultured for long periods as undifferentiated cells, and still retain their potential to give rise to cell types representing all three germinal layers. Given their unique properties, hESCs are expected to serve as an invaluable tool for basic and applied research. However, to exploit their remarkable potential, the development of effective strategies for genetic modification of hESCs is required. Lentiviral-based vectors offer an attractive system for efficient gene delivery into hESCs. These vectors are derived from lentiviruses, a group of complex retroviruses that cause slow chronic immunodeficiency diseases in humans and animals. Gene delivery into hESCs by vectors derived from lentiviruses has the following advantages: (1) lentiviral-vectors efficiently transduce hESCs, (2) they integrate into the host-cell genome, thus promoting stable transgene expression, (3) transgene expression is not significantly silenced in hESCs, and (4) transduced hESCs retain their self-renewal and pluripotent potential. In recent years, we and others have developed protocols for efficient transduction of hESCs by advanced modified replication-defective lentiviral-based vectors. Transduction of hESCs by these vectors resulted in high and stable transgene expression that was maintained over long periods of undifferentiated cultivation, as well as following differentiation. This chapter focuses on methods for the use of lentiviral-based vectors for gene delivery into hESCs.

1. Introduction

Human embryonic stem cells (hESCs) are pluripotent cells derived from the inner cell mass of preimplantation embryos (Reubinoff *et al.*, 2000; Thomson *et al.*, 1998). These cells can self-renew for prolonged periods in culture, and yet retain their potential to differentiate into cells representing all three germinal layers both *in vivo* and *in vitro*. Given their unique properties, hESCs are expected to serve as an invaluable tool for the study of early human development. Moreover, these cells could be potentially used as an unlimited source of transplantable cells for cell replacement therapies.

However, to exploit the potential of hESCs, the development of effective methods for genetic modification of these cells is required. Several strategies have been used to introduce exogenous genes into hESCs. These strategies included the use of nonviral delivery methods such as transfection or electroporation, and viral vectors such as lentiviruses. While nonviral systems are considered safer than viral systems, they mostly promote gene transfer at low efficiencies and with transient gene expression. In contrast, lentiviral-based vectors were reported to promote highly efficient stable gene transfer into hESCs (see review by Menendez and colleagues (Menendez *et al.*, 2005)).

In this chapter, we will focus on the use of lentiviral-based vectors as gene delivery tools into hESCs. More specifically, we will concentrate on the design of lentiviral-based vectors suitable for transduction of hESCs, and on the methods for production and delivery of recombinant virus particles into hESCs.

1.1. Development of lentiviral-based vectors

Since the first report twelve years ago, on the development of a gene delivery system based on the human immunodeficiency virus type I (HIV-1) (Naldini *et al.*, 1996a,b), vectors derived from lentiviruses have proved to be efficient gene delivery tools into a variety of cell types both *in vitro* and *in vivo*.

Lentiviral-based vectors are derived from lentiviruses, a group of complex retroviruses that cause slow chronic diseases in humans and animals. This group includes the HIV-1, which causes acquired immunodeficiency syndrome (AIDS) in humans, as well as lentiviruses which are nonpathogenic to humans, such as the simian immunodeficiency virus (SIV) and the feline immunodeficiency virus (FIV). All lentiviruses share a common infection mechanism: The virus binds to specific receptors on the host-cell membrane and enters the cell. In the cytoplasm, the viral RNA is uncoated and reverse transcribed by its own reverse transcriptase into double-stranded proviral DNA. The proviral DNA then enters the nucleus and integrates into the host cell genome. The virus remains permanently integrated and uses the host-cell cellular machinery for its replication and expression. A detailed description of the biology of lentiviruses can be found elsewhere (Buchschacher, 2001; Goff, 2001; Palu *et al.*, 2000).

Lentiviral-based vectors were designed to take advantage of the unique characteristics of lentiviruses: (1) efficient infection of both dividing and nondividing cells; (2) integration of the virus into the genome of the infected cell, enabling stable expression and germline transmission of the transgene; (3) lack of immunogenicity; and (4) the unique sequence arrangement of the viral genome, which allows easy genetic manipulation of the regulatory and the coding sequences.

1.2. Design of HIV-1-based vectors

The current advanced HIV-1-based vectors were designed to improve viral performance. Moreover, since HIV-1 causes a severe disease in humans, great effort was invested in the design of safer HIV-1-based vectors. Accordingly, several steps were carried out to improve vector biosafety and performance: First, all pathogenic coding sequences were deleted from the vector, resulting in a replication-defective vector containing only the transgene and several essential regulatory viral sequences, such as the encapsidation signal and the viral LTR. Second, the proteins necessary for the early steps of viral infection (entering into the host cell, reverse transcription and integration) were provided *in trans* by two additional plasmids: a packaging plasmid expressing the *gag*, *pol*, and *rev* genes, and an envelope plasmid expressing a heterologous envelope glycoprotein of the vesicular stomatitis virus (VSV-G). The use of the heterologous envelope prevented generation of wild-type virus. It was also beneficial because it broadened the viral host range and increased the stability of the viral particles (Naldini *et al.*, 1996a,b). Third, a large deletion was introduced into the U3 region of the viral LTR abolishing the viral promoter/enhancer activity. The self-inactivating (SIN) vector that was generated contained a heterologous internal promoter driving the expression of the transgene (Miyoshi *et al.*, 1998; Zufferey *et al.*, 1998). The steps described earlier resulted in a vector that could only go one round of infection and integration, a process termed "transduction." Moreover, they minimized the risk of generation of wild-type HIV-1 by recombination.

The HIV-1-based vector was further improved by the addition of two regulatory sequences: (1) the central polypurine tract (cPPT) sequence, that is found within the *pol* gene of HIV-1, was reinserted into the vector upstream of the internal promoter. cPPT is necessary for the nuclear import of HIV-1 into the nucleus of the host cell. Insertion of this element into the vector greatly increased transduction efficiencies (Follenzi *et al.*, 2000); (2) the post transcriptional regulatory element of the woodchuck hepatitis virus (WPRE) was introduced into the vector downstream of the transgene. WPRE was shown to increase transgene expression by a mechanism that is still unclear, and might involve increasing the number of nuclear RNA transcripts or enhancing mRNA export into the cytoplasm (Zufferey *et al.*, 1999).

To date, the advanced recombinant viral particles are generated by transient cotransfection into producer cells of three plasmids: (1) the advanced modified replication-defective, SIN transfer vector, expressing the transgene from an internal promoter; (2) a packaging plasmid expressing the *gag, pol,* and *rev* genes; and (3) an envelope plasmid expressing VSV-G (see Fig. 25.1).

1.3. Transduction of hESCs by lentiviral-based vectors

Vectors based on retroviruses such as the Moloney murine leukemia virus (MoMLV) were used for transduction of mouse ES cells. However, it was found that the expression of the transgene was silenced over time (Niwa *et al.*, 1983; Pannell and Ellis, 2001). The recent reports on the development of advanced HIV-1-based vectors, that could efficiently transduce a variety of cell types, promoted the study of these vectors as potential gene transfer vectors into ESCs. Several groups have reported that advanced lentiviral-based vectors pseudotyped with VSV-G could efficiently transduce mouse (Hamaguchi *et al.*, 2000; Pfeifer *et al.*, 2002), and cynomolgus monkey (Asano *et al.*, 2002) ESCs. Furthermore, transgene expression was not silenced throughout undifferentiated proliferation as well as following differentiation *in vivo* and *in vitro*. These results indicated that lentiviral-

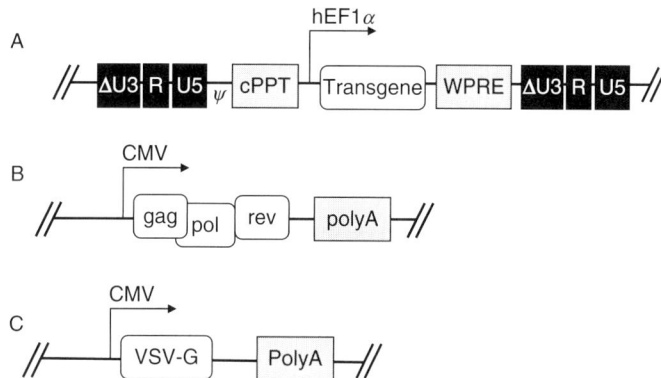

Figure 25.1 The advanced modified HIV-1-based vector system. The figure depicts a schematic representation of the three plasmids used to generate the recombinant HIV-1 based viral particles. (A) The transfer vector is a replication-defective advanced HIV-1 based vector, expressing the transgene from an internal constitutive promoter such as the human elongation factor 1α (hEF1α) promoter. This vector is self-inactivating, due to a large deletion in the 3′UTR region of its LTR, which abolishes the viral promoter/enhancer activity. The performance of the transfer vector is increased by the addition of two regulatory elements: the central polypurine tract (cPPT) sequence that is found within the *pol* gene of HIV-1, which increases viral integration, and the post transcriptional regulatory element of the woodchuck hepatitis virus (WPRE) which increases transgene expression. (B) The packaging plasmid expresses the three viral genes *gag, pol,* and *rev,* from a constitutive cytomegalus (CMV) promoter. The proteins provided by this plasmid are essential for the early steps of viral infection (entry into the host cell, reverse transcription, and viral integration into the genome of the host cell). (C) The envelope plasmid expresses the heterologous envelope glycoprotein of the VSV-G from the CMV promoter. The use of the heterologous envelope prevents generation of wild-type virus, broadens the viral host range and increases the stability of the viral particles.

based vectors may prove advantageous over MoMLV vectors for transduction of ES cells, thus encouraging the use of these vectors for genetic modifications of hESCs.

In recent years, we and others have developed protocols for efficient transduction of hESCs by advanced modified HIV-1-based vectors pseudotyped with VSV-G (Clements *et al.*, 2006; Gropp *et al.*, 2003; Ma *et al.*, 2003; Pfeifer *et al.*, 2002; Rodriguez *et al.*, 2007; Suter *et al.*, 2005; Xiong *et al.*, 2005; Zaehres *et al.*, 2005). Transduction of hESCs by these vectors resulted in high and stable transgene expression that was maintained over long periods of undifferentiated cultivation, as well as following differentiation (Gropp *et al.*, 2003).

Thus, HIV-1-based vectors have proved to be efficient gene delivery tools into hESCs. Moreover, these vectors confer stable transgene expression within hESCs that is sustained throughout prolonged proliferation of undifferentiated hESCs, as well as following differentiation of the cells *in vitro* and *in vivo*. Most importantly, the transduced hESCs retain their potential for self-renewal and their pluripotency.

1.4. Potential applications of gene delivery into hESCs by lentiviral-based vectors

Transduction of hESCs by lentiviral-based vectors could be employed for various applications, both in regenerative medicine and basic research.

Lentiviral-based vectors over-expressing specific transcription factors could direct the differentiation of hESCs toward a desired cell lineage, or cell type. In addition, lentiviral-based vectors expressing a reporter or a selectable marker under the control of a tissue-specific or cell-type specific promoter will enable to select and thus enrich a specific cell type (Gallo *et al.*, 2008; Huber *et al.*, 2007). The ability to direct the differentiation of hESCs or to derive enriched populations of specific types of differentiated cells, may allow the exploitation of hESCs as an unlimited source of cells for transplantation.

Transduced hESCs could be used not only for cell therapy but also for gene therapy. For this purpose, hESC transduced by lentiviral-based vectors expressing genes that are defective in specific genetic diseases, will be transplanted and could potentially correct these diseases. Inherited mutations can also be corrected using gene-editing lentiviral vectors. In this system hESCs are transduced with two lentiviral vectors: the first vector expresses an engineered zinc-finger nuclease, designed to bind and excise a specific mutated genomic sequence, while the second vector provides the template DNA for mutation correction. Transduction of cells with these two vectors stimulates homologous recombination that leads to repair of a specific mutation (Lombardo *et al.*, 2007).

Transduction of hESCs by lentiviral-based vectors constitutively expressing a reporter gene (GFP, RFP, or luciferase) will generate reporter hESC cell lines, allowing easy detection and monitoring the fate of transplanted hESCS within the animals (Gropp *et al.*, 2003; Li *et al.*, 2008; Rodriguez *et al.*, 2007).

Finally, lentiviral-based vectors could be employed to silence genes of interest in hESCS (Clements *et al.*, 2006; Gropp and Reubinoff, 2007; Rodriguez *et al.*, 2007; Suter *et al.*, 2005; Xiong *et al.*, 2005; Zaehres *et al.*, 2005). Thus, they might prove invaluable for studying the role of specific genes involved in early human development.

2. Design of HIV-1 Based Vectors for Transduction of hESCs

2.1. Choice of HIV-1 based vector

Thus far, only HIV-1-based lentiviral vectors were reported to transduce hESCs. We will therefore concentrate on the design and production of these vectors. Most groups have used the safer advanced SIN transfer HIV-1-based vectors, containing cPPT and WPRE, for transduction of hESCs.

The simple monocistronic HIV-1-based vectors express a single transgene from an internal promoter. However, for various applications, it is desired to coexpress two genes. In vectors coexpressing two genes, usually the first gene is the gene of interest, while the second gene is the reporter/selection gene. This design allows monitoring of transduction efficiencies and transgene expression, or selection for cells expressing the transgene.

2.1.1. Monocistronic lentiviral–based vectors:

These vectors are suitable for generation of reporter hESC lines. Transduction of hESCs with HIV-1 based vectors expressing the reporter gene from a constitutive promoter is valuable for monitoring the fate of hESCs following transplantation into animals (Gropp *et al.*, 2003; Li *et al.*, 2008). Lately, two groups have reported on the generation of a lineage-specific reporter hESC line using HIV-1 based vectors expressing eGFP from cardiac-specific promoters (Gallo *et al.*, 2008; Huber *et al.*, 2007).

2.1.2. Lentiviral–based vectors coexpressing two genes:

2.1.2.1. Bicistronic vectors

Two strategies were employed thus far to coexpress two genes in hESCs. In these vectors, the two genes are transcribed from a single promoter, but are translated separately. This is achieved by insertion of an internal ribosome entry site (IRES) between the two genes (Fig. 25.2A). This strategy ensures a coordinated expression of the two genes. Several groups have successfully used bicistronic vectors containing viral IRES elements for transduction of hESCs (Ben-Dor *et al.*, 2006; Clements *et al.*, 2006; Xiong *et al.*, 2005). Ben-Dor *et al.* have compared two IRES elements from the encephalomyocarditis virus (EMCV) and the poliovirus (PV) and found that the EMCV–IRES conferred higher expression levels of the

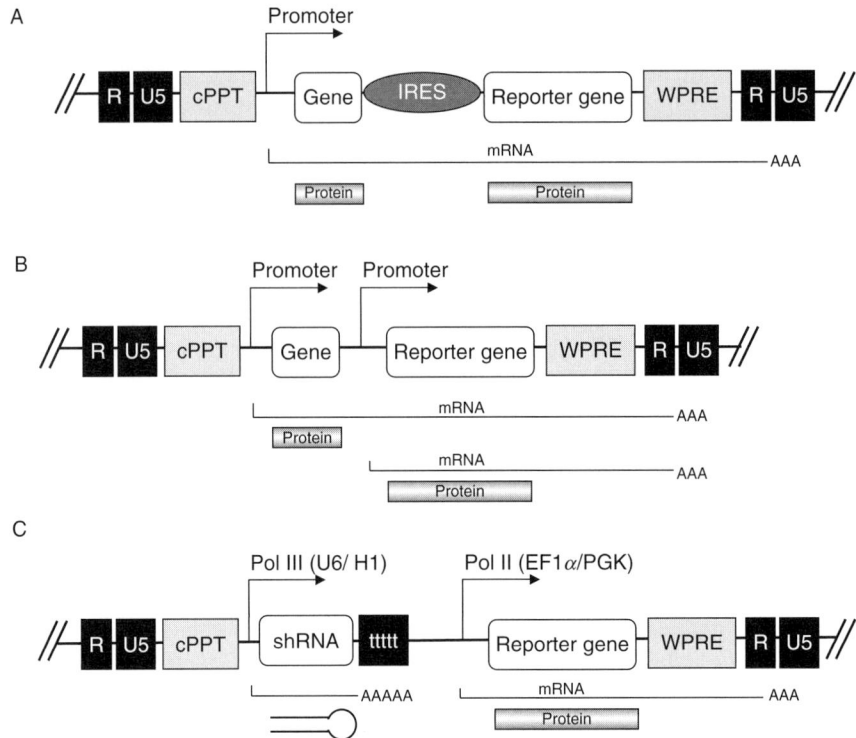

Figure 25.2 HIV-1 transfer vectors coexpressing two genes. The figure depicts a schematic representation of two types of advanced HIV-1 based vectors coexpressing two genes. These vectors enable the expression of a gene of interest together with a reporter/selection gene. (A) Bicistronic vectors: In these vectors the coordinated expression of the two genes is achieved by transcription of the two genes from a single promoter. An internal ribosome entry site (IRES) inserted between the two genes allows separate translation of their transcripts. (B) Dual promoter vectors: These vectors contain two separate expression cassettes, each harboring a gene under the control of its own promoter, thus enabling an uncoupled expression of transgenes (C) Dual promoter vectors for silencing specific genes *via* RNA interference (RNAi): These vectors contain two cassettes, each harboring a different type of promoter: (1) a silencing cassette expressing a specific short hairpin RNA (shRNA) from a pol III promoter (H1 or U6) and harboring a unique transcription termination signal (ttttt), and (2) a second cassette harboring a reporter/selection gene expressed from a constitutive pol II promoter.

downstream gene (Ben-Dor *et al.*, 2006). Clements *et al.* have further shown that separation of the upstream gene from the EMCV–IRES by a 50-bp linker allowed comparable expression levels of both genes (Clements *et al.*, 2006). However, a major drawback of bicistronic vectors is that the expression of the gene downstream to the IRES, may be low and inconsistently dependent on the first gene sequence (Yu *et al.*, 2003). In addition, IRESs do not allow uncoupled expression of the two genes.

2.1.2.2. Dual promoter vectors

These vectors contain two separate expression cassettes, thus each gene is expressed from its own promoter (Fig. 25.2B). The uncoupled expression of transgenes is required for a variety of applications, such as the development of genetically modified hESCs that allow tracing of differentiated cells of a specific lineage. For this purpose constitutive expression of a selectable marker is required in combination with the expression of a reporter gene under a tissue-specific promoter. So far, dual promoter lentiviral vectors were used for the transduction of hESCs by several groups (Ben-Dor *et al.*, 2006; Suter *et al.*, 2005; Vieyra and Goodell, 2007). Ben-Dor *et al.* have developed and optimized dual-promoter vectors harboring a reporter and a selectable marker under the regulation of human elongation factor 1α (hEF1α) and phosphoglycerate (PGK1) promoters, respectively. Optimal efficiency in hESCs was obtained when: (1) the reporter cassette was upstream rather than downstream of the selectable marker cassette, (2) the puromycin rather than the neomycin resistance gene was used, (3) a 5' deletion (314 bp) was created in the PGK promoter, and (4) two copies of a 120-bp element derived from the hamster *Aprt* CpG island were introduced upstream of the hEF1α promoter (Ben-Dor *et al.*, 2006). The major disadvantage of this strategy is that the expression driven by one promoter may be disrupted by interference from the other promoter.

Dual promoter vectors are very efficient tools for silencing specific genes *via* RNA interference (RNAi). In this case, the silencing cassette contains a specific short hairpin RNA (shRNA) under the control of a pol III promoter (H1 or U6) while the second cassette harbors a reporter/selection gene expressed from a constitutive pol II promoter (Fig. 25.2C). Since the expression of the shRNA is directed from a pol III promoter and the silencing cassette includes a unique transcription termination signal, these dual-promoter vectors do not encounter promoter interference. Several groups have recently used dual promoter vectors to efficiently silence specific genes in hESCs (Clements *et al.*, 2006; Suter *et al.*, 2005; Xiong *et al.*, 2005; Zaehres *et al.*, 2005).

However, complete silencing of specific genes may be incompatible with the survival of hESCs in culture. In addition, genes that are involved in ESCs pluripotency and differentiation may act in a concentration-dependent mode. We, therefore, developed a lentiviral–RNAi-based system that allowed the derivation of hESC-clones with different homogenous levels of silencing. It also facilitated simple monitoring of the relative levels of silencing within these hESC clones. In this system, dual-promoter lentiviral vectors coexpressing an RNAi cassette and a reporter gene were initially used for efficient induction of heterogeneous levels of gene silencing in hESCs. This step was further combined with the isolation of transduced clones with different homogenous levels of gene silencing. The level of silencing in each of the clones correlated and could be monitored by the level of expression of the vector's reporter transgene. Thus, this system allows easy identification of clones with relatively different homogenous levels of gene silencing (Gropp and Reubinoff, 2007).

To summarize this section, various lentiviral vector types can be employed for transduction of hESCs. The choice of which vector to use depends on the desired application. It should be noted that the efficiency of various lentiviral vectors may vary, depending on the vector construct, the specific promoters, and the transgenes that are expressed.

2.2. Choice of internal promoter

2.2.1. Constitutive transgene expression

Constitutive expression of the transgene in hESCs requires the use of promoters active in undifferentiated hESCs and their differentiated derivatives. So far, several constitutive promoters were used to drive lentiviral mediated transgene expression in hESCs. We and others have found that the hEF1α promoter promoted high transgene expression in hESCs that was not silenced over time, or after differentiation (Gropp *et al.*, 2003; Kim *et al.*, 2007; Ma *et al.*, 2003; Suter *et al.*, 2005; Xia *et al.*, 2007; Xiong *et al.*, 2005). Other promoters that were used are the human PGK promoter (Zaehres *et al.*, 2005), the SV40 promoter (Suter *et al.*, 2005), the hybrid CAG promoter composed of the cytomegalovirus (CMV) immediate early enhancer, chicken β-actin promoter and a rabbit β-globin intron (Pfeifer *et al.*, 2002), the spleen focus forming virus (SFFV) promoter (Clements *et al.*, 2006), and the ubiquitin-C (Ubi-C) promoter (Li *et al.*, 2008).

Notably, the commonly used promoter from CMV was suppressed in hESCs following transduction (Xia *et al.*, 2007).

2.2.2. Regulated transgene expression

Two groups have recently reported on the development of conditional lentiviral mediated gene expression systems in hESCs, using two components TET–ON systems (Vieyra and Goodell, 2007; Zhou *et al.*, 2007). Zhou *et al.* have first transduced hESCs with a vector constitutively expressing the tetracycline repressor fused to a transcriptional suppression domain (tTS). The cells were then transduced with a second vector containing a transgene controlled by a constitutive promoter (EF1α or UbiC) adjacent to a high affinity tTs-binding site (*tet* operator-tetO). In this system, in the absence of tetracycline, transgene expression from the EF1α promoter is suppressed by binding of tTS to tetO. In the presence of the tetracycline analog doxycycline (Dox), tTs is released from tetO and transgene expression is restored (Zhou *et al.*, 2007). A potential disadvantage of this system is that the constitutive expression of tTS may be toxic to hESCs. An alternative TET–ON system was reported by Vieyra and Goodell. hESCs were first transduced with a vector constitutively expressing the tetracycline-controlled transactivator rtTA2SM2. The cells were then transduced with a second vector containing a reporter gene controlled by a minimal CMV promoter flanked by a *tet* responsive element (TRE) containing tetO sequences. In this system only in the presence of tetracycline or its analog DOX, the transactivator binds to tetO and thus promotes activation of the CMV promoter (Vieyra and Goodell, 2007). In contrast to constitutive expression of tTS, the continuous expression of rtTA2SM2 was well tolerated by hESCs and did not interfere with their pluripotent potential.

2.3. Virus pseudotyping

Most groups have used the VSV-G protein for viral pseudotyping. However, a major drawback of the use of viruses pseudotyped with VSV-G is that they efficiently transduce the mouse embryonic fibroblast (MEF) feeders, used by many groups to support the propagation of hESCs. To overcome this problem Jang *et al.* have pseudotyped lentiviral vectors with envelopes from either the gibbon ape leukemia virus (GALV) or the RD114 feline endogenous virus (RD114). Lentiviral vectors pseudotyped with these envelopes transduced hESCs as efficiently as vectors pseudotyped with VSV-G, but did not transduce the MEF feeders (Jang *et al.*, 2006).

3. Generation of Recombinant Viral Particles

In general, recombinant viral particles are generated by transient cotransfection of the transfer vector, the packaging plasmid, and the envelope plasmid, into the highly transfectable human embryonic kidney (HEK) 293T cell line. The supernatant containing the viral particles is collected, concentrated, and used to transduce hESCs.

3.1. Plasmid constructs

The transfer vectors used by most groups are advanced modified HIV-1-based vectors, expressing the transgene from an internal promoter (Gropp *et al.*, 2003). The envelope plasmid expressing the heterologous envelope protein is pMDG (Naldini *et al.*, 1996a,b), and the packaging plasmid expressing the *gag*, *pol*, and *rev* genes is pCMVΔR8.91 (Zufferey *et al.*, 1997). Note that some groups use a more advanced packaging system, in which the *rev* gene is expressed from a separate promoter (Follenzi and Naldini, 2002). Plasmid DNA should be pure, and can be purified by commercial kits.

3.2. Production of viral particles

Viral particles are produced by transient cotransfection into 293T cells. We routinely use FuGene-6 transfection reagent (Roche Molecular Biochemicals, Mannheim, Germany) or TransIT-LT1 transfection reagent (Mirus, Madison, USA). These reagents yield high reproducible transfection efficiencies, which lead to high viral titers. An alternative economical strategy is the calcium–phosphate transfection method (see Follenzi, 2002 no. 18). In general, for production of high viral titers the transfection method chosen should be very efficient. The transfection protocol described in the next section is suitable for both FuGene-6 and TransIT-LT1 transfection reagents. The components of the transfection are calculated for 10-cm tissue culture dishes. However, transfection can be performed in larger or smaller tissue culture dishes, provided that all the other components of the transfection (DNA, transfection reagent, and medium) are scaled proportionally.

3.3. FuGene-6/TransIT-LT1 transfection protocol

1. 293T cells are routinely cultured in 90% Dulbecco's modified eagle's medium (DMEM, Gibco-BRL, Gaithersburg, MD) supplemented with 10% Fetal Calf Serum (FCS, Biological Industries, Beit Haemek, Israel), 1 mM L-glutamine, 100 units g/ml penicillin, and 50 μg/ml streptomycin. Note that 293T cells are highly susceptible to mycoplasma infection, which leads to low viral titers. Therefore, it is recommended to ascertain prior to transfection that the cells are mycoplasma-free.

2. 24 h prior to transfection, plate $1.4–2 \times 10^6$ 293T cells on a 10-cm tissue culture dish, in 10 ml 293T tissue culture medium. The confluence of the cells on the day of transfection should be ~70%.

3. Incubate the cells overnight at 37 °C.

4. The next day, perform the transfection in a total volume of 600 μl.

5. Prepare the plasmids mixture: in a sterile Eppendorf tube combine 10 μg of the transfer vector, 6.5 μg of the packaging plasmid, and 3.5 μg of the envelope plasmid (a total of 20 μg plasmid DNA).

6. In a second Eppendorf tube, prepare the transfection reagent solution: place a serum-free medium (Optimem, Gibco) in the tube. The volume of the Optimem medium is determined according to the volumes of the other components of the transfection (transfection reagent and plasmid DNA). Add 55 μl of the FuGene-6/TransIT-LT1 transfection reagent (a ratio of 2.7 μl reagent per 1 μg DNA) directly to the Optimem medium, in a drop wise manner. Mix completely by gently flicking the tube. Incubate at room temperature for ~5 min.

7. Add the reagent solution to the plasmids mixture in a dropwise manner. Mix by gently flicking the tube. Incubate at room temperature for 15–45 min, to enable reagent/DNA complex formation.

8. Add the reagent/DNA complex mixture drop wise to the 293T cells. Gently rock the tissue culture dish to ensure even distribution of the mixture. Incubate for 16–20 h at 37 °C.

3.4. Collection of recombinant viral particles

1. 16–20 h after transfection, replace the medium with 10 ml fresh 293T culture medium.

2. After additional incubation for 24 h at 37 °C, collect the supernatant containing the viral particles. Store the supernatant in the dark at 4 °C.

3. Add 10 ml fresh medium and incubate the transfected cells further for 24 h. Collect the supernatant.

4. Combine the supernatants that were collected 48 and 72 h after transfection (a total of 20 ml). Filter them through a 0.45 μM filter (we use Sartorius, Goettingen, Germany). Proceed directly to concentration of the virus. Save a small volume for determination of viral titer.

3.5. Concentration of viral particles

High viral titers (10^7–10^8 TU/ml) are vital for efficient transduction of hESCs. Therefore the collected viral particles should be concentrated. The use of the VSV-G envelope generates a stable virus that can be concentrated by ultracentrifugation without a significant loss of viral titer.

1. Concentrate the virus by centrifugation at 50,000 × g at 4 °C for 2 h. We use a Sorvall ultracentrifuge model Discovery 100, with a Surespin 630 swinging bucket rotor.

2. After centrifugation resuspend the pellet containing the viral particles immediately in 0.1 volume of the collected supernatant (2 ml from each 10-cm tissue culture dish), in the desired hESC culture medium. Proceed directly to transduction of hESCs. Save a small volume for determination of viral titer. Note that the viral pellet is very loose and mostly invisible. Therefore, mark the centrifugation tubes in advance and aspirate the supernatant carefully using a pipette.

3. Optionally, the concentrated viral pellet can be resuspended in a minimal volume of freezing buffer and stored in small aliquots at −80 °C. The freezing buffer is composed of 19.75 mM Tris–HCl buffer, 40 mg/ml lactose, 37.5 mM sodium chloride, 1 mg/ml human or bovine serum albumin, and 5 μg/ml protamine sulphate. Avoid freeze–thaw cycles.

4. Transduction of hESCs

Transduction of hESCs can be performed using several strategies. The strategy chosen is dependent on the culture method of the hESCs. Each of the three protocols described below is suitable for a different culture method.

4.1. Transduction of hESCs cultured in serum-containing medium and passaged as cell clusters

hESCs are routinely cultured in 80% high glucose DMEM (Gibco) supplemented with 20% Fetal Bovine Serum (FBS, Hyclone, Logan, UT), 1 mM L-glutamine, 50 units g/ml penicillin, 50 μg/ml streptomycin, 1% nonessential amino acids, and 0.1 mM β-mercaptoethanol. The cells grow on MEF feeders, and are passaged weekly as clusters of 100–200 cells (Reubinoff *et al.*, 2000).

1. At the time of routine passage, isolate small clusters of undifferentiated hESCs by mechanical slicing of undifferentiated sections of the hES colonies, followed by treatment with 10 mg/ml dispase (Gibco).
2. Resuspend the viral pellet in 2 ml hESC culture medium. Incubate the hESC clusters with 1 ml of the concentrated virus, in the presence of 5 μg/ml polybrene (Sigma, St. Louis, MO), in a 35 mm petri dish (BD Falcon, Franklin Lakes, NJ), at 37 °C for 2 h. At time intervals, gently rock the dish to prevent adherence of the clusters, and increase exposure of the clusters to the virus.
3. Add 1 ml fresh concentrated virus and continue the incubation for 1–2 h.
4. Wash the transduced hESC clusters with PBS and plate them on fresh MEF feeders, on a 35-mm tissue culture dish.
5. After 1 week determine transduction efficiency.

This method generally promotes maximal transduction efficiencies of 30%. However, transduced hESC populations can be further enriched.

4.2. Transduction of hESCs cultured in serum-free medium on feeders and passaged as single cells

hESCs are routinely cultured in 85% Knockout (KO) DMEM medium (Gibco) supplemented with 15% KO-Serum replacement (SR, Gibco), 1 mM L-glutamine, 50 units g/ml penicillin, 50 μg/ml streptomycin, 1% nonessential amino acids, and 4 ng/ml basic fibroblast growth factor (bFGF, Cytolab, Rehovot, Israel).The cells grow on human foreskin feeders, and are passaged weekly as single cells (using Trypsin/EDTA or EDTA). Note that due to the high frequency of chromosomal abnormalities observed after extended passaging of hESCs as single cells, it is recommended to culture the cells in this method for not more than 12–15 passages (Mitalipova *et al.*, 2005). This protocol is also suitable for hESCs cultured on human feeders as described here and passaged routinely as small clusters using collagenase type IV. In this case, 1–2 passages prior to transduction, split the cells as single cells. 1–2 passages after transduction the cells can be passaged again with collagenase.

Two optional strategies can be used for transduction of hESCs:

A. Transduction of hESCs on the day of cell passage:

1. At the time of routine passage, dissociate the cells into a single cell suspension by digestion with 0.05% trypsin/0.53 mM EDTA (Gibco), or 0.05% EDTA solution (Biological Industries, Beit-Haemek, Israel).

2. Resuspend the viral pellet in 2 ml hESC culture medium (described earlier).

3. Combine 1×10^5 trypsinized hESCs with the virus containing culture medium, in the presence of 5 μg/ml polybrene.

4. Plate the cells in the virus containing culture medium on fresh foreskin feeders, on 35-mm tissue culture dishes. Incubate overnight at 37 °C.

5. The next day replace the medium with fresh hESC culture medium.

6. Since this method promotes transduction of not only the hESCs but also the feeders, one or two passages are required to eliminate the transduced feeders. By this time, the transduced hESCs regain normal growth and transduction efficiency may be determined. This method generally promotes high transduction efficiency of over 90%.

B. Transduction of hESCs 2 days after passage and plating:

1. At the time of routine passage, dissociate the cells into a single cell suspension and plate them on fresh foreskin feeders on 35-mm tissue culture dishes (~10^5 cells per dish).

2. Culture the cells for 2 days, allowing them to adhere and expand.

3. Resuspend the concentrated virus in 2 ml hESC culture medium (described earlier).

4. Replace the culture medium of the hESCs with the virus containing culture medium, supplemented with 5 μg/ml polybrene, and incubate overnight at 37 °C.

5. The next day replace the medium with fresh hESC culture medium.

6. Determine transduction efficiency 2 weeks after transduction.

Transduction of hESCs 2 days after passage, generally promotes lower efficiencies (80%–85%) compared to transduction of single cells at the time of passage. However, it is less toxic to the hESCs.

4.3. Transduction of hESCs cultured in serum-free medium without feeders, and passaged enzymatically as small clusters

hESCs are grown in a feeder-free culture system. In this system, the cells are routinely cultured on Matrigel coated plates in MEF-conditioned medium. The cells are passaged weekly as small clusters using collagenase type IV (a detailed culture protocol is described by Xu and colleagues (Xu *et al.*, 2001).

1. At the time of routine passage, dissociate the cells into small clusters by digestion with 200 u/mg collagenase type IV (Gibco).
2. Resuspend the viral pellet in 2 ml MEF-conditioned medium containing 5 μg/ml polybrene.
3. Combine the hESC clumps with the virus containing medium.
4. Plate the cells in the virus containing medium on fresh Matrigel coated 35-mm tissue culture dishes. Incubate overnight at 37 °C.
5. The next day, replace the medium with fresh conditioned medium.
6. Determine transduction efficiency 1 week after transduction.

This method enables transduction of pure populations of hESCs (no feeders). Therefore, high efficiencies can be obtained, even though the cells are transduced as small clusters.

5. Measurement of Transduction Efficiency

The method described in this section for determining transduction efficiency is based on FACS analysis of the expression of a reporter gene (GFP or RFP) by the transduced cells. The percentage of undifferentiated cells among the transduced hESCs is determined by the analysis of the percentage of transduced cells that are also immunoreactive with antibodies against stem cell-specific markers, such as GCTM2 (Pera *et al.*, 1988), SSEA-4 or TRA-1–60 (both are available commercially).

1. Depending on the method of hESCs culture separate the transduced hESCs from the feeders.
2. Dissociate the transduced hESCs cells into a single cell suspension using 0.05% trypsin/0.53 mM EDTA.
3. Wash with cold PBS and spin for 5 min at 1500 rpm.
4. Incubate 10^5 cells with the stem cell-specific antibody on ice for 30 min.
5. As a control, incubate 10^5 cells with the appropriate isotype control.
6. Wash the cells with cold PBS and spin again for 5 min at 1500 rpm.
7. Incubate the cells with the secondary fluorescent antibody on ice for 30 min.
8. Wash the cells with cold PBS and spin for 5 min at 1500 rpm. Resuspend the pellet in cold FACS buffer (1 × PBS supplemented with 1% BSA, and 0.1% sodium azide) containing propidium iodide (PI), to gate out dead cells.
9. Analyze the percentage of GFP-expressing cells by FACS. Determine the percentage of the undifferentiated transduced hESCs by analyzing the percentage of GFP expressing cells that are immunoreactive with the stem cell-specific antibody.

6. Enrichment for Transduced hESCs Expressing High Levels of the Transgene

Transduction of hESCs by lentiviral-based vectors without selection leads to a population of cells with varied transgene expression levels. However, high and homogenous transgene expression is preferred for many applications. Here we describe several strategies to obtain homogenous and high expression levels of transgenes within hESCs transduced by lentiviral-based vectors.

6.1. Mechanical enrichment of transduced hESC colonies expressing high levels of a reporter gene

This method is suitable for enrichment of hESCs transduced with a lentiviral-based vector expressing a reporter gene that are maintained on feeders and passaged mechanically as small clusters.

1. At the time of routine passage, examine the transduced hESC colonies by fluorescence microscopy and mark colonies expressing high levels of the reporter gene.
2. Selectively passage the marked colonies by mechanical slicing of undifferentiated sections of the hESC colonies, followed by treatment with 10 mg/ml dispase (Gibco).
3. Continue propagating the transduced colonies. Repeat the enrichment protocol for three to four additional passages, until the majority of the transduced hESC population expresses high levels of the transgene.
4. Determine transgene expression levels in the enriched transduced hESC population by FACS.

6.2. Isolation of transduced hESC clones expressing high levels of a reporter gene by limiting dilution

This method can be used to obtain transduced hESC clones with high and relatively homogenous expression of a reporter gene. It is suitable for hESCs maintained and transduced as single cells.

1. One day before passage seed several 96-wells tissue culture dishes with fresh feeders.
2. At the time of passage, dissociate the transduced hESCs into a single cell suspension using 0.05% trypsin/0.53 mM EDTA (Gibco)
3. Count viable cells and make serial dilutions in hESC culture medium, to final dilutions of 100, 50, and 10 cells/ml.

4. For each dilution, prepare 2–3 96-well culture dishes. Seed in each well of the first group of culture dishes, 100 μl of the 100 cells/ml dilution (10 cells/well). In the second group seed 100 μl of the 50 cells/ml dilution (5 cells/well), and in the third group 100 μl of the 10 cells/ml dilution (1 cell/well).

5. Culture the cells for 1 week.

6. After 1 week, monitor cells growth and mark wells containing single colonies. Examine the transduced hESC colonies by fluorescence microscopy and mark colonies expressing desired levels of the reporter gene.

7. Pick single colonies using a micropipette, and replate them on fresh feeders in separate wells, in a 48-well tissue culture dish. Since each transduced hESC colony evolved from a single transduced cell, the clones should express homogenous transgene levels.

8. One week after cloning, select clones expressing desired levels of the reporter gene.

9. Continue propagation of the selected clones.

10. Determine transgene expression levels in individual clones by FACS.

6.3. Enrichment of transduced hESCs expressing high levels of a reporter gene by FACS sorting

This protocol can be used to obtain clonal or nonclonal transduced hESC populations, with high levels of reporter gene expression. It is suitable for hESCs that can be maintained, at least for several passages as single cells. The protocol below is adapted from Hewitt et al. (Hewitt et al., 2006) and describes clonal and nonclonal sorting of transduced hESCs using a FACSAria cell sorter (Becton Dickinson Immunocytometry System [BD], UK). Alternative protocols for FACS sorting of hESCs were described by Eiges et al. (Eiges et al., 2001) and Nicholas et al. (Nicholas et al., 2007).

A. **Nonclonal population sorting:**

1. At the time of passage, dissociate the transduced hESCs into a single cell suspension using 0.05% trypsin/0.53 mM EDTA (Gibco). Resuspend the cells in hESC culture medium.

2. To ensure single-cell suspension filter the cells through a 0.2 μM filter (Sartorius, Goettingen, Germany).

3. Transfer the cells to a sterile FACS tube.

4. For sorting, use a standard filter set, with a 70 μm nozzle, the sheath pressure set at 70 PSI, and a flow rate of 1. This should give a sorting rate of less than 3600 events/second, using the sort precision mode set for purity.

5. Exclude dead cells based on their forward and side scatter profile.

6. Apply sort gate to define cells expressing the reporter gene.

7. Collect the sorted cells into 5 ml polystyrene sterile tubes (Falcon, BD) primed with hESC culture medium. Centrifuge the cells and resuspend them in a fresh hESC culture medium.

8. To determine the recovery and viability of sorted cells count cells, and plate them on fresh feeders.

B. **Clonal single cell sorting:**

1. Perform steps 1–7 as described earlier. Use a FACSAria cell sorter equipped with an automated cell deposition unit (ACDU) and a 488-nm laser light.

2. Prepare 96-well tissue dishes plated with fresh feeders, and primed with hESC culture medium.

3. Collect sorted cells and deposit in each well of the 96-well dish 50, 25, 10, 5, or 1 cells.

4. Continue propagation of sorted cells, allowing them to recover for 10–14 days.

5. Screen the plates for wells containing single colonies.

6. Expand selected clones and analyze them for reporter gene expression.

6.4. Enrichment of transduced hESC cells expressing high levels of the transgene using antibiotic selection

This enrichment protocol can be employed for hESCs transduced by lentiviral vectors coexpressing a gene of interest, together with an antibiotic resistance gene. It allows robust and simple development of nonclonal transduced hESC populations, with high levels of transgene expression.

1. Prior to transduction, test the sensitivity of the hESCs to the specific antibiotic. We found that different hESC lines exhibit varied sensitivities to specific antibiotics.

2. Three days after transduction, start antibiotic selection. Antibiotic selection can be performed on hESCs transduced on Matrigel or on feeders. For cells maintained on feeders, the feeders should be resistant to the specific antibiotic used.

3. Continue antibiotic selection until the majority of the transduced hESCs express high homogenous levels of the gene of interest.

7. Determination of Viral Titer

Since efficient transduction of hESCs is dependent on high viral titers it is important to determine viral titers before transduction. The method described below is based on measurement of the expression of the reporter gene

(GFP or RFP). However, if the lentiviral construct does not contain a reporter gene, other methods can be employed to determine viral titers (see, (Follenzi and Naldini, 2002)).

1. One day prior to transduction, plate 5×10^4 293T cells per well in a 12-well tissue culture dish, in 2 ml 293T culture medium. The number of cells per well at the time of transduction should be 1×10^5. Incubate overnight at 37 °C.

2. The next day, transduce the 293T cells with dilutions of the virus samples that were collected before and after concentration. Use dilutions that will promote transduction efficiencies of less than 15%, to avoid transductions with multiple integration sites. We usually transduce the cells with 200 μl, 100 μl, and 50 μl of the nonconcentrated virus, and 20μl, 10 μl, and 5 μl of the concentrated virus. The transduction is performed in a total volume of 1 ml in the presence of 5 μg/ml polybrene. Incubate the transduced cells overnight at 37 °C.

3. The next day, replace the medium with fresh 293T culture medium and continue growing the cells for 2–3 days, until they reach 80%–90% confluence.

4. Dissociate the cells into a single cell suspension using 0.25% trypsin/1 mM EDTA.

5. For each dilution, analyze the percentage of GFP/RFP expressing cells by FACS.

6. The viral titer is represented as transducing units per ml (TU/ml). Calculate the titer as follows: the number of cells at the time of transduction (1×10^5) multiplied by the percentage divided by 100 of the GFP/RFP expressing cells, multiplied by the fraction of 1 ml of the virus used for transduction, and multiplied by the final dilution in the culture medium. For example, if you use 200 μl out of 1 ml virus for transduction in a total volume of 1 ml culture medium, and the percentage of GFP expressing cells is 15%, then the viral titer is $10^5 \times 15/100 \times 5 \times 5 = 3.75 \times 10^5$ TU/ml.

7. Average the titers from the three dilutions and determine viral titer.

References

Asano, T., Hanazono, Y., Ueda, Y., Muramatsu, S., Kume, A., Suemori, H., Suzuki, Y., Kondo, Y., Harii, K., Hasegawa, M., Nakatsuji, N., and Ozawa, K. (2002). Highly efficient gene transfer into primate embryonic stem cells with simian lentivirus vector. *Mol. Ther.* **6,** 162–168.

Ben-Dor, I., Itsykson, P., Goldenberg, D., Galun, E., and Reubinoff, B. E. (2006). Lentiviral vectors harboring a dual-gene system allow high and homogeneous transgene expression in selected poly-clonal human embryonic stem cells. *Mol. Ther.* **14,** 255–267.

Buchschacher, G. L., Jr. (2001). Introduction to retroviruses and retroviral vectors. *Somat. Cell Mol. Genet.* **26,** 1–11.

Clements, M. O., Godfrey, A., Crossley, J., Wilson, S. J., Takeuchi, Y., and Boshoff, C. (2006). Lentiviral manipulation of gene expression in human adult and embryonic stem cells. *Tissue Eng.* **12,** 1741–1751.

Eiges, R., Schuldiner, M., Drukker, M., Yanuka, O., Itskovitz-Eldor, J., and Benvenisty, N. (2001). Establishment of human embryonic stem cell-transfected clones carrying a marker for undifferentiated cells. *Curr. Biol.* **11**, 514–518.

Follenzi, A., Ailles, L. E., Bakovic, S., Geuna, M., and Naldini, L. (2000). Gene transfer by lentiviral vectors is limited by nuclear translocation and rescued by HIV-1 pol sequences. *Nat. Genet.* **25**, 217–222.

Follenzi, A., and Naldini, L. (2002). Generation of HIV-1 derived lentiviral vectors. *Methods Enzymol.* **346**, 454–465.

Gallo, P., Grimaldi, S., Latronico, M. V., Bonci, D., Pagliuca, A., Ausoni, S., Peschle, C., and Condorelli, G. (2008). A lentiviral vector with a short troponin-I promoter for tracking cardiomyocyte differentiation of human embryonic stem cells. *Gene Ther.* **15**, 161–170.

Goff, S. P. (2001). Retroviridae; The retroviruses and their replication. *In* "Fields Virology" (D. M. Knipe, and P. M. Howley, Eds.), Vol. **2**, pp. 1871–1939. Lippincott Williams and Wilkins, Phildelphia.

Gropp, M., Itsykson, P., Singer, O., Ben-Hur, T., Reinhartz, E., Galun, E., and Reubinoff, B. E. (2003). Stable genetic modification of human embryonic stem cells by lentiviral vectors. *Mol. Ther.* **7**, 281–287.

Gropp, M., and Reubinoff, B. E. (2007). Lentiviral-RNA-interference system mediating homogenous and monitored level of gene silencing in human embryonic stem cells. *Cloning Stem Cells* **9**, 339–345.

Hamaguchi, I., Woods, N. B., Panagopoulos, I., Andersson, E., Mikkola, H., Fahlman, C., Zufferey, R., Carlsson, L., Trono, D., and Karlsson, S. (2000). Lentivirus vector gene expression during ES cell-derived hematopoietic development *in vitro*. *J. Virol.* **74**, 10778–10784.

Hewitt, Z., Forsyth, N. R., Waterfall, M., Wojtacha, D., Thomson, A. J., and McWhir, J. (2006). Fluorescence-activated single cell sorting of human embryonic stem cells. *Cloning Stem Cells* **8**, 225–234.

Huber, I., Itzhaki, I., Caspi, O., Arbel, G., Tzukerman, M., Gepstein, A., Habib, M., Yankelson, L., Kehat, I., and Gepstein, L. (2007). Identification and selection of cardiomyocytes during human embryonic stem cell differentiation. *FASEB J.* **21**, 2551–2563.

Jang, J. E., Shaw, K., Yu, X. J., Petersen, D., Pepper, K., Lutzko, C., and Kohn, D. B. (2006). Specific and stable gene transfer to human embryonic stem cells using pseudotyped lentiviral vectors. *Stem Cells Dev.* **15**, 109–117.

Kim, S., Kim, G. J., Miyoshi, H., Moon, S. H., Ahn, S. E., Lee, J. H., Lee, H. J., Cha, K. Y., and Chung, H. M. (2007). Efficiency of the elongation factor-1alpha promoter in mammalian embryonic stem cells using lentiviral gene delivery systems. *Stem Cells Dev.* **16**, 537–545.

Li, Z., Suzuki, Y., Huang, M., Cao, F., Xie, X., Connolly, A. J., Yang, P. C., and Wu, J. C. (2008). Comparison of reporter gene and iron particle labeling for tracking fate of human embryonic stem cells and differentiated endothelial cells in living subjects. *Stem Cells* **26**, 864–873.

Lombardo, A., Genovese, P., Beausejour, C. M., Colleoni, S., Lee, Y. L., Kim, K. A., Ando, D., Urnov, F.D, Galli, C., Gregory, P. D., Holmes, M. C., and Naldini, L. (2007). Gene editing in human stem cells using zinc finger nucleases and integrase-defective lentiviral vector delivery. *Nat. Biotechnol.* **25**, 1298–1306.

Ma, Y., Ramezani, A., Lewis, R., Hawley, R.G, and Thomson, J. A. (2003). High-level sustained transgene expression in human embryonic stem cells using lentiviral vectors. *Stem Cells* **21**, 111–117.

Menendez, P., Wang, L., and Bhatia, M. (2005). Genetic manipulation of human embryonic stem cells: A system to study early human development and potential therapeutic applications. *Curr. Gene Ther.* **5**, 375–385.

Mitalipova, M. M., Rao, R. R., Hoyer, D. M., Johnson, J. A., Meisner, L. F., Jones, K. L., Dalton, S., and Stice, S. L. (2005). Preserving the genetic integrity of human embryonic stem cells. *Nat. Biotechnol.* **23**, 19–20.

Miyoshi, H., Blomer, U., Takahashi, M., Gage, F. H., and Verma, I. M. (1998). Development of a self-inactivating lentivirus vector. *J. Virol.* **72**, 8150–8157.

Naldini, L., Blomer, U., Gage, F. H., Trono, D., and Verma, I. M. (1996a). Efficient transfer, integration, and sustained long-term expression of the transgene in adult rat brains injected with a lentiviral vector. *Proc. Natl. Acad. Sci. USA* **93**, 11382–11388.

Naldini, L., Blomer, U., Gallay, P, Ory, D., Mulligan, R., Gage, F. H., Verma, I. M., and Trono, D. (1996b). *In vivo* gene delivery and stable transduction of nondividing cells by a lentiviral vector. *Science* **272**, 263–267.

Nicholas, C. R., Gaur, M., Wang, S., Pera, R. A., and Leavitt, A. D. (2007). A method for single-cell sorting and expansion of genetically modified human embryonic stem cells. *Stem Cells Dev.* **16**, 109–117.

Niwa, O., Yokota, Y., Ishida, H., and Sugahara, T. (1983). Independent mechanisms involved in suppression of the Moloney leukemia virus genome during differentiation of murine teratocarcinoma cells. *Cell* **32**, 1105–1113.

Palu, G., Parolin, C., Takeuchi, Y., and Pizzato, M. (2000). Progress with retroviral gene vectors. *Rev. Med. Virol.* **10**, 185–202.

Pannell, D., and Ellis, J. (2001). Silencing of gene expression: Implications for design of retrovirus vectors. *Rev. Med. Virol.* **11**, 205–217.

Pera, M. F., Blasco-Lafita, M. J., Cooper, S., Mason, M., Mills, J., and Monaghan, P. (1988). Analysis of cell-differentiation lineage in human teratomas using new monoclonal antibodies to cytostructural antigens of embryonal carcinoma cells. *Differentiation* **39**, 139–149.

Pfeifer, A., Ikawa, M., Dayn, Y., and Verma, I. M. (2002). Transgenesis by lentiviral vectors: Lack of gene silencing in mammalian embryonic stem cells and preimplantation embryos. *Proc. Natl. Acad. Sci. USA* **99**, 2140–2145.

Reubinoff, B. E., Pera, M. F., Fong, C. Y., Trounson, A., and Bongso, A. (2000). Embryonic stem cell lines from human blastocysts: Somatic differentiation *in vitro*. *Nat. Biotechnol.* **18**, 399–404.

Rodriguez, R. T., Velkey, J. M., Lutzko, C., Seerke, R., Kohn, D. B., O'Shea, K. S., and Firpo, M. T. (2007). Manipulation of OCT4 levels in human embryonic stem cells results in induction of differential cell types. *Exp. Biol. Med. (Maywood)* **232**, 1368–1380.

Suter, D. M., Cartier, L., Bettiol, E., Tirefort, D., Jaconi, M. E., Dubois-Dauphin, M., and Krause, K. H. (2006). Rapid generation of stable transgenic embryonic stem cell lines using modular lentivectors. *Stem Cells* **24**, 615–623.

Thomson, J. A., Itskovitz-Eldor, J., Shapiro, S. S., Waknitz, M. A., Swiergiel, J. J., Marshall, V. S., and Jones, J. M. (1998). Embryonic stem cell lines derived from human blastocysts. *Science* **282**, 1145–1147.

Vieyra, D. S., and Goodell, M. A. (2007). Pluripotentiality and conditional transgene regulation in human embryonic stem cells expressing insulated tetracycline-ON transactivator. *Stem Cells* **25**, 2559–2566.

Xia, X., Zhang, Y., Zieth, C. R., and Zhang, S. C. (2007). Transgenes delivered by lentiviral vector are suppressed in human embryonic stem cells in a promoter-dependent manner. *Stem Cells Dev.* **16**, 167–176.

Xiong, C., Tang, D. Q., Xie, C. Q., Zhang, L., Xu, K. F., Thompson, W. E., Chou, W., Gibbons, G. H., Chang, L. J., Yang, L. J., and Chen, Y. E. (2005). Genetic engineering of human embryonic stem cells with lentiviral vectors. *Stem Cells Dev.* **14**, 367–377.

Xu, C., Inokuma, M. S., Denham, J., Golds, K., Kundu, P., Gold, J. D., and Carpenter, M. K. (2001). Feeder-free growth of undifferentiated human embryonic stem cells. *Nat. Biotechnol.* **19**, 971–974.

Yu, X., Zhan, X., D'Costa, J., Tanavde, V. M., Ye, Z., Peng, T., Malehorn, M. T., Yang, X., Civin, C. I., and Cheng, L. (2003). Lentiviral vectors with two independent internal promoters transfer high-level expression of multiple transgenes to human hematopoietic stem-progenitor cells. *Mol. Ther.* **7**, 827–838.

Zaehres, H., Lensch, M. W., Daheron, L., Stewart, S. A., Itskovitz-Eldor, J., and Daley, G. Q. (2005). High-efficiency RNA interference in human embryonic stem cells. *Stem Cells* **23**, 299–305.

Zhou, B. Y., Ye, Z., Chen, G., Gao, Z. P., Zhang, Y. A., and Cheng, L. (2007). Inducible and reversible transgene expression in human stem cells after efficient and stable gene transfer. *Stem Cells* **25,** 779–789.

Zufferey, R, Donello, J. E., Trono, D., and Hope, T. J. (1999). Woodchuck hepatitis virus posttranscriptional regulatory element enhances expression of transgenes delivered by retroviral vectors. *J. Virol.* **73,** 2886–2892.

Zufferey, R., Dull, T., Mandel, R. J., Bukovsky, A., Quiroz, D., Naldini, L., and Trono, D. (1998). Self-inactivating lentivirus vector for safe and efficient *in vivo* gene delivery. *J. Virol.* **72,** 9873–9880.

Zufferey, R., Nagy, D., Mandel, R. J., Naldini, L., and Trono, D. (1997). Multiply attenuated lentiviral vector achieves efficient gene delivery *in vivo*. *Nat. Biotechnol.* **15,** 871–875.

CHAPTER 26

Engineering Embryonic Stem Cells with Recombinase Systems

Frank Schnütgen,★ **A. Francis Stewart,**† **Harald von Melchner,**★ **and Konstantinos Anastassiadis**†

★Department of Molecular Hematology
University of Frankfurt Medical School
Theodor-Stern-Kai 7
60590 Frankfurt am Main, Germany

†Genomics
Technische Universitaet Dresden
BioInnovationZentrum
Am Tatzberg 47
01307 Dresden, Germany

Abstract
1. Introduction
2. Site-Specific Recombination
    2.1. Cre and FLPe
    2.2. Other tyrosine recombinases
    2.3. Large serine recombinases
3. Designing Substrates for Site-Specific Recombination
    3.1. Characteristics of loxP and FRTs
    3.2. Factors affecting Cre and FLPe recombination
    3.3. Characteristics of ΦC31 recombination
4. Generation of Conditional Alleles
    4.1. Strategies employing excisions
    4.2. Strategies employing inversions
    4.3. Recombinase delivery to ES cells
    4.4. Recombinase delivery to mice
5. Recombinase Mediated Cassette Exchange (RMCE)
6. Molecular Switches
7. Protocols
    7.1. Recombinase delivery into ES cells (DNA transfection)
    7.2. Low-density seeding of ES cells

7.3. Picking of colonies
7.4. Molecular confirmation of excisions
7.5. Tamoxifen treatment of primed ES cells
7.6. *In vitro* verification of conditional constructs
7.7. Verification of gene trap cassette inversions
7.8. RMCE of gene trap cassettes with CreER(T2)
References

Abstract

The combined use of site-specific recombination and gene targeting or trapping in ES cells has resulted in the emergence of technologies that enable the induction of mouse mutations in a prespecified temporal and spatially restricted manner. Their large scale implementation by several international mouse mutagenesis programs will lead to the assembly of a library of ES cell lines harboring conditional mutations in every single gene of the mouse genome. In anticipation of this unprecedented resource, this article will focus on site-specific recombination strategies and issues pertinent to ES cells and mice. The upcoming ES cell resource and the increasing sophistication of site-specific recombination technologies will greatly assist the functional annotation of the human genome and the animal modeling of human disease.

1. Introduction

The combined use of site-specific recombination and gene targeting or trapping in mouse ES cells has resulted in the emergence of a precise technology to manipulate the mouse genome (Glaser *et al.*, 2005). The preeminent application of this technology is conditional mutagenesis, which permits the controlled mutagenesis of specific genes in the mouse or in cells in culture for functional studies. Because conditional mutagenesis presents an unrivalled accuracy for mammalian functional studies, international efforts are underway to generate conditional alleles for every protein coding gene in the mouse genome. Based on the development of high throughput methodologies, this task is now feasible and an international cooperation has been initiated to build a complete set of mutated ES cell lines as a readily available resource (Austin *et al.*, 2004; Auwerx *et al.*, 2004; Collins *et al.*, 2007a,b). Because this resource will contain alleles that can be modified in various ways by site-specific recombination, here we will focus on issues and methods relevant to the generation and use of site-specific recombination strategies in ES cells and mice.

2. Site–Specific Recombination

Site-specific recombinases (SSRs) mediate DNA rearrangements by breaking and joining DNA molecules at two specific sites, termed recombination targets (RTs). These enzymes fall into two main classes termed tyrosine or serine recombinases according to the amino acid which becomes transiently covalently linked to DNA during recombination. They have been found in diverse prokaryotes and lower eukaryotes, however only three so far have been shown to work efficiently in ES cells. These are the tyrosine recombinases, Cre and FLPe and the large serine recombinase, ΦC31 integrase. We will first consider the characteristics of these proteins and their expression for engineering, then characteristics of their RTs.

2.1. Cre and FLPe

Cre and FLPe are presently the recombinases most widely used in ES cells. Cre recombinase was found in the *Escherichia coli* P1 phage (Sternberg *et al.*, 1986). In contrast to several other prokaryotic enzymes that have been used in mammalian cells, the prokaryotic codon bias present in the original Cre coding region appears to be satisfactory for expression in mammalian cells. Altering the codon bias from prokaryotic to mammalian to create a version termed iCre, only slightly improved expression levels and performance (Shimshek *et al.*, 2002). Therefore, most applied studies are still performed with the original prokaryotic coding sequence. For Cre recombination in ES cells, the early expression vector generated by Gu *et al.*, MC1Cre (Gu *et al.*, 1993), routinely delivers high levels of recombination (>25%) upon transient expression in ES cells. Consequently, it is still widely used. However, for difficult recombination exercises or recombination in mice (see below) more potent expression vectors have been used, such as pOG231 (O'Gorman *et al.*, 1997; Zheng *et al.*, 2000), pBS185 (Li *et al.*, 1996), pBS500 (Schlake *et al.*, 1999), pICCre (Madsen *et al.*, 1999), and pCAGGSCre (Okabe *et al.*, 1997).

FLP recombinase was found in the yeast 2 circle. Not surprisingly, this enzyme's temperature optimum is 30 °C and it is quite inefficient at 37 °C (Buchholz *et al.*, 1996). Therefore, molecular evolution was applied to develop an enzyme, termed FLPe, with a better temperature optimum (Buchholz *et al.*, 1998). However, FLPe is still significantly less efficient than Cre as evaluated by transient expression in ES cells. It was, therefore, important to use the best FLPe expression vector, pCAGGS-FLPe-IRES-puro (Schaft *et al.*, 2001) for transient expression experiments. Even with this vector, no one has yet achieved more than 10% recombination in ES cells for any allele. Consequently, this vector was designed to express also puromycin resistance, which can be used to enhance recombination frequencies by transient selection pressure (see below), or stable selection in difficult cases. Recently, a codon-optimized version of FLPe termed FLPo was published (Raymond and Soriano, 2007). Our tests with FLPo indicate that it is about

4 times more active than FLPe in ES cell excision experiments, so we recommend pCAGGS-FLPo-IRES-puro for future experiments (KA, AFS, unpublished).

Both Cre and FLPe are relatively small enzymes, which appear to be able to enter the mammalian nucleus. While Cre has its own weak, cryptic, nuclear localization signal (nls) (Le *et al.*, 1999), addition of an nls to FLPe increases the efficiency of recombination about 3-fold (Schaft *et al.*, 2001).

For experiments in mice, it is usually advisable to establish a mouse line from the engineered ES cell as soon as possible, because germ line transmission can be compromised by multiple handling of the cells. This is particularly important for ES cells derived from the C57BL/6 strain, which are less robust in tissue culture than ES cells derived from other strains, such as, 129Sv. Consequently, two methods for site-specific recombination in the germ line have been developed, which avoid ES cell handling. First, recombination can be induced in the zygote by injection of an expression plasmid. For pCAGGS-FLPe, recombination in one third of surviving zygotes has been reported (Schaft *et al.*, 2001). Second, recombination can be induced by crossing to a "deleter" mouse line, which expresses either Cre or FLPe in the germ line so that the progeny of the cross carry the recombination event. Deleters for both Cre and FLPe work perfectly for small events (<3 kb). Some Cre deleter strains possibly express dangerously high levels of the enzyme, which provoke unwanted mutagenesis by mediating recombination between cryptic RTs in the genome (Loonstra *et al.*, 2001; Schmidt *et al.*, 2000). A recent bioinformatics study estimated cryptic RTs to occur with a frequency 1.2 per megabases of mouse genome (Semprini *et al.*, 2007). Although the affinity of Cre for these recognition sites is much lower than for the consensus RTs, recombination between cryptic RTs following prolonged exposure to high levels of Cre has been shown to cause DNA damage activated apoptosis (Baba *et al.*, 2005; Loonstra *et al.*, 2001). Toxicity can be avoided by transient Cre expression in ES cells or mice by using inducible Cre proteins (see below) or self deleting retroviruses for delivery (Pfeifer *et al.*, 2001; Russ *et al.*, 1996b; Silver and Livingston, 2001). However, in the absence of a systematic study of Cre deleters, we can only comment that the numerous reported successes with Cre recombination in mice indicate that the window between adequate and excessive levels of Cre expression is rather wide.

Two FLPe deleter lines have been published thus far, and both work very well (Farley *et al.*, 2000; Rodriguez *et al.*, 2000). The first of these is a β-actin FLPe transgenic (Rodriguez *et al.*, 2000), which appears to express FLPe more strongly than the second, which is a ROSA26 FLPe knock-in (Farley *et al.*, 2000). With the β-actin FLPe deleter, in addition to deletion of small cassettes in all expected progeny, we have observed deletion of a 150 kb interval in nearly all expected progeny (G. Testa, KA, AFS; unpublished). It therefore appears that the FLPe in the mouse performs better than expected from transient expression experiments in ES cells.

The efficiency difference between Cre and FLPe provokes a simple consequence. Cre is the enzyme of choice for demanding applications, such as conditional mutagenesis or long distance chromosomal engineering, whereas FLPe is used for simpler tasks, such as removal of selectable cassettes.

For conditional mutagenesis, the recombination event needs to be spatially and/ or temporally regulated. Spatial regulation is achieved by expressing Cre recombinase in a cell-type specific manner. Consequently, recombination only occurs in those cells that express Cre. A variety of mouse lines have been developed which express Cre in cell-type specific patterns. Information regarding these lines is available from the CreXMice database assembled by Andras Nagy (http://www. mshri.on.ca/nagy). It is worth mentioning again that the use of Cre recombinase presents some potential risk, which entails careful controls. The interested reader may wish to consider a recent commentary on the vagaries of conditional gene targeting (Schmidt-Supprian and Rajewsky, 2007).

The most popular method for temporal regulation of recombination is based on expression of the SSR fused to a steroid receptor ligand binding domain (LBD) (Logie and Stewart, 1995). In the absence of ligand binding to the LBD, the fusion protein is cytoplasmic. Binding of a ligand triggers recombination because it permits translocation to the nucleus. Consequently, SSR–LBD constructs should not contain a nls added to the SSR because this partially circumvents regulation and increases recombination before ligand activation. In addition to temporal control, the SSR–LBD strategy reduces the chances of unwanted mutagenesis by limiting the activity of the SSR to the experimental need. The most popular application of the SSR–LBD strategy in the mouse uses the fusion of a mutated estrogen receptor LBD to Cre, called CreER(T2), because the mutated ER(T2) LBD is insensitive to endogenous estrogens yet can be activated by the synthetic estrogen antagonist, 4-hydroxytamoxifen (Feil *et al.*, 1996). The CreER(T2) fusion protein is superior to the earlier version CreERT (Feil *et al.*, 1997; Schwenk *et al.*, 1998) because it requires lower concentrations of 4-hydroxytamoxifen for activation (Feil *et al.*, 1997). It should be noted that the actual ligand is 4-hydroxytamoxifen, however, tamoxifen, which is cheaper, can be used *in vivo* because it is rapidly metabolized in the liver to the 4-hydroxy form. Experiments with cells in culture must use 4-hydroxytamoxifen. Commercially available 4-hydroxytamoxifen is an equal mixture of two enantiomers, only one of which is active. Although pure preparations of each enantiomer can be obtained at considerable expense, no benefit to experiments with cultured cells was observed (AFS, unpublished). Several other SSR–mutant LBD fusions have been described, for example, Cre-PR (Wunderlich *et al.*, 2001); however, none have yet emerged with good properties to match the CreER(T2).

Conditional mutagenesis has also been reported using tetracycline regulation of Cre transgene expression (Saam and Gordon, 1999). Possibly because the tetracycline system is more complicated to establish, or still entails certain unpredictabilities, this approach for conditional mutagenesis has not found widespread application. However, the fact that synthetic steroids, like tamoxifen, are not neutral ligands in mammalian cells whereas tetracycline is, recommends further work with tetracycline and other neutral ligand strategies (Cronin *et al.*, 2001). This point is particularly relevant for conditional mutagenesis during differentiation studies with ES cells in culture. Whereas treatment of ES cells with either 4-hydroxytamoxifen, dexamethasone, RU486, or mibolerone for 5 days in normal LIF culture did not noticeably promote differentiation or loss of germ line transmission (K. Vintersten, F. van der Hoeven, and AFS, unpublished), the prominent roles that steroid hormones play in ES cell differentiation protocols make it unlikely that these ligands can be used without consequence. Similarly, the use of synthetic steroids to induce conditional mutagenesis *in utero* is possible, but problematic (Danielian *et al.*, 1998). Therefore, the development of a tetracycline, or similar, strategy to avoid steroid ligand usage will be particularly relevant for studies *in utero* or with ES cells *in vitro* (Mao *et al.*, 2005).

2.2. Other tyrosine recombinases

Two other tyrosine recombinases, R and Kw, have been examined for eukaryotic genetic engineering applications (Araki *et al.*, 1992; Ringrose *et al.*, 1997). Both originate from variations of the 2 circle found in divergent yeasts. Although both recognize distinct sequences, they share several features with FLP including reduced efficiencies at 37 °C. Recently, a search for Cre-like enzymes amongst P1 phages uncovered a new candidate, Dre, which could be the long sought after tool that works as well as Cre (Sauer and McDermott, 2004). Initial studies in ES cells show that Dre works with a similar efficiency as Cre (KA, AFS; unpublished).

2.3. Large serine recombinases

Until the discovery of an unexpected subset of the serine recombinases, termed the large serine recombinases (Smith and Thorpe, 2002), it had been assumed that serine recombinases were unlikely to be useful for genetic engineering. ΦC31 was the first large serine recombinase described (Thorpe and Smith, 1998) and has subsequently been used in mammalian cells (Andreas *et al.*, 2002; Belteki *et al.*, 2003; Olivares *et al.*, 2002). However, it is still not clear how useful the ΦC31 and other large serine recombinases will prove to be or how best they may be employed. A codon-optimized version of ΦC31 has been reported to deliver more efficient recombination in ES cells (Raymond and Soriano, 2007). Another large serine recombinase, ΦBT1, has been reported to function in mammalian cells (Xu *et al.*, 2008).

3. Designing Substrates for Site–Specific Recombination

Upon expression, the SSR acts upon two or more recombinase target sites (RTs) that have been introduced into the genome. Designing how the RTs are deployed in the genome is the first step. The design choices are influenced by the objective and also by certain constraints that affect the efficiency of recombination.

For both tyrosine and large serine recombinases, the site of recombination is a short crossover region of hybrid DNA composed of one strand from each of the substrate RTs. Because the recombination product must have complementarity in the crossover regions, and because none of the RTs usually employed for genetic engineering are palindromic in the hybrid region, it follows that RTs have a directionality.

3.1. Characteristics of loxP and FRTs

In general, tyrosine recombinases have RTs that are centered on a palindromic recombinase binding site, which flanks the recombination site. For FLP and Cre, these minimal RTs are termed FRT and loxP respectively (Fig. 26.1A). They are both composed of inverted 13 bp binding sites flanking an 8 bp spacer, which includes the 8 or 6 bp hybrid region. For FLP and other 2 micron recombinases, the minimal RT is usually accompanied by an additional 13 bp binding site. This extra binding site appears to convey only a very moderate effect for genetic engineering (Ringrose et al., 1999) and is usually not employed. Recombination is usually employed to occur between two identical RTs and is inherently reversible (Ennifar et al., 2003; Van Duyne, 2001).

Recombination will still occur between RTs that carry limited sequence variations; however, the sequence in the crossover regions must be identical. Consequently, two classes of sequence variations can be defined.

First, the binding sites can be altered, which will reduce the binding affinity of the recombinase for the binding site. Because binding of both FLP and Cre to their palindromic RTs is cooperative (Ringrose et al., 1998), it is possible to mutate one half site without greatly affecting recombination efficiency as long as the other half site remains wild type. The compensatory effect of cooperativity has been employed to impose asymmetry upon Cre recombination (Albert et al., 1995; Araki et al., 1997). In this strategy, two mutant loxP sites, termed lox66 and lox71, carry mutations in the left or right 13 bp binding sites, respectively (Fig. 26.1A). After recombination, one product RT will contain both wild type binding half sites, and the other will contain both mutant binding half sites (Fig. 26.2A). The double mutant product RT will have a greatly reduced ability to bind Cre, hence the reverse recombination reaction will be disfavored. Second, the crossover region in the spacer can be altered. Because the sequence in the crossover region must be identical for productive recombination, a variety of

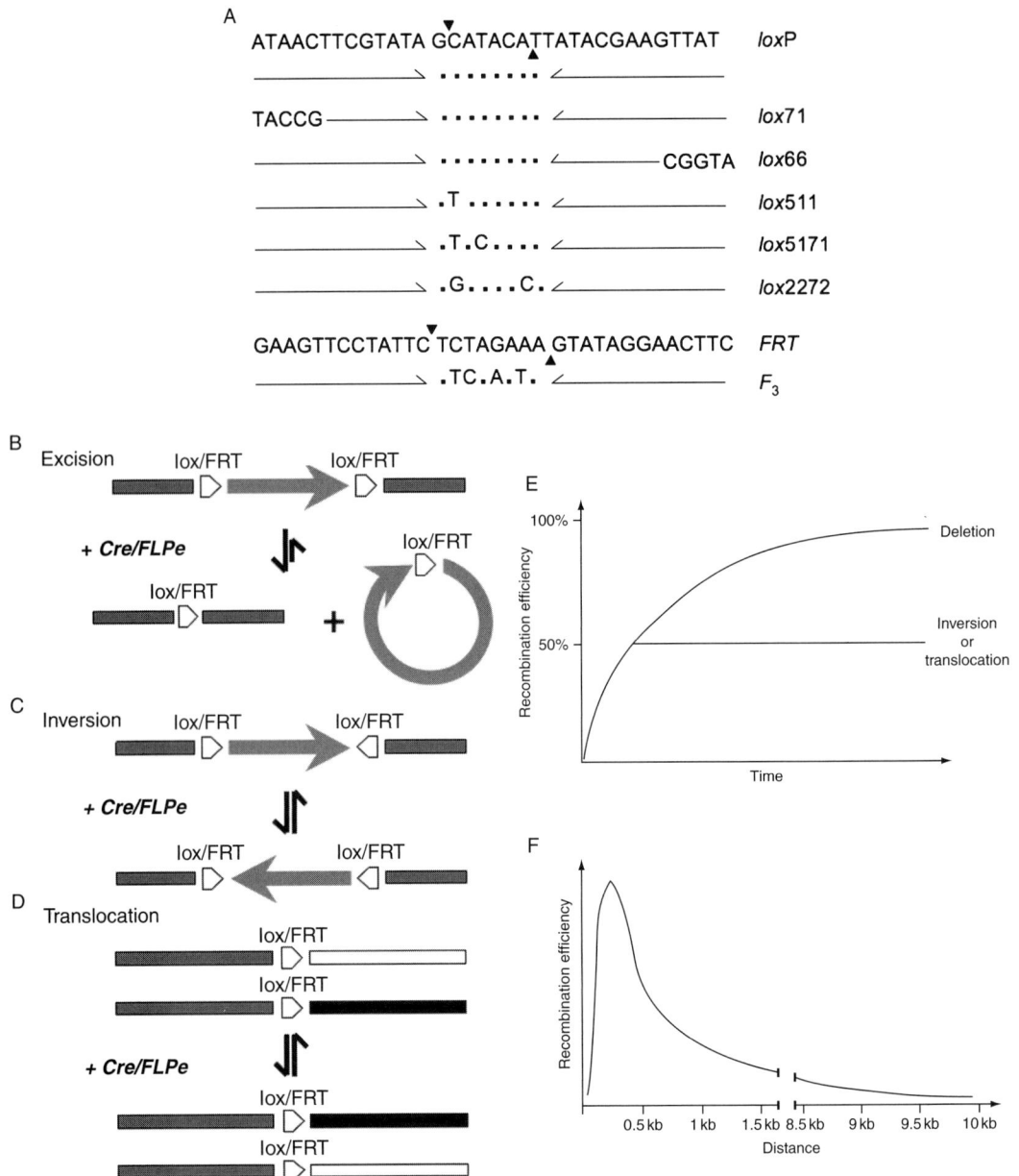

Figure 26.1 Characteristics of Cre and FLP site-specific recombination substrates. (A) Cre or FLP recombinations targets (RTs) known as loxP or FRT, respectively, are depicted along with several useful variations. The 13 bp inverted binding sites are illustrated by inverted arrows flanking the spacer sequences, which contain the region of recombination. The recombination region is denoted by arrow heads above and below the sequence at each end of the spacer. For the variant RTs, only the sequence differences are shown. (B) Excision recombination between two directly repeated RTs. The excised

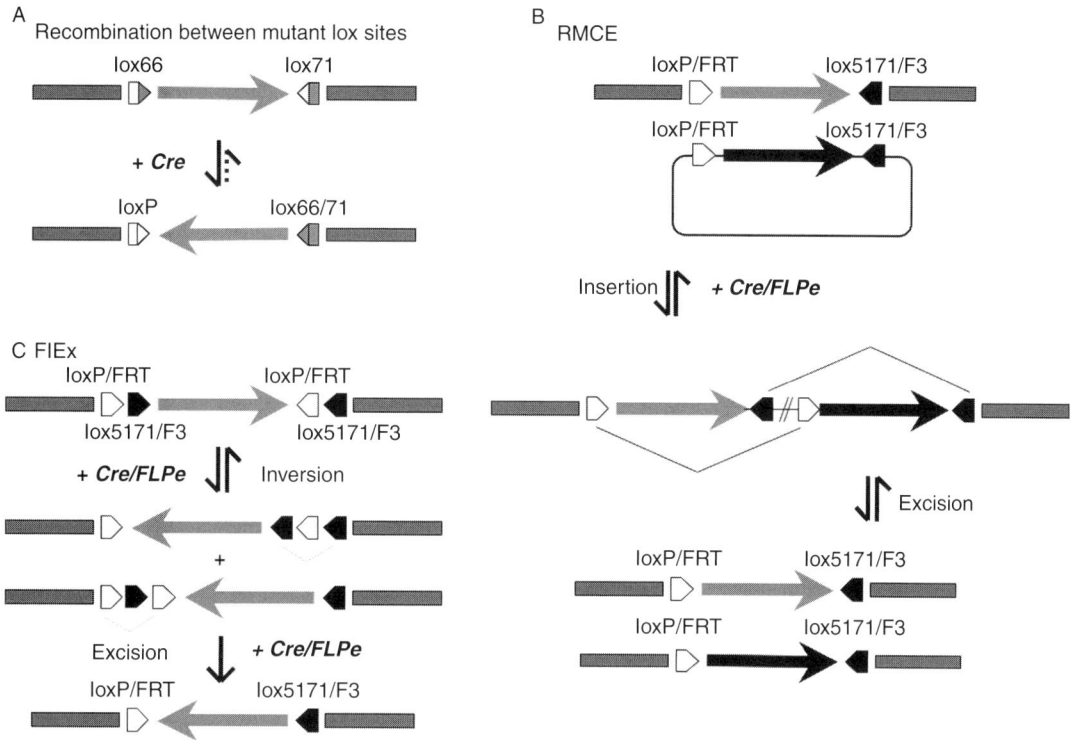

Figure 26.2 Directional recombination using Cre and FLP. (A) Inversion between mutant lox66 and lox71 generates a wild type loxP site and a double mutant lox66/lox71 site which cannot recombine. (B) Recombinase mediated cassette exchange (RMCE). Intermolecular recombination occurs between two different (heterotypic) RTs, here shown as wild type (white pentagons) and mutant (black pentagons). Intramolecular recombination between heterotypic RTs is not possible. The RTs flank the exchange cassettes on a linear recipient molecule (in gray) and on a circular donor molecule (in black). The insertion of the donor cassette is followed by the excision of the recipient's cassette (gray) in 50% of cases. (C) Flip-Excision (FLEx). Inversion of the RT flanked fragment occurs either at wild-type (white pentagons) or mutant (black pentagons) RTs. After either inversion, pairs of homotypic RTs in direct orientation flank a heterotypic RT. Excision between the directly repeated homotypic RTs excises the heterotypic RT, hence locking the recombination product. The final product is flanked by heterotypic RTs which cannot recombine.

DNA is released as a covalent circle. Reintegration is possible but disfavored. (C) Inversions occur when RTs are arranged as indirect repeats. (D) Recombination between RTs on different molecules results in a reciprocal translocation. The orientation of the translocated fragments is dictated by the orientation of the RTs. White pentagons indicate RTs and their orientation. (E) Idealized kinetics for excision, inversion, and translocation. Note that all three reactions approach 50% with equivalent kinetics. Thereafter, excisions approach 100% whereas inversions and translocations equilibriate at 50% (See Logie and Stewart, 1995, for supporting data). (F) Site-specific recombination frequency as a product of the distance between two RTs, as determined for naked DNA (Ringrose et al., 1999). Note the maximum recombination efficiency at 400 bp.

"heterotypic" RTs have been generated, which permit different recombination events directed by one recombinase. Intensive work on heterotypic lox sites has revealed that the loxP crossover region cannot be freely mutated and only certain mutations are possible (Langer *et al.*, 2002; Lee and Saito, 1998). Early work with heterotypic lox sites employed singly mutated crossover regions, for example, lox511 (Fig. 26.1A; Lauth *et al.*, 2000). However, a low but significant frequency of recombination between loxP and lox511 sites has been observed so that doubly mutated lox sites, such as lox5171 or 2272, are now preferred (Fig. 26.1A; Kolb, 2001; Lauth *et al.*, 2002). Work on the FRT crossover region indicates that it is more freely mutatable than loxP, and is mainly constrained by the GC content of the 8 bp crossover region (Umlauf and Cox, 1988). Increasing the GC content decreased the recombination efficiency. The increased mutability of the FRT compared to loxP correlates with the looser synapse implicit in an 8 bp compared to a 6 bp crossover region, as revealed by the crystal structure (Chen and Rice, 2003; Chen *et al.*, 2000). The most commonly used heterotypic FRT is termed F3 (Fig. 26.1A; Baer and Bode, 2001; Schlake and Bode, 1994).

When two RTs are placed in a DNA molecule so that their spacers are arranged as direct repeats, the DNA interval between the RTs is excised by the appropriate recombinase and released as a covalent circle (Fig. 26.1B). Because the reactions are inherently reversible, the covalent circle can reintegrate into the site it was excised from. However, intramolecular excision is favored over intermolecular integration because the two RTs can be separated by an infinite distance during intermolecular recombination but cannot separate by more than the distance between them when on the same DNA molecule. Furthermore in replicating cells, the excised circles are progressively diluted with each cell division because they are not replicated whereas the chromosomal excision product is (Logie and Stewart, 1995). When two RTs are placed in a DNA molecule so that their spacers are arranged as indirect repeats recombination results in the inversion of the DNA interval between the two RTs. Because the reaction is reversible, the end product of recombination will be 50% in one orientation and 50% in the other with an ongoing frequency of inversion if the recombinase continues to act (Fig. 26.1C). Finally, when RTs are placed in two different linear DNA molecules, a translocation occurs whose orientation is dictated by the direction of the spacers (Fig. 26.1D). Again, reversibility imposes an end product comprising a 50:50 mixture of the untranslocated and translocated molecules. The expected kinetics for these three reactions is depicted in Fig. 26.1E.

In addition to these three basic ways to deploy RTs, several more complicated designs have been described. Most notable is the use of pairs of RTs on two DNA molecules. Usually this deployment involves a heterotypic RT pair and is the basis for RMCE (recombinase mediated cassette exchange; Fig. 26.2B; Schlake and Bode, 1994). RMCE can also be based on two pairs of inverted homotypic RTs (Feng *et al.*, 1999). Also notable is the deployment of interwoven heterotypic RTs in the FlEx (Flip Excision) strategy (Fig. 26.2C; Schnutgen *et al.*, 2003). These applications are explored in more detail below.

3.2. Factors affecting Cre and FLPe recombination

After integration into a genome, the rate at which two RTs recombine in a population of cells is influenced by several factors, including:

i. The levels of active recombinase. Studies in ES cells employing cell permeable Cre protein have shown that the recombination rate between two RTs inserted into the genome directly correlates with the amount of added enzyme. At high concentrations recombination proceeds very quickly and is complete within a few hours (Peitz *et al.*, 2002).

ii. The physical distance between the two RTs. Site-specific recombination depends on random collision between two RTs. When the two RTs are positioned on the same, naked, DNA molecule, the rate of random collision is determined by the inherent flexibility of DNA, known as the persistence length, which is 50 nm (Ringrose *et al.*, 1999). This is an inherent property of DNA as a polymer, which yields maximal random collision frequencies between two sites 400 bp apart. Hence, maximum recombination rates between two RTs are observed at a distance of 400 bp. Increased distance leads to a predictable and asymptotic decrease in recombination frequencies, whereas two RTs cannot randomly collide when the distance is too short (less than approximately 120 bp) because of the inherent stiffness of DNA (Fig. 26.1F; Ringrose *et al.*, 1999).

iii. The position of the RTs within chromosomes. DNA in cells is not naked but embedded in chromatin. Chromatin is not a uniform polymer but includes fundamental regional variations, such as hetero- or euchromatin, as well as local, nucleosomal specific variations and probably distinct local tertiary structures. These variations produce an influence of chromatin known as "position effect," which has been reported to influence the recombination efficiency between two RTs (Vooijs *et al.*, 2001). However, it is still unclear to what extent the chromatin environment affects recombination. The influence of variables such as gene expression levels, whether an RT is in a transcribed region, whether the two RTs are separated by a chromatin boundary and many other considerations all remain unresolved. Nevertheless, the relationship between distance and recombination frequencies in active chromatin have been observed to conform to polymer predictions, albeit with an unexpected decrease of the persistence length to 25 nm (Ringrose *et al.*, 1999).

Although the absence of understanding about position effects makes predictable design of recombination substrates impossible at this stage, the best guideline for design involves placing the RTs as close to each other as practically possible because increased distance will decrease the rate of collision between the RTs, which is the rate limiting step (Ringrose *et al.*, 1998).

3.3. Characteristics of ΦC31 recombination

In principle, the above considerations for Cre and FLPe also apply to the large serine recombinases. However, the serine recombinases employ a fundamentally different recombination mechanism (Li *et al.*, 2005; Oram *et al.*, 1995), which may

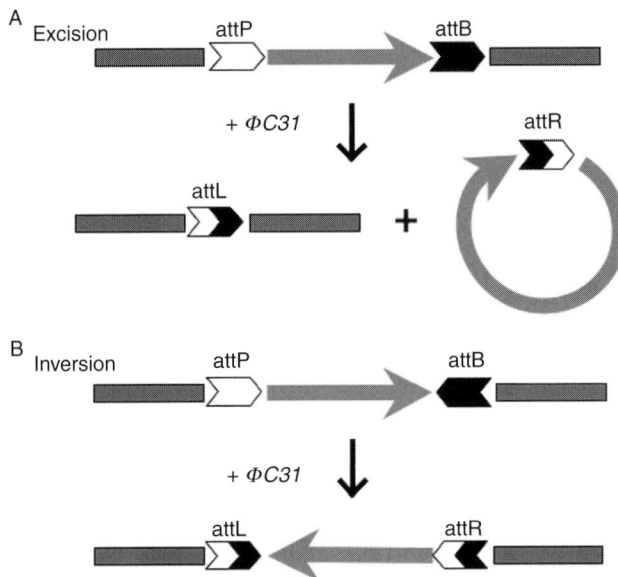

Figure 26.3 Products of ΦC31 recombination. The ΦC31 specific RTs, attP and attB are shown in excision (A) or inversion (B) orientations. After recombination, the RTs are changed to attL and attR and cannot be recombined by ΦC31 in the reverse reaction.

impose a limitation on how they can be applied. ΦC31 mediates recombination between two different RTs, attP and attB, to produce attL and attR (Fig. 26.3). The minimal attP and attB sites, 39 and 34 bp, respectively, include imperfect inverted repeats and a three nucleotide recombination site which here again imposes directionality. Like Cre and FLPe, ΦC31 excises and inverts DNA intervals flanked by attP and attB in direct and indirect orientation, respectively. However, unlike with Cre and FLPe, ΦC31 reactions are irreversible, presumably because reversibility requires a cofactor which is not present in mammalian cells. While directionality is an advantage in certain applications (see below), ΦC31's performance is presently no match for Cre and FLPe, despite early indications (Andreas *et al.*, 2002). ΦC31 has been successfully applied for insertional mutagenesis (Sclimenti *et al.*, 2001) and RMCE (Belteki *et al.*, 2003). Both of these applications involve intermolecular recombination. Interestingly, the Belteki *et al.* study revealed an unusual asymmetry in reaction products that appears to reflect inefficient intramolecular recombination. This could be due to the fact that serine recombinases need to topologically wrap one of the substrates before recombination (Li *et al.*, 2005; Oram *et al.*, 1995). Although it is not known whether the large serine recombinases also need to wrap the substrate, ΦC31 inefficiency with intramolecular chromosomal recombination could be due to its inability to wrap a chromatin template, as opposed to an incoming transgenic template, prior to recombination.

4. Generation of Conditional Alleles

Two major gene driven strategies have been used to induce mutations in ES cells, gene targeting and gene trapping. While gene targeting requires targeting vectors for each gene selected for inactivation, gene trapping induces mutations by inserting a generic gene disruption cassette throughout the genome. A combination of the two, called "targeted trapping" inserts a gene trap style cassette into a gene of interest by homologous recombination (Friedel *et al.*, 2005).

Each of these strategies takes advantage of the ability of ES cells to pass mutations induced *in vitro* to the germ line of transgenic offspring *in vivo*, which provides a unique way to analyze gene function in living organisms. Consequently, over 4000 mouse strains with functionally inactivated genes ("knock out" mice) have been generated using these technologies and many have proven useful as models for human disease (Zambrowicz and Sands, 2003). However, for most of these mutant strains, the significance for human disease remains uncertain because germline mutations can reveal only the earliest, nonredundant role of a gene. Moreover, about 30% of the genes targeted in ES cells are required for development and cause embryonic lethal phenotypes when passaged to the germline. These factors preclude accurate functional analysis in the adult.

Because most human disorders are the result of a late onset gene dysfunction, strategies of conditional mutagenesis have been developed in ES cells that combine gene targeting or trapping with site-specific recombination. They fall into two main categories—strategies employing excisions and strategies employing inversions.

4.1. Strategies employing excisions

The classic conditional allele is the so called "floxed" allele. In a floxed allele, one or more exons of a target gene are flanked by loxP sites inserted into introns in a direct orientation. Introns are preferred insertion sites because loxP sites in these positions are unlikely to have a mutagenic effect before recombination. As already mentioned, it is useful to place the loxP sites as close together as practically possible and so that Cre recombination deletes an essential section of the protein coding sequence. Alternatively, exons can be floxed whose deletion results in a shift of the protein translational reading frame (Shibata *et al.*, 1997).

Figure 26.4 illustrates a common strategy for generating conditional alleles. In the targeting vector, loxP sites flank the exon(s) of the target gene and FRT sites flank the expression cassette of a selectable marker gene. After selecting for homologous recombinants in ES cells, the selectable marker is removed by FLPe (Rodriguez *et al.*, 2000; Schaft *et al.*, 2001). Until now, FLPe has been used for this task to reserve the more potent Cre for conditional mutagenesis in the mouse. Presently, codon optimized FLPo offers a more potent alternative (see above) (Raymond and Soriano, 2007). Although deletion of the selectable marker gene is not essential, most investigators remove it because it can sometimes interfere

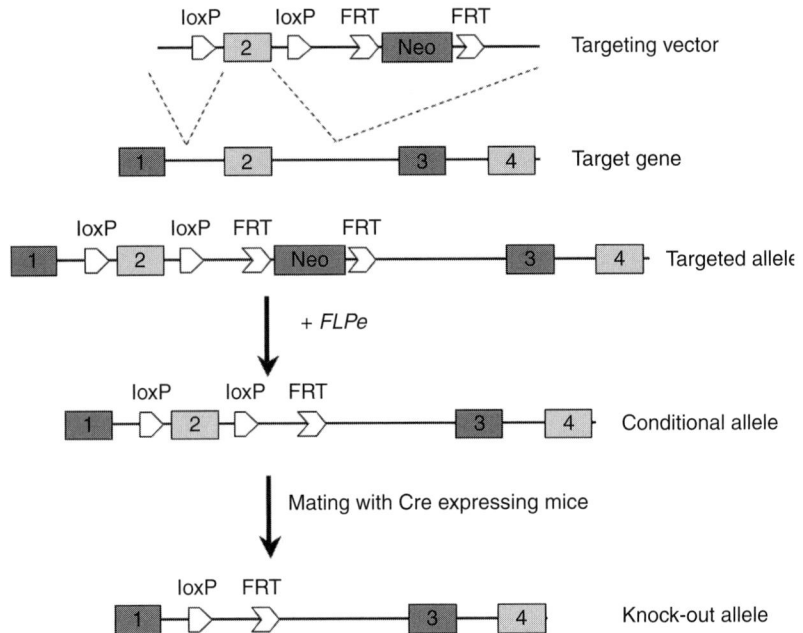

Figure 26.4 A common strategy for creating conditional alleles in the mouse. In the targeting vector, an exon (here exon 2) of the target gene is flanked by loxP sites in direct (excision) orientation (white pentagons). The targeting vector contains a selection cassette (here neomycin; neo), which is flanked by FRT sites in direct orientation (white polygons). After introducing the vector into ES cells and selecting for homologous recombinants, the neomycin cassette is removed using FLPe, either in ES cells, mice or oocytes. Homozygous mice for the conditional allele are crossed to mice expressing Cre in a spatially and/or temporally restricted manner. This deletes the loxP flanked exon 2 from the target gene and causes a mutation.

with gene expression in the mouse (Kaul *et al.*, 2000). This can result in a hypomorphic mutation of the target gene, which can however have a merit of its own (Meyers *et al.*, 1998; Nagy *et al.*, 1998).

4.2. Strategies employing inversions

Inversion strategies rely on insertional mutagenesis performed with cassettes designed to block gene expression when inserted into a gene. Originally developed for gene trapping, these cassettes consist of a reporter/selectable marker gene flanked by a splice acceptor sequence and a transcriptional termination signal (polyA). When inserted into an intron of an expressed gene, the cassette is transcribed from the endogenous promoter in the form of a fusion transcript in which the exon(s) upstream of the insertion site is spliced in frame to the reporter/selectable marker gene. Because transcription is terminated prematurely at the inserted polyadenylation site, the processed fusion transcript encodes a truncated and nonfunctional version of the cellular protein and the reporter/selectable marker or just the reporter/selectable marker if the insertion is upstream of a translation initiating methionine. Thus, to cause a mutation the cassette needs to insert in the same transcriptional orientation relative to the target gene (Stanford *et al.*, 2006). Cassette insertions in reverse orientations should not interfere with gene expression hence cause no mutation (Schnutgen *et al.*, 2005). Alternatively, conditional mutations can be induced by inserting an artificial intron into an exon of a target gene. Although the intron splits the exon in two, its sense orientation is innocuous because it is removed by splicing, which reconstitutes the exon. However, in reverse orientation on the noncoding strand the intron cannot be removed by splicing and causes a mutation (A. Economides and G.D. Yancopoulos, personal communication).

As we have discussed earlier, both Cre and FLPe invert DNA fragments positioned between RTs with spacers arranged as indirect repeats. However, inversions induced by recombination between loxP or FRT sites are inherently reversible and equilibrate at an equal mixture of the two orientations. This is of limited utility in mutagenesis and has been superceded by strategies that enable directional inversions.

The mutant loxP sites, lox66 and lox71 were the first to be used for this purpose (Fig. 26.2A; Oberdoerffer *et al.*, 2003). However, they recombine about 3 times less efficiently than the wild type loxP sites. The lox66/71 strategy has been used in ES cells for inversions of immunoglobulin genes (Oberdoerffer *et al.*, 2003), of gene trap cassettes (Xin *et al.*, 2005), and of artificial introns inserted into exons (A. Economides and G.D. Yancopoulos, personal communication). However, directionality with the half site mutants is not absolute as recombination can still occur between the loxP and double mutant lox66/71 sites at a low rate in both plants and ES cells (Albert *et al.*, 1995; Oberdoerffer *et al.*, 2003).

A second strategy termed FlEx, uses pairs of inversely oriented heterotypic RTs to flank a cassette in inverse orientations. Recombination between either pair of homotypic RTs inverts the cassette and places the other homotypic RT pair near to each other in a direct orientation. Recombination between this pair of directly repeated RTs excises one of the other heterotypic RTs, hence "locking" the recombination product against reinversion to the original orientation (Fig. 26.2B). The FlEx design was originally developed to monitor Cre recombination at cellular level in transgenic mice (Schnutgen *et al.*, 2003). A targeting construct with an inversely oriented splice acceptor-β-galactosidase-polyA (SAβgalpA) gene trap cassette was introduced into the RARγ gene by homologous recombination. In this construct, a RARγ exon and the SAβgalpA cassette were jointly flanked ("flexed") by pairs of heterotypic loxP/lox511 sites. When exposed to Cre, the gene trap cassette was inverted into a sense orientation on the coding strand whereas the exon was placed into an antisense orientation on the noncoding strand. This caused a mutation and simultaneously activated the β-gal reporter. Studies in mice carrying two flexed RARγ alleles showed that β-gal accurately reported recombination at the cellular level and mirrored the expression pattern of the RARγ gene (Schnutgen *et al.*, 2003).

FlEx arrays have been used recently in gene trap vectors developed for high throughput conditional mutagenesis in ES cells (Schnutgen *et al.*, 2005). These vectors rely on gene trap cassettes flexed by heterotypic FRT/F3 and loxP/lox5171 sites, which enable two directional inversions if exposed to Cre and FLPe in succession. As exemplified in Fig. 26.5A for a SAβgeopA cassette, FLPe inverts

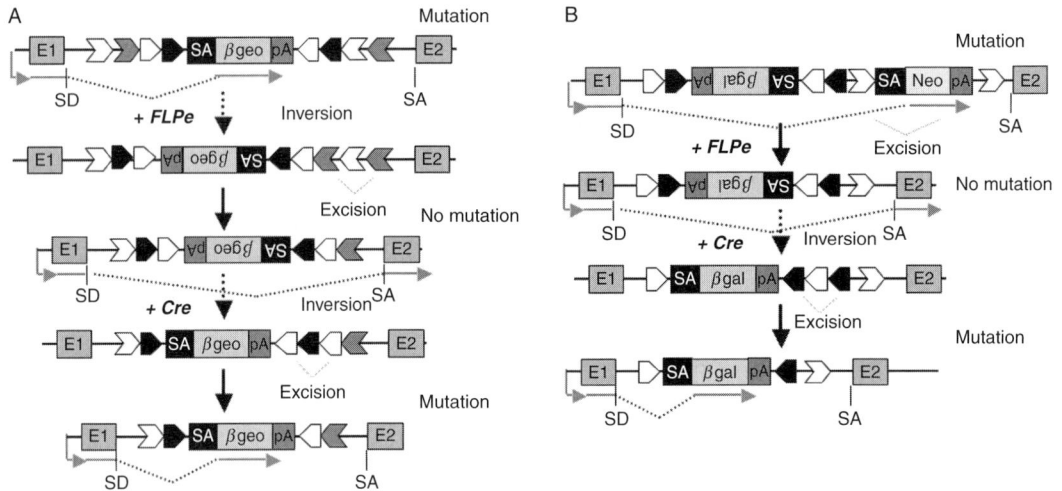

Figure 26.5 Conditional gene inactivation by gene trap vectors. (A) Single cassette strategy. A flexed SAβgeopA gene trap cassette is illustrated after integration into an intron of an expressed gene. White polygons, FRT sites; white polygons, F3 sites; grey pentagons, loxP sites; black pentagons, lox5171 sites; βgeo, β-galactosidase/neomycinphosphotransferase fusion gene. Transcripts (shown as gray arrows) initiated at the endogenous promoter are spliced from the splice donor (SD) of an endogenous exon (here exon 1) to the splice acceptor (SA) of the SAβgeopA cassette. Thereby the βgeo reporter gene is expressed and the endogenous transcript is captured and prematurely terminated at the cassette's polyadenylation sequence (pA) causing a mutation. FLPe inverts the SAβgeopA cassette onto the antisense, noncoding strand at either FRT (shown) or F3 (not shown) RTs and positions FRT and F3 sites between direct repeats of F3 and FRT RTs, respectively. By simultaneously excising the heterotypic RTs, the cassette is locked against reinversion as the remaining FRT and F3 RTs cannot recombine. This reactivates normal splicing between the endogenous splice sites, thereby repairing the mutation. Cre mediated inversion repositions the SAβgeopA cassette back onto the sense, coding strand and reinduces the mutation (Schnutgen *et al.*, 2005, 2008). (B) Double cassette strategy. Two gene trap cassettes in opposite orientation relative to each other are illustrated after integration into an intron of an expressed gene. Neo, neomycinphosphotransferase; βgal, β-galactosidase. The sense cassette (SAneopA) is flanked by wild type FRT sites in direct orientation and is amenable to excision by FLPe. The antisense cassette (SAβgalpA) is flexed by heterotypic lox sites, hence invertible by Cre. Transcripts initiated at the endogenous promoter are spliced from the SD of the endogenous exon to the splice acceptor site of the SAneopA cassette, causing a mutation (see above). FLPe repairs the mutation by excising the SAneopA cassette and reactivating normal splicing between the endogenous splice sites. Cre mediated inversion places the flexed SAβgalpA cassette onto the sense, coding strand and reinduces the mutation.

the gene trap cassette into an antisense orientation on the noncoding strand from where it is removed by endogenous splicing. This repairs the original mutation by restoring the normal gene expression. A second inversion induced by Cre places the cassette back onto the coding strand, which reinduces the mutation. Thus, the gene trap vectors allow (i) high throughput recovery of gene trap lines by selecting in G418, (ii) inactivation of gene trap mutations before ES cell conversion into mice by blastocyst injection, and (iii) reactivation of the mutation at prespecified times in selected tissues of the resulting mice. Alternatively, the gene trap mutations can be directly passaged to the germ line and then repaired and reinduced by two successive breeding cycles to FLPe and Cre *deleter* mice.

Essentially similar results have been obtained with vectors containing two gene trap cassettes in opposite orientation relative to each other, one of which is flexed and the other floxed (Fig. 26.5B; J. Altschmied, FS, HvM unpublished). In this strategy, the floxed cassette is used for trapped cell line selection and the flexed cassette for inducing the conditional mutation.

Any of the gene trap cassettes described earlier, insert heterologous RTs into the genome, opening the possibility for a large variety of postinsertional modifications using RMCE. Examples include replacing the gene trap cassettes with Cre recombinase genes to expand the number of tissue specific deleter strains (see below), or point mutated minigenes to study point mutations. Further options are the induction of gain of function mutations or the ablation of specific cell lineages by inserting gain of function cassettes or toxin genes, respectively.

4.3. Recombinase delivery to ES cells

Recombinases are routinely delivered to cultured cells by transient transfection of recombinase expression plasmids. For ES cells, electroporation is the method of choice. Coexpression of a selectable marker gene can be useful because it helps to enrich for recombinants. Promoter elements that are highly active in ES cells include the PGK and CMV/chicken β-actin (CAGGS) promoters. Usually, between 10% and 40% of the electroporated cells express the transduced recombinase at levels high enough to cause recombination. This frequency increases if a short selection is applied. However, to avoid genome integration of expression plasmids, selection should not exceed 48 h.

For Cre an alternative delivery method exists in the form of cell permeable derivatives of the Cre enzyme (Jo *et al.*, 2001; Nolden *et al.*, 2007; Peitz *et al.*, 2002). Of the several cell permeable Cre proteins described, the most widely used is HTN-Cre (HTNC). HTNC contains an N-terminal cell permeation domain derived from the HIV-TAT protein fused to a nuclear localization (NLS) sequence (Peitz *et al.*, 2002). Incubation of ES cells reporting Cre activity (see below) for 20 h with 2 μM of recombinant HTNC induced recombination in over 95% of the cells without any noticeable toxicity (Peitz *et al.*, 2002). This procedure is simple, efficient, and fast but does involve a period of cellular stress that may be deleterious for sensitive ES cell lines such as C57BL/6 lines. While large quantities of recombinant HTNC protein can be produced in *E. coli* and stored frozen without loss of activity (Nolden *et al.*, 2007; Peitz *et al.*, 2002), FLPe has thus far resisted conversion to a cell permeable protein and further efforts are in progress to achieve this goal.

4.4. Recombinase delivery to mice

To induce mutations in mice with conditional alleles, Cre is usually delivered by one of the following methods:

i. Spatial: crossing to a transgenic strain in which Cre is expressed from a tissue specific promoter. Only tissues with Cre expression develop a target gene mutation in the double transgenic offspring and thus reveal its tissue specific function.

ii. Temporal: crossing to a transgenic strain containing a ubiquitously expressed, ligand inducible Cre. Recombination occurs in all cells of double transgenic offspring after ligand induction and thus the cell types requiring the target gene at the time of induction can be identified.

iii. Spatio-temporal: crossing to a transgenic strain containing a tissue specific, ligand regulated Cre transgene. In double transgenic offspring, recombination occurs only in cell types expressing the transgene and after ligand induction. Depending on the inducible system the ligand can be applied orally (tetracyline), as a topical or by IP injection (tamoxifen).

iv. Local delivery of Cre, notably by adenoviruses, to specific cells and tissues. Unlike (i)–(iii), this method does not require extensive breeding.

Major limitations encountered with conditional mutagenesis are presently (i) a relatively low number of tissue specific promoters available for exclusive recombinase expression in somatic cells, (ii) the leakiness of most inducible SSR systems, and (iii) chromatin position effects affecting recombination efficiency between RTs as well as the expression of the SSRs. Chromatin position effects can be weakened and tissue specificity improved by inserting the SSRs genes into the mutant locus by gene targeting or RMCE (see below) to ensure an SSR expression pattern which is similar if not identical to that of the disrupted gene.

5. Recombinase Mediated Cassette Exchange (RMCE)

RMCE is a gene knock-in strategy that provides a simple alternative to gene replacement by homologous recombination. RMCE was first described by Bode and colleagues using FLP recombinase but has found wider use with Cre (Baer and Bode, 2001; Schlake and Bode, 1994). RMCE in ES cells is performed in two steps.

First, a pair of inversely oriented heterotypic RTs is inserted into a genomic region of interest by homologous recombination. This step can be circumvented if a suitable conditional gene trap ES cell line is available. As we discussed earlier, conditional gene trap vectors insert indirect repeats of heterotypic RTs across the genome. Hence, trapped ES cell lines can be used directly for RMCE. Over 5000 ES cell lines with conditional mutations in single genes have been already assembled by the International Gene Trap Consortium (IGTC) (Nord *et al.*, 2006) and the European Conditional Mouse Mutagenesis (EUCOMM) project (Collins *et al.*, 2007b) and it is anticipated that a further 8000 will be available in the near future. All cell lines can be ordered from the IGTC or EUCOMM (http://www. genetrap.org, http://www.eucomm.org) and are freely available to the scientific community.

Second, a circularized DNA fragment containing a replacement cassette between the same pair of inversely oriented heterotypic RTs (incoming construct) is electroporated into ES cells with a recombinase expression plasmid. Recombination inserts the incoming construct into the locus in a two step mechanism (Fig. 26.2B). Incoming cassettes almost always contain a positive selectable marker so that selection for the replaced locus can be applied. Selection can be improved by avoiding constitutively expressed selectable marker genes in the incoming constructs. For replacing conditional gene trap cassettes, for example, promoterless selectable marker genes are advisable to reduce background levels. Frequencies of up to 70% of correctly replaced loci have been reported with this method (Cobellis *et al.*, 2005). Best results are obtained with preengineered docking sites that activate or reconstitute an incoming promoterless or truncated selectable marker gene. Because only correct cassette insertions confer drug resistance the success rate with this approach is nearly 100% (Hitz *et al.*, 2007; Seibler *et al.*, 2007).

A potential limitation of RMCE relates to the length of the regions that can be exchanged. We could replace 5 kb of gene trap sequence by an SA-CreERT2-IRES-puro-pA cassette with a success rate of 28%, suggesting that RMCE with DNA fragments up to this size works reasonably well (FS, HvM, unpublished). Nevertheless, the relationship between the distance separating the genomic RTs and RMCE efficiencies has not been systematically explored. Similarly, the relationship between the distance separating RTs on the incoming construct and the efficiency of RMCE to insert or exchange long stretches of DNA has not been published.

6. Molecular Switches

Recombinase mediated DNA excision can be applied as a molecular switch to achieve permanent gene activation after transient exposure to a recombinase activity (Angrand *et al.*, 1998; Russ *et al.*, 1996a; Thorey *et al.*, 1998). The core element of a SSR molecular switch is a DNA fragment flanked by directly orientated RTs. The fragment, commonly referred to as a "STOP" sequence, is used to block gene expression; hence, it either disrupts the coding sequence of a gene or keeps the gene at a distance from an active promoter. STOP sequences used for spacing are frequently selectable marker or reporter genes, which are usually expressed before recombination. Recombination deletes the STOP sequence from the vector, which activates gene expression (Fig. 26.6A). The most widely used reporter gene encodes for β-galactosidase, which is an easily detectable histochemical marker and quantifiable in solution by spectrophotometry. Other popular reporter genes encode the firefly luciferase (LUC) or the enhanced green fluorescent protein (EGFP), which are both detectable in living cells and useful for *in vivo* imaging (Agah *et al.*, 1997; Constien *et al.*, 2001; Kolb *et al.*, 1999; Mao *et al.*, 2001; Pasqualetti *et al.*, 2002).

In their simplest application, switch vectors are used to monitor recombinase activity in cultured cells and mice. Because reporter activation is irreversible,

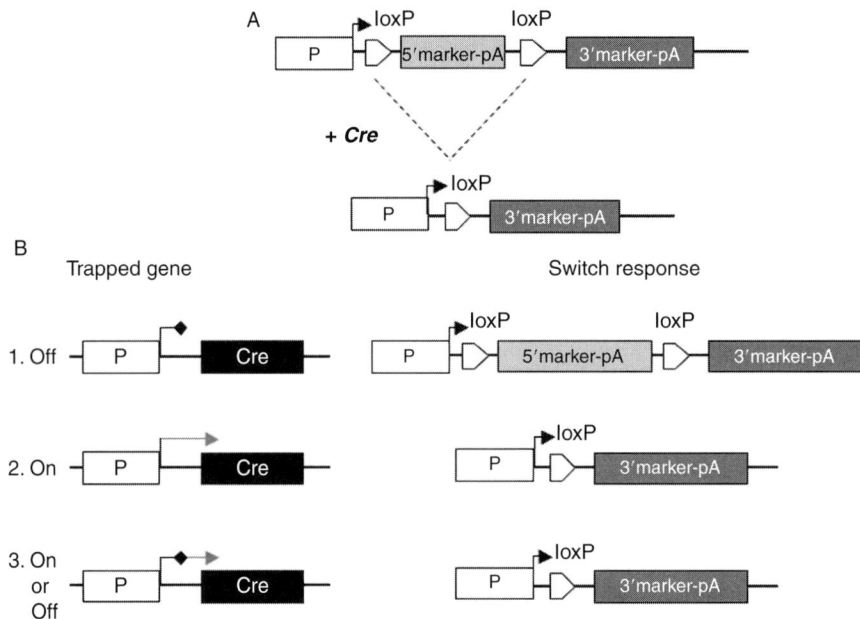

Figure 26.6 Gene expression switch. (A) A selectable marker (5′marker-pA), flanked by loxP sites (white pentagons) in direct orientations, is expressed from a constitutively active promoter (P) and simultaneously blocks the expression of a downstream gene (3′marker-pA) by polyadenylation (pA). Cre deletes the upstream cassette, which activates the downstream gene. (B) Use of the gene expression switch in a genetic screen for inducible genes. A switch cassette is integrated into ES cells by selection for the upstream selection marker, which is flanked by loxP sites (white pentagons). The downstream marker is not expressed because of premature polyadenylation (pA). These cells are transduced with a Cre gene trap vector. Vector integrations into active genes express Cre so that these are eliminated by continued selection for the upstream marker. Hence surviving cells do not express Cre, which must be integrated into silent genomic sites (1. OFF). Then, cells are treated with an inducer, which may be a cytokine or a hormone or a differentiation protocol. If this change activates a gene trap insertion site to express Cre, recombination ensues and selection applied for the activation of the downstream marker can be used to isolate these cells (2. ON). Regardless of whether Cre expression stays on or is subsequently turned off, the expression of the downstream marker continues (3. ON or OFF), thus facilitating the identification of transient sites of activation of Cre expression (Thorey *et al.*, 1998).

switch vectors are well suited for tracing recombinase activity in transgenic mice and are applied routinely in pretesting the ability of Cre deleter strains to induce tissue specific mutations.

The best reporter mouse lines have been obtained using vectors knocked into ubiquitously transcribed chromosomal regions, such as ROSA26 or HPRT (Akagi *et al.*, 1997; Lallemand *et al.*, 1998; Soriano, 1999; Thorey *et al.*, 1998; Tsien *et al.*, 1996). From such locations, the expression of the reporter gene after recombination accurately reflects the expression of the Cre transgene. If the cells that express Cre are known, the expression pattern of the reporter gene charts the progenitor/progeny relationship (lineage tracing) and the pattern of cell migration (fate mapping) during mouse development.

Recently, a switch vector enabling the combinatorial expression of fluorescent proteins (XFPs) in the mouse brain was published (Livet *et al.*, 2007). The vector contains three XFP-encoding genes (red, yellow, and cyan) flanked by alternating heterotypic lox sites such that Cre excision between either pair of identical lox sites removes the other pair ensuring the expression of only one XFP per cell. Since recombination is stochastic, coexpression of the vector with Cre in mammalian cells labeled each individual cell with a distinct color. In Cre expressing transgenic mice carrying multiple copies of the vector, the combinatorial expression of XFPs generated as many as 90 different colors each uniquely labeling individual neurons and their ramifications (Livet *et al.*, 2007).

Switch vectors have also been used for conditional knock-downs in mice employing short hairpin RNAs (shRNAs) for posttranscriptional gene silencing of gene expression. Switch vectors supporting PolIII expression of shRNAs have been successfully applied to inactivate the antiapoptotic bcl2 family protein A1 in B-cells (Oberdoerffer *et al.*, 2005) and the p53 tumor suppressor protein in mouse embryonic fibroblasts (MEFs) (Ventura *et al.*, 2004). Both strategies employed a U6-lox-STOP-lox-shRNA switch inserted into the ES cell genome. Cell type specific knock-downs in U6-lox-STOP-lox-shA1 and U6-lox-STOP-lox-shp53 transgenic mice were achieved either by crossing to B-cell specific Cre deleter mice (Oberdoerffer *et al.*, 2005) or by infecting cultured fibroblasts with adenoviral Cre expression vectors (Ventura *et al.*, 2004).

More recently, lox-STOP-lox cassettes have been used in combination with shRNAs to knock down two genes simultaneously in a tissue specific manner. The strategy involved inserting two specific shRNAs whose strands were separated by lox-STOP-lox cassettes into the ROSA26 locus. Excision of the lox-STOP-lox cassettes by Cre recombinase enabled strand assembly and shRNA formation, which effectively silenced the target genes in ES cells and mice (Steuber-Buchberger *et al.*, 2008).

Because activation of switch vectors requires only transient Cre expression, they have been used in combination with gene trapping to screen for genes induced during ES cell differentiation (Chen *et al.*, 2004; Thorey *et al.*, 1998). The strategy uses an ES cell line, which constitutively expresses a switch vector containing two selectable marker genes, of which the upstream one is flanked by loxP sites and blocks the expression of the downstream one (Fig. 26.6B). When this cell line is treated with gene trap vectors that insert a Cre gene randomly throughout the genome, insertions into active genes will express Cre, leading to deletion of the upstream gene by recombination. Selection for the deleted gene eliminates those ES cells with expressed gene trap integrations. The surviving ES cells are now enriched for insertions into "silent" genes and can be used to screen for genes that are activated by signals. Various signals can be employed, such as signal transduction pathways activated by differentiation, cytokines, or hormones. If the signal activates the site of Cre insertion, Cre is expressed and the upstream gene is deleted. Selection for the downstream gene, therefore, identifies the signal-responsive genes (Fig. 26.6B). Applied to retinoic acid responsive genes, this strategy effectively selected for developmentally regulated genes (Chen *et al.*, 2004).

7. Protocols

General protocols for growing ES cells in tissue culture are described elsewhere in this issue. As a rule, if manipulated ES cells are to be converted into mice they must be prevented from differentiating in the culture dish. In most cases, this is achieved by growing the cells in the presence of leukemia inhibitory factor (LIF) on mitotically inactivated feeder layers prepared with MEFs. If selection is required, MEFs need to be drug resistant. Feeder layers can be avoided by using ES cells adapted to feeder free culture conditions.

7.1. Recombinase delivery into ES cells (DNA transfection)

This protocol is used for removing targeted DNA fragments flanked by FRT or loxP sites using site-specific FLPe- or Cre-recombinases, respectively. For higher efficiency of recombination (or reducing work load) the recombinases are bicistronically expressed with a puromycin resistance gene (*pac*) under the control of the CAGGs promoter. Thereby, transient selection for 48 h can be used to reduce background level (Taniguchi *et al.*, 1998). The expression vector, pCAGGs-Flpe-IRES-puro-pA has been described elsewhere (Schaft *et al.*, 2001). Similar vectors for Cre and ΦC31 integrase (pCAGGs-Cre-IRES-puro-pA and pCAGGs-ΦC31IRES-puro-pA) are available. Usually, the flanked fragments include selection cassettes, so that the efficiency of recombination can be checked afterwards by sensitivity assays to the selection pressure used previously.

1. Split confluent ES cells (usually 1×10^7 cells per 10 cm TC dish) and start with 2×10^6 cells per 10 cm TC dish so that after 24 h they are in the log phase of growth and have a density of $4–5 \times 10^7$ cells per dish.
2. After 24 h, wash cells with PBS and trypsinize with 1 ml 0.5% v/v trypsin solution for 5 min at 37 °C.

3. Inactivate trypsin with 9 ml ES medium and pipette the suspension using a 10 ml pipette (8–10 times) in order to get single cells.

4. Take an aliquot of 10 μl and count cells.

5. Centrifuge in 15 ml Falcon tube at 1000 rpm for 5 min at RT.

6. Resuspend the pellet in PBS at 1×10^7 cells in 900 μl PBS.

7. Mix the 900 μl suspension with the DNA (40 μg, circular) and transfer to an electroporation cuvette (4 mm gap).

8. Leave on ice for 5–10 min.

9. Electroporate using the following conditions: 250 V, 500 μF (Gene Pulser X-cell; BioRad).

10. After electroporation immediately flick the cuvette to stabilize the pH.

11. Place the cuvette on ice.

12. Transfer the electroporated suspension using 1 ml pipette into a 15 ml Falcon tube containing 9.5 ml ES medium. Total volume is around 10.4 ml.

13. Distribute the cells in 10 cm TC dishes (1 ml per dish–10 dishes) that are already filled with 10 ml ES medium each.

14. After 24 h, replace the ES medium with one that contains 1 μg/ml puromycin (Sigma P-8833).

15. Keep the selection for 48 h without changing the medium.

16. Remove the selection medium and replace with normal ES medium and change every day.

17. 10–12 days after electroporation colonies become visible.

Other methods of recombinase delivery use transfection of *in vitro* transcribed RNA (Van den Plas *et al.*, 2003) or incubation with recombinase proteins fused to transduction domains (Nolden *et al.*, 2007). Detailed protocols can be found in the cited literature.

7.2. Low-density seeding of ES cells

If selection with puromycin is not possible or desired, the colonies will usually display a mosaic genotype for the recombination event. To purify the genotype, the cells have to be replated for growing clonal colonies.

1. Proceed as described above until step 14. Instead of step 14, let the cells grow for 2 days.

2. Wash cells with PBS and trypsinize with 1 ml trypsin for 5 min at 37 °C.

3. Inactivate trypsin with 9 ml ES medium and pipette the suspension using a 10 ml pipette (8–10 times) in order to get single cells.

4. Take an aliquot of 10 μl and count cells.

5. Seed 500 cells onto gelatinized 6 cm dishes containing feeder layers.

6. Grow cells for 10–12 days with daily change of medium until colonies reach an appropriate size for picking.

7.3. Picking of colonies

1. Seed feeder cells onto the appropriate number of freshly gelatinized 96-well plates one day before picking.
2. Aspirate culture medium and add prewarmed PBS. If colonies from more than one plate have to be picked, process them one by one. Break up the colonies using a Pipetman (P200) and sterile tips and collect the cells in 30 μl of PBS. Transfer the cell clumps to a 96-well plate and repeat until all colonies are picked.
3. Add 30 μl of 2 $\times$ trypsin solution to each well and incubate until cells can be easily dispersed by tapping the plate. Add 100 μl of medium to each well and transfer cells onto 96-well plates containing feeder cells.
4. When cells approach confluence, trypsinize, and expand for freezing and further analysis.

In cases where a selection cassette has been deleted, you can screen the colonies by applying selection pressure. For this reason, make duplicates or triplicates (depending if you want to see loss of selection markers in both alleles). One plate should contain normal medium and the second medium containing antibiotic corresponding to the deleted selection marker. Colonies that grow in normal medium but not in antibiotic containing medium have been efficiently recombined.

7.4. Molecular confirmation of excisions

To confirm the deletion of an FRT- or loxP-flanked fragment, DNA should be prepared and analyzed by PCR or Southern blot. A detailed protocol for preparing genomic DNA from ES-cells growing in 96-well plates has been described by Ramirez-Solis *et al.*(1993).

A schematic presentation of the Southern blot and PCR strategies is shown in Fig. 26.7. For the Southern blot, the DNA should be digested using an enzyme with recognition sequences upstream and downstream of the FRT-flanked selection cassette (enzyme x in Fig. 26.7B). It has become common practice to introduce the same restriction site in the targeting cassette. For our purpose, a restriction site immediately downstream of the FRT-site is ideal. After Southern blot, the membranes can be hybridized using a probe situated between the sites (probe 1 in Fig. 26.8B). This probe can distinguish between wt-allele, targeted-allele (unrecombined), and recombined allele (Fig. 26.7B). If both unrecombined and recombined alleles are detectable, the recombination has been partial (colonies with a mixed population of cells). In addition, the DNA of the colonies should be

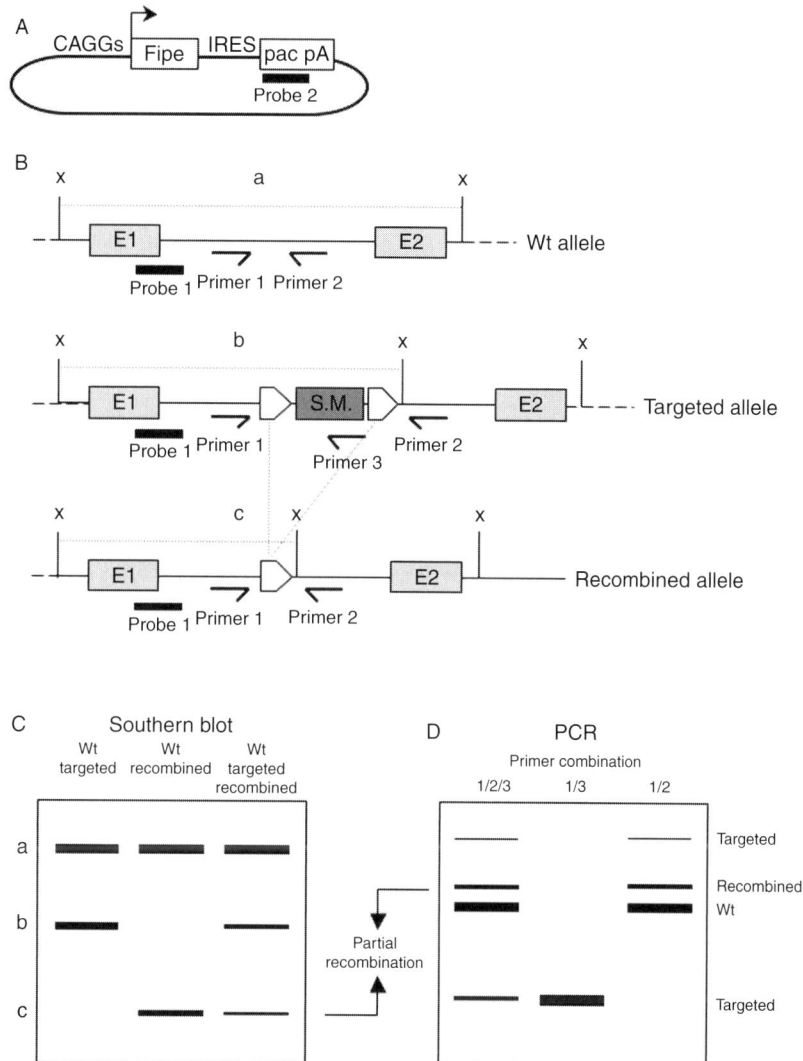

Figure 26.7 Recombination in ES cells. (A) The expression vector pCAGGs-Flpe-IRES-puro-pA that is used for transfection in ES cells. (B) Schematic presentation of an imaginary gene in all three possible conformations: wt-allele, targeted allele, and recombined allele. Depicted are only exons 1 and 2 (gray boxes). The selection marker (S.M.) flanked by FRTsites (white pentagons) has been integrated in the first intron. Vertical arrows show positions of the hypothetical restriction site (x). Horizontal arrows represent the PCR-primers used for testing recombination efficiency. Dashed horizontal lines that connect the restriction sites (x) represent the different fragments (a, b, c) that can be detected after hybridization with probe 1 (black bar). (C) Schematic presentation of Southern blot and the resulting fragments using probe 1. (D) Schematic presentation of PCR using different primer combinations.

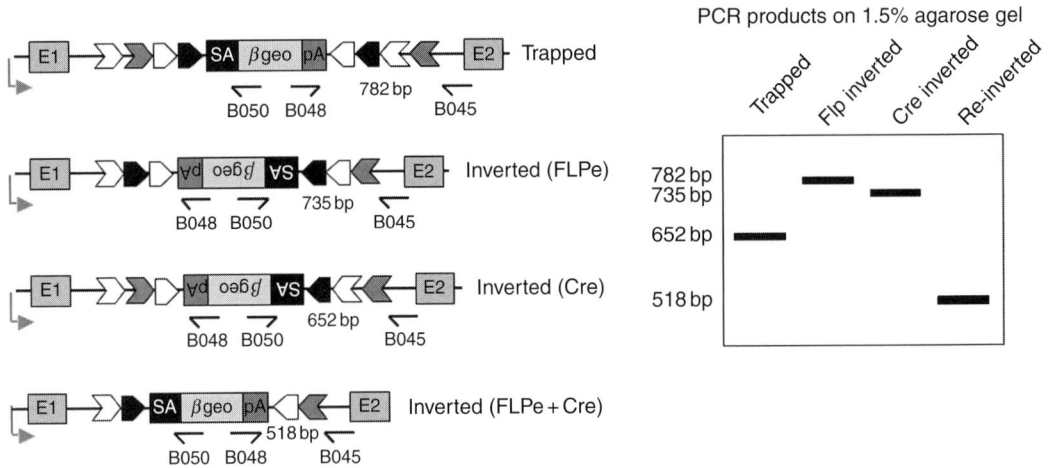

Figure 26.8 Multiplex PCR strategy for the verification of gene trap cassette inversions in ES cells and mice. Positions of primers (A) and expected amplification products (B) from the three possible postrecombination alleles of a FlipRosaβgeo gene trap locus.

analyzed for undesired integration of the recombinase expression plasmid (Southern blot using a puro-probe (probe 2 in Fig. 26.7A). For the PCR analysis, one should design the primers so that all possible variations can be distinguished. Usually a set of 3 primers can be used. Primers 1 and 2 amplify the wt- and the recombined-allele. Differences in size occur because of the recombinase target site left after recombination or additional sequences introduced with the targeted vector downstream of the cassette. One can also use the introduced restriction site for digesting the PCR product (only the recombined allele and not the wt will be digested). Primers 1 and 2 can also amplify the targeted allele (dashed line in Fig. 26.7D) but usually they lie far apart because of the introduced cassette so that they will rarely be amplified under standard PCR conditions. Therefore primer 3, which anneals in the targeting cassette, can be used for detecting residual unrecombined alleles.

For relatively small deletions (less than 4 kb), the usual rate of recombination using pCAGGs FLPe-expression vector is around 5%. In the case where both alleles contain a cassette flanked by FRT sites, colonies carrying a deletion of one allele were observed about three times more often than colonies with the deletion of both alleles. The puromycin selection protocol includes the risk that the genomic integration of pCAGGs-FLPe-IRES-puro plasmid will be selected. In our experience, we observed integration in up to half of all transiently puromycin selected colonies. It is quite likely that this frequency of undesired integration of the expression plasmid will be promoted by DNA breaks in the plasmid DNA preparation, so the use of supercoiled expression plasmid is recommended.

7.5. Tamoxifen treatment of primed ES cells

This approach is used in ES cells that contain a recombinase-LBD fusion integrated in the genome. The most widely used fusion is CreER(T2) (Feil *et al.*, 1997). Integration of Cre-ER(T2) in a locus that is ubiquitously expressed, such as the *Rosa26* has been described (Seibler *et al.*, 2003). In the inactive state, Cre-ER(T2) expressed in ES cells is associated with the Hsp90 in the cytoplasm. Induction occurs upon administration of a ligand, such as 4-hydroxytamoxifen. Cre-ER(T2) disassociates from Hsp90 and translocates to the nucleus. 4-Hydroxytamoxifen (4-OHT; Sigma H-6278) is dissolved at a concentration of 10^{-2} M in 100% ethanol and stored at $-20\,^{\circ}$C. An intermediate stock with a concentration of 10^{-4} M in 100% ethanol is also stored at $-20\,^{\circ}$C. The working concentration is 5×10^{-7} M in ES medium. Recombination efficiencies of almost 100% (for deletion of fragments around 1 kb) have been observed within a day of tamoxifen induction. Factors affecting the efficiency of recombination are the levels of recombinase, the distance between the loxP sites, and their site of integration in the genome. Therefore, the efficiency of recombination for different loxP-flanked fragments should be experimentally tested. Ideally, one should set a time-course and collect tamoxifen treated cells at different time points. For each time point, use 2 dishes so that you can analyze the efficiency of recombination at the DNA level and the effects at the RNA or protein level. Calculate the amount of plates you need (in duplicates) for your time course. A time course using uninduced, 6, 12, 24, 36, 48, 72, and 96 h induction will give you an idea of the kinetics of recombination. Do the inductions in reverse order, that means induce the cells for the 6 h time-point 6 h before collecting. This will facilitate your experiment because you will have the same amount of cells for each different time point and you can collect the cells at the same time.

1. Grow ES cells to confluence on 10 cm TC dishes in normal medium.
2. Split confluent ES cells (usually 1×10^7 cells per 10 cm TC dish) by trypsinization and start with 1×10^6 cells per 10 cm TC dish in medium containing 5×10^{-7} M 4OHT. Continue growing 4 plates of cells in normal medium (uninduced).
3. Change medium (4-OHT) after 24 h and add 4-OHT in 2 of the uninduced plates (time point 72 h).
4. After 48 h, the ES cells usually reach confluence and should be split by trypsinization as described in step 2. Split the induced plates and continue culturing in 4-OHT containing medium. Split the uninduced plate in at least 12 plates. Two of them already contain 4-OHT (time point 48 h). To the other 8 plates add 4-OHT 36 h, 24 h, 12 h, and 6 h before collecting (2 plates each time point). The last 2 plates remain uninduced.
5. After additional 48 h (96 h from the beginning of induction), collect cell-pellets for DNA, and protein extraction.
6. Wash twice with PBS.

7. Add 1 ml cold **PBS** per TC dish and scrap the cells using a rubber policeman. Transfer the scraped cells into 1.5 ml Eppendorf tubes, on ice.

8. Centrifuge at 4000 rpm for 4 min at 4 °C.

9. Aspirate supernatant and either freeze the cell pellet at −80 °C, or proceed with DNA and protein extraction, using standard methods.

7.6. *In vitro* verification of conditional constructs

Since subsequent steps often represent a considerable investment of time and labor, it is desirable to verify at an early stage that introduced recombination target sites are competent for recombination. Therefore, we suggest pretesting conditional constructs before integration into embryonic stem cells.

Pretesting conditional constructs in bacterial strains

1. Generate competent 294-Cre or 294-Flp bacteria (Buchholz *et al.*, 1996) that contain stably integrated expression cassettes for Cre recombinase or FLPe recombinase, respectively.

2. Transform constructs to be tested into the appropriate bacterial strain and inoculate 5 ml of LB medium containing antibiotics directly with the transformed bacteria without plating them onto LB agar plates. Grow overnight at 37 °C.

3. Isolate the plasmid DNA from these bacteria by standard procedures.

4. Transform appropriate lab strain (DH5α, XL1 blue) with 1 μl of miniprep DNA and plate onto LB agar plates containing antibiotic and incubate overnight.

5. Pick colonies and check for recombination events by restriction digest or PCR screening.

Pretesting conditional constructs in mammalian cells

1. Seed 8×10^5 293T cells into a 6 cm TC dish and allow to adhere overnight.

2. Cotransfect the conditional construct (2 μg) with a 4-fold excess of a recombinase expression plasmid (e.g., pCAGGs-Cre-IRES-puro-pA) using standard calcium phosphate precipitation.

3. Grow cells for 48 h with one medium change after 24 h.

4. Isolate DNA by standard procedures.

5. Transform *E. coli* (DH5α, XL1 blue) with 50 ng of DNA and inoculate 5 ml LB medium.

6. Incubate overnight at 37 °C in a bacterial shaker.

7. Miniprep plasmid DNA.

8. Analyze miniprep DNA by PCR, as described in Fig. 26.7.

7.7. Verification of gene trap cassette inversions

Conditional gene trap alleles contain gene disruption and selection cassettes that can be inverted twice by the successive application of FLPe and Cre recombinases. While this sequence can be reversed for certain *in vitro* applications, it is advisable to use FLPe for the first inversion in ES cells and the more potent Cre for the second inversion in the mouse. For recombinase delivery *in vitro*, follow the transfection protocol described above. Recombinase delivery *in vivo* requires breeding of the gene trap line derived transgenic mice to a FLPe or Cre deleter strain. Because the gene trap cassette is generic, the verification protocol applies to all ES cell lines and mice produced with a particular gene trap vector. Here, we describe the protocol for the FlipRosaβgeo vector and its derivatives (Schnutgen *et al.*, 2005), which have been used for the vast majority of the ES cell lines.

To monitor inversions, we use multiplex PCR with one forward primer complementary to the 3′ end of βgeo (B048; 5′-CCT CCC CCG TGC CTT CCT TGA C-3′) and two reverse primers complementary to the 5′end of βgeo and to viral sequences downstream of the 3′ RTs, respectively (B050; 5′-TTT GAG GGG ACG ACG ACA GTA T-3′ and (B045; 5′-CTC CGC CTC CTC TTC CTC CAT C-3′) (Fig. 26.8). The size of the amplification product is diagnostic for the orientation of the gene trap cassette (Fig. 26.8).

1. Isolate DNA from ES cells or mouse tails by standard procedures (Laird *et al.*, 1991; Ramirez-Solis *et al.*, 1995).

2. For each reaction pipette 23 μl of a PCR-master mix containing of 2.5 μl 10 × PCR buffer (Invitrogen), 0.75 μl of 2 mM MgCl$_2$, 0.5 μl dNTPs (10 μM each), 1 μl of each B045, B048, and B050 primers (10 pmol/μl), 0.5 μl DMSO, and 0.2 μl Taq polymerase (Invitrogen) in H$_2$O into 0.2 ml PCR tubes.

3. Add approximately 500 ng of DNA in 2 μl to each tube and place the tubes into an appropriate thermocycler.

4. Allow amplification reactions to proceed for 30 cycles using 94 °C for 45 s for denaturation, 61 °C for 45 s for annealing, and 72 °C for 60 s for elongation. Terminate the reaction with a final elongation cycle at 72 °C for 7 min.

5. Resolve amplification products on 1.5% agarose gels.

7.8. RMCE of gene trap cassettes with CreER(T2)

As described above, FlipRosaβgeo trap insertions are amenable to cassette exchange by RMCE. Here, we describe a method for replacing the SA-βgeo-pA gene trap cassette by a similar SA-P2A-CreERT2-IRES-Puro-pA cassette

(Fig. 26.9). Inserting CreERT2 into the gene trap locus ensures a CreERT2 expression pattern which is similar if not identical to that of the disrupted gene.

For RMCE follow the recombinase delivery into ES cells protocol from step 1–6. At step 7, add 30 μg of circular donor vector (Fig. 26.9) and 100 μg of pgkFLPo expression plasmid (Raymond and Soriano, 2007) to the 900 μl of cell suspension. Proceed with steps 8–17 as described. Pick puromycin resistant clones after 10 days in selection. Due to massive cell death during selection, change medium daily to remove the accumulating toxic products. Screen puromycin resistant colonies for correct cassette exchange as follows:

1. Stain for β-galactosidase (LacZ) by incubating the cells overnight at 37 °C in X-Gal staining solution (1 mg/ml X-Gal; 5 mM potassiumferricyanice; 5 mM potassiumferrocyanide; 50 mM MgCl$_2$ in PBS). If the original gene trap clone is LacZ+ (blue), clones with correct cassette exchange should be LacZ- (white). However, keep in mind that in presence of FLPo, LacZ-clones can also arise from gene trap cassette inversions.

2. Check LacZ-clones for correct insertions using RT-PCR. For this, design a forward primer complementary to exon sequences sequences immediately upstream of the insertion site and two reverse primers complementary to βgeo (B34: 5′-TGT AAA ACG ACG GGA TCC GCC-3′) and CreERT2 (K38: 5′-TCC CTG AAC ATG TCC ATC AG-3′) Set up RT-PCR reactions as follows:

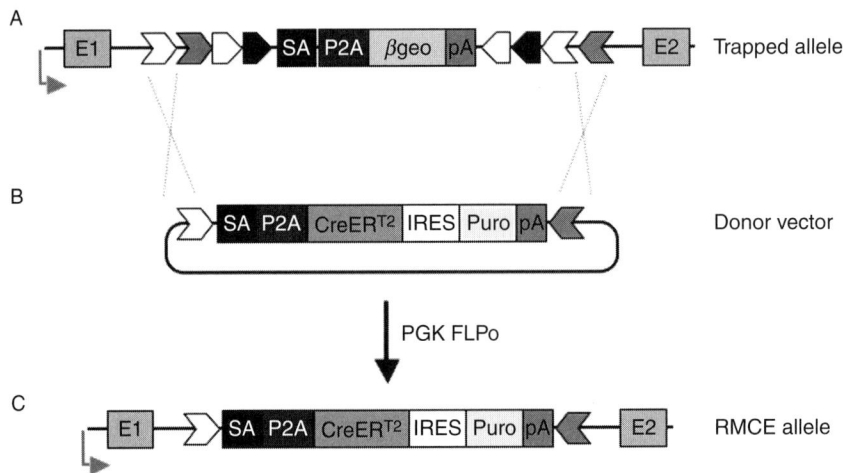

Figure 26.9 RMCE strategy for replacing a FlipRosaβgeo gene trap insertion with CreERT2. (A) FlipRosaβgeo gene trap insertion into the first intron of a gene. (B) Incoming CreERT2 gene trap vector consisting of a SA, a polyprotein cleavage sequence (P2A), the CreERT2 cDNA, an internal ribosomal entry site (IRES), a puromycin resistance gene (puro) and a polyadenylation sequence (pA). (C) FLPo mediated recombination replaces SAbgeopA with the CreERT2 exchange cassette in the same transcriptional orientation and uses the same translational reading frame. FRT sites, white polygons; F3 sites, gray polygons.

- Isolate RNA from ES cells by standard procedures (RNA Isolation kit. Sigma Aldrich, Hamburg, Germany).
- Reverse transcribe the 1 μg RNA in 11 μl using the Transcriptor First Strand cDNA Synthesis kit (Roche Applied Sciences, Mannheim, Germany) and the manufacturers' instructions.
- Set up two PCR reactions by pipetting into each of two 0.2 ml PCR tubes 23 μl of a PCR-master mix containing 2.5 μl 10 $\times$ PCR buffer (Invitrogen), 0.75 μl of 2 mM MgCl$_2$, 0.5 μl dNTPs (10 μM each), and 1 μl of gene specific forward primer (10 pmol/μl) in H$_2$O. Add 1 μl of the B034 reverse primer to one tube and 1 μl of the K038 primer to the other from a stock solution of 10 pmol/μl. Finally, add 2 μl first strand product and 0.2 μl of Taq polymerase (Invitrogen) to each tube.
- Place tubes into a thermocycler and allow amplification reactions to proceed for 30 cycles using 94 $°$C for 45 s for denaturation, 57 $°$C for 30 s for annealing, and 72 $°$C for 60 s for elongation. Terminate the reaction with a final elongation cycle at 72 $°$C for 7 min.
- Resolve amplification products on 1.5% agarose gels. Correct cassette exchanges should generate amplification products only with the KO38 primer.

To test whether CreER(T2) is active in clones with molecularly verified cassette exchanges, we transiently transfect the Cre activity reporter plasmid-pgk-lox-puropA-lox-βgeopA-into the cassette exchanged cells and stain with X-Gal following exposure to 4OHT.

1. Seed 6×10^4 ES cells per well onto 24 well plates and incubate for 24 h.
2. Prepare transfection complexes containing of 0.8 μg DNA and 2 μl Lipofectamine 200 according to the manufacturer's instructions.
3. Add the complexes dropwise to the cells and mix by gently shaking the plate.
4. Incubate cells overnight.
5. Trypsinize and seed one sixth of the cells from one well into each of two new wells of 24-well plate.
6. Add 4OHT to one of the wells at a final concentration of 5×10^{-7} M and incubate for 48 h.
7. Wash cells twice with PBS.
8. Fix the cells for 5' minutes at room temperature in 2% formaldehyde.
9. Wash again to remove the formaldehyde.
10. Add X-Gal staining solution (see above) and incubate overnight at 37 $°$C.

Acknowledgments

This work was supported by grants from the Deutsche Forschungsgemeinschaft to HvM and from the European Union to AFS and HvM.

References

Agah, R., Frenkel, P. A., French, B. A., Michael, L. H., Overbeek, P. A., and Schneider, M. D. (1997). Gene recombination in postmitotic cells. Targeted expression of Cre recombinase provokes cardiac-restricted, site-specific rearrangement in adult ventricular muscle *in vivo*. *J. Clin. Invest.* **100**, 169–179.

Akagi, K., Sandig, V., Vooijs, M., Van der Valk, M., Giovannini, M., Strauss, M., and Berns, A. (1997). Cre-mediated somatic site-specific recombination in mice. *Nucleic Acids Res.* **25**, 1766–1773.

Albert, H., Dale, E. C., Lee, E., and Ow, D. W. (1995). Site-specific integration of DNA into wild-type and mutant lox sites placed in the plant genome. *Plant J.* **7**, 649–659.

Andreas, S., Schwenk, F., Kuter-Luks, B., Faust, N., and Kuhn, R. (2002). Enhanced efficiency through nuclear localization signal fusion on phage PhiC31-integrase: Activity comparison with Cre and FLPe recombinase in mammalian cells. *Nucleic Acids Res.* **30**, 2299–2306.

Angrand, P. O., Woodroofe, C. P., Buchholz, F., and Stewart, A. F. (1998). Inducible expression based on regulated recombination: A single vector strategy for stable expression in cultured cells. *Nucleic Acids Res.* **26**, 3263–3269.

Araki, K., Araki, M., and Yamamura, K. (1997). Targeted integration of DNA using mutant lox sites in embryonic stem cells. *Nucleic Acids Res.* **25**, 868–872.

Araki, H., Nakanishi, N., Evans, B. R., Matsuzaki, H., Jayaram, M., and Oshima, Y. (1992). Site-specific recombinase, R, encoded by yeast plasmid pSR1. *J. Mol. Biol.* **225**, 25–37.

Austin, C. P., Battey, J. F., Bradley, A., Bucan, M., Capecchi, M., Collins, F. S., Dove, W. F., Duyk, G., Dymecki, S., Eppig, J. T., Grieder, F. B., Heintz, N., *et al.* (2004). The knockout mouse project. *Nat. Genet.* **36**, 921–924.

Auwerx, J., Avner, P., Baldock, R., Ballabio, A., Balling, R., Barbacid, M., Berns, A., Bradley, A., Brown, S., Carmeliet, P., Chambon, P., Cox, R., *et al.* (2004). The European dimension for the mouse genome mutagenesis program. *Nat. Genet.* **36**, 925–927.

Baba, Y., Nakano, M., Yamada, Y., Saito, I., and Kanegae, Y. (2005). Practical range of effective dose for Cre recombinase-expressing recombinant adenovirus without cell toxicity in mammalian cells. *Microbiol. Immunol.* **49**, 559–570.

Baer, A., and Bode, J. (2001). Coping with kinetic and thermodynamic barriers: RMCE, an efficient strategy for the targeted integration of transgenes. *Curr. Opin. Biotechnol.* **12**, 473–480.

Belteki, G., Gertsenstein, M., Ow, D. W., and Nagy, A. (2003). Site-specific cassette exchange and germline transmission with mouse ES cells expressing phiC31 integrase. *Nat. Biotechnol.* **21**, 321–324.

Buchholz, F., Angrand, P. O., and Stewart, A. F. (1996). A simple assay to determine the functionality of Cre or FLP recombination targets in genomic manipulation constructs. *Nucleic Acids Res.* **24**, 3118–3119.

Buchholz, F., Angrand, P. O., and Stewart, A. F. (1998). Improved properties of FLP recombinase evolved by cycling mutagenesis. *Nat. Biotechnol.* **16**, 657–662.

Chen, Y. T., Liu, P., and Bradley, A. (2004). Inducible gene trapping with drug-selectable markers and Cre/loxP to identify developmentally regulated genes. *Mol. Cell. Biol.* **24**, 9930–9941.

Chen, Y., Narendra, U., Iype, L. E., Cox, M. M., and Rice, P. A. (2000). Crystal structure of a Flp recombinase-Holliday junction complex: Assembly of an active oligomer by helix swapping. *Mol. Cell* **6**, 885–897.

Chen, Y., and Rice, P. A. (2003). New insight into site-specific recombination from FLP recombinase-DNA tructures. *Annu. Rev. Biophys. Biomol. Struct* **32**, 135–159.

Cobellis, G., Nicolaus, G., Iovino, M., Romito, A., Marra, E., Barbarisi, M., Sardiello, M., Di Giorgio, F. P., Iovino, N., Zollo, M., Ballabio, A., and Cortese, R. (2005). Tagging genes with cassette-exchange sites. *Nucleic Acids Res.* **33**, e44.

Collins, F. S., Finnell, R. H., Rossant, J., and Wurst, W. (2007). A new partner for the international knockout mouse consortium. *Cell* **129**, 235.

Collins, F. S., Rossant, J., and Wurst, W. (2007). A mouse for all reasons. *Cell* **128**, 9–13.

Constien, R., Forde, A., Liliensiek, B., Grone, H. J., Nawroth, P., Hammerling, G., and Arnold, B. (2001). Characterization of a novel EGFP reporter mouse to monitor Cre recombination as demonstrated by a Tie2 Cre mouse line. *Genesis* **30**, 36–44.

Cronin, C. A., Gluba, W., and Scrable, H. (2001). The lac operator-repressor system is functional in the mouse. *Genes Dev.* **15,** 1506–1517.

Danielian, P. S., Muccino, D., Rowitch, D. H., Michael, S. K., and McMahon, A. P. (1998). Modification of gene activity in mouse embryos *in utero* by a tamoxifen-inducible form of Cre recombinase. *Curr. Biol.* **8,** 1323–1326.

Ennifar, E., Meyer, J. E., Buchholz, F., Stewart, A. F., and Suck, D. (2003). Crystal structure of a wild-type Cre recombinase-loxP synapse reveals a novel spacer conformation suggesting an alternative mechanism for DNA cleavage activation. *Nucleic Acids Res.* **31,** 5449–5460.

Farley, F. W., Soriano, P., Steffen, L. S., and Dymecki, S. M. (2000). Widespread recombinase expression using FLPeR (flipper) mice. *Genesis* **28,** 106–110.

Feil, R., Brocard, J., Mascrez, B., LeMeur, M., Metzger, D., and Chambon, P. (1996). Ligand-activated site-specific recombination in mice. *Proc. Natl. Acad. Sci. USA* **93,** 10887–10890.

Feil, R., Wagner, J., Metzger, D., and Chambon, P. (1997). Regulation of Cre recombinase activity by mutated estrogen receptor ligand-binding domains. *Biochem. Biophys. Res. Commun.* **237,** 752–757.

Feng, Y. Q., Seibler, J., Alami, R., Eisen, A., Westerman, K. A., Leboulch, P., Fiering, S., and Bouhassira, E. E. (1999). Site-specific chromosomal integration in mammalian cells: Highly efficient CRE recombinase-mediated cassette exchange. *J. Mol. Biol.* **292,** 779–785.

Friedel, R. H., Plump, A., Lu, X., Spilker, K., Jolicoeur, C., Wong, K., Venkatesh, T. R., Yaron, A., Hynes, M., Chen, B., Okada, A., McConnell, S. K., *et al.* (2005). Gene targeting using a promoterless gene trap vector ("targeted trapping") is an efficient method to mutate a large fraction of genes. *Proc. Natl. Acad. Sci. USA* **102,** 13188–13193.

Glaser, S., Anastassiadis, K., and Stewart, A. F. (2005). Current issues in mouse genome engineering. *Nat. Genet.* **37,** 1187–1193.

Gu, H., Zou, Y. R., and Rajewsky, K. (1993). Independent control of immunoglobulin switch recombination at individual switch regions evidenced through Cre-loxP-mediated gene targeting. *Cell* **73,** 1155–1164.

Hitz, C., Wurst, W., and Kuhn, R. (2007). Conditional brain-specific knockdown of MAPK using Cre/loxP regulated RNA interference. *Nucleic Acids Res.* **35,** e90.

Jo, D., Nashabi, A., Doxsee, C., Lin, Q., Unutmaz, D., Chen, J., and Ruley, H. E. (2001). Epigenetic regulation of gene structure and function with a cell-permeable Cre recombinase. *Nat. Biotechnol.* **19,** 929–933.

Kaul, A., Koster, M., Neuhaus, H., and Braun, T. (2000). Myf-5 revisited: Loss of early myotome formation does not lead to a rib phenotype in homozygous Myf-5 mutant mice. *Cell* **102,** 17–19.

Kolb, A. F. (2001). Selection-marker-free modification of the murine beta-casein gene using a lox2272 [correction of lox2722] site. *Anal. Biochem.* **290,** 260–271.

Kolb, A. F., Ansell, R., McWhir, J., and Siddell, S. G. (1999). Insertion of a foreign gene into the beta-casein locus by Cre-mediated site-specific recombination. *Gene* **227,** 21–31.

Laird, P. W., Zijderveld, A., Linders, K., Rudnicki, M. A., Jaenisch, R., and Berns, A. (1991). Simplified mammalian DNA isolation procedure. *Nucleic Acids Res.* **19,** 4293.

Lallemand, Y., Luria, V., Haffner-Krausz, R., and Lonai, P. (1998). Maternally expressed PGK-Cre transgene as a tool for early and uniform activation of the Cre site-specific recombinase. *Transgenic Res.* **7,** 105–112.

Langer, S. J., Ghafoori, A. P., Byrd, M., and Leinwand, L. (2002). A genetic screen identifies novel non-compatible loxP sites. *Nucleic Acids Res.* **30,** 3067–3077.

Lauth, M., Moerl, K., Barski, J. J., and Meyer, M. (2000). Characterization of Cre-mediated cassette exchange after plasmid microinjection in fertilized mouse oocytes. *Genesis* **27,** 153–158.

Lauth, M., Spreafico, F., Dethleffsen, K., and Meyer, M. (2002). Stable and efficient cassette exchange under non-selectable conditions by combined use of two site-specific recombinases. *Nucleic Acids Res.* **30,** e115.

Le, Y., Gagneten, S., Tombaccini, D., Bethke, B., and Sauer, B. (1999). Nuclear targeting determinants of the phage P1 cre DNA recombinase. *Nucleic Acids Res.* **27,** 4703–4709.

Lee, G., and Saito, I. (1998). Role of nucleotide sequences of loxP spacer region in Cre-mediated recombination. *Gene* **216**, 55–65.

Li, W., Kamtekar, S., Xiong, Y., Sarkis, G. J., Grindley, N. D., and Steitz, T. A. (2005). Structure of a synaptic gammadelta resolvase tetramer covalently linked to two cleaved DNAs. *Science* **309**, 1210–1215.

Li, Z. W., Stark, G., Gotz, J., Rulicke, T., Gschwind, M., Huber, G., Muller, U., and Weissmann, C. (1996). Generation of mice with a 200-kb amyloid precursor protein gene deletion by Cre recombinase-mediated site-specific recombination in embryonic stem cells. *Proc. Natl. Acad. Sci. USA* **93**, 6158–6162.

Livet, J., Weissman, T. A., Kang, H., Draft, R. W., Lu, J., Bennis, R. A., Sanes, J. R., and Lichtman, J. W. (2007). Transgenic strategies for combinatorial expression of fluorescent proteins in the nervous system. *Nature* **450**, 56–62.

Logie, C., and Stewart, A. F. (1995). Ligand-regulated site-specific recombination. *Proc. Natl. Acad. Sci. USA* **92**, 5940–5944.

Loonstra, A., Vooijs, M., Beverloo, H. B., Allak, B. A., van Drunen, E., Kanaar, R., Berns, A., and Jonkers, J. (2001). Growth inhibition and DNA damage induced by Cre recombinase in mammalian cells. *Proc. Natl. Acad. Sci. USA* **98**, 9209–9214.

Madsen, L., Labrecque, N., Engberg, J., Dierich, A., Svejgaard, A., Benoist, C., Mathis, D., and Fugger, L. (1999). Mice lacking all conventional MHC class II genes. *Proc. Natl. Acad. Sci. USA* **96**, 10338–10343.

Mao, J., Barrow, J., McMahon, J., Vaughan, J., and McMahon, A. P. (2005). An ES cell system for rapid, spatial and temporal analysis of gene function *in vitro* and *in vivo*. *Nucleic Acids Res.* **33**, e155.

Mao, X., Fujiwara, Y., Chapdelaine, A., Yang, H., and Orkin, S. H. (2001). Activation of EGFP expression by Cre-mediated excision in a new ROSA26 reporter mouse strain. *Blood* **97**, 324–326.

Meyers, E. N., Lewandoski, M., and Martin, G. R. (1998). An Fgf8 mutant allelic series generated by Cre- and Flp-mediated recombination. *Nat. Genet.* **18**, 136–141.

Nagy, A., Moens, C., Ivanyi, E., Pawling, J., Gertsenstein, M., Hadjantonakis, A. K., Pirity, M., and Rossant, J. (1998). Dissecting the role of N-myc in development using a single targeting vector to generate a series of alleles. *Curr. Biol.* **8**, 661–664.

Nolden, L., Edenhofer, F., Peitz, M., and Brustle, O. (2007). Stem cell engineering using transducible Cre recombinase. *Methods Mol. Med.* **140**, 17–32.

Nord, A. S., Chang, P. J., Conklin, B. R., Cox, A. V., Harper, C. A., Hicks, G. G., Huang, C. C., Johns, S. J., Kawamoto, M., Liu, S., Meng, E. C., Morris, J. H., *et al.* (2006). The International Gene Trap Consortium Website: A portal to all publicly available gene trap cell lines in mouse. *Nucleic Acids Res.* **34**, D642–D648.

Oberdoerffer, P., Kanellopoulou, C., Heissmeyer, V., Paeper, C., Borowski, C., Aifantis, I., Rao, A., and Rajewsky, K. (2005). Efficiency of RNA interference in the mouse hematopoietic system varies between cell types and developmental stages. *Mol. Cell. Biol.* **25**, 3896–3905.

Oberdoerffer, P., Otipoby, K. L., Maruyama, M., and Rajewsky, K. (2003). Unidirectional Cre-mediated genetic inversion in mice using the mutant loxP pair lox66/lox71. *Nucleic Acids Res.* **31**, e140.

O'Gorman, S., Dagenais, N. A., Qian, M., and Marchuk, Y. (1997). Protamine-Cre recombinase transgenes efficiently recombine target sequences in the male germ line of mice, but not in embryonic stem cells. *Proc. Natl. Acad. Sci. USA* **94**, 14602–14607.

Okabe, M., Ikawa, M., Kominami, K., Nakanishi, T., and Nishimune, Y. (1997). 'Green mice' as a source of ubiquitous green cells. *FEBS Lett.* **407**, 313–319.

Olivares, E. C., Hollis, R. P., Chalberg, T. W., Meuse, L., Kay, M. A., and Calos, M. P. (2002). Site-specific genomic integration produces therapeutic Factor IX levels in mice. *Nat. Biotechnol.* **20**, 1124–1128.

Oram, M., Szczelkun, M. D., and Halford, S. E. (1995). Recombination. pieces of the site-specific recombination puzzle. *Curr. Biol.* **5**, 1106–1109.

Pasqualetti, M., Ren, S. Y., Poulet, M., LeMeur, M., Dierich, A., and Rijli, F. M. (2002). A Hoxa2 knockin allele that expresses EGFP upon conditional Cre-mediated recombination. *Genesis* **32**, 109–111.

Peitz, M., Pfannkuche, K., Rajewsky, K., and Edenhofer, F. (2002). Ability of the hydrophobic FGF and basic TAT peptides to promote cellular uptake of recombinant Cre recombinase: A tool for efficient genetic engineering of mammalian genomes. *Proc. Natl. Acad. Sci. USA* **99**, 4489–4494.

Pfeifer, A., Brandon, E. P., Kootstra, N., Gage, F. H., and Verma, I. M. (2001). Delivery of the Cre recombinase by a self-deleting lentiviral vector: Efficient gene targeting *in vivo. Proc. Natl. Acad. Sci. USA* **98**, 11450–11455.

Ramirez-Solis, R., Davis, A. C., and Bradley, A. (1993). Gene targeting in embryonic stem cells. *Methods Enzymol.* **225**, 855–878.

Ramirez-Solis, R., Liu, P., and Bradley, A. (1995). Chromosome engineering in mice. *Nature* **378**, 720–724.

Raymond, C. S., and Soriano, P. (2007). High-efficiency FLP and PhiC31 site-specific recombination in mammalian cells. *PLoS ONE* **2**, e162.

Ringrose, L., Angrand, P. O., and Stewart, A. F. (1997). The Kw recombinase, an integrase from *Kluyveromyces waltii. Eur. J. Biochem.* **248**, 903–912.

Ringrose, L., Chabanis, S., Angrand, P. O., Woodroofe, C., and Stewart, A. F. (1999). Quantitative comparison of DNA looping *in vitro* and *in vivo*: Chromatin increases effective DNA flexibility at short distances. *EMBO J.* **18**, 6630–6641.

Ringrose, L., Lounnas, V., Ehrlich, L., Buchholz, F., Wade, R., and Stewart, A. F. (1998). Comparative kinetic analysis of FLP and cre recombinases: Mathematical models for DNA binding and recombination. *J. Mol. Biol.* **284**, 363–384.

Rodriguez, C. I., Buchholz, F., Galloway, J., Sequerra, R., Kasper, J., Ayala, R., Stewart, A. F., and Dymecki, S. M. (2000). High-efficiency deleter mice show that FLPe is an alternative to Cre-loxP. *Nat. Genet.* **25**, 139–140.

Russ, A. P., Friedel, C., Ballas, K., Kalina, U., Zahn, D., Strebhardt, K., and von Melchner, H. (1996). Identification of genes induced by factor deprivation in hematopoietic cells undergoing apoptosis using gene-trap mutagenesis and site-specific recombination. *Proc. Natl. Acad. Sci. USA* **93**, 15279–15284.

Russ, A. P., Friedel, C., Grez, M., and von Melchner, H. (1996). Self-deleting retrovirus vectors for gene therapy. *J. Virol.* **70**, 4927–4932.

Saam, J. R., and Gordon, J. I. (1999). Inducible gene knockouts in the small intestinal and colonic epithelium. *J. Biol. Chem.* **274**, 38071–38082.

Sauer, B., and McDermott, J. (2004). DNA recombination with a heterospecific Cre homolog identified from comparison of the pac-c1 regions of P1-related phages. *Nucleic Acids Res.* **32**, 6086–6095.

Schaft, J., Ashery-Padan, R., van der Hoeven, F., Gruss, P., and Stewart, A. F. (2001). Efficient FLP recombination in mouse ES cells and oocytes. *Genesis* **31**, 6–10.

Schlake, T., and Bode, J. (1994). Use of mutated FLP recognition target (FRT) sites for the exchange of expression cassettes at defined chromosomal loci. *Biochemistry* **33**, 12746–12751.

Schlake, T., Schupp, I., Kutsche, K., Mincheva, A., Lichter, P., and Boehm, T. (1999). Predetermined chromosomal deletion encompassing the Nf-1 gene. *Oncogene* **18**, 6078–6082.

Schmidt, E. E., Taylor, D. S., Prigge, J. R., Barnett, S., and Capecchi, M. R. (2000). Illegitimate Cre-dependent chromosome rearrangements in transgenic mouse spermatids. *Proc. Natl. Acad. Sci. USA* **97**, 13702–13707.

Schmidt-Supprian, M., and Rajewsky, K. (2007). Vagaries of conditional gene targeting. *Nat. Immunol.* **8**, 665–668.

Schnutgen, F., De-Zolt, S., Van Sloun, P., Hollatz, M., Floss, T., Hansen, J., Altschmied, J., Seisenberger, C., Ghyselinck, N. B., Ruiz, P., Chambon, P., Wurst, W., *et al.* (2005). Genomewide production of multipurpose alleles for the functional analysis of the mouse genome. *Proc. Natl. Acad. Sci. USA* **102**, 7221–7226.

Schnutgen, F., Hansen, J., De-Zolt, S., Horn, C., Lutz, M., Floss, T., Wurst, W., Ruiz Noppinger, P., and von Melchner, H. (2008). Enhanced gene trapping in mouse embryonic stem cells. *Nucleic Acids Res.* doi:10.1093/nar/gkn603.

Schnutgen, F., Doerflinger, N., Calleja, C., Wendling, O., Chambon, P., and Ghyselinck, N. B. (2003). A directional strategy for monitoring Cre-mediated recombination at the cellular level in the mouse. *Nat. Biotechnol.* **21,** 562–565.

Schwenk, F., Kuhn, R., Angrand, P. O., Rajewsky, K., and Stewart, A. F. (1998). Temporally and spatially regulated somatic mutagenesis in mice. *Nucleic Acids Res.* **26,** 1427–1432.

Sclimenti, C. R., Thyagarajan, B., and Calos, M. P. (2001). Directed evolution of a recombinase for improved genomic integration at a native human sequence. *Nucleic Acids Res.* **29,** 5044–5051.

Seibler, J., Kleinridders, A., Kuter-Luks, B., Niehaves, S., Bruning, J. C., and Schwenk, F. (2007). Reversible gene knockdown in mice using a tight, inducible shRNA expression system. *Nucleic Acids Res.* **35,** e54.

Seibler, J., Zevnik, B., Kuter-Luks, B., Andreas, S., Kern, H., Hennek, T., Rode, A., Heimann, C., Faust, N., Kauselmann, G., Schoor, M., Jaenisch, R., *et al.* (2003). Rapid generation of inducible mouse mutants. *Nucleic Acids Res.* **31,** e12.

Semprini, S., Troup, T. J., Kotelevtseva, N., King, K., Davis, J. R., Mullins, L. J., Chapman, K. E., Dunbar, D. R., and Mullins, J. J. (2007). Cryptic loxP sites in mammalian genomes: Genome-wide distribution and relevance for the efficiency of BAC/PAC recombineering techniques. *Nucleic Acids Res.* **35,** 1402–1410.

Shibata, H., Toyama, K., Shioya, H., Ito, M., Hirota, M., Hasegawa, S., Matsumoto, H., Takano, H., Akiyama, T., Toyoshima, K., Kanamaru, R., Kanegae, Y., *et al.* (1997). Rapid colorectal adenoma formation initiated by conditional targeting of the Apc gene. *Science* **278,** 120–123.

Shimshek, D. R., Kim, J., Hubner, M. R., Spergel, D. J., Buchholz, F., Casanova, E., Stewart, A. F., Seeburg, P. H., and Sprengel, R. (2002). Codon-improved Cre recombinase (iCre) expression in the mouse. *Genesis* **32,** 19–26.

Silver, D. P., and Livingston, D. M. (2001). Self-excising retroviral vectors encoding the Cre recombinase overcome Cre-mediated cellular toxicity. *Mol. Cell* **8,** 233–243.

Smith, M. C., and Thorpe, H. M. (2002). Diversity in the serine recombinases. *Mol. Microbiol.* **44,** 299–307.

Soriano, P. (1999). Generalized lacZ expression with the ROSA26 Cre reporter strain. *Nat. Genet.* **21,** 70–71.

Stanford, W. L., Epp, T., Reid, T., and Rossant, J. (2006). Gene trapping in embryonic stem cells. *Methods Enzymol.* **420,** 136–162.

Sternberg, N., Sauer, B., Hoess, R., and Abremski, K. (1986). Bacteriophage P1 cre gene and its regulatory region. Evidence for multiple promoters and for regulation by DNA methylation. *J. Mol. Biol.* **187,** 197–212.

Steuber-Buchberger, P., Wurst, W., and Kuhn, R. (2008). Simultaneous Cre-mediated conditional knockdown of two genes in mice. *Genesis* **46,** 144–151.

Taniguchi, M., Sanbo, M., Watanabe, S., Naruse, I., Mishina, M., and Yagi, T. (1998). Efficient production of Cre-mediated site-directed recombinants through the utilization of the puromycin resistance gene, pac: A transient gene-integration marker for ES cells. *Nucleic Acids Res.* **26,** 679–680.

Thorey, I. S., Muth, K., Russ, A. P., Otte, J., Reffelmann, A., and von Melchner, H. (1998). Selective disruption of genes transiently induced in differentiating mouse embryonic stem cells by using gene trap mutagenesis and site-specific recombination. *Mol. Cell. Biol.* **18,** 3081–3088.

Thorpe, H. M., and Smith, M. C. (1998). *In vitro* site-specific integration of bacteriophage DNA catalyzed by a recombinase of the resolvase/invertase family. *Proc. Natl. Acad. Sci. USA* **95,** 5505–5510.

Tsien, J. Z., Chen, D. F., Gerber, D., Tom, C., Mercer, E. H., Anderson, D. J., Mayford, M., Kandel, E. R., and Tonegawa, S. (1996). Subregion- and cell type-restricted gene knockout in mouse brain. *Cell* **87,** 1317–1326.

Umlauf, S. W., and Cox, M. M. (1988). The functional significance of DNA sequence structure in a site-specific genetic recombination reaction. *EMBO J.* **7,** 1845–1852.

Van den Plas, D., Ponsaerts, P., Van Tendeloo, V., Van Bockstaele, D. R., Berneman, Z. N., and Merregaert, J. (2003). Efficient removal of LoxP-flanked genes by electroporation of Cre-recombinase mRNA. *Biochem. Biophys. Res. Commun.* **305,** 10–15.

Van Duyne, G. D. (2001). A structural view of cre-loxp site-specific recombination. *Annu. Rev. Biophys. Biomol. Struct.* **30,** 87–104.

Ventura, A., Meissner, A., Dillon, C. P., McManus, M., Sharp, P. A., Van Parijs, L., Jaenisch, R., and Jacks, T. (2004). Cre-lox-regulated conditional RNA interference from transgenes. *Proc. Natl. Acad. Sci. USA* **101,** 10380–10385.

Vooijs, M., Jonkers, J., and Berns, A. (2001). A highly efficient ligand-regulated Cre recombinase mouse line shows that LoxP recombination is position dependent. *EMBO Rep.* **2,** 292–297.

Wunderlich, F. T., Wildner, H., Rajewsky, K., and Edenhofer, F. (2001). New variants of inducible Cre recombinase: A novel mutant of Cre-PR fusion protein exhibits enhanced sensitivity and an expanded range of inducibility. *Nucleic Acids Res.* **29,** E47.

Xin, H. B., Deng, K. Y., Shui, B., Qu, S., Sun, Q., Lee, J., Greene, K. S., Wilson, J., Yu, Y., Feldman, M., and Kotlikoff, M. I. (2005). Gene trap and gene inversion methods for conditional gene inactivation in the mouse. *Nucleic Acids Res.* **33,** e14.

Xu, Z., Lee, N. C., Dafhnis-Calas, F., Malla, S., Smith, M. C., and Brown, W. R. (2008). Site-specific recombination in *Schizosaccharomyces pombe* and systematic assembly of a 400 kb transgene array in mammalian cells using the integrase of Streptomyces phage phiBT1. *Nucleic Acids Res.* **36,** e9.

Zambrowicz, B. P., and Sands, A. T. (2003). Knockouts model the 100 best-selling drugs–will they model the next 100? *Nat. Rev. Drug Discov.* **2,** 38–51.

Zheng, B., Sage, M., Sheppeard, E. A., Jurecic, V., and Bradley, A. (2000). Engineering mouse chromosomes with Cre-loxP: Range, efficiency, and somatic applications. *Mol. Cell. Biol.* **20,** 648–655.

CHAPTER 27

Tissue Engineering Using Embryonic Stem Cells

Shahar Cohen,★ Lucy Leshanski,★ and Joseph Itskovitz-Eldor★,†

★Stem Cell Center
Bruce Rappaport Faculty of Medicine
Technion—Israel Institute of Technology
Haifa, Israel 31096, Israel

†Department of Obstetrics and Gynecology
Rambam Medical Center
Haifa, Israel 31096, Israel

Abstract
1. Introduction
2. Special Considerations when using hESCs as the Cell Source for TE
3. Growing hESCs in Defined Animal-free Conditions
4. Obtaining the Desired Cell Population
5. Choosing the Right Scaffold
6. Scaling-Up a Regulable Bioprocess
7. hESC-Derived Connective Tissue Progenitors for TE
8. Culture and Maintenance of hESC on MEF Feeders
    8.1. Preparation of growth media
    8.2. Preparation of MEF feeder layers
    8.3. Starting hESC Culture
    8.4. Passaging hESCs
    8.5. Passaging protocol
    8.6. hEB formation
    8.7. Changing hEB medium
    8.8. Derivation and propagation of connective tissue progenitors
    8.9. Choosing the right scaffold for connective TE
    8.10. Scaffold fabrication and cell seeding
    8.11. CTP seeding protocol

9. Harvesting Samples for Analyses
    9.1. Immunofluorescent staining
    9.2. Electron microscopy
    9.3. Histological analysis
References

Abstract

The possibility of using stem cells for tissue engineering has greatly encouraged scientists to design new platforms in the field of regenerative and reconstructive medicine. Stem cells have the ability to rejuvenate and repair damaged tissues, and can be derived from both embryonic and adult sources. Among cell types suggested as a cell source for tissue engineering (TE), human embryonic stem cells (hESCs) are one of the most promising candidates. Isolated from the inner cell mass of preimplantation stage blastocysts, they possess the ability to differentiate into practically all adult cell types. In addition, their unlimited self-renewal capacity enables the generation of sufficient amount of cells for cell-based TE applications. Yet, several important challenges are to be addressed, such as the isolation of the desired cell type and gaining control over its differentiation and proliferation. Ultimately, combing, scaffolding, and bioactive stimuli, a newly designed bioengineered constructs could be assembled and applied to various clinical applications. Here we define the culture conditions for the derivation of connective tissue lineage progenitors, design strategies and highlight the special considerations when using hESCs for TE applications.

1. Introduction

Tissue engineering (TE) is an evolving interdisciplinary area that combines biological and engineering principles, aimed at mimicking the natural processes of tissue formation, providing with transplantable substitutes for the field of reconstructive and regenerative medicine (Vacanti and Langer, 1999).

The essentials of TE involve cells, scaffolds and bioactive factors, such as chemical substances and mechanical stimuli. For cells to be viable for TE, they must be easily isolated, sufficient in numbers and have an appropriate defined and controlled phenotype. A number of cell types has already been used for TE applications, including fully matured cells derived from adult tissues, and stem cells derived from either embryonic, fetal, or adult tissues (Sharma and Elisseeff, 2004).

Designing the right strategy for engineering tissues using stem cells is especially challenging, since it requires plastic cells to form tissues while differentiating. The formation of stem cell-derived neotissues integrates several dynamic processes occurring simultaneously. Primarily, the basic building block, a stem cell, is a very powerful and flexible unit. Its differentiation and proliferation characteristics have an inverse relationship; the more it differentiates the less it proliferates.

Additionally, the surrounding matrix builds and degrades at the same time, enabling cell migration and spatial organization. Thus, the microenvironment continuously remodels, and in turn affects basic cellular processes.

Ultimately, aiming to higher order engineered tissues to be suitable for graftment into patients, it is essential to have a vascular network assimilated, enabling proper nutrition supply and waste disposal. This can be achieved through either stem cell-derived vascular progenitors setting up the infrastructure for functional blood vessels, or promoting host derived vasculature invading and integrating into the graft. Possibly, stem cell-based tissue engineered grafts, being growing and differentiating biological elements, will continue remodeling once transplanted *in vivo*, adjusting to the specific host requirements.

As for having the ultimate cell source to accomplish these goals, human embryonic stem cells (hESCs) hold a great promise to be just that. Ever since they were first isolated (Thomson *et al.*, 1998), they have been shown to possess a developmental potential to differentiate into cells representing all three embryonic germ layers (Itskovitz-Eldor *et al.*, 2000), including neurons (Reubinoff *et al.*, 2001), cardiomyocytes (Kehat *et al.*, 2001), hematopoietic cells (Kaufman *et al.*, 2001), endothelial cells (Levenberg *et al.*, 2002), and more. In addition, their practically unlimited self-renewal capacity can provide a sufficient amount of cells needed for TE applications. Nevertheless, these unique properties bring up several critical issues. Since undifferentiated hESCs have the potential to form tumors *in vivo*, it is crucial to study how to direct and control their differentiation and proliferation.

2. Special Considerations when using hESCs as the Cell Source for TE

Previous studies have provided protocols for differentiating hESCs along various lineages. Cell phenotype and functionality are usually tested to ensure that genuine differentiation has occurred. The following is an outline of the general principles and special considerations when using hESCs for TE applications.

3. Growing hESCs in Defined Animal-free Conditions

Traditionally, hESCs are cocultured with feeder layers made of mouse embryonic fibroblasts (MEFs) and fed with serum-containing media (Pera *et al.*, 2003); they can therefore be considered as xenografts. While MEF feeders are essential to support the growth of undifferentiated hESCs, the use of animal products is associated with the risk of pathogen transmission, thus major advantages were made in this field. Alternatives include several human-derived feeder layers, feeder-free conditioned medium, and defined medium supplemented with growth factors and matrix proteins (Amit *et al.*, 2003, 2004; Mallon *et al.*, 2006).

Ultimately, new hESC lines should be derived in defined, animal-free culture conditions to be suitable for clinical applications. Guidelines for setting defined, animal-free culture systems are beyond the scope of this article and may be found elsewhere (Ludwig *et al.*, 2006).

4. Obtaining the Desired Cell Population

hESCs are the most potent stem cells available for TE, therefore, controlling their differentiation into the desired cell type is of great challenge. In general, hESCs can be induced to differentiate once removed from the MEF feeders and introduced with bioactive signals. This is done either directly or through the formation of embryoid bodies (EBs). EBs are small clumps of hESC colonies grown in suspension, which form 3D spheroid bodies representing a differentiation model with the widest spectrum of cell types that can be achieved *in vitro*. Differentiating cells within the EBs enjoy cell–cell interaction and paracrine effects of the three embryonic germ layers, in a developing 3D microenvironment that mimics to the closest extent the temporal and spatial processes taking place in the developing embryo (Dvash and Benvenisty, 2004). Although direct differentiation is possible, most differentiation systems rely on EB formation.

Once cells start to differentiate, a variety of bioactive manipulations can be applied in order to control and direct the differentiation journey. In general, these include:

- Soluble signals, such as growth factors, hormones, and cell-conditioned media
- Genetic modifications, such as overexpression of transcription factors known to derive stem cells into the desired cell type
- Direct or indirect coculturing with other developing or mature somatic cell populations
- Physical stimuli, such as mechanical forces, temperature, and oxygenation changes

Whether differentiation is allowed to occur spontaneously or in a directed manner, the resultant cell population in most known protocols is still heterogeneous, which limits its potential to be used as a cell source for TE. Thus, the next challenge is the isolation of the desired cell type. Methods for isolating a specific cell type from heterogeneous populations include either positively selecting the desired cells or negatively removing the undesired cells. Both can be done using several approaches, including fluorescence- and magnetic-activated cell sorting, or genetic modification and selection using antibiotics. Defining the target cell population and its phenotype is crucial when utilizing these strategies and planning the next stages and appropriate time points for seeding cells onto scaffolds and transplanting them into animal models.

5. Choosing the Right Scaffold

Scaffolds utilized in TE are designed to provide cells with a solid 3D framework, allowing cells to attach, migrate, grow, and differentiate, while meeting their nutritional and biological needs (Lavik and Langer, 2004). They could be used as means of delivering cells to the patient, or support cell growth and *ex vivo* tissue formation, prior to transplantation.

Ideally, scaffolds should imitate the chemical and physical properties of the native extracellular matrix and provide the cells with the most "homey" environment. They can be made of either natural materials, such as collagen, hydoxyapatite, alginate, and silica, synthetic materials, such as polyesters, or both. While natural materials are more biocompatible and recognizable by cells, synthetic materials offer more control over properties, such as degradation rate, permeability, specific architecture, and mechanical properties. Their architecture can be processed in various techniques, attaining controlled porous structures, different-scale fibrous matrices, and hydrogels, which are either chemically or physically cross-linked water-soluble polymers that can be mixed with cells and potentially be injected in a variety of clinical situations. Additionally, the surface properties of a scaffold have a crucial effect on its biocompatibility. A surface can be modified in a range of physical and chemical ways, including applying biological molecules and binding peptides, such as RGD—an ubiquitous peptide found in many ECM proteins (such as fibronectin and laminin) which binds to cells through integrins. In summary, the properties of a scaffold can be specifically tailored and tightly controlled, to meet the many biological and physical requirements.

6. Scaling-Up a Regulable Bioprocess

Large scale production of functional tissues to be suitable for biomedical applications requires that bioprocesses are scalable, tightly controlled, and easily regulated. Standard, "investigative science"–scale, static culture systems for growing hESCs offer limited control over the culture conditions.

While different hESC lines show diversity in growth kinetics, phenotype, and differentiation potential, the variety of culture protocols and laboratory skills result in lack of consistency and different desirable yields. Each step of hESC-based systems should be ultimately scaled-up in a regulated and controlled manner. Perhaps the most challenging process to control is the formation and cultivation of EBs. EB remains the preferable approach in many differentiation systems. Made of a small aggregate of hESCs grown in suspension, EBs are independently growing and differentiating units, heterogeneous in size and in spatial arrangement and may aggregate in-between themselves to form agglomerates. In addition, cells within the growing EB respond differently and sometimes unpredictably to bioactive stimuli such as growth factors and physical cues. Methods to obtain some control over these processes include encapsulation of EBs in beads made of specific material and in specific size. Ultimately, tissue culture bioreactors, such as spinner flasks and rotating vessels, can be scaled in size to meet with specific production needs and offer control over culture conditions such as nutrients and growth factors concentrations, pH, oxygen, and carbon dioxide levels.

7. hESC-Derived Connective Tissue Progenitors for TE

Defining the desired target cell population derived from hESCs, one could aim to either somatic cell type or earlier committed progenitor cell. The objective of the following protocols developed in our laboratory was to direct hESC progeny along the mesenchymal lineage, and to achieve a progenitor cell population that is committed to connective tissue derivatives, and meets the basic requirements of being viable for cell-based TE applications. The general principles and special considerations can be applied to other cell types and differentiation assays, with appropriate modifications.

8. Culture and Maintenance of hESC on MEF Feeders

8.1. Preparation of growth media

8.1.1. MEF growth medium

- High-glucose Dulbecco's modified eagle's medium (DMEM), supplemented with:
- 10% fetal bovine serum (FBS)

8.1.2. hESC growth medium

Knockout DMEM, supplemented with:

- 20% knockout serum replacement
- 1 mM glutamine
- 1% non essential amino acids
- 0.1 mM 2-mercatoethanol
- 4 ng/ml basic fibroblast growth factor (bFGF)

8.1.3 EB growth mediumKnockout DMEM, supplemented with:

- 20% FBS
- 1 mM glutamine
- 1% non essential amino acids

8.1.4. CTP growth medium

Minimum Essential Medium-Alpha (MEM-α), supplemented with:

- 15% FBS (selected lots)
- 50 μg/ml ascorbic acid
- 10 mM beta-glycerophosphate
- 10–7 M dexamethasone

Use low protein binding, 22 μm pore size filters for sterilizing media components.

8.2. Preparation of MEF feeder layers

The procedure of deriving MEFs of 13-day old ICR mouse embryos is described elsewhere (Pera *et al.*, 2003). Once derived, MEFs should be subcultured and used for feeder preparation at passages 3–4. Lower passages use is possible but wasteful.

1. Inactivate MEFs by incubating with 8 g/ml mitomycin C for 2 h.
2. Wash with Dulbecco's phosphate buffered saline (PBS).
3. Harvest cells by trypsinization.
4. Plate 40,000 cells/cm^2 on gelatin-pretreated 6-well plates in MEF growth medium.

   - Over-night incubation is recommended before plating hESCs.
   - The fresher the better—Use plates within a week, keep for 2 weeks only

8.3. Starting hESC Culture

1. Make sure MEF feeders are intact and healthy.
2. Change feeders' medium to hESC medium 1 h before plating.
3. Thaw out a frozen hESC vial and resuspend in fresh hESC medium.
4. Gently centrifuge and resuspend pellet in final volume not exceeding 1 ml per feeder well.
5. Plate on feeders, place inside the incubator and shake plates for evenly distributing hESCs on feeders.
6. Feed cells with fresh medium on a daily basis.

8.4. Passaging hESCs

Frequency of splitting hESC cultures changes between different cell lines. Generally, timing and splitting ratio should be determined according to the following two principles:

8.4.1. Morphology of the colonies

A high quality colony is round, has well-defined edges and shows no signs of differentiation. Cells within the colony are small, with high nucleus/cytoplasm ratio (Fig. 27.1).

Poor colonies could be selectively taken out, mechanically, using a sterile needle or pipette tip. Alternatively good colonies could be saved in the same manner.

8.4.2. Density of the colonies

hESC colonies favor a crowded environment, where they support each other. At the same time, avoid overly crowded cultures. Grow ~40–60 medium size colonies per well (Fig. 27.2).

8.5. Passaging protocol

1. Prepare feeders by removing MEF medium and incubating with hESC medium 1 h before hESC seeding.
2. Incubate hESCs with type 0.1% type IV collagenase for 20–40 min.

Wait for the colonies edges to lift off the feeders before continuing.

3. Add fresh medium.

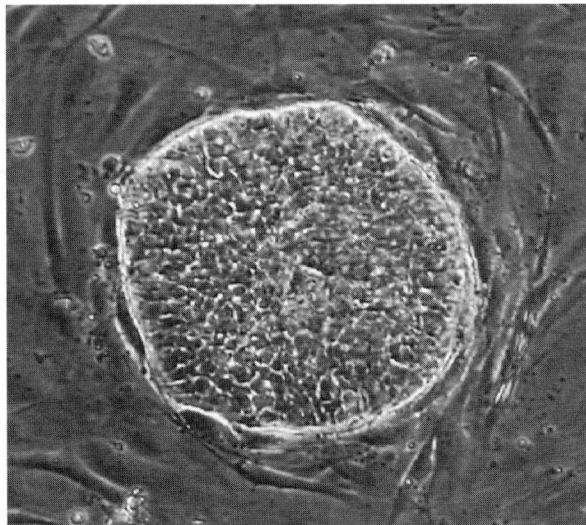

Figure 27.1 The appearance of a high quality hESC colony. Note round-shaped, well-defined edges, and high nucleus/cytoplasm ratio of cells within the colony.

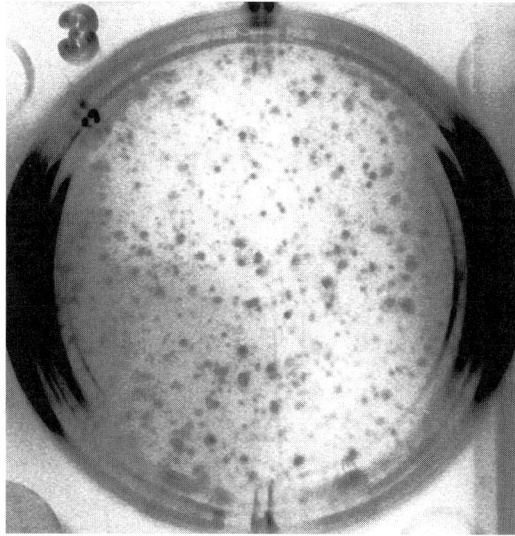

Figure 27.2 The optimal density of hESC colonies.

4. Use a pipette tip and thoroughly scratch out the colonies.
5. Collect the scratched material into a conical tube.
6. Gently centrifuge and resuspend pellet in final volume not exceeding 1 ml per feeder well.
7. Plate on feeders, place inside the incubator, and shake plates to evenly distribute the hESCs on feeders.
8. Feed cells with fresh hESC medium on a daily basis.

8.6. hEB formation

1. Prepare 60 mm Petri dished (bacterial grade, nontissue culture treated).
2. Repeat steps 2–5 of passaging protocol.
3. Gently centrifuge and resuspend pellet in fresh hEB medium.
4. Plate into Petri dishes.

   - hEBs are considered to be independently growing units, thus density of culture is of lesser importance. In general, plate one well of a 6-well plate content into one 60 mm Petri dish.

5. Place inside the incubator.
6. Feed cells with fresh hEB medium every 3–4 days.

8.7. Changing hEB medium

1. Collect the content of the Petri dish into a conical tube.
2. Gently centrifuge and resuspend pellet in fresh hEB medium.
3. Plate into new Petri dishes.

Alternatively, the following method can be used:

1. Place the Petri dish inside the working hood, topless.
2. Gently swirl the dish until all hEBs are centered.
3. Aspirate off the medium from the edges of the dish.
4. Add fresh hEB medium and place inside the incubator.

8.8. Derivation and propagation of connective tissue progenitors

1. Collect 10 day old hEBs growing in suspension into a conical tube.
2. Wash with PBS.
3. Trypsinize.
4. Thoroughly pipette up and down, and pass through a 40 m mesh cell strainer.
5. Resuspend in CTP medium, centrifuge, and resuspend again.
6. Plate 5×10^4 cells per cm^2 on tissue culture-treated flask.
7. Upon reaching subconfluence, incubate with 0.1% type IV collagenase for 40–60 min.
8. Wash with PBS.
9. Harvest cells by trypsinization.
10. Resuspend in fresh CTP medium, centrifuge, and resuspend again.
11. Split 1:3 onto new flasks.

8.9. Choosing the right scaffold for connective TE

In contrast to parenchymal organs, which are mainly cellular and function via their cells, most of the volume of connective tissues consists of their functional element—the extracellular matrix (ECM).

Connective tissue ECMs cope with tensile and compressive mechanical stresses. Tension is transmitted and resisted by nanoscaled fibrous proteins such as collagen and elastin, while compression is opposed by water-soluble proteoglycans, such as chondroitin sulphate (Scott, 2003). The proteoglycan part form a highly hydrated, gel-like "ground substance" in which the fibrous proteins are embedded.

So that cells would enjoy the most suitable 3D surrounding environment resembling the native ECM, we have postulated that nanoscaled fabricated surface topography of a synthetic scaffold would be best one to utilize. Electrospining is the most common and practical way to fabricate polymeric nanofiber matrix (reviewed by Ma *et al.*, 2005). We hypothesized that electrospun nanofiber biodegradable polymer scaffolds would support hESC-derived CTPs' organization into complex 3D tissues, as we show in the following sections.

8.10. Scaffold fabrication and cell seeding

Electrospun nanofiber mash scaffolds were made of a 1:1 blend of polycaprolactone (PCL) and poly lactic acids (PLA) by a process described elsewhere (Ma *et al.*, 2005). The average thickness of the prepared scaffold was 500 μm, fiber diameter ranging between 200 and 450 nm, with porosity of 85%.

For preparation for cell seeding we recommend the following procedure:

1. Cutting scaffold mat into 0.5 × 0.5 cm^2 squares, making them fit into 24-well plates.
2. Gas-sterilizing with ethylene oxide.
3. Immersing in 5M sodium hydroxide and washing in PBS to increase surface hydrophilicity.

8.11. CTP seeding protocol

1. Incubate subconfluent CTP cultures with 0.1% type IV collagenase for 40–60 min.
2. Rinse with PBS.
3. Harvest cells by trypsinization.
4. Resuspend in fresh CTP medium, centrifuge, and resuspend again.
5. Seeding volume should be minimal: count cells and resuspend to obtain 5 × 10^5 cells per 10 μl.
6. Seed 10 μl on each scaffold and allow cells to attach for 30 min inside the incubator.
7. Gently add fresh medium and change medium every 3–4 days.

9. Harvesting Samples for Analyses

9.1. Immunofluorescent staining

To avoid misinterpretation of the staining results, assessing autofluorescence prior to staining is highly recommended.

1. Prewash samples with PBS.
2. Soak in 4% paraformaldehyde fixative.

3. Rinse with PBS.

4. Apply primary antibody. Optimal dilution should be calibrated individually.

5. Rinse with PBS.

6. Apply appropriate secondary antibody.

7. Counter-stain nuclei with appropriate dye, such as DAPI.

8. Mount cells and view under fluorescent light microscope (Fig. 27.3A and B).

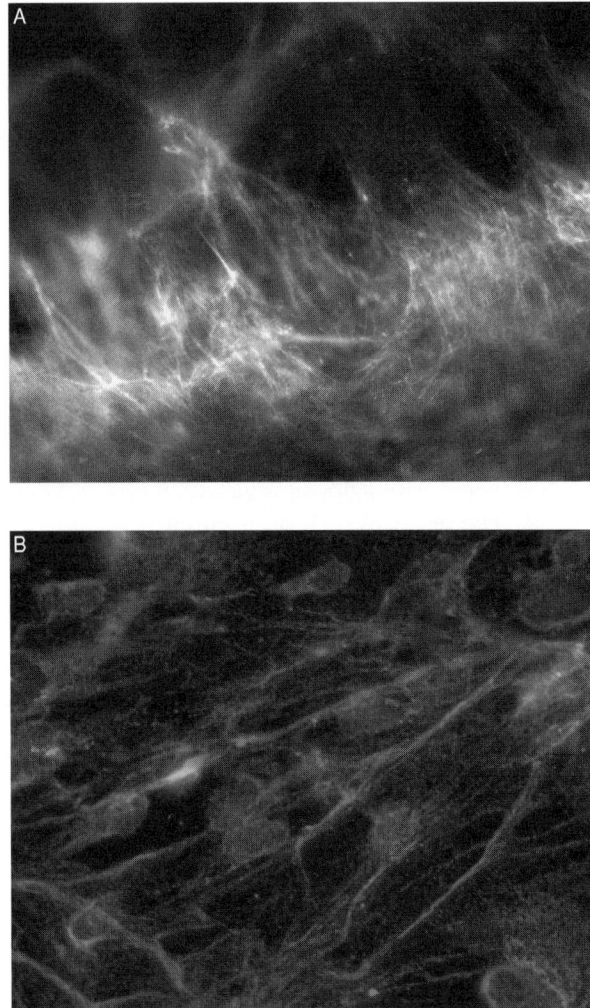

Figure 27.3 CTPs stained with antibody against type I collagen (A) and type II collagen (B). DAPI was used for nuclear visualization. Note cells embedded in self-produced extracellular matrix.

9.2. Electron microscopy

1. Prewash samples with PBS.
2. Soak in 2.5% glutaraldehyde in 0.1 M sodium cacodylate buffer.
3. Gradually dehydrate in ethanol followed by soaking in hexamethyldisilazane (HMDS).
4. Coat with carbon and view under scanning electron microscope (Fig. 27.4A–D).

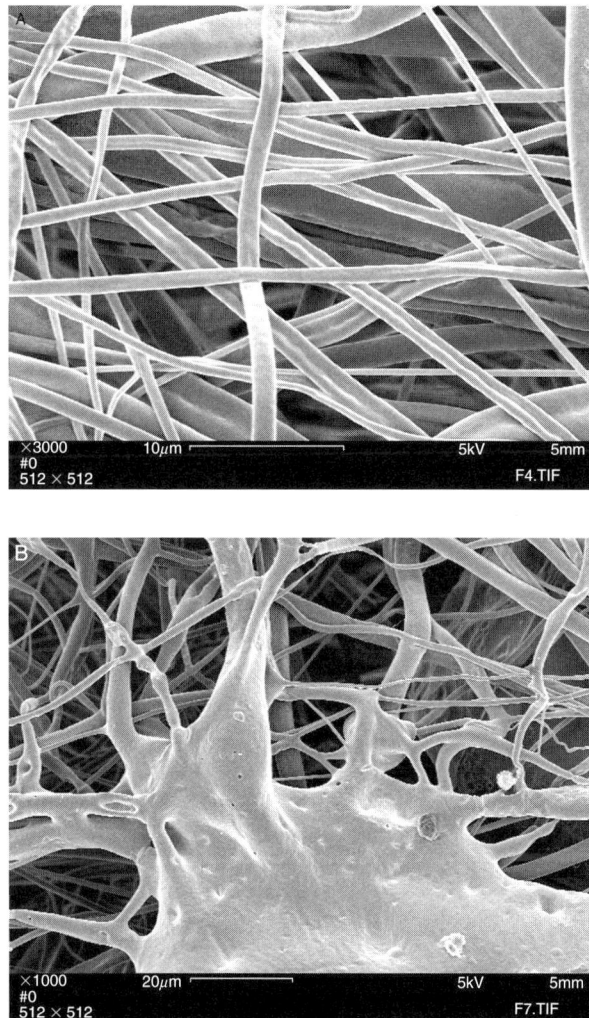

Figure 27.4 (Continued)

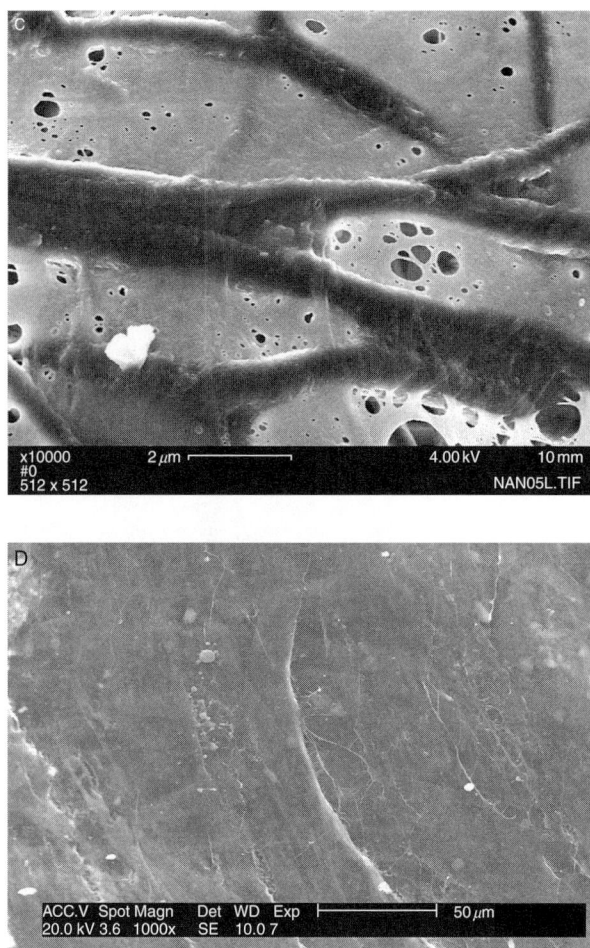

Figure 27.4   Scanning electron micrograph of electrospun PCL/PLA nanofiber mash scaffold alone (A) and of seeded CTPs (B) producing extra-cellular matrix (C) and eventually forming 3D sheet-like tissue completely covering the scaffold (D).

9.3. Histological analysis

1. Prerinse samples with PBS.
2. Soak in 10% natural buffered formalin fixative.
3. Gradually dehydrate in ethanol and embed in paraffin.
4. Cut sections and stain with hematoxylin and eosin (H and E).
5. View under light microscope (Fig. 27.5A and B).

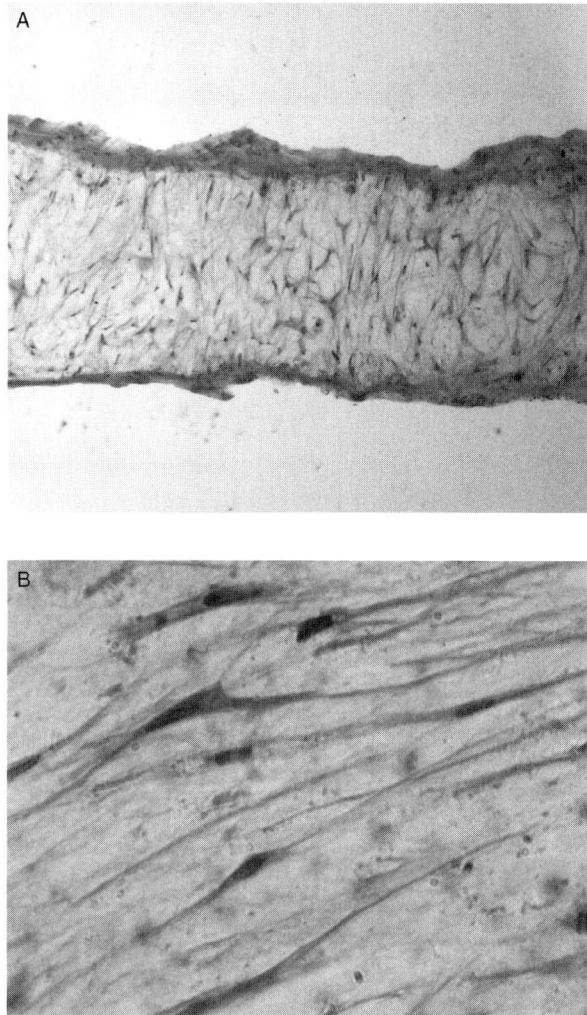

Figure 27.5 H and E stained histological images of cross-sectioned CTPs seeded on nanofiber scaffolds at low (A) and high (B) power magnifications.

References

Amit, M., Margulets, V., Segev, H., Shariki, K., Laevsky, I., Coleman, R., and Itskovitz-Eldor, J. (2003). Human feeder layers for human embryonic stem cells. *Biol. Reprod.* **68,** 2150–2156.

Amit, M., Shariki, C., Margulets, V., and Itskovitz-Eldor, J. (2004). Feeder layer- and serum-free culture of human embryonic stem cells. *Biol. Reprod.* **70,** 837–845.

Dvash, T., and Benvenisty, N. (2004). Human embryonic stem cells as a model for early human development. *Best Pract. Res. Clin. Obstet. Gynaecol.* **18,** 929–940.

Itskovitz-Eldor, J., Schuldiner, M., Karsenti, D., Eden, A., Yanuka, O., Amit, M., Soreq, H., and Benvenisty, N. (2000). Differentiation of human embryonic stem cells into embryoid bodies compromising the three embryonic germ layers. *Mol. Med.* **6,** 88–95.

Kaufman, D. S., Hanson, E. T., Lewis, R. L., Auerbach, R., and Thomson, J. A. (2001). Hematopoietic colony-forming cells derived from human embryonic stem cells. *Proc. Natl. Acad. Sci. USA* **98,** 10716–10721.

Kehat, I., Kenyagin-Karsenti, D., Snir, M., Segev, H., Amit, M., Gepstein, A., Livne, E., Binah, O., Itskovitz-Eldor, J., and Gepstein, L. (2001). Human embryonic stem cells can differentiate into myocytes with structural and functional properties of cardiomyocytes. *J. Clin. Invest.* **108,** 407–414.

Lavik, E., and Langer, R. (2004). Tissue engineering: Current state and perspectives. *Appl. Microbiol. Biotechnol.* **65,** 1–8.

Levenberg, S., Golub, J. S., Amit, M., Itskovitz-Eldor, J., and Langer, R. (2002). Endothelial cells derived from human embryonic stem cells. *Proc. Natl. Acad. Sci. USA* **99,** 4391–4396.

Ludwig, T. E., Levenstein, M. E., Jones, J. M., Berggren, W. T., Mitchen, E. R., Frane, J. L., Crandall, L. J., Daigh, C. A., Conard, K. R., Piekarczyk, M. S., Llanas, R. A., and Thomson, J. A. (2006). Derivation of human embryonic stem cells in defined conditions. *Nat. Biotechnol.* **24,** 185–187.

Ma, Z., Kotaki, M., Inai, R., and Ramakrishna, S. (2005). Potential of nanofiber matrix as tissue-engineering scaffolds. *Tissue Eng.* **11,** 101–109.

Mallon, B. S., Park, K. Y., Chen, K. G., Hamilton, R. S., and McKay, R. D. (2006). Toward xeno-free culture of human embryonic stem cells. *Int. J. Biochem. Cell Biol.* **38,** 1063–1075.

Pera, M. F., Filipczyk, A. A., Hawes, S. M., and Laslett, A. L. (2003). Isolation, characterization, and differentiation of human embryonic stem cells. *Methods Enzymol.* **365,** 429–446.

Reubinoff, B. E., Itsykson, P., Turetsky, T., Pera, M. F., Reinhartz, E., Itzik, A., and Ben-Hur, T. (2001). Neural progenitors from human embryonic stem cells. *Nat. Biotechnol.* **19,** 1134–1140.

Scott, J. E. (2003). Elasticity in extracellular matrix 'shape modules' of tendon, cartilage, etc. A sliding proteoglycan-filament model. *J. Physiol.* **553,** 335–343.

Sharma, B., and Elisseeff, J. H. (2004). Engineering structurally organized cartilage and bone tissues. *Ann. Biomed. Eng.* **32,** 148–159.

Thomson, J. A., Itskovitz-Eldor, J., Shapiro, S. S., Waknitz, M. A., Swiergiel, J. J., Marshall, V. S., and Jones, J. M. (1998). Embryonic stem cell lines derived from human blastocysts. *Science* **282,** 1145–1147.

Vacanti, J. P., and Langer, R. (1999). Tissue engineering: The design and fabrication of living replacement devices for surgical reconstruction and transplantation. *Lancet* **354**(Suppl. 1), SI32–SI34.

INDEX

A

ABCG2 expression, in HSC, 240–242
A2B5 positive cells, 26
Acini, 151–152, 155
β-actin FLPe transgenic, 544
Activin A, 427, 478, 487, 493, 496
Adipose tissue-derived stemcells, 290
ADSCs. *See* Adipose tissue-derived stemcells
Adult lung
    anatomical and cellular diversity of, 115–117
    stem cell phenotypes and niches in, 117–121
    in vivo injury models of, 121–128
Adult stem cells, 156–157
AEC I. *See* Alveolar type I cells
AEC II. *See* Alveolar epithelial type II
Affymetrix GeneChip, principle of, 252–253
Aggrecan genes, 305
Air–liquid interface, 136
Airway epithelial cells, culture of, 133
Airway epithelium, CFE role in, 136–140
Airway SMG, stem/progenitor cells of, 141–143
Akt activation, 31, 40
ALI. *See* Air–liquid interface
Alkaline phosphatase (AP), 76, 83, 129, 167, 196–197, 307, 381
    detection of, 383
    undifferentiated hESC colonies, 380
Alveolar epithelial type II, 114
Alveolar type I cells, 115
Alzheimer's disease, 26
Amniocentesis, for fetal abnormalities, 193
Amniotic and placental-derived progenitor cells, differentiation of, 195–197
    hepatocytes, 197–198
    myocytes, 198
    neuronal, 198–199
    osteocytes and endothelial cells, 197
Amniotic cell, for therapy, 193–194
Amniotic fluid and placenta, in developmental biology, 192–193
Angiogenic factors, 286
Animal-free culture system, 396, 582
Antibody staining, of Hoechst-stained cells, 245
Apaf1, 31

Apoptosis, 27, 30–31, 36, 308, 544
Appendicular skeletogenesis
    mesenchymal–epithelial interactions, 298
A⊂pr and A⊂al spermatogonia, 89, 100
Array based comparative genomic hybridization, 260–261. *See also* Microarray technology
Array CGH. *See* Array based comparative genomic hybridization
Arteriosclerosis, 289
Ascl1. *See* bHLH transcription factor
Astrocytes, 5, 17, 24, 26, 30, 35, 41, 420
Astrocytic differentiation, 35
    and Hes, 38
Autoclave method, 466

B

BADJ. *See* Bronchioalveolar duct junctions;
    Bronchoalveolar-duct junction
BASCs. *See* Bronchioalveolar stem cells
Basic fibroblast growth factor, 5, 28, 40, 42, 44, 107, 399–400, 427, 464, 506, 530, 584
Bayesian network model, importance of, 269
β-catenin, 28
B-cell development, 429
bcl2 expression, 31
Beta cells
    dedifferentiation of, 157
    in postnatal life, origins of, 153
Beta-3-tubulin, 412
bFGF. *See* Basic fibroblast growth factor
BFU-E. *See* Burst forming unit-erythroid
BFU-E/CFU-GEMM. *See* Burst forming unit-
    Erythroid/Colony forming unit-
    Granulocyte/ Erythroid/ Macrophage/
    Megakaryocyte colony forming assay
β–galactosidase activity, 98, 129, 131, 138, 140, 555–556, 560, 570
bHLH transcription factors, 36, 54
β1 intergrin, 31
Bioluminescence imaging (BLI), 449–451
Blastocysts, 343, 363
    outgrowth, 364
    types of, 344

Bleomycins, role in pulmonary toxicity, 124
bmi1, polycomb group of genes, 38
BMP. *See* Bone morphogenetic protein
BMPCs. *See* Bone marrow progenitor cells
BMP signaling, 58
BMSCs. *See* Bone marrow stromal cells
Bone marrow cells, 156
Bone marrow content
    expansion of, 300–302
    isolation of, 300
Bone marrow-derived cells, 299
Bone marrow HSCs, 238
    cell surface markers of, 238–239
    from different purification schemes,
        features of, 242
    fluorescent dye effux in, 239–242
    sorting with Hoechst 33342 staining, 243–247
Bone marrow mesenchymal stem cells
    (BMMSCs), 77
Bone marrow progenitor cells, 298
Bone marrow stromal cells, 298
Bone morphogenetic protein, 26, 37, 58, 76, 273,
    288, 303, 307, 399, 414, 431, 440, 444
Bovine serum albumin, 79, 184, 422, 489, 492, 529
Brain disorders, cellular treatment of, 291
BrdU. *See* Bromodeoxyuridine
BrdU proteins, 65–66
Bromodeoxyuridine, 117
Bronchioalveolar duct junctions, 120
Bronchioalveolar stem cells, 120
Bronchoalveolar-duct junction, 117
BSA. *See* Bovine serum albumin
Burst forming unit-erythroid, 207
Burst forming unit-Erythroid/Colony forming
    unit-Granulocyte/ Erythroid/ Macrophage/
    Megakaryocyte colony forming assay,
    214–215
Busulfan, 93–95, 108

C

CAFC. *See* Cobblestone area-forming cell
Calcitonin gene-related peptide, 120
Cardiomyocyte. *See also* ES (D3) cells into
    cardiomyocytes *in vitro* differentiation
    differentiation, 470–471
    enrichment of, 471–473
Cartilage tissue, 289
Caspase 3, 31
Cath5, 56
Cavitated blastocyst, 344
C57BL mES cell lines, 324

CCAAT/enhancer-binding proteins (C/EBP), 309
CCSP. *See* Clara cell secretory protein
CD150 marker, in HSC, 239
cDNA microarrays, 251–252. *See also*
    Microarray technology
CDX4 expression, 434–435
Cell-adhesion substrate, 284
Cell-based regenerative medicine techniques
    applying, limitations of, 283
Cell death, 31, 36, 75, 96, 104, 368, 403, 409, 412,
    415, 419, 421, 465, 487, 570
Cell population, 582
    for TE, 582
Cells helping cells, concept of, 347
Cell surface markers, of HSC, 238–239
Cell transplantation, principles of, 282
Cellular and anatomical diversity, of adult lung,
    115–117. *See also* Adult lung
Central polypurine tract (cPPT), 520
Ceramide, 33–34
CeramideGM1, 28
CFC. *See* Colony forming cells
CFE. *See* Colony-forming efficiency
CGRP. *See* Calcitonin gene-related peptide
cHairy, 56
ChIP-chip analysis. *See* Chromatin
    immunoprecipitation followed by
    microarray analysis
5-Chloro-2-deoxyuridine (CldU), 119
Chorionic mesenchyme, 438
Chromatin immunoprecipitation followed by
    microarray analysis, 261–262. *See also*
    Microarray technology
Chx10, transcription factors, 56
Ciliary marginal zone (CMZ), 55–57
c-kit, expression of, 90
c-kit ligand, 169
c-Kit$^+$Sca-1$^+$Lineage$^-$, 238
Clara cell secretory protein, 120
CLSM. *See* Confocal laser scanning microscope
CMV. *See* Cytomegalovirus
CMZ in fish and frogs, 56
CMZ progenitors, 57
CNTF-mediated MAPK activation, 34
Cobblestone area-forming cell, 217
Collagenase
    hES dissociation with, 367–368
    type IA, for digestion, 349
    type IV dissociation, 385, 414, 436, 530, 588
Colonies of blast morphology (BL-CFC), 426
Colonies of definitive erythroid (BFU-E), 427
Colonies of primitive erythroid (EryP) cells, 427

Colony forming cells, 217
Colony-forming efficiency, 136
Colony-forming unit-fibroblasts (CFU-F), 77
Colony forming unit-granuloctye, 207, 212–213
Colony forming unit-granuloctye, macrophage (CFU-GM), 207–208, 212, 215, 427, 431, 435, 440
Colony forming unit-granulocyte, erythroid, macrophage, megakaryocyte (CFU-GEMM), 207–208
Compaction, 343–344
Computer modeling, 39
Conditional alleles, generation of, 553
    inversion strategies, 554–557
    recombinase delivery, to ES cells, 557–558, 562–563
    recombinase delivery, to mice, 558
    strategies employing excisions, 553–554
Confocal laser scanning microscope, 489
Connective tissue progenitor cells, 298
Cord blood transplantation, 204–206
    EPC, 209–211
    hematopoietic stem and progenitor cells, 206–209
Coronary heart disease, 460
Craniofacial morphogenesis
    mesenchymal–epithelial interactions, 298
Cre and FLPe recombinases, 543–546
CreER(T2) fusion protein, 545
Cre-lox system, characteristics of, 159
CreXMice database, 545
CTPCs. See Connective tissue progenitor cells
Cultured mouse SSCs, siRNA transfection of, 105–106. See also Spermatogonial stem cells (SSCs)
    cultured germ cells, transient transfection of, 106–107
    gene knockdown, determining level of, 108
    transplanting transfected cells, to quantify SSC numbers, 108–109
Cy3 channel labeling, 254
Cytochrome C, 31
Cytokine-supplemented cultures, 440–441
Cytomegalovirus (CMV), 441, 502, 526

D

Days post coitum (dpc), 167
Definitive erythroid (EryD) precursors, 427
Dendritic cell (DC) development, 430
Dental pulp stem cells (DPSCs), 75–77
    differentiation of, 79–82
    isolation of, 77–79
    transplantation, protocol for, 82–83
Dentin phosphoprotein (DPP), 75
Dentin sialophosphoprotein (DSPP), 75
Dentin sialoprotein (DSP), 75
Diabetes mellitus, causes of, 478
Differentiation capacity, hESCs
    potential applications of, 386
    in vivo differentiation
        through embryoid bodies, 387–388
        through teratoma formation, 387
Dimethyl sulfoxide, 170
Dispase, 78, 367–368, 461, 464, 530, 533
DMEM. See Dulbecco's modified Eagle's medium
DMSO. See Dimethyl sulfoxide
Donor tissue, source of, 282
Doublecortin, 26, 35
Double-stranded RNA, 503
DPBS. See Dulbecco's phosphate-buffered saline
Drosophila lethal giant larvae gene (Lgl1), 32
dsRNA. See Double-stranded RNA
Dulbecco's modified Eagle's medium, 140, 244, 349, 351, 423, 449, 505, 584
Dulbecco's phosphate-buffered saline, 171

E

EBD. See EB-derived
EB-derived CD34+ cells into immunodeficient mice
    in vivo transplantation of, 448
EB-derived cell formation, 183–185
EB formation, and analysis, 182
ECC. See Embryonal carcinoma cells
ECM. See Extracellular matrix
EDTA/PBS. See Ethylenediaminetetraacetic acid/phosphate buffered saline
EGC. See Embryonic germ cells
EGC derivation, 168–170
    feeder layer, 175–176
    protocol for deriving, 170–175
EG cultures, characterization of, 177–182
EGF. See Epidermal growth factor
EGF-responsive murine neural stem cells, 12
EGF-responsive stem cells, 55
ELISA assay, 181, 305, 441, 487, 492, 494
Embryo biopsy, 345–347
Embryo culture system, and embryo development in vitro, 342
Embryoid bodies
    formation of, 387–388

Embryoid bodies (*cont.*)
  out growths, immunocytochemistry analysis
    of, 388–389
Embryoid bodies (EBs), 166, 398, 426,
    478, 582. *See also* hESC
  formation and hematopoietic differentiation,
    442–444
    aggregation methods for, 444
    hanging drop EB cultures, 442–444
    spin technique for, 444
  generation of, 461 (*see also* ES (D3)
    cells into cardiomyocytes
    in vitro differentiation)
Embryonal carcinoma cells, 166
Embryonic germ cell markers, profile of, 177
Embryonic germ cells, 166
Embryonic stem cells, 153, 166, 283, 501
  retrovirus expression vectors and, 502–503
  and RNA interference, 503–504
Embryonic stem (ES) cells, 408, 477
Enamel, 74
Encephalomyocarditis virus (EMCV), 523
Endocrine pancreas, 151
Endoglin marker, in HSC, 239
Endometrial epithelial cells culture, 350–352
Endothelial progenitor cells, 204
Engraftment, *in vivo*, 433–434
Enhanced green fluorescent protein
    (EGFP), 560
Enzymatic dissociation with trypsin, hES cells
  materials for, 365
  trypsinization, 365–366
EPC. *See* Endothelial progenitor cells
Eph receptors, 30
ephrinA/EphA receptor, 30
EphrinB ligands, 30
Ephrins, 26
Ephs, 26
Epidermal growth factor, 4, 6, 26, 28, 31, 55, 57,
    99, 418, 420, 423
Epigenetic variation, 413
EryP precursors, 427
Erythroid, 206–207, 428, 434, 445
Erythroid differentiation, 445–446
Erythropoiesis, 428, 431, 445
Erythropoietin (Epo), 207
ES cell-derived NS cell lines, 408
ES cell media, 422
ES cells and NS cells, similarities and differences
    between, 416
ES cells, conditional mutagenesis strategies in,
    553–557

ES cells engineering, with recombinase systems,
    542. *See also* Conditional alleles, generation
    of; Site-specific recombination substrates,
    designing of
  protocols for
    conditional constructs, pretesting of, 568–569
    gene trap cassette inversions, verification of,
      569
    low-density seeding of ES cells, 563–564
    molecular confirmation of excisions, 564–566
    picking of colonies, 564
    recombinase delivery into ES cells, 562–563
    RMCE of gene trap cassettes, with
      CreERT2, 569–571
    tamoxifen treatment, of primed ES cells,
      567–568
129 ES cells strains, 324
Escherichia coli P1 phage, 543
ESC lines, hematopoietic differentiation of.
    See mESC differentiation
ESCs. *See* Embryonic stem cells
ES (D3) cells into cardiomyocytes, *in vitro*
    differentiation, 462
  EBs in bioreactor, scalable production of,
    465–470
  hanging drop method, 463–464
Ethylenediaminetetraacetic acid/phosphate
    buffered saline, 506
European Conditional Mouse Mutagenesis
    Project (EUCOMM), 559
Exocrine pancreas, 151
Expanded blastocyst, 344
Extracellular matrix (ECM), 25, 141, 479
Extracellular matrix proteins, 282
Eye-field transcription factors (ETFs), 54

F

Facial vein injection (FVI), 448–449
FACS. *See* Fluorescence activated cell sorter;
    Fluorescent-activated cell sorting
FACS analysis, 99–100, 447, 449, 505, 532–533,
    536
Fallopian tube epithelium, 396
False discovery rate, 263
Fanconi anemia, 204
FAS ligand, 31
FBS. *See* Fetal bovine serum
FCS. *See* Feeder cells and sera; Fetal calf serum
FDR. *See* False discovery rate
Feeder cells and sera, 487

Feeder layer-free culture of hESCs, methods, 401
  cell splitting, 402–403
    splitting medium, 402
  culture medium, 402
  fibronectin coating of plates, 401–402
  freezing cells, 403
  thawing cells, 403–404
Fetal bovine serum, 325, 396, 437, 505, 584
Fetal calf serum, 18–19, 244, 420, 463, 485, 528
Fetal hemoglobin, 432, 446
FGF receptor, 28
Fibroblast growth factor-2 (FGF2), 56–57, 59–60, 68, 169, 301
Fibronectin, 398, 413
Fibronectin stock solution, 401
FISH. *See* Fluorescent *in situ* hybridization
FKBP12 rapamycin associated protein (FRAP), 34
Flip-Excision (FlEx) strategy, 549–550, 555
Flk1-/-embryonic stem cell chimeras, 427–428
Flow cytometric enrichment, 19
Flow cytometry, 4, 300, 384, 446, 448
  analysis, of surface markers, 382–383
Floxed allele, 553
FLPo, 543–544
Flt3 gene expression, 435
Flt-3-ligand (FL), 207
Fluorescence activated cell sorter, 505
Fluorescence-labeled secondary antibodies, 491
Fluorescent-activated cell sorting (FACS), 99, 194, 238
Fluorescent *in situ* hybridization, 181
Focal adhesion kinase (FAK), 31
Forebrain restricted orphan nuclear receptor TLX, 26
Foreskin fibroblasts, 397
Fox class, 55
Foxn4, 55
Freezing ratio, 403
Functional tissue replacement, 282
Fuzzy clustering method, 275

G

Gas-exchanging airspaces, 113
GATA-1 null ES cells, 427
G-CSF. *See* Granulocyte colony stimulating factor
GDNF-dependent Thy1+germ cell cultures, 106
GDNF family receptor α1 (GFRα1), 100
Gene module analysis, in microarray experiment, 266–268

Gene set enrichment analysis, 267
Gene targeting, 553
Genetic/epigenetic variation, 413
Genetic pulse-chase, 154
Gene trapping, 553
Germ cell development, 167–168
GFAP. *See* Glial fibrillary acidic protein
GFAP expression, 420
GFAP promoter, 38
GFP. *See* Green fluorescent protein
GFP/luciferase fusion gene, 449
GFRα1, 100, 103–104, 106
Gibbon ape leukemia virus (GALV), 527
Glial fibrillary acidic protein, 198
Gli1 and Gli3, 56
GM3 ceramide, 31
GM-CSF. *See* Granulocyte-macrophage colony stimulating factor
G1 phase, 27
G-protein signals, 33, 36
Granulocyte colony stimulating factor, 207
Granulocyte-macrophage colony stimulating factor, 207
Green fluorescent protein, 503
Growth factors, 34
GSEA. *See* Gene set enrichment analysis

H

Haemocytometer, 42, 61
hAFS cells. *See* Human amniotic fluid stem cells
Harvesting samples. *See also* Tissue engineering
  for analyses
    CTPs seeded on nanofiber scaVolds, 593
    electron microscopy, 591
    histological analysis, 592
    immunofluorescent staining, 589–590
    PCL/PLA nanofiber mash scaffold, SEM of, 592
Hatched blastocyst, 344–345
hEGCs. *See* Human EGCs
Hemangioblast formation, 428
Hematopoiesis, 217, 427–428, 432, 434, 506
Hematopoietic colonies, 440
Hematopoietic growth factors, 429
Hematopoietic stem cells, 117, 206–209, 238, 272, 506. *See also* Cord blood transplantation
Hemoglobin (Hb), 446
HES1, neural progenitor cells, 26
hESC. *See also* mESC differentiation
  gene delivery by lentiviral-based vectors in, 522–523

hESC. *See also* mESC differentiation (*cont.*)
  hematopoietic differentiation, using stromal
      conditioned medium, 441–442
  hematopoietic differentiation with stromal cell
      lines, 438–441
    stromal plus cytokine-supplemented
        cultures, 440–441
    stromal supported hESC differentiation,
        438–440
  lymphoid differentiation from, 447
  maintenance of, 436
  morphology of colony, 397
  stromal cell lines, maintenance of, 437–438
  transduction of, 529–532 (*see also*
      Recombinant viral particles, generation of)
    enrichment for transduced hESCs, 533–535
    transduction efficiency, measurement of, 532
hESCs. *See* Human embryonic stem cells; Human
    ES (hES) cells
hES derivation media, 362–363
Hierarchical clustering, in microarray data
    analysis, 265
High proliferative potential-endothelial colony
    forming cells, 211
High proliferative potential primitive
    hematopoietic progenitors (HPP-CFCs), 427
Histone H2BGFP, expression, 119
HIV-1 based vectors. *See also* Recombinant viral
    particles
  design of, 520–521
  regulatory sequence for improvement of, 520
  for transduction of hESCs, 523
    choice of HIV-1 based vector, 523–526
    choice of internal promoter, 526–527
    virus pseudotyping, 527
HLA. *See* Human leukocyte antigen
hMSCs. *See* Human mesenchymal stem cells
Hoechst SP cells, FACS analysis for, 245–247
Hoechst 33342 staining, in bone marrow HSCs,
    243–247
HOXB4, expression, 434–435
HPAP. *See* Human placental alkaline
    phosphatase
hPGCs. *See* Human PGCs
HPP-ECFC. *See* High proliferative potential-
    endothelial colony forming cells
HSCs. *See* Hematopoietic stem cells
hTERT gene, 398
HTNC protein, 557–558
Human amniotic fluid stem cells, 291
Human bone marrow stroma, 438
Human bronchial xenograft protocol, 129–132

Human cord blood, hematopoietic progenitor
    and stem cells in, 211
  BFU-E/CFU-GEMM, 214–215
  CFU-GM, 212–213
  colony replating, 215–216
  human hematopoietic stem cells, methods for,
      216–227
  umbilical cord blood, methods in, 212
Human DSP protein, 80
Human EBD cells, protocols for, 184–185
Human EGCs, 167
Human embryo culture, 347–351
  coculture system, 348–351
  sequential culture, 347–349
Human embryo development, 342–345
Human embryonic kidney (HEK), 527
Human embryonic stem cells, 396, 342, 400,
    408, 503
  animal-free conditions, 581–582
  cell morphology, 379
  change cultures, 585
  changing hEB medium, 588
  coculture of irradiated OP9 cells and, 439
  colonies, morphology of, 586
  colony, optimal density of, 587
  connective TE, scaffold for, 588
  connective tissue progenitors
      derivation and propagation of, 588
  conversion to neural cells in adherent
      monolayer, 414
  CTP seeding protocol, 589
  culture and maintenance of
    CTP growth medium, 585
    Dulbecco's modified Eagle's medium
        (DMEM), 584
    MEF growth medium, 584
  derivation of, 360
    hES derivation medium preparation for,
        362–363
    from inner cell mass (*see* Inner cell mass)
    MEFs in (*see* Mouse embryo fibroblast)
    planning and considerations for, 361
  enzymatic dissociation with trypsin
    materials for, 365
    trypsinization, 365–366
  freezing of, 366–367
  growth medium, 584
  hEB formation, 587
  high quality colony, 586
  isolation of, 374
  lentiviral-based vectors for gene delivery into
      HIV-1-based vectors (*see* HIV-1-based vectors)

long-term feeder-free cultures of, 376–378
MEF feeder layers, preparation of, 581, 585
methods for culture of
    dissociation with collagenase/dispase,
        367–368
    with human feeder cells, 368
passaging protocol, 586
potent stem cells, 582
proliferation of, 360
scaffold fabrication and cell seeding, 589
scalable bioprocesses, 583–584
TE, cell source, 581
thawing of, 367
transduction of, 450–451
undifferentiated
    cardinal features of, 375
    differentiation capacity, 386–389
    karyotype analysis, 386
    markers for (*see* Markers for
        Undifferentiated hESCs)
    proliferation, 384–386
    self-renewal of, 374
Human ESC lines
    hematopoietic differentiation of, 431–433
Human feeder cells, hES culture with, 368
Human gonads and EGCs, immunostaining,
    180–181
Human hematopoietic stem cells, assessment, 216
    EPC in umbilical cord blood, 224–227
    in vitro hematopoietic stem cell assays for cord
        blood, 217–219
    in vivo animal models for cord blood-derived
        hematopoietic stem cells assay, 219–224
Human immunodeficiency virus type I (HIV-1),
    519
Human leukocyte antigen, 204
Human mesenchymal stem cells
    isolation of, 301
    osteoblasts, 301
Human NSCs, culture of, 43–45
Human PGCs, 168
Human placental alkaline phosphatase, 154
Human serum albumin (HSA), 347–348, 350
Human-specific alu cDNA probe, 81
4-Hydroxytamoxifen, 545
Hypoxia inducible factor 1 (HIF1), 30, 36

I

Id and Hes protein, 38
Ideal discriminator method, in microarray data
    analysis, 265

IGF1. *See* Insulin-like growth factor 1
IGFBP7 gene, 76
IgorPro image analysis, 451
IHC. *See* Immunohistochemistry
Immune cell generation, 433
Immunocytochemistry (ICC), analysis, 41
    of EB out growths, 388–389
    markers of undifferentiated hESCs, 381
        alkaline phosphatase activity, 383
        flow cytometry analysis, 382–383
        live cell staining of surface marker
            expression, 381–382
        RT-PCR analysis, 384
    of markers to identify stem cells and
        differentiation products, 45
Immunofluorescent analysis, of stem/progenitors,
    65–66
Immunohistochemistry, 77, 127, 182, 271,
    305, 308
Induced pluripotent stem cells, 502, 509
Inner cell mass
    dispersion of, 363–364
    isolation of, 363
    outgrowth, 364
Inner cell mass (ICM), 343, 345–346, 396
In situ hybridization (ISH), 271
Insulin, 31, 56
Insulin deficiency, and diabetes mellitus, 478
Insulin-like growth factor 1, 57, 99, 183–184, 374
Insulin receptor, 31
Insulin, transferrin, sodium selenite, fibronectin,
    nicotinamide (ITSFn), 479
ϕC31 integrase, 543, 546, 562
Integrin receptors, 31
Integrins, 31, 39, 583
Internal ribosome entry site (IRES), 411,
    523–524, 570
International Gene Trap Consortium
    (IGTC), 559
Intrafemoral bone marrow transplantation
    technique (IBMT), 448
Intrahepatic transplant method (IHT), 448
In vitro fertilization (IVF), 342, 351, 353–354
In vivo differentiation, hESCs
    through embryoid bodies, 387–388
    through teratoma formation, 387
5-Iodo-2-deoxyuridine (IdU), 119
iPS. *See* Induced pluripotent stem cells
γ−irradiation, of MEF cells, 362
Iscove's modified Dulbecco's medium (IMDM),
    441–442
Islets of Langerhans, 150, 478

J

JAK/tyk kinases, 36
Janus kinase/tyrosine kinase (JAK/TYK), 34

K

Kainic acid, 57
Karyotype abnormalities, 401
Karyotype analysis, of hESC cultures, 386
Karyotyping, 182, 331
Knockout-Dulbeccos Modified Eagle Medium
    (ko-DMEM), 402
Krebs' Ringer bicarbonate hepes (KRBH), 492
KSL. *See* c-Kit⁺Sca-1⁺Lineage⁻

L

Label-retaining cells, 114
Laminin, 40, 374, 381, 398, 413, 415, 487, 583
Large serine recombinases. *See* Serine
    recombinases
Lentihair vector system, 504
Lentilox vector system, 504–505, 509
Lentiviral-based vectors, 518
    coexpressing two genes
        bicistronic vectors, 523–524
        dual promoter vectors, 525–526
    for gene delivery, 519–520
    transduction of hESCs by, 521–522
Lentiviral gene transfer, in mouse and human
    ESCs, 506–507
Lentiviral–RNAi-based system, 525
Lentivirus-mediated gene transfer, 503
Leukemia inhibitory factor (LIF), 30, 99, 168,
    398, 507, 562
LIF receptor (LIFR), 29, 37, 169
Ligand binding domain (LBD), 545, 567
Lipid signaling, 28
Lipoprotein lipase (LPL), 310
LivingImage (Xenogen) software, 451
Lkb1, protein kinase, 32
Long-term culture-initiating cell, 217
Long-term HSC, 242
Long terminal repeat, 502
Lox-stop-lox, 160
LRCs. *See* Label-retaining cells
LSL. *See* Lox-stop-lox
LTC-IC. *See* Long-term culture-initiating cell
LT-HSC. *See* Long-term HSC
LTR. *See* Long terminal repeat
Luciferase imaging, *in vivo*, 451
Luciferase (LUC), 560

M

Macrophage, 207, 212–213
    colony stimulating factor, 207
    differentiation, 447
Magnetic enrichment, of Hoechst-stained cells,
    245
Male fertility, 109
Mammalian neural stem cells
    isolation and characterization of, 5
        adult C57Bl/6 mouse brain, ventral
            view of, 14
        adult NSCs, flow cytometric enrichment of,
            19–20
        equipment for, 5
        growth factors and media, 5–7
        neurosphere cultures, differentiation of,
            16–19
        passaging neurosphere cultures, 15–16
        primary adult neurosphere cultures,
            establishment of, 11–15
        primary embryonic cultures, establishing,
            10–11
        primary embryonic neurosphere cultures,
            establishment of, 8–10
        stock solutions, 7–8
Mann–Whitney Rank Sum test. *See* Wilcoxon
    Rank Sum test, in microarray data analysis
MAPK pathway, 28, 31
Markers, for undifferentiated hESCs, 379
    alkaline phosphatase activity, 380
    expression of, 381
    immunocytochemical analysis of
        alkaline phosphatase activity, 383
        flow cytometry analysis, 382–383
        live cell staining of surface marker
            expression, 381–382
        RT-PCR analysis, 384
MASH1, 26
Mash1, proneural genes, 55, 65, 412
Matrigel, 374
    feeder-free cultures of hESCs, 374, 376–380
mcl-1, 31
MDR-1. *See* Multidrug resistance-1
MEF. *See* Mouse embryo fibroblast
MEF-CM, 374–376, 380
Megakaryocyte, 207
    differentiation, 446
Megakaryopoiesis, 429
MEM. *See* Modified Eagles medium
2-Mercaptoethanol (2ME), 229
mESC. *See* Mouse embryonic stem (ES) cells;
    Murine embryonic stem cells

mESC differentiation, 426
  into cardiomyocyte, 460
  early ESC development, 426–428
  EB differentiation systems, 428
  hematopoiesis and homeobox gene expression,
    in ESC, 434–435
  hematopoietic differentiation, of human ESC
    lines, 431–434
  lymphoid differentiation, of ESC, 429–430
  mESC-derived hematopoietic cells, *in vitro*
    engraftment of, 430
  mESC-stroma coculture systems, 429
  primate ESC-derived hematopoiesis, 431
Mesenchymal progenitor cells, 298
Mesenchymal stem cells, 194, 287, 298
  adipogenic differentiation
    adipose tissue, 308
    Oil Red-O staining, 310
    protocol, 309–310
    stimulants, 309
  autologous tissue grafts, 299
  bone marrow stroma, 287
  chondrocytes, 289, 301
  chondrogenic differentiation
    chondrogenic stimulants, 303–304
    human chondrogenic media, 305
    rat chondrogenic, 304
    SOX family, 305
    tissue-engineered cartilage, 306
  definition of, 298
  isolation protocol
    human bone marrow, 302
  multilineage differentiation of
    chondrogenic differentiation, 303–306
  osteogenic differentiation
    bone morphogenetic proteins (BMPs), 307
    markers, 307–308
    protocol, 307
    stimulants, 306–307
  progenitor cells, 311
  vertebrate species, 299
Methylases, 38
MGSC. *See* Multipotent germ-line stem cells
Mice, lineage analysis in, 158–160
Microarray hybridization, 259–260. *See also*
    Microarray technology
Microarray technology
  cDNA microarray, 251–252
  confirmation studies for, 270–271
  experimental design for, 253–255
    array CGH, 260–261
    ChIP–chip, 261–262

microarray data analysis, 263–270
microarray hybridization, 259–260
MicroRNA arrays, 262–263
RNA amplification, 256–259
total RNA isolation, 255–256
  future perspectives of, 275
  oligonucleotide arrays, 252–253
  in stem cell biology and differentiation, 271–275
Microinjection, 96–97
Microinjection pipette, 96
MicroRNA (miRNA) arrays, 262–263. *See also*
    Microarray technology
Minimum essential medium-alpha (MEM-α), 585
Mitogen activated protein kinase (MAPK)
    pathway., 28
ML-IC. *See* Myeloid-lymphoid initiating cell
MLV. *See* Murine leukemia virus
Modified Eagles medium, 135
MOI. *See* Multiplicity of infection
Molecular switches, 559–562
Moloney murine leukemia retrovirus, 502
MoMuLV. *See* Moloney murine leukemia
    retrovirus
Mononuclear bone marrow cells, 290
Mononuclear cells (MNC)
  preparation and culture of, 224–226
Mouse embryo fibroblast, 169, 396, 506, 581
  dissection and primary culture of, 362
  γ−irradiation and plating of, 362
  protocol for preparing, 361
Mouse embryonic stem (ES) cells, establishment,
    322. *See also* Mouse ES cells
  blastocyst outgrowth development phases, 327
  characterization of, cell lines, 330–331
  chimeric embryos, factors affecting, 325
  events during
    attachment, 328
    cell colonies in early phase, 329–330
    embryos placement and hatching, 326–328
    first disaggregation timing, 328
    outgrowth formation, 328
    source embryos, 326
  factors affecting the effciency of, 324–325
  freezing of, ES cell lines, 330
  historical overview, 322–324
  protocols
    culture of initial colonies, 335–336
    disagregation of outgrowth, 334–335
    establishment, 333–334
    maintainance, 336–337
    mouse embryonic fibroblasts feeder layer
      preparation, 332–333

Mouse embryonic stem (ES) cells, establishment, 322. *See also* Mouse ES cells (*cont.*)
    other alternate approaches, 336
    plating, 334
Mouse ES cells, 169, 322, 502
    conversion to neural cells in adherent monolayer, 410–411
    conversion to neural progenitors in adherent monolayer, 409–410
    into insulin-producing cells, 477–479, 493–494
        activin A-induction, 487–488
        culture of undifferentiated mES cells, 485
        differentiated phenotypes, analysis of, 488–493
        generation of mES cells, 485–486
        pancreatic differentiation, induction of, 487
        protocols and parameters analyzed, 480–483
    monitoring neural and neuronal differentiation, 411–412
    troubleshooting, during conversion, 412–413
Mouse pluripotent ES cells, 415–416
    conversion to tissue-specific NS cells, 415–421
        differentiation of NS cells, 420
        expansion, freezing, and thawing of NS cells, 419
        NS cell derivation from ES cells (*see* NS cell derivation)
        troubleshooting, 421
Mouse tracheal epithelial cells CFE assay protocol, *in vitro*, 138–140
Mouse tracheal epithelial cell seeding, 143
MPCs. *See* Mesenchymal progenitor cells
MSC-based therapies
    autologous, schematic diagram of, 311
    clinical translation of, 310
    total joint replacement, 312
MSCs. *See* Mesenchymal stem cells
MUC-18 marker, 77
Mller glia, 54–55, 59–60
Multidrug resistance-1, 240
Multiplicities of infection (MOI), 441, 506
Multipotent germ-line stem cells, 166
Multipotent retinal progenitors, 54
Murine embryonic stem cells, 426
Murine leukemia virus, 502
Musashi, 26
Muscle-derived stem cells, 289
Myeloid-lymphoid initiating cell, 217
Myogenic differentiation, in amniotic fluid, 198

N

Nanog, transcription factors, 398
Naphthalene and clara cells, 125

Naphthalene injury, and BrdU labeling protocol, 126–127
Natural killer cell, 447
N2B27 media, 410–411, 417–418, 422
N-cadherin substrates, 64
NCoR in glial differentiation, 38
NEB. *See* Neuroepithlial bodies
Nestin, 33, 56, 65, 80, 198, 479, 496
Neural crest stem cells (NCSCs), 24–25
Neural induction of human ES cells, in adherent monolayer, 413–415
    monitoring induction, 414–415
    to neural cells, 414
    troubleshooting, 415
Neural lineage markers, 414
Neural stem cells (NSCs), 24
    culture and characterization of, 40
    in vitro manipulation of, 27–39
        apoptosis, central role in, 31
        cell division and regulation, 32–34
        differentiation and regulation, 34–39
        proliferation and survival, 27–28
        signals regulating proliferation of, 29
Neurobasal media supplemented with B27 + BDNF, 420
NeuroD1, 54
Neuroepithlial bodies, 114
Neurogenesis, 24
    from RPE cells, 58
Neurogenin2, proneural genes, 412
Neurosphere assay (NSA), 4–5
Neutrophil differentiation, 446–447
Ngn2, 54–55
NK cell development, 429–430
N-myc, 39
NOD/SCID mouse hematopoietic engraftment, 448–449
Noggin, 26, 35, 399
Noncollagenous proteins (NCPs), 75
Non-neural differentiation, 413
Non obese diabetic (NOD), 206
Nonparametric t-test, in microarray data analysis, 263–264
Nonpigmented epithelium (NPE), 57–58
Notch1, 56
NS basal media, 420
NS cell derivation
    from any ES cell line, 418–419
    using Sox1-lineage selection, 417
NSC medium, 11
N-2 supplement, 422–423
Nuclear localization signal, 544–545
Nuclear transfer ES (NT-ES) cells, 460

O

Oct 4, transcription factors, 398
Oligodendrocytes, 5, 17, 24, 34–35, 40, 420
Oligonucleotide arrays, 252–253. *See also*
    Microarray technology
Olig2, proneural genes, 412
Oncogenes, 39–40
Organ replacement, cell-based regenerative
    medicine techniques, 283
Osteocalcin, 76, 197, 306, 308
Osteogenic markers, 76
Osteopontin, 76, 308
Ovum donation patients, 353–354

P

PA6 cell line, 410
p53activation, 31
Pancreas development, 151–153
Pancreatic ducts, 155
Paraformaldehyde (PFA), 489
Pathogen testing, in mouse ES cell cultures, 331
Pax6, 56
Pax6/Chx10-expressing cells, 57
Pax6, transcription factors, 56
PBS. *See* Phosphate buffered saline
PCNA gene, 56
PCR analyses, 448
Peanut agglutinin (PNA), 19
PECAM-1. *See* Platelet endothelial cell adhesion
    molecule 1
Periodontal ligament stem cells (PDLSCs), 75
Peroxisome proliferator-activated receptor g
    (PPARg), 309
PEV. *See* Position-effect-variegation
PGC. *See* Primordial germ cell
PGK. *See* Phosphoglycerate kinase
Phosphate buffered saline, 6, 18, 43, 45, 66, 79,
    127, 130, 226, 302, 366, 382, 388, 423, 436,
    448, 489, 505, 563, 585, 588, 590, 592
Phosphoglycerate kinase, 502
PKCξ inhibitor, 33
Placental cell, for therapy, 193–194
Platelet-derived growth factor (PDGF), 31, 37,
    39, 41, 420
Platelet endothelial cell adhesion molecule 1, 197
Plating density, 415
Plating, of MEF cells, 362
Pluripotent cells, 166, 177, 380, 396, 519
Pluripotent epiblast cells, 192
PMEF. *See* Primary MEF; Primary mouse
    embryonic fibroblast

PMNs. *See* Polymorphonuclear leukocytes
PNECs. *See* Pulmonary neuroendocrine cells
Polidocanol injury and BrdU labeling
    protocol, 128
Polidocanol, role in proximal airway injury, 125
Pollutants, in lung injury models, 123
Polycomb repressive complex 1 (PRC1), 38
Polycomb transcription factor bmi1, 26
Poly-d-lysine (PDL), 61
Poly(ethylene glycol) diacrylate (PEGDA), 304
Polymorphonuclear leukocytes, 123
Poor plating effciency, 412
Position effect, 551
Position-effect-variegation, 509
Postnatal progenitors, 55
Preexisting beta cells, 154
Primary airway epithelial cells, retroviral
    infection of, 133
Primary MEF, 175
Primary mouse embryonic fibroblast, 141
Primate ESC-derived hematopoiesis, 431
Primordial germ cell, 166
Progenitor and pancreatic cell types, antibodies
    to characterize, 491
Progenitor cell, 206–209. *See also* Cord blood
    transplantation
    of airway SMG, 141–143
    isolation and characterization of, 194–195
Proliferative potential primitive hematopoietic
    progenitors (HPP-CFCs), 427
Proneural genes, 412
Proximal airway
    ex vivo epithelial tracheal xenograft model for,
        128–136
    of lung, 114
p300/SMAD complex, 38
Pulmonary neuroendocrine cells, 116

Q

Quantitative real-time (QRT), 505–506

R

Rb protein, 34, 38
RCR. *See* Replication-competent retrovirus
Reactive NO species (RNOS), 36
Real-time reverse transcriptase-polymerase chain
    reaction (RT-PCR) analysis, 381
    of Oct3/4, hTERT, Nanog, and Cripto, 384
    primers and probes for, 385

Receptor tyrosine kinases (RTKs), 39
Recombinant viral particles, generation of
   FuGene-6/TransIT-LT1 transfection, 528
   plasmid construction, 527
   viral particles
      collection of, 529
      concentration of, 529
      production, 527–528
Recombinase mediated cassette exchange,
   549–550, 552, 558–559, 569–570
Recombination targets (RTs), 543
Replication-competent retrovirus, 505
REST, in glial maintenance, 38
Retinal progenitor cells, 54
Retinal stem cells, 24, 55–57
   sources of, 57
      ciliary epithelium, 57–58
      Mller glia, 59–60
      pigmented epithelium, 58
Retinal stem/progenitors cells
   analysis of culture of, 64–67
   harvest and culture of, 60–61
      culture medium, 61
      N-cadherin substrates, 64
   possible relationships, among types of, 59
   primary cell culture of, 60
   substrate for adherent culture of, 61–63
   in vivo methods for study of, 67
      dissection and fixation of tissues, 68
      intraocular injections, 67–68
Retinoic acid (RA), 484
Retroviral gene transfer, in mouse and human
   ESCs, 506–507
Retroviral/lentiviral vector technology, usages of,
   511–512
Retrovirus, production of, 505–506
Reverse transcriptase-polymerase chain reaction
   (RT-PCR), 41
rev gene, 527
Rex 1, transcription factors, 398
RMCE. See Recombinase mediated cassette
   exchange
RNA interference (RNAi), 503, 525
RNA labeling, for microarray hybridization,
   257–259. See also Microarray technology
Rodent embryonic stem cells, 26
Rodent/murine NSCs, culture of, 41–43
   dishes preparation, 41
   freezing and thawing cells, 42–43
   NSCs isolation, 41–42
   NSCs, passaging of, 42
   reagents for, 43

Rostral-migratory stream (RMS), 25
Rotating cell culture systems (RCCS), 465
Runx1-deficient EBs, 427
Rx, transcription factors, 56

S

SAE. See Surface airway epithelium
SAG. See Serine/asparagine/glutamine
SAM. See Significant analysis of microarray
SAM, for microarray data analysis, 265
Sandoz thioguanine and ouabain, 169
SCF. See Stem cell factor
SCID-repopulating cells, 208
SDF-1. See Stromal cell derived factor-1
Self-organizing maps, 265
Serine/asparagine/glutamine, 228
Serine recombinases, 546
Sertoli cells, 88
Serum-free culture, 102–103
Serum replacement (SR), 325
Severe combined immunodeficiency (SCID), 206
Shh, role in NSCs proliferation, 28, 30
Short interfering RNAs, 503
shRNAs. See Small hairpin RNAs
Sialoprotein, 76
Side population cells, 115
Signals regulation for
   differentiation of NSCs, 37
   proliferation of NSCs, 29
Significant analysis of microarray, 265
Simian immunodeficiency virus (SIV), 519
Singular value decomposition, 266
SIN transfer vector, 520
siRNA expression
   in mouse and hESCs, 507–511
   vector design of, 504
siRNAs. See Short interfering RNAs; Small
   interfering RNAs
siRNA transfection, 107
Site-specific recombinases (SSRs), 543
Site-specific recombination substrates, designing
   of, 547
   Cre and FLPe recombination, factors
      affecting, 551
   φC31 recombination, characteristics of, 551–552
   loxP and FRTs, characteristics of, 547–550
SLAM family marker, in HSC, 239
slit1 and slit2, 26
Slow turning lateral vessel (STLV) system,
   465–467
   manipulation of, 468–470

SMAD 1/5 signaling, 400

Small hairpin RNA (shRNA), 323, 503

Small interfering RNAs, 503

SMGs. *See* Submucosal glands

Smooth muscle α-actin marker, 77

Soluble amyloid precursor protein (sAPP), 26

Soluble stem cell factor, 169

SOM. *See* Self-organizing maps

Somatic cell nuclear transfer (SCNT), 460

SOM, in microarray data analysis, 265–266

Sonic hedgehog (Shh), 26, 57

Southern blot, 448

Sox1–GFP reporter cell line, 411

Sox$^+$/Sox2$^+$/BLBP$^+$ cultures, 416

SP cells. *See* Side population cells

Spermatogonial stem cells (SSCs), 89–91

    culture, 98–101

        isolation from mouse testis, 101–102

        serum-free culture, on feeder cell

            monolayers, 102–104

        subculturing practices, 104–105

    transplantation, 91–93

        donor animals, choice of, 93–94

        donor cell suspension, microinjection of,

            96–97

        donor-derived spermatogenesis, analysis,

            97–98

        recipient mice, preparation of, 94–96

Spermatogonial stem cell transplantation

    technique, in mice, 92

Splice acceptor-β-galactosidase-polyA

    (SAβgalpA) gene trap cassette, 555–557

SRC. *See* SCID-repopulating cells

sSCF. *See* Soluble stem cell factor

SSEA. *See* Stage-specific embryonic antigens

SSEA 3 and SSEA4, surface marker, 398

SSR–LBD strategy, 545

Stage-specific embryonic antigens, 177

Stage-specific embryonic antigen-3 (SSEA-3), 379

Stat3 activity, 34, 38–39

Stem cell factor, 168, 207

Stem cell, of airway SMG, 141–143

Stem cell phenotypes and niches, in adult lung,

    117–121. *See also* Adult lung

Stem cells from human exfoliated deciduous teeth

    (SHED), 75–76

Stem cells *in vivo*

    adult niches for, 25–27

Stem cells, microarray technology for

    differentiation of, 272–273

    features of niches, 273–275

    identification of, 271–272

STO. *See* Sandoz thioguanine and ouabain

STO feeder layer, 176

Streptozotocin toxin, in beta cells, 156

Stromal cell derived factor-1, 206

STRO-1 marker, 77

Subgranular zone (SGZ), 25

Submucosal glands, 114

Subventricular zone (SVZ), 25, 32

Sulphur dioxide (SO$\subset$2), role in respiratory

    disease, 124

Surface airway epithelium, 116

Surface markers. *See* Markers for

    Undifferentiated hESCs

SVD. *See* Singular value decomposition

SVD, in microarray data analysis, 266

Symmetric and asymmetric division of NSCs,

    regulation of, 33

T

Tamoxifen, 151, 159, 545–546, 567

Targeted trapping", 553

T-cell development, 429

Teleost fish, 55

Telomerase activity, 398

Terasaki microwells, 419

Teratoma assay, 331

Terminally differentiated (TD), 115

Tetraploid complementation assay (TCA),

    323, 331

    and ES cell lines, 324

tet responsive element (TRE), 526

TGFα. *See* Transforming growth factor alpha

TGF-β. *See* Transforming growth factor beta

Therapeutic cloning

    nuclear transplantation, 283

Thrombopoietin (TPO), 207

Thy1 (CD90), expression of, 90

Thy1+germ cells for transfection, 107

Thyroid hormone T3, 34

Tie-2 marker, in HSC, 239

Tissue culture polystyrene, 300

Tissue engineering

    adult stem cells for

        bone tissue, engineering of, 288–289

        cardiac tissue, 289–290

        cartilage tissue, 289

        embryonic stem cells, compared with, 287

        fetal stem cells, 291–292

        mesenchymal stem cells (MSCs), 287

        neural tissue, 290–291

        tissue-specific progenitor cells, 287

Tissue engineering (*cont.*)
  biomaterials
    acellular tissue matrices, 285–286
    cell-adhesion substrate, 284
    naturally derived materials, 285
    spatial tissue reorganization, 285
    synthetic polymers, 286
  cell population, 582
  hESC-derived connective tissue progenitors,
    584
  regenerative medicine, components of, 282–283
  scaffolds utilization, 583
Tissue nonspecific alkaline phosphatase (TnAP),
  171–172
Tooth structure, 74–75
Total RNA isolation, 255–256. *See also*
  Microarray technology
TRA. *See* Tumor rejection antigen
TRA-1–60, and TRA-1–81, surface markers, 379,
  398
Transcription factor SMAD2/3, 400
Transduced hESCs expressing transgene,
  enrichment. *See also* Lentiviral-based vectors
  by FACS sorting, 534–535
  isolation by limiting dilution, 533–534
  mechanical enrichment, 533
  using antibiotic selection, 535
Transduction efficiency, 532
Transforming growth factor alpha, 26, 28–29, 55
Transforming growth factor beta, 76, 99, 399, 484
Transgene expression, 526–527. *See also*
  Lentiviral-based vectors
Transgene expression, in mouse and hESCs, 507–
  511
Transient amplifying (TA), 114
Transitin, 56, 65, 67
Trizol extraction method, in total RNA isolation,
  255
Trophectoderm (TE), 343–346
Trophoblast cells
  bubbling of, 364
Trypsinization, 365–366, 413
Tubulin, 17, 33, 35, 66, 388, 412, 489
TuJ1 antibody, 66
Tumor rejection antigen, 177
Type I and II diabetes, 150
Type I collagen, 76

Tyrosine kinase, 30, 34, 238, 275
Tyrosine recombinases
  Cre and FLPe, 543–546
  Dre, 546
  R and Kw, 546

U

Umbilical cord blood (UCB)
  collection of, 224
  importance of, 204
  methods in, 212
Urothelial strain, 284

V

Vascular cell adhesion molecule, 197
Vascular endothelial growth factor, 26, 183, 286,
  426–427, 431, 441, 444–445
VCAM. *See* Vascular cell adhesion molecule
VCAM-1 marker, 77
VEGF. *See* Vascular endothelial growth factor
Vesicular stomatitis virus surface protein, 505
Viral titer, determination of, 535–536. *See also*
  Lentiviral-based vectors
Virus pseudotyping, 527
VSV-G. *See* Vesicular stomatitis virus surface
  protein

W

White matter precursor cells (WMPCs), 26
Wilcoxon Rank Sum test, in microarray data
  analysis, 264–265
wnt proteins, 35–36
Woodchuck hepatitis virus (WPRE), 520

Y

Yeast 2 circle, 543

Z

Zona pellucida (ZP), 326, 344–345, 363